T0393030

FRONTIERS OF CIVIL ENGINEERING AND DISASTER PREVENTION AND
CONTROL VOLUME 1

Frontiers of Civil Engineering and Disaster Prevention and Control is a compilation of selected
papers from The 3rd International Conference on Civil, Architecture and Disaster Prevention and
Control (CADPC 2022) and focuses on the research of architecture and disaster prevention in
civil engineering. The proceedings features the most cutting-edge research directions and achievements related to construction technology and prevention and control of disaster. Subjects in this
proceedings include:

Construction Technology
Seismicity in Civil Engineering
High-Rise Building Construction
Disaster Preparedness and Risk Reduction
Smart Post-Disaster Rescue

These proceedings will promote development of civil engineering and risk reduction, resource
sharing, flexibility and high efficiency. Moreover, promote scientific information interchange
between scholars from the top universities, research centers and high-tech enterprises working all
around the world.

PROCEEDINGS OF THE 3RD INTERNATIONAL CONFERENCE ON CIVIL, ARCHITECTURE AND DISASTER PREVENTION AND CONTROL (CADPC 2022), WUHAN, CHINA, 25–27 MARCH 2022

Frontiers of Civil Engineering and Disaster Prevention and Control

Volume 1

Edited by

Yang Yang
Chongqing University, China

Ali Rahman
Southwest Jiaotong University, China

CRC Press
Taylor & Francis Group
Boca Raton London New York Leiden

CRC Press is an imprint of the
Taylor & Francis Group, an **informa** business

A BALKEMA BOOK

First published 2023
by CRC Press/Balkema
4 Park Square, Milton Park, Abingdon, Oxon OX14 4RN, UK
e-mail: enquiries@taylorandfrancis.com
www.routledge.com – www.taylorandfrancis.com

CRC Press/Balkema is an imprint of the Taylor & Francis Group, an informa business

© 2023 selection and editorial matter, Yang Yang and Ali Rahman; individual chapters, the contributors

The right of Yang Yang and Ali Rahman to be identified as the authors of the editorial material, and of the authors for their individual chapters, has been asserted in accordance with sections 77 and 78 of the Copyright, Designs and Patents Act 1988.

Although all care is taken to ensure integrity and the quality of this publication and the information herein, no responsibility is assumed by the publishers nor the author for any damage to the property or persons as a result of operation or use of this publication and/or the information contained herein.

ISBN: 978-1-032-39108-3 (SET hbk)
ISBN: 978-1-032-39111-3 (SET pbk)

ISBN Volume 1: 978-1-032-31200-2 (hbk)
ISBN Volume 1: 978-1-032-31201-9 (pbk)
ISBN Volume 1: 978-1-003-30857-7 (ebk)

DOI: 10.1201/9781003308577

ISBN Volume 2: 978-1-032-39103-8 (hbk)
ISBN Volume 2: 978-1-032-39106-9 (pbk)
ISBN Volume 2: 978-1-003-34843-6 (ebk)

DOI: 10.1201/9781003348436

Typeset in Times New Roman
by MPS Limited, Chennai, India

Frontiers of Civil Engineering and Disaster Prevention and
Control – Yang & Rahman (Eds)
© 2023 The Editor(s), ISBN: 978-1-032-31200-2

Table of contents

Building material properties and highway bridge construction

Frontiers of Civil Engineering and Disaster Prevention and Control – Yang & Rahman (Eds)
© 2023 The Editor(s), ISBN: 978-1-032-31200-2

Preface

Due to recent pandemic, the 3rd International Conference on Civil, Architecture and Disaster Prevention and Control (CADPC 2022) which was planned to be held in Guangzhou, China, was held virtually online during March 25–27, 2022. The decision to hold the virtual conference was made in compliance with many restrictions and regulations that were imposed by countries around the globe. Such restrictions were made to minimize the risk of people contracting or spreading the COVID-19 through physical contact. There were 120 individuals who attended this online conference, represented many countries including Malaysia, Iran, India and China.

CADPC 2022 focused on construction technology, disaster prediction, disaster prevention and control, and post-disaster reconstruction. The conference provided a platform for scholars from universities at home and abroad, researchers and engineers to share research results and cutting-edge technologies. All attendants get the chance to know about updated academic trends, develop new way of thinking, improve academic research and discussions, and promote industrialization of research findings.

During the conference, the conference model was divided into three sessions, including oral presentations, keynote speeches, and online Q&A discussion. In the first part, some scholars, whose submissions were selected as the excellent papers, were given about 5-10 minutes to perform their oral presentations one by one. Then in the second part, keynote speakers were each allocated 30-45 minutes to hold their speeches. In the second part, we invited four professors as our keynote speakers. Dr. Mohammadreza Vafaei, School of Civil Engineering, Faculty of Engineering, Universiti Teknologi Malaysia. His field of specialty and interest: Seismic design and rehabilitation, Structural health monitoring, Signal processing, Neural networks and Wavelet Transforms. And then we had Distinguished Professor. Zhifeng Xu. He delivered a wonderful speech: *Analytical approximations for the frequency response, localization factor and attenuation coefficient of one-dimensional periodic foundations*. Professor. Xinlin Wan, College of Civil Engineering, Anhui Jianzhu University. His research area: Geotechnical Engineering, Geophysical Exploration, Engineering Seismic, etc. Lastly, we were glad to invite Associate Professor. Deng Peng, College of Civil Engineering, Hunan University. He performed a speech on *Experimental Investigation of Seismic Uncertainty Propagation through Shake Table Tests*. Their insightful speeches had triggered heated discussion in the third session of the conference. Every participant praised this conference for disseminating useful and insightful knowledge.

The proceedings are a compilation of the accepted papers and represent an interesting outcome of the conference. Topics include but are not limited to the following areas: Civil, Construction Technologies, Disaster Prevention and Control and more related topics. All the papers have been through rigorous review and process to meet the requirements of international publication standard.

We would like to acknowledge all of those who supported CADPC 2022. The help and contribution of each individual and institution was instrumental in the success of the conference. In particular, we would like to thank the organizing committee for its valuable inputs in shaping the conference program and reviewing the submitted papers.

We sincerely hope that the CADPC 2022 turned out to be a forum for excellent discussions that enabled new ideas to come about, promoting collaborative research. We are sure that the proceedings will serve as an important research source of references and knowledge, which will lead to not only scientific and engineering findings but also new products and technologies.

The Committee of CADPC 2022

Committee members

Conference General Chair
Prof. Fuyou Xu, *Dalian University of Technology, China*

Local Organizing Chair
Dr. Yiding Zhao, *Yancheng Institute of Technology, China*

Technical Program Committee
Prof. Tetsuya Hiraishi, *Kyoto University, Japan*
Prof. Fauziah Ahmad, *Universiti Sains Malaysia, Malaysia*
A. Prof. Mohammadreza Vafaei, *Universiti Teknologi Malaysia, UTM, Malaysia*
Dr. Sadegh Rezaei, *Shargh-e Golestan Institute of Higher Education, Iran*
Dr. Haytham F.A. Isleem, *Tsinghua University, China*

Civil construction technology and structural
seismic reinforcement

Frontiers of Civil Engineering and Disaster Prevention and Control – Yang & Rahman (Eds)
© 2023 The Author(s), ISBN: 978-1-032-31200-2

A layered finite element method for transient heat conduction of functionally graded concrete beams

Qiang Liu
School of Civil Engineering and Architecture, Nanchang Hangkong University, Nanchang, Jiangxi, P.R. China
Poly Changda Engineering Co. LTD, Guangzhou, Guangdong, P.R. China

Huihua Zhang*, Shangyu Han & Xiaolei Ji*
School of Civil Engineering and Architecture, Nanchang Hangkong University, Nanchang, Jiangxi, P.R. China

ABSTRACT: Functionally graded concrete (FGC) is a type of new concrete composite whose functions and properties change with the spatial position. In this paper, a finite element model of the FGC beam is established with the idea of the layered model and the finite element software Ansys. A benchmark example is used to verify the feasibility and accuracy of the model for transient heat conduction, and the results show that the present solutions are in good agreement with the reference ones. Furthermore, the transient heat conduction of an FGC beam is analyzed and the influence of the gradient parameter on the temperature field is obtained.

1 INTRODUCTION

It is well known that traditional concrete has several defects, such as heavy self-weight, poor crack resistance, and complex field construction. In addition, the design idea of FGC is derived from functionally graded materials (i.e., FGMs, whose components and properties change continuously in space (Koizumi 1997). Compared with traditional concrete, FGC is superior in many aspects, e.g., mechanical properties, impermeability, fire resistance, and corrosion resistance (Evangelista et al. 2009; Gao et al. 2007). Due to excellent performance, it has been used in some practical projects. Considering the fact that there is not only mechanical load but also temperature load (temperature gradient generated by hydration heat, drastic change of ambient temperature, etc.), it is of great scientific and practical significance to study the thermal behavior of FGC, especially in the transient state.

The current studies on FGC mainly focus on the mechanical properties (Mastali et al. 2015, Moghadam & Omidinasab 2020), while the thermal behavior is rarely involved. Some representative work about thermal analysis are as follows: Wang et al. (2017) experimentally obtained the variation law of temperature, stress, and strain of FGC beam with the layer thickness of ultra-high performance concrete, based on concrete hydration heat. Zhang et al. (2018) used the finite element method (FEM) to analyze the thermal buckling and post-buckling of FGC plates.

To the authors' knowledge, there is no report on the transient heat conduction of FGC beams. Further, with regard to the advantages of FEM (i.e., high applicability, low cost, and easy parametric analysis), this paper combines FEM with the layered model (Huang et al. 2005) to explore the influence of material gradient parameters on the transient heat conduction behavior of FGC beams.

---

*Corresponding Author: hhzhang@nchu.edu.cn

2 TRANSIENT HEAT CONDUCTION EQUATIONS OF FGC

Consider the physical domain Ω composed of 2D isotropic FGC shown in Figure 1. Ω is enclosed by the contour $\Gamma = \Gamma_1 \cup \Gamma_2$, with Γ_1 and Γ_2, respectively, the temperature boundary and the heat flux boundary. The differential governing equation of this problem is (Wang 2003):

$$\frac{\partial}{\partial x_1}\left(k(\mathbf{x})\frac{\partial T(\mathbf{x},t)}{\partial x_1}\right) + \frac{\partial}{\partial x_2}\left(k(\mathbf{x})\frac{\partial T(\mathbf{x},t)}{\partial x_2}\right) + Q = \rho(\mathbf{x})c(\mathbf{x})\frac{\partial T(\mathbf{x},t)}{\partial t} \tag{1}$$

where ∂ denotes partial derivative; k, ρ, and c are the thermal conductivity, the density, and the specific heat, respectively, at a constant pressure of the FGC, and may vary spatially with $\mathbf{x} = (x_1, x_2)$; T, t, and Q are the temperature, the time, and the heat source, respectively.

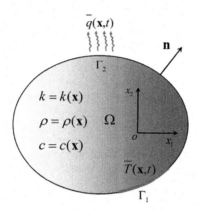

Figure 1.   Transient heat conduction in a domain composed of isotropic FGC.

The corresponding boundary conditions are:

$$T(\mathbf{x},t) = \overline{T}(\mathbf{x},t) \quad \text{on } \Gamma_1 \tag{2}$$

$$-k(\mathbf{x})\frac{\partial T(\mathbf{x},t)}{\partial x_1}n_1 - k(\mathbf{x})\frac{\partial T(\mathbf{x},t)}{\partial x_2}n_2 = \overline{q}(\mathbf{x},t) \quad \text{on } \Gamma_2 \tag{3}$$

where $\overline{T}$ and $\overline{q}$ are the given temperature on Γ_1 and the heat flux on Γ_2, respectively; $(n_1, n_2) = \mathbf{n}$ is the outward unit normal to Ω. Moreover, the initial condition is $T(\mathbf{x},t)|_{t=0} = T_0(\mathbf{x})$.

3 FINITE ELEMENT EQUATIONS FOR FGC

By virtue of the Galerkin method and Equations (1)–(2), the following formula can be obtained (Wang 2003):

$$\int_{\Omega}\left[\begin{array}{l}\delta T(\mathbf{x},t)\left(\rho(\mathbf{x})c(\mathbf{x})\frac{\partial T(\mathbf{x},t)}{\partial t}\right) + \frac{\partial\delta T(\mathbf{x},t)}{\partial x_1}\left(k(\mathbf{x})\frac{\partial T(\mathbf{x},t)}{\partial x_1}\right) \\ +\frac{\partial\delta T(\mathbf{x},t)}{\partial x_2}\left(k(\mathbf{x})\frac{\partial T(\mathbf{x},t)}{\partial x_2}\right) - \delta T(\mathbf{x},t)\rho(\mathbf{x})Q\end{array}\right]d\Omega - \int_{\Gamma_2}\delta T(\mathbf{x},t)q\,d\Gamma = 0 \tag{4}$$

where δ is the variational symbol.

When modeled by the FEM, the physical domain Ω is first discretized into finite elements, and then the temperature in the element is approximately interpolated by the node temperature T_i of this element, i.e.,

$$\widetilde{T} = \sum_{i=1}^{n_e} N_i(\mathbf{x})T_i(t) \tag{5}$$

where N_i is the shape function at node i and n_e is the node amount attached to the element.

By substituting Equation (5) into Equation (4), and further considering the arbitrariness of the variation, the FEM global equations for 2D transient heat conduction problems in the FGC can be deduced as (Wang 2003):

$$C\dot{T} + KT = P \tag{6}$$

where C is the heat capacity matrix, K is the heat conduction matrix, $\dot{T} = dT(\mathbf{x},t)/dt$ is the derivative array of node temperature with respect to time, and P is the array of temperature load. The details of C, K, and P can be found in Wang (2003).

4 LAYERED MODEL OF FGC BEAM

Because the finite element software Ansys finds it difficult to directly represent the continuous gradient change in FGC, the layered model (Huang et al. 2005) is used. Taking Figure 2 as an example, in the layered model, the FGC is divided into N parts, in which the material parameters of each layer are different, but those in the same layer are constant; i.e., FGC is regarded as a combination of N-layer-homogeneous materials.

Assume that the material properties of FGC are exponential along y-axis as shown in Figure 2, i.e.,

$$S(y) = S_0 e^{\beta_S y} \quad (S = k, \rho, c) \tag{7}$$

where S_0 is constant and β_S is the gradient parameter. From the idea of the layered model, and taking the center of each layer as the benchmarking, the material parameters of any homogeneous layer can be expressed by

$$S(i) = S_0 e^{\frac{\beta_S l(2i-1)}{2N}} \quad (i = 1, 2, \ldots, N) \tag{8}$$

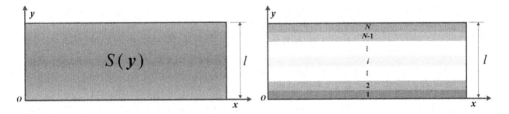

Figure 2.   Theoretical (left) and layered model (right) of FGC.

Theoretically, the more the number of layers, the closer it is to the actual material. However, many layers will lead to lower calculation efficiency. Therefore, Section 5 discusses the determination of layer number through a typical example.

5 NUMERICAL EXAMPLE

In this section, a benchmark problem is first examined to verify the feasibility and accuracy of the method, and then a parametric study is carried out around the transient heat conduction in an FGC beam.

5.1 Transient heat conduction in an FGMs plate

As shown in Figure 3, transient heat conduction in a square FGM plate of edge size L is considered. The material parameters are $k(\mathbf{x}) = k_0(1 + \gamma x_1)^2$, $c(\mathbf{x}) = c_0(1 + \gamma x_1)^2$ and $\rho(\mathbf{x}) = \rho_0$. The temperature on the left side of the plate is $\overline{T}(\mathbf{x}, t) = 0$, while that on the right side is $\overline{T}(\mathbf{x}, t) = T_r H(t)$, where $H(t)$ represents the Heaviside step function). Moreover, the other edges are adiabatic and the initial temperature is 0.

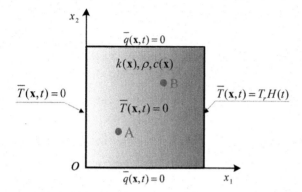

Figure 3.   Transient heat conduction of square FGM plate.

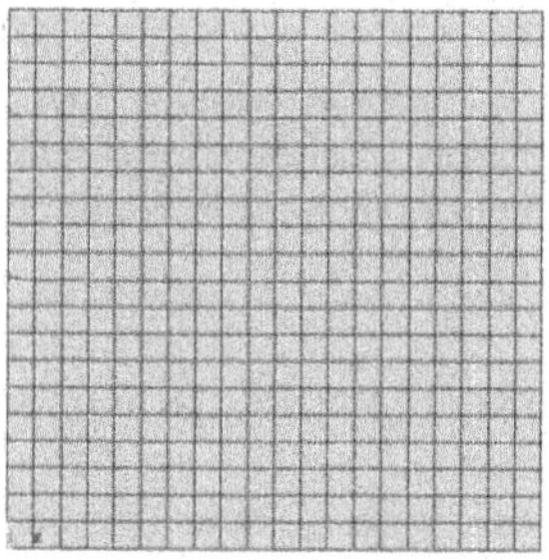

Figure 4.   The finite element mesh corresponding to $h = 0.05, N = 20$.

According to the work by Sutradhar and Paulino (2004), the analytical solution of temperature field is represented as follows:

$$T(\mathbf{x}, t) = \frac{T_1 x_1}{\sqrt{k}L} + \frac{2T_1}{\sqrt{k}} \sum_{n=1}^{\infty} \frac{\cos(n\pi)}{n\pi} \sin\left(\frac{n\pi x_1}{L}\right) \exp\left(-\frac{n^2\pi^2}{L^2}\kappa t\right) \tag{9}$$

where $T_1 = \sqrt{k_0}(1 + \gamma L)T_r$ and $\kappa = k/(\rho c)$.

When modeling, we take $L = 1.0$, $k_0 = 5.0$, $c_0 = 1.0$, $\rho_0 = 1.0$, $\gamma = 2.0$, and $T_r = 100$. In the Ansys platform, the plate is discretized by Plane55 four-noded elements. The element size (edge length of a finite element) is, respectively, taken as $h = 0.2, 0.1, 0.05, 0.025$ and the corresponding number of divided layers is $N = 2, 10, 20, 40$. Figure 4 shows the finite element mesh when $h = 0.05, N = 20$ (the total number of elements is 400 and the total number of nodes is 441). In the analysis, the time step Δt is taken as 0.001. Tables 1 and 2, respectively, provide the simulated

Table 1.   Temperatures of point A at different time, layer number, and element size.

N	$t \backslash h$	0.07	0.08	0.09	0.1
2	0.2	41.436	42.443	43.072	43.465
	0.1	41.469	42.475	43.099	43.486
	0.05	41.506	42.503	43.119	43.500
	0.025	41.515	42.509	43.124	43.504
10	0.1	52.777	54.056	54.850	55.342
	0.05	52.824	54.092	54.875	55.360
	0.025	52.836	54.100	54.882	55.364
20	0.05	52.895	54.466	54.951	55.437
	0.025	52.907	54.174	54.957	55.441
40	0.025	52.925	54.193	54.977	55.461
Exact solutions (Sutradhar & Paulino 2004)		53.198	54.387	55.112	55.555

Table 2. Temperatures of point B at different time, layer number, and element size.

N	$t \setminus h$	0.07	0.08	0.09	0.1
2	0.2	75.712	76.881	77.611	78.066
	0.1	75.709	76.892	77.626	78.081
	0.05	75.753	76.925	77.649	78.098
	0.025	75.764	76.933	77.655	78.102
10	0.1	78.877	79.973	80.652	81.073
	0.05	78.917	80.003	80.674	81.088
	0.025	78.927	80.010	80.679	81.092
20	0.05	78.957	80.044	80.715	81.131
	0.025	78.967	80.051	80.721	81.135
40	0.025	78.977	80.062	80.731	81.145
Exact solutions (Sutradhar & Paulino 2004)		79.209	80.225	80.846	81.224

temperatures of two sampling points, i.e., A: (0.3, 0.3) and B: (0.6, 0.6), as well as the analytical solutions by Equation (9). It can be found that with increase in the layer number and the refinement of the elements, the simulated results tend to be stable and closer to the exact solutions. However, the effect of the increasing layer number is more significant than that of element refinement on the improvement of solution accuracy. In addition, when the layer number is very small (e.g., $N = 2$), the simulated results are very unsatisfactory since the difference between the layered model and the actual one is very large.

5.2 Transient heat conduction in an FGC beam

Consider the transient heat conduction in an FGC beam as shown in Figure 5. The beam size is $L \times H$, and its material components are ordinary concrete and polypropylene fiber concrete (PPC). The bottom and top of the beam, respectively, correspond to pure PPC and pure ordinary concrete. The temperature at the bottom of the beam is $\overline{T_0}(x, y, t) = 0$, the temperature at the top is $\overline{T_1}(x, y, t) = T_R H(t)$, and the initial temperature is 20°C. The remaining boundary is insulated. It is assumed that the material parameters change exponentially along the y-axis, as shown in Equation (7).

In the simulation, we take $L \times H = 0.7 \text{ m} \times 0.15 \text{ m}$, $T_R = 50°C$, thermal conductivity of PPC as $k_0 = 1.28 \text{ W}/(\text{m} \cdot \text{k})$ (Wang et al. 2014). and the thermal conductivity for ordinary concrete as $k_1 = 1.453 \text{ W}/(\text{m} \cdot \text{k})$ (Wang et al. 2014). Plane55 element is still used, the element size is 5 mm and the number of layers is $N = 30$. The corresponding finite element mesh is shown in Figure 6. The total number of elements and nodes are, respectively, 4,200 and 4,371. Because the density and specific heat of PPC and ordinary concrete are almost similar, the influences of density and specific heat between them can be ignored, and the unified value is $c = 913 \text{ J}/(\text{kg} \cdot \text{K})$ and $\rho = 2300 \text{ kg/m}^3$. Further, we take time step $\Delta t = 50$ s.

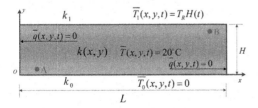

Figure 5. Physical model of FGC beam.

Figure 6. Finite element mesh in Ansys.

In this example, we mainly inspect the influence of the thermal conductivity gradient parameter on the temperature field of the FGC beam. Accordingly, the parameter $\beta_k H$ is taken as $\ln(1.453/1.28), \ln(1.679/1.28)$, and $\ln(1.942/1.28)$, respectively. Figure 7(a) and (b) show the simulated temperatures of sampling points A: $(0.1, 0.02)$ and B: $(0.6, 0.12)$ at different times. It can be seen that the temperature of point A decreases, while that of point B increases with the increase in k. We can also view that during the initial period, the variation of temperature at different gradients of point A is undistinguishable, while that of point B is more obvious. Moreover, the temperature of both points gradually tends to be stable with the increase in time.

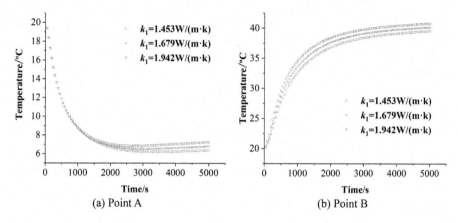

Figure 7. Simulated temperatures of sampling points at different time and material gradient.

6 CONCLUSIONS

In this paper, FEM in combination with the layered model is used to analyze the transient heat conduction of FGC. With the governing equation and boundary conditions of the problem, the corresponding finite element formula is derived by using the Galerkin method. A typical example is used to verify the feasibility, convergence, and accuracy of the proposed method. Based on this, the influence of material gradient parameters on the transient heat conduction of the FGC beam was further studied. The main conclusions were as follows:

(1) The proposed method can effectively address the transient heat conduction problem of FGC. In terms of solution accuracy, the more layers and elements are used, the higher calculation accuracy can be achieved. Moreover, the contribution of layer number is more significant than that of element refinement.
(2) The gradient change of thermal conductivity affects the temperature distribution, which indicates that the results given in this paper can provide some support for the component design of FGC under temperature loading.

ACKNOWLEDGMENTS

This work was supported by the National Natural Science Foundation of China (Grant No. 12062015 and 52068054) and the Provincial Natural Science Foundation of Jiangxi, China (Grant No.: 20212BAB211016).

REFERENCES

Evangelista, F., Roesler, J., Paulino, G.H. (2009) Numerical simulations of fracture resistance of functionally graded concrete materials. Transport. Res. Rec., 2113: 122–131.

Gao, Y.L., Ma, B.G., Wang, X.G., et al. (2007) Development and properties study on functionally graded concrete segment used in shield tunneling. Chinese Journal of Rock Mechanics and Engineering, 11: 2341–2347.

Huang, G.Y., Wang, Y.S., Yu, S.W. (2005) A new multi-layered model for in-plane fracture analysis of functionally graded materials (FGMs). Theoretical and Applied Mechanics, 01: 1–8.

Koizumi, M. (1997) FGM activities in Japan. Compos. Part B-Eng., 28(1–2): 1–4.

Mastali, M., Naghibdehi, M.G., Naghipour M., et al. (2015) Experimental assessment of functionally graded reinforced concrete (FGRC) slabs under drop weight and projectile impacts. Constr. Build. Mater., 95: 296–311.

Moghadam, A.S., Omidinasab, F. (2020) Assessment of hybrid FRSC cementitious composite with emphasis on flexural performance of functionally graded slabs. Constr. Build. Mater., 250: 118904.

Sutradhar, A., Paulino, G.H. (2004) The simple boundary element method for transient heat conduction in functionally graded materials. Comput. Method. Appl. M., 193(42): 4511–4539.

Wang, J.F., Zhang, Q., Du, H.X., et al. (2014) Thermal diffusivity research on polypropylene fiber reinforced high strength concrete (HSC) after high temperature. China Concrete and Cement Products, 12: 56–59.

Wang, K., Zhu, E., Li, B.D., et al. (2017) Effect of UHPC hydration heat on early age shrinkage cracking of functionally graded composite beams. Bulletin of the Chinese Ceramic Society, 36(01): 38–42.

Wang, X.C. (2003) Finite element method (second edition). Tsinghua University Press, Beijing.

Zhang, H.Q., Zhang, Z., Wu, H.L., et al. (2018) Thermal buckling and post-buckling analysis of functionally graded concrete slabs with initial imperfections. Int. J. Struct. Stab. Dy., 18(11): 1850142.

Frontiers of Civil Engineering and Disaster Prevention and
Control – Yang & Rahman (Eds)
© 2023 The Author(s), ISBN: 978-1-032-31200-2

Research on the application of BIM technology in the design of prefabricated buildings and green construction

Xiaoqiang Tang*

Sichuan Aerospace Vocational College, Chengdu, Sichuan, China

ABSTRACT: With regard to "high energy consumption, high pollution, high waste, and low efficiency" caused by traditional construction methods used in different construction stages, this paper proposes "BIM technology as the support and core of prefabricated building." The "integrated" construction method realizes digital innovation in all aspects of construction engineering planning, design, construction, management, etc. Prefabricated building is an important innovation in the construction industry. It not only has the characteristics of specialization, standardization, and scale, but also meets the requirements of China's industrial structure adjustment and green energy-saving building evaluation standards, and has become the only way for the transformation and development of the construction industry. Prefabricated building technology can realize the integrated application and whole-process management of prefabricated buildings, improve operation and management efficiency, effectively reduce costs, save resources, and reduce risks; thereby, providing ensuring smooth implementation of prefabricated buildings. The digital upgrade of the construction industry has led to healthy and sustainable development of the industry.

1   INSTRUCTION

Currently, China has entered a new era of social development. The internal and external environment has undergone profound changes under this new historical reform. All industries and fields of the society are leveraging new opportunities and situations for development, but they will also be faced with new problems and new challenges. In an environment where economic development is shifting from a high-speed growth stage to a high-quality development stage, as an important industry in China's national economy, the construction industry actively seeks breakthroughs, conforms to the development trend of the industry, accelerates integration and innovation, and comprehensively realizes transformation and upgrading of high-quality development (Gao 2021). For a long time, China's construction industry has been affected by the application level of science and technology and the quality of employees, and most construction projects follow the traditional construction method of in situ pouring. Traditional construction methods have several shortcomings, such as high energy consumption, high pollution, high waste, low efficiency, and large errors in the process of practical application, and they are faced with the pressures of ecological environmental protection, labor costs, and rising building material costs, making it challenging to adapt to the construction. Developments in the new era of the industry call for new technologies and models to open up new directions for the traditional construction industry and rebuild new industry ecology. Prefabricated buildings are based on the prefabricated production of component factories and on-site prefabricated installation as an important form of modernization of the construction industry. They are characterized by standardized design, factory production, prefabricated construction, integrated decoration, and information management. Research and development, design, manufacturing, on-site assembly, and other business fields to achieve a new sustainable development of building products, energy-saving, environmental protection, and maximizing the full-cycle value of building production methods. (Wen 2020). Increase in prefabricated buildings not only has

---

*Corresponding Author: 304071943@qq.com

DOI 10.1201/9781003308577-2

unparalleled advantages in engineering construction quality, construction work efficiency, cost consumption control, energy-saving, and environmental protection (as shown in Tables 1 and 2), but also brings a change in the overall craftsmanship and thinking mode of the industry. Since the "Twelfth Five-Year" Green Building and Green Ecological Regional Development Plan was issued by the Ministry of Housing and Urban-Rural Development in 2013, clear requirements of the current *"14th Five-Year" Development Plan for the Construction Industry* were mentioned for the first time, and prefabricated buildings entered the development stage rapidly. However, it can be seen that, as of 2021, the national prefabricated building penetration rate is at 14%, far lower than the mature markets in major countries in the world. As shown in Figure 1, the prefabricated building penetration rate in Japan and the U.S. is 90%. Fundamentally, weak technical support is an important reason behind the hindrance to the development of prefabricated buildings in China. Therefore, this paper believes that introduction of BIM technology into the design, production, construction, operation, and maintenance stages of prefabricated buildings can effectively improve the efficiency of collaborative designs of prefabricated buildings, reduce design errors, optimize the production process of prefabricated components, and improve the inventory management of prefabricated components. In addition, it may also help simulate and optimize the construction process, realize quality management and energy consumption management in the operation and maintenance stage of prefabricated buildings, and become the core for the application development of prefabricated buildings. It has made important contributions to the transformation and upgrading of the city and to healthy and green development (Yu 2021).

Table 1. Comparison of the advantages of prefabricated buildings and traditional buildings.

	Prefabricated building	Traditional building
Quality	The prefabricated construction method can better avoid quality problems such as water leakage, cracking, and large dimensional errors.	In the cast-in-place concrete structure, it is difficult to set up the formwork, the scaffolding is dense, and the forming quality is difficult to guarantee.
Energy-saving and environmental protection	Factory prefabrication can significantly minimize the high-altitude painting operation of workers' hanging baskets while also lowering safety risks. Prefabricated structures use less energy.	In traditional buildings, a lot of plastering, leveling, and wet work are required, which wastes materials and causes harsh construction environments.
Duration	Most of the components are completed in the engineering assembly line and are not affected by the weather. The overall delivery time is generally 30%–50% faster than the traditional one.	A more mature construction team can achieve one floor in five days for a structural project, but it also requires secondary structures such as bricklaying and plastering.

Table 2. Comparison of energy-saving and environmental protection data between prefabricated buildings and traditional buildings.

	Energy consumption per square meter (kg standard coal)	Water consumption per square meter (t)	Amount of garbage per square meter (t)
Prefabricated building	14.70	0.314	0.002
Traditional building	19.10	1.495	0.022

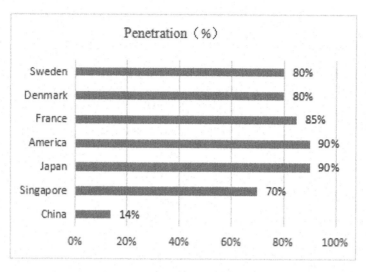

Figure 1. The penetration rate of prefabricated buildings in major countries in the world.

2 PREFABRICATED BUILDINGS AND BIM TECHNOLOGY

2.1 *Prefabricated building*

Prefabricated buildings refer to the transfer of a large number of on-site pouring operations in the traditional construction method to the factory, that is, to prefabricate various components or accessories for construction in advance, such as beams, columns, shear wall panels, stairs, etc. After being transported to the construction site, a suitable connection method is selected for different components or accessories to complete the assembly of the building. Prefabricated buildings can be classified according to different categories, including steel-framed buildings, wood-framed buildings, concrete-framed buildings, and composite buildings.

In terms of practical application, the concrete building type in the structural classification is the most common and widely used. Concrete structure buildings are mainly prefabricated concrete components (PC components for short), and the design, production, and construction of PC components are completed according to the shear wall structure system, frame structure system, or frame-shear wall structure system. Different structural systems have different characteristics, and their prefabricated components and scope of application also exhibit certain differences. Different building structures require different structural technical systems (He 2020). The selection of each structural technical system will include the determination of planning and design schemes, production standards for prefabricated components, on-site construction processes, and connection methods for PC components. It reflects the technical integration, refinement, and humanization of prefabricated buildings. Only by applying the appropriate technical system to the projects suitable for the use of prefabricated buildings can the advantages of prefabricated buildings be maximized and the purpose of improving construction quality, accelerating project progress, and improving the overall quality of construction projects. The benefits of prefabrication are fully reflected in the advantages of construction, energy savings, and environmental protection, preventing potential safety problems and enhancing the market link between supply and demand.

The prefabricated building integrates multiple links and processes, such as technical system research and development, overall design planning, factory manufacturing, prefabricated construction, operation and maintenance management, etc. It gradually improves and forms a set of overall technical solutions for prefabricated buildings in the development process. The prefabricated construction industry is in the process of transition from extensive to fine (Liao 2019). However, we can see that the development of prefabricated buildings is still unsatisfactory under the current country's vigorous promotion and publicity of prefabricated buildings: Due to the rough design requirements in the general environment in the past 30 years in China. The Chinese people do not have higher requirements due to the mining-type construction method for refined and humanized building environment; key technologies need to be broken through, and the integration of design-production-construction-management is not enough. The adaptability is low and the molds and components between different projects are not common. The lack of modular design increases the project construction cost; lack of overall planning, single consideration of the design and development of each component or component, ignoring in-depth design, interior decoration design, and collaborative design. It is easy to have a huge impact on professional fields, such as structure, electromechanical decoration, and pipelines. Therefore, development of prefabricated buildings needs to achieve breakthroughs in key technologies on the basis of previous work, promote the application of networked and digital technologies in prefabricated buildings, master the key technical points of prefabricated components, and ensure that each process is completed. Smooth implementation to meet the specification requirements of the construction project.

2.2 *BIM technology*

Building Information Modeling is abbreviated as BIM, which is translated into Building Information Model in Chinese. The core of BIM is to use digital technology to complete the integration of data and information models of construction projects by establishing virtual three-dimensional models of construction projects and enabling models to be transmitted and shared with various links or processes in the entire life cycle of construction projects to facilitate the engineering and technical personnel and project managers to make accurate understanding and efficient processing of building information. At the same time, it also provides a platform for the exchange and sharing of engineering information among all participants in the construction project, thus laying the foundation for the realization of collaborative work in the construction project. As shown in Figure 2, the digital expression of the physical and functional characteristics of a construction project by BIM technology covers all stages of the construction project and runs through the entire life cycle of the construction project. As a result, BIM technology can be viewed as a comprehensive solution for realizing spatial digital informatization of construction projects, with data information serving as the core, an application model serving as the result, collaborative operation serving as the focal point, and a software program serving as the tool. BIM technology has entered a new stage (Zhang 2020).

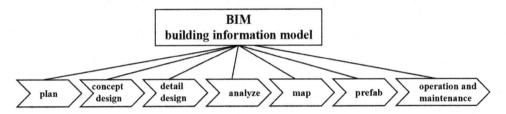

Figure 2.   Features of BIM software.

The digital simulation of construction projects by BIM technology relies on three-dimensional digital technology. The content of a digital simulation is not only the building height, floor area, and external shape of the construction project, but also includes a large amount of auxiliary information, such as material strength, performance, security level, purchasing information, etc. In the actual operation process, 3D digital technology needs the support of software tools to build digital information models for successful engineering construction. The commonly used BIM modeling tools include the modeling software and related functional modules of Autodesk, Bentley, Nernetschek Graphisoft, and Gery Technology Dassault, as shown in Figure 3. Among them, Autodesk Revit series software is commonly used, which is also the BIM model in China's construction industry. One of the most widely used software systems in the world is Autodesk Revit is an application that combines the capabilities of Autodesk Revit Architecture, Autodesk Revit MEP, and Autodesk Revit Structure software. After relying on the modeling software to establish the corresponding BIM model of the building structure, the BIM data model is analyzed and calculated in combination with the corresponding structural quality simulation analysis software. The commonly used simulation analysis software include PKPM, IES, Echotect, and Green Building Studio.

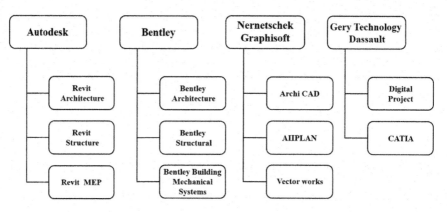

Figure 3.   The most commonly used modeling software in BIM software.

Through BIM technology, a comprehensive database based on the three-dimensional digital information model can be constructed to realize the dynamic modification, adjustment, and storage of all data information for the construction project, thus realizing the coordination between the whole and each part of the engineering design process. Operation 3D digital information model under BIM technology can also help in the construction management stage of the project in the form of visualization. For example, through a large number of collision experiments, unreasonable places in the design can be quickly found and modified, to reduce the probability of safety accidents in actual construction and the probability of rework. Through the construction of the BIM digital information model, it is possible to make a statistical information table of the engineering quantity, the required quantity of each component, and the construction progress during the construction process, to further refine the control of the construction process of the construction project and realize the real environmental protection and energy-saving green construction, to avoid unnecessary waste of resources.

3   APPLICATION OF BIM TECHNOLOGY IN PREFABRICATED BUILDINGS

Prefabricated buildings have introduced new changes and developments in the construction industry, which not only significantly improve the construction efficiency but also promote the integration of the production of accessories into the construction industry (Qu 2019). However, prefabricated buildings also have problems and defects that need to be solved urgently. Prefabricated building technology dates much earlier than that of BIM technology, and the BIM technology, with its application characteristics of visibility, simulation, and coordination, has produced a strong degree of fit between the two, which can be combined with prefabricated building technology. The organic blending of construction technologies not only fulfills the complementing benefits of the two, but also breathes fresh life into prefabricated structures and even the whole construction sector, becoming an essential future growth trend.

3.1 The application of BIM technology in the design stage

In the planning and design stage of the prefabricated building, BIM technology is used to complete the architectural design. Compared with traditional CAD drawing, BIM software combines different data and complete simulation and quantification to form a three-dimensional digital model, which can provide designers with a detailed and scientific basis to complete the relevant design of construction projects. It can not only integrate multiple types of design drawings to improve design efficiency but also avoid subsequent repairs or repeated construction due to design errors during the project operation process and ensure design quality while controlling project costs. The BIM technology design process of prefabricated buildings is shown in Figure 4.

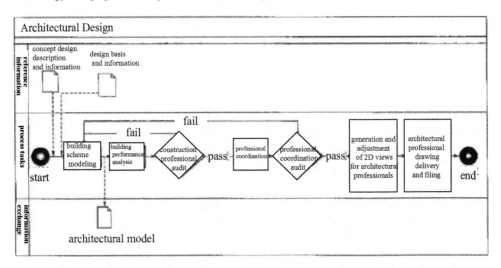

Figure 4. Architectural design phases based on BIM technology.

3.1.1 Application of BIM technology in the early planning stage

In the preliminary planning stage, after the data model of the overall building is constructed using BIM software, the corresponding analysis software is used to complete the relevant analysis of the construction project site in combination with the local environment, climate, and hydrological data. For example, in the planning and construction process of People's Hospital in a city along the southern coast, the solar and climate data of the place are collected with the help of relevant software, and based on the BIM model data, the relevant analysis software are used to conduct climate analysis, and the plan is then carried out, which include environmental impact assessment, comprising the impact of sunshine, wind, thermal environment, and acoustic environment (You 2019). Figure 5 shows the architectural model of the hospital building project, and Figure 6 shows the wind environment simulation of the main building of the hospital, which realizes the simulation and prediction of the normal distribution of indoor and outdoor airflow. The left side is for indoor ventilation, and the right side is for outdoor ventilation. According to various data, the building form and main structural system are optimized and adjusted in different aspects so that the architectural design scheme can achieve the best standard.

3.1.2 Application of BIM technology in the detailed design stage

After the prefabricated building design scheme is determined, it enters the deepening design stage. At this stage, BIM technology focuses on the processing of each node in the structural model—that is, to generate a component model that contains the data information attributes required for specific manufacturing. The detailed design of the component model is the key to the design of the prefabricated building. Professional designers collect design demand information, such as

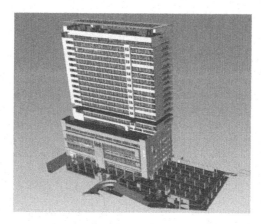

Figure 5. The main building model of the hospital.

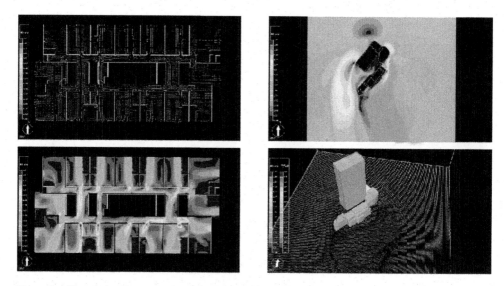

Figure 6. Site wind environment simulation test.

structure, water supply and drainage, HVAC, and power supply, and rely on the BIM technology platform to integrate the data and information to complete the solid design of the components. Compared with the plane drawing of CAD, BIM technology can complete the reinforcement work of components more intuitively, effectively reduce the error rate, and realize control of construction project cost. In addition, the three-dimensional model perspective under BIM technology supports the three-dimensional view of the construction, realizes the systematic design management of components, and improves the design completion and accuracy of prefabricated components.

3.1.3 Application of BIM technology in collaborative design stage

The design of prefabricated buildings is a comprehensive task that integrates multiple departments and disciplines. Various professionals in different departments need to carry out their work based on design drawings (models) and documents. This has resulted in a strong dependence between departments and personnel. On the one hand, various departments cooperate with each other to complete the design of the overall architectural plan; on the other hand, any modification in any department affects all departments, making the problem more complex. The unified BIM platform built under the BIM technology ensures information exchange among various departments and majors. Using BIM technology and a BIM server, collaborative design is carried out on the same BIM model. Through collaborative design and visual analysis, it can be timely. Solving the inconsistency in the above design ensures smooth progress of the later construction (Yang 2016). In addition, in the collaborative design stage, the key aspect of the application of BIM technology is collision detection. That is, the contradictions and errors existing in the design of various departments and majors are automatically marked to facilitate the adjustment and modification of designers. After importing the BIM model into the collision-checking software, Navisworks, for analysis, it is discovered that there is a cross-connection between the water heating pipes and the air conditioning ducts in the hospital's main building, as shown in Figure 7, and the system will automatically identify and mark them to form the collision check result.

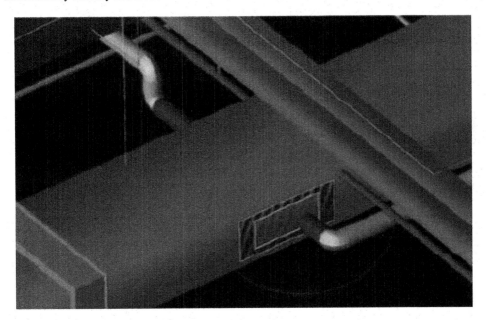

Figure 7. Display of the collision check result.

3.2 *Application in green construction*

Green construction is the general guiding concept that involves using prefabricated structures; however, in the actual construction process, issues such as insufficient construction management and substantial resource waste frequently have a direct impact on prefabricated building green construction evaluation outcomes. The aforesaid challenges may be successfully solved by applying BIM technology in prefabricated structures, and the needs of green construction can be really met.

3.2.1 *BIM technology to solve the problem of resource and energy waste*
First, in the construction stage of prefabricated buildings, construction managers can rely on BIM technology to generate a bill of quantities to achieve control over building resources and energy utilization. After the integration of the data and information models of the construction project, the attribute parameters of each component are automatically recorded by the BIM software system, and the corresponding family comprehensive database is automatically generated, and the shared parameters are established to generate the bill of quantities. Modifications and changes in the architectural design automatically change the bill of quantities. Second, it can clarify the details

and actual application of the materials required in each link of the construction and realizes the material consumption statistics according to the material list, to reduce the waste caused by not picking up materials as required. Finally, the application of BIM technology can also realize cost control in the construction process of prefabricated buildings. The established digital information model is combined with construction time, construction procedures, and other factors to form a multi-dimensional comprehensive information database, so that different departments and different professionals can update and collect the progress and cost in real time according to the actual project progress. Combined with the summary analysis function of BIM software, the cost analysis needs of the project can be met, and the summary and analysis of costs can be more convenient and rapid (Yuan 2021).

3.2.2 *BIM technology to solve the problem of comprehensive construction management*

The comprehensive management of the prefabricated building construction process ensures smooth completion of the construction project. Through the overall planning of construction projects, we can try our best to be environmentally friendly and finely control our progress to achieve the purpose of green construction of prefabricated buildings. The application of BIM technology can not only complete the design scheme of the prefabricated building, but also complete the simulation of the construction scheme and optimize the design of the construction scheme through continuous adjustments and modifications. Hence, BIM technology can add the construction schedule plan to the digital model to dynamically analyze the construction process and simulate the site conditions, as well as investigate potential problems in advance, arrange the construction site reasonably, do a good job of dispatching equipment and personnel, and ensure adequate construction safety measures (Tian 2017). For example, in the planning and construction process of the People's Hospital in a city along the southern coast above, after using BIM software to complete the architectural design, Aotodesk Naviswork was used to add the specific construction schedule to the digital model to form a 4D construction model, and using BIM to carry out construction simulation to enhance the construction management and control of the main building of the People's Hospital as a whole. As shown in Table 3, the grid plan for the construction progress of the main building of the People's Hospital can intuitively reflect the node relationship and the duration of each process in

Table 3. Plans for the construction of the main building of the people's hospital.

	Construction details (day)							
	1	2	3	4	5	6	7	8
I construction section wall hoisting	–							
II construction section wall hoisting		–						
I construction end sleeve simple grouting		–						
II Construction end sleeve grouting			–					
I construction section wall hoisting			–					
IV construction section wall hoisting				–				
Sleeve grouting in construction section II				–				
Sleeve grouting in construction section IV					–			
Post-casting with steel bar binding	–	–	–	–				
Post-casting with support formwork reinforcement		–	–	–	–			
Support frame erection		–	–	–	–			
Stacked beam hoisting					-			
Laminated board hoisting								
Pre-buried water and electricity pipelines						–	–	
Reinforcement of hanging formwork and anti-sill support Formwork at the drop plate							–	
Concrete pouring								–
Stair hoisting					–			

the construction process, further improving the control of the construction progress. In addition, it is also possible to find out possible problems in the construction through simulation and propose corresponding feasible adjustment plans to avoid potential safety hazards in time and increase the safety of project construction.

4 CONCLUSION

This paper proposes a comprehensive solution to the problems of "high energy consumption, high pollution, high waste and low efficiency" existing in the traditional construction mode of the current construction industry. Supported by BIM technology, the "integrated" construction method with prefabricated building technology as the core realizes digital innovation in all aspects of construction project planning, design, construction, management, etc., with the management covering the entire life cycle of construction projects as the foundation, focusing on improving the operation efficiency in the planning and design stage and improving the detail control in the construction stage, to achieve the standards of green design and green construction. The combination of BIM technology and prefabricated buildings not only realizes the standardization, systematization, and industrialization transformation and upgrading of the traditional construction industry, but also signals that the final development of the construction industry moves toward a green and low-carbon direction.

REFERENCES

Gao Fan. (2020). During the "14th Five-Year Plan" period, the transformation and upgrading of traditional buildings to smart buildings. Informatization of China Construction.

He Qiang. (2019). Research on new-type prefabricated building construction technology based on PC components. Construction & Design for Project.

Liao Liping.(2019). Development status and strategies of green prefabricated buildings. Enterprise Economy.

Qu Lipeng. (2019). Application exploration of BIM technology in prefabricated building design and construction management. China University of Mining and Technology.

Tian Dongfang. (2017). Research on the application of BIM technology in the construction management of prefabricated housing. Hubei University of Technology.

Wen Linfeng, Liu Meixia. (2020). Actively promote prefabricated buildings and promote high-quality development of the construction industry. Construction Science and Technology.

Yang Qilin. (2016). Research on the application of visual collaborative design based on BIM. Southwest Jiaotong University.

You Qian. (2019). On the role of BIM in various stages of the project. Jushe.

Yu Jing. (2021). Independent BIM technology promotes the transformation, upgrading and high-quality development of the construction industry. Informatization of China Construction.

Yuan Yuan. (2021). Application analysis of BIM in the design of prefabricated residential buildings for green energy saving. Building Technology Development.

Zhang Juxian. (2020). Research on the application of BIM technology in the whole life cycle of prefabricated buildings. Chongqing Architecture.

Frontiers of Civil Engineering and Disaster Prevention and Control – Yang & Rahman (Eds)
© 2023 The Author(s), ISBN: 978-1-032-31200-2

Research on ventilation ecological building design under regional influence

Zhang Yin* & Zhou Qi
School of Art and Design, Wuhan Institute of Technology, Wuhan, Hubei, China

ABSTRACT: Natural ventilation is one of the most commonly used methods in ecological build-ing design. This paper cites several typical cases to illustrate the application of natural ventilation for green buildings in various regions, and finally proposes several measures for natural ventila-tion design. Its design process should be combined with geographical factors, humanistic history, economic conditions, etc.

1 INTRODUCTION

A building has a long service life because it is a consumable item built with a lot of of time and effort. Therefore, when planning a construction project, a rigorous design must be planned, considering the long-term problems that could arise in the future, with ecological concepts incorporated. Buildings used to be regarded as being energy intensive. The designs of diversified energy-saving devices have gradually gained people's attention (Wang 2020). During the 19th National Congress of the Communist Party of China, new requirements were laid out for constructing ecological civilizations, as well as the timetable and roadmap for building a beautiful China in the new era. To realize ecological civilization and build new-age China, it is necessary to establish the fundamental concept of harmonious coexistence between man and nature, adhere to strict environmental protection policies, and accelerate the reform of the ecological civilization system (Wang 2020). This indicates that the ecological problems in architecture industry are closely connected to the political, economic, and other issues that contribute to the development of a nation.

The ecological theme of buildings involves addressing the problem of building energy con-sumption. Buildings have diverse styles based on geographical conditions, but their core functions largely remain the same. Given this context, it is critical to explore the concepts and methods of building and natural ventilation design. It is envisaged that the findings of this study would inspire the construction of natural ventilation ecological buildings.

2 THE NEED FOR NATURAL VENTILATION

High-rise buildings thrive as cities grow, but some architects seek high-rise building designs with particular aesthetics, and natural ventilation has been a vital technological method for the betterment of people and the environment since long. Many ancient Chinese structures, such as ventilation, inner courtyards, and quadrangle courtyards, reflect its shadow (Li 2006). Natural ventilation, as a measure to minimize indoor heat, may accelerate the circulation of air within the building and disperse the temperature of the human body surface, thereby cooling the inside of the building at an era when electric power technology was not completely utilized.

---

*Corresponding Author: 809895605@qq.com

                                                      DOI 10.1201/9781003308577-3

The basic principle of today's existing advanced ecological building strategies is to cut the building from the external space, and use advanced technology and equipment to solve the problem of building energy consumption. However, this so-called ecological strategy increases the engineering cost at the beginning of construction. The cost would be high, and if this building is not used for a long time in the future, this strategy will be "putting the cart before the horse".

Compared with other ecological technologies, natural ventilation is a low-cost and relatively mature technology system. Natural ventilation is a technology that exchanges interior and outdoor air by depending on natural air pressure or temperature differences rather than mechanical energy (Chen & Ye 2021). The purpose of achieving ecological protection with the least number of resources applies to the ecological environmental protection concept of the national policy and is in line with the national sustainable development strategy. Under the national conditions, under the "ecological environment" policy, the issue of architecture and ecology has been given much attention by architects, and the ecological architecture strategy utilizing natural ventilation is widely used by designers.

3 APPLICATION CASE OF VENTILATION BUILDING UNDER REGIONAL INFLUENCE

3.1 *Arizona state university*

3.1.1 *Background factor research*

Arizona State University (ASU) is located in the Phoenix metropolitan area. It is a top research university. The school is highly valued by the government as well as school leaders. Therefore, the teaching building will be used for a long time in the future. A point to consider is the significance of campus strength has prompted an increase in the number of students on campus, which in turn raises important questions about how the campus can develop sustainably in the future (Figure 1).

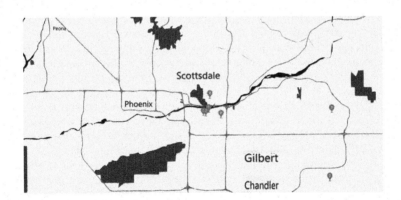

Figure 1.   School location (created by the authors).

Geographical Background Facts: Arizona State University is located in the urban area of Phoenix, which belongs to the tropical desert area. It exhibits a typical subtropical continental arid and semi-arid climate. The rainy season is short, and the most common green plant is cactus. The tropical desert climate is characterized by sunshine, especially strong sunshine, a dry and hot summer, a large temperature difference between day and night, and more wind and sand. There are no abundant clouds in the sky, and the sun shines directly on the ground all year round, which makes part of the sun unstoppable and not very friendly. Because the weather is hot, dry, and with less rainfall, there is no water in the ground, and it is impossible to evaporate and cool down. Therefore, the

21

temperature in summer is particularly high, and the temperature difference between day and night is particularly significant, which poses a potential threat to the living (Figure 2).

Figure 2.   Natural environment (source: network).

3.1.2   *Ventilation and eco-design applications*

Strategy 1: Ecological sustainability is a key design driver throughout the process. Tooker House at Arizona State University is a brand new seven-story 4.58 million sq. ft living and learning facility space for students. The architectural design layout of the teaching building is in the form of "one divided into two". The entire building complex is composed of two irregular rectangular volumes. The connecting bridges between the two buildings connect the circulation of people. So as to ensure the convenience of passage. And the connecting corridor bridge is made of perforated material. This special material can promote the passage of natural wind inside the building and form a "window through the hall". Compared with the traditional closed two-building form, this kind of ventilation corridor ensures that the building will not be damaged. It will block the natural wind flow, reduce the wind resistance of the building, and achieve the effect of allowing the smooth passage of tropical sand.

In addition, the composition of the building volume of the teaching building has been carefully calculated and planned by the designers. The entire building complex faces the east-west direction and is placed in a parallel position. The east-west high walls provide maximum shade for the site and its large openings. It has excellent lighting and powerful ventilation on the south and north sides, so it can also have sufficient sunshine in winter, which is helpful in maintaining the inside temperature (Figure 3).

Strategy 2: The design of the teaching building itself also promotes the movement of wind. People are accustomed to the traditional ventilation method of opening windows. In terms of simple ventilation, one of the buildings makes use of an algorithm to combine with the position of each window on the outer wall to form a unique curvilinear and wavy appearance. Opening and closing with changing weather conditions ensures proper control of sunlight entering the room, thus avoiding considerable energy consumption caused by strong sunlight. On the other hand, the perforated metal panels of this structure can drive the passage of natural wind and promote the improvement of indoor space. Air flow reduces indoor thermal effects, reduces the use of mechanical energy for ventilation and air conditioning, and saves building energy consumption. The materials used in the building walls are all local materials, which are like the roots of a tree deeply rooted in the soil. They are original and regional, giving people a visual sense of security and intimacy.

Affected by natural factors, strong winds will be blocked by the building, and the acceleration of the wind speed will lead to negative pressure in the building. The building adopts the overhead mode of the first floor to effectively relieve this pressure and allows the airflow to pass through the planned route, which is beneficial to the overhead layer. The evacuation of harmful gases from the parking lots from both sides, while providing the function of ventilation and noise reduction for the entire building complex in summer, can improve the impact of the desert climate (see Figure 4).

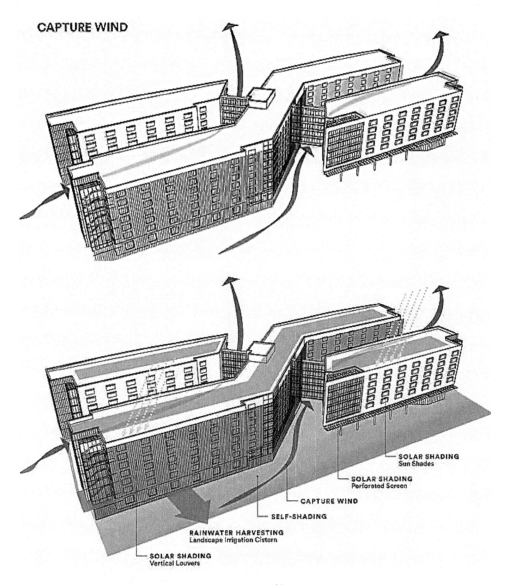

CAPTURE WIND

SOLAR SHADING
Sun Shades

SOLAR SHADING
Perforated Screen

CAPTURE WIND

SELF-SHADING

RAINWATER HARVESTING
Landscape Irrigation Cistern

SOLAR SHADING
Vertical Louvers

Figure 3.   Buildings with two sky bridges (source: network).

3.2   *Star city shopping center, Wuhan, China*

3.2.1   *Background factor research*

Social background factors: Wuhan Qunxingcheng Shopping Center is located on Xudong Street, Wuchang District, Wuhan City, Hubei Province. Wuhan has nine provinces. Wuhan's superior geographical location created a prosperous atmosphere in Wuhan and promoted the development of commercial groups. People have an innate closeness to nature and yearn for a free blue sky and white clouds, mountains, rivers, streams, green trees, and flowers. The designer used this concept into the design of the building. At the same time, based on this group of star city shopping malls, they have rich visual impact (Figure 5).

Figure 4.   A model for building facades (source: network).

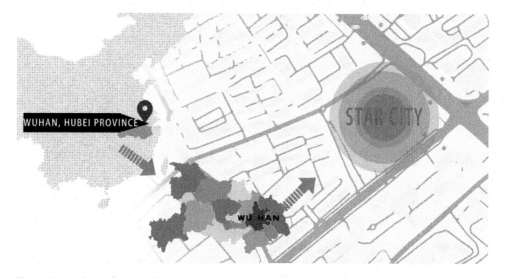

Figure 5.   Analysis of geographic location (created by the authors).

Humid subtropical monsoon (humid) climate with abundant rainfall and sufficient heat in summer, rain and heat in the same season, and light and heat in the same season is typical of Wuhan.

The end of 2019 to the beginning of 2020 was a difficult period for people in facing the challenges of the pandemic. Although the spread of the disease has been effectively prevented and controlled, people are still urged to reduce crowd gatherings and confined spaces. Due to the relatively airtight environment in the building, coupled with various human activities, the air quality in the building is poor. Based on this, major business circles have implemented measures to switch off the central air conditioner to prevent the spread of virus that can aggravate by the circulating air. This goes back to the building itself, whether it can shoulder its mission and play a huge role in the epidemic.

3.2.2   *Ventilation and eco-design applications*
During the initial design stages of the architecture, the idea of nature was taken as the central motif, with the "canyon" as its design form. A "garden-style" shopping center has been created by integrating natural elements with the architectural environment. At the same time, it is important to consider ecological effects.

In the design of the façade, Wuhan's Qunxingcheng Square combines the concept of mountains and canyons to show the rock formation curves of the canyon through wooden textures (Figure 6). The open atrium is designed with reference to the topography of the canyon, creating an external feature of mountains, introducing the stream trail in the canyon to create a vertical "waterfall" landscape of the building, and the layers of plants on the platform are the air. A three-dimensional vertical atrium space designed to supplement oxygen and reduce atmospheric particles is surrounded by green plants and flowing water, creating a pure natural atrium exhaust system.

Figure 6.   External of Qunxingcheng Square (Source: network).

In the ventilation system of the atrium, the natural wind can penetrate into the indoor space of each floor of the building, so that the air is circulated inside the originally airtight building, which is a natural method to realize the conversion of indoor and outdoor air. After the epidemic, natural ventilation is an important protective measure to effectively alleviate the retention and accumulation of indoor bacteria and harmful gases. It may be possible to reduce the virus concentration in the air by implementing protective ventilation measures, thereby minimizing the overall exposure of users to viruses (Figure 7).

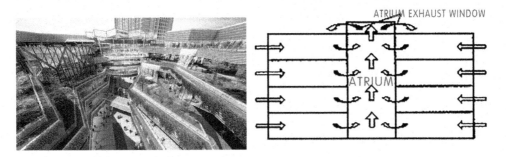

Figure 7.   Atrium space of Qunxingcheng Square (source: network).

3.3   *Chengdu dayi agricultural science base exhibition center*

3.3.1   *Background factor research*

Chengdu Pastoral Resort is home to the Dayi Agricultural Science Base Exhibition Center. Under the background of national rural revitalization, breaking the dual structure of urban and rural areas, and tracing the roots of the city to the countryside, we create a western Sichuan Linpan with pastoral hotels (homestays) as the core and popular science, children, theme parks, and health care cultural experiences as the content of activities (Xie & Ruan 2020). This pastoral complex, to inherit and carry forward the traditional culture of western Sichuan, continues the architectural style of the famous residences in western Sichuan. It is paved with small green tiles and combined with modern technology to construct a unique Dayi Agricultural Science Base Exhibition Center (Figure 8).

In Chengdu, there is a subtropical humid monsoon climate zone. The days are short throughout the year, and the weather is hot and humid. Chengdu is known as "Rainy City". According to

25

Figure 8.   Dayi Agricultural Science Base Exhibition Center (picture source network).

climate conditions and functional requirements, the ancient historical period in Chengdu used a traditional architectural style of western Sichuan: dry-column construction (Figure 9).

Figure 9.   Dry-column buildings (source: network).

3.3.2   *Ventilation and eco-design applications*

To inherit the national culture, the building of the exhibition center continues with the biggest feature of traditional buildings in western Sichuan: the ground floor is overhead, and its architectural shape is square, in the form of a "thousand-footed house," inheriting the "top to avoid heat and the bottom to avoid moisture" (Song 2020). This type of building increases the flow between the building and the natural wind, acts as thermal and moisture insulation, and can isolate the pests on the ground.

The overall volume of the building is lifted, and the vast space on the first floor becomes the activity space for wind and people. The open space is used for public resting seats and for planting and greening, which fully depicts the effectiveness of the space. The microclimate of the building is further adjusted by utilizing the transpiration of plants. In terms of materials, the exterior of the building is composed of glass curtain walls and steel structure rods, and large glass windows are installed. There are openings on the windward and the windward. There are no obstacles between the indoor openings and the air can flow smoothly. The formation of ventilation is beneficial to the air circulation and lighting inside and outside the building and effectively relieves indoor humidity (see Figure 10).

4   REALIZATION OF NATURAL VENTILATION IN BUILDINGS

Natural ventilation is an important factor affecting the building's design. The benefits of natural ventilation on the building are affected by factors such as structure dislocation, front-to-back

Figure 10.   Void spaces of buildings (source: network).

relationship, location direction, size and number of windows, and material. The space layout should be carefully planned. However, in design, it is necessary to fully combine the local natural climate, environmental conditions, human history, economic conditions, and more, as well as formulate technical measures in line with regional characteristics to ensure natural ventilation with good ecological benefits and can be regarded as a real design.

4.1   *Chengdu dayi agricultural science base exhibition center*

The opening and closing of windows in buildings requires a rigorous knowledge of doors. In Feng Shui, the shape and orientation of windows are related to the five elements. If used properly, it will help strengthen the energy absorption of indoor space. Natural ventilation depends on the location, size, and shape of the windows in the building. The position of the opening affects the distribution of natural wind. The size of the cornice, the method of opening, and the use of materials in the window structure have different impact on the natural wind. Generally, the ventilation is used as the basic air circulation method, but the design aspects of the building should be respected and specific problems should be analyzed in detail with regard to whether it is suitable for the use of hall ventilation (Figure 11).

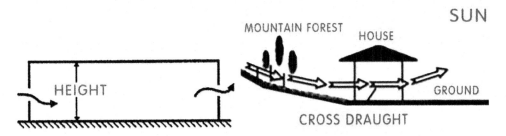

Figure 11.   Schematic diagram of the ventilation in the hall (source: network).

4.2   *Realization of natural ventilation in building facades*

Reasonable building layout is one of the important ways in achieving natural ventilation. When designing the building scheme, the orientation of the building group and the structure between the buildings should be considered. Different structural composition methods have different blocking factors for wind force. Generally, the "bridge-connected" building composition method is adopted

27

to ensure that the building does not block the circulation of natural wind and reduce the building's pressure.

The building layout can be combined with large spaces, such as atriums and stairwells, to adjust the indoor air. For example, the Villa Savoye, designed by Le Corbusier, and the dry-style buildings at the Hemudu site in Yuyao all adopt the form of building facades on the ground floor. This structure is both artistic and practical and can effectively promote natural wind flow (Figure 12).

(a) The Villa Savoy   (b) Schematic diagram of a dry building

Figure 12.   (source: network).

In addition, in the atrium space of a large building, the vertical air density difference should be considered to design the atrium ventilation of the building. For example, the Qunxingcheng shopping mall in this article mainly strengthens the air circulation through the difference in densities between the high-temperature air and the low-temperature air. The air temperature in the atrium or the airshaft is higher. The small airflow tends to increase, and the air with a lower ambient temperature tends to decline due to high density, thus forming an up and down air circulation, which is the so-called atrium wind (Figure 13).

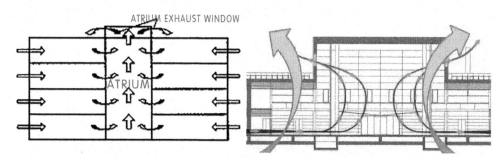

Figure 13.   Ventilation in atrium building (source: network).

4.3   *The Realization of natural ventilation in other aspects of the building*

Natural ventilation is one of the key components of building ventilation. The roof is not only used as the top of the building, which acts as a thermal insulation and rain shelter, but can also be used as an independent part to promote natural ventilation of the building. For example, setting up lighting and ventilation, skylights, and roof insulation are the most common methods in modern technology. When the building cannot make use of the natural wind, some special materials can be comprehensively considered for realization of energy-saving ventilation, such as double-layer glass curtain walls, perforated metal plates, perforated aluminum plates, etc., as shown in Figure 14.

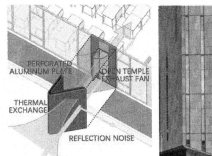

Figure 14.   Ventilation material (source: network).

5   CONCLUSIONS

Urbanization has resulted in an increase in the number of high-rise buildings. However, some architects use the construction of high-rise buildings to develop their art and skills, continually destroying the original appearance of a city that is both unconventional and eye-catching (Wang et al. 2021). As a result, the relationship between the ecological benefit and the spatial environment of the building is gradually ignored. Based on this background, it is particularly important to trace the origin and study the relationship between architecture, nature, and people.

The development of the digital age has brought about many technologies to achieve ecological benefits, and there are many methods and means to implement ecological buildings. However, while natural ventilation is beneficial to human health, it also focuses on the concept of "energy-saving and environmental protection" and other ecological technologies. In contrast, natural ventilation is a low-cost and relatively mature technical system that helps save energy with fewer labor and materials. Natural ventilation systems play an essential role in energy savings and environmental protection in buildings, and are long-lasting.

To attain a more effective design, it is necessary to comprehensively consider the regional features of a building, its surroundings, and its own characteristics, as well as adapt design strategies appropriate to the actual design requirements. We must exhibit respect for nature to optimize the performance of natural ventilation in architectural design so that regional environmental belts can be alleviated. Natural ventilation serves the needs of people, allowing them to live in harmony with nature.

REFERENCES

Chen Jing, Ye Xin. Research on natural ventilation mode of Hong Kong public buildings [J]. Interior Design and Decoration, 2021(02): 118–119.

Li Min, Yang Zugui. Discussion on natural ventilation design in ecological architecture [J]. Sichuan Architecture, 2006(04): 29–30.

Song Zuomei. Architectural design of dry-column dwellings [J]. Industrial Architecture, 2020, 50(12): 230.

Wang Hui. Review and prospect of China's ecological civilization construction since the 18th National Congress of the Communist Party [J]. Comparative Research on Cultural Innovation, 2020, 4(12): 136–137.

Wang Yang, Chen Yongwen, Xu Ke. Research on the regional design mode of contemporary high-rise building complexes [J]. Architecture and Chemistry, 2021(05): 249–251.

Wang Zijia. Analysis of Building Energy Efficient Ventilation Design [J]. China Building Metal Structure, 2020(12): 86–87.

Xie Xiaolu, Ruan Guoshi. The modern translation of the architectural style characteristics of traditional residential houses in western Sichuan—the exhibition center of Dayi agricultural science base [J]. Architecture and Culture, 2020(04): 261–262.

*Frontiers of Civil Engineering and Disaster Prevention and
Control – Yang & Rahman (Eds)*
© 2023 The Author(s), ISBN: 978-1-032-31200-2

Interaction between double-track tunnels when slurry shield passes through water-rich sand layer

Yilei Zhang & Jianxun Ma*

School of Human Settlements and Civil Engineering, Xi'an Jiaotong University, Xian, China

ABSTRACT: Taking the phase III project of Xi'an Metro Line 1 as the research background, the finite element model of single-track tunnel construction and double-track tunnel construction at the same time is established, and the relationship between the vertical displacement of soil and segment and the internal force of segment is obtained. The results show that: 1) when the double-track tunnel is constructed at the same time, the maximum vertical displacement of soil and segments is less than that of the single-track tunnel; 2) when the double-track tunnel is constructed at the same time, the internal force of the segment is greater than that of the single-track tunnel.

1 INTRODUCTION

The main features of slurry balance shield construction are that a diaphragm is set behind the cutterhead at the front end of the shield, a slurry pressure tank is formed between the cutterhead and the diaphragm, and the pressurized slurry is pumped to the slurry tank. When the mud tank is filled with mud at a certain pressure, a mud film will be formed on the excavation surface. The pressure of the mud tank acts on the soil on the excavation surface through the mud film, so as to maintain the stability of the excavation surface. Mud film formation and shield tunneling form a dynamic balance process. The cutterhead installed on the front of the shield rotates continuously to cut the soil containing mud film, and the mud film will be quickly formed on the new excavation surface. After the soil cut by the cutterhead is mixed with the mud water in the mud sump, it will be stirred by the mixing device to form a high-concentration mud water that will be transported to the ground by the mud pump and mud discharge pipeline.

Whether in the excavation stage or segment assembly stage, there is always a layer of mud film in order to ensure the stability of the excavation surface. When the mud film is cut by the cutter head, a new mud film will be formed rapidly and repeated again and again, so as to achieve the effect of maintaining the stability of the excavation surface. The mud film has good support effect on the excavation surface and little impact on the environment, which makes the mud water balance shield show strong vitality in various tunnel construction methods. At present, the research methods for shield construction stability mainly include the following: theoretical calculation method, numerical simulation method, and indoor model test method.

Theoretical calculation methods mainly include the limit analysis method and the limit equilibrium method. The limit analysis method is based on the upper and lower limit theorems of the extreme value theorem, also known as the upper and lower limit solution. The key to finding the upper bound solution is to determine the velocity field, including the assumed sliding surface. Usually, the surface passing through a point is selected as the possible sliding surface to obtain the approximate solution of the surface failure mode (Murakami 1995). The key to solving the lower bound solution is to build a stress field that includes the assumed sliding surface in order to stabilize the soil. The required ultimate load can be obtained directly by using the stress field. The limit equilibrium method considers that the failure of soil occurs on the sliding surface, and the soil meets the

---

*Corresponding Author

                                            DOI 10.1201/9781003308577-4

yield condition along the sliding surface. Assuming that the form of the sliding surface is known, such as a plane, circular arc surface, logarithmic spiral surface, or other irregular surfaces, the ultimate load or safety factor when the soil along the sliding surface is in the limit state is determined by considering the static balance and moment balance of each isolator in the sliding body.

The numerical simulation method has been widely used in shield construction. Wang Minqiang et al. (Wang & Chen 2002) used three-dimensional nonlinear finite elements to simulate the shield propulsion process and proposed a calculation model that can be used to investigate excavation surface stability. Xu Ming (Xu et al. 2012) did a three-dimensional numerical analysis on the stability of the excavation surface of the super large diameter shield in the sandy soil layer, focusing on the instability mode and ultimate support pressure of the excavation surface under different friction angles, and revealed the importance of the uneven distribution of support pressure caused by the unit weight of mud and water for the prediction results.

The indoor model test method is a research method carried out in the laboratory by reducing the actual project according to a certain proportion. It can more truly restore the actual situation of the site and predict the possible problems on the site in advance.

Taking the phase III project of Xi'an Metro Line 1 as the research object, this paper establishes a finite element model through Midas GTS NX to study the variation law of vertical displacement and internal force of soil and segments during the construction of single-track tunnels and double-track tunnels.

2 ENGINEERING BACKGROUND

The project is located in the middle of the alluvial plain of the Weihe River. The line crosses the floodplain of the Weihe River between the first and second stage areas. The Weihe River is a perennially flowing river. The water volume is greatly affected by seasonality. It is the largest tributary of the Yellow River. The geological conditions are mainly sandy soil with strong permeability. During shield construction, the diameter of the cutterhead is 6.28 m, the diameter of the segment is 6 m, and the length is 1.5 m. Every two segments are a construction step, and the buried depth of the tunnel is about 29 m.

3 FINITE ELEMENT SIMULATION

The finite element model consistent with the actual construction is established by using Midas GTS NX software. The length, width, and height of the soil in the model are 426 m, 81 m, and 75 m, respectively, to ensure the accuracy of the results. According to the geological survey report, the soil layers are miscellaneous fill, loess, sandy soil, and silty clay from top to bottom. The modified Moore-Coulomb model is adopted. See Table 1 for specific parameters. The shield shell is made

Table 1. Soil parameters.

Soil name	Poisson's ratio	Bulk density (g/cm^3)	Elastic modulus (MPa)	Cohesion (kPa)	Friction angle (°)	Dry density (g/cm^3)
Loess	0.30	1.85	8.70	33.00	25.30	1.53
Medium sand 1	0.35	2.20	30.00	0.00	34.50	1.77
Silty clay 1	0.30	2.00	7.00	30.80	18.20	1.64
Medium sand 2	0.25	2.00	30.00	5.60	32.87	1.74
Silty clay 2	0.29	2.02	7.00	47.00	27.30	1.67
Fine sand	0.28	2.00	20.00	0.00	30.48	1.74
Artificial fill	0.33	1.90	8.00	33.00	25.30	1.53

of Q345 steel, the segment and grouting layer are made of C50 concrete, the top surface of the model is a free surface, the bottom surface is a fixed surface, the normal displacement is restrained, and the water depth is set to 5 m. The activation and passivation of soil and segment are used to simulate the dynamic construction. The finite element model is shown in Figure 1.

Figure 1.   Finite element model.

4   NUMERICAL SIMULATION RESULTS

4.1   *Model reliability verification*

In order to verify the reliability of the finite element model, 8 points are taken every 3 m from the tunnel axis to both sides, and the displacement results are compared with the calculation results of Peck formula, as shown in Figure 2.

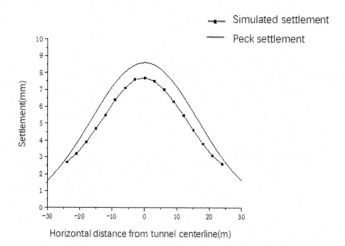

Figure 2.   Comparison between numerical simulation results and calculation results.

As can be seen from the above figure, the simulation results and calculation results show a normal distribution, and the difference between the values is very small, which can verify the reliability of the model.

4.2 Influence of double-track tunnel construction on vertical displacement of soil and segments

In the process of shield construction, the vertical displacement of soil and segments during the simultaneous construction of two tunnels is different from that of a single tunnel. It is of great significance to study the interaction of double-track tunnels to select appropriate construction parameters to guide the construction.

Take a node every 30m on the model as the research object, and compare the change of vertical displacement during the construction of a single-track tunnel and a double-track tunnel at the same time. The change of soil vertical displacement is shown in Figure 3, and the change of segment vertical displacement is shown in Figure 4. It can be seen that the maximum vertical displacement of soil and segments during the construction of a double-track tunnel is less than that of a single-track tunnel.

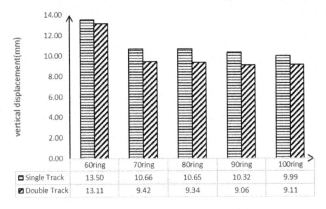

Figure 3.   Relationship between the number of tunnels and the vertical displacement of soil mass.

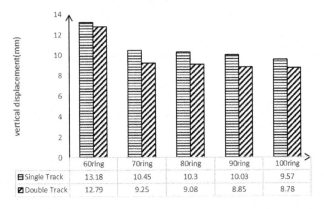

Figure 4.   Relationship between the number of tunnels and the vertical displacement of segments.

4.3 Influence of double-track tunnel construction on segment internal force

Take segments every 30 m on the model, and take the segment apex as the research object to analyze the changes in internal force during single-track construction and double-track construction. The results are shown in Figure 5. It can be seen from the figure that the internal force of the segment during the construction of a double-track tunnel is greater than that of a single-track tunnel.

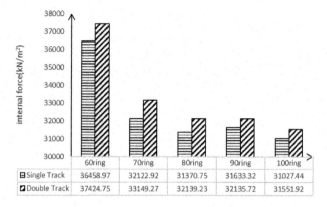

Figure 5. Relationship between the number of tunnels and segment internal force.

4.4 *Comparison between numerical simulation results and field monitoring data*

The statistics of the numerical simulation results of riverbed settlement are shown in Figure 6, and the field monitoring data are shown in Table 2.

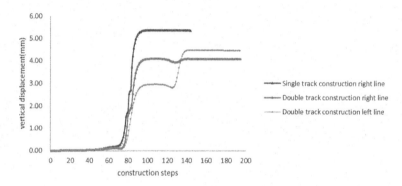

Figure 6. Riverbed settlement value.

Table 2. Settlement value of monitoring results.

	Monitoring point	Settlement value (mm)
Right line	2	4.87
	4	5.00
	6	5.40
	8	1.68
	10	3.95
	12	3.67
Left line	1	4.84
	3	3.10
	5	1.46
	7	2.37
	90	1.95
	11	4.38

The monitoring data shows that the maximum settlement value of the riverbed above the tunnel during single-track construction is about 5 mm. During double-track construction, the maximum settlement value of the right line is about 4 mm and the maximum settlement value of the left line is about 4.5 mm, which is consistent with the simulation results.

5 CONCLUSIONS

In this paper, a finite element model consistent with the actual project is established by Midas GTS NX to study the vertical displacement of soil and segments and the change of segment internal force during the construction of single-track tunnels and double-track tunnels when the slurry shield passes through the water-rich sand layer. The conclusions are as follows:

(1) During the construction of a double-track tunnel, the maximum vertical displacement of soil and segments is less than that of a single-track tunnel.
(2) During the construction of a double-track tunnel, the internal force of each segment is greater than that of a single-track tunnel.

6 OUTLOOK

This paper mainly studies the influence of a double-track tunnel on the vertical displacement of soil and segment and the internal force of the segment when a slurry shield passes through a water-rich sand layer. There are still some problems that can be further improved in the follow-up research:

(1) In the highly permeable stratum, the seepage force of water will have a certain impact on the results. For technical and time reasons, the seepage effect is not considered in this paper. In future research, the seepage conditions can be added to the finite element model for analysis.
(2) Surface settlement is influenced by the hardening time of the grouting layer. In this paper, we focus only on the onset and completion of hardening, and future research can be conducted to examine the hardening time of the grouting layer.

REFERENCES

M H, Murakami E. Stability and failure mechanisms of a tunnel face with a shallow depth: proceedings of the Proceedings of the 8th Congress of the International Society for Rock Mechanics, F, 1995 [C]. International Society for Rock Mechanics.
Wang Minqiang, Chen Shenghong. Three-dimensional nonlinear finite element simulation of shield driven tunnel structure [J]. Journal of rock mechanics and engineering, 2002(02): 228–232.
Xu Ming, Zou Wenhao, Liu Yao. Stability analysis of excavation face of super large diameter slurry shield in sand [J]. Journal of civil engineering, 2012, 45(03).

Frontiers of Civil Engineering and Disaster Prevention and
Control – Yang & Rahman (Eds)
© 2023 The Author(s), ISBN: 978-1-032-31200-2

Seismic reliability analysis of pipeline in underground powerhouse of Baihetan hydropower station

Jie Fang* & Shengbing Li
Huadong Engineering Corporation Limited, Hangzhou, Zhejiang, China

Baoshan Zhu, Zhigang Zuo & Shuhong Liu
State Key Laboratory of Hydro Science and Engineering, Department of Energy and Power Engineering, Tsinghua University, Beijing, China

Chunjian Cao
Huadong Engineering Corporation Limited, Hangzhou, Zhejiang, China

ABSTRACT: In this paper, a simplified calculation model of the underground workshop pipeline system has been established to study the seismic reliability of the pipeline system in the underground powerhouse of Baihetan Hydropower Station. The structural dimensions and material properties of the pipeline and bracket pillar are treated as random variables with a Gaussian distribution. A pipeline sampling method is used to analyze the pipeline system. The seismic reliability of the pipeline system under actual conditions has been proven. Sensitivity analysis shows that the material density and elastic modulus of the pipeline are the most important factors affecting the seismic reliability of the pipeline system. This study provides theoretical support for seismic reliability and structural optimization for similar pipeline systems.

1 INTRODUCTION

Baihetan Hydropower Station is located in Ningnan County, Sichuan Province and Qiaojia County, Yunnan Province in the lower reaches of the Jinsha River. It is the second-stage hydropower station for cascade development in the lower reaches of the Jinsha River, with Wudongde Hydropower Station in the upper part and Xiluodu Hydropower Station in the lower part. The controlled basin area is 430,300 km^2, accounting for 91.0% of the Jinsha River basin area. The main tasks of power station development are to generate electricity, consider flood control and shipping, and promote local economic and social development (Dai et al. 2006; Fan et al. 2019; Ma et al. 2020). Because the hydropower station is located in an earthquake-prone area of China, it is particularly important to analyze the seismic reliability of each pipeline system in the underground powerhouse of the hydropower station. Many papers (Ai & Li 2007; Dong et al. 2008; Gazis & Nikolaos 2011; Han & Jame, 2014; Wang & Hong 1993; Wang et al. 2005, 2019; Zhang et al. 2010) have studied the dynamic characteristics and seismic response of submarine pipelines and buried pipelines based on numerical or experimental methods. Chen (Chen 1996) has studied the reliability of above-ground pipelines under earthquakes. However, the research on the reliability of the above-ground suspended pipeline under an earthquake is rarely seen in the open literature. Based on this, this paper adopts the PDS module in ANSYS finite element software, which is widely used to preliminarily study the safety and reliability of the suspended pipeline of the underground powerhouse of Baihetan Hydropower Station under an earthquake.

The ANSYS-PDS module combines finite element technology and probability design theory (Ye et al. 2004; Qin 2006). By setting random variables that conform to certain distribution rules as

---

*Corresponding Author: fang_j@hdec.com

                                                      DOI 10.1201/9781003308577-5

input parameters, after limited cycle sampling, it cannot only get more accurate failure probability values to evaluate customers' design satisfaction with product quality and reliability, but also get the sensitivity analysis results of each input parameter to output parameters, so as to properly adjust and control each input parameter according to the sensitivity degree, and achieve the purpose of improving product safety and reliability. He Shuanghua et al. (2012) applied the ANSYS-PDS module to earthquake damage prediction models and random reliability analyses of underground pipelines, which provided a scientific basis for optimization of underground pipelines. Li Wenzhen et al. (Li et al. 2018) used the ANSYS-PDS function to analyze the reliability of a pressure vessel and obtained the sensitivity of design parameters such as stress probability distribution characteristics, pressure load, and wall thickness to stress distribution. Zhu Yuankun and others (Zhu et al. 2017) applied the ANSYS-PDS module to analyze the reliability of the tubing hanger structure, which provided theoretical support for the reliability and structural optimization of other underwater complex structures such as the tubing hanger. Zhang Aihua et al. (Zhang & Ren 2010) applied ANSYS-PDS module design to analyze the anti-resonance reliability of a high-speed motorized spindle. Rong Zhixiang and others (Rong & Lin 2011) applied the ANSYS-PDS module to analyze the reliability of the connecting rod.

In this paper, the reliability of suspended pipeline systems under earthquakes is mainly analyzed, and the reliability analysis model under normal service limit state is established. Through the Monte Carlo method and Latin Hypercube Sampling technology (Gao et al. 2008; Wang et al. 2006; Zhao et al. 2015), the related design parameters of the pipeline system (pipeline and support column) are randomly sampled, and the seismic reliability analysis of the pipeline system is completed on the basis of finite element analysis.

2 NUMERICAL ANALYSIS METHOD AND SIMULATION MODEL

2.1 Reliability related theory

The reliability of the structure refers to the probability that the structure will complete the predetermined function within the specified time and under the specified conditions (normal service limit state and bearing capacity limit state) (Li & Zhou 2001; Pan 2005; Wu 2009; Yu et al. 2017, 2016, 2008; Yan et al. 2007).

The basic variables of the structure consist of X1, X2, ..., Xn, and the structural function Z is a function of the basic variables, then the functional function (limit state function) of the structure can be expressed as:

$$Z = g(X_1, X_2, \cdots X_n) \tag{1}$$

In the probabilistic limit state design theory, the limit state equation is:

$$g(X_1, X_2, \cdots X_n) = 0 \tag{2}$$

Therefore, in the structural design of probability limit state, the following conditions must be met:

$$Z = g(R, S) = R - S \geq 0 \tag{3}$$

According to the reliability theory, the reliability of a structure is the probability of the limit state function $g(X_1, X_2, \cdots X_n) \geq 0$, so the reliability of the structure can be obtained by using ANSYS probability analysis of the probability calculated by PDS module function.

2.2 Case overview

Select a section of pipeline in the turbine layer of Baihetan underground powerhouse, with a length of L = 12.525 m. Poisson's ratio γ of pipe material is 0.3; Poisson's ratio γ of support column material is 0.3; initial thickness r of pipe wall is r = 5 mm; pipe material is stainless steel; yield strength is 205 Mpa; support column material is ordinary carbon steel Q235; and yield strength is 235 MPa.

2.3 Finite element analysis model

In the ANSYS finite element model of the pipeline system, the pipeline and the support column are set as rigid connections. All the models adopt Beam188 beam elements, and parametric modeling is carried out by Ansys Parametric Design Language (APDL). The model building process is as follows:

- The displacement constraints, i.e., UX=0, UY=0 and UZ=0, are imposed on the nodes of X =-75 $\sim$ 0, Y = 0 and Z = 0;
- The displacement constraints, i.e., UX=0, UY=0 and UZ=0, are imposed on the nodes with X =-75 $\sim$ 0, Y=4500$\sim$4575 and Z = 0;
- The displacement constraints, i.e., UX=0, UY=0 and UZ=0, are imposed on the nodes with X =-75 $\sim$ 0, Y=7850$\sim$7925 and Z = 0;

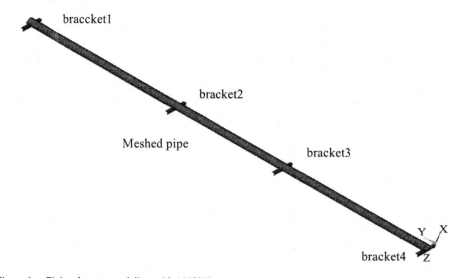

Figure 1. Finite element modeling with ANSYS.

- The displacement constraints, i.e., UX=0, UY=0 and UZ=0, are imposed on the nodes with X =-75 $\sim$ 0, Y=12450$\sim$12525 and Z = 0;
- The displacement constraints, i.e., UX=0 and UZ=0, are applied to the nodes with X = 110 $\sim$ 220, Y=25 and Z=340$\sim$560;
- The displacement constraints, i.e., UX=0 and UY=0, are imposed on the nodes with x = 0 $\sim$ 110, Y=4525 and Z=340$\sim$560;
- A displacement constraint, i.e., UX=0, is imposed on the nodes with X = 0 $\sim$ 110, Y=7875 and Z=340$\sim$560;
- The displacement constraints, i.e., UX=0 and UZ=0, are applied on the nodes with X=110$\sim$220, Y=12475, Z=340$\sim$560.

Finally, the three-dimensional finite element model of simplified pipes and supports is established, as shown in Figure 1.

2.4 Variable setting for seismic reliability calculation of pipeline system

In reality, there are randomness such as processing size errors and $Z1 = DS - UMAX \geq 0$ material parameters of pipes and support columns, which obey certain distribution. Therefore, in this paper, only 11 parameters of the sizes and material parameters of pipes and support columns are selected as

38

random variables. See Table 1 for the random variable parameters and their distribution characteristics of the selected pipe and support column materials. Enter the PDS analysis module of ANSYS software, specify the completed reliability analysis documents, and define the random input variables, the output variables of the limit displacement failure state function $Z1 = DS - UMAX \geq 0$ and the output variables of the limit strength failure state function $Z1 = DS - UMAX \geq 0$.

Table 1. Distribution of random input variables for pipeline and bracket pillar parameters.

Random variable name	Parameter meaning	Distribution pattern	Average/mean value	Standard deviation
W1	Column section length	Normal distribution	75 mm	2 mm
W2	Column section length	Normal distribution	75 mm	3 mm
T1	Column section width	Normal distribution	8 mm	0.2 mm
T2	Column section width	Normal distribution	8 mm	0.35 mm
RT	Inner diameter of pipe	Normal distribution	105 mm	1 mm
YOUNG1	Elastic modulus of column	Normal distribution	2.06E5 N/mm^2	2.06E3 N/mm^2
YOUNG2	Elastic modulus of pipeline	Normal distribution	1.95E5 N/mm^2	1.95E3 N/mm^2
DENSITY1	Column density	Normal distribution	7.8E-9 t/mm^3	1E-10 t/mm^3
DENSITY2	Pipeline density	Normal distribution	7.9E-9 t/mm^3	1E-10 t/mm^3
DS	Pipeline failure check displacement	Normal distribution	11 mm	0.5 mm
CH	Pipeline failure checking strength	Normal distribution	164 Mpa	12 Mpa

2.5 Seismic reliability analysis of pipeline system

The TH4TG045 synthetic seismic wave has been selected, and its peak acceleration is 2,048 mm/s^2 at 2.6 s over the duration of an action lasting 10 seconds. The Monte Carlo method (Monte Carlo method) and Latin Hypercube Sampling (U-IS sampling method) were used to randomly sample the geometric finite element models of pipes and support columns, and the sampling times were 500 times, and the sampling trends of the ultimate displacement failure state function Z1 and the ultimate strength failure state function Z2 shown in Figure 2 were obtained. Among the three sampling curves, the upper and lower curves represent the upper and lower limits of the confidence interval, and the middle curve represents the change of the sampling mean.

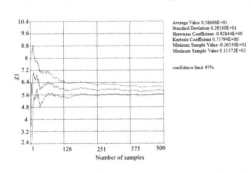
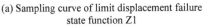
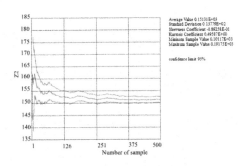

(a) Sampling curve of limit displacement failure state function Z1

(b) Sampling curve of ultimate strength failure state function Z2

Figure 2. Sampling result curve.

As can be seen from Figure 2, the three curves tend to be horizontal in the later stages, and the average values of the sampling curves of limit displacement failure state function Z1 and limit strength failure state function Z2 gradually converge, which indicates that 500 sampling times can meet the accuracy requirements of reliability analysis of pipelines and support columns. In order to verify whether the sampling times of the probability analysis (PDS) simulation results are enough, the histogram of random variables can be drawn as a reference. Because there are many random input variables, only the distribution histograms of random variables W1 and YOUNG2 are given in Figure 3.

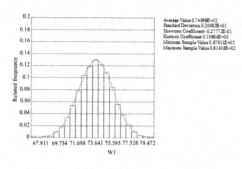 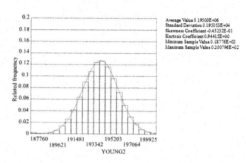

(a) Input variable of section length of support column	(b) Input variable of pipeline elastic modulus

Figure 3.   Random input variable histogram.

It can be seen from Figure 3 that the distribution histograms of W1 and YOUNG2 are close to the normal distribution curve, and the curve is smooth, which shows that the sampling times are enough to ensure the accuracy of the analysis results.

3   NUMERICAL CALCULATION RESULTS AND DISCUSSION

3.1   *Seismic reliability analysis results of pipeline system*

List probability (possibility): PDS provides a function using the command PDPROB to determine the value of any point of the cumulative distribution function on the axis of probability design variable, including interpolation function, so that the data between sample points can be evaluated. This property is very effective for evaluating the failure probability of structures at given data. The results show that the probability of Z1<0 and Z2<0 is 0.05872 and 0, respectively, when the confidence level of this pipeline system is 95%, and the reliability of these two failure state functions is 94.128% and 100%, as shown in Figure 4 and Figure 5.

3.2   *Sensitivity analysis results*

Sensitivity analysis is an important part of probability analysis in the ANSYS-PDS module. Figure 6 shows the results of sensitivity analysis for the ultimate displacement failure state function Z1 and the ultimate strength failure state function Z2 for the pipeline system under the influence of seismic waves.

As can be seen from Figure 6, the main factors that have significant influence on the ultimate displacement failure function Z1 are the material density, elastic modulus, and check displacement of the pipeline, and the input variables for the material density and check displacement of the pipeline are positive (the positive sign indicates that the change of this random variable has a positive effect on the structural reliability), which indicates that the reliability of the system increases with the increase of this random variable. The elastic modulus of the pipeline is negative (the negative sign indicates that the change of this random variable has a negative effect on the structural reliability), indicating that the reliability of the system decreases with the increase of the change. The main

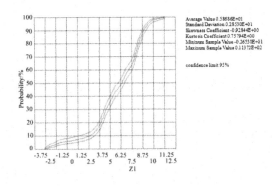

Average Value 0.58686E+01
Standard Deviation 0.28530E+01
Skewness Coefficient -0.92844E+00
Kurtosis Coefficient 0.75794E+00
Minimum Sample Value -0.26530E+01
Maximum Sample Value 0.11372E+02

confidence limit 95%

Probability Result of Response Parameter Z1

Solution Set Label   = BRACKETDESIGNPDS
Simulation Method   = Monte Carlo with Latin Hypercube Sampling
Number of Samples   = 500
Mean {Average} Value = 5.8685847e+000
Standard Deviation   = 2.8529502e+000
Skewness Coefficient = -9.2844455e-001
Kurtosis Coefficient = 7.5794182e-001
Minimum Sample Value = -2.6530114e+000
Maximum Sample Value = 1.1372010e+001

The probability that Z1 is smaller than 0.0000000e+000 is :

Probability [Lower Bound, Upper Bound]
5.87224e-002 [4.03684e-002, 8.15743e-002]

NOTE: The confidence bounds are evaluated with a confidence level of 95.000%.

(a) Cumulative distribution function of limit displacement failure state function Z1

(b) Probability of failure state function Z1 of limit displacement

Figure 4.   Reliability analysis results of ultimate displacement failure state function Z1.

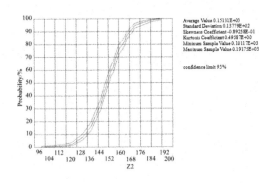

Average Value 0.15131E+03
Standard Deviation 0.13779E+02
Skewness Coefficient -0.89255E-01
Kurtosis Coefficient 0.49587E+00
Minimum Sample Value 0.10117E+03
Maximum Sample Value 0.19175E+03

confidence limit 95%

Probability Result of Response Parameter Z2

Solution Set Label   = BRACKETDESIGNPDS
Simulation Method   = Monte Carlo with Latin Hypercube Sampling
Number of Samples   = 500
Mean {Average} Value = 1.5130605e+002
Standard Deviation   = 1.3779039e+001
Skewness Coefficient = -8.9258286e-002
Kurtosis Coefficient = 4.9586804e-001
Minimum Sample Value = 1.0117135e+002
Maximum Sample Value = 1.9174783e+002

The probability that Z2 is smaller than 0.0000000e+000 is :

Probability [Lower Bound, Upper Bound]
0.00000e+000 [0.00000e+000, 0.00000e+000]

NOTE: The confidence bounds are evaluated with a confidence level of 95.000%.

(a) cumulative distribution function of ultimate strength failure state function Z2

(b) Probability of failure state function Z2 of ultimate strength

Figure 5.   Reliability analysis results of ultimate strength failure state function Z2.

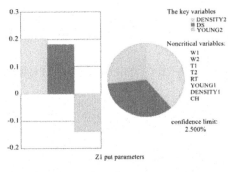
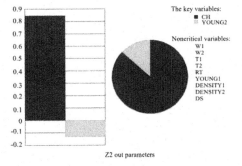

(a) Output parameters of the limit displacement failure state function Z1

(b) Output parameter of ultimate strength failure state function Z2

Figure 6.   Sensitivity analysis results.

variables that have significant influence on the ultimate strength failure function Z2 are the elastic modulus and check stress of the pipeline, with the check stress input variable being positive, whereas the elastic modulus input variable of the pipeline is negative.

The influence degree of 11 random input variables selected on the reliability of the pipeline system is shown in Table 2 and Table 3. In order to ensure the structural reliability of the pipeline system, the related random design variables should be adjusted according to the results of sensitivity analysis. For example, the variables that have significant influence on reliability should be strictly controlled in the design and manufacturing process. In order to reduce the maximum displacement and the maximum stress of the pipeline system structure, the following measures should be taken first: optimize the structure size instead of considering other unimportant random design variables.

Table 2. Influence degree of random variables on reliability of ultimate displacement failure state function Z1 of pipeline system.

W1	W2	T1	T2	RT	
−0.058	0.011	−0.055	−0.043	−0.054	
YOUNG1	YOUNG2	DENSITY1	DENSITY2	DS	CH
−0.025	−0.139	0.069	0.203	0.181	−0.001

Table 3. Influence degree of random variables on reliability of ultimate strength failure state function Z2 of pipeline system.

W1	W2	T1	T2	RT	
−0.023	−0.038	0.003	0.033	−0.034	
YOUNG1	YOUNG2	DENSITY1	DENSITY2	DS	CH
0.001	−0.130	−0.006	0.094	0.061	0.845

4 CONCLUSIONS

In this paper, the APDL modeling-analysis command flow file of a pipeline system structure of the underground powerhouse of Baihetan Hydropower Station is compiled by using the ANSYS analysis platform, and the PDS analysis module is used to analyze the seismic reliability of the pipeline system structure. The main conclusions are as follows:

- It provides a simulation analysis method for seismic reliability analysis of pipeline system structure;
- The analysis results show that the limit displacement failure state function Z1 and the limit strength failure state function Z2 of the pipeline system are both highly reliable at 94.128% and 100%, respectively;
- Through sensitivity analysis, it can be seen that the elastic modulus, material density, and check displacement of the pipeline are the key factors affecting the structural reliability of the ultimate displacement of the pipeline system;
- The elastic modulus and check strength of the pipeline are the key factors that affect the structural reliability and ultimate strength of the pipeline system.

Through the numerical simulation analysis of a pipeline example in the hydraulic turbine layer of the underground powerhouse of Baihetan Hydropower Station, it is proved that it is feasible to use the ANSYS-PDS module function to analyze the failure probability or reliability of this pipeline system, which provides a basis for the subsequent optimization design of other pipeline systems in the underground powerhouse of Baihetan Hydropower Station.

REFERENCES

Chen Huai. (1996) Seismic reliability analysis of above-ground pipelines. J. Industrial Architecture, 26(011): 24–27.

Dai Huichao, Cai Zhiguo, He Wenshe, et al. (2006) Review on the research and practice of cascade operation of cofferdam of Three Gorges-Gezhouba Water Control Project during power generation. J. Journal of Hydropower, 25(6): 8–15.

Dong Rubo, James Zhou, Feng Xin. (2008) Seismic response analysis of partially suspended submarine pipeline under multi-input. J. Journal of Vibration Engineering, 21(2):146–151.

Fan Qixiang, Zhang Chaoran, Chen Wenbin, et al. (2019) Key technologies of intelligent construction of ultra-high arch dams in Wudongde and Baihetan. J. Journal of Hydropower, 38(2): 22–35.

Gao Juan, Luo Qifeng, Che Wei. (2008) Monte Carlo theory and its realization in ANSYS. J. Journal of Qingdao University of Technology, 29(004): 18–22.

GAZIS, NIKOLAOS. (2011) Monte Carlo-based response analysis of subsea free spanning pipeline systems subjected to non-stationary random seismic excitations. Oceans'11 MTS/IEEE KONA.: IEEE, pp. 1–8.

Han Wenhai, James Zhou. (2014) Seismic reliability analysis of suspended span of corroded submarine pipeline. J. Chemical Equipment and Pipeline, 51(5): 75–78.

He Shuanghua, Song Can. (2012) Seismic damage prediction model and random reliability analysis of underground pipelines. Journal of North China Institute of Water Resources and Hydropower, 33(2): 4.

Li Dianqing, Zhou Jianfang. (2001) Back analysis method of reliability in mechanical design. J. Mechanical design, 18(3): 33–35.

Li Wenzhen, Huang Si, Xu Zhengnan. (2018) Reliability analysis of pressure vessel based on PDS module of ANSYS software. J. Machinery Manufacturing, 56(012): 14–16, 30.

Ma Bin, She Xin, Guo Yiliang. (2020) Experimental study on hydroelastic model of flood discharge and vibration reduction in Wudongde hydropower station. J. Journal of Hydropower, 39(1): 110–120.

Pan Xueguang. (2005) Causes and prevention of submarine pipeline suspension. J. China Ship Inspection, (10): 68–69.

Qin Quan. (2006) Stochastic finite element method of structural reliability: theory and engineering application. Tsinghua University Press, Beijing.

Rong Zhixiang, Lin Shaofen. (2011) Reliability analysis of connecting rod based on stochastic finite element method. J. Naval Science and Technology, 33(9): 68–70.

Wang Qianxin, Hong Feng. (1993) Seismic displacement response and seismic reliability analysis of offshore pipeline bridge. J. Earthquake Engineering and Engineering Vibration, 013(003): 43–53.

Wang Xiaoling, Li Xiao, Zhu Xiaobin, et al. (2019) Reliability analysis of dam foundation anti-sliding stability based on PLS-ELM response surface method. J. Journal of Hydropower, 38(4): 224–233.

Wang Yan, Yin Haili, Dou Zaixiang. (2006) Application of Monte Carlo Method. J. Journal of Qingdao University of Technology, 27(2): 111–113.

Wang Yuchuan, Zhao Changjun, Huang Gang, et al. (2014) Quantitative evaluation of failure probability of oil and gas pipelines with cracks. J. Safety and Environmental Engineering, 21(3): 126–129.

Wang Zhiping, Li Xia, Zhan Liu, et al. (2005) Reliability analysis and calculation of buried pipeline corrosion. J. Mechanical strength, 27(3): 339–341.

Wu Shiwei. (2009) Structural reliability analysis. People's Communications Press.

X.Q.AI, J.H.LI. (2007) Seismic reliability analysis of underground pipelines based on probability density evolution method. The first international symposium on geotechnical engineering safety and risk (proceedings of the first international symposium on geotechnical safety & risk is GSR 2007).

Yan Hongsheng, Yu Jianxing, Hu Yunchang, et al. (2007) Research on reliability analysis method of large-scale structural system. J. Ship Mechanics, 11(3): 444–452.

Ye Yong, Hao Yanhua, Zhang Changhan. (2004) Structural reliability analysis based on ANSYS. J. Mechanical Engineering and Automation, (6): 63–65.

Yu Jianxing, Fu Mingxi, Yang Yi, et al. (2008) Reliability analysis of vortex-induced vibration fatigue of submarine pipelines. J. Journal of Tianjin University, 41(11): 1321–1325.

Yu Jianxing, Guo Shuai, Yu Yang, et al. (2016) Influence of crack parameter on fatigue life of deep-water submarine pipeline. J. Journal of Tianjin University, 49(9): 889–895.

Yu Jianxing, Liu Xiaoqiang, Yu Yang, et al. (2017) Riser fatigue reliability calculation based on improved response surface method. J. Journal of Tianjin University, 50(10): 1011–1017.

Zhang Aihua, Ren Gongchang. (2010) Reliability analysis of anti-resonance of high-speed motorized spindle based on probability design of ANSYS. J. Mechanical Design and Manufacturing, 7: 112–114.

Zhang Lisong, Yan Xiangzhen, Xiujuan Yang. (2010) Application research of parameter reliability back analysis method in X80 steel pipeline design. J. Pressure Vessel, 27(8): 19–23.

Zhao Huigan, Guo Mingzhu, Zhai Changda, et al. (2015) Reliability analysis of seismic connectivity of urban gas pipeline network based on Monte Carlo method. J. Earthquake Research, 038(002): 292–296.

Zhu Yuankun, Liu Jian, Qin Haozhi, et al. (2017) Reliability analysis of tubing hanger structure based on ANSYS. J. Oil Field Machinery, 46(4): 28–31.

*Frontiers of Civil Engineering and Disaster Prevention and
Control – Yang & Rahman (Eds)*
© *2023 The Author(s), ISBN: 978-1-032-31200-2*

Study on the distribution law of sidewall earth pressure during the sinking stage of a super large open caisson

Zhewen Chen

School of Civil, Architectural and Environmental Engineering, Hubei University of Technology, Wuhan, China
State Key of Geomechanics and Geotechnical Engineering, Institute of Rock and Soil Mechanics, Chinese
Academy of Science, Wuhan, China

Tiechui Yang

Power China Henan Electric Power Engineering Co. Ltd, Zhengzhou, Henan, China

Gaojie Lan

State Key of Geomechanics and Geotechnical Engineering, Institute of Rock and Soil Mechanics, Chinese
Academy of Science, Wuhan, China
School of Civil Engineering and Architecture, Anhui University of Science and Technology, Huainan, China

Mingwei Guo*

School of Civil, Architectural and Environmental Engineering, Hubei University of Technology, Wuhan, China
State Key of Geomechanics and Geotechnical Engineering, Institute of Rock and Soil Mechanics, Chinese
Academy of Science, Wuhan, China

ABSTRACT: Considering the open caisson of the Changtai Yangtze River Bridge project as the research object, based on the monitoring data of the sidewall soil pressure during the sinking process of the caisson foundation, the distribution characteristics of the sidewall soil pressure are analyzed. A new distribution model for the sidewall soil pressure was proposed in comparison with the existing distribution model. The analysis results show that the soil pressure of the sidewall increases linearly in a specific range with the sinking depth of the caisson and then decreases with the depth after reaching the peak, and finally remains constant. Furthermore, the location corresponding to the peak value of the sidewall soil pressure moves down with the increase in the sinking depth.

1 INTRODUCTION

The large caisson foundations are increasingly being used with the increasing construction of large bridge projects in China. It is critical for expanding research on the sinking of large caissons (Guo et al. 2021; Shi et al. 2019). It was found that with the increasing size of the caisson foundation, the proportion of side friction resistance decreases in the total resistance of caisson sinking, and the proportion of Changtai caisson is 0.37 (Qin et al. 2019). Zhou Hexiang et al. (2018) and Zhou, Ma, Zhang, et al. (2018) carried out centrifugal simulation experiments with the local models of straight-arm open caisson and stepped open caisson. They proposed the calculation model of sidewall friction resistance according to the experimental results of the centrifugal simulation experiments. Zhang Kai et al. (Zhang et al. 2019,) studied the sinking process of the caisson by centrifugal simulation experiments with the main pier of the Hutong Yangtze River Bridge. They concluded that the sidewall soil pressure of the caisson first increases and then decreases, and the sidewall soil pressure reaches extreme when the depth of entry is two-thirds of the final sinking depth. They proposed a calculation model of the sidewall soil pressure related to the frictional resistance. He Qiaoling et al. (He et al. 2020) monitored the sidewall earth pressure during the

---

*Corresponding Author: 562913072@qq.com

sinking of the south anchor caisson foundation of Taizhou Bridge in real time and analyzed the data to derive the distribution form of the frictional resistance. Jiang Bingnan et al. (Jiang et al. 2019) considering the construction process of Hutong Bridge-29 # caisson as their research object, monitored the whole construction process. Post eliminating the influence of tilt on the caisson, they derived the distribution form of sidewall earth pressure along the depth of the open caisson and proposed that the calculation model of sidewall earth pressure be used as the basis for the calculation of sidewall friction resistance.

Chen Xiaoping et al. (Chen et al. 2005) monitored the whole process of soil extraction and sinking of the main pier caisson of the Haikou Century Bridge and derived the calculation model of sidewall earth pressure based on the monitoring data obtained. However, at present, there is no uniform distribution model of sidewall soil pressure during the sinking process of a super large caisson foundation.

Based on the large caisson foundation of the Changtai Yangtze River project, this paper deeply analyzes the distribution characteristics of sidewall soil pressure based on the measured data during the sinking process of the open caisson. Furthermore, a new distribution model of the sidewall soil pressure was proposed based on the existing calculating model.

2   CHANGTAI YANGTZE RIVER BRIDGE PROJECT OVERVIEW

As seen from Figure 1, the caisson foundation of the main tower of the Changtai Changjiang River Bridge is in the form of a circular end-shaped plane. The sidewall earth pressure sensors are divided into six layers, as shown in Figure 2.

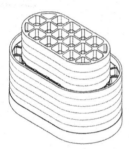

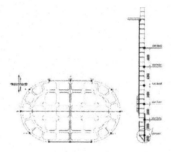

Figure 1.   Elevation drawing of open caisson.   Figure 2.   Monitoring point layout map.

The caisson foundation of the main pier is located in geological conditions with a sediment thickness higher than 170 m, as shown in Table 1.

Table 1.   Open caisson stratigraphic parameters.

Soil layer	Top altitude (m)	Bottom altitude (m)	γ (kN/m³)	c' (kPa)	φ' (°)
Loose silty sand	−14.7	−17.0	2	2	29.5
Hard plastic silty clay	−17.0	−24.4	19.8	29.3	10.6
Loose silty sand	−24.4	−25.6	20.6	5.6	31.8
Hard plastic silty clay	−25.6	−27.7	19.8	29.3	10.6
Slightly dense fine sand	−27.7	−32.7	19.8	2.5	33.39
Silt silty clay	−32.7	−34.0	19	29.7	4.8
Medium dense silty sand	−34.0	−39.4	20.6	5.6	31.4
Medium dense fine sand	−39.4	−50.2	19.8	2.5	33.39

(*continued*)

Table 1. Continued.

Soil layer	Top altitude (m)	Bottom altitude (m)	γ (kN/m³)	c' (kPa)	φ' (°)
Soft plastic silty clay	−50.2	−51.3	19	27.8	7.2
Dense medium sand	−51.3	−54.2	20	2	34.47
Dense coarse sand	−54.2	−70.0	20.8	2	36.55

3 MONITORING DATA OF SIDEWALL SOIL PRESSURE

After floating the caisson into the target position, it was sunk with water injection, and sinking was carried out until the elevation of the outer edge of the caisson reached −26.8 m, finally reaching the sinking depth of −65 m. Because of the caisson's better attitude during the fourth construction (Wang et al. 2021), data from the fourth stage of the sinking were chosen in this study, as shown in Figure 3.

The sidewall earth pressure at the depths from 33.74 m to 37.94 m is shown in Figure 4. The effect of the caisson step on the sidewall soil pressure starts to decrease with the increase in caisson depth.

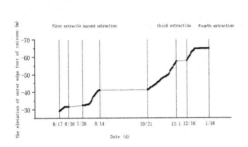

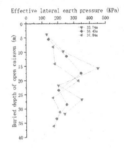

Figure 4. Curve of sidewall soil pressure with sinking depth.

Figure 3. Elevation curve of caisson.

4 A NEW CALCULATION MODEL OF SIDEWALL SOIL PRESSURE

On the basis of the geotechnical engineering and monitoring data of the caisson foundation of the Changtai Yangtze River Bridge, a new distribution model of sidewall soil pressure was proposed, with reference to the existing calculating model of sidewall soil pressure. Furthermore, this model considered the stress relaxation effect at the bottom of the caisson, as shown in Equations (1) and (2) and Figure 5.

$$\left.\begin{aligned} E &= \gamma h K_Z & h \leq \tfrac{1}{2}H \\ E &= \gamma K_Z H - h & \tfrac{1}{2}H < h \leq \tfrac{2}{3}H \\ E &= \gamma \tfrac{1}{3} H K_Z & \tfrac{1}{2}H < h \leq H \end{aligned}\right\} \tag{1}$$

$$\left.\gamma = \frac{\sum_{i=1}^{n} \gamma_i h_i}{\sum_{i=1}^{n} h_i}\right\} \tag{2}$$

Where H is the total sinking depth, γ is the average floating weight, H_t is the depth of the footing step into the earth, and α is the discount factor of the sidewall friction above the step of the well wall, which is related to the nature of the stratum soil. Based on the measured data and compared with the calculated values of static and passive earth pressure, K_Z was selected as 0.5 times the

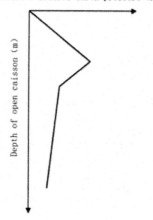

Figure 5. Sidewall soil pressure of the proposed model.

passive earth pressure. The effective earth pressure at the sidewall calculated from the above model is compared with the measured earth pressure. From Figure 6, it can be seen that the calculated and measured values of the sidewall soil pressure are well consistent.

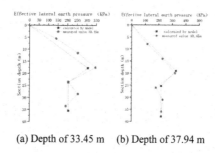

(a) Depth of 33.45 m   (b) Depth of 37.94 m

Figure 6. Comparison between measured and calculated values.

5 CONCLUSIONS

1. The measured data of the sidewall soil pressure of the caisson foundation showed that the steps of the caisson foundation exhibit significant reduction effect on the sidewall soil pressure above and below the step.
2. In addition, the relaxation effect and the influence range are increased considerably as the sinking depth increases, and both the caisson step effect on soil pressure and the relaxation effect on the pressure of the ground are considered.
3. A new sidewall soil pressure calculation model was proposed in line with the existing sidewall soil pressure distribution model, which provides an important reference for the sidewall soil pressure of similar large caisson foundations.

REFERENCES

Chen, X.P, Qian, P.Y, Zhang, Z.Y. (2005) Study on penetration resistance distribution characteristic of sunk shaft foundation[J]. Chinese Journal of Geotechnical Engineering, 27(2). 148–152 (in Chinese)

Chen, X.Z, Ye, J. Geotechnical Engineering[M]. Tsinghua University publishing house Co., Ltd.

Guo, M.W, Dong X.C, Shen, K.J et al. (2021) Study on the variation of the bottom resistance during sinking stage of super large caisson foundation[J]. Chinese Journal of Rock Mechanics and Engineering, 40(1). 2977–2985 (in Chinese)

He, Q.L, Wei, R.M, Xiong, H et al. (2020) Calculation and analysis of the side friction of caisson foundation[J]. Construction Technology, 49.1715–1719 (in Chinese)

Jiang, B.L, Ma, J.L, Chu, J.L et al. (2019) On-site monitoring of lateral pressure of ultra-deep large and subaqueous open caisson during construction[J]. Rock and Soil Mechanics, 40(4). 1551–1560 (in Chinese)

Qin, S.Q, Tan, G.H, Lu, Q.F et al. (2019) Research on Design and Sinking Method of Super large Caisson Foundation[J]. Bridge Construction, 50(5).1–9 (in Chinese)

Shi, Z, Li, S.Y, Yang, S.L et al. (2019) Study on the characteristics of friction resistance and the mechanism of sudden sinking in the middle and late sinking stages of super large caisson foundation[J]. Chinese Journal of Rock Mechanics and Engineering, 38(2). 3894–3904 (in Chinese)

Wang, Z.C, Yang, Q, Chen, J.R et al. (2021) Monitoring data analysis of super large underwater steel open caisson construction[J]. Journal of China & Foreign Highway. (in Chinese)

Zhang, K, Ma, J.L, Zhou, H.X et al. (2019) Seismic response law of large span railway cable-stayed bridge under non-uniform excitation[J]. Railway Engineering, 59(6). 28–32 (in Chinese)

Zhou, H.X, Ma, J.L, Zhang, K et al. (2018) Experimental study of distribution characteristics of friction resistance on sidewall of open caisson[J]. Bridge Construction, 48(5). 27–32 (in Chinese)

Frontiers of Civil Engineering and Disaster Prevention and
Control – Yang & Rahman (Eds)
© 2023 The Author(s), ISBN: 978-1-032-31200-2

Experimental research on local outburst prevention effect of coal road driving

Zhonghua Wang*

National Key Laboratory of Gas Disaster Detecting, Preventing and Emergency Controlling, Chongqing, China
China Coal Technology Engineering Group Chongqing Research Institute, Chongqing, China

ABSTRACT: In order to study the local outburst prevention effect of coal roadway driving, the cuttings index method is used to predict the outburst risk of coal roadway driving and the effect is tested in the process of coal roadway driving. At the same time, the gas concentration after the blasting of coal roadway excavation is analyzed statistically, which verifies the accuracy of outburst prediction and effect inspection of coal roadway excavation so as to ensure the safe excavation of coal lanes.

1 INTRODUCTION

Shangzhuang Minefield is a part of the Wusheli mining area in Fengcheng Hexi Coalfield, located in Fengcheng City in the middle and lower reaches of the Ganjiang River. The mine is about 5.1 km long, about 2.65 km wide, and has an area of about 13.288 km². The approved production capacity in 2005 was 350 kt/a, and the annual output has stabilized at 350,000 to 400,000 tons in recent years. The main geological factors affecting the zoning of Shangzhuang Coal Mine gas are: coal seam burial depth, coal seam thickness, coal seam gas storage conditions, regional geological structure, etc. The gas pressure test results are shown in Table 1. This paper aims to test the local sensitivity index of coal roadway precisely, which is of great importance to the safe excavation of coal roadway.

Table 1. Measurement result of gas pressure.

Measurement location	See coal elevation (m) Gas pressure (MPa)	See coal elevation (m) Gas pressure (MPa)
710 Floor Roadway	−622.5	6.0
505 Floor Roadway	−783	2.0

2 OUTBURST PREDICTION METHOD OF COAL ROAD EXCAVATION

2.1 *Prediction of outburst hazard in coal road excavation*

After the outburst prevention measures are implemented on the coal roadway driving face, the drill cuttings index method is used to predict the outburst risk of the coal roadway driving face. At least

---

*Corresponding Author: boaidajia2007@126.com

    DOI 10.1201/9781003308577-7

three boreholes with a diameter of 42 mm and a hole depth of 8–10 m should be constructed on the coal seam face toward the front coal body (Chen et al. 2019; Li et al. 2020). Determine the drill cuttings' gas desorption index and the amount of drill cuttings. Drill holes should be arranged in soft layers as much as possible. One hole is located in the middle of the tunnel section and parallel to the direction of excavation. The final hole points of other drill holes should be located 4 m outside the contour line on both sides of the tunnel section. The total amount of cuttings S of the 1 m section is measured every 1 m of the borehole, and the value of the gas desorption index of drill cuttings K1 is measured at least once every 2 m of the drilling. When the drilling cuttings index method is used to predict the outburst hazard of a coal tunnel driving face, the outburst hazard critical value of each index is shown in Table 2.

Table 2. Critical value of drilling cuttings index method to predict the outburst risk of coal roadways.

Drill cuttings gas desorption index K1 (mL/g·min$^{1/2}$)	Drilling cuttings S (kg/m)
Dry coal sample < 0.5 Wet coal sample < 0.4	<6 (Aperture Φ42 mm)

2.2 Outburst prevention effect test of coal tunnel driving

In the event that the working face is predicted to be one free of outburst hazards, the driving drill cutting index method is used to test the effectiveness of coal tunnel driving measures.

2.2.1 Indicators and critical values

Normal coal seam critical index K1 value is 0.5 mL/g·min$^{1/2}$, Smax value is 6 kg/m; fault structure zone (20 m before and after the structure), coal seam thickness change zone, stress concentration zone K1 critical value is 0.4 mL/g·min$^{1/2}$, Smax value is 6 kg/m.

2.2.2 Inspection steps

(1) After the discharge drilling is completed, the effect inspection will be carried out. Three effect inspection holes are constructed in the middle of the measure holes. The hole depth is 10 m, the diameter is 42 mm, and the control area is 4 m outside the contour line of the roadway, which is arranged in a fan shape (Mao et al. 2019; Li. & Jiang 2019).

(2) When K1 < 0.5 mL/g·min$^{1/2}$ and S < 6 kg/m, the prevention and control measures are considered effective. According to the design of intensive drilling, ensure an advance distance of 5 m in front of the roadway (calculated by projection), and the allowable footage is 4.5 m.

(3) In the 20 m before and after the fault structure zone and the geological structure change zone, when K1 = 0.4 mL/g·min$^{1/2}$ and S = 6 kg/m, only a small loop footage can be allowed, i.e., a maximum of 1.5 m per sub-shift. The leading distance must be kept at 7 m.

(4) If K1 = 0.5 mL/g·min$^{1/2}$ or S = 6 kg/m, the inspector must drill a number of counter holes beside the inspection hole of the supercritical index according to the inspection drilling situation, and then an effect inspection is carried out between the supplementary drilling measures. If the effectiveness inspection holes meet the requirements of the hole layout and the three sets of effectiveness inspection indicators are not supercritical, the prevention and control measures are considered effective; otherwise, the regional outburst prevention measures must be taken again (Wang et al. 2017).

(5) When the inspection result measures are effective, if the projected length of the inspection hole and the anti-outburst drill hole in the direction of roadway driving (referred to as the projected hole depth) is equal, the conditions that allow sufficient anti-outburst measures to advance and take safety protection measures during down digging At the same time, when the depth of the projection hole of the inspection hole is less than the hole of the anti-outburst measure, it shall

be implemented after taking safety protection measures under the condition that the required advance distance of the anti-outbreak measure is reserved and at the same time there is at least 2m of the inspection hole projection hole depth.

3 OUTBURST PREVENTION EFFECT TEST OF COAL TUNNEL DRIVING

3.1 Analysis of coal tunnel driving index

The outburst hazard prediction in Shangzhuang Coal Mine's coal roadway was tested during the tunneling process. The inspection results of K1 and S indicators are shown in Table 3, and the indicator change curve is shown in Figure 1.

Table 3. K1 and S statistics of coal roadway driving efficiency inspection index.

Total footage (m)	Drilling diameter (mm) Size	Drill cuttings desorption index K1 (mL/g·min$^{1/2}$)		Cuttings S (kg/m)	
		Excess (number)	Size		Excess (number)
176	75	0.08–0.43	0	2–4.7	0

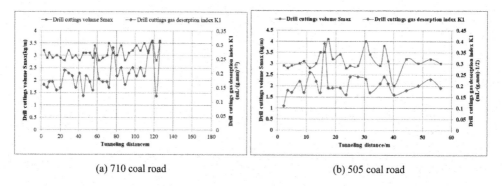

(a) 710 coal road   (b) 505 coal road

Figure 1. K1 and S prediction indexes of coal road excavation.

It can be seen from Figure 1 that after the coal roadway adopts the regional measures of floor pressure relief rock roadway combined with cross-layer borehole pre-drainage coal roadway strip gas, the K1 and S prediction indexes are both small, and the drilling cuttings gas desorption index K1 value is 0.08–0.43 mL/g·min$^{1/2}$, which is less than the critical value 0.5 mL/g·min$^{1/2}$, and the drill cuttings S value is 2–4.7 kg/m, which is less than the critical value of 6 kg/m, realizing safe excavation of coal roads.

3.2 Analysis of gas concentration after blasting in Coal roadway driving

In order to ensure the safety of coal tunnel excavation, in the process of coal tunnel excavation, the gas concentration of Shangzhuang Coal Mine's test roadway after blasting was carried out to verify the accuracy of the coal tunnel effect. The gas concentration change curve is shown in Figure 2.

It can be seen from Figure 2 that the gas concentration in the coal roadway is between 0.12% and 0.73% after the blasting. No gas has appeared since the coal roadway adopted the regional measures of floor pressure relief rock roadway combined with through-bed drilling and pre-draining coal road strip gas. Over-limit phenomenon. The coal tunnel demonstrates a safe excavation method in order to verify the danger of the coal tunnel excavation effect test.

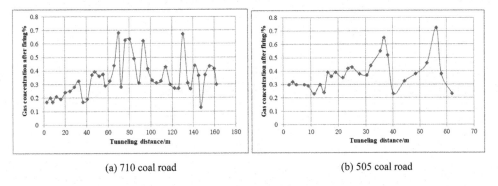

(a) 710 coal road   (b) 505 coal road

Figure 2. Concentration change curve after blasting in coal roadway excavation.

4 CONCLUSION

Following the adoption of the regional measures of pressure relief rock roadway and drilling through the layer to pre-drain the coal roadway strip gas, both the K1 and S prediction indices are small, and both are less than the critical value of outburst risk from coal roadway excavation.

The gas concentration in the coal roadway was between 0.12% and 0.73% after the blasting, and there was no gas over-limit phenomenon, which verified the danger of the coal roadway driving effect test and the coal roadway realized safe driving.

ACKNOWLEDGMENTS

This work was financially supported by the Tiandi Technology Co., Ltd. Special Project of Science and Technology Innovation and Entrepreneurship Fund (Grant No. 2021-2-TD-ZD008), and Chongqing Natural Science Foundation General Project (Grant No. cstc2021jcyj-msxmX1149).

REFERENCES

Chen X.X., Zhang P.G., Hou F.J. (2019) Chen Guangjian. Determination of sensitive index and critical value for coal seam outburst prediction in a mine of Xin'an Coal Field[J]. Journal of North China Institute of Science and Technology, 16(06): 1–6.

Li Y., Zhang Y., Qin Z. (2020) Research on determination of sensitive indexes for coal and gas outburst prediction in Hexi coal mine[J]. Energy Technology and Management, 45(04): 143–145.

Li Y.H., Z.K., Jiang H. (2019) Experimental research on sensitive indicators of local outburst prediction in coal mines[J]. Energy and Environmental Protection, 41(09): 1–4+9.

Mao J.N., Wang S.L., Jiang Q.F. Chen Chunyuan. (2019) Determination of the critical value of outburst prediction sensitivity index for coal roadway driving face[J]. China Mining Industry, 28(S1): 150–153.

Wang J.S., Zhang Y.M., Fan H.M. (2017) Determination of sensitive indexes and critical values of driving faces[J]. China Coal, 43(07): 131–133.

Frontiers of Civil Engineering and Disaster Prevention and Control – Yang & Rahman (Eds)
© 2023 The Author(s), ISBN: 978-1-032-31200-2

Research on design and experimental application of flexible pavement base

Jianping Su*

China Urban and Rural Holding Group Co., Ltd., Beijing, China

ABSTRACT: In recent years, with the increase in traffic volume and heavy vehicles, asphalt pavement has prematurely shown problems such as fatigue cracking and insufficient rutting resistance. The commonly used mixture structure of asphalt pavement cannot adapt to the growing volume of traffic and heavy traffic situations. It has become an urgent and important task to study new pavement structures. This paper presents the design and experimental performance evaluation of flexible base pavements, analyses of the relevant performance indicators of different flexible pavement structures, as well as construction suggestions that could be applied to practical applications. The research result shows that the engineering effect is good.

1 INTRODUCTION

In recent years, with the increase of traffic volume, heavy vehicles and the influence of traffic channelization, asphalt pavement has shown shortcomings such as premature fatigue cracking and insufficient anti-rutting ability, which reduce the performance of the pavement, shorten the service life of the pavement, and greatly increase the cost of reconstruction and maintenance. The commonly used mixture structure of asphalt pavement cannot adapt to the growing volume of traffic and heavy traffic situations. It has become an urgent and important task to study new pavement structures (Hardy & Cebon 1994; Liu 2002; Zaghloul & White 1993). Meanwhile, it is of great practical significance to adopt a flexible base structure to improve the high and low-temperature performance of asphalt pavement and intensify the water loss and fatigue resistance of the asphalt mixture. In this paper, the design and experimental application of the flexible pavement base of the Qingdao-Yinchuan Expressway from Jilujie to Shijiazhuang are studied and analyzed. Additionally, the paving of the road has been tested on the flexible base of the road section, and the engineering application has demonstrated a good effect.

2 PROJECT OVERVIEW

The Qingdao-Yinchuan Highway from Jilujie to Shijiazhuang is an organizational part of the "five north-to-south longitudinal superhighways and seven east-to-west transverse superhighways" national highway trunk lines planned by the Ministry of Communications. According to the current traffic volume forecast results and considering the status and role of the project in the road network, it is planned to adopt the standard construction of expressways. The driving speed is calculated at 120 km/h, and the design load of the bridge and culvert adopts the automobile-super-20 level and the hanging-120. The roadbed is 28 meters wide, with four lanes in both directions, which are fully enclosed and fully interchangeable. The total length of the expressway from Jilujie to the Shijiazhuang section of the Qingdao-Yinchuan Highway is 183.551 kilometers, and the total length of the research section is 18.79 kilometers.

---

*Corresponding Author: sujianping0104@163.com

DOI 10.1201/9781003308577-8

3 RESEARCH ON FLEXIBLE BASE DESIGN

The design idea of the flexible base comes from the American practice of using asphalt stabilized gravel or graded gravel as the base on the semi-rigid base and then laying the asphalt concrete surface. Generally, in the pavement structure, the upper layer of the pavement is defined as the functional layer, and the middle and lower layers, as well as the base layer, are defined as the load-bearing layers of the structure. Moreover, flexible base pavement usually has two typical structures. One is a full-thickness flexible pavement, that is, the upper, lower base, and surface layers are asphalt mixture, and the total thickness of the asphalt layer is thick, which is also called a permanent pavement structure, and the other type is a hybrid flexible pavement, which has two types. One is to pave the flexible transition base on the semi-rigid base, and the other is to take the graded gravel as the lower base, and then to pave the asphalt surface, which is also known as the inverted structure or sandwich structure (Lv 2006).

Combined with the traffic characteristics, regional characteristics, and research needs of the project, coarse-grained asphalt macadam is used as the base material in this experiment, and full-thickness and hybrid structures are adopted in the structural form. The length of a hybrid pavement structure is 8.96 km (half), and the length of a full-thickness pavement structure is 5.769 km (half). The specific structural design is shown in Table 1.

Table 1. Structure design of flexible base pavement.

Structure type	Hybrid structure	Full-thickness structure
Stationary paragraph	K129+320-K138+280 left	K119+490-K125+259 right
Upper layer	4 cm fine-grained asphalt concrete SAC-13 SBR modified emulsified asphalt sticky layer	4 cm fine-grained asphalt concrete SAC-13 SBR modified emulsified asphalt sticky layer
Middle layer	6 cm medium grain asphalt concrete AC-20 SBR modified emulsified asphalt sticky layer	6 cm medium grain asphalt concrete AC-20 SBR modified emulsified asphalt sticky layer
Lower layer	6 cm coarse-grained asphalt concrete AC-25	6 cm coarse-grained asphalt concrete AC-25
Up-sealing layer	Thermal spray modified asphalt plus crushed stone seal	SBR modified emulsified asphalt sticky layer
Upper level	12 cm coarse-grained asphalt stabilized crushed stone LSM-30 SBR modified emulsified asphalt sticky layer transparent layer	16 cm coarse-grained asphalt stabilized crushed stone BL1-30 SBR modified emulsified asphalt sticky layer 17 cm coarse-grained asphalt stabilized crushed stone BL1-30
Lower level	18 cm cement stabilized and crushed stone	5-6 mm emulsified asphalt slurry seal
Sub-base	18 cm cement lime stabilized soil	15 cm lime-fly ash stabilized aggregate treatment soil foundation 18 cm cement lime soil treatment soil foundation

4 TEST ANALYSIS OF FLEXIBLE BASE

4.1 *Test ratio*

For the flexible base of the test road, there are two types of structures: BL1-30 and LSM-30. The asphalt is selected from No. 70 petroleum asphalt, and the aggregate is preliminarily crushed into 2,050 cm semi-finished gravel, both of which are secondarily processed and shaped into 4 m to 6 cm, 2 m to 4 cm, 1 m to 3 cm, 1 m to 2 cm, and 0.5 m to 1 cm. Mineral powder made of crushed stone of 1,020 cm processed by a powder concentrator is ordinary mineral powder without adding slaked lime. The US Bailey method is used to adjust the mix ratio design. The synthetic gradation of the flexible base layer is shown in Table 2, the CA ratio parameters of each synthetic gradation are shown in Table 3, and the cutting photo of the specimen is shown in Figure 1.

Table 2. Synthetic gradation of asphalt-stabilized crushed stone mixture.

Gradation type	Percentage of mass passing through the following sieves (square sieves) (%)														
	53	37.5	31.5	26.5	19.0	16.0	13.2	9.5	4.75	2.36	1.18	0.6	0.3	0.15	0.075
LSM-30	100	99.5	94.3	80.8	61.8	56.4	49.7	41.2	29.3	24.6	18	11.6	8.1	6.3	4.2
BL1-30	100	98.9	95.5	85.7	74.3	65.7	50.2	35	27.2	18	9.2	6.6	6.2	5.1	3.8

Table 3. CA parameter values of synthetic gradation test.

Parameter	LSM-30	BL1-30
Thickness (cm)	12	33
CA	0.539	1.529

Figure 1. Photo of specimen cutting of BL1-30 and LSM-30.

4.2 *Road performance test*

4.2.1 *High-temperature stability test results*

The 300 mm × 300 mm × 100 mm specimens are formed by the wheel mill of the site mixing station, and the rutting test results are shown in Table 4.

It can be seen from Table 4 that the high-temperature stability of coarser and thicker BL1-30 gradation is significantly higher than that of flimsier and thinner LSM-30 gradation, which indicates

Table 4. The rutting test results of the asphalt stabilized gravel base of the test road.

Gradation type	45 min Deformation (mm)	60 min deformation (mm)	Dynamic stability (time/mm)	Standard deviation σ	Coefficient of variation Cv (%)
LSM-30	4.037	4.519	1321	167	12.7
BL1-30	3.321	3.603	2239	101	7.7

that appropriately increasing the thickness of the asphalt stabilized base and strengthening the embedding effect of the coarse aggregate can improve the ability of the asphalt mixture to resist shear deformation.

4.2.2 *Water stability test results*
The test pieces are formed by the large Marshall compactor at the site mixing station. The test results of water stability residual stability and freeze-thaw splitting strength in the laboratory are shown in Table 5.

Table 5. The test results of Marshall residual stability and freeze-thaw splitting strength.

Gradation type	Optimal asphalt content (%)	Relative density of bulk volume	Percentage of void (%)	Residual stability (%)	Freeze-thaw splitting strength ratio TSR (%)
LSM-30	3.4	2.452	4.0	110.0	91.3
BL1-30	3.2	2.420	5.5	84.2	87.9

It can be seen from Table 5 that the water stability of the mixture with a good compact structure is better than that of the mixture with a poor compact structure. For the asphalt stabilized gravel base mixture, when the properties of asphalt and stone are not much different, the larger the porosity is, the worse the water damage resistance will be, and the smaller the porosity is, the better the water damage resistance will be. Therefore, controlling the porosity is the key to ensuring the water damage resistance of the asphalt stabilized base.

4.2.3 *Low-temperature crack resistance test results*
The 300 mm×300 mm×100 mm specimen is formed by the wheel mill of the site mixing station and then cut into 40 mm×40 mm×250 mm trabeculae. The low-temperature bending test results are shown in Table 6.

Table 6. Low-temperature bending test results of asphalt-stabilized gravel base of the test road.

Gradation type	flexural tensile strength (MPa)	Failure strain ($\mu\varepsilon$)	Failure stiffness modulus (MPa)	Strain energy (J)
LSM-30	5.384	1750	3147	0.171
BL1-30	7.488	1719	4459	0.165

It can be seen from the test results that the low-temperature bending strain energy of graded LSM-30 with a strong compact structure is greater than that of graded BL1-30 with a weak compact structure, which indicates that a looser skeleton structure is beneficial to improve the low-temperature performance of the asphalt mixture, and the formation of a tight skeleton should not be overemphasized.

5 FLEXIBLE BASE PAVEMENT APPLICATION

5.1 *Construction advice*

1) During mixing, due to the good high-temperature stability, large construction thickness, and slow cooling of the flexible base layer, there is enough compaction time, so the temperature requirements for the asphalt-stabilized crushed stone mixture are reduced. The recommended mixing time is 30 to 35 seconds for clean mixing and 5 seconds for dry mixing. The uniformity of the asphalt mixture should be checked at any time. If the phenomenon of whitening occurs, the reasons should be analyzed for improvement.

2) When paving, due to the large thickness of the flexible base layer, the unit length needs a lot of material, and the paver needs to wait for the material. Therefore, the paving speed should be controlled between 1.0–1.5 meters to keep an even, continuous, uninterrupted paving operation on the paver. During the paving process, the paver should be kept half full, and the number of buckets should be minimized.

3) When compacting, a rubber wheel roller over 26 tons is used for initial pressure, and then a drum roller is used for re-compression, which is conducive to the compaction effect of the drum vibratory roller from top to bottom. It should be noted that before the initial pressure, the rubber wheel should be heated to the same temperature as the mixture to avoid sticking to the wheel, which can be achieved by driving on the paved road for 5 to 15 minutes.

4) When the distance from the end is 6 to 8 meters, the vibration will be turned off, and the static pressure will exceed the end. The vibration will be turned on when returning to prevent the end from being too low and cutting the overlong part. The front row paver that has been paved part of the 40 cm wide temporarily does not roll, which is used as the reference surface for the subsequent part. Then roll across the seam to eliminate wheel marks.

5.2 *Performance index detection*

After the completion of the project construction, the relevant flexible base pavement shall be tested for indicators to verify the relevant parameters, as shown in Table 7. The test results show that all technical indicators can meet the design requirements (Institute of highway science, Ministry of communications of China, 2004).

Table 7.   Test results of compaction degree and flatness.

Gradation type	LSM-30 (upper level)	BL1-30 (lower level)	BL1-30 (upper level)
Degree of compaction (%)	99.27	99.24	99.3
Flatness σ (mm)	1.082	1.147	0.942

6 CONCLUSIONS

In this paper, the design and experimental performance test analysis of flexible base pavement are carried out, the relevant performance indicators of different flexible pavement structures are analyzed, and construction suggestions that can be applied to practical projects are given. The engineering effect is good. The specific conclusions are as follows:

(1) Analyzing the composition and structure of the asphalt mixture, increasing the amount of coarse aggregate to form a skeleton is conducive to improving the high-temperature stability of the asphalt mixture. However, if the skeleton effect is too strong, it will also affect and weaken other properties of the asphalt mixture. Boosting the compactness of the asphalt mixture is beneficial to improve its low-temperature crack resistance and water stability.

(2) Due to its own characteristics, the construction of asphalt-stabilized flexible base has certain particularities, especially the segregation problem, compaction process, and flatness control, which should be paid attention to during construction.

(3) The use of flexible base pavement structure increases the project cost compared with the ordinary semi-rigid base, and its excellent performance is better, which is conducive to reducing diseases such as rutting, shear damage, and cracks of asphalt pavement, improving the service performance of the pavement, and prolonging the service life of the pavement. Therefore, the flexible base pavement structure has broad application prospects.

REFERENCES

Hardy M, Cebon D. Importance of speed and frequency in flexible pavement response[J]. Journal of Engineering Mechanics, 1994, 120(3): 83–98.

Institute of highway science, Ministry of communications of China. Technical code for construction of highway asphalt pavement [M]. People's Communications Press, 2004.

Liu Z. L. New technology of asphalt concrete pavement for high-grade highway [M]. People's Communications Press, 2002.

Lv W. M. Theoretical analysis of shear performance of asphalt concrete pavement with asphalt stabilized base [J]. Highway, 2006 (4): 220–224.

Zaghloul S M, White T D. Use of a three-dimensional, dynamic finite element program for analysis of flexible pavement[J]. Transportation Research Record Journal of the Transportation Research Board, 1993(1388): 60–69.

*Frontiers of Civil Engineering and Disaster Prevention and
Control – Yang & Rahman (Eds)
© 2023 The Author(s), ISBN: 978-1-032-31200-2*

Study on energy system distribution of high-temperature damaged rock under uniaxial loading

Shuang Yang, Lan Qiao, Ming Zhou & Qingwen Li*
Department of Civil Engineering, University of Science and Technology Beijing, Beijing, China

ABSTRACT: With the rapid development of energy exploitation, chemical waste and radioactive nuclear waste underground burial, energy underground storage and other fields, a large number of tunnels, stopes and mine surrounding rocks are in the environment of high-temperature and high in-situ stress for a long time, and the macroscopic fracture caused by rock damage caused by high-temperature has become a great threat to the development of these fields. In this paper, uniaxial compression tests on intact rocks and rocks with different initial thermal damage were carried out based on rock triaxial testing machine, PCI-II acoustic emission system and tubular high-temperature furnace. By comparing the variation trend of the dissipated energy ratio of intact rocks and rocks with different initial thermal damage in the process of uniaxial compression failure, it can be found that: the decrease rate of the dissipated energy ratio of the rock sample with higher fracture degree is slower, indicating that the internal fracture degree can slow down the release of the elastic energy accumulated in the rock mass. Therefore, artificial cracks can be manufactured to reduce the energy storage capacity of rock mass and transfer the accumulated elastic energy to the depth of rock mass, so as to consciously control the on-site energy and reduce the risk of disaster.

1 INTRODUCTION

Rock mass is a complex geological body formed after a long period of geomechanically action, which has the characteristics of non-homogeneity, discontinuity, and anisotropy. Under the influence of external disturbance, cracks in rock mass are often activated to open or expand, and these joint cracks often become the "breakthrough" of rock mass engineering instability failure because of their extremely low strength, small stiffness, large deformation capacity and strong permeability. Therefore, from the perspective of energy, considering the influence of stress environment and internal fracture of micro-rock mass in different positions away from the surface comprehensively, analyzing the influence mechanism of fracture on energy evolution during rock deformation under different stress environment, and then avoiding the influence of macroscopic fracture caused by fracture reasonably and effectively is great significance.

A great deal of research work has been carried out on the evolution of rock deformation energy (Liu et al. 2019; Meng et al. 2019, 2016; Song et al. 2020; Zhang et al. 2017, 2021). Meng et al. (2018) used MTS 815 rock mechanics test system to conduct cyclic uniaxial compression tests on 180 red sandstone specimens under 36 different loading and unloading schemes in order to reveal the characteristics of energy accumulation and dissipation in the process of rock deformation and failure. Based on thermodynamic theory and synthetic stress-strain curve analysis, a method to calculate rock energy density based on the characteristics of loading and unloading curves is proposed. Li (Li et al. 2017) et al. proposed a coal mine rock mass energy calculation method by using a self-designed test device and the SANS-CMT5305 testing machine, and proposed that rock

---

*Corresponding Author: qingwenli@ustb.edu.cn

DOI 10.1201/9781003308577-9

particle size has an important influence on the energy evolution of rock fragmentation. Diyuan Li et al (2017). conducted triaxial compression tests on fine-grained and medium-grained granite specimens with initial confining pressures of 10, 20, 40 and 60 MPa respectively under different loading and unloading stress paths. The energy evolution characteristics of granite samples in a quarry during triaxial deformation and failure process are studied. The ratio of the dissipative strain energy to the total strain energy can be used to describe the deformation and degree of damage to rock specimens during the triaxial loading and unloading processes. These theoretical studies (Jia et al. 2019; Li et al. 2021; Pan et al. 2020; Zhang et al. 2019, 2021, 2014; Zhao et al. 2020) analyze the influence of loading and unloading paths, particle size and other factors on the energy evolution mechanism from the perspective of energy, providing theoretical basis for the study of this paper.

In the study of fracture rock damage, Baghbanan (Baghbanan & Jing 2008) used discrete element method (DEM) to study the influence of fracture aperture distribution and fracture length on permeability and permeability mode of fractured rock mass. Sainoki et al. (2019) studied the influence of fractures on seismic activity intensity from the perspective of energy. Yang (2020) et al carried out a series of tests to reveal the influence of anchorage method on the evolution mechanism of rock mass fractures with non-continuous joints. However, in the study of rock with fractures, scholars tend to simplify the fractures as single, double, or multiple visible artificial cracks, ignoring the influence of the interaction between disordered fractures on the evolution of rock mass deformation energy completely. Considering the occurrence environment of deep rock mass with high-temperature and high in-situ stress, and the cracks induced by thermal stress are more random than those induced by load stress, which have a strong directional distribution. Therefore, in this paper, samples with different initial thermal damage were prepared at high-temperature for subsequent tests.

In this paper, uniaxial compression tests of rocks with different initial thermal damage were carried out based on rock triaxial testing machine, PCI-II acoustic emission system and tubular high-temperature furnace. The energy system distribution law of fractured granite under uniaxial loading is studied, and the influence of stress and fracture coupling on energy evolution mechanism is discussed, so as to provide theoretical basis for on-site energy regulation and reduction of disaster risk in practical engineering.

2 PREPARATION OF ROCK DAMAGED BY INITIAL HIGH-TEMPERATURE

The granite samples were taken from Sanshandao gold deposit in Shandong Province. The main mineral components of granite samples include quartz, plagioclase, alkaline feldspar, biotite, and vermicide. The granite has a density of 2.72 g/cm^3, porosity of 0.59% and p-wave velocity of 4,587 m/s. The size of the rock sample used in the test is φ 50×100 mm. The prepared standard sample is shown in Figure 1.

Figure 1.   Granite sampling and sample making pictures.

Before the uniaxial compression test, the specimen was preloaded axially to 3 MPa and then relieved to 0MPa at a compression rate of 0.5 MPa/s, so as to reduce different degrees of damage caused by manual sample preparation. In conventional uniaxial test, the specimen was compressed to failure at a compression rate of 0.5 MPa/s. The main test instruments used in the test are TAW-2000 microcomputer controlled electro-hydraulic servo rock triaxial testing machine, PCI-II acoustic emission system and tubular high-temperature furnace.

The granite specimens with the size of φ 30 mm × 60 mm were heated in a programmable tube furnace to prepare the granite specimens with cracks. In order to obtain granite samples with different degrees of fracture, four target temperatures were set during the test, namely 300°C, 400°C, 500°C and 600°C, and three samples were heated at each target temperature. The heating path is as follows: the sample is heated from room temperature to the target temperature at a heating rate of 55°C/min, and then the target temperature is kept constant for 1 h, and finally the sample is cooled to room temperature at a cooling rate of 5°C/min.

Nonmetallic ultrasonic instrument RS-ST01C was used to measure the p-wave velocity of rock samples before and after high-temperature treatment. The p-wave velocity of samples treated at different temperatures is shown in Figure 2.

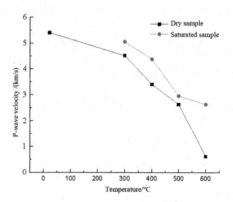

Figure 2. Variation of p-wave velocity of granite with temperature.

Referring to Lemaitre's definition of damage variable, the damage variable was established according to the change of p-wave velocity after high-temperature:

$$D = 1 - \frac{V_i}{V_0} \tag{1}$$

Where, D is the equivalent damage factor; V_0 is the p-wave velocity of the sample at room temperature; V_i is the p-wave velocity of the sample after high-temperature.

According to Formula 1, the equivalent damage factors of specimens under various temperature gradients from 23°C to 600°C are 0, 0.16, 0.36, 0.51 and 0.89, respectively.

3 ACOUSTIC EMISSION TEST RESULTS AND ANALYSIS OF GRANITE UNDER UNIAXIAL LOADING

3.1 *Analysis of failure modes and stress-strain curves of rock under uniaxial compression*

3.1.1 *Total stress-strain curves of intact rocks*

Uniaxial compression tests were conducted on three standard granite samples, as shown in Figure 3. It can be seen from the stress-strain curve that the specimens made of the same rock sample have slight differences in compressive strength. The average uniaxial compressive strength is 202.4 MPa, the peak strain is 0.307%, and the elastic modulus is 647.36 GPa.

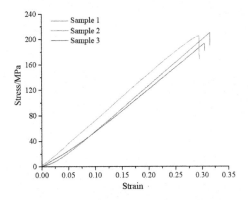

Figure 3.   Stress-strain curves of intact granite under uniaxial condition.

Figure 3 shows that the deformation and failure of granite under uniaxial compression can be divided into four stages: (I) compaction stage. In this stage, with the increase of axial stress, the axial strain increases rapidly, the internal micro-cracks or joints are compressed, and the rock sample stiffness gradually increases. (II) elastic stage, the deformation conforms to Hooke's law, and the axial stress-strain increases linearly. Most of the axial deformation of unloading rock samples will recover at this stage. (III) the strain hardening stage, the strain increases slowly with the continuous increase of axial stress, showing an up convex shape on the curve. In this stage, the development of old and new cracks changes from slow to rapid growth, and the micro-cracks initiate, expand, and finally coalesces to form macro-failure cracks. (IV) the post-peak failure stage, the internal structure of the rock sample has been basically destroyed, and obvious macroscopic cracks appear at this stage.

3.1.2   Total stress-strain curves and failure modes of rock with initial thermal damage

Figure 4 shows the stress-strain curve, ringing count and failure pattern of the initial thermal damage granite specimen during uniaxial compression. In order to accurately describe the effect of equivalent damage factors on rock deformation and failure, quantitative and qualitative analyses were carried out from the characteristic mechanical parameters, stress-strain morphology and ringing count. It can be seen from the figure that the characteristic parameters and stress-strain morphological changes of rock samples with equivalent damage factor D of 0–0.89 show gradual change.

From Figure 4 peak strength, strain corresponding to peak strength and elastic modulus of rock samples with different initial thermal damage can be obtained. It can be seen that the peak strength of rock samples with equivalent damage factor D from 0 to 0.89 is 166.29 MPa, 161.55 MPa, 157.84 MPa, 150.18 MPa and 102.48 MPa respectively, which has a good linear correlation before D = 0.89. When D = 0, the strain corresponding to the peak strength of the sample is 0.0046, which is 1.18, 1.37, 1.58 and 2.35 times of that of the sample with D = (0.16–0.89), respectively, indicating that the deformation of rock failure time changes greatly with the increase of equivalent damage factor. The elastic moduli of rock samples with equivalent damage factor D from 0 to 0.89 are 44.8 GPa, 36.78 GPa, 32.55 GPa, 28.77 GPa and 14.37 GPa, respectively. From the changes of the three characteristic parameters mentioned above, it can be found that with the increase of equivalent damage factor D, brittle weakening, and plastic strengthening of rock samples. It can also be seen from the figure that when the equivalent damage factor D reaches a certain critical point, mechanical strength parameters will weaken significantly, which is consistent with the changes in the ringing count rate and the total count rate observed, namely, a sudden increase in the macroscopic failure ringing count rate occurs near the peak strength. This is due to the initiation and expansion of cracks in the sample to peak failure after early loading, and the initiation of slip on the macroscopic fracture surface produces a large number of ring-number impacts.

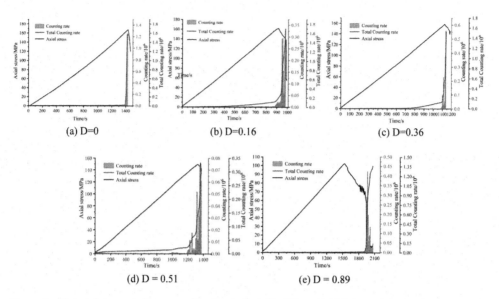

(a) D=0 (b) D=0.16 (c) D=0.36

(d) D = 0.51 (e) D = 0.89

Figure 4. Uniaxial stress-strain curves of rock samples with different initial thermal damage.

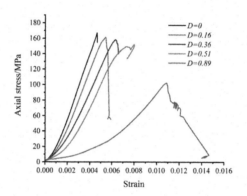

Figure 5. Stress-strain curves of samples with different initial thermal damage.

As shown in Figure 4, with the increase of equivalent damage factor D, the failure modes of rock under uniaxial loading can be roughly divided into three types: One macroscopic fracture surface (e.g., D = 0), one main macroscopic shear surface with multiple sub-fracture surfaces (e.g., D = 0.16, 0.36 and 0.51), and multiple fracture surfaces with a large amount of debris (e.g., D = 0.89). It can be seen that under the same external compression conditions, the failure modes of sample rock mass with different internal fracture densities are significantly different.

In order to compare the effects of different equivalent damage factors D on rock stress-strain under uniaxial compression, each stress-strain is plotted in Figure 5.

It can be seen from Figure 5, stress-strain curves of samples with different initial thermal damage in uniaxial compression have all gone through the compaction stage, elastic stage, crack development stage and strain softening stage of cracks or holes and other defects. However, with the increase of the equivalent damage factor, the performance is different in each stage. For example, in the consolidation stage, with the increase of equivalent damage factor on the stress-strain curve response is different, intact rock sample (D = 0) has less original crack, the external stress compression closing time is short, and with the increase of equivalent damage factor, closed micro-cracks need more external stress, therefore on the stress-strain curve under present convex more significantly; In the fracture development stage, under the influence of the internal initial fracture, the stress concentration will occur at the fracture tip when the rock is stressed. The more developed the initial fracture is, the more uniform the stress inside the rock is, and the stronger the pre-peak plasticity is displayed on the stress-strain curve. In the strain softening stage, the rate of stress decline slows down with the increase of equivalent damage factor, indicating that the brittleness of rock mass can be reduced by controlling fracture development in rock mass.

3.2 *Energy evolution analysis of rock deformation failure under uniaxial compression*

3.2.1 *Energy evolution of intact rock under uniaxial compression*

For uniaxial compression test, according to the law of energy conservation

$$W = U^d + U^e \tag{2}$$

$$u_i^e = \int_{\varepsilon_b}^{\varepsilon_c} \sigma_i d\varepsilon_i \tag{3}$$

$$u_i^d = \int_0^{\varepsilon_c} \sigma_i d\varepsilon_i - \int_{\varepsilon_b}^{\varepsilon_c} \sigma_i d\varepsilon_i \tag{4}$$

$$U_i^e = \sigma_i^2 / 2E_0 \tag{5}$$

Where, W is the energy input to rock sample by press; U^e is the elastic energy accumulated in the rock; U^d is mainly damage energy used for crack initiation and propagation during rock deformation and plastic energy used for rock particles to produce plastic deformation. ε_b and ε_c are the strain values at the point where the stress is 0 after unloading and at the beginning of unloading respectively. In this paper, the elastic modulus E_0 of the elastic segment before the peak is taken as the unloading elastic modulus E_i in the calculation of energy. For uniaxial compression test, the elastic energy U_i^e can be released at stress σ_i.

According to Equations (2) to (5), the input energy density, elastic energy density and dissipated energy density of rock samples in the process of deformation and failure under uniaxial compression were calculated, as shown in Figure 6.

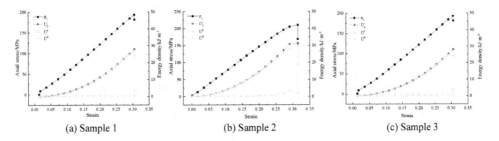

(a) Sample 1	(b) Sample 2	(c) Sample 3

Figure 6. Energy evolution of intact granite under uniaxial compression.

As can be seen from Figure 6, with the continuous increase of axial load, the evolution of total input energy and elastic energy density is similar, showing a trend of "slow growth – steady growth – cliff decline", and the elastic energy density reaches its maximum at peak strength. With the increase of axial load, the dissipated energy density presents a trend of "small increase – rapid increase", and reaches the maximum value at peak strength. In addition, it can also be seen from the figure that the elastic energy density curve is close to the total input energy curve during the whole deformation process, that is, the elastic energy density accounts for more than 95% of the total input energy. The rock samples are mainly characterized by energy accumulation, and the elastic energy releases rapidly at the peak failure stage. After entering the plastic strengthening stage, the proportion of dissipated energy density in the total input energy increases gradually due to the steady increase of dissipated energy generated by the initiation and expansion of new and old fractures in the rock sample, until the macroscopic cracks are formed by the confluence and connection of the cracks at peak failure. At this time, the dissipated energy increases rapidly, and the overall instability failure of rock mass structure occurs.

3.2.2 *Energy evolution of initial thermal damage rock under uniaxial compression*

Figure 7 shows the variation curves of input total strain energy, elastic energy, and dissipated energy of granite with different initial thermal damage during deformation and failure process under uniaxial compression. It can be seen from the figure that the energy characteristic curves of the five samples have similar evolution, which can be roughly divided into pre-peak stage and post-peak stage. In the pre-peak stage, with the increase of axial stress, the increase rate of absorbed total strain energy gradually accelerates. The evolution of elastic energy is similar to that of the total input energy. With the increase of axial stress, the growth rate of strain energy accelerates gradually, but with the accumulation of damage, the growth rate of strain energy slows down near the peak strength, and the larger the equivalent damage factor D is, the more significant this trend is. This is because the sample with large equivalent damage factor D under the same compressive stress value has more developed cracks. Before the formation of macroscopic main crack, the rock mass with relatively developed internal cracks will produce more short wing cracks near the stress concentration area at the crack tip, and absorb a large amount of surface energy, so that the elastic energy accumulated in the rock mass is less, the dissipated energy is reversed. In the post-peak stage, the energy absorbed by rock sample increases slowly with the increase of equivalent damage factor. However, the elastic energy decreases rapidly at the peak strength, while the dissipated energy increases rapidly, indicating that when the accumulated elastic energy in the rock mass exceeds the energy storage limit, the stored elastic energy in the rock mass is released rapidly, and the remaining energy is mainly used to start the kinetic energy of ejection of fragments.

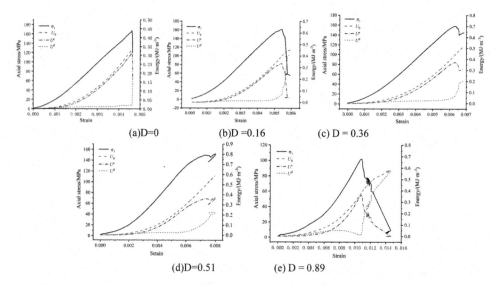

Figure 7. Energy evolution curves of rock samples with different initial thermal damage under uniaxial compression.

According to the existing research results, the energy absorbed by specimens under uniaxial compression is mainly transformed into elastic energy and dissipated energy. And the essence of rock compression failure is the accumulation of internal damage. Therefore, in order to observe the influence of fracture development degree in rock mass on energy characteristic evolution more directly. The ratio of the dissipated energy to the total input energy in each stage of rock deformation and the conversion rate of dissipated energy will be studied emphatically.

Figure 8 shows the ratio of elastic performance and dissipated energy to total input energy of different samples in the process of deformation and failure. As the rock mass of the sample is greatly affected by the equivalent damage factor when D = 0.89, the results obtained are quite different from those of other samples, which is not conducive to the rejection of the study on the evolution of energy characteristics by fracture degree. It can be seen from Figure 8(b) that in the pre-peak stage, the proportion of dissipated energy increases with the increase of fracture degree in the sample, and the average values are 0.23, 0.25, 0.3 and 0.36, respectively. This is related to the energy dissipated by the initial fracture closure and the plastic energy formed by stress concentration at the crack tip. In the post-peak stage, the decrease rate of dissipated energy ratio is significantly different. For the intact rock sample (D = 0), the dissipated energy ratio suddenly and rapidly decreases after the peak, while the decrease rate of the dissipated energy ratio of the rock sample with higher fracture degree is slower, indicating that the internal fracture degree can slow down the release of the elastic energy accumulated in the rock mass. In engineering, artificial cracks can be made to reduce the energy storage capacity of rock mass and transfer the accumulated elastic energy to the depth of rock mass. In this way, on-site energy regulation can be consciously carried out to reduce the risk of disaster.

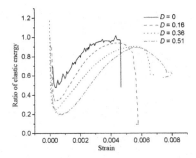 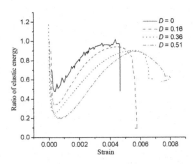

(a) Relationship between elastic energy ratio and strain (b) Relationship between dissipated energy ratio and strain

Figure 8.   Proportion of elastic energy and dissipated energy in each deformation stage.

Figure 9 shows the change curve of dissipation energy conversion rate of samples under uni-axial compression, which is used to reflect the damage rate of rock samples during deformation. The incremental dissipation energy per unit time was defined as the dissipation rate, and the U^d conversion rate was used to represent the dissipation rate.

As can be seen from Figure 9, the conversion rate of dissipated energy is low in the pre-peak stage, and the mechanical strain energy is mainly stored in the rock in the form of elastic energy at this stage. In the post-peak stage, the dissipation energy conversion rate accelerates obviously, and the maximum value occurs when the stress drop is large. With the increase of equivalent damage factor, the maximum dissipation energy conversion rates of the specimens were 0.050 MJ·m^{-3}·h^{-1}, 0.18 MJ·m^{-3}·h^{-1}, 0.027 MJ·m^{-3}·h^{-1}, 0.045 MJ·m^{-3}·h^{-1}, and 0.06 MJ·m^{-3}·h^{-1}, respectively. The dissipative strain energy conversion rate is very sensitive to the initial fracture damage of rock. That is to say, the dissipative strain energy conversion rate of the specimen is the minimum at room temperature, and the damage rate is the maximum when the equivalent damage factor is 0.16, and then increases with the increase of the equivalent damage factor. It shows that the deformation damage rate of rock mass is not only related to the accumulated energy before the peak, but also closely related to the development degree of fractures in rock mass.

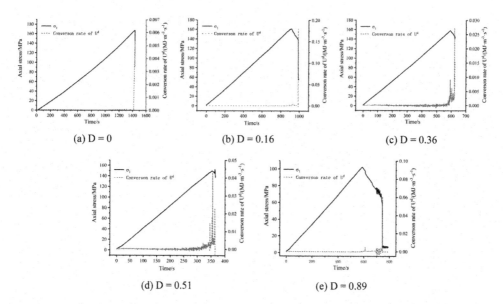

(a) D = 0   (b) D = 0.16   (c) D = 0.36

(d) D = 0.51   (e) D = 0.89

Figure 9. Evolution curve of dissipative energy conversion rate of fractured rock samples under uniaxial compression.

4 CONCLUSIONS

Based on the results and discussions presented above, the conclusions are obtained as below:

(1) For intact rock, with the continuous increase of axial load, the evolution of the total energy input from the outside and the elastic energy density is similar, showing a trend of "slow growth – stable growth – cliff decline", and the elastic energy density reaches the maximum at the peak strength. As the axial load continues to increase, the dissipated energy density presents a trend of "small increase – rapid increase", and reaches its maximum value at peak strength.

(2) With the increase of equivalent damage factor, the maximum transformation rate of rock mass exhibits a trend of "increase-decrease-increase". When D=0.16, the maximum transformation rate is 0.18 MJ·m^{-3}·h^{-1}, and there is an optimal equivalent damage factor to maximize the damage rate of rock mass.

(3) For the intact rock sample (D=0), the dissipated energy ratio suddenly and rapidly decreases after the peak, while the decrease rate of the dissipated energy ratio of the rock sample with higher fracture degree is slower, indicating that the internal fracture degree can slow down the release of the elastic energy accumulated in the rock mass.

ACKNOWLEDGMENTS

This work was supported by the National Natural Science Foundation of China and Shandong Province joint program (Grant No. U1806209) and the Basic Scientific Research Operating Expenses of Central Universities (Grant No. FRF-TP-19-021A3 and FRF-IDRY-19-002).

REFERENCES

Baghbanan A, Jing L. (2008) Stress effects on permeability in a fractured rock mass with correlated fracture length and aperture[J]. International Journal of Rock Mechanics and Mining Sciences, 45(8): 1320–1334.

Jia Z, Li C, Zhang R, et al. (2019) Energy evolution of coal at different depths under unloading conditions[J]. Rock Mechanics and Rock Engineering, 52(11): 4637–4649.

Li D, Sun Z, Xie T, et al. (2017) Energy evolution characteristics of hard rock during triaxial failure with different loading and unloading paths[J]. Engineering Geology, 228: 270–281.

Li M, Zhang J, Zhou N, et al. (2017) Effect of particle size on the energy evolution of crushed waste rock in coal mines[J]. Rock Mechanics and Rock Engineering, 50(5): 1347.

Li P, Cai M, Wang P, et al. (2021) Mechanical properties and energy evolution of jointed rock specimens containing an opening under uniaxial loading[J]. International Journal of Minerals, Metallurgy and Materials, 28(12): 1875–1886.

Liu W, Zhang S, Sun B. (2019) Energy evolution of rock under different stress paths and establishment of a statistical damage model[J]. KSCE Journal of Civil Engineering, 23(10): 4274–4287.

Meng Q, Zhang M, Han L, et al. (2016) Effects of acoustic emission and energy evolution of rock specimens under the uniaxial cyclic loading and unloading compression[J]. Rock Mechanics and Rock Engineering, 49(10): 3873–3886.

Meng Q, Zhang M, Zhang Z, et al. (2018) Experimental research on rock energy evolution under uniaxial cyclic loading and unloading compression[J]. Geotechnical Testing Journal, 41(4): 717–729.

Meng Q, Zhang M, Zhang Z, et al. (2019) Research on non-linear characteristics of rock energy evolution under uniaxial cyclic loading and unloading conditions[J]. Environmental Earth Sciences, 78(23): 1–20.

Pan J, Wu X, Guo Q, et al. (2020) Uniaxial experimental study of the deformation behavior and energy evolution of conjugate jointed rock based on AE and DIC methods[J]. Advances in Civil Engineering, 2020.

Sainoki A, Mitri H S, Chinnasane D, et al. (2019) Quantitative energy-based evaluation of the intensity of mining-induced seismic activity in a fractured rock mass[J]. Rock Mechanics and Rock Engineering, 52(11): 4651–4667.

Song S, Liu X, Tan Y, et al. (2020) Study on failure modes and energy evolution of coal-rock combination under cyclic loading[J]. Shock and Vibration, 2020.

Yang S Q, Chen M, Huang Y H, et al. (2020) An experimental study on fracture evolution mechanism of a non-persistent jointed rock mass with various anchorage effects by DSCM, AE and X-ray CT observations[J]. International Journal of Rock Mechanics and Mining Sciences, 134: 104469.

Zhang L M, Gao S, Ren M Y, et al. (2014) Rock elastic strain energy and dissipation strain energy evolution characteristics under conventional triaxial compression[J]. Journal of China Coal Society, 39(7): 1238–1242.

Zhang L, Cong Y, Meng F, et al. (2021) Energy evolution analysis and failure criteria for rock under different stress paths[J]. Acta Geotechnica, 16(2): 569–580.

Zhang M, Meng Q, Liu S. (2017) Energy evolution characteristics and distribution laws of rock materials under triaxial cyclic loading and unloading compression[J]. Advances in Materials Science and Engineering, 2017: 1–16.

Zhang Y, Feng X T, Yang C, et al. (2021) Evaluation method of rock brittleness under true triaxial stress states based on pre-peak deformation characteristic and post-peak energy evolution[J]. Rock Mechanics and Rock Engineering, 54(3): 1277–1291.

Zhang Z, Xie H, Zhang R, et al. (2019) Deformation damage and energy evolution characteristics of coal at different depths[J]. Rock Mechanics and Rock Engineering, 52(5): 1491–1503.

Zhao K, Yu X, Zhou Y, et al. (2020) Energy evolution of brittle granite under different loading rates[J]. International Journal of Rock Mechanics and Mining Sciences, 132: 104392.

Frontiers of Civil Engineering and Disaster Prevention and Control – Yang & Rahman (Eds)
© 2023 The Author(s), ISBN: 978-1-032-31200-2

Analysis of shear bearing capacity of assembled monolithic concrete frame joints strengthened with CFRP

Yan Cao

School of Arts Design, Wuchang University of Technology, Wuhan, China

Zhao Yang*

School of Urban Construction, Wuhan University of Science and Technology, Wuhan, China

ABSTRACT: In order to grasp the shear performance of the seismically damaged assembled monolithic concrete frame joints reinforced by carbon fiber reinforced polymer (CFRP), the paper carried out finite element simulation analysis on the basis of the previous experimental research, and studied the shear performance of the reinforced seismically damaged joints under different axial compression ratios. Parameters such as skeleton curve, yield load, yield displacement, initial stiffness and peak load are analyzed. The research shows that CFRP reinforcement can effectively improve the shear performance of seismically damaged joints. The effect of axial compression ratio on the shear performance of seismically damaged frame joints after CFRP reinforcement is obvious. Increasing the axial compression ratio can improve the stiffness and bearing capacity of the reinforced joint, but it will reduce the ultimate displacement and ductility of the joint. The increase in the degree of seismic damage will reduce the reinforcement effect of the CFRP. The research results provide a reference for the reinforcement of seismically damaged prefabricated frame joints.

1   INTRODUCTION

Prefabricated concrete structure has the advantages of high level of industrialization, convenient winter construction, reducing the amount of wet work on the construction site, reducing material consumption, reducing dust and construction waste (Wang et al. 2021). As a large number of new generations assembled concrete structures are put into use in China, it is of great value to carry out research on seismic damage and reinforcement. For the widely used assembled monolithic concrete frame structure, beam-column joint is an important part to ensure effective force transmission and structural integrity. Due to the construction interface between precast components and post-cast concrete at the joint, the seismic damage of such joints is different from that of cast-in-situ concrete frame joints (Hu et al. 2021).

The FRP reinforcement method has the advantages of high material strength, light quality and simple process. Many scholars have carried out experimental research and theoretical analysis on FRP reinforced concrete structures. Abu Tahnat et al. (2018) used FRP sheet wraps to improve the ductility of R.C joints, results show that the using CFRP wraps around beam converts the brittle failure to ductile failure. Attari et al. (2019) used carbon and fiberglass fabric, and a hybrid braided FRP fabric to strengthen concrete beam-column joints. Yang et al. (2018) studies the seismic behavior of RC beam-column joints strengthened with sprayed FRP; the results show that sprayed FRP strengthening can improve the seismic behavior of RC beam-column joints effectively. Llkhani et al. (2019) used the data collected from the existing standards and studies regarding the FRP strengthened RC joints to develop an artificial neural network model to predict the shear strength contribution of FRP jacket to the joints. In addition, scholars have also studied

---

*Corresponding Author: yzwh77@163.com

                                        DOI 10.1201/9781003308577-10

the effects of reinforcement forms (Ha et al. 2013), FPR materials (Li et al. 2019), dry and wet environments (Chotickai P. & Somana S. 2018), and other factors on FRP-reinforced reinforced concrete structures.

In general, the current research on FRP-reinforced concrete structures is basically used in cast-in-place structures, while the research on the reinforcement of seismically damaged assembled monolithic structures is relatively rare. It is necessary to carry out special research on the seismic damage reinforcement and bearing capacity of such joints. Based on the previous test data, this paper uses the finite element software ABAQUS to establish an assembled monolithic beam-column joint model with seismic damage strengthened by carbon fiber, and analyzes the shear performance and influencing factors of the joint.

2 TEST INTRODUCTION

Four assembled monolithic reinforced concrete frame joints with the same size and reinforcement are made with the scale ratio of 1 : 2, and the number is KJ-0, KJ-1, KJ-2 and KJ-3, respectively. The size of each component of the specimen and the reinforcement are shown in Figure 1, and the specimen after assembly is shown in Figure 2. The main materials of the specimens are concrete and steel bars. The design strength grade of concrete is C30. The average value of the measured concrete cube compressive strength is 34.3 MPa and the elastic modulus is 3×104 MPa. The longitudinal reinforcement is HRB400 steel bar with 12 mm diameter, and the yielding strength is 423 MPa. The stirrup is 6 mm HPB300 steel bar and the yielding strength is 342 MPa. The elastic modulus of these two kinds of steel bars is 2.01×105 MPa. The tensile strength of CFRP is 3,216 MPa and the elastic modulus is 2.2×105 MPa. The specimen pre-damage and reinforcement scheme is shown in Table 1 and Figure 3.

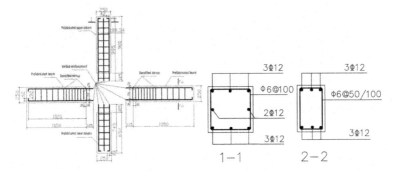

Figure 1.   Dimensions and Reinforcement Diagrams of Components.

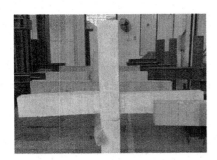

Figure 2.   Picture of specimens.

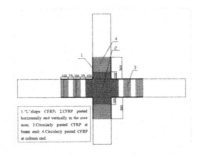

Figure 3.   Reinforcement specimens.

Table 1. Pre-damage and reinforcement scheme of specimen.

Specimen number	Axial compression ratio	Damage degree	Pre-damaged interlayer displacement angle (displacement)	reinforcement method
KJ-0	0.4	No pre-damage	–	–
KJ-1	0.4	No pre-damage	–	CFRP strengthening
KJ-2	0.4	Mild pre-damage	1/150 (10 mm)	CFRP strengthening
KJ-3	0.4	Moderate pre-damage	1/75 (20 mm)	CFRP strengthening

3 FINITE ELEMENT MODEL OF BEAM-COLUMN JOINTS

3.1 *Establishment of finite element model*

C3D8R element in Abaqus was used to simulate concrete, using T3D2 element to simulate steel bar, m3D4R element was selected to simulate CFRP. In this paper, the finite element model of the test specimen is established. Three-dimensional concrete and steel models are created by stretching, and CFRP models are created by using three-dimensional shell elements. Then the properties of concrete and steel bar are defined according to the relevant constitutive model. In order to make the model more consistent with the mechanical characteristics of the core area of the fabricated structure, a 10 mm weak area is set at the connection between the post pouring area and the beam column. After the components are created, they are assembled in the general coordinate system, as shown in Figure 4.

a) Unreinforced model   b) Reinforced model

Figure 4. Assembly schematic of finite element model.

3.2 *Loading and failure criteria*

The loading steps of the finite element model are divided into two types according to the direct loading and the pre-damage loading. Under direct loading, the bottom of the column and the beam are firstly restrained, and then the vertical constant load is applied at the top of the column, and then the horizontal displacement is applied on the upper column. When pre-damaged reinforcement is loaded again, the bottom end of the lower column and the two ends of the beam are first constrained, and then the vertical constant force is applied at the top of the column, and then the pre-damaged horizontal displacement is applied on the upper column, and then the carbon fiber element is activated, and then the horizontal displacement is applied on the upper column.

Relative slip is not considered between steel and concrete, carbon fiber and concrete. The friction coefficient between concrete components is set to 0.85. The bottom end of the lower column is set as hinge, and the beam end is set as sliding hinge support. The compressive stress of 10 MPa is applied on the top of the upper column, and the horizontal displacement is applied. The applied load and boundary conditions are shown in Figure 5. The mesh size of concrete and steel elements is 25 mm, and that of CFRP elements is 10 mm. After the mesh is completed, as shown in Figure 6.

When the finite element model appears one of the following phenomena, determine the model failure : 1)In the current cycle, the maximum load is less than 85 % of the maximum load in the previous cycle; 2)Serious global or local instability occurred in the specimen; 3)Due to material failure, the finite element iteration does not converge, and the calculation cannot continue.

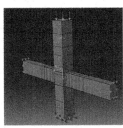

a) Reinforced model

b) Unreinforced model

Figure 5. Assembly schematic of finite element model.

Figure 6. Meshing of reinforcement model and unreinforced model.

4 ANALYSIS OF FINITE ELEMENT SIMULATION RESULTS

4.1 *Comparison of finite element and experimental results*

As shown in Figure 7, the skeleton curve obtained by finite element simulation is close to that obtained by experiment. In addition, the stress distribution program obtained by the finite element simulation is in good agreement with the failure phenomenon of the specimen in the test. Taking the KJ-1 specimen as an example, it can be seen that the maximum stress of the concrete and carbon fiber cloth in the specimen is mainly concentrated in the joint core area and the beam-column ends, and the cracks in the test specimen are mainly concentrated in this area as well. The above comparison proves that the finite element model established in this paper can accurately reflect the mechanical properties of actual joint specimens.

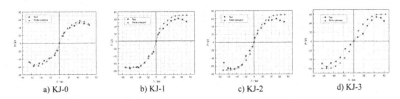

a) KJ-0　　　　　　b) KJ-1　　　　　　c) KJ-2　　　　　　d) KJ-3

Figure 7. Comparison of skeleton curves simulated by each specimen test.

a) Stress distribution of concrete

b) Stress distribution of CFRP

Figure 8. Stress distribution of KJ-1 obtained by the finite element simulation.

4.2 Influence analysis of axial compression ratio

The skeleton curves of each specimen under 0.2, 0.4 and 0.6 axial compression ratios were obtained by simulation, as shown in Figure 9. The related performance parameters are shown in Table 2. According to the above chart:

1) As the axial compression ratio increases, the limit displacement of each simulated specimen gradually decreases, indicating that increasing the axial compression ratio reduces the deformation capacity of the specimen;

2) As the axial compression ratio increases, the initial stiffness of each simulated specimen increases, the yield displacement decreases, the yield strength increases, and the ultimate bearing capacity increases, indicating that increasing the axial compression ratio can effectively improve the stiffness and bearing capacity of the specimen. force;

3) The axial compression ratio increases from 0.2 to 0.4 and 0.6, and the ultimate bearing capacity of KJ-0 is increased by 7.9% and 19.5%; the ultimate bearing capacity of KJ-1 is increased by 10.9% and 22.7%; the ultimate bearing capacity of KJ-2 is increased 7.3% and 19.8%; the ultimate bearing capacity of KJ-3 increased by 9.6% and 18.1%. Compared with the control specimen KJ-0, the effect of axial compression ratio on the bearing capacity of the carbon fiber reinforced specimen KJ-1 is more significant; but as the degree of seismic damage deepens, the bearing capacity enhancement effect also decreases.

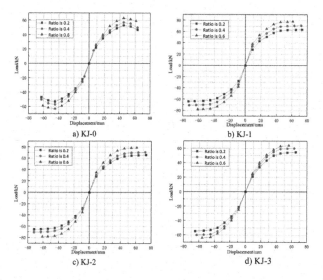

a) KJ-0   b) KJ-1

c) KJ-2   d) KJ-3

Figure 9.   Skeleton curve of simulated specimen.

Table 2.   Performance parameters of simulated specimens under different axial compression ratios.

Ratio	Parameter	Yield load/kN	Yield displacement/mm	Initial stiffness/kN/mm	Peak load/kN
KJ-0	0.2	22.4	9.3	2.4	53.2
	0.4	25.3	9.3	2.7	57.4
	0.6	26.5	9.2	2.9	63.6
KJ-1	0.2	28.5	9.5	3	63.8
	0.4	34.7	9.5	3.7	70.8
	0.6	38.6	9.4	4.1	78.3
KJ-2	0.2	28.1	9.4	3.0	65.5
	0.4	30.3	9.4	3.2	70.3
	0.6	35.8	9.2	3.9	78.5
KJ-3	0.2	21.4	9.3	2.3	54.7
	0.4	23.2	9.3	2.5	60
	0.6	25.2	9.2	2.7	64.6

5 CONCLUSIONS

In this study, the effect of the axial compression ratio on the shear performance of the seismic-damaged assembled monolithic frame joints reinforced with carbon fiber sheets was studied by finite element simulation of the test. The research shows that:

1) Strengthening the seismic damaged frame joints with carbon fiber sheets can effectively improve the shear resistance of the joints.
2) It is effective to simulate the shear performance of seismically damaged frame joints by considering the concrete strength reduction.
3) The greater the axial compression ratio, the greater the ultimate bearing capacity of the reinforced joint, but the decrease in ductility.
4) This study only considers one influencing factor, and there is only one type of joint, so the results of the study have certain limitations. In the future, more influencing factors and different types of nodes will be considered for research.

ACKNOWLEDGEMENTS

This research is funded by the Science and Technology Research Project of Education Department of Hubei Province (Project No. B2019289) and Philosophy and Social Science Research Project of Education Department of Hubei Province (Project No. 21G149). Their support is gratefully acknowledged.

REFERENCES

Abu Tahnat, Y. B., Dwaikat, M. M. S., & Samaaneh, M. A. (2018). Effect of using CFRP wraps on the strength and ductility behaviors of exterior reinforced concrete joint. Composite Structures, 201, 721–739.

Attari, N., Youcef, Y. S., & Amziane, S. (2019). Seismic performance of reinforced concrete beam-column joint strengthening by FRP sheets. Structures, 20, 353–364.

Chotickai P. & Somana S. (2018). Performance of CFRP-strengthened concrete beams after exposure to wet/dry cycles. Journal of Composites for Construction, 22(6): 1090–0268.

Ha G.J., Cho C.G., Kang H.W., et al. (2013). Seismic improvement of RC beam column joints using hexagonal CFRP bars combined with CFRP sheets. Composite Structures, 95(1): 464–470.

Hu Z., Shah Y. I. & Yao P. (2021). Experimental and numerical study on interface bond strength and anchorage performance of steel bars within prefabricated concrete. Materials(13).

Li Y., Liu X.F., Wang Z., et al. (2019). Experimental study on reinforcement and chloride extraction of concrete column with MPC-CFRP composite anode. Kscf Journal of Civil Engineering, 23(4): 1766–1775.

Llkhani, M. H., Naderpour, H., & Kheyroddin, A. (2019). Soft computing-based approach for capacity prediction of FRP-strengthened RC joints. SCIENTIA IRANICA, 26(5), 2678–2688.

Wang X., Li L., Deng B., Zhang Z. & Jia L. (2021). Experimental study on seismic behavior of prefabricated RC frame joints with T-shaped columns. Engineering Structures.

Yang, Z., Liu, Y., & Li, J. J. (2018). Study of Seismic Behavior of RC beam-column joints strengthened by sprayed FRP. Advances in Materials Science and Engineering, 2018.

Frontiers of Civil Engineering and Disaster Prevention and
Control – Yang & Rahman (Eds)
© 2023 The Author(s), ISBN: 978-1-032-31200-2

Structural design and analysis of recycled aggregate concrete high-rise building

Jia-Sen Lu*
Shanghai Urban Construction Design & Research Institute (Group) Co., Ltd., Shanghai, China

ABSTRACT: This paper studies the key issues of recycled aggregate concrete in the structural design and analysis of high-rise buildings. Firstly, the design strength of recycled aggregate concrete is determined based on the partial safety factor. Secondly, the static elastic-plastic analysis model of recycled aggregate concrete high-rise structures and the plastic hinge rotation limit of recycled aggregate concrete members is proposed, and the seismic analysis method of recycled aggregate concrete high-rise buildings and the design method of recycled aggregate concrete members are suggested. Finally, the basic principles of structural design and analysis of recycled aggregate concrete high-rise buildings and the strengthening measures for earthquake-resistant structures are proposed. The reasonability of the seismic analysis method in this paper is proven, and the world's first recycled aggregate concrete high-rise building is successfully designed.

1   GENERAL INSTRUCTIONS

The application of recycled aggregate concrete (RAC) can effectively solve the environmental and ecological problems caused by the growing amount of waste concrete in cities. Scholars at home and abroad have carried out a lot of experimental research and theoretical analysis on the performance of RAC, which provides a foundation for the use of RAC structures in practical engineering.

Scholars have carried out experimental research on the stress-strain curve of RAC. Henrichshen & Jensen (1989) found that the shape of the stress-strain curve of RAC is similar to that of normal concrete (NC). Bairagi et al. (1993) found that the curvature of the stress-strain curve increases as the replacement rate of recycled aggregate increases. Topçu & Günçan (1995) found that as the replacement rate of recycled aggregate increased, the peak stress of RAC decreased and the elastic modulus decreased. Rqhl & Atkinson (1999) found that the peak strain of RAC increased as the replacement rate of recycled aggregate increased. Xiao et al. (2005, 2008) found that with the increase of the replacement rate of recycled aggregate, the curvature of the ascent part of the stress-strain full curve gradually increased, and leading to a decrease of the elastic modulus and more brittle than NC. Belen et al. (2011) found that the presence of recycled aggregate increases the peak and ultimate strain of RAC. It can be seen that these stress-strain relationships are the stress-strain relationships of unconfined RAC, and there is no research on the stress-strain relationship of confined RAC.

Scholars have completed some quasi-static tests of beam-column joints of RAC frames (Corinaldesi et al. 2011; Tian 2011; Xiao & Zhu 2005), and the test results show that the failure process of RAC joints is similar as that of NC joints, but the brittleness characteristics of the failure process are more obvious, and the seismic performance does not decrease significantly, which meets the seismic design requirements.

There are few studies on the seismic performance of RAC structures abroad, but in recent years, scholars at home have conducted research on this structures (Cao & Zhang 2011; Min et al. 2011; Liu et al. 2014; Xiao et al. al. 2006), the seismic characteristics of the specimens, such as failure mechanism, force characteristics and failure form, are not significantly different from those of NC frame structures, and the seismic performance is slightly reduced, but still meets the

---

*Corresponding Author: Lujiasen99@sina.com

DOI 10.1201/9781003308577-11

seismic requirement of the current code. At present, the codes (ASCE 2007, GB50011-2010 2016) lack the plastic hinge rotation limit of the performance-based design of RAC structures, and the performance-based analysis and design cannot be carried out.

Cao et al. (2020) pointed out that the use of RAC mainly focuses on pavement and low-rise building. Most of the existing researches focus on medium and low-strength RAC, and there is a lack of research on high-strength RAC. The measures for the seismic performance of RAC structures and the numerical simulation based on experimental research and theoretical analysis need to be improved.

This paper studies the key issues of RAC in the structural design and analysis of high-rise building based on a project. The design strength of different grades, including high-strength RAC (such as RAC50, that is the cube compressive strength of RAC is 50 Mpa), is determined through the partial safety factor. Confined RAC stress-strain curve, elastic-plastic analysis method and the plastic hinge rotation limit of RAC were given, and the seismic analysis method of RAC high-rise building and RAC member design were suggested, and finally put forward the basic principles of structural design and analysis of RAC high-rise buildings and the strengthening measures of seismic structures. The reasonability of the seismic analysis method in this paper is proven by the comparative analysis of numerical analysis and experimental results, and the world's first RAC high-rise building is successfully designed by using the proposed method in this paper.

2 RAC MATRIAL PARAMETER

2.1 Partial safety factor

In order to determine the bearing capacity of RAC structural members, the characteristic value of RAC strength f_{Rk} should be determined, then the design strength f_R of RAC is determined based on the partial safety factor γ_R.

$$f_R = \frac{f_{Rk}}{\gamma_R} \tag{1}$$

Xiao & Li (2005) found that the conversion relationship between the characteristic value of the axial compressive strength of RAC and that of the cube compressive strength is the same as that of NC. Considering the brittleness of RAC and lack of engineering application in high-rise building, the partial safety factor of RAC should be considered to improve the safety reserve of the structure, and to achieve the reliability similar to the NC.

The replacement rate of recycled aggregate is controlled at a lower level of 30 % and below for high-rise building, the partial safety factor of RAC40 and below is increased from 1.4 to 1.5, that is, the design strength of RAC is 93% that of the same grade NC. For RAC50, because of less research at home and no current code support, its partial safety factor increases to 1.55, and its design strength is 90% of that of the same grade NC.

10,000 members are selected to calculate the bearing capacity of RAC members, and the ratio of the bearing capacity between RAC members and NC members with the same specified concrete compressive strength, as shown in Table 1.

Table 1. Ratio of bearing capacity between RAC and NC members*.

Design method	Normal section bending	axial compression	Shear
DB11/T803-2011	0.9949	0.9705	0.9643
DBJ61/T88-2014	0.9826	0.9	0.9643
Proposed method (RC40, RC30)	0.9930	0.9607	0.9762
Proposed method (RC50)	0.9865	0.9429	0.9655

*the replacement rate of recycled aggregate is controlled at a lower level of 30% and below?

Table 2. Elastic Modulus (MPa).

Grade	RAC30	RAC40	RAC50
Elastic Modulus	2.70	2.92	3.10

It can be seen from Table 1 that the bearing capacity ratio of proposed method is basically consistent with that of DB11/T803-2011and DBJ61/T88-2014, and the error is basically within 5%, which verifies the reliability of the proposed design method for RAC members.

2.2 Elastic modulus

According to Xiao (2008) experimental and theoretical analysis, and referring to the current RAC codes, the elastic modulus of RAC is 90% of the same specified concrete compressive strength of NC, as shown in Table 2.

3 RAC PERFORMANCE-BASED DESIGN

3.1 Nonlinear stress-strain curve

Unconfined NC stress-strain curve is based on the Chinese concrete code (GB50010-2010, 2015). Unconfined RAC stress-strain curve is based on Xiao (2008). The coordinates of the four control points (Figure 1) of the unconfined NC normalized simplified stress-strain curve are: $A_c(0.6f_{c,r}/Ec/\varepsilon_{c,r},0.6)$, $B_c(0.85,1)$, $C_c(1.15,1)$, $D_c(\varepsilon_{cu0}/\varepsilon_{c,r},0.5)$, where $f_{c,r}$ is the characteristic value of the uniaxial compressive strength of NC, i.e. $f_{c,r}=f_{ck}$, $\varepsilon_{c,r}$ is the strain peak at $f_{c,r}$, ε_{cu0} is the compressive strain when the uniaxial compressive strength in NC descent part of stress-strain curve is $0.5f_{c,r}$, Ec is the elastic modulus of NC. The coordinates of the four control points (Figure 1) of the unconfined RAC normalized simplified stress-strain curve are: $A_{Rc}(1.2f_{Rc,r}/ERc/\varepsilon_{Rc,r},0.6)$, $BRc(0.95,1)$, $C_{Rc}(1.05,1)$, $D_{Rc}(\varepsilon_{Rcu0}/\varepsilon_{Rc,r},0.5)$, where $f_{Rc,r}$ is the characteristic value of the uniaxial compressive strength of RAC, i.e. $f_{Rc,r}=f_{Rck}$, $\varepsilon_{Rc,r}$ is the strain peak at $f_{Rc,r}$, ε_{Rcu0} is the compressive strain when the uniaxial compressive strength in RAC descent part of stress-strain curve is $0.5f_{Rc,r}$, E_{Rc} is the elastic modulus of RAC.

Confined NC stress-strain curve is based on the Mander concrete stress-strain curve, and the coordinates of the four control points (Figure 2) of the confined NC normalized simplified stress-strain curve are: $A_{cc}(0.6f_{cc}/E_c/\varepsilon_{cc},0.6)$, $B_{cc}(0.75,1)$, $C_{cc}(1.25,1)$, $D_{cc}(\varepsilon_{cu}/\varepsilon_{cc},0.6)$, where f_{cc} is the characteristic value of the compressive strength of confined NC, ε_{cc} is the strain peak at f_{cc}, ε_{cu} is the concrete spalling strain for confined NC.

Confined RAC stress-strain curve is based on the modified Mander concrete stress-strain curve by reducing the gradient of the descent part to consider the reduction of the ductility of RAC, and the coordinates of the four control points (Figure 2) of the confined RAC normalized simplified stress-strain curve are: $A_{Rcc}(1.2f_{Rcc}/ERc/\varepsilon_{Rcc},0.6)$, $B_{Rcc}(0.85,1)$, $C_{Rcc}(1.15,1)$, $D_{Rcc}(\varepsilon_{Rcu}/\varepsilon_{Rcc},0.6)$, where f_{Rcc} is the characteristic value of the compressive strength of confined RAC, ε_{Rcc} is the strain peak at f_{Rcc}, ε_{Rcu} is the concrete spalling strain for confined RAC and calculates as follows.

$$\varepsilon_{Rcu}=\frac{\varepsilon_{Rcu0}-\varepsilon_{Rc,r}}{\varepsilon_{cu0}-\varepsilon_{c,r}} \cdot (\varepsilon_{cu}-\varepsilon_{cc}) +\varepsilon_{Rcc} \tag{2}$$

3.2 Static elastic-plastic analysis modal

A static elastic-plastic finite element model is established based on SAP2000. The beams and columns use the centralized plastic hinge element, the beams use M hinges, and the columns use fiber PMM hinges. Shear Walls use layered Shell element. The bearing capacity of the M hinges

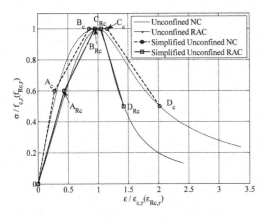

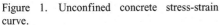

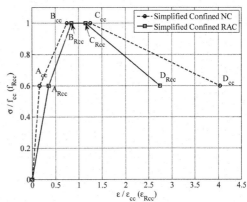

Figure 1. Unconfined concrete stress-strain curve.

Figure 2. Confined concrete stress-strain curve.

of the beams is calculated according to the Chinese concrete code (GB50010-2010, 2015). Fibers of column section and layers of shear wall use nonlinear stress-strain curve shown in 3.1.

Unconfined concrete material models are used for the simulation of shear wall without confined boundary elements. Confined concrete material models are used for the simulation of confined boundary elements of shear wall, confined fiber elements of columns and coupling beams.

3.3 Seismic performance objective

Seismic performance objective includes the plastic hinge rotation limit and seismic performance level. Based on Xiao (2008) experimental results, the ductility coefficient of recycled concrete is decreased by 89%. When the replacement rate of recycled aggregate is 30 % and below, the plastic hinge rotation limit of RAC members is put forward by reducing to 85% of the plastic hinge rotation limit of NC members in ASCE (2007). The story drift of RAC high-rise building should be less than 1/100, the seismic performance level of RAC high-rise building meets the D level of JGJ 3-2010, 2011.

3.4 Comparison between theoretical analysis and experiments

The ultimate push load of 3 RAC frames in Xiao (2008) and 1 RAC frame-shear-wall structure in Cao (2011) are numerically simulated using the proposed static elastic-plastic analysis model in 3.2. The comparison between theoretical analysis and experiments are shown in Table 3.

Table 3. The ultimate push load of RAC structures (Unite: KN).

No.		F30*	F50*	F100*	KJQ-1**
Ultimate push load	Experiment	93.44	88.98	87.90	267.15
	Proposed method	91.8	92.2	89.8	260.9
	Error	-1.8%	3.6%	2.2%	-2.3%

* Xiao (2008) experiments, ** Cao (2011) experiment.

Table 3 shows that the error is basically within 5%, which verifies the reliability of the proposed static elastic-plastic finite element model for RAC structures.

4 RAC STRUCTURE DESIGN METHOD

4.1 *Structural design and analysis principle*

Earthquake action reduction is reduced due to the elastic modulus reduction of RAC. Considering the measured data of RAC elastic modulus are higher than the value of NC elastic modulus, elastic modulus reduction factor is 1.0 in seismic elastic analysis. When the displacement of RAC high-rise building is analyzed, the elastic modulus is determined according to Table 2, the period ratio, displacement ratio and story drift of the structure can meet the requirements of the Chinese seismic code GB50011-2010 (2016).

By modifying the design parameters of RAC (such as material design strength and elastic modulus), and using the component design method provided by the concrete specification GB50011-2010 (2016), the design of RAC structures can be realized on the main structural design software (such as YJK and MIDAS).

4.2 *Details of seismic design*

In order to further improve the safety of RAC structures, details of seismic design are put forward as follows:

(1) Control the application scope of RAC materials. Except for the plastic hinge areas at the bottom of the vertical structures (mainly 1st and 2nd story) and below ±0.000, RAC materials can be used.
(2) Control the replacement rate of recycled aggregate in RAC materials at a low level (30% and below). The replacement rate of RAC50 recycled aggregate is 10%, and the replacement rate of RAC40 and below recycled aggregate is 30%.
(3) The limit value of the axial compression ratio of RAC structures is 0.05 lower than that of NC structures.
(4) The seismic level of RAC structure is increased by one level.
(5) The recycled aggregate should meet the requirements of fixed site, fixed point, and fixed proportion, that is, single source.

5 THE WORLD'S FIRST RAC HIGH-RISE BUILDING DESIGN AND ANALYSIS

5.1 *Project overview*

Wu Jiao Chang Town 340 block commercial office building 2A is located in Shanghai Yang Pu District China, and adopts RAC frame-shear-wall structure system with the height of 45.6 m, and is the first high-rise recycled concrete building in the world (Figure 3). The plane size is 46.2 m × 29.4 m (Figure 4), and the distribution of RAC material is shown as Table 4.

5.2 *Analysis of frequent earthquake and wind load*

The mode decomposition response spectrum method and elastic time history analysis method is adopted in frequent earthquake analysis. The software YJK and MIDAS are used, and the main results are shown in Table 5.

5.3 *Seismic performance evaluation under rare earthquake*

The static elastic-plastic analysis (i.e., pushover) of the structure is carried out, by looking for the corresponding performance point, the elastic-plastic story drift, top displacement, and bottom shear of the structure under rare earthquake are expressed, as shown in Table 6, and the distribution diagram of plastic hinge at rare earthquake performance point is shown in Figure 8.

Figure 3.   Site photograph.

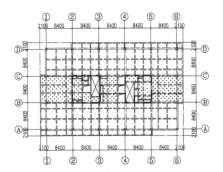

Figure 4.   The typical plane diagram.

Table 4.   The distribution of concrete material.

Floor	Level	Height	Shear Wall, Column		Beam, Slab	
	(m)	(m)	Grade	r*	Grade	r
Roof	49.60	–	–	–	RAC30	30%
12	45.60	4.0	RAC40	30%	RAC30	30%
7–11	26.05–41.65	3.9	RAC40	30%	RAC30	30%
4–6	14.35–22.15	3.9	RAC50	10%	RAC30	30%
3	9.85	4.5	RAC50	10%	NC30	–
2	5.35	4.5	NC50	–	NC30	–
1	–0.05	5.4	NC50	–	NC35	–
B1	–5.50	5.45	NC50	–	NC30	–
B2	–9.00	3.50	NC50	–	–	–

* r is the replacement rate of recycled aggregate.

Table 5.   Analysis of frequent earthquake and wind load.

Software		YJK	MIDAS
Total Weight/t		41147	41253
1,2 horizontal period/s		T1=1.47,T2=1.33	T1=1.49,T2=1.33
1torsion period/s		T3=1.31	T3=1.32
Period ratio		0.891	0.886
Base shear/KN	X earthquake	10968	11265
	Y earthquake	11897	12317
Shear weight ratio(1.2%)	X	4.87%	4.78%
	Y	5.28%	5.22%
the proportion of overturning	X earthquake	27.7%	21.5%
moment of frame	Y earthquake	18.9%	14.1%
Max story drift	X wind	1/7871	1/8474
	Y wind	1/6049	1/6389
	X earthquake	1/892	1/866
	Y earthquake	1/1016	1/990
Max story drift of 1F	X earthquake	1/2021	1/2018
	Y earthquake	1/2081	1/2145
Displacement Ratio	X	1.22	1.06
	Y	1.32	1.23

Table 6. Performance point under rare earthquake.

Earthquake direction	mode	Load step	Max Layer drift		Top Displacement mm	Base shear KN
X	uniform	68	1/173	7	220.6	44617
	modal	63	1/192	6	203.4	49661
Y	uniform	60	1/140	12	205.6	45568
	modal	56	1/152	12	191.5	50053

At the performance point, the structural damage is mainly concentrated on the shear wall, most of the reinforcement of the coupling beams yield, some tension and compression cracks appear in the bottom shear wall, some end columns of the bottom shear wall are pulled out, the frame beam has plastic hinges, the state is between B and IO, and the frame columns in the structure do not have plastic hinges.

6 CONCLUSIONS

This paper studies the key issues of RAC material in the structural design and analysis of high-rise buildings, and puts forward a seismic analysis method for RAC high-rise buildings and a design method for RAC components based on main structural design software.

(1) The replacement rate of recycled aggregate is controlled at a low level of 30% and below. For RAC40 and below, the partial safety factor γRc is increased from 1.4 of NC to 1.5. For RAC50, the partial safety factor γRc is further increased to 1.55.
(2) Based on the modified Mander confined concrete model, the confined RAC model is suggested by reducing the gradient of the descent part to consider the reduction of the ductility of RAC. The plastic hinge rotation limit of RAC members is put forward by reducing to 85% of the plastic hinge rotation limit of NC members in ASCE (2007). The static elastic-plastic analysis model of RAC structure is established, and the reasonability of the seismic analysis method in this paper is proven.
(3) By modifying the design parameters of RAC (such as material design strength and elastic modulus), and using the component design method provided by the concrete specification GB50011-2010 (2016), the design of RAC structures can be realized on the main structural design software (such as YJK and MIDAS). The reliability of the bearing capacity design for RAC members is proven
(4) In order to further improve the safety of RAC structures, details of seismic design are put forward, such as the strength control of RAC materials, the elastic modulus reduction factor, the scope control of RAC application, the replacement rate of recycled aggregate, the limit of axial compression ratio, the seismic level, and the requirement of single source.

As the seismic performance of RAC is widely concerned, the performance-based design of RAC will be studied in the future, especially the dynamic elastic-plastic analysis method of recycled concrete structures will be further improved.

REFERENCES

ASCE (2007), *Seismic Rehabilitation of Existing Buildings*, American Society of Civil Engineers.
Bairagi, N.K., Ravande, K., Pareek, V.K. (1993). Behavior of concrete with different proportions of natural and recycled aggregates. *Resources conservation and recycling.* 9(1): 109-126.

Belén, G., Fernando, M., Diego, C.L. et al. (2011) Stress–strain relationship in axial compression for concrete using recycled saturated coarse aggregate. *Construction and Building Materials*. 25(5): 2335-2342.

Cao, W.L., Xiao, J.Z., Ye, T.P., Li, L.L. (2020). Research progress and engineering application of reinforced RAC structure. *Journal of Building Structures*, 41(12): 1-16

Cao, W.L., Yin H.P., Zhang J.W. (2011). Seismic Behavior Experiment of Recycled Concrete Frame Structures. *Journal of Beijing University of Technology*, 37(2): 191-198.

Cao, W.L., Yin, H.P., Zhang, J.W. (2010) Experiment study on the seismic behavior of recycled concrete frame-shear-wall structures. *Engineering Mechanics*, 27(2): 135–141.

Corinaldesi V., Letelier V., Moriconi G. (2011) Behavior of beam-column joints made of recycled-aggregate concrete under cyclic loading. *Construction and Building Materials*, 2011(25): 1877-1882.

DB11/T803-2011, *Code for design of recycled concrete structures*. Beijing: China Architecture press.

DBJ61/T88-2014, *Technical specification for recycled concrete structures*. Xi'an: Shaanxi architectural standard design office.

DG/TJ08-2018-2007, *Technical specification for application of recycled concrete*. Shanghai: Tongji University.

GB50010-2010. (2015). *Code for design of concrete structures*. Beijing: China Construction Industry Press.

GB50011-2011. (2016). *Code for seismic design of buildings*. Beijing: China Construction Industry Press.

Herinchsen, A., Jensen, B. (1989). *Styrkeegenskaber for beton med genanvendelsesmaterialer*. Internal report, only available in Danish.

JGJ 3-2010. (2011) *Technical specification for concrete structures of high-rise buildings*. Beijing: China Construction Industry Press.

Liu, X.L., Zhang, C.Q., Zhou, Y., (2014) Seismic performance of reinforced totally-recycled concrete frame. Journal of Central South University(Science and Technology), 45(6): 1932-1942.

Min, Z., Sun, W.M., Guo, Z.G. (2011) Experimental research on seismic behavior of recycled concrete frames. *World Earthquake Engineering*, 27(1): 22-27.

Rqhl, M., Atkinson, G. (1999). The influence of RAC on the stress-strain relation of concrete. *Darmstadt concrete*. 26(14): 36-52.

Tian Y., (2011). *Research on Seismic Behavior of RAC Frame Knee Joints*. Hefei University of Technology.

Topcu, I.B., Güncan, N.F. (1995). Using waste concrete as aggregate. *Cement and Concrete Research*. 25(7): 1385-1390.

Xiao J. Z., Sun Y. D., Falkner H. (2006) Seismic performance of frame structures with RAC. *Engineering Structures*, 28 (1): 1-8.

Xiao, J.Z. (2008). *Recycled concrete*, Beijing: Construction Industry Press.

Xiao, J.Z., Du, J.T. (2008). Complete Stress-Strain Curve of Concrete with Different Recycled Coarse Aggregates under Uniaxial Compression. Journal of Building Materials. 11(1): 111-115. (in Chinese)

Xiao, J.Z., Li, J.B. (2005). Study on Relationships between Strength Indexes of Recycled Concrete. Journal of Building *Material*. 8(2): 197-201.

Xiao, J.Z., Li, J.B., Zhang, C. (2005). Mechanical properties of RAC under uniaxial loading. *Cement and Concrete Research*. 35(6): 1187-1194.

Xiao, J.Z., Zhu X.H., (2005). Study on Seismic Behavior of Recycled Concrete Frame Joints. *Journal of Tongji University (Natural Science Edition)*, 33 (4): 436-440.

Frontiers of Civil Engineering and Disaster Prevention and
Control – Yang & Rahman (Eds)
© 2023 The Author(s), ISBN: 978-1-032-31200-2

Seismic reinforcement effect of civil structure buildings

Yilei Du*
Xian University of Science and Technology, Xian, China

ABSTRACT: Civil structure is the most common form of residential structure in rural areas in Yunnan, and its seismic performance is poor in earthquakes. Seismic reinforcement structures designed for civil buildings can help mitigate damage to buildings caused by earthquakes. In this paper, we investigate the reinforcement situation and damage form of civil structure houses in the earthquake-stricken areas of magnitude 6 in Yunnan. Further analysis is made from the comparison of the damage of earthquakes of similar magnitude in Yunnan, the comparison of reinforced and unreinforced houses, the strong vibration records, and the comparison of the relationship with the intensity attenuation. The results show that seismic reinforcement can reduce the damage of civil structures. It can even reduce its macro-intensity by 1 to 2 degrees, and evaluate the earthquake-proof and disaster-reduction effect of seismic reinforcement; put forward suggestions on seismic reinforcement and intensity assessment.

1 INTRODUCTION

The occurrence of earthquakes is highly random and instantaneous. Even the most advanced measuring instruments cannot accurately predict the arrival of earthquakes. Therefore, in the process of strengthening earthquake resistance in the design of civil engineering structures, it is necessary to analyze the seismic performance and response status of civil structures according to the seismic theory of civil structures and the actual experience of civil structure design, in order to solve a series of problems in the design of civil structures. In the design process, it is necessary to consider not only the overall layout of the civil structure, but also the connection position of the structure, so as to fundamentally improve the seismic performance of the civil structure.

Yunnan Province is located in the eastern margin of the collision zone between the Indian plate and the Eurasian plate, and in the southern section of the north-south seismic zone. There are 49 active faults developed in the province (Guang et al. 2006), with strong tectonic movement and significant seismic activity, it is one of the areas with the highest earthquake frequency in the world (Hua et al. 2015). According to the earthquake records since the 20th century, earthquakes of magnitude 5.0-5.9 in Yunnan average twice a year, earthquakes of magnitude 6.0 to 6.9 occur twice every three years, and earthquakes of magnitude 7 and above occur about every eight years on average. 1 time, the occurrence frequency of earthquakes of all magnitudes is very high.

Civil structure is the most common form of residential structure in rural areas in Yunnan, and has the characteristics of poor seismic performance (Cheng et al. 2015). The seismic performance of civil structures can be increased due to seismic reinforcement (Liu et al. 2016). Previous scholars' evaluation of the effect of seismic strengthening measures for civil structures is mostly based on shaking table tests, and has not experienced the actual test of earthquakes (Deng et al. 2020; Liu et al. 2021; Peng et al. 2019; Su et al. 2021). This paper mainly studies the seismic reinforcement effect of civil structure houses based on the field survey results of the magnitude 6 earthquake in Yunnan Province in May 2021 and the data recorded by strong earthquake stations.

---

*Corresponding Author: 13103512146@163.com

                                                                    DOI 10.1201/9781003308577-12

2 ANALYSIS OF SICHUAN CIVIL STRUCTURE

The most widely used form of civil structure in Yunnan Province is the cross-bucket timber frame. During this investigation, no civil structure houses with beam-lifting type, wooden post and wooden roof truss and other structural forms were found. The bucket-piercing timber frame is a light-weight frame, mainly composed of beams, columns, piers, and other components, with good integrity. At the same time, the mortise and tenon joints are used to connect the nodes of the timber frame, which can absorb a certain amount of seismic energy during an earthquake, so it has better seismic performance than other structural forms of timber frame houses. According to the survey, the civil structure houses in Yangbi County and surrounding districts and counties are relatively old, the houses are basically over 30 years old, and some old houses are over 50 years old.

In recent years, the huge losses caused by earthquake disasters to civil structures have gradually attracted the attention of the government and scholars, and the research on seismic reinforcement of wooden structures has been gradually deepened. A large number of scholars in China have carried out performance tests with different reinforcement methods (Dönmez et al. 2021; İşleyen et al. 2021). According to the field investigation, it is found that the seismic reinforcement measures of civil buildings in Yunnan Province mainly include the following:

(1) According to the earthquake damage of civil structure houses in the multiple earthquakes in Yunnan Province, it can be found that the performance of beam-column joints is the key to the seismic performance of houses. collapse. There are two common joint reinforcement methods in the disaster area: nail reinforcement and steel plywood reinforcement shown in Figure 1(a).
(2) The integrity of a timber-framed house is crucial to its performance in earthquakes—especially in strong earthquakes. For houses with small cross-sections or poor construction techniques, which lead to insufficient integrity, it is generally necessary to use longitudinal bracing to increase the connection between adjacent horizontal timber frames to enhance the integrity of the timber frames. The main longitudinal bracing method used for civil structure houses in the disaster area is scissor bracing shown in Figure 1(b).

3 ANALYSIS OF THE ETHQUAKE

3.1 *Typical destructions analysis*

The intensity of the earthquake's meloseismic zone was assessed as VIII degree, and the damage of civil structures in the VIII-degree, VII degree and VI-degree areas were different. In the VIII-degree area, individual column bases were displaced, a few walls collapsed, some partially collapsed, most of them were cracked, the walls flashed out, and tiles were generally lost; In the area, a few walls cracked and tiles fell off. Typical destructions are shown in Figure 2.

3.2 *Comparison with previous earthquakes*

In 2014, three earthquakes of magnitude 6-7 occurred in Yunnan Province, namely the Yingjiang MS6.1 earthquake, the Jinggu MS6.9 earthquake, and the Ludian MS6.6 earthquake, all of which were comparable in magnitude to the Yangbi earthquake. A number of scholars have conducted a comparative analysis of the damages of the three earthquakes. Because the civil structure houses in the earthquake-stricken area of Ludian are mainly "walls and beams", the seismic performance of the buildings in the earthquake-stricken areas of Yangbi, Yingjiang and Jinggu is quite different, and the damage caused by the secondary disasters of the earthquake is relatively large. Therefore, this paper only compares the Yingjiang earthquake and the Jinggu earthquake with the Yangbi earthquake. Compared with the other two earthquakes, the Yangbi earthquake disaster area has the following characteristics: (1) the epicenter is the shallowest; (2) the peak acceleration (PGA) is the largest; (3) the number of casualties is the least (with the same population density); (4) per

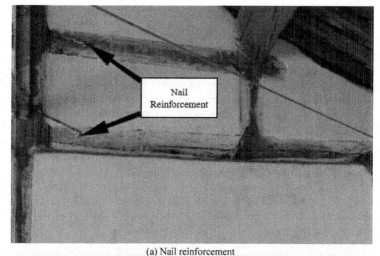

(a) Nail reinforcement

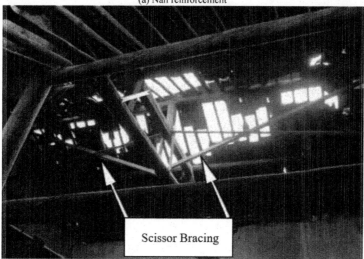

(b) Scissor bracing

Figure 1.   Examples of reinforcement measures.

capita GDP is the lowest; (5) Isoseismal areas of degrees VIII and VII are the smallest. In addition, according to the post-earthquake disaster loss assessment survey of the three earthquakes, except for the county seat located on the terrace of the Shunbi River, the Yangbi earthquake-stricken area is basically mountainous, and the slope amplification effect of the earthquake is significant; Over the years, the seismic performance is relatively poor.

3.3   *Comparison of reinforced and unreinforced structures*

In the post-earthquake disaster damage assessment survey, it was found that Xishan Township, Tielian Township and Fengyu Township in the south of Eryuan County bordering Yangbi County did not carry out seismic reinforcement of civil structures (passenger-wood structures), and the south side was closer to the epicenter. Seismic reinforcement was carried out in Yangjiang Town. 5 and 7 natural villages (investigation sites) were selected in the above-mentioned areas respectively

Figure 2. Examples of reinforcement measures.

to investigate the damage of civil structures, and according to the "China Earthquake Intensity Scale" (GB/T17742-2008). Seismic damage index and macroscopic intensity. It can be seen from Table 3 that the distances from the seven measuring points in Yangjiang Town, Yangbi County to the epicenter are 11.6-17.2 km, and the macroscopic intensity is generally VI degree; the five measuring points in Eryuan County are 20.7-22.4 km from the epicenter. Although the 7 survey points in Yangjiang Town, Yangbi County were relatively close to the epicenter, the damage was less severe. Considering that the survey points are not far apart, the terrain slopes are similar, and the building ages are similar, the analysis believes that the main reasons for the difference in earthquake damage are the civil structures and houses in Yangjiang Town, and the seismic reinforcement in Yangbi County. It can be seen that the civil structure buildings in Yangbi County all adopt the seismic structure. It can be seen from the data in Table 1. Compared with Eryuan County, the number of civil constructions in Yangbi County is 0. At the same time, the buildings in damage in Yangbi County are far less than those in Eryuan County. This shows that the reinforcement of seismic structures can effectively reduce the damage of civil buildings in earthquakes. Further, we analyze the effect of different seismic structures in actual earthquakes. As shown in Table 2, the number of damages received by the civil structure using Nail reinforcement in Yangbi County is 8. Compared with the number 40 using the Scissor bracing structure, it is much less. This inspires us to apply Nail reinforcement as a seismic structure in practical engineering.

Table 1. Comparison of damage to civil structures in Yangbi county and Eryuan county, Yunnan province.

District	Epicenter distance/km	Number of households	Destruction	Damage	Intact
Eryuan	21.6	215	11	153	51
Yangbi	17.2	241	0	48	193

Table 2. Comparison of different seismic structures in Yangbi county.

Structures	Number of households	Destruction	Damage	Intact
Nail reinforcement	135	0	8	91
Scissor bracing	106	0	40	66

4 CONCLUSIONS

Through the investigation and analysis of the damage to civil and structural buildings caused by the M6 earthquake in Yunnan Province, the following conclusions are initially drawn:

(1) Yunnan Province adopts steel plywood, steel shear bracing and other methods to reinforce the civil structure building, which increases the integrity and stability of the civil structure and improves the seismic capacity of the civil structure building.

(2) The data of earthquake damage comparison, strong earthquake records, intensity attenuation comparison and other data show that seismic reinforcement has effectively reduced the damage of civil and structural buildings in the disaster area, and the macroscopic intensity has been reduced by 1 to 2 degrees, which greatly reduces its impact under the action of earthquakes. The degree of damage has reduced economic losses and casualties.

The seismic reinforcement work in Yunnan Province played a role in this earthquake, greatly reducing casualties and economic losses, further proving the rationality and effectiveness of the seismic reinforcement work, and providing valuable experience for our earthquake prevention and disaster mitigation work. Suggestions Further promote the seismic strengthening of old residential buildings in high-earthquake-risk areas, improve the seismic performance of buildings, reduce earthquake risks, and reduce earthquake losses and casualties. In the seismic intensity assessment, due to the large difference in seismic performance between the two, the buildings that have been seismically strengthened and those that have not been seismically strengthened should be distinguished. The earthquake damage index of the former can be appropriately increased so as not to underestimate the damage caused by the earthquake. To sum up, this paper analyzes the effect of seismic reinforcement technology in the earthquake of magnitude 6 in Yunnan Province through field data, and provides guidance for the seismic reinforcement of civil structures in actual scenarios. Future work could take samples from more seismic records and compare the effects of different seismic structures in earthquakes of different sizes.

REFERENCES

Cheng, Y. Z., Tang, J., Chen, X. B., Dong, Z. Y., Xiao, Q. B., & Wang, L. B. (2015). Electrical structure and seismogenic environment along the border region of Yunnan, Sichuan, and Guizhou in the south of the North-South seismic belt. Chinese Journal of Geophysics, 58(11), 3965–3981.

Deng, S., Zhang, W., Yu, X., Song, Q., & Wang, X. (2020). Analysis on crustal structure characteristics of southern Sichuan-Yunnan by regional double-difference seismic tomography. Chinese Journal of Geophysics, 63(10), 3653–3668.

Dönmez, A., Rasoolinejad, M., & Bažant, Z. P. (2020). Size effect on FRP external reinforcement and retrofit of concrete structures. Journal of Composites for Construction, 24(5), 04020056.

Gang, H. U. A. N. G. F. U., & Jia, Z., Q. I. N. (2006). Study of the seismicity of strong earthquakes in Yunnan area. Seismology and Egology, 28(1), 37.

Hua, W., Fu, H., Chen, Z., Zheng, S., & Yan, C. (2015). Reservoir-induced seismicity in high seismicity region—a case study of the Xiaowan reservoir in Yunnan province, China. Journal of Seismology, 19(2), 567–584.

İşleyen, Ü. K., Ghoroubi, R., Mercimek, Ö., Anil, Ö., & Erdem, R. T. (2021). Behavior of glulam timber beam strengthened with carbon fiber reinforced polymer strip for flexural loading. Journal of Reinforced Plastics and Composites, 40(17–18), 665–685.

Liu, C., Yang, H., Wang, B., & Yang, J. (2021). Impacts of reservoir water level fluctuation on measuring seasonal seismic travel time changes in the Binchuan basin, Yunnan, China. Remote Sensing, 13(12), 2421.

Liu, X., Ma, J., Du, X., Zhu, S., Li, L., & Sun, D. (2016). Recent movement changes of main fault zones in the Sichuan-Yunnan region and their relevance to seismic activity. Science China Earth Sciences, 59(6), 1267–1282.

Peng, C., Jiang, P., Chen, Q., Ma, Q., & Yang, J. (2019). Performance evaluation of a dense MEMS-based seismic sensor array deployed in the Sichuan-Yunnan border region for earthquake early warning. Micromachines, 10(11), 735.

Su, J., Liu, M., Zhang, Y., Wang, W., Li, H., Yang, J., & Zhang, M. (2021). High resolution earthquake catalog building for the 21 May 2021 Yangbi, Yunnan, M S 6.4 earthquake sequence using deep-learning phase picker. Chinese Journal of Geophysics, 64(8), 2647–2656.

Frontiers of Civil Engineering and Disaster Prevention and
Control – Yang & Rahman (Eds)
© 2023 The Author(s), ISBN: 978-1-032-31200-2

Scientific discussion on the construction process of rammed earth walls of Hakka dwellings

L.H. Zhang, Z.M. Xiao*, M. Ouyang*, Y. Yu, C.Y. Zhu & Y.X. Zhao
Jiaying University, School of Civil Engineering, Meizhou, Guangdong, China

ABSTRACT: The strength and construction influencing factors of rammed earth wall materials for Hakka dwellings are scientifically explored. Based on the research of improving the accuracy of the test, the working principle of rebound meter and the relationship between the strength of rammed earth walls in different periods and parts were fully considered, and the strength test curve of rammed earth walls was obtained by increasing the displacement of the elastic rod end of the rebound meter. Through field research and on-site measurement, the applicability of the improved rebound meter was verified, and it was also shown that the strength of rammed earth walls of Hakka traditional dwellings is low, ranging from 0.44 to 0.55 Mpa, and the strength and durability of the walls can be improved by improving the formula and increasing the number of plate rams on both sides. The method also provides a basis for the daily inspection of the protection of the walls of Hakka dwellings and helps to further improve the protection measures of the dwellings.

1   BACKGROUND

Meizhou is one of the main gathering places of Hakka people, and the in-depth study of Meizhou Hakka dwellings is typical for the study of Hakka dwellings all over the country. In the course of the development of Meizhou's history and culture for more than 2,000 years, many historical celebrities have been nurtured, leaving behind a large number of Hakka dwellings and the former residences of celebrities with education, strong cultural color and full of historical value. However, over time, many factors, such as changes in the natural environment and human damage, have caused varying degrees of damage to these precious Hakka dwellings, greatly weakening the safety of the architectural structure as well as the historical value of the building (Fang 2017).

With the development of modern construction and the speed of urbanization, residential rammed earth architecture has been impacted, including the construction process and other intangible cultural heritage relying on the transmission of the word is rapidly disappearing, many traditional skills are on the verge of extinction, the traditional building construction process is facing the influence of modern architectural thinking and construction methods and increasingly shrinking situation, and has even reached the verge of losing the situation (Li 2019). Therefore, strengthening the study of intangible cultural heritage such as rammed earth construction techniques can help the preservation and transmission of traditional architecture. At present, the research on rammed earth buildings mainly focuses on the seismic performance of building structures, the mechanical properties of rammed earth materials, and the reinforcement and weatherproofing of rammed earth walls (Shi 2014).

Based on the reliability evaluation of the existing Hakka traditional rammed earth building structures, the actual compressive strength of the rammed earth walls was tested. In order not to destroy

---

*Corresponding Author: xzmyy0226@163.com

   DOI 10.1201/9781003308577-13

the existing intact Hakka dwelling buildings, the strength data of the tested rammed earth walls of Hakka dwelling houses were analyzed by non-destructive testing of Hakka dwelling buildings using rebound meters from the aspects of economic applicability and heritage conservation, and on the basis of this experimental data, the similarities and differences of each group of experimental data were analyzed to establish the strength curves of the rammed earth walls of Hakka dwelling houses in different periods and different parts of the buildings to ensure the originality and authenticity of the experimental data to the greatest extent.

2   REBOUND METER ROD END SELECTION

In the case of protecting the rammed earth walls of residential houses, a nondestructive testing method applicable to the strength evaluation of rammed earth walls of traditional residential houses in Meizhou is sought. Modern non-destructive testing methods for walls include radiographic testing, ultrasonic, rebound method, etc. Among them, the rebound method is often used in concrete and mortar strength testing, which has the advantages of being more economical, faster, easier to use and having considerable testing accuracy compared to other testing methods. Based on the comprehensive evaluation, the rebound method was experimentally selected as the nondestructive testing method for the rammed earth walls of residential houses.

2.1   *Principle of rebound meter*

Rebound method is to use the spring-driven hammer, through the bouncing rod (transmission rod), bounce the concrete surface, and measure the distance the hammer was bounced back to the rebound value (rebound distance to the ratio of the impact length of the bouncing hammer) as a strength-related indicators, to presume a method of concrete strength. Since there is a certain phase relationship between the compressive strength of concrete and its surface hardness, and the rebound hammer of the rebound meter is struck by a certain elastic force on the concrete surface, the rebound height (rebound value read by the rebound meter) and the concrete surface hardness into a certain proportional relationship. The compressive strength of concrete can be deduced from the surface hardness.

2.2   *Rebound meter detection improvement*

Rammed earth materials are closer to concrete materials in terms of mechanical properties, but their compressive strength is still relatively low compared to concrete. Therefore, the mortar rebound meter with a kinetic energy of 0196 J was selected for the rammed earth wall detection test. Relative to the rammed earth wall, the impact energy acting on the unit area is too high or accidental phenomenon is more, so the rebound value reading on the rebound meter is too small or too large, thus causing the problem of low accuracy of the reading. Therefore, in order to improve the reading accuracy of the rammed earth wall and to keep the rebound value reading stable, the contact area of the rammed earth wall is increased in such a way that a more stable rebound value is obtained by the ramming hammer (Qiu 2021).

From the experimental study on the effect of changing the spherical radius R value of the front section of the bouncing rod on the rebound value, it can be seen that with the change of the spherical radius R value, the rebound value measured on the examined test block also changes. In order to improve the rebound meter at the same time, without too much impact on the other components of the rebound meter and damage its detection performance, and taking into account the strength and appearance of different rammed earth walls will be different, can adapt to the detection of many rammed earth walls, so the use of the same material as the rod material processing and production of different R value can be set in the head of the steel sleeve of the rod, and in the front of the sleeve corresponding to the front of the original rod spherical radius and grinding a certain degree of curvature, in order to make the rebound meter in the installation of the sleeve will not be due to the rebound meter is not absolutely perpendicular to the surface of the inspected components and cannot be good contact with the surface of the rammed earth materials.

Through field tests and analysis in different areas, as the spherical radius R value of the front section of the bouncing rod becomes larger, the test value is small and the reading error increases. The broken and collapsed walls were sampled and then tested and compared using a laboratory press, and a steel head with closer test data was selected. The results show that it is more appropriate to choose a set head with R of 25 mm for the inspection of rammed earth walls from 50 to 240 years. Steel sockets of different diameters, as shown in Figure 1.

Figure 1.   Steel sleeve with different diameters.

3   RAMMED EARTH WALL REBOUND TEST COMPRESSIVE STRENGTH EXPERIMENT

The rammed earth buildings in Meizhou were taken as the test object, and the rammed earth buildings in different periods in Meizhou city and in different periods in the suburbs of Meizhou were taken as the research object, and the surface of the rammed earth buildings were cleaned at different locations to remove their surface lime layer and to level them (Zeng 2008). After the previous cleanup work is completed, the rebound meter is then used to delineate the area for testing, taking a random 250×250 mm plane at different locations and dividing it into 5×5 squares for testing.

3.1   Rammed earth buildings in Meizhou city

3.1.1   Rammed earth architecture of Yuehmei village
Yuemei Village is in the old city of Meizhou. The houses inside the village are still preserved more than 100 years ago, and the houses are relatively well protected. Take 500 mm from the ground, half height (950 mm from the ground), 500 mm from the eaves three locations for testing, try to ensure that the test to the rain on the wall of different locations erosion, and reduce the rebound value is not affected by accidental factors in the appearance of the wall.

Use chalk to draw a good test plane of 250×250 mm at each position, align the head of the rebound meter bullet rod to the center of the square and make its axis always perpendicular to the surface of the rammed earth wall, slowly apply pressure to the end of the rebound meter, fast reset for one time. Since the surface properties of rammed earth walls vary greatly, three consecutive bounces are required to ensure that the surface conditions of rammed earth walls are as uniform as possible (Liu 2021). When the pressure is applied after the 3rd time, the reading is recorded and reset to exactly 1 scale before ending the measurement at 1 point here. During the test, care should be taken to control the rebound meter to always be perpendicular to the test block surface. Repeat the above steps for each of the remaining measurement points separately.

Due to the 500 mm from the ground, when it rains, the rain falls low on the ground after dripping splash and wind carries raindrops to hit the wall, so that the wall is eroded by rain, and it does not receive enough light for sun drying, resulting in serious dampness of the wall and loose top soil. When using the rebound meter for testing, the rebound meter has no value display.

The distribution of rebound values at two locations, half height of the rammed earth wall (950 mm from the ground) and 500 mm from the eaves, were analyzed and counted in a scatter plot of rebound values, as shown in Figure 2.

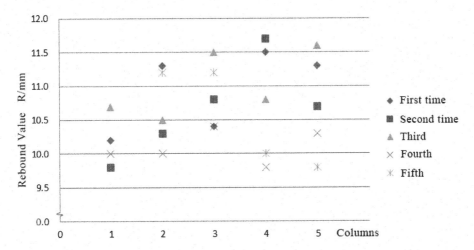

a) Scatter diagram of rebound value at half height of rammed earth wall (950 mm from the ground)

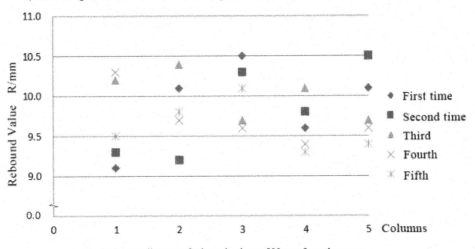

b) Scatter diagram of rebound value at 500 mm from the eaves

Figure 2.   Scatter plot of rebound value.

3.1.2 *Rammed earth architecture of Xintian village*

Xintian Village is at the border of the urban area of Meizhou, where many architecturally representative ancestral halls are still preserved, and the study chose the Jiang ancestral hall, which is more than 240 years old, for testing. It was measured at 500 mm from the ground, half height (950 mm from the ground), and 500 mm from the eaves at three locations.

After testing and analysis, 500 mm from the ground also due to rain erosion of the wall, the wall is seriously damp, the surface soil is soft, the rebound meter does not show the value.

93

The distribution of rebound values at two locations, half height of the rammed earth wall (950 mm from the ground) and 500 mm from the eaves, were analyzed and counted in a scatter plot of rebound values, as shown in Figure 3.

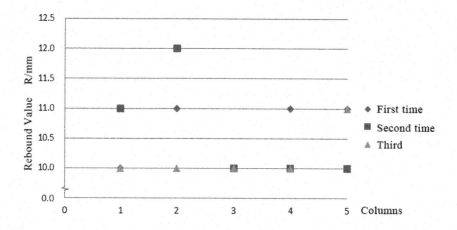

a) Scatter diagram of rebound value at half height of rammed earth wall (950 mm from the ground)

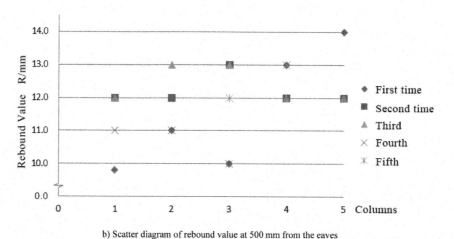

b) Scatter diagram of rebound value at 500 mm from the eaves

Figure 3.   Scatter plot of rebound value.

3.2   *Rammed earth buildings on the outskirts of Meizhou*

3.2.1   *Rammed earth architecture of Chashan village*

Located in the Meixian District of Meizhou, Chashan Village has a history of over 200 years and is a nationally protected cultural village with relatively complete and intact rammed earth buildings. The rammed earth wall of Chashan Village is characterized by the fact that the part within 1540 mm from the ground is built by rammed earth one board after another, and its height of one board is 280 mm to 300 mm; The upper part of the building is made of clay bricks, which are 300×200×110mm in size.

As the building here belongs to the municipal cultural relics' protection unit, only the exposed part is conditionally tested, and 250×250 mm form measurement is not used. Instead, it was replaced with an unspecified test on the same plate ramming wall, and 10 points were selected as the test data for that plate.

Due to the high terrain, the walls are at a certain distance from the ground and are basically not subject to rain erosion. The average rebound value of the rammed earth wall is 10.0 to 11.0, and the rebound value is relatively stable. The rebound value of the first row of earthen brick walls is around 5.0 to 6.0, and the rebound value of the second row is around 5.0 to 5.5. The detailed drawing of the test wall and the rebound values of each location are shown in Figure 4.

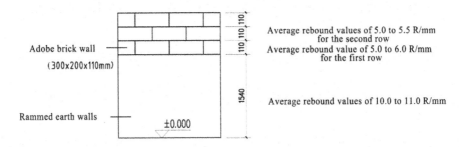

Figure 4. Detailed drawing of the test wall in Chashan Village and the rebound values at each location.

3.2.2 *Rammed earth architecture in Chengtan village*

3.2.2.1 *Renchang building*

Inchang House is more than 100 years old and is relatively fully protected, and still meets the daily needs of villagers today. The characteristic of rammed earth wall is that the part within 1,500 mm from the ground is built by rammed earth one board after another, and its height of one board is 280 mm to 300 mm; The upper part is made of clay bricks, which are $300 \times 200 \times 110$ mm in size.

Since there are still people living in this building cash and there are many debris and household items, we cannot test the internal data and clean the exterior, therefore, we also use the method of conducting indeterminate test in the rammed earth wall of the same slab and take 10 points as the test data of this slab.

The corners of the walls at the sampling site have been subjected to rain erosion for many years, and the surface layer of the walls has been loosened, making it impossible to measure accurate data. The average rebound value of the rammed earth wall at 600 mm from the ground was 10.0 to 11.0; The rebound values of the earthen brick walls were around 5.0 to 5.5. The rebound values for each location of the test wall details are shown in Figure 5.

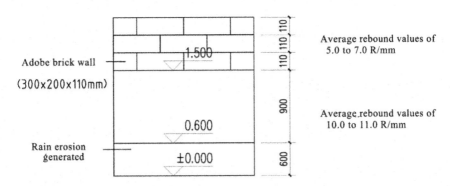

Figure 5. Detailed drawing of the test wall of Renchang Building and the rebound value of each position.

3.2.2.2 *Chengxu building*

Chengxulou is more than 150 years old, and it is different from other architectural styles in this area, not in the style of a sirocco house, but in the style of a courtyard-like building. The rammed earth wall is characterized by the fact that the part within 1,200 mm from the ground is built of rammed earth and the height of a slab is 280 mm to 300 mm; The upper part is made of earthenware bricks with the specifications of 300×250×240 mm, and the eaves are 500 mm from the wall and 800 mm from the eaves, so the wall is not subject to rain erosion.

Average rebound value of 9.5 R/mm for rammed earth walls at 500 mm from the ground; The rebound value of the earthen brick wall is in the range of 5.0-5.5 R/mm. A detailed drawing of the test wall, as shown in Figure 6.

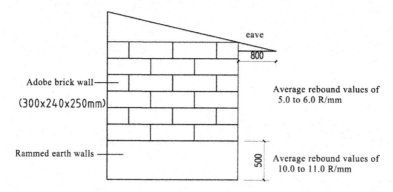

Figure 6.   Detailed drawing of the test wall of the Chengxu Building and the rebound values at each location.

4   ESTABLISHMENT OF COMPRESSIVE STRENGTH CURVE OF RAMMED EARTH WALL

Through the analysis and statistics of the test rebound value, we can learn that the foundation of the rammed earth building in Meizhou city is basically flush with the ground, and the foundation of the rammed earth building outside Meizhou city is higher from the ground, resulting in the Meizhou city 500 mm from the ground is seriously eroded by rainwater, and the rebound meter value cannot be tested. The average rebound value of the rammed earth wall at 950 mm from the ground for the rammed earth building in the Meizhou city area is 9-10 R/mm for the rammed earth wall in Yue Mei village and 10-12 R/mm for the rammed earth wall in Xin Tian village. The average rebound value of a rammed earth wall at 500 mm from the eaves is 10-11 R/mm. There is a gap between the wall composition of rammed earth buildings outside Meizhou city and inside the city. The average rebound value of rammed earth slab wall position is 10-11 R/mm, and the average rebound value of earth embryo brick wall is 5-7 R/mm.

For different regions, different periods of rammed earth wall rebound test and compressive strength test to establish the relationship between the compressive strength of the wall under inspection and the rebound value applied numerical analysis (Zhang 2020), the field measured wall rebound value R substituted into the measured strength curve formula can be converted to wall strength, from the rebound meter manual can be obtained, around 100 years of wall strength formula see (1) and around 200 years of wall strength formula see (2), to establish a correlation compressive strength curve. The average rebound values and compressive strength curves for different periods and regions are shown in Table 1.

$$f_{cu}^e = 0.1306\, R_m^{0.6521} \tag{1}$$

$$f_{cu}^e = 0.0612\, R_m^{0.7979} \tag{2}$$

Table 1. Average rebound value and compressive strength of rammed earth wall.

Average rebound value R/mm		Compressive strength f_{cu}/MPa	
Around 100 years	Around 200 years	Around 100 years	Around 200 years
10	9	0.586	0.353
10	10	0.586	0.384
9	9	0.547	0.353
11	9	0.624	0.353
10	11	0.586	0.415
9	10	0.547	0.384
12	9	0.66	0.353
10	10	0.586	0.384
13	10	0.696	0.384
9	9	0.547	0.353
10	9	0.586	0.353
11	10	0.624	0.384
9	9	0.547	0.353
12	11	0.66	0.415
10	10	0.586	0.384

5 CONCLUSIONS

In this paper, based on the reliability evaluation of the rammed earth building structure of the existing Hakka houses, the non-destructive testing method of the rebound meter is mainly adopted. By adding steel heads of different diameters at the rod end of the rebound meter and combining the field test and laboratory press test data after sampling, a set of heads with R of 25 mm is finally selected for the testing of rammed earth walls from 50 to 240 years, which truly realizes the non-destructive testing of the compressive strength of rammed earth walls and greatly To a great extent, the original appearance of the rammed earth walls of the existing Hakka houses was preserved, laying the foundation for the evaluation of the construction techniques of the traditional Hakka houses.

The authors' team verified the applicability of the improved rebound meter through field research and on-site measurements. On top of analyzing a large number of detected experimental data on the strength of rammed earth walls of Hakka houses, they also analyzed the similarities and differences of each group of experimental data and established the strength measurement curves of rammed earth walls of Hakka houses at different periods and different parts of the walls to ensure the reliability of the experimental conclusions to the greatest extent. Based on the analysis of the experimental data, the following experimental conclusions are mainly drawn in this paper.

(1) Rammed earth walls that have been eroded by rain over the years and have not been dried by the sun in time will have severely reduced compressive strength and loose surface soil.
(2) The stability of the building wall is related to its survival time, construction method, material selection and changes in the surrounding environment, etc. The long survival time and insufficient number of ramming passes on the side plates will cause the rammed earth wall to be less strong.
(3) The strength of rammed earth walls of Hakka traditional houses is low, between 0.44 and 0.55 Mpa. The strength and durability of the walls can be improved by improving the formula and increasing the number of plate rams on both sides.

(4) In the conservation and inheritance of residential houses, the construction of walls should be based on construction techniques suitable for local natural and social development as far as possible, but attention should also be paid to the need to meet local humanistic requirements and living customs.

However, there are certain limitations of this experiment, which are mainly reflected in the following aspects.

(1) The experiments were conducted in a narrow range of areas and with relatively few types of building walls, and some of the experimental findings may be contingent or show regional characteristics of Meizhou.
(2) Some of the Hakka residential buildings selected for the experiments were still occupied. In addition, for the consideration of heritage building conservation, individual experimental walls were not tested for the compressive strength of the inner side of the rammed earth walls.
(3) The environment inside and outside the building is different, and the influencing factors such as light, temperature and rainwater produce chemical reactions on the wall materials, resulting in differences in compressive strength on both sides of the wall, which may lead to slight errors in the experimental data. In individual experiments, the rebound meter showed no numerical feedback.
(4) The accuracy of the rebound meter selected for the experiment is not enough and the readings may have certain deviations. The principle of rebound meter is to measure different data according to the hardness of the measured material. However, in the actual construction of rammed earth walls, it is not possible to achieve uniformity and the same hardness everywhere inside and outside the wall, so the results obtained by bouncing on different surfaces may also be different.

Overall, this paper provides a scientific and useful exploration of the rammed earth wall construction process of the Hakka houses in Meizhou. It is of positive and critical significance for the further conservation and repair of Hakka houses in Meizhou. In the future, as the economy continues to develop and science and technology continues to change, NDT technology and related experimental instruments will also be further improved and perfected. By then, the authors' team will further expand the scope of experiments and enrich the types of experimental walls, further reduce the experimental errors based on a large number of extensive experiments, and use more scientific and reasonable experimental analysis methods as a comparison with the existing experimental data, so as to draw more accurate experimental conclusions and provide more important reference and reference for the protection of rammed earth walls in Meizhou and even in the whole country.

REFERENCES

Fang, L. D., J. Wang & Y. B. Jin (2017). Research on the construction technology and optimized inheritance of rammed earth walls in Zhuangkou, Qinghai region. Huazhong Architecture. 35(03), 117–121.
Li, G. L. (2019). The development and application of rammed earth process in traditional Chinese raw earth camping process. Nature and Cultural Heritage Inquiry. 4(10), 94–99.
Liu Y. J., & Z. H. Yue (2021). Exploration on the technology of concrete strength testing by rebound method. Inner Mongolia Water Resources. 11, 48–49.
Qiu, J., D. C. An & Z. Y. Li (2021). A study on modern rammed earth wall construction techniques–the example of slab construction wall. Art Education. 11, 243–246.
Shi, W. J., X. Q. Peng & Y. H. Chen (2014). Erosion measurements of rammed earth walls of Tulou in natural environment. Journal of Natural Hazards. 23(01), 138–143.
Zeng, J. Z., Y.X. Fu, J. Guo & B. Zhou (2008). Exploration of typical rammed earth walls of residential houses in southeast Sichuan. Sichuan Architectural Science Research. 5, 182–184.
Zhang, Y.Q. (2020). Study on out-of-plane bearing capacity of masonry wall based on numerical analysis. Green Building. 12(04), 119–122.

Frontiers of Civil Engineering and Disaster Prevention and
Control – Yang & Rahman (Eds)
© 2023 The Author(s), ISBN: 978-1-032-31200-2

Study progress and prospect of bridge erecting machine monitoring technology

Jia Chi, Li Wei* & Liu Wei
Research Institute of Highway, Ministry of Transport, Beijing, China

ABSTRACT: In the process of prefabricated beams installation, erecting machine accidents often lead to serious loss of life and property, and erecting machine monitoring technology is an important technical way to prevent accidents. In order to understand the development and application effect of erecting machine monitoring technology, the research on erecting machine safety monitoring system and structural damage identification technology in China in the past decade are reviewed. According to the different monitoring emphases, it is divided into state-based erecting machine safety monitoring and structural damage-based erecting machine health monitoring. At the end of the article, the future development direction of erecting machine monitoring technology is prospected, which can be used as a reference for the study and application of erecting machine monitoring technology.

1 INTRODUCTION

Bridge erecting machine is a kind of lifting special machinery with complex structure, which is the key equipment to ensure the safety construction of bridges, and its main function is to lift the prefabricated beams to the designated position. The erecting machine is supported on the built bridge structure and can move longitudinally along the slide rail to change the support position so that the prefabricated beams can be installed at the specified position. The working conditions of the bridge erector are more demanding and riskier because it needs to move itself longitudinally through the holes on the erected girders, and more accidents occur in this segment. (Liu 2018; Ma et al. 2018).

There are many forms of erecting machine, but according to the function, they all include the bearing structure (beam and legs), lifting structure (girder trolley), walking system, transverse system, electric control system and hydraulic system (Zhang 2011). The current national standards mainly include 'general technical conditions of erecting machine (GB/T 2011)' and 'safety regulations of erecting machine (GB 2011)'.

Entrusted by the Special Equipment Bureau of the State Administration of Market Supervision and Administration, our research group carried out a research project on the optimization and evaluation of the safety use and regulatory measures of gantry cranes (erecting machines). In this project, the construction site investigation and market investigation were conducted. The questionnaire investigation of erecting machine was also carried out, in which the construction units, inspection units and supervision departments are included. The questionnaire data were statistically analyzed, it was found that the following problems were obvious:

(1) Over-reliance on the operator's experience in the operation and management of the construction operation is significant, in this case, inability to fully grasp the running state of the bridge machine, inattentiveness and other reasons will lead to disoperation, resulting accidents.

---

*Corresponding Author: w.li@rioh.cn

DOI 10.1201/9781003308577-14

(2) Self-launching operation and beam erection are the most dangerous working conditions. Due to the limited stability design of the erecting machine, even if the hydraulic, electronic control, brake, and other aspects are normal, there is still a risk of forward tilt and left, right tilt collapse during self-launching operation and beam erection. If the operation is affected by wind, this risk may be greatly increased. In addition, it is prone to overweight in the process of beam erection, which causes damage to the crane and steel wire rope, and may aggravate the risk of tilting when shifting the beam left and right.

(3) Due to the increase of service life, poor working environment, high frequency of use and transfer assembly, resulting in material corrosion, structural aging, so it is difficult to accurately judge the health status of the structure. At present, the safety assurance of bridge erecting machines in China still depends on the maintenance and repair of maintenance units and the regular verification of quality supervision departments. But there has actually been little research on active monitoring and early warning of damage conditions for erecting machines (Wang et al. 2013).

Go into seriously from reason, management factors are notable, but we also should pay attention to the lack of monitoring system function. In recent years, bridge erecting machine accidents have occurred from time to time, resulting in significant losses of life and property. With the continuous development of information technology, the research on bridge erecting machine monitoring technology has gradually become a hot spot (Huang & Yu 2015a, 2015b, Liu 2010).

2 RESEARCH STATUS OF SAFETY MONITORING TECHNOLOGY

The monitoring system of bridge erecting machine uses sensors installed on the structure to collect real-time data of key systems such as mechanical structure, transmission system and drive system. Combined with signal processing method, the characteristic parameters of the structure are extracted, so as to warn and control potential safety hazards, to achieve remote state monitoring and fault self-examination. Besides, it provides technical means for operators to take reasonable emergency measures for sudden accidents.

Since the 1980s, some foreign companies have developed relatively mature systems in the research of crane condition monitoring and fault diagnosis systems (Yen 2003). Such as the crane monitoring system based on wireless network communication technology developed by Anchuan Company, Japan. The system developed by Sumitomo Company, Japan, using high-speed route switching technology, which can carry out online condition monitoring of global lifting equipment. The electronic computer monitoring system developed by Caterpillar Company, the United States, which can view monitoring parameters in real time and has fault alarm function. The remote monitoring system for excavator satellite data transmission was developed by P & K Company, Germany. The fault monitoring and maintenance platform of lifting equipment was built by KONES company in Finland. However, on the whole, there are few studies abroad on the safety monitoring system of bridge erecting machine.

The process of safety of China's hoisting machinery is relatively slow. It was not until the 1980s that the equipment diagnosis technology was initially contacted. Until the 21st century, the domestic hoisting machinery industry has not entered the application stage of computer state monitoring and fault diagnosis technology, and relevant regulations have been promulgated (Wang & Zhao 2006). The State Administration of Quality Inspection requires that since July 1,2013, the manufacturing and installation units should install the safety monitoring and management system on the newly-manufactured bridge erecting machine according to the national standard 'crane machinery safety monitoring and management system (GB 2012), otherwise they cannot leave the factory. Since January 1,2014, the erecting machine with no safety monitoring and management system cannot be used anymore.

According to the different monitoring emphases, this paper divides the monitoring system of bridge erecting machine into the safety monitoring system based on the state and the health monitoring system based on structural damage.

2.1 Study progress of safety monitoring system based on erecting machine state

The safety monitoring system based on the state of the erecting machine is the main type of monitoring system used currently in engineering, it is mainly based on the specified monitoring content for the erecting machine in the 'crane safety monitoring and management system (GB 2012). The system is composed of hardware and software, its functional units include information acquisition unit, information processing unit, control output unit, information storage unit, information display unit, information output interface unit, etc. It mainly involves the monitoring of the working environment, operating parameters and electrical system of the erecting machine and the video monitoring of the dangerous area. However, there is no specific requirement for monitoring parameters, and it does not involve the monitoring of mechanical parameters. The system structure is shown in Figure 1.

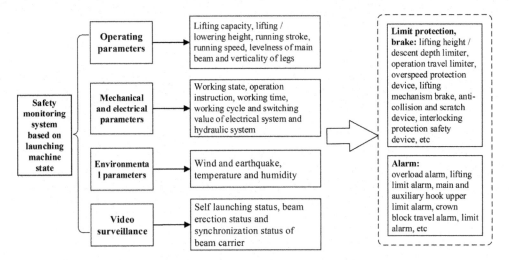

Figure 1. Safety monitoring system based on erecting machine state.

Due to the research and development of universities and enterprises, especially the outstanding contribution of Shijiazhuang Tiedao University, remarkable achievements have been made in the design and development of this kind of erecting machine monitoring system in China.

(1) In terms of the safety analysis theory of erecting machine, Yang et al.(2013) proposed a three-dimensional safety analysis system of large construction machinery, as shown in the Figure 2., the system is based on the construction process as the main line. According to the functional categories of machinery, a hierarchy is developed to systematically analyze the mechanical operation status, risk factors and consequences corresponding to each subsystem and process state, and the theory provides theoretical guidance for bridge erector safety design and safety monitoring. From the perspective of human-machine-environment system engineering, Xie (2015) established a qualitative analysis method of safety monitoring and a three-dimensional safety monitoring analysis system. The safety factor quantitative analysis method was used to construct the safety monitoring system structure, and the safety monitoring program was written. It provides a new idea for the development of safety monitoring system.

(2) There are many studies and practices on the design of monitoring system. Li et al. (2012) took TLJ900 bridge erecting machine as the object and designed a monitoring system. Through the collection of PLC status and sensor information, the system uploads data to the remote data center through GRPS network to realize real-time monitoring. Xing (2014) sent the monitoring data to the cloud server by 3G communication mode. After obtaining the data, the remote monitoring system drives the 3D model action of the erecting machine in the monitoring

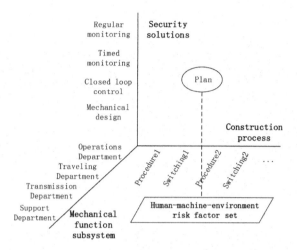

Figure 2. Graph of three-dimensional security analysis system. Adapted from 'Safety analysis of large construction machinery driven by construction technology' by Yang S.P., 2013.

system developed based on GENESIS64. The previous monitoring system only contains two-dimensional information, this study innovatively uses a new generation of three-dimensional configuration monitoring software to realize the remote virtual monitoring of the working process and operation parameters of the erecting machine. Xu (2017) carried on the intelligent transformation of the bridge erecting machine. The highlight is that through the video surveillance to expand the field of vision and a variety of sensors to collect real-time data, the erecting machine drivers and other operators can fully grasp the state of the bridge erecting machine at any time, and it can alarm in the critical state, which greatly reduces the degree of operator dependence on experience. Thus, the cross-integration of monitoring system platform development and other disciplines such as computers, automatic control, and artificial intelligence have been strengthened. Figure 3 is the structure of the security monitoring system developed by a company, which represents the type of monitoring system widely used in current projects.

(3) Applications of new sensors in monitoring are still emerging. Guo et al. (2013) used vibration sensors to monitor the vibration state of the bottom wheel of the erecting machine during operation in real time. The collected data were processed by the algorithm of the most advanced architecture kernel, so as to judge the operation state and warn the vibration state. In addition to the commonly used strain gauges, fiber grating sensors, piezoelectric sensors, shape memory alloy sensors are also developed in recent years, and show advantages in practical applications. However, sensors always have to receive the impact of the operating environment, so how to improve the signal-to-noise ratio is a topic that cannot be bypassed by the development of sensor technology. For special sensors, strict use conditions and life problems are also the limiting factors for the development of sensor technology.

(4) In the way of information transmission, the traditional security monitoring system used to lay cables, the electromagnetic interference is easy to occur between strong and weak electricity, and the reliability of the system is poor. In order to improve this situation, Zhao (2013) used ZigBee wireless network sensor to realize real-time monitoring of erecting machine structure, and used GPRS as data transmission mode to upload data, and designed a monitoring system with three layers of C/S. Zheng and Zhang (2019) proposed a security monitoring data acquisition and transmission mode based on 433 MHz wireless communication technology, and developed the hardware system architecture and PC software interface. Liang (2019) designed a main beam safety monitoring system based on STM32 micro controller and 4G communication module for data transmission. At present, the monitoring technology based on wireless

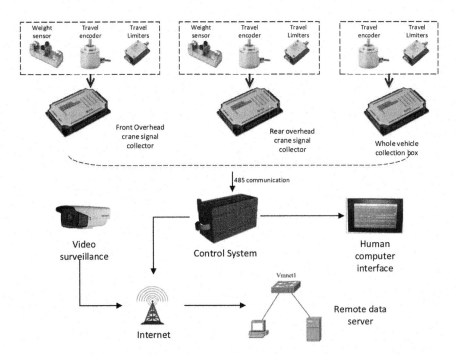

Figure 3.   Product structure diagram of a company's security monitoring system.

sensor has achieved a lot of practical application results, the information transmission mode has gone through the integration of Wi-Fi, ZigBee, Bluetooth, GPRS/CDMA, radio frequency communication technology and other wireless communication technology for data acquisition and transmission, so that the monitoring technology from wired to wireless, from low speed to high speed. And the improvement of the speed has laid a solid foundation for the real-time transmission of a large number of monitoring data and video.

(5) Due to the complexity of the mechanical structure of large construction machinery, a large number of points are needed to monitor potential safety hazards. For the problem of the location and number of sensors monitoring, Chen (2016) believed that the setting of safety monitoring points of large construction machinery should not only consider the needs of measuring points, but also control the cost of equipment and the efficiency of diagnosis. Therefore, the effectiveness evaluation strategy of measuring points of mechanical equipment safety monitoring system was proposed, which suggested that the sensitivity or effectiveness of each measuring point should be measured according to the ability of different monitoring points to perform working conditions. On this basis, the evaluation model of the ability of monitoring data to perform working conditions was established. However, this study does not deeply study the optimization of sensor measurement point layout. Actually, there are some studies on the measurement point layout of hoisting machinery (Wang 2015, Zhang 2013). Practice has proved that as the front end of monitoring, the layout scheme of sensors directly affects the accuracy and effectiveness of monitoring results. In comparison, the layout scheme of measuring points optimized by optimization algorithm has certain advantages. Unfortunately, there is no research and practice in the field of erecting machine monitoring.

(6) For the monitoring of the running attitude of the erecting machine, He (2017) proposed the monitoring and early warning technology of the running attitude of the erecting machine with the inclination angle of the front leg and the deviation between the two guide beams as the monitoring indexes, the principle of measuring the relative offset between guide beams is shown in Figure 4. The monitoring data were sent to the remote server, and the running state

was displayed and alerted by means of raw data, time course curves and hazard classification indicators. Du et al. (2020) found through research and practice that although the detection accuracy and test effect of the front leg verticality of the erecting machine monitoring system are good, the horizontality detection of the left and right beams needs to be studied. Due to the deflection of the bridge erecting machine itself, the selection of horizontality measurement points has a great influence on the results. Du et al. (2021) analyzed the bridge erecting machine due to the wind speed forward or rollover accident, and found that the middle main beam of the middle leg and the front leg were the dangerous parts of the whole machine, the transient dynamic analysis showed that the bridge erecting machine was more prone to tilt forward. The monitoring based on the running attitude of the erecting machine is essential, which can more directly reflect the state of the bridge erecting machine and provide early warning. However, the effect is not good as expected in practical application, mainly because the operation environment of the bridge erecting machine is severe and the working condition is complex, and it is often closed by the staff due to false alarm. In addition, whether the monitoring index of running posture is related to the stress state of the structure, the degree of correlation and the selection of early warning value need further study.

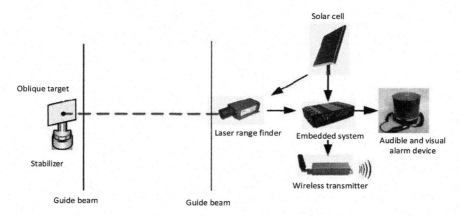

Figure 4. Graph of measurement principle of relative offset between guide beams. Adapted from 'Research on Bridge Erection Machine Monitoring System Based on Operation Attitude Monitoring' by He, J.W., 2017.

At present, the safety monitoring system based on the state of erecting machine is supported by standard specifications. Besides, there are many in-depth academic research and practice, and such monitoring system is widely installed in practical engineering. However, from the use and feedback of each unit, the effect is not as good as expected, which will be discussed below.

2.2 Study progress of health monitoring system based on erecting machine structural damage

The health monitoring system based on erecting machine structural damage monitors and evaluates the damage and health status from the perspective of structure, mainly covering the monitoring of static and dynamic parameters of erecting machine. The system structure is shown in Figure 5.

2.2.1 Static monitoring

The monitoring system with static parameters as the monitoring object mainly monitors the stress, strain, displacement, and inclination angle, so as to master the healthy state of the structure in operation.

In general, the monitoring system including electrical and mechanical parameter monitoring and video monitoring has met the requirements of the standard specification, just a few monitoring systems will monitor and warn the mechanical parameters of the structure. In the safety monitoring

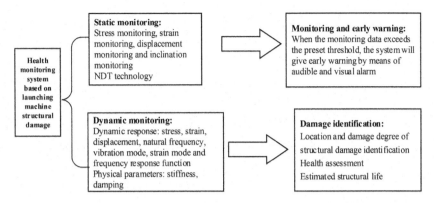

Figure 5.   Health monitoring system based on erecting machine structural damage.

system designed by Liang (2019), monitoring contents such as stress and deflection deformation of the main beam were added. Li et al. (2020) developed a construction safety monitoring system for super-large equipment in their project, shown in the Figure 6. During the assemble of erecting machine, stress strain gauges were installed at the corresponding positions to monitor and warn the pressure and horizontal thrust of the supporting legs, and the inclination of the body due to the force. The innovation of the monitoring system is that the monitoring of mechanical data, on-site video monitoring and the action monitoring of the main operating components of the erecting machine are set on one platform to achieve full dynamic real-time monitoring. It is a successful attempt of the existing monitoring system combined with mechanical data monitoring, there are many similar experiences. The stress and deformation values calculated by finite element or theoretical calculation are used as monitoring thresholds. When the monitoring value exceeds the threshold, the structure will be likely to damage, resulting in accidents. However, the resistance provided by the overall structure or local components of the erecting machine is reduced due to fatigue, corrosion, and other factors, so the collapse accident often occurs in the project without reaching the monitoring threshold.

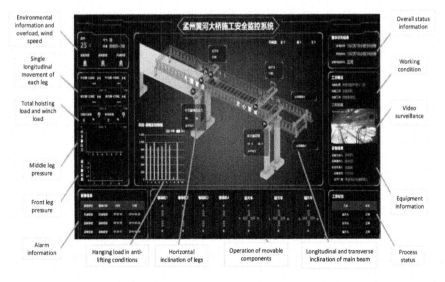

Figure 6.   Graph of monitoring system display interface. Adapted from 'Application of Super Large Equipment Safety Monitoring System in Construction' by Li, K. K. et al. 2020.

In addition to this method of using sensors fixed on important components to collect information for damage identification, electromagnetic detection techniques such as ultrasonic method, X-ray method and γ-ray method can also be used to detect the local damage of the structure (Zhao et al. 2020), namely nondestructive testing (NDT).

2.2.2 Dynamic monitoring

The main parameters used for dynamic monitoring are frequency, mode shape, mode curvature, strain mode, frequency response function and so on, also known as dynamic fingerprint. When the structure is damaged, the structural parameters such as mass, stiffness and damping will change, resulting in changes in the corresponding dynamic fingerprint. Through the monitoring of these dynamic fingerprints, the structural damage can be identified to achieve the purpose of monitoring the structural health status and estimating the structural life (Hearn 1911; Kim 1975). At present, many scholars in China have studied the damage identification of the main beam and the front leg of the erecting machine.

The identification method based on structural dynamic characteristics is one of the most common methods. Zhang (2014) established parametric models of different damage conditions based on dynamic characteristics, and carried out modal analysis and transient analysis. The obtained modal frequency was used as the damage identification quantity, and the support vector machine classification algorithm and regression algorithm were used to locate the damage location and predict the damage degree. Su (2013) carried out strength, deformation, stability checking and modal analysis for the front leg. The element stiffness reduction method was used to simulate the structural damage, and the influence of damage on the modal parameters of the front leg was studied.

The emergence of heuristic algorithm greatly enriched the damage identification method. Zhao (2013) systematically studied the location and degree of crack damage of bridge erecting machine based on the multi-distribution principle of neural network. The research results show that the multi-distribution damage identification theory based on PNN network and BP network is feasible to identify the hidden crack of bridge erecting machine in the form of unit node disconnection, and can better determine the location and degree of damage. Wang (2014) regarded the damage identification problem as a mathematical problem of constraint optimization, improved the cuckoo search algorithm, established the structural damage model, identified the damage of the structure, and predicted the damage. The results showed that the improved cuckoo search algorithm and the proposed kernel fuzzy C-clustering algorithm based on kernel and multi-resolution wavelet kernel correlation vector machine could better determine the location and degree of the damage of the main beam structure.

In addition to the above methods, the damage identification can also be carried out based on the deflection influence line curvature, which is a new and emerging approach. Chen (2005) found in the study that the static displacement curvature confidence factor is very sensitive to the local damage of the structure, and the curvature of the structure is inversely proportional to the structural stiffness, which is particularly obvious in large structures. By using the principle of virtual work and the coordination condition of structural deformation, Cheng (2018) deduced the analytical expressions of the deflection influence line and the curvature curve of the deflection influence line at the identification point of the erecting machine. It was pointed out that the deflection influence line at different identification points, the deflection rate curve after damage, and the difference curve of the deflection influence line before and after damage can be effectively used to identify the damage location and degree of the main beam structure. But this method also has certain limitations, when the damage location is too much or the interval is too small, the identification effect will be affected. Influence line damage identification method is a method for damage identification of bridges and other structures, which has certain advantages over other methods, such as no health state parameters before damage, less sensors needed, and good anti-noise performance of displacement deflection instrument. This method is a hot research topic in recent years.

In the complex service environment, the erecting machine not only bears the design load, but also bears various sudden external factors. In this case, the change of component state and the accumulation of damage will be increasingly serious, which will threaten the safety of the structure and cause the sudden failure of the structure (Ou 2002). The main beam is the most important bearing part of the erecting machine, and the front leg is relatively weak due to the construction needs and support point structure constraints, besides, the unsafe factors also include fatigue and instability. Through the accident analysis in recent years, it can be seen that the highest proportion is the tilt accident of the erecting machine. Among these accidents, the instability of the front leg and the insecurity of the main beam accounted for the majority (Li et al. 2012). Therefore, it is of great practical significance to carry out mechanical monitoring and damage identification of the main beam and the front leg of the bridge erecting machine to determine its health status. In other fields, there are many health monitoring technologies based on dynamic or static parameters, and good practical results have been achieved. However, at present, the research on this type of erecting machine is very limited.

3 DISCUSSION

At present, the monitoring technology of bridge erecting machine has made remarkable achievements. Below are some views on its limitations.

(1) In the early development process of erecting machine, most of the attention was paid to its erection function, but less attention was paid to safety monitoring. Monitoring technology involves a wide range of fields, including sensing and testing technology, electronics, signal processing, computer technology and artificial intelligence. In addition, there are many types of erecting machines, so safety monitoring means cannot be unified. Due to all the above factors, so far in China, erecting machine safety monitoring still has not formed a scientific and systematic theoretical system. And the provisions of 'crane safety monitoring and management system' for bridge erecting machine are not detailed enough.

(2) Effectiveness of detection tools is still insufficient. Taking the stress monitoring as an example, the stress monitoring compares the theoretical value with the monitoring value under certain mechanical assumptions. The assumption itself has a certain deviation from the actual situation. Moreover, due to the unknown damage of the material, the vibration of the bridge erecting machine in the movement, the influence of wind, temperature and humidity on the environment and other factors, it is difficult for the monitoring personnel to distinguish what specific reasons caused the difference between the theoretical calculation value and the measured value. For the damage identification technology, the structural vibration mode and frequency are indeed related to the structural damage, but the natural frequency is not sensitive to the change of stiffness and mass. By the way, it is difficult to measure the structural vibration mode clearly and the measurement results are also accidental (Reflections on bridge health monitoring, 2012).

(3) In general, the existing monitoring technology has made remarkable achievements, but now many technologies still remain in the research stage, or its use conditions are very harsh, so it is unable to achieve good application effect in the harsh environment of erecting operation. In particular, the damage identification technology is still in the stage of simple laboratory model or numerical simulation, and there is no monitoring system to incorporate structural damage into the monitoring content. However, in fact, the phenomenon that damage accumulation leads to structural failure is very common.

(4) According to the survey results, most of the existing monitoring systems in the market just apply the standards mechanically, and often do not consider the particularity of the type and structure of erecting machine. Therefore, the design of monitoring system has become to choose the monitoring content options recommended by the standard specification according to the budget. As the erecting machine market managers reflected, to a certain extent, the bridge erecting machine monitoring system perform practically no function, some studies (Li 2013) also support this view. And the problem of weak usage level leads to the loss of trust in the

monitoring system by the operation or management staff, which is the most urgent problem that needs to be solved for the further development.

(5) The survey also found that in addition to the need for high requirements for monitoring systems, some problems also rely on strict management. Accident risk often occurs in the weakest link in the system. Usually, we attach great importance to the main components of the erecting machine, like main beam and supporting legs, spending a lot of manpower and material resources for monitoring. However, the risk is often hidden in places that are not easy to notice, such as the bolts of the supporting legs and the sleeper. For example, on July 18, 2019, the steel cylinder support pin shaft of the left front leg of an erecting machine in Mianxian County, Shaanxi Province fell off, resulting in the overturning and disintegration of the overall structure, resulting in 5 deaths and 7 serious injuries. Actually, the monitoring system of erecting machine is just one way to obtain the management information of bridge erecting machine operation, rather than the monitoring system can avoid accidents. Strict management of beam erection operation should always be the top priority.

4 PROSPECT

(1) The monitoring market of erecting machine is mixed, and the level of professional construction team is uneven. The special standards or guidelines for erecting machine monitoring should be issued as soon as possible.

(2) At the technical level, more advanced sensing technology should be developed to achieve more precise measurement and improve the noise resistance of the sensor. The intelligent safety monitoring system will be needed. At the same time, the automation degree of erecting operation is supposed to be improved, the manual participation links should be reduced. The fault self-test ability and fault diagnosis technology level need to be improved, and the early warning and alarm information must be given in time for the real-time monitoring.

(3) Attention should be paid to the damage accumulation of erecting machine, and the damage identification technology should be developed to avoid the difficulty in repairing the large parts and the interacting parts due to the local damage accumulation. Another direct cause of Mianxian bridge erecting machine accident was that the service life of the accident erecting machine exceeded 8 years, the structure was seriously damaged, the overall safety performance was poor, and it did not satisfy the basic safety technical requirements. Therefore, damage identification and health monitoring are suggested to be included in the monitoring scope of the existing monitoring system.

(4) As the front end of the monitoring, the layout of the sensor directly affects the monitoring results. The optimization of the sensor measuring point layout of the erecting machine will be one of the future research directions.

(5) Strengthening the operation management is also an important part of ensuring the safe operation of bridge erecting machine. It also requires us to improve the management system and safety working mechanism of electrical and mechanical equipment, improve the professional level of technicians and operators, increase the construction site hidden danger investigation, and standardize the site operation.

REFERENCES

Chen, N. (2016). Research on Effectiveness Evaluation and Diagnostic Methods of Safety Monitoring of Mechanical Equipment [D]. Beijing Jiaotong University.

Chen, X.Z. (2005). Damage identification of bridge structures based on static measurement data [D]. Huazhong University of Science and Technology.

Cheng, Y. (2008). Research on girder damage identification of bridge erector based on deflection influence line [D]. Shijiazhuang Tiedao University.

Du, C., Tian, Z., & Yi, Y. K. (2020). Research on safety monitoring system of bridge erector [J]. Engineering construction and design, 2020 (18): 233–234.

Du, H., Cheng, Y., Wu, S.J., & Wang, X.S. (2021). Research on dynamic response characteristics of JQ900A bridge erector structure under wind load [J]. Mechanical science and technology, 40 (06): 934–940.

GB 26469-2011. (2011). Bridge erector safety regulations [S].

GB/T 26470 - 2011. (2011). General technical conditions for bridge erectors [S].

GB/T 28264 - 2012. (2012). Safety monitoring and management system of hoisting machinery [S].

Guo, P., Peng, C., & Song, Y.W. (2013). Vibration safety monitoring of bridge erector based on STM32 [J]. Industrial control computer, 26 (09): 14–15.

He, J.W. (2017). Research on Bridge Erection Machine Monitoring System Based on Operation Attitude Monitoring [J]. Highway Traffic Technology (Applied Technology Edition), 13(09): 283–285.

Hearn, G., & Testa, R.B. (1911). Modal analysis for damage detection in structures [J]. Journal of structural engineering, ASCE, 1911, 117(10): 3042–3063.

Huang, Y.Y.,&YU, C.H. (2015a). On China's large tonnage bridge erector from the start to the world leading road (I) [J]. Railway construction technology, 2015 (02): 1-13 + 24.

Huang, Y.Y.,&YU, C.H. (2015b). On China's large tonnage bridge erector from the beginning to the world leading road (II) [J]. Railway construction technology, 2015(04): 1-13 + 23.

Kim, V. J. (1975). Detection of structural failure on fixed platforms by measurement of dynamic response [C]. Proceedings of the 7th annual offshore technology conference, 243–252.

Li, C.Y., Zhao, X.H., Yang, S.P., & Ma, Q. (2012).Design of bridge erector remote condition monitoring system based on GPRS [J]. National defense traffic engineering and technology, 10 (01): 12–15.

Li, F.S. (2013). Test verification of safety monitoring and management system of bridge erecting machine [J] Hoisting and transportation machinery, (08): 107–109.

Li, L.M. (2018). Current situation and development trend of domestic bridge erecting machine [J] Shanxi architecture, 44 (27): 218–219.

Li, X., Jing, Z., Jiang, F. (2012). Cause Analysis of Bridge Erecting Machine Tipping Accident based on Fault Tree and the Corresponding Countermeasures [J]. Procedia Engineering, 45(3):43–46.

Li, K. K., Wang, C.W., Wang, J.Y., &Yang, J.S. (2020). Application of Super Large Equipment Safety Monitoring System in Construction [J]. Highway Traffic Technology (Application Technology Edition), 16 (08): 194-195 + 198.

Liang, L.H. (2019). Research on safety monitoring system of bridge erector girder [J]. Mechanical and electrical engineering technology, 48 (11): 109–111.

Liu, Patriotic. (2010). Safety center of bridge construction-safety operation of bridge erector [J]. Journal of China Safety Science, 20 (03): 70–75. DOI: 10.16265/j.cnki.issn1003-3033.2010.03.028.

Ma, J., Sun, S.Z., Rui, H.T., Wang, L., Ma, Y., Zhang, W.W., Zhang, W., Liu, H., Chen, H.Y., Liu, J., &Dong, Q.Z. (2018). Review on China's road construction machinery research progress:2018 [J]. China highway Journal, 31(06):1–164.

Ou, J.P. (2002). Intelligent Monitoring and Health Diagnosis of Major Engineering Structures [C]. The 11th National Symposium on Structural Engineering, 2002:59–68.

Reflections on bridge health monitoring. (2012). https://mp.weixin.qq.com/s/fEiTqp-ESx 9laJp0oXM0vQ.

Su, Y. (2013). Safety analysis of front leg of high-speed railway bridge erector [D]. Shijiazhuang Tiedao University, 2013.

Wang, F.M., & Zhao, X. (2006). Safety and management of hoisting machinery [J]. Safety of special equipment in China, 22 (03): 10–13.

Wang, L.Y. (2014). Research on damage identification of bridge girder structure based on safety factor set [D]. Beijing Jiaotong University.

Wang, X.H., Yang, Z.H., Huang, G.J., Liu, J., & Wang, D.H. (2013). Research progress on safety inspection, monitoring and risk management of special electromechanical equipment [J]. Automation and information engineering, 34 (01): 1–5.

Wang, Y.Q. (2015). The optimal layout of gantry crane sensors for risk identification [D]. Southeast University.

Xie, W.L. (2015). Research on safety monitoring system of large construction machinery [D]. Shijiazhuang Tiedao University.

Xing, P.F. (2014). Design and implementation of bridge erector remote 3D monitoring system based on GENESIS64 [D]. Zhengzhou University.

Xu, K. (2017). Application of Intelligent Safety Monitoring System in Highway Bridge Erecting Machine [J]. Jiangxi Building Materials, 2017 (11): 179–180.

Yang, S.P., Xing, H.J., Guo, W.W., Wang, J.X., & Cai, W.D. (2013). Safety analysis of large construction machinery driven by construction technology [J]. China Mechanical Engineering, 2013, 24 (09): 1169–1173.

Yen, G. G., & Ho, L.W. (2003). Online multiple-model-based fault diagnosis and accommodation. IEEE Transactions on Industrial Electronics, 50(2), 296–312.

Zhang, C. (2011). Foreign highway bridge erector research and enlightenment [J]. Lifting transport machinery, 2011 (04): 5–9.

Zhang, C. (2014). Research on girder structural damage of bridge erector based on dynamic characteristics [D]. Shijiazhuang Tiedao University.

Zhang, H., Ni, W.J., Qin, X.R., Sun, Y.T. (2013). Optimal sensor configuration for low-order modal measurement of a certain crane [J]. Journal of China Construction Machinery, 11 (03): 263–266. DOI: 10.15999

Zhao, W.J. (2013). Multi-step damage identification of bridge erector based on neural network [D]. Shijiazhuang Tiedao University.

Zhao, X.H. (2013). Research on safety monitoring and management system of bridge erector [D]. Shijiazhuang Tiedao University.

Zhao, Y.M., Gong, M.S., & Yang, Y. (2020). Review of structural damage identification methods [J]. World Earthquake Engineering, 36 (02): 73–84.

Zheng, X.W., & Zhang, H.S. (2019). Safety monitoring system of bridge erector based on wireless communication [J]. Construction machinery technology and management, 32 (09): 68–70.

Frontiers of Civil Engineering and Disaster Prevention and
Control – Yang & Rahman (Eds)
© 2023 The Author(s), ISBN: 978-1-032-31200-2

Research on the influence of the temperature difference inside and outside the structure of the open-cut tunnel with Cofferdam at sea

Guangcai Yang*, Wenchang Li*, Wen Liu* & Xiangchuan Yao*
CCCC Fourth Harbor Engineering Bureau Co., Ltd., Guangzhou, Guangdong, China

ABSTRACT: Considering the fact that the construction of submarine tunnels has become more frequent in recent years, and the research on the influence of temperature difference on the internal force of open-cut tunnels in the sea is relatively scarce, a finite element numerical model is established in conjunction with the tunnel project of the University of Macau to analyze the causes of temperature difference in tunnels and the effect of temperature difference on the internal force of the structure. The results show that the effect of temperature change on the bending moment of the tunnel structure is obvious. When the tunnel section is large, the bending moment increment caused by the temperature difference cannot be ignored, but the shear force and axial force of the tunnel structure are insensitive to the temperature difference inside and outside the structure. In the construction of open-cut tunnels under the sea, it is required that temperature control calculations should be carried out before construction to theoretically budget the temperature stress caused by temperature difference and temperature drop, to predict the stability of the structure volume, and to provide scientific guidance for construction.

1   INTRODUCTION

Since the second half of the 20th century, with the development of human society, growing population, and expanding cities, the spaces for human production activities have expanded from the land to the sea, and these effective connections between land, sea, and air are made in two ways: bridges and tunnels. At present, there are more than 20 undersea tunnels that have been built or are planned to be built around the world, mainly in Japan, the United States, Western Europe, Hong Kong, China, and other regions. There are also many undersea tunnels under construction at home and abroad.

The undersea tunnel is a very safe all-weather channel passage as a solution for transportation between the straits and bays without taking up space, obstructing navigation, or affecting the ecological environment. However, when the undersea tunnel construction is carried out by the open-cut method, a large cofferdam needs to be built, then the deep foundation pit is dewatered, next, the deep foundation pit is excavated, and finally, the main structure of the tunnel is constructed. The heat of hydration caused by tunnel construction and the cold environment of the sea will form a large temperature difference between the internal and external tunnel structure, which in turn will affect the tunnel structure. At present, the research on the temperature at home and abroad mainly focuses on the influence of temperature on the deformation and internal force of deep and large foundation pits. Zheng (2002) pioneered an iterative calculation method for temperature stresses in horizontal supports using an improved elastic resistance method; Lu (2008) coupled the temperature field to the stress field for numerical analysis of the temperature effect of foundation pit support, and obtained the law that the support stress changes linearly with temperature; based on the Winkel model, Fan (2014) considering the multi-layer foundation soil, the interaction of multi-channel horizontal support and retaining wall, established the calculation method of temperature

*Corresponding Authors: yguangcai@cccc4.com, lwenchang2@cccc4.com, 379047775@qq.com and 605866412@qq.com

DOI 10.1201/9781003308577-15

internal force under multi-channel horizontal support, and concluded that the stiffness of the soil layer, the stiffness of the enclosure piles, and the length of the supports are the three main factors affecting the temperature internal force. Xiang (2014) proposed temperature control measures by monitoring the stress and deformation of the support structure in the deep foundation pit. In terms of temperature research of tunnel structure, the main focus is on the effect of tunnel fire on the tunnel structure. Tajima (2005) used the RABT heating curve to conduct fire tests and numerical simulations on the tunnel segment structure, and analyzed the change law of the temperature field of the tube sheet structure with the fire time; Yan (2012) conducted a full-scale fire test on the shield tunnel segment structure, and discussed the bursting, deformation and bearing capacity of the shield tunnel segment lining structure under fire; in response to the large uncertainty in the full-scale fire experiment, Chen (2022) used the temperature distribution in the tunnel under fire scenarios obtained from multiple repetitions of experiments under the same working conditions and adopted statistical methods to quantitatively assess the error of the longitudinal section temperature data in different areas of the tunnel and at different times.

However, there are few studies on the influence of temperature difference on the internal force of open-cut tunnel structure in sea cofferdam. This paper will study the effect of temperature difference on the internal force of the tunnel structure, and discuss the reasons for the temperature difference in the tunnel, the influence of the temperature difference on the internal force of the structure, the mechanism of the additional stress generated by the temperature difference, and the analysis method of the temperature effect, thus providing a reference for the undersea tunnel project.

2 PROJECT OVERVIEW

The cross-sea tunnel of the University of Macau starts from the planning road of the Hengqin campus of the University of Macau in the west and ends at the Lianhua Waterfront Road in the east of the Coloane, Macau. The linearity of the project is a "Z" shape. The total length of the tunnel is 1570 m, of which the length of the fully enclosed section of the tunnel is about 1030m, as shown in Figure 1.

Figure 1. Schematic diagram of the tunnel crossing the sea on the Hengqin campus of the university of Macau.

The tunnel project of the University of Macau adopts the open-cut method in the sea cofferdam, and the main structure is made of reinforced concrete. After building two cofferdams in the cross-gate waterway to pump the middle seawater dry, the 530m-long open cut tunnel section in the sea will be constructed by dry operation. The excavation depth of the foundation pit is 18 ~ 24m, with 5 ~ 6 supports; the first support will use reinforced concrete support, and the rest will set steel support or reinforced concrete support according to the structural calculation; the steel support

112

will be Φ800mm Q235 steel pipe with a 16mm-thick wall; a row of continuous Φ800@600 will be laid out, close to the pile, with the same depth of the enclosure pile.

3 NUMERICAL ANALYSIS OF TEMPERATURE EFFECTS

3.1 *Establishment of numerical model*

This chapter mainly studies the effect of temperature difference on the internal force of the structure, which inevitably involves the coupling effect of temperature and structure, and its parameters also need to consider two aspects, one is the physical and mechanical parameters, and the other is the thermodynamic parameters. Its mechanical parameters are also divided into two parts, one is the mechanical parameters of the soil layer, and the other is the mechanical parameters of the structure. The main geological and material parameters are selected according to the geological exploration data and analog engineering examples, as shown in Table 1.

Table 1. Main geological and material parameters.

Material name	Elastic modulus E (Pa)	Weight γ (kN/m^3)	Thermal conductivity Kxx (W/(m·K))
2- Silt	2.64×10^7	16	1.08
3-1 Clay	6.48×10^7	19.4	1.12
3-2 Silty soil	3.9×10^7	17.8	1.13
3-4 Clay	3.972×10^7	19.3	1.12
3-5 Silty soil	2.892×10^7	16.4	1.13
3-6 Medium-coarse sand	2.4×10^8	20.5	1.20
3-8 Silty soil	3.072×10^7	17.7	1.13
Concrete	3.35×10^{10}	25	1.74
Steel support	2.06×10^{11}	78.5	—

ANSYS finite element software was selected to establish a numerical calculation model. Considering the temperature field analysis, the stratigraphic structure method was used for calculation. The relevant specifications and analogous engineering examples were consulted, and the model range was preliminarily determined. The range of twice the tunnel width was taken on both sides of the tunnel, and the range of twice the width of the tunnel was taken on the bottom of the tunnel, as shown in Figure 2. Soil and structural elements are solid plane elements, both of which can be considered as linear elastic materials; enclosure piles and supports use beam3 elements. Among them, the expansion coefficient of concrete can be regarded as not changing with temperature.

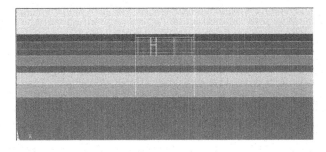

Figure 2. Analysis model diagram.

3.2 Analysis of working conditions

The changes in site temperature during construction and operation are shown in Figure 3. It can be seen from the temperature change curve that the seasonal temperature difference in Zhuhai is not too large. The highest temperature in summer can reach 35°C, and the average temperature is about 28.6°C; the annual average temperature in Zhuhai over the past few decades was about 22°C. Besides, considering the temperature change on the soil side caused by the heat of hydration, since there is no actual monitoring data for the temperature of the heat of hydration in the early stage, the temperature change trend cannot be obtained intuitively. Here mainly based on the later monitored temperature data and analogy with reference to other engineering examples, the temperature change trend can be analyzed. According to the monitoring data and analogy to similar projects, it can be known that the temperature inside the structure can reach up to 36°C–48°C.

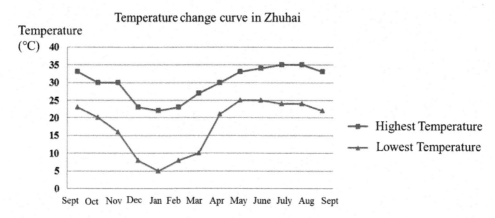

Figure 3.   Temperature change curve in Zhuhai.

Regarding the analysis of temperature stress, it mainly involves the influence of temperature difference on the internal force of the tunnel. Considering the coupling of temperature field and stress field, this coupled analysis may increase or decrease the structural stress compared with simple stress analysis. According to the simple discussion of the above two aspects of temperature, it can be seen that the extreme working conditions in the tunnel construction stage are: when the heat of hydration is the largest and when the outside temperature is the lowest. In the operation stage, the temperature inside and outside the structure changes a lot, which is mainly divided into two aspects for analysis. One is that the temperature inside the structure is lower than the outside, and the other is that the temperature inside the structure is higher than the outside. The specific effects of the above working conditions on the stress of the structure itself will be analyzed in detail in the following sections on temperature field and stress field.

In the ANSYS numerical analysis, the specific construction process is simulated, timely support is adopted, and a layer of soil is dug for immediate support. The specific working conditions are as follows:

Step 1: Activate the fencing pile and the first support;
Step 2: Excavate the first layer of soil and activate the second support;
Step 3: Excavate the second layer of soil and activate the third support;
Step 4: Excavate the third layer of soil and activate the fourth support;
Step 5: Excavate the fourth layer of soil and activate the third support;
Step 6: Excavate the fifth layer of soil and activate the sixth support;
Step 7: Excavate the sixth layer of soil;
Step 8: Activate the tunnel lining structure;
Step 9: Backfill the soil.

3.3 *Analysis of temperature field*

Before calculating and analyzing the temperature stress, a simple temperature field analysis is carried out to study the specific distribution of the temperature field inside the tunnel structure. Since the temperature parameters of the material can be considered not to change with the temperature change, under different working conditions, that is, different temperature conditions, the internal temperature of the structure is different but the distribution should be similar, so one of the sections is used for research and analysis. The measured results of atmospheric temperature and soil temperature corresponding to the construction period are input into the analysis model as the boundary conditions of the surface temperature of the structure, and relevant parameters are set to analyze the temperature field. The analysis results of the temperature field inside the tunnel structure are shown in Figure 4 and Figure 5.

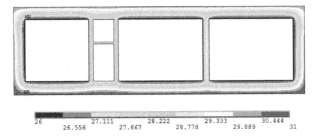

Figure 4.   Temperature field of the tunnel structure.

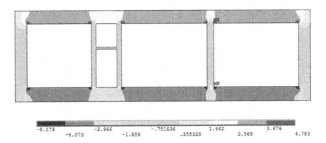

Figure 5.   Temperature gradient of the tunnel structure.

From the temperature field diagram of the tunnel structure, it can be seen that the temperature difference inside and outside the structure produces a relatively uniform temperature field inside the structure, and the temperature of the intermediate partition wall is slightly higher than the air temperature because of the heat of hydration and its small thickness; the bottom plate is thicker, the heat of hydration is large, and the temperature is significantly higher than that of the top plate and sidewall; from the depth direction, the temperature of the sidewall near the top plate is obviously lower than that of the part near the bottom plate, and the temperature in the middle of the sidewall is slightly raised. The reason for this phenomenon can be attributed to the release of the hydration heat on the one hand, and the faster dissipation of temperature on the other hand, because the open-cut method is used, the upper part is in more contact with the air.

From the temperature gradient diagram of the tunnel structure, it is obvious that the gradient is the largest at the corners of each tunnel structure. In other words, the temperature changes are relatively large at each corner. At the same time, from the design point of view, the corners are also the stress concentration points, which are the structural weak link, so the corner should be strengthened.

3.4 Coupling field analysis

The coupled field analysis is divided into three working conditions, one is without applying temperature load, the other is applying temperature load but the temperature outside the structure is higher than the temperature inside the structure, and the third is applying temperature load but the temperature outside the structure is lower than the temperature inside the structure. Calculations are carried out for these three working conditions respectively, and the results of the internal force and deformation of the tunnel structure are obtained. Due to the large span of the tunnel, the maximum span is as long as 10 meters, the water and soil pressure and temperature difference have a greater influence on its deformation. In addition, the height of the tunnel is also relatively large, nearly 9 meters high. However, owing to the existence of enclosure piles, the exterior of its sidewalls is less subject to soil and water pressure, and the change of the longitudinal burial depth causes the sidewall to be significantly affected by the temperature difference. Generally speaking, the top plate and sidewalls are most affected by the peripheral water and soil pressure and temperature difference of the entire tunnel structure. The top plate and sidewalls of the structure are analyzed intensively.

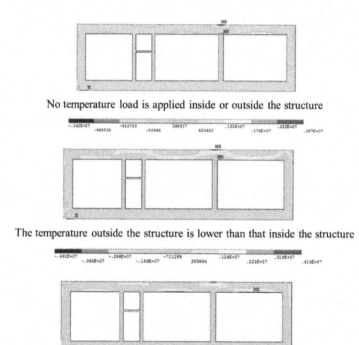

No temperature load is applied inside or outside the structure

The temperature outside the structure is lower than that inside the structure

The temperature outside the structure is higher than that inside the structure

Figure 6. The principal stress cloud diagram of the tunnel structure.

It can be clearly seen from Figure 6 that the effect of temperature difference on the structure is obvious. When the temperature load is applied, the principal stress increases significantly compared to that without applying temperature load, and when the temperature outside the structure is lower than the temperature inside the structure, the effect is more significant when the temperature outside the structure is higher than the temperature inside the structure. From the scope of influence, the top plate of the tunnel lining is the most affected, the sidewalls are second, and the bottom plate is the least affected.

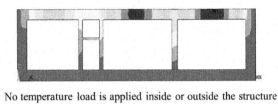

No temperature load is applied inside or outside the structure

The temperature outside the structure is lower than that inside the structure

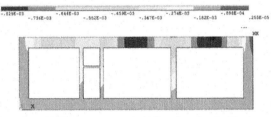

The temperature outside the structure is higher than that inside the structure

Figure 7.   Vertical displacement cloud map of the tunnel structure.

For the vertical displacement nephograms of the three working conditions, it can be clearly noted that different temperature differences have different effects on structural deformation (as shown in Figure 7). When the temperature outside the structure is lower than the temperature inside the structure, the overall vertical displacement of the structure is slightly reduced, mainly due to the effect of thermal expansion and contraction, and the direction of temperature expansion is opposite to the direction of stress deformation, so it is slightly offset. When the temperature outside the structure is higher than the temperature inside the structure, the overall vertical displacement of the structure increases slightly, mainly due to the effect of thermal expansion and contraction, and the direction of temperature expansion is the same as the direction of stress deformation, so it increases slightly.

Since the finite element software used in this section is ANSYS, and plane elements are used in the above-mentioned temperature field and stress field analysis modeling for lining simulation, plane elements can obtain good results for stress and deformation, but for bending moment, axial force and shear force cannot be described intuitively. For this reason, the model was partially adjusted, the lining structure was simulated by beam elements, and the soil body was simulated by plane elements.

It can be observed from Figure 8 that when the temperature outside the structure is lower than the temperature inside the structure, the maximum bending moment of the top plate increases significantly, especially at the mid-span, but it decreases slightly at the junction of the partition wall and the top plate, and for the sidewall, the bending direction of part of the bending moment has changed; when the temperature outside the structure is higher than the temperature inside the structure, the maximum bending moment of the top plate decreases significantly. The bending

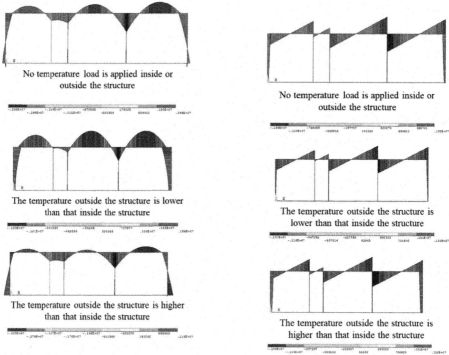

No temperature load is applied inside or
outside the structure

The temperature outside the structure is lower
than that inside the structure

The temperature outside the structure is higher
than that inside the structure

Figure 8. Comparison of structural bending moments under the influence of temperature difference.

No temperature load is applied inside or
outside the structure

The temperature outside the structure is
lower than that inside the structure

The temperature outside the structure is
higher than that inside the structure

Figure 9. Comparison of structural shear force under the influence of temperature difference.

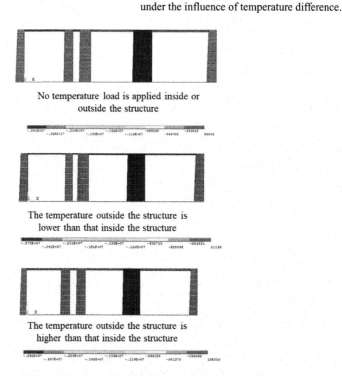

No temperature load is applied inside or
outside the structure

The temperature outside the structure is
lower than that inside the structure

The temperature outside the structure is
higher than that inside the structure

Figure 10. Comparison of structural axial force under the influence of temperature difference.

moment decreases in the middle of the span and increases at the junction of the partition wall and the top plate, while for the sidewall, the position of the point where the bending moment is zero is obviously moved up, and the bending moment also increases slightly.

Figure 9 reflects the effect of the temperature inside and outside the structure on the sheer force of the tunnel structure. The overall trend of the shear force under the three working conditions does not change significantly, and the value only increases slightly. In general, the effect of temperature differences on the shear force is not great, relatively speaking, the shear force on the sidewall is slightly more affected by the temperature than the top plate. Similarly, the temperature difference inside and outside of the structure does not have a large change in the axial force (as shown in Figure 10). When the temperature outside the structure is lower than the temperature inside the structure, the maximum positive axial force increases, and the negative axial force with the largest absolute value decreases slightly. When the temperature outside the structure is higher than the temperature inside the structure, the maximum positive axial force increases, and the negative axial force with the largest absolute value increases slightly. In general, the temperature difference has little effect on the axial force of the structure, but the axial force of the structure will change greatly locally.

4 CONCLUSIONS

In light of the current construction status of the open-cut method of undersea tunnels, relying on the tunnel project of the University of Macau, the research and analysis of the influence of temperature difference on the internal force of the open-cut method of the sea cofferdam tunnel structure was carried out, and the formed technical achievements were applied to the project practice.

(1) Through analysis and comparison, it is considered that the temperature gradient is the largest at the corner of the tunnel structure, and the resulting temperature stress is correspondingly the largest. The corner is also the stress concentration point and the weak link of the structure, so the corner should be strengthened.
(2) When the temperature outside the structure is lower than the temperature inside the structure, the effect is more significant than that when the temperature outside the structure is higher than the temperature inside the structure. The effect on the top plate of the tunnel lining is the largest, the sidewalls are the second, and the bottom plate is the least affected.
(3) When the temperature outside the structure is lower than the temperature inside the structure, the maximum bending moment of the top plate increases significantly, the bending moment at the junction of the middle partition wall and the top plate decreases, and the structural shear force and the structural axial force are insensitive to the influence of temperature.

REFERENCES

Chen J F, Cheng H H, Wei X et al. (2022). Error analysis of temperature measurement in a tunnel full-scale fire experiment. J. Tsinghua University (Sci & Technol), 3: 614–618.
Fan J Y. (2014). A deep big hole excavated for building foundation level support temperature endogenic force and distortion computational method. J. Shanxi Architecture, 40(18): 59–62.
Lu P Y, Han L J, YU Y. (2008). Finite element analysis of temperature stress in strut of foundation pit. J. Rock and Soil Mechanics, 29(5): 1290–1294.
Tajima H, Kishida M, Kanda T, et al. (2005). Study on the deformation and load-bearing capacity of TBM shield tunnel lining in fire. J. Underground Space Use: Analysis of the Past and Lessons for the Future, 2: 793–799.
Xiang Y. (2014). Influence of temperature stress on internal force and deformation of retaining structures for deep excavations. J. Chinese Journal of Geotechnical Engineering, 37(S2): 64–69.
Yan Z G, Zhu H H, Ju J W, et al. (2012). Full-scale fire tests of RC metro shield TBM tunnel linings. J. Construction and Building Materials, 36: 484–494.
Zheng G, Gu X L. (2002). A simple method for calculating temperature stress in horizontal strut of foundation pit considering strut-pile-soil interaction. J. China Civil Engineering Journal, 35(3): 87–89, 108.

Frontiers of Civil Engineering and Disaster Prevention and Control – Yang & Rahman (Eds)
© 2023 The Author(s), ISBN: 978-1-032-31200-2

Practice of research ideas and framework for improving energy consumption performance of building envelope system in existing residential areas in North China based on BIM technology

Ye Wen* & Guangmei Zhang*

School of Architectural Engineering, City Institute, Dalian University of Technology, Dalian, China

ABSTRACT: With the improvement of energy-saving renovation of existing residential buildings and the corresponding requirements, in order to ensure the energy consumption level of the building envelope system of existing residential areas in the north, this paper analyzes the building envelope system of existing residential areas in the north by BIM technology, so as to know the specific energy consumption of the envelope system. And using BIM technology, we deeply explore the energy-saving and consumption-reducing methods of building envelope system in existing residential areas in north China, and improve the energy-saving benefits of envelope system structure.

1 INTRODUCTION

With the introduction of China's policies, different regions are carrying out the renovation of existing residential buildings. However, there are some problems in the process of renovation, such as single form, poor effect and low efficiency, which will have a direct impact on the improvement of the quality of existing residential buildings in China. Therefore, based on BIM technology, it is necessary to analyze the specific situation of existing residential buildings in northern China in the process of envelope transformation, master the performance of envelope system, and fully understand the energy consumption performance of existing residential building envelope system by using BIM technology, so as to ensure the energy consumption performance of existing residential building envelope system.

2 OVERVIEW OF BUILDING ENVELOPE SYSTEM IN EXISTING RESIDENTIAL AREAS

2.1 *Classification of building envelope system in existing residential areas*

In the process of classifying the building envelope system of existing residential areas, it mainly includes the following three types of building residential areas (as shown in Figure 1): First, the location classification of residential areas. This classification mainly divides the existing residential areas from the perspective of urban development and evolution, including the original residential areas of urban residents, reserved residential areas in urban centers and special residential areas in urban sub-centers. The original residential areas of urban residents are formed by the centralized development of the overall planning of the city in the initial development process, which is characterized by scale and openness, with certain supporting facilities (Zhang 2016). The reserved residential area in the urban center is developed and evolved on the basis of the original residential area, because in the process of development of some residential areas, lots with better location will be transformed into commercial land, while large-scale residential areas will be scattered, and only some of the lots with low commercial value will be reserved, which will form relatively scattered residential areas and become reserved residential areas in the urban center. Special residential areas

---

*Corresponding Authors: 532680192@qq.com and 20446777@qq.com

DOI 10.1201/9781003308577-16

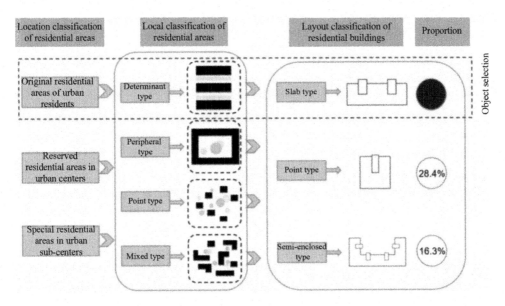

Figure 1. Classification of existing residential building types.

in urban sub-centers are usually formed with characteristic industries as ties, and the core residents are mainly specific occupational groups, including industrial parks, industrial parks and so on.

Second, local classification of residential areas. The layout patterns of this residential area in the process of urban planning and development are mainly determinant, enclosing, point and mixed. Based on the investigation of the existing residential areas in a city, it is found that because of the particularity of the region itself, the terrain is sloping, and the height difference, the orientation, spacing and other factors are considered in the planning process. The layout modes of residential buildings mainly include different types such as peripheral type, point type, mixed type and determinant type, and various layout forms of residential areas are important references for selecting the simulation objects of the building envelope system of existing residential areas.

Third, the layout classification of residential buildings. This paper investigates the existing residential areas in a city, and completes the copying and drawing of residential building drawings, and summarizes several architectural layouts of existing residential buildings in 1980s and 1990s. The differences in layout forms, envelope systems and locations will have an impact on the comprehensive performance of residential building envelope systems. There are more than 300 survey drawings of existing residential buildings in the city, which classify the layout types of residential buildings, including point type, semi-enclosed type and slab type, among which slab type accounts for a large proportion, exceeding 50%, and point type and semi-enclosed type account for about 28% and 16% respectively (Wei 2020). Different residential buildings have some differences in information such as building layout and shape coefficient, which will lead to differences in energy consumption performance of building envelope system. Therefore, it is necessary to classify the layout of residential buildings to facilitate targeted research and analysis.

2.2 *Demand change of building envelope system performance improvement in existing residential areas*

In the process of urban planning and development, to study and analyze the performance of building envelope system in existing residential areas, it is necessary to understand the demand for improving the performance of envelope system. Since the completion of housing, after a long period of application and change, its quality has some deviation from the development demand of social times. The main reasons for this gap are the deterioration of housing and historical performance, etc. In addition, with the continuous revision and optimization of national standards and specifications, the performance of residential buildings has also declined (Wu 2016). In this case, the evolution and optimization of standards can reflect the performance of existing residential envelope system in different periods on a relatively broad level, thus showing the evolution of performance.

There will be simple to complex processes in the process of standard formulation: (1) In 1980s, because of the limitations of China's economic level and construction level, the national standard only designed some safety-related physical key performance indicators and structural performance indicators for different systems, and the number of standards was not much, and more emphasis was placed on seismic and fire-proof standards and specifications, and the content of formulation was to summarize the experience of practical projects (Ji 2021). (2) In the mid-1980s, commercial housing began to appear in China. People gradually pursued the aesthetics of architecture, and the breadth and depth of its building standards also changed greatly. Ventilation and thermal standards became the main contents of the building standards at that time. In the process of determining the standard content, in addition to referring to the practical experience of the project, more attention is paid to the study of the related theories of architecture. (3) At the beginning of this century, the development of China's building standards and norms has formed an engineering construction standardization system with comprehensive standards, basic standards, general standards and professional standards as levels. Especially under the background of continuous improvement of living standard, economic level and construction level, there are more and more requirements for special specifications. For example, fire-resistance and moisture-proof performance indexes of material structure, and some codes also have recommended provisions of envelope system, which promotes the diversification and personalized development of Chinese architecture and helps to improve the living quality of residents (Dong 2017).

With the increasing demand for improving the performance of residential building envelope system, the original residential buildings cannot meet the living needs of residents. Therefore, it is necessary to reform and design the building envelope system of existing residential areas. The main problem in the practice of reforming the building envelope system of existing residential areas is the deterioration of the performance of existing houses. When carrying out the renovation, it has gone through the renovation of dilapidated buildings, energy saving, water saving, comprehensive renovation, green renovation and safety renovation, etc. Under the background of rapid economic and technological development, new features have emerged in the content and methods of housing renovation. At present, in the process of transformation, China mainly adopts three ways of comprehensive, energy-saving and green transformation (Liu 2019). Because the northern region is a severe cold region, the original envelope structure is more seriously damaged, and the harsh environment requires higher performance of residential buildings. Therefore, it is necessary to improve the overall performance of existing residential buildings in the north as much as possible in this respect. At present, China's large-scale renovation and design of old and existing residential areas in northern cities has made some achievements, and it has greatly promoted the renovation process of existing residential buildings and improved the performance of residential buildings to a certain extent.

3 ENERGY CONSUMPTION ANALYSIS OF BUILDING ENVELOPE SYSTEM IN EXISTING RESIDENTIAL AREAS IN NORTH CHINA BASED ON BIM TECHNOLOGY

3.1 *Application value of BIM technology*

BIM technology refers to the technology of digital expression and information management of all information in the whole life cycle of buildings or facilities. BIM technology can be applied as a virtual substitute model of physical buildings or facilities, and a shared information database can be built during the application process, which is convenient for users to access the database to obtain corresponding information. And the information can be updated in real time in the process of building BIM database. In addition, BIM technology can establish an information model based on open standards and communication operation standards, and dynamically manage all links in the construction process of buildings or facilities. BIM model is an important technical means to ensure the quality of dynamic management. All participants can use BIM model to extract their own information and update the information. BIM modeling can improve the efficiency and quality of building management to a certain extent (Ye 2020).

3.2 *Simulation analysis of energy consumption of building envelope in existing residential areas in North China based on BIM technology*

When building energy consumption analysis is carried out based on BIM model data, it is necessary to optimize and adjust the reconstruction scheme based on specific analysis results, and finally determine the reconstruction scheme. In the process of operation, the following steps are required: First, import the gbxml file exported from the Revit model, complete the call of model monitoring data, and check and sort out the model at the same time. Second, adjust, extract and integrate the building information of existing residential areas, and set the environmental parameters that affect the energy consumption performance. Among them, the external environmental information needs to be scientifically set based on the external meteorological conditions of the area in China's "Special Meteorological Database for Building Thermal Environment Analysis in China" (Yang 2020). In the reference process of internal factors, the new environmental design is mainly based on the energy-saving design standards of residential buildings in cold and cold areas. It is required to complete the setting of environmental information, mainly including structural data, thermal parameters of envelope structure, activities of staff, use time of room and equipment, etc. Third, simulate the energy consumption of different promotion schemes, and compare and analyze the simulation results. The thickness of newly added insulation layer can be used as the main parameter of renovation design, analyze the influence of insulation layer material thickness required for renovation and promotion of envelope system on energy consumption, and carry out verification test. Externally, the combination of different parts' lifting modes should be taken as the transformation design parameter, so as to verify the reduction degree of overall building energy consumption under different combination schemes of various components' molding and lifting modes, and improve the reliability of scheme selection. The specific simulation process is shown in Figure 2.

4 METHOD AND GENERAL PROCESS FOR IMPROVING ENERGY CONSUMPTION PERFORMANCE OF ENVELOPE SYSTEM BASED ON BIM TECHNOLOGY

In order to comprehensively analyze the energy consumption performance of building envelope system in existing residential areas, BIM technology can be used for modeling and energy consumption simulation analysis, so as to take targeted measures to solve the problems. In this process, we can give full play to the application advantages of BIM technology. In this research process, mainly based on BIM technology, when analyzing the energy consumption performance improvement of existing building envelope system, the methods used mainly include the following contents: (1) Complete the acquisition and integration of existing residential building information. After the information is obtained, it can be input into BIM information database, and the data model can be built by BIM technology to ensure the comprehensiveness of building envelope system information, and at the same time, the consistency and correlation coordination of information should be improved. (2) In the process of model application, information integration attribute and model visualization attribute can greatly improve the intuition and efficiency of the transformation process. Visualization can directly show the roof, doors and windows, exterior wall, material structure, etc. to the constructors, Party A and designers, which is convenient for all participants to simulate the envelope system. (3) In the process of BIM model construction, it is necessary to separately analyze the original situation of the envelope system and the performance of the envelope system construction in each reconstruction scheme, and to form a selective reconstruction and promotion scheme by changing the combination method of different components that make

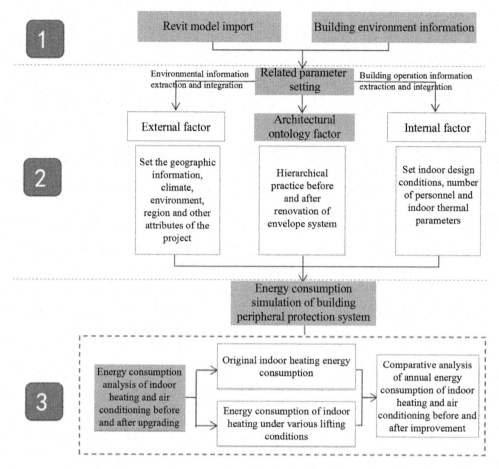

Figure 2. Schematic diagram of energy consumption simulation process.

up the envelope system. (4) In the simulation calculation of natural energy consumption, using the visual data monitoring and analysis function of BIM technology, we can get the specific values of energy consumption of different envelope systems, and evaluate the importance of objective data to prevent unrealistic problems in the process of upgrading scheme selection. Using BIM technology can not only improve the efficiency of transformation and upgrading and the accuracy of scheme decision-making, but also ensure the integrity, relevance and consistency of information (Wang 2019).

In the application process of BIM platform, different majors, different stages and different softwares can carry out information conversion. Using effective connection interfaces is beneficial to improve the rate of information import, and eliminate various problems such as information duplication, contradictions and omissions that may exist when inputting information from multiple ports. This advantage can fundamentally improve the accuracy of data, and ensure the effective sharing of information among different majors and different stages.

In the application process of the envelope system, BIM should be combined to build the energy consumption performance improvement process of the envelope system of existing residential buildings (see Figure 3).

The first stage is the information integration stage, the main purpose of which is to maintain the appropriate level of building environment. Different information between buildings and environment affects the energy consumption performance of buildings. Therefore, it is necessary to acquire

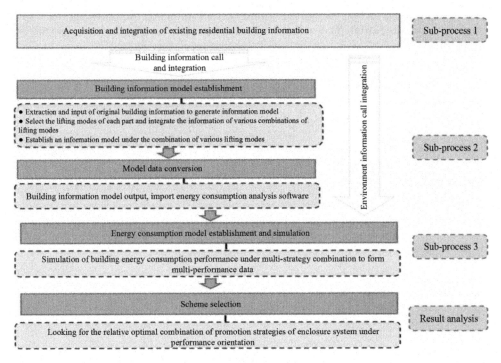

Figure 3. Energy consumption performance improvement process of building envelope system in existing residential areas based on BIM.

and integrate this part of information when carrying out the work of improving the energy consumption performance of building envelope system in existing residential areas. So as to accurately evaluate various factors influencing building energy consumption performance. Because the existing residential building itself has certain complexity, at this stage, it is necessary to fully grasp the information types and different information indicators needed in the process of improving energy consumption performance, and to know the sources of all information indicators. After sorting out the information indicators and sources, we can use the integration methods and integration platforms for in-depth research.

The second stage mainly completes the construction of information integration platform, which can carry out data integration and application from two aspects of model building software Revit and energy consumption performance simulation software Ecotect, and complete the whole process of information call from information acquisition to information integration. After the information integration in the first stage, the data of the first stage of the envelope system can be reflected in the model construction stage, and the BIM software can be used to complete the model construction of building information related to the envelope system of existing residential buildings. This is an important means to integrate indoor information. In this process, it is necessary to combine the hierarchical analysis method cited by the envelope system, use the visual attributes of Revit software to complete the hierarchical feature analysis of the envelope system, and build a new hierarchical visualization model. In this process, the attribute information of every building level of the building is recorded, and the information can be called at any time according to the demand. In addition, in the process of building the 3D information model, the 2D diagram of the scheme diagram can be shown in other views, and the modified building components in the plan view need to be presented. Corresponding changes will automatically occur in this model, which can improve the work efficiency. Revit has high compatibility, can cooperate with various BIM softwares, can transform the model into IFC files, and can transmit the model information to the next stage of performance simulation, which brings great convenience to the whole workflow and can reduce various problems such as information loss or omission in the process of information transmission (Zhang 2019).

The third stage is the performance simulation stage. In this stage, it is mainly combined with the information integration in the first stage to effectively integrate the environmental information inside and outside the building and provide basic data for environmental simulation. In the specific simulation process, it is necessary to export Revit information model to Ecotect energy consumption performance simulation software first, and then integrate it with indoor and outdoor environmental information parameters by setting the design variable parameters of the envelope system, so as to calculate the annual energy consumption report. The Ecotect simulation software used in this study, as an auxiliary design software for technical performance analysis, can fully play a role in the design and optimization of building schemes. The main function of this software is to complete the analysis of many physical performance parameters of buildings, such as thermal and energy-saving design, building manufacturing and building lighting. The most remarkable feature in the application process of the software is the real-time analysis feature. After changing a window of the model, we can immediately see the changes of indoor thermal effect and light environment, and can efficiently analyze the energy consumption performance improvement scheme of the envelope system.

The fourth stage is the conclusion stage. After all the work in the first, second and third stages is completed, it is necessary to analyze the annual energy consumption report before and after the transformation, so as to obtain the energy consumption results and select the optimal scheme.

5 CONCLUSIONS

All in all, in the process of studying the energy consumption performance of building envelope system in existing residential areas, it is necessary to introduce various data information of energy consumption of building envelope system in existing residential areas by using BIM technology from the perspective of energy consumption system upgrading. At the same time, it is necessary to complete the simulation of energy consumption in the renovation of envelope system, and select and optimize the upgrading scheme.

In this study, the author did not build an energy consumption performance improvement system of envelope system based on BIM technology comprehensive utilization of many scientific means such as field investigation, standard consultation and expert consultation, so he could not fully grasp the application effect of BIM platform in energy consumption analysis of envelope system.

In the subsequent research, researchers need to use BIM technology to verify the improvement of energy consumption performance of typical existing residential building envelope system in specific areas, so as to scientifically grasp the feasibility of the process system.

REFERENCES

Dong Mei. Discussion on the Practice of Building Engineering Design Management Based on BIM. Architectural Engineering Technology and Design, 000(015):3421–3421.

Ji Weiwei, Wang Dapeng, Lan Chengjie, et al. Building House Sunshine Analysis and Simulation Device Based on BIM Technology: CN212365323U.

Liu Yingchuan. Three-dimensional Model Construction and Energy Consumption Model Research of Green Renovation of Existing Residential Buildings in Cold Areas. Shenyang Jianzhu University.

Wang Yijing, Wang Runsheng, Liu Yuan. Analysis of Passive Ultra-low Energy Consumption and Energy-saving Renovation of Existing Public Buildings Based on BIM Technology. Architecture & Culture, (10):2.

Wei Ting. Application Advantage Analysis of BIM Technology in Green Smart Residential Area Construction. Technology Innovation and Application, (26):2.

Wu Qiang, Qin Chao. Energy Consumption Evaluation of Green Buildings Based on BIM Technology. Chinese Sci-tech Periodical Database (Citation Edition) Engineering Technology, (9):00005-00005.

Yang Liqin. Research on Energy Consumption Evaluation of Green Building Based on BIM Technology. Famous Scenery, (6):1.

Ye Qing, Fang Min, Zhao Qiang, et al. Research on Multi-objective Comprehensive Optimization Method of Green Transformation of Existing Urban Residential Areas Based on Differential Evolution Method. Urban Development Studies, 27(11):6.

Zhang Chunhuan. Research on Energy Consumption Performance Improvement of Building Envelope System of Existing Residential Areas in North China Based on BIM Technology. Dalian University of Technology.

Zhang Xueshun, Fang Tingyong, Xu Jun, et al. Design of Energy-saving Renovation Scheme for Existing Residential Buildings Based on BIM Technology. Engineering and Construction, 030(003):379–381,390.

Frontiers of Civil Engineering and Disaster Prevention and
Control – Yang & Rahman (Eds)
© 2023 The Author(s), ISBN: 978-1-032-31200-2

Experimental research on the connection node of square steel tube column and T-shaped piece

Zhang Jianke*
School of Civil Engineering, Henan polytechnic University, Jiaozuo, Henan, China

Wang Xinwu*
International Joint Laboratory for New Type Civil Engineering Structures of Henan Province, Luoyang Institute of Science and Technology, Luoyang Henan, China

Fan Lidan
School of Civil Engineering, Henan polytechnic University, Jiaozuo, Henan, China

Liu Huanhuan
International Joint Laboratory for New Type Civil Engineering Structures of Henan Province, Luoyang Institute of Science and Technology, Luoyang Henan, China

Zhao Junyang
School of Civil Engineering, Henan polytechnic University, Jiaozuo, Henan, China

ABSTRACT: To study the mechanical performance and failure mode of the square steel pipe column-T-shaped connection joint, the unidirectional static loading test was conducted. The study involves the unilateral high strength bolt connection – The tensile bearing capacity and failure mode of T-piece square steel tubular column connection joints are analyzed as bolt type and bolt diameter. The size of the T-shaped member and other mechanical properties of the stiffener in the column determine that the steel tube column and T-shaped member is a connection node. The results show that plastic deformation is mainly distributed at the junction of flange and web of T-shaped steel tube, and the deformation develops most rapidly with the increase in displacement. Bolt type and bolt aperture have little influence on the ultimate bearing capacity of specimens. The welded stiffeners in the column enhance the stiffness of the column and reduce the plastic deformation capacity of the specimen. By increasing the thickness of the flange and web of T-shaped parts, the bearing capacity can also be improved obviously.

The steel structure is a natural prefabricated structure. Its material has the advantages of lightweight, high reliability, and good seismic performance. The construction process of steel structure buildings is fast and involves less wet construction. It is easy to implement industrialization with standardized construction in steel-structured buildings (Qu 2020). The use of steel structures is becoming more extensive and the joint is an important stress-bearing part connecting beam-column members in the structure. Its structural form and mechanical performance affects the safety of the entire structure, hence the study of joint performance is of great significance.

In the previous research on the square steel tube column-H-beam joint, there was no operating space while using ordinary high-strength bolts to fasten, and it was necessary to open installation hand holes, which reduced the construction efficiency. The unilateral high-strength bolts have the characteristics of unilateral tightening, reliable connection, and convenient construction. This new type of fastener can effectively solve the problem of difficulty in the application of ordinary high-strength bolts to closed-section members (Li 2015). In recent years, scholars at home and abroad have studied the steel tube column-T-joint joints, mostly steel beams are simplified by connecting the joints with flush end plates or overhanging end plates, and focusing on exploring

---

*Corresponding Authors: 1695401712@qq.com, and lywxw518@163.com

DOI 10.1201/9781003308577-17

the influence of self-locking unilateral bolts on the joints. Z.Y. Wang (2010) et al. found that, compared with ordinary bolts, unilateral bolts increased the plastic deformation capacity of joints by using unilateral bolts and ordinary bolts to pull out T-shaped pieces. Wang Yan et al. (2018) studied the self-locking unilateral high-strength bolt T-joint joints and conventional high-strength bolted joints, and found that the failure modes of unilateral high-strength bolts and conventional high-strength bolts are different, and the ultimate tensile load of the two is also different. The force and initial stiffness are the same. The edge distance and bolt distance of unilateral high-strength bolts have little effect on their ultimate tensile capacity, and the bolt distance has little effect on the tensile capacity of T-shaped joints. Walid Tizani (2010) et al. simplified the steel beam and end plate part of the CFST column-steel beam flat end plate joint into T-shaped pieces, carried out the tensile test of the T-shaped joint of the CFST column, and studied the common high-strength bolts, common unilateral tensile capacity of bolts, and unilateral bolts with nuts in square steel tubes with and without concrete. Studies have shown that joints have the greater bearing capacity with concrete, and unilateral bolts with nuts have greater tensile properties. J. Lee (2010) et al designed the square steel tube column to connect the test piece with the T-shaped piece in the tension zone and used the unilateral screw Ajax ONESIDE connection to conduct monotonic tensile and compression tests on the test piece. The test results are consistent with the existing theoretical model of the yield line. The maximum load that the flex cylinder can bear in the tension zone before yielding is compared, and a good agreement is obtained. In the experiments with more simplified models, Jia Shushuo (2020) et al. studied the seismic performance of the unilateral bolted connection between the rectangular steel tube column and the H-shaped steel beam, and found that the seismic performance of the self-locking unilateral bolt can meet the requirements of the joint design. Currently, many scholars at home and abroad have done little research on the connection joints between steel pipe columns and H-beams by splitting T-section steel as connectors and unilateral bolts as fasteners. The foundation is laid by the in-depth study of the connection nodes of steel beams through T-pieces.

In this paper, the unidirectional static loading test was carried out on the square steel tubular column T-shaped joint connected by a nested unilateral high-strength bolt. The failure mode and tensile capacity of T-type joint of square steel tubular columns connected with single side high strength bolt are studied. The effects of different bolt types, bolt diameters, T-piece sizes, and stiffeners in the column on the mechanical properties of the T-piece joint are analyzed. The research content can provide a reference for the engineering application of square steel tubular T-shaped joint.

1 THE EXPERIMENT

1.1 *Specimen design*

According to GB50017-2017 *Standard for design of steel structures* (Ministry of Housing and Urban-Rural Development of the People's Republic of China 2017) and GB50011-2010 *Code for seismic design of buildings* (Ministry of Housing and Urban-Rural Development of the People's Republic of China 2010), 7 full-foot square steel tubular column-T-shaped joint specimens were designed in this test. The specimen is simplified from the beam-column joint connected by T-shaped parts, as shown in Figure 1. T-shaped web and end welding connecting plate are connected and fixed with the loading device. A triangle plate is welded on the side of the T-shaped web and connecting plate to prevent deformation, as shown in Figure 2. All the specimens were Q235B steel, and the square steel column was 250*250*12. The T-shaped web of JD-2 was connected with the square steel column by ordinary high-strength bolts of Class 10.9 M20, and the other specimens were connected by nested unilateral high-strength bolts of class 10.9 M20 (Figure 3). Due to the existence of the outer sleeve of the unilateral bolt, 30 mm bolt holes are required for both the steel pipe column and the T-shaped parts. Specimen JD-4 changed the length of the bolt sleeve, and 22 mm and 30 mm bolt holes were opened for T-shaped parts and columns, respectively (Figure 4), forming special-shaped holes. Jd-3 and JD-7 specimens consider the influence of welding stiffeners in a column under different T-shaped parts sizes on specimens. See Table 1 for sample numbers and T-shaped parts size.

Figure 1. Simplified node mode.

Table 1. Specimen parameters.

Specimen number	T-piece section size	Bolt type	Bolt diameter (mm)	The column ribbed
JD-1	270*200*9*14	Unilateral bolt	30	no
JD-2	270*200*9*14	High strength bolt	22	no
JD-3	270*200*9*14	Unilateral bolt	30	yes
JD-4	270*200*9*14	Unilateral bolt	30 (pillars), 22 (T)	no
JD-5	270*200*8*12	Unilateral bolt	30	no
JD-6	270*200*10*16	Unilateral bolt	30	no
JD-7	270*200*8*12	Unilateral bolt	30	yes

1.2 Material test

Before the test installation, the steel required by the specimen was sampled with different thicknesses (Steel and steel products—Location and preparation of test pieces for mechanical testing 2018), and the material property test was carried out according to GB/T288.1-2010 (Method of the test at room temperature 2011) Standard for Tensile Test of Metallic Materials at Room temperature. All samples were subjected to INSTRON5587 hydraulic universal testing machine for tensile testing, as shown in Figure 5. The following related indexes were detected: material yield strength (fy), tensile strength (fu), elastic modulus (E), and shrinkage rate (A). The test results were shown in Table 2.

Figure 2. Test specimen.

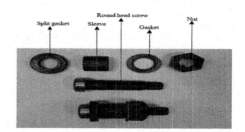

Figure 3. Nested single side bolt.

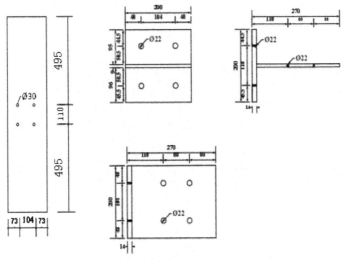

(a) Column opening size   (b) T-piece size

Figure 4. Specimen size diagram.

Table 2. Material test results.

Steel	t mm	MPa	MPa	GPa	(%)
Q235B	7	286.9	448.2	198.0	36.3
	9	313.0	464.0	206.0	32.1
	11	265.5	423.4	201.0	38.7
	12	283.0	456.0	203.0	30.0
	14	280.6	453.0	208.6	33.5
	16	290.7	450.6	202	33.3

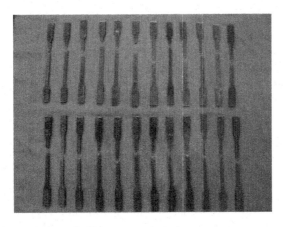

Figure 5. Material property test sample diagram.

1.3 *Test loading device and measuring scheme*

The test was carried out at the structural laboratory of Luoyang Institute of Science and Technology. A 100T electro-hydraulic servo control system was used to apply unidirectional static load on the specimens, and the vertical displacement of the specimens was measured accurately using a displacement meter. The two ends of the square steel tube column are fixed by pressing beams to prevent the column from mis-moving. The test loading method is shown in Figure 6.

Figure 6.   Test loading device.

To accurately record the test measurement data, the test load and displacement were controlled. The specimens were loaded with 15 kN/s before yielding, and 0.02 mm/s after yielding. The test ended when the component was damaged or the bearing capacity decreased to less than 85% of the ultimate load.

2   EXPERIMENTAL PHENOMENON AND FAILURE PATTERN

2.1   *Experimental phenomenon*

2.1.1   *JD-1 (Basic test)*
The failure mode of JD-1, a square steel tubular t-shaped joint connected with unilateral high-strength bolts, is shown in Figure 7(a). In the initial loading stage, the T-shaped flanges are closely attached to the cylinder. When the load of 120.5 kN was applied and the displacement was 2.3 mm, the T-shaped part yielded and the flange plate of the T-shaped part began to crack. Following this, displacement control was used to load step-by-step with an integer multiple of yield displacement. When the displacement was 16.1 mm, the gap between the flange plate and the cylinder of the T-shaped part reached 12 mm, and the paint peeling of the column and the joint of the T-shaped part swelled slightly. When the displacement was up to 23 mm, small cracks appeared at the junction of flange and web of T-shaped parts. When the displacement was 32.2 mm, the gap between the flange plate and the cylinder of T-shaped parts was 30 mm, the paint of the web of T-shaped parts was peeled and the cracks at the junction with the flange were increased, and the bolt gaskets were squeezed and deformed. With increase in load displacement, T piece of flange and web junction fracture extend unceasingly. When the displacement load increase to 51.1 mm(as the ultimate displacement load), it applied to T-shaped web and causes flange junction fracture (left), with relatively neat cross section. And bolt gasket extrusion deformation is relatively serious, tilt the bolts to the side, the web is bent to the unbroken side (right).

2.1.2   *JD-2*
Specimen JD-2 was connected with high-strength bolts, and the failure mode is shown in Figure 7(b). When the load of 140 kN is applied and the displacement is 3.25 mm, the T-shaped parts

yield and the flange plate and cylinder surface of the T-shaped parts begin to crack. When the displacement loading reached 16.25 mm, the gap between the flange plate and the cylinder of T-shaped parts reached 12 mm, the column slightly swelled, and the bolts moved sideways. When the displacement load reaches 22.75 mm, small cracks appear at the junction of flange and web of T-shaped parts. When the displacement was loaded to 49.3 mm, deep discontinuous cracks appeared at the junction between the web and flanges of T-shaped parts. The paint on the web of T-shaped parts was peeled and slightly bent in a large area. The pillars were swollen severely, and the bolts tilted significantly.

2.1.3 *JD-3*
Compared with specimen JD-1, stiffener ribs were welded inside the column of specimen JD-3 to increase the column stiffness. The test phenomenon and failure mode were similar to specimen JD-1, but the column did not bulge obviously. As shown in Figure 7(c), when the displacement loading reached 31.4 mm, the junction between the web and flange of T-shaped parts (right side) broke, the web slightly bent, and the bolts slightly tilted.

2.1.4 *JD-4*
Specimen JD-4 cut the outer sleeve of a unilateral high-strength bolt with the same thickness as the column to form different holes. During the test process, there was no obvious extrusion deformation of gasket and nut, and other test phenomena were similar to specimen JD-1. Finally, when the displacement loading reached 50.9 mm, the junction between the web and flange of the T-shaped part (left) broke, as shown in Figure 7(d).

2.1.5 *JD-5*
Based on JD-1, the thickness of the flange and web of the T-shaped part was reduced for specimen JD-5. When the displacement load reaches 20 mm, cracks appear at the junction of flange and web of T-shaped parts. When the displacement load reaches 26 mm, the bolt gasket is squeezed into the bolt hole. With increasing displacement, the nut begins to enter the bolt hole slowly, and the cracks at the junction between the flange and the web of the T-shaped parts increase continuously, forming a crack. Finally, when the displacement loading reached 60 mm, the junction (right side) of the flange and web of the T-shaped part broke (see Figure 7(e)).

2.1.6 *JD-6*
Based on JD-1, the thickness of the flange and the web of T-shaped piece was increased in JD-6. When the displacement loading reached 26.4 mm, the bolts slipped and small cracks appeared at the junction of flange and web of T-shaped piece. When the displacement load reaches 46.2 mm, the gasket is squeezed into the bolt hole. When it was finally loaded to 59.4 mm, the bolt was suddenly pulled off, as shown in Figure 7(f). This was because the thickness of T-flange and web increased, and the strength of T-shaped parts increased so that the bolt was pulled off first, and the bolt tilt became more obvious.

2.1.7 *JD-7*
Specimen JD-7 welded stiffeners inside the column based on the JD-5, and the test phenomenon was the same as THAT of JD-5, as shown in Figure 7(g).

2.2 *Failure pattern*

During the initial stage of loading, cracks appeared at the junction between flange and web of all T-shaped pieces, and gaps appeared between flange and column of T-shaped pieces. With increase in displacement, the crack gradually forms a bigger crack, , and the web starts bending. The bolts of specimens JD-1, JD-3, JD-4, JD-5, and JD-7T were slowly dented into the bolt holes, and the joints of flange and web of T-shaped parts were finally broken, and the bolts were tilted. The joint

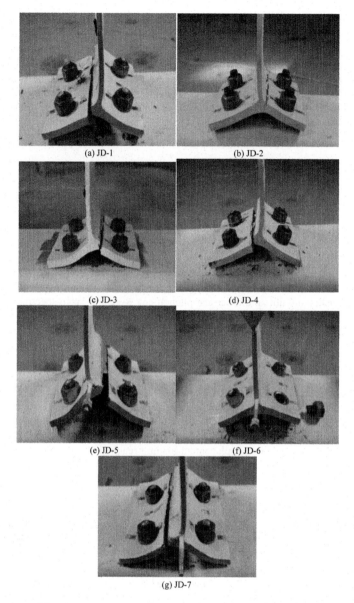

(a) JD-1   (b) JD-2

(c) JD-3   (d) JD-4

(e) JD-5   (f) JD-6

(g) JD-7

Figure 7. Specimen failure pattern.

between flange and web of T-shaped part of specimens JD-2 was not broken, but the column bulged severely and the bolt tilted obviously. Jd-5 specimen increased the thickness of T-shaped parts, and the bolts became weaker, which eventually led to bolt fracture.

3 TEST ANALYSIS

Figure 8 shows the load-displacement curves of 7 specimens. By analyzing the load-displacement curves, the tensile capacity, ductility, and other mechanical properties of the connection joints of square steel tubular T-shaped parts can be solved. The yield displacement, yield load, ultimate displacement, ultimate load, and initial stiffness of each specimen are shown in Table 3.

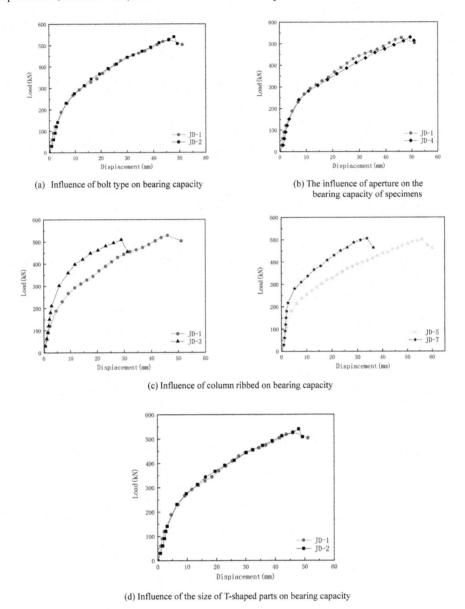

(a) Influence of bolt type on bearing capacity

(b) The influence of aperture on the bearing capacity of specimens

(c) Influence of column ribbed on bearing capacity

(d) Influence of the size of T-shaped parts on bearing capacity

Figure 8.   Load-displacement curve.

It can be seen from Figure 8 and Table 3 that:

(1) The load-displacement curves of the seven specimens were the same. All the specimens passed through the elastic stage, elastic-plastic stage, and strengthening stage during the test loading process, showing strong plastic deformation ability.

(2) It can be seen from Figure 8(a) that the load-displacement curves of the joint connecting steel tubular column T-shaped parts using the same type of unilateral high-strength bolt and ordinary

Table 3. Test results.

Specimen number	Yield capacity (kN)	Yield displacement (mm)	Limit-bearing capacity (kN)	Limit displacement (mm)	Initial stiffness (kN/mm)
JD-1	120.5	2.3	528.5	46.1	52.4
JD-2	141.4	3.3	540.1	48.0	42.8
JD-3	211.1	2.9	509.6	29.0	72.8
JD-4	152.2	3.6	530.2	49.4	42.3
JD-5	108.6	2.0	500.9	56.1	54.3
JD-6	168.2	3.3	619.4	58.5	51.0
JD-7	217.3	2.6	506.1	33.9	83.6

high-strength bolt have a very high coincidence degree, especially after specimen yield. The ultimate bearing capacity of SPECIMEN JD-1 and JD-2 is 528.5 kN and 540.1 kN, respectively. Compared with JD-2, the ultimate bearing capacity of specimen JD-1 is only reduced by 2.1%, while the initial stiffness is increased by 22.4%. It can be seen that the bolt type has little influence on the ultimate bearing capacity of the joint, but the initial stiffness of the joint is improved by using a single-side high-strength bolt connection.

(3) Figure 8(b) shows that during the elastic stage, pore size has little influence on bearing capacity. In the elastic stage, the load-displacement curves of specimens JD-1 and JD-4 were separated and no longer overlapped. This may be because the specially shaped hole (JD-4) makes the bolt and T-piece more closely connected, which limits the bolt displacement and deformation, thus reducing the plastic deformation capacity of the joint. The ultimate bearing capacity of specimens JD-1 and JD-4 is 528.5 kN and 530.2 kN, respectively, and the initial stiffness is 52.4 and 42.3, respectively. The ultimate bearing capacity of the two specimens is almost the same, but the initial stiffness of JD-1 is 23.9% higher than that of JD-4.

(4) Figure 8(b) shows the comparison of the two groups of in-column ribbed, and the load-displacement curves of the two groups have the same development trend. In both cases, the displacement of the specimens with in-column ribbed (JD-3 and JD-7) is small, which is because the joint stiffness is improved by welding the stiffeners in the column, which weakens the plastic deformation capacity of the specimens. It can be seen from Table 3 that the limit displacement of JD-3 and JD-7 specimens decreased by 37.1% and 39.6%, respectively, while the initial stiffness increased by 38.9% and 54.0%, respectively. The ultimate bearing capacity of the two groups of experiments was almost the same.

(5) As shown in Figure 8(b), in the elastic stage, load-displacement curves of the three specimens almost coincide; in the plastic stage, the bearing capacity increases with the increase in flange and web thickness. The ultimate bearing capacity of specimens JD-6 is 619.4 kN. Compared with specimen JD-5, the ultimate bearing capacity of JD-1 and JD-6 is increased by 5.5% and 23.7%, respectively. The initial stiffness of the three specimens remains the same.

4 CONCLUSIONS

Based on the results and discussions presented above, the conclusions obtained are as follow:

(1) The plastic deformation of the joint between square steel tubular column and T-shaped parts is mainly distributed at the junction of the flange and web of T-shaped parts. Except for JD-6, the bolt was pulled off due to the excessive thickness of the T-shaped parts, all the other specimens were damaged at the junction of the flange and web of T-shaped parts.

(2) The type of bolt has little influence on the ultimate bearing capacity of the specimens, which indicates that the single-side high-strength bolt connection can maintain the better bearing

capacity of the joints, and the initial stiffness is significantly improved. It can be seen that the use of unilateral high-strength bolt connection has better advantage.

(3) The size of the aperture has little effect on the bearing capacity, and the special-shaped hole closely connects the bolt and T-shaped parts, which limits the displacement and deformation of the bolt, thus reducing the plastic deformation capacity of the joint.

(4) The welded stiffeners in the column enhance the stiffness of the column and reduce the plastic deformation capacity of the specimen, but the bearing capacity is not significantly improved.

(5) By increasing the thickness of the flange and web of T-shaped parts, the bearing capacity can also be improved, but the thickness of flange and web of T-shaped parts is very large, resulting in the failure of bolt being pulled off. It is indicated that the matching T-piece size and bolt type should be selected during design.

REFERENCES

Ataei A., Bradford M.A., Valipour H.R. Moment-Rotation Model for Blind-Bolted Flush End-Plate Connections in Composite Frame Structures [J]. Journal of Structural Engineering, 2014, 141(9): 04014211.

GB/T2975—2018.Steel and steel products—Location and preparation of test pieces for mechanical testing [S]. Beijing: Standards Press of China, 2018.

Jia Shushuo, WANG Yan, WANG Xiujun, LIU Xiuli. Seismic behavior and restoring force model of connections between rectangular tubular columns and H-shaped beams using single direction bolts [J]. Journal of Building Structures, 2020, 41(05):168–179.

Lee J, Goldsworthy H M, Gad E F. Blind bolted T-stub connections to unfilled hollow section columns in low rise structures[J]. Journal of Constructional Steel Research, 2010, 66(8): 981–992.

Li De-shan, TAO Zhong, WANG Zhi-bin. Experimental Investigation of Blind-bolted Joints to Concrete Filled Steel Columns [J].Journal of Hunan University(Natural sciences), 2015, 42(03):43–49. DOI:10.16339/j.cnki.hdxbzkb.2015.03.007.

Ministry of Housing and Urban-Rural Development of the People's Republic of China. Standard for design of steel structures GB 50017-2017 [S]. Beijing: China Architecture & Building Perss, 2017.

Ministry of Housing and Urban-Rural Development of the People's Republic of China. Code for seismic design of buildings GB50011-2010[S].Beijing: China Architecture & Building Perss, 2010.

Qu Shengyu. Analysis of the advantages of prefabricated steel structure buildings [J]. Urban Dwelling, 2020, 27(07):227–228.

State General Administration of the People's Republic of China for Quality Supervision and Inspection and Quarantine. Metallic materials-Tensile testing-Part 1: Method of test at room temperature. GB/T228.1-2010 [S]. Beijing: Standards Press of China, 2011.

Wang Yan, Jia Shushuo, Chai Wenjuan. Experimental Study and Numerical Analysis of T-Stub Connections with Single Direction High Strength Bolts [J]. Journal of Tianjin University(Science and Technology), 2018, 51(S1):78–85.

Wang Z Y, Tizani W, Wang Q Y. Strength and initial stiffness of a blind-bolt connection based on the T-stub model[J]. Engineering Structures, 2010, 32(9):2505–2517.

Frontiers of Civil Engineering and Disaster Prevention and
Control – Yang & Rahman (Eds)
© 2023 The Author(s), ISBN: 978-1-032-31200-2

Deformation behaviour of embankment on soft soils heightened with light soil replacement

Taiping Mu
Guangdong Provincial Freeway Co., Ltd., Guangzhou, China

Tianyi Chen*
School of Civil Eng., Tsinghua Univ., Beijing, China

Yangguang Sun
Guangdong Provincial Communications Planning and Design Institute Co., Ltd., Guangzhou, China

ABSTRACT: The light soil replacement technology can effectively decrease the settlement in embankment heightening projects and avoid the pavement cracking. A series of finite element analysis were conducted on the deformation of embankment on soft soil that was heightened with different light soil replacement schemes. The results show that the light soil replacement effect is related to the replacement areas. The settlement of the pavement is reduced and distributes more uniformly if the new embankment is constructed by the light soil in place of common soil with a depth of 2 m. Whereas, differential settlement of the pavement may be increased if the old part is replaced of the light soil. The effect of the light soil on the embankment deformation includes two aspects. On the one hand, the light soil replacement leads to self-weight reduction, which significantly reduces the settlement of the embankment. On the other hand, the light soil replacement changes underlying surface for the heightening part and thus changes the stress distributions under the same load.

1 INTRODUCTION

Large settlement of embankment leads to pavement cracking and affects the traffic safety (De Sarno et al. 2020; Siavash et al. 2019). The light soil is widely used in embankment heightening project because the settlement of embankment can be effectively decreased by replacing the soil with the light soil to reduce the load (Xu et al. 2022). Lots of studies have been carried out on the settlement control of the light soil, most of which mainly focus on the properties of it. Kim et al. conducted the tests about the lightweight air-mixed soil (LWS) for expanded road construction, showing that the numerous tiny pores inside the lightweight soil changed the density of the LWS (Kim et al. 2013). Zhu et al. developed an equipment used to measure the strength of the foamed mixture lightweight soil (FMLS) along the embankment depth during construction, and advanced the method to determine the optimum pouring thickness of the FMLS (Hwang et al. 2010). Hwang et al. found that the stress-strain behaviour of the light air mixed soil was affected by water and its density would also change (Zhu et al. 2016). Hou and Yang researched the earth pressure of the light weight soil mixed with expandable polystyrene (EPS) particles behind a retaining wall through model tests, which reflected the buffering effect of EPS particles could effectively absorb and disperse the vertical pressure (Hou & Yang 2021).

---

*Corresponding Author: chen-ty19@mails.tsinghua.edu.cm

 DOI 10.1201/9781003308577-18

However, there is a lack of research on the deformation behavior of embankment with light soil replacement. Therefore, a series of numerical analysis was carried out to investigate the deformation of embankment on soft soils heightened with light soil replacement.

2 NUMERICAL MODEL

2.1 *Model description*

The model is established based on a widening and heightening expressway project. The finite element numerical software ABAQUS is used for analysis, as shown in Figure 1.

The top width of the old embankment is 24.5 m and the height is 8 m. The new embankment is widened on both sides with 11.75 m on each side. The top of the embankment is heightened by 2 m using the light soil. The gradients of the new and old embankments are both 1:1.5. There is a soft soil layer with a thickness of 15 m under the embankment that is located on a bearing stratum of 5-m-thickness.

The soft soil layer is reinforced by six piles on each side. The spacing between piles is 2.1 m, with the outermost pile located at the toe of the new embankment. The pile is 18 m long and inserted into the bearing stratum of 3 m. The pile cap is 1.2 m in length and 0.36 m in height. A layer of geotextile with a thickness of 5 mm is laid on the top of the piles.

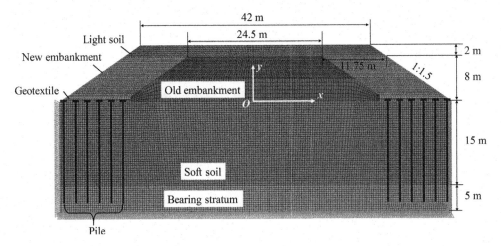

Figure 1. The numerical model.

2.2 *Materials*

The soil is simulated by Mohr-Coulomb model. The pile and geotextile are simulated by linear elastic model. The calculation parameters are shown in Table 1.

2.3 *Calculation conditions*

The four-node bilinear plane strain quadrilateral elements (CPE4) are applied considering that the calculations are performed on the basis of the plane strain assumption. The numerical model includes 19725 nodes and 17731 elements. The stress is considered positive by pressure while negative by tension.

As for the boundary conditions, the bottom of the model is fixed for the horizontal and vertical displacement and the lateral sides are fixed only for the horizontal displacement (Figure 1).

Table 1. The calculation parameters.

Material	ρ (g/cm^3)	E (MPa)	ν	c (kPa)	φ (°)
Light soil	0.8	150	0.25	40	40
Embankment	1.9	15	0.30	40	15
Soft soil	1.6	1.5	0.38	15	20
Bearing stratum	1.8	5.5	0.40	30	20
Geotextile	2.0	200	0.33	/	/

A plane rectangular coordinate system is established for further analysis as shown in Figure 1. The intersection of the central axis of the old embankment and the foundation surface is taken as the origin point. The positive direction is vertically upward of the y-axis, and for the x-axis is horizontally to the right.

2.4 *Schemes*

The numerical models are simulated according to the actual construction process. Firstly, the geo-stress is calculated for the old embankment, the soft soil, and the bearing stratum. Secondly, pile construction is carried out and the geotextile is laid on the top of the piles. Next, the new embankment is widened and then the light soil is used for heightening.

Based on the model scheme in Figure 1, different light soil replacement schemes of the new and old embankments are considered. The model schemes of 2-m-depth light soil replacement of the new and the old embankments are shown in Figure 2, respectively.

For convenience of expression, the model scheme in Figure 1 is called *Case-Basic*. The model scheme in Figure 2a is called *Case-New* and Figure 2b is called *Case-Old*.

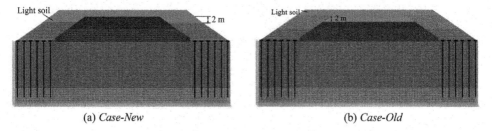

(a) *Case-New*   (b) *Case-Old*

Figure 2. Model schemes of different replacement areas by the light soil.

3 DISPLACEMENT RESPONSE

3.1 *Settlement*

Figure 3 shows the cloud contours of the settlement of the embankment and soil base under different light soil replacement schemes. It could be observed that the settlement mainly occurs inside the soil base. The settlement is comparatively small and distributes more evenly inside the pile reinforced area. It is obvious that the settlement of *Case-Old* is the largest and that of *Case-New* is the smallest, indicating that the light soil replacement effect is related to the replacement areas. Replacement of the new embankment by the light soil with 2-m-depth reduces the settlement. However, replacement of the old embankment with the same depth might increase the settlement.

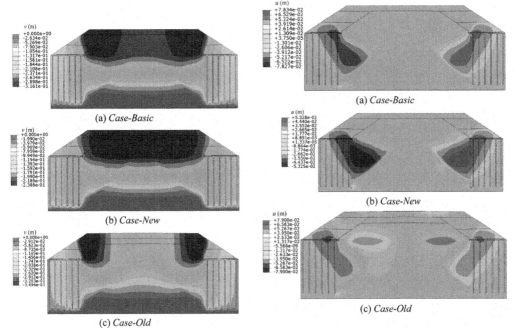

Figure 3.   Cloud contours of settlement.

Figure 4.   Cloud contours of horizontal displacement.

Figure 3a shows that the settlement of the new embankment is larger than that of the old area, and the settlement concentrates near the boundary of the new and the old embankments of *Case-Basic*. That leads to large uneven settlement on the surface of the new and old embankments where cracks will easy to appear. However, the settlement of different areas can be adjusted through the replacement by the light soil due to its light weight. As could be seen from Figure 3b, the settlement of the new embankment decreases because of the light soil replacement of the new area. The embankment settlement distributes relatively uniformly of *Case-New* and the settlement concentration does not occur, implying that the light soil replacement of the new embankment could effectively reduce the uneven settlement. However, in Figure 3c, the settlement near central axis of the old embankment decreases obviously when the old embankment is replaced by the light soil, resulting in more distinct settlement concentration and uneven settlement on the pavement.

3.2 *Horizontal displacement*

Figure 4 shows the cloud contours of the horizontal displacement of the embankment and soil base under different light soil replacement schemes. It can be seen that the embankment has little horizontal displacement. Significant horizontal displacement occurs only near the innermost pile inside the soft soil base, where the soil moves horizontally to the pile reinforced area. The area range of that develops larger as the new embankment is replaced by 2-m-depth light soil, while the range becomes smaller as the old embankment is replaced.

There is also a horizontal movement towards the new embankment shoulder within the certain area on the pile top. The horizontal displacement becomes smaller as closer to the new embankment shoulder. In *Case-Old* (Figure 4c), the horizontal displacement appears at the bottom of the old embankment near the surface of the soil base. The horizontal movement at the bottom of the old embankment may relate to the reduction of the old embankment weight.

141

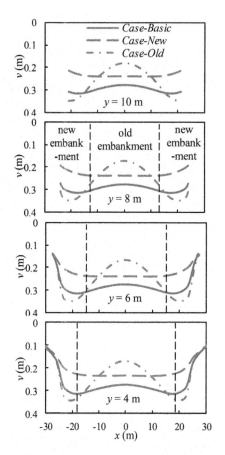

Figure 5. Horizontal distributions of settlement at different elevations of the embankment. v, settlement.

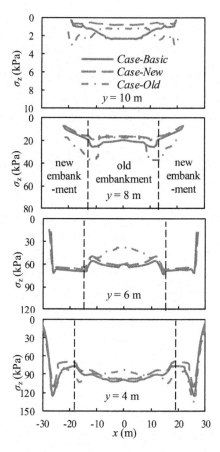

Figure 6. Horizontal distributions of vertical stress at different elevations of embankment. σ_z, vertical stress.

4 DEFORMATION CHARACTERISTICS

4.1 Embankment deformation

Figure 5 shows the horizontal distributions of settlement at different elevations of the embankment. The dotted lines are used to divide the new and old embankments. It can be seen that the displacement distribution on the embankment surface of *Case-New* is approximately a straight line, demonstrating that there is basically no uneven deformation distribution of the pavement. The uneven deformation of *Case-Old* is the most obvious because the settlement of the old embankment is significantly reduced due to the replacement by the light soil with a depth of 2 m.

Figure 6 compares the horizontal distribution of vertical stress at different elevations. The vertical stress of the embankment becomes smaller as the elevation goes higher. It is obvious that the vertical stress at the central axis of *Case-Old* decreases significantly, due to the self-weight reduction at the central axis caused by the light soil replacement the old embankment. It can also be found that the stress distributions of different light soil replacement schemes at the same elevation are diverse. The diversity is most obvious at the upper surface of the old embankment ($y = 8$ m), which is exactly the underlying surface for the heightening part. Thus, it can be inferred that underlying surface changes due to light soil replacement, and therefore the stress distributions develop differently under the same heightening load.

142

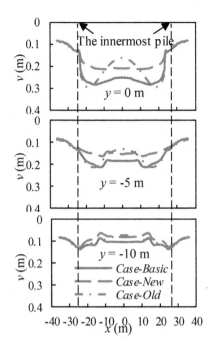

Figure 7. Horizontal distributions of settlement of soil base. v, settlement.

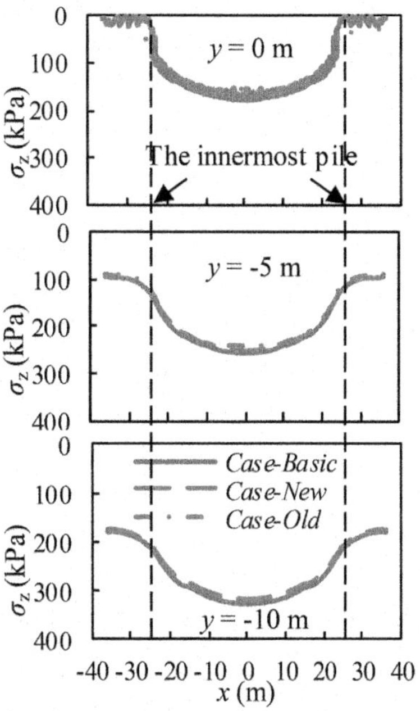

Figure 8. Horizontal distributions of vertical stress of soil base. σ_z, vertical stress.

4.2 Soil base deformation

The results in Figure 6 also show that the settlement of the embankment at different elevations are similar, indicating that the settlement mainly occurs inside the soil base. The horizontal distributions of the settlement at different depths of the soil base are drawn in Figure 7. The positions of the innermost pile are marked with dotted lines. Results show that the settlement of the soil base decreases gradually with increasing depth of the soil base. The settlement on the soil base surface at the central axis of *Case-Old* is relatively small due to the light soil replacement. The settlements of three cases are similar at the depth of 10 m, meaning that the light soil replacement has a small effect on a certain depth of the soil base. The settlement decreases obviously in the pile reinforced area and is hardly affected by the light soil replacement.

Figure 8 shows the horizontal distributions of vertical stress at different depths of the soil base. It can be found that the light soil replacement has little impact on the stress distribution of the soil base. And the vertical stress in the pile reinforced area is also significantly smaller.

5 CONCLUSIONS

Based on the results and discussions presented above, the conclusions are obtained as follows:

(1) The settlement mainly occurs inside the soil base. The settlement is comparatively small and distributes more evenly inside the pile reinforced area.
(2) The light soil replacement effect is related to the replacement areas. Replacement of the new embankment significantly decreases the unevenness of the pavement.
(3) The light soil effect includes two aspects. On the one hand, the light soil replacement leads to self-weight reduction, which significantly reduces the settlement of the embankment. On the

other hand, the light soil replacement changes underlying surface for the heightening part and thus changes the stress distributions under the same load.

(4) The light soil replacement has a small effect on the deformation of the soil base.

It should be noted that the density of the light soil is mainly considered. The effect of the light soil mixed with different properties particles with the same density still needs further research.

REFERENCES

De Sarno, D., Vitale, E., Nicotera, M.V., et al. (2020) Lightweight Cemented Soils: Mix Design, Production and Control. LNCE., 40: 743–752.

Hou, T.S., Yang, K.X. (2021) Model test on earth pressure at rest of light weight soil mixed with EPS particles behind a retaining wall. Rock and Soil Mechanics, 42: 3249–3259.

Hwang, J.H., Ahn, Y.K., Kim, T.H. (2010) Effect of Water on the Lightweight Air-Mixed Soil Containing Silt Used for Road Embankment. J. Korean Geo. Soc., 26: 23–32.

Kim, T.H., Kim, T.H., Kang, G.C. (2013) Performance evaluation of road embankment constructed using lightweight soils on an unimproved soft soil layer. Eng. Geol., 160: 34–43.

Siavash, S., Yaser, J., Alborz, H. (2019) Bearing Capacity and Uneven Settlement of Consecutively Constructed Adjacent Footings Rested on Saturated Sand Using Model Tests, Int J Civ Eng, 17: 737-749.

Xu, J.B., Wang, Y.Z., Qi, Y., et al. (2022) Deformation characteristics of fiber-reinforced foam lightweight soil under cyclic loading and unloading. J. Zhejiang Univ., 56: 111–117.

Zhu, J.J., Liu, X., Hong, B.N., et al. (2016) A method for determining optimum casting thickness of foamed mixture lightweight soil embankment. Rock Mech. Rock Eng., 37: 3642–3649.

Frontiers of Civil Engineering and Disaster Prevention and Control – Yang & Rahman (Eds)
© 2023 The Author(s), ISBN: 978-1-032-31200-2

Finite element analysis of concrete circular pole substation structures under coupling action of carbonization-corrosion-load

Liu Yong* & Wei Zhenzhong*
Shandong Electric Power Engineering Consulting Institute Corp, Jinan, China

Sui Bin*, Wan Jia* & Jiao Jinfeng*
School of Civil Engineering, Taiyuan University of Technology, Taiyuan, China

ABSTRACT: With regard to a full open-air environment and load coupling effect, it is very necessary to study the safety assessment of existing concrete circular pole substation structures. In this paper, based on the on-site inspection and numerical simulation, the parametric theoretical analysis of the single-span non-lateral support frame is carried out for typical concrete substation structures in service, and the influence of four factors, including concrete strength, protective layer thickness, steel corrosion, and carbonation depth, on the bearing capacity of the frame is mainly discussed. The research results show that the substation structures with different damage levels meet the requirements but are associated with considerable safety risks. The research helps us understand the damage mechanism of the circular pole substation structures and the broken ring of the cross beam possesses a typical brittle characteristic; the frame bearing capacity is positively related to concrete strength, protective layer thickness, reinforcement diameter, and there is fluctuation and little difference in carbonization depth. The order of structural sensitivity is Reinforcement corrosion → outer diameter dimension (thickness of the protective layer) → concrete strength.

1 INTRODUCTION

The structural safety of the concrete substation structures (Figure 1), as an important component of the substation, is directly related to the normal operation of the substation. Under coupling adverse effects of the full open-air environment conditions and the service loads of the substation structures, factors such as the concrete strength hardening of the existing substation frame, the thickness reduction of the protective layer, the corrosion of the steel rebar, the crack of the components (Figure 2) and the carbonation depth may lead to the attenuation of the bearing capacity (Feng 2012; Lian 2004; Sun 2004). This may result in serious hidden dangers with regard to safety and normal operation. Therefore, it is necessary to study the safety evaluation of the existing concrete substation structures.

Zeng. 2010, Liu. 2010 & Liu. 2011 considered the concrete poles in service for nearly 50 years as the research object, carried out the bending test and theoretical calculation of three annular poles and six steel plate joints, revealed the failure mechanism of the poles, and obtained the corresponding important mechanical properties, such as ultimate bearing capacity. Xia. 2011 selected three members of concrete poles in service for 30 years to carry out a flexural loading test, and revealed its failure mechanism and corresponding bearing capacity performance. Based on experimental research and theoretical analysis, Li (2006) researched the bearing capacity of the concrete pole with defects. The influence of concrete damage and steel bar corrosion on its safety was further discussed. Based on the variable angle space truss theory, Wang (2017), while studying the calculation model of ultimate bearing capacity of reinforced concrete ring

---

*Corresponding Authors: liuyong@sdepci.com, weizhenzhong@sdepci.com, 492429657@qq.com, adam2ftcttt@163.com and jiaojinfeng@tyut.edu.cn

Figure 1.   Concrete substation structures.   Figure 2.   The crack of the components.

section members under the combined stress of tension, compression, bending, shear, and torsion, obtained the geometric expression of failure section and the calculation formula of ultimate bearing capacity. Steel corrosion caused by concrete carbonation is one of the important factors affecting the durability of reinforced concrete structures. In addition to conventional factors, concrete structures under different stress states also affect the carbonation performance (Ma 2019). At the same time, the corrosion of steel bars in concrete also affect the carbonation of concrete, thereby aggravating the corrosion of concrete, and the cracks on the surface of the concrete also affect this process (Rita Maria Ghantous 2017; Valérie L'Hostis 2020). Xu (2016) studied the influence of carbonation on the strength, stiffness, and deformation capacity of concrete using the loading test of rapid carbonation concrete. Chen (2018) proposed a new analysis method for evaluating the development of concrete cracks that could estimate the degradation degree of bond strength of reinforcement. For the crack propagation caused by steel corrosion, Chen, H., & Alani, A. (2013) proposed a strategy for optimal maintenance. Using numerical simulation, Chen (2016) discussed the influence of concrete strength on the dynamic performance of ring concrete rod under impact load, and put forward corresponding solutions.

Vadlūga, R., & Kliukas, R. (2012) proposed a universal and suitable calculation method for the bearing capacity of in-service annular concrete members based on theoretical analysis. Hashimoto, T. (2015)studied the causes of ring concrete pole cracks, and proposed that graded fly ash can inhibit the development of cracks. Dhakal (2002) explored the general method to determine the buckling strength of reinforced concrete structures, and predicted the damage of its protective layer. Kioumarsi, M., Benenato, A., Ferracuti, B., & Imperatore, S. (2021) sum-marized the previous experimental studies on the influence of reinforcement corrosion on the residual bearing capacity of prestressed reinforced concrete structures. The analysis shows that low corrosion level would lead to gradual decrease in ductility without causing serious loss of bearing capacity, while high corrosion level leads to lower bending strength and changes in failure mechanism.

In summary, the relevant literature shows that the research on the in-service concrete ring rod mainly focuses on the material and component, especially the durability of materials or components and the bearing capacity of components. However, research on the existing structure with the concrete ring rod is relatively less. Therefore, it is necessary to perform field measurement and theoretical analysis on the exciting concrete circular pole substation structures.

In this paper, the field investigation on 37 typical concrete substation structures in five cities and counties in Shandong and Shanxi Province was conducted. The typical concrete

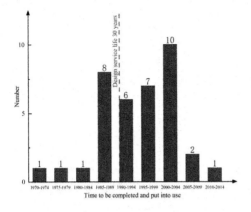

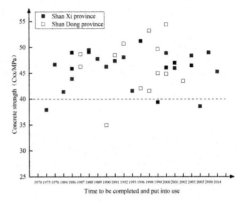

Figure 3. Cylindrical chart of built-up age statistics.

Figure 4. Concrete strength and service life.

substation structures with single-span non-lateral support frame were selected as the research object, considering different service life and different substation scales. The influences of concrete strength, protective layer thickness, steel corrosion, and carbonation depth on the residual bearing capacity of the substation frame were discussed, which provided an evaluation basis for the safety evaluation, modification, and the expansion of the substation frame.

2 FIELD INVESTIGATION AND RESULT EVALUATION

To accurately evaluate the bearing capacity performance of the existing concrete circular pole substation structures, considering the objective factors such as substation selection, different regions, and different service life, 37 substations were investigated in Shandong Province and Shanxi Province. Considering the purpose of the research, the main work done is as follows: the strength of the concrete was evaluated using rebound method; the number of steel rebar and the concrete protective layer was checked using the steel bar scanner; the crack width, direction, and appearance were inspected based on the crack caliper; and the damage and corrosion were surveyed and classified. The field investigation results of the concrete circular substation structures are shown in Figures 3–5.

As seen in Figure 3, the service life of 11 concrete circular pole substation structures has exceeded the specified design service life (30 years), and the longest service life is nearly 50 years, but all of them operate normally. The transformed electric frame is within the normal service life range, which indicates that the concrete annular pole transformation frame still has a large market. In Figures 4 and 5, it can be seen that the actual detection strength of 33 concrete ring rods of the substation frame (89.2%) is higher than the original design strength C40, and the actual detection strength of 4 concrete ring rods (10.8 %) is C35–C40, which basically reflects the characteristics that concrete strength decreases with the increase in service time. The number of reinforcing steels meets the design requirements, but the thickness of a concrete protective layer is unevenly distributed along the circumferential direction of the component, and the difference between the maximum protective layer thickness and the minimum protective layer thickness of concrete is about 2.5–5 times. It is proved that the thickness of the protective layer also decreases with the increase in service time, and the decrease of the same component is uneven.

In addition, the average service life of the concrete substation frame investigated in this paper is approximately 20 years, and the concrete members have been carbonized to different degrees.

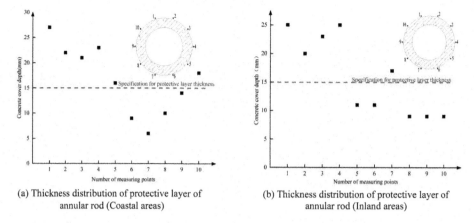

(a) Thickness distribution of protective layer of annular rod (Coastal areas)

(b) Thickness distribution of protective layer of annular rod (Inland areas)

Figure 5. Protection layer distribution of concrete circular pole of representative substation structures.

The thickness of the concrete ring bar is only 50 mm; therefore, the influence of the concrete carbonization on its bearing capacity must be considered.

3 FINITE ELEMENT ANALYSIS OF SUBSTATION STRUCTURES

Based on the field investigation on the concrete circular pole substation frame structure system, the typical substation with frame-single-span non-lateral support is selected as the analysis object, and the influence of concrete strength, protective layer thickness, steel corrosion, and carbonization depth on the ultimate bearing capacity is mainly discussed. The concrete analysis process is as follows:

3.1 *Model establishment and static analysis results*

The finite element analysis software, ABAQUS, was used to establish the model of the selected substation frame. The geometric dimensions of the structure and members, and the reinforcement of the concrete ring member, are shown in Tables 1 and 2. The concrete grade is C40, and the longitudinal reinforcement, stirrup, and spiral reinforcement are HPB235 steel bar. The three-dimensional solid element C3D20R is used for concrete, and the truss element (T3D2) is used for longitudinal reinforcement, stirrup, and spiral reinforcement. The constitutive relationship between concrete and truss is based on the curve proposed in Wang (2017). The finite element model of the structure is shown in Figure 6. The mesh element of concrete circular pole adopts hexahedral shape, the mesh size is 0.06 m, and the element type is standard linear three-dimensional stress (C3D8R). While the elements of the spiral reinforcement, stirrup, and longitudinal reinforcement are in the form of standard linear truss, with mesh size 0.05 m. Three-point load was adopted, and the horizontal load is applied at the coupling reference point.

Through the analysis of static ultimate bearing capacity, the results are shown in Figures 7–9. It can be seen from the figures that: 1) When the crack width of the component is 0.2 mm for the normal service limit state, the tensile damage of the concrete beam extends along the loading point to both sides, and there is a small amount of damage between the loading points. 2) When the steel bar in the tensile zone of the component yields, the compressive damage occurs at the loading position of the concrete beam, and the tensile damage of the concrete between the loading points begins to expand and the tensile damage of the concrete column begins to appear. 3) When half of the steel bars yield, there is a lot of damage in the structure, and plastic hinges develop in the beam. The concrete has been completely destroyed under the load, and the new load is completely

148

Table 1. Geometric dimensions of structure and component.

Constructional dimension (m)			Size of concrete circular pole (mm)		
Span	Foot distance	Height	Outer diameter/r1	Inside diameter/r2	Circular radius of longitudinal reinforcement/rs
6	4	10	150	100	125

Table 2. Section table of reinforced concrete ring bar.

Longitudinal reinforcement	Stirrup	Spiral reinforcement
16Φ12	Φ6@500	Φ4@100

Figure 6. Finite element model of concrete substation frame.

borne by the steel bars. The structure has completely lost its bearing capacity and the broken ring form is similar to the "less reinforced beam" failure. The analysis results demonstrated above are consistent with the failure mode and failure mechanism in the field investigation in Liu (2010) and Liu (2011), which verifies the feasibility of the model.

3.2 Single factor analysis

3.2.1 Influence of concrete strength

Through field investigation, it can be seen in Figure 4 that the concrete strength of ring pole substation frame shows certain degree of fluctuation. To obtain the influence of strength grades of concrete on the ultimate bearing capacity of substation frame, the concrete strength grades of C20, C25, C30, C35, and C40 were discussed in this paper. The analysis results are shown in Figure 10, wherein it can be seen that the concrete strength is positively correlated with the bearing capacity of the frame. As the concrete strength decreases from C40 to C20, the corresponding frame bearing capacity decreased by 1.2%, 4.6%, 9.8%, and 13.3%.

Figure 7.   Analysis results of frame with allowable crack width of 0.2 mm.

Figure 8.   Reinforcement yield in tensile zone of member.

3.2.2 *Influence of protective layer thickness*

Figure 5 shows that the thickness distribution of concrete circular pole protective layer is not uniform. In this paper, the influence of protective layer thickness on the bearing capacity of the structure is simulated by changing the outer diameter of the component. Considering that both ends of the concrete ring rod are closed and the inner wall surface of the component cannot be measured, it is assumed that the thickness of the inner wall protective layer of the component is evenly distributed and meets the structural requirements. It can be seen from Figure 11 that the ultimate bearing capacity of the concrete ring bar substation frame is positively correlated with the

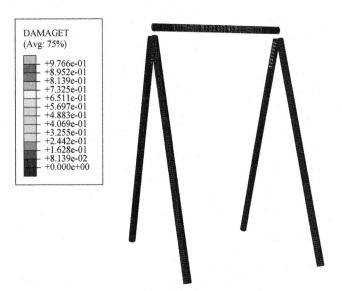

Figure 9. Yield of half steel bar of component.

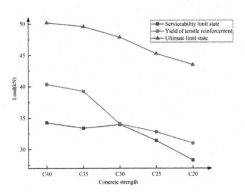

Figure 10. Influence of concrete strength on bearing capacity of frame.

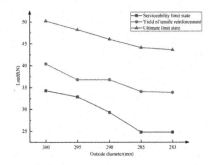

Figure 11. Influence of outer diameter of ring rod on bearing capacity of frame.

thickness of the protective layer. When the protective layer thickness decreases from 19 mm to 10.5 mm, the corresponding bearing capacity decreases by 4%, 8.4%, 12.2%, and 13.1%, respectively.

3.2.3 *Influence of steel corrosion*

The corrosion of the reinforced concrete members may mainly results in the loss of the cross-section of steel bars, the degradation of mechanical properties of steel bars, and the decrease in bond strength between steel bars and concrete; thereby, affecting the overall bearing capacity of the frame (Kioumarsi, M. 2021), causing serious safety hazards to the structure and component. It is assumed that the longitudinal reinforcement of the ring bar is corroded uniformly, and the influence of reinforcement corrosion is simulated by changing the section area of reinforcement.

It can be seen from Figure 12 that the bearing capacity of the concrete ring bar substation frame is positively correlated with steel corrosion. As the diameter of reinforcement gradually decreases from 12 mm to 10.0 mm, the corresponding bearing capacity of the frame decreases by -1.2%, 5.8%, 11.2%, and 16.9%, respectively.

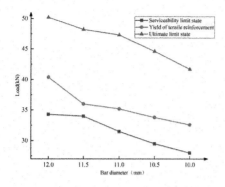

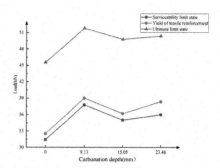

Figure 12. Influence of diameter of longitudinal reinforcement on bearing capacity of frame.

Figure 13. Influence of carbonization depth on bearing capacity of frame.

3.2.4 *Influence of carbonization depth*

The concrete carbonation could influence the constitutive relation to some extent, as proposed in by Xu (2016). The difference between the ideal and actual constitutive relation curve is that the peak strain increases, but the decline section of the curve becomes steeper after the peak strain thus the deformation ability decreases. To investigate the influence of the constitutive relation of concrete after carbonation on the bearing capacity of its structure, the bearing capacity and deformation performance are analyzed based on the material properties in Xu (2016). Table 3 shows the carbonation rate of normal operation pole concrete in Zhang (1999).

Table 3. Carbonation rate of normal operation pole concrete.

Service life (year)	1	2	3	5	10
Carbonation depth(mm)	0.2~0.6	0.4~1	0.7~1.5	1.5~3	3.3~7

It can be seen from Figures 13 and 14 that concrete carbonation can improve the bearing capacity of the structure to a certain extent. Compared with the ideal concrete, when the carbonization depth of concrete increases from 9.13 mm to 23.48 mm, the ultimate bearing capacity of the corresponding structure increases by 13.7%, 9.3%, and 10.6%, respectively, but the ductility of the structure decreases after carbonation. The depth of carbonation has little effect on the bearing capacity, and the difference is mainly reflected in the carbonation and non-carbonization of concrete.

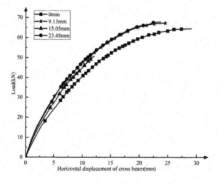

Figure 14. Relationship curve between frame bearing capacity and horizontal displacement.

3.3 Orthogonal analysis of bearing capacity sensitivity of concrete circular pole substation structures

To investigate the influence of concrete strength, protective layer thickness, and steel bar diameter on the bearing capacity of the frame, the orthogonal analysis method was applied to investigate the bearing capacity of single-span annular rod (Tables 4 and 5), and the range and variance analysis were used to explore the sensitivity of the factors above the bearing capacity. It can be seen from Tables 3 and 6 that the sensitivity of the three factors above the bearing capacity of the concrete ring bar substation frame can be arranged as follows: steel corrosion → outer diameter size (protective layer thickness) → concrete strength.

Table 4. $L_{25}(5^3)$ orthogonal list.

Number	Concrete strength	Outside diameter (mm)	Bar diameter (mm)	Ultimate bearing capacity (kN)
LS-1	C40	300	12	52
S-2	C40	295	11	44.6
S-3	C40	290	10	38
S-4	C40	285	11.5	41.8
S-5	C40	283	10.5	37.4
S-6	C35	300	10	38.3
S-7	C35	295	11.5	48
S-8	C35	290	10.5	43
S-9	C35	285	12	46
S-10	C35	283	11	38
S-11	C30	300	10.5	43.4
S-12	C30	295	12	46
S-13	C30	290	11	38.8
S-14	C30	285	10	33.4
S-15	C30	283	11.5	37.8
S-16	C25	300	11	43.2
S-17	C25	295	10	36.3
S-18	C25	290	11.5	38.9
S-19	C25	285	10.5	33.4
S-20	C25	283	12	38
S-21	C20	300	11.5	41.4
S-22	C20	295	10.5	35.9
S-23	C20	290	12	38.3
S-24	C20	285	11	33
S-25	C20	283	10	29.6

Through range analysis and variance analysis results of single-span substation frame bearing capacity, it can be obtained that the sensitivity order of the three factors on the influence of ring concrete pole substation frame is: steel corrosion → outer diameter size (protective layer thickness) → concrete strength.

4 CONCLUSIONS

Based on the field investigation of 37 substation structures, the parametric theoretical analysis of the typical single-span concrete circular pole substation structures was conducted using the numerical simulation software ABAQUS. The influence of four factors, namely, concrete strength,

Table 5. Range analysis of bearing capacity of concrete circular pole substation structures.

Index	Ultimate bearing capacity (kN)		
Factor	Concrete strength	Outside diameter	Bar diameter
K_1	213.8	218.3	220.3
K_2	213.3	210.8	207.9
K_3	199.4	197	197.6
K_4	189.8	187.6	193.1
K_5	178.2	180.8	175.6
$\overline{K}$	198.9	198.9	198.9
$\overline{K1}$	42.8	43.7	44.1
$\overline{K2}$	42.7	42.2	41.6
$\overline{K3}$	39.9	39.4	39.5
$\overline{K4}$	38.0	37.5	38.6
$\overline{K5}$	35.6	36.2	35.1
R	7.2	7.5	9.0

Note: K_1, K_2, K_3, K_4, K_5 represent the sum of indicators at each factor level; $\overline{K}$ represents the average value of the sum of indicators under the same factor; $\overline{K1}, \overline{K2}, \overline{K3}, \overline{K4}, \overline{K5}$ are the average values of each factor level; R represents the range of indicators under different levels of the same factor.

Table 6. Analysis of variance of bearing capacity of concrete circular pole substation structures.

Index	Factor	Squares	Freedom	Mean square
Ultimate bearing capacity	Concrete strength	940.9	4	235.2
	Outside diameter	976.9	4	244.2
	Bar diameter	1117.2	4	279.3

protective layer thickness, steel corrosion, and carbonation depth, on the bearing capacity of the frame was mainly discussed. The conclusions are as follows:

(1) The concrete circular pole substation structures are still widely used and meet the requirements, but the durability of the structure or component is reduced to varying degrees, and the potential safety hazard is large.

(2) The bearing capacity analysis of concrete circular pole substation structures shows that the failure mechanism of substation frame can be divided into three stages: normal service state (with the allowable crack width being 0.2 mm), yielding of one longitudinal tensile steel, and yielding of half of the longitudinal steel; while the beam failure presents brittle failure of "inadequate steel."

(3) The bearing capacity of a concrete circular pole substation frame is positively correlated with concrete strength, protective layer thickness, and steel bar diameter. Concrete strength grade decreases from C40 to C20, the ultimate bearing capacity decreased by 1.2%–13.3%; the protection layer decreases from 19 mm to 10.5 mm, the bearing capacity decreased by 4%–13.1%; the diameter of steel bars decreases from 12 mm to 10.0 mm, the bearing capacity decreased by -1.2%–16.9%.

(4) The carbonation of concrete leads to the change of its constitutive relationship. Compared with ideal concrete, the bearing capacity of concrete frames with carbonation depths of 9.13 mm, 15.05 mm, and 23.48 mm increases by 13.7%, 9.3%, and 10.6%, respectively. The carbonation depth has little effect on the bearing capacity of the frame, while mainly influencing the deformation capacity of the frame.

(5) Using orthogonal analysis, range analysis, and variance analysis, the sensitivity order of influencing factors on the bearing capacity of substation frame is steel corrosion → outer diameter size (protective layer thickness) → concrete strength.

ACKNOWLEDGMENTS

This paper was financially supported by Fund Program for the Scientific Activities of Selected Returned Overseas Professionals in Shanxi Province, China through grant no. DC1900000602.

REFERENCES

Chen, F., Tang, C., & Zhang, J. (2016, 3). A Study on the Impact between Car to Circular Concrete Pole. 2016 Eighth International Conference on Measuring Technology and Mechatronics Automation (ICMTMA). 2016 Eighth International Conference on Measuring Technology and Mechatronics Automation(ICMTMA).

Chen, H., & Alani, A. (2013). Optimized Maintenance Strategy for Concrete Structures Affected by Cracking due to Reinforcement Corrosion. ACI Structural Journal, 110(2). https://doi.org/10.14359/51684403.

Chen, Hua-Peng., & Nepal, Jaya. (2018). Modeling Residual Flexural Strength of Corroded Reinforced Concrete Beams. Aci Structural Journal 115 (6), pp.1625-1635.

Feng Ruimin. (2012, 5)Study on Disease and Durability Evaluation in Reinforce Concrete Power Poles of Ring-shaped Cross Section by Ultrasonic Diagnosis[D].Central South University,

Hashimoto, T., Kanai, S., Hirono, S., & Torii, K. (2015). Aconsiderationon Improvement in Durability Aspects of Concrete Poles. Cement Science and Concrete Technology.

Kioumarsi, M., Benenato, A., Ferracuti, B., & Imperatore, S. (2021). Residual Flexural Capacity of Corroded Prestressed Reinforced Concrete Beams. Metals,11(3),442.

Li Feng, Xia Kaiquan & Zhang Zifu. (2006)Research on Bearing Capability of Concrete Poles with Defects[J].Northeast Electric Power Technology.(9):17-18+42.

Lian Hui, Fu Wei& Zhu Yundong. (2004.8,22) Analysis of structural style of 220kV frames in 500kV transformer station[J]. Henan Sciences (4):536–538.

Liu Siyuan, Xia Kaiquan & Chen Zongping. (2011, 11) Study on residual carrying capacity of existing reinforced concrete poles [J]. Concrete, (6):60–63+72.

Liu Siyuan, Xia Kaiquan, Chen Zongping & Su Yisheng. (2010) Study on residual carrying capacity of existing reinforced concrete poles[J]. Concrete.(4):48–50+53.

Ma Hongyuan, Yu Yanfei, Liang Chaofeng, He Zhihai & Zhao Jiangxia. (2019). Research progress on carbonation performance of concrete under different stress states[J]. China Concrete and Cement Products. (06),20-24. doi:10.19761/j.1000-4637.2019.06.020.05.

Rajesh Prasad Dhakal & Koichi Maekawa.(2002)Reinforcement Stability and Fracture of Cover Concrete in Reinforced Concrete Members[J]. Journal of Structural Engineering, 128(10): 1253–1262.

Rita Maria Ghantous et al. (2017). Effect of crack openings on carbonation-induced corrosion. Cement and Concrete Research, 95pp. 257–269.

Sun Zengshou. & Ying. (2004.12,22) The structural appraisal and strengthen of substation framework[J].Henan Sciences,(6):825–828.

Vadlūga, R. & Kliukas, R. (2012). The Use Of Annular And Circular Cross -Section Members In Transport And Power Engineering Building Construction. The Baltic Journal of Road and Bridge Engineering, 7(2), 77–83.

Valérie L'Hostis.(2020).Long-term corrosion of rebars submitted to concrete carbonation. Materials and Corrosion (5), doi:10.1002/maco.201911401.

Wang Meng, Wang Pu. & Huang Zhen. (2017, 4)Ultimate bearing capacity of annular section RC members subjected to compound forces[J]. Journal of Hebei University of Technology, 46(2):99–104.

Wang Meng. (2017) Unified model of reinforced concrete members with annular section under combined loads[D]. Shanghai Jiao Tong University.

Xia Kaiquan, Zhang Xianggang, Chen Zongping & Xiao Hua. (2012, 2)An experimental study on the ultimate bending capacity of in-service prestress concrete pole[J]. Journal of Guangxi University (Natural Science Edition), 37(1):29–33.

Xu Shanhua, Li Anbang, Cui Huanping & Liu Xiaowei. (2016.3) Experimental investigation of the stress-strain relationship of carbonated concrete under monotonic loading[J]. Building Structure, 46(6):81–85.

Zeng Xianle, Xu Fei & Xia Kaiquan. (2010.12) Study on Bend Bearing Capacity of Reinforced Concrete Pole after Extended Service[J]. Guangxi Electric Power, 33(6):1–3+24.

Zhang Sudong & Tao Juan. (1999) Detection and diagnosis of defective poles after operation [J]. Electric power construction. (10):52–54.

Frontiers of Civil Engineering and Disaster Prevention and
Control – Yang & Rahman (Eds)
© 2023 The Author(s), ISBN: 978-1-032-31200-2

Strain energy response of asphalt pavement under high temperature and wheel load

Y.F. Lin & J.Q. Gao*
Nanjing University of Aeronautics and Astronautics, Nanjing, China

ABSTRACT: In order to study the shear damage effect of asphalt pavement structure under high temperature, heavy traffic and horizontal loading conditions, the maximum strain energy value and the most unfavorable point of shear resistance at each depth of asphalt pavement under different temperature, heavy load and horizontal load are studied by using asphalt pavement mechanics calculation software. The results show that the temperature affects the asphalt mixture material, but has little effect on the strain energy at each depth, with the maximum strain energy occurring at the middle surface layer. Heavy loads and large horizontal loads cause a large amount of strain energy to be generated at the top layer of the asphalt pavement, with the most unfavorable point of shear resistance occurring at the rear edge of the tire.

1 INTRODUCTION

Heavy traffic sections such as urban intersections are particularly prone to distress in the summer and the strain energy generated by these distressed sections is often high (Song 2021). This is because asphalt pavements, under the action of high temperatures and wheel loads, generate a large amount of elastic deformation (strain energy) due to the elastic deformation of the structural layer, some of which is converted into plastic dissipation energy and creep dissipation energy through the plastic deformation of the structural layer and creep deformation, and this energy will damage the structural layer of the pavement in the process of dissipation, which is the main reason for permanent deformation of the pavement (Tian 2013).

Traditional shear strength theory suggests that rutting damage in asphalt pavement structures occurs because the shear stresses generated in the pavement under load exceed the internal shear strength (Wang 2014). Related studies have shown that the maximum shear stress in asphalt pavements increases significantly with the increase in horizontal loading, and the location of the maximum shear stress moves towards the road surface; under vertical loading, the maximum shear stress generally occurs in the middle surface layer (Qiao 2015). Shear strength can be calculated using Moore-Coulomb strength theory, but Moore-Coulomb strength theory, which does not consider the second principal stress, has some drawbacks.

The fourth strength theory introduces a second principal stress based on the traditional shear strength theory and considers shape change to be the main factor that causes yielding damage in plastic materials. Road researchers at home and abroad have conducted relevant studies on the crack expansion of asphalt mixtures under load, the selection of asphalt layer thickness and the shear test of pavements based on the strain energy method, and have achieved certain results (Chen 2018; Huang 2015; Roque 2002). The strain energy index comprehensively considers the failure of materials under various stress states, so the theory is closer to the actual working situation of pavement (Li 2005; Wang 2015). Taking strain energy as the research object, this paper analyzes

---

*Corresponding Author: junqi_gao@nuaa.edu.cn

                    DOI 10.1201/9781003308577-20

the maximum strain energy response of asphalt pavement structure under different temperatures and wheel loads.

2 ANALYTICAL MODEL

2.1 *Pavement structure*

The strain energy represents the damage that the mixture in the asphalt pavement structure can withstand. The greater the strain energy, the easier the material is to be damaged. Under the action of vehicle load, the strain energy generated at different positions in the asphalt pavement structure must be different, and the area with the largest strain energy is generally the area with the most unfavorable shear resistance.

This paper refers to the domestic typical semi-rigid base pavement structure, and combined with the code for design of highway asphalt pavement, selects the structural parameters as shown in Table 1 for calculation.

Table 1. Structure composition and thickness of semi-rigid base pavement (Zhu 2016).

Structural composition	Thickness/ m	Temperature/ °	Modulus of elasticity/MPa	Poisson's ratio
SMA-13	4	30	620	0.30
		40	554	0.35
		50	530	0.40
		60	526	0.45
AC-20	6	30	752	0.30
		40	600	0.35
		50	440	0.40
		60	380	0.45
AC-25	8	30	900	0.30
		40	910	0.35
		50	500	0.40
		60	390	0.45
Cement stabilized macadam base	40	—	1500	0.2
lime-ash soil	20	—	500	0.3
Soil base	—	—	45	0.4

The asphalt pavement structure calculation software is applied to the discrete nodal analysis calculations relying on the assumptions plus the meshless method under the elastic laminated half-space system in the bicircular loading mode. The infinite boundary problem of the model is dealt with using the infinite element method of coupled calculations. The standard axle load BZZ-100 is used to load the pavement structure model, where the vertical load is 0.7MPa, the bicircular uniform load equivalent circle radius is 10.65cm, and the contact condition between the pavement structure layers is completely continuous, the computational model is shown in Figure 1.

2.2 *Point layout*

The x-axis direction is the transverse direction of the pavement, the y-axis direction is the longitudinal direction along the pavement, the downward direction of the road surface is the z-direction, and the horizontal force direction is the positive direction of the y-axis. The calculation points are selected as follows, extending from the coordinate origin to 3 along the plus direction of the x-axis δ, extend the plus or minus directions along the y-axis to $\pm$ 1.5 δ, 0.1 per interval δ (δ=1.065 cm) set a calculation point, as shown in Figure 2.

Figure 1. Computational model.

For the depth z direction, there are 10 planes: road surface z=0, z=0.004m, upper layer internal z=0.01m, upper layer internal z=0.02m, upper layer bottom z=0.04m, middle layer internal z=0.07m, middle layer bottom z=0.1m, lower layer internal z=0.14m, lower layer bottom z=0.18m and base layer z=0.19m. The state of bonding between the layers is completely continuous and strain energy is calculated for each point on each plane.

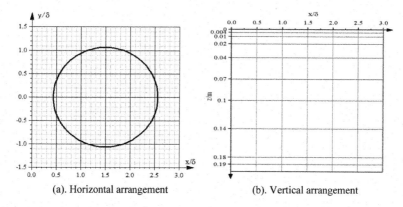

(a). Horizontal arrangement   (b). Vertical arrangement

Figure 2. Layout of calculation points.

3 CALCULATION RESULTS AND ANALYSIS

The coordinates of the shear test points in the 1990 version of the Urban Road Design Code are $(1.5\delta, -1.0\delta, 0)$; the shear test points in the 2012 version of the Urban Road Pavement Design Code are $(2.4\delta, 0, 0)$ and $(2.5\delta, 0, 0.1h1)$. Compare this with the calculated strain energy maximum points.

3.1 Calculation and analysis of strain energy at different temperatures

The maximum value and coordinates of strain energy at each depth when the asphalt surface temperature is 30°C, 40°C, 50°C and 60°C respectively are calculated. The results are shown in the table below:

From Table 2, it can be seen that the strain energy maximum at different temperatures at each depth show a certain regularity, with the strain energy first increasing and then decreasing, and

Table 2. Distribution of maximum strain energy at different depths at different temperatures.

	30°C		40°C		50°C		60°C	
z/m	U/J	(x, y)/δ	U/J	(x, y)/δ	U/J	(x, y)/δ	U/J	(x, y)/δ
0	64.85	(2.3,0)	70.79	(2.3,0)	79.7	(2.3,0)	89.64	(2.3,0)
0.004	118.1	(2.2,0.7)	131.8	(2.2,0.7)	139.6	(2.2,0.7)	144.1	(2.3,0.6)
0.01	115.2	(2.4,0.3)	127.6	(2.4,0.4)	147.1	(2.3,0.5)	153.8	(2.4,0.4)
0.02	120.8	(2.4,0.1)	134.7	(2.4,0.2)	147	(2.4,0.3)	159.7	(2.4,0.3)
0.04	126	(2.3,0.1)	137.3	(2.3,0.1)	171.3	(2.3,0)	188.1	(2.2,0)
0.07	136.7	(1.6,0)	139.5	(1.7,0)	186.3	(1.6,0)	208.9	(1.6,0)
0.10	121.6	(1.6,0)	101.5	(1.6,0)	140.4	(1.6,0)	170.1	(1.6,0)
0.14	44.69	(1.6,0)	54.52	(1.6,0)	73.36	(1.6,0)	83.78	(1.6,0)
0.18	19.21	(1.6,0)	19.08	(1.6,0)	20.34	(1.6,0)	23.01	(1.6,0)
0.19	15.4	(1.6,0)	16.34	(1.6,0)	17.6	(1.6,0)	19.74	(1.6,0)

the strain energy maximum of the pavement structure appear in the middle surface layer, and the coordinates are located in the center of the load circle action. With the increase of temperature, the strain energy of road surface is transferred from the center of load circle to the edge of load circle, and its value increases with the increase of temperature; At the same time, it is found that the maximum strain energy changes little when the surface temperature does not exceed 40°C. When the temperature rises to 50°C, the maximum strain energy increases from 139.5j at 40°C to 186j, with an increase of 33%, while the maximum strain energy increases by 12% from 50°C to 60°C.

The change of temperature directly affects the change of the properties of asphalt mixture, resulting in the change of strain energy value. Under low temperature, all energy values are small, and the asphalt pavement tends to elastic structure. When the temperature rises by 30°C, the strain energy of each depth begins to increase, and the asphalt layer enters the low viscoelastic stage; When the temperature rises to 50°C, the asphalt layer enters the high viscoelastic stage, which shows that the value of strain energy increases, but the increase decreases.

3.2 Calculation and analysis of strain energy under vehicle load overload

When the asphalt surface is at 60°C, the double circle vertical uniform load p is 0.9MPa and 1.1MPa respectively. The calculation results of strain energy under vehicle load are shown in Figure 3:

As can be seen from the calculations in Figure 2, the strain energy maxima of the pavement structure increase with increasing vehicle loads at each depth. The maximum point of strain energy is (1.5δ, 1.0δ, 0), and the strain energy at 0.7MPa, 0.9mpa and 1.0MPa points are 89.64j, 589.4j and 693.2j respectively, an increase of 657.4% and 773%. The impact of overload on pavement structure cannot be ignored. When subjected to vertical load only, the maximum point of strain energy should be (1.5δ, 1.0δ/-1.0δ, 0), which is consistent with the shear stress test point of the 1990 design code; with the increase of vertical load, the maximum point of strain energy at z=0.004m is (2.5δ, 0), which is consistent with the shear test point of the 2012 version of the road design code.

Comparing the data for a vehicle load of 0.7MPa shows that the vertical load overload not only increases the strain energy of the pavement structure at all depths, especially causing a significant increase in the strain energy of the upper layer, but also causes the maximum strain energy at each depth to move away from the center of the load circle towards the edge, with the unfavorable point position of the road surface shifting from the outer edge to the horizontal side edge.

3.3 Calculation and analysis of strain energy under horizontal load

In general, the horizontal load factor is usually in the range of 0 to 0.3. A horizontal force factor of 0 means that no horizontal force is considered, a horizontal force factor of 0.2 characterizes a

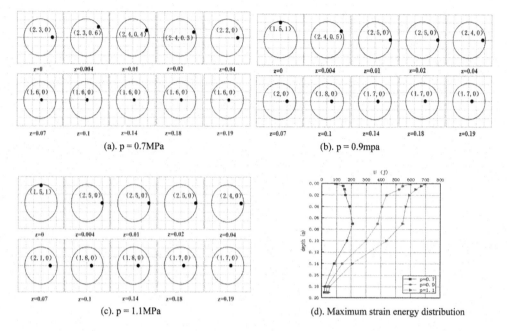

(a). p = 0.7MPa

(b). p = 0.9mpa

(c). p = 1.1MPa

(d). Maximum strain energy distribution

Figure 3.　Distribution of maximum strain energy under different vertical loads.

vehicle in normal driving conditions and a factor of 0.5 characterizes a vehicle under emergency braking etc.

Taking the vertical mean load as 0.7MPa, the asphalt surface layer is at 60°C, the horizontal load coefficient μ is 0.1, 0.3 and 0.5 respectively, the strain energy coordinate distribution and calculation results under the horizontal load are shown in Figure 4.

By analyzing Figure 4, it can be seen that the influence of horizontal load coefficient on strain energy has the following characteristics:

(1) When horizontal load coefficient is small ($\mu < 0.1$), it is found that the horizontal load has little effect on the maximum strain energy and coordinates in the vertical direction, and the calculated strain energy is far less than the action result of heavy load, so the influence of horizontal load can be ignored in general.

(2) As the coefficient increases , the maximum strain energy appears on the road surface, and the characteristic point of the maximum strain energy on the road surface is (1.5 δ,-1.0 δ,0), which is consistent with the shear stress test point of the 1990 design code, the maximum strain energy of the upper layer of asphalt appears at the edge of the load circle, and the maximum strain domain of the middle and lower layer appears at the center of the load circle.

(3) As the coefficient increases from 0.1 to 0.3 and 0.5, the maximum strain energy of pavement structure increases from 211.3j to 685.7j and 1892j respectively. With the increase of horizontal load coefficient, the peak value of strain energy in the upper layer increases significantly, and has little effect on the maximum value of strain energy in the middle and lower layers. This is similar to the development pattern of shear stresses in pavements.

It is inferred that the action range of horizontal load mainly occurs in the upper layer and has little impact on the middle and lower layers. When the horizontal load coefficient is small, the strain energy produced by horizontal load is smaller than that produced by vertical load. Therefore, in general, the influence of horizontal load on the strain energy of pavement structure can be ignored. While the coefficient increases from 0.3 to 0.5, the maximum strain energy increases significantly.

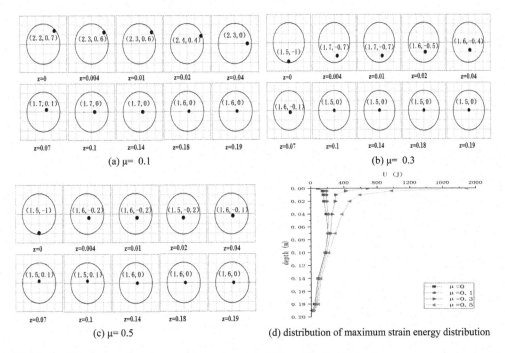

(a) μ= 0.1

(b) μ= 0.3

(c) μ= 0.5

(d) distribution of maximum strain energy distribution

Figure 4.   Distribution of maximum strain energy under different horizontal loads.

The influence of horizontal load on the road surface should be carefully considered in special sections such as intersections.

4   CONCLUSION

(1) Temperature has little effect on the strain energy of the asphalt pavement structure, with the maximum strain energy occurring mainly in the middle surface layer and the most unfavorable point of shear resistance located at the center of the load circle. Temperature variations change the nature of the asphalt mixture, and the increase in strain energy maxima between 40°C and 50°C is much greater in the middle surface layer of asphalt than in other temperatures.

(2) Heavy load will not only increase the strain energy of the pavement structure at all depths, especially making the strain energy of the upper layer increase significantly, the strain energy maximum from and will cause the maximum strain energy point at each depth to move away from the center of the load circle towards the edge, the unfavorable point of the road surface shifts from the outer edge to the horizontal side edge.

(3) The range of action of horizontal loads occurs mainly in the upper layers and has little effect on the middle and lower layers. When the horizontal load factor is small, the strain energy generated by the horizontal load is smaller than that generated by the vertical load, and its effect on the strain energy of the pavement structure is negligible. As the horizontal load factor gradually increases, the maximum strain energy of the pavement increases significantly. Therefore, the effect of horizontal loads on the pavement should be carefully considered at special sections such as intersections.

(4) The strain energy is the object of study and its response to the pavement under horizontal loading is consistent with the shear stress response. Under heavy loads the strain energy is considered to be mainly concentrated in the middle and upper layers and not only in the middle surface layer. The maximum strain energy points under vertical and horizontal loads are generally consistent

with the shear stress calculation points given in the CJJ 30-90 Urban Road Design Code and the CJJ 169-2012 Urban Road Pavement Design Code.

REFERENCES

Chen D. & Huang L.K. (2018). Strain energy response analysis of CRC+AC composite pavement. *Highway Engineering* (02), 253–256.

Huang Z.Y. (2015). *Determining the Minimum Thickness of Asphalt Pavement Based on Energy Dissipation Principle*. (Doctoral dissertation, Harbin Institute of Technology).

Li J., &Yuan J. (2005). Research on low temperature cracking resistance of asphalt mixture. *Highway Traffic Science and Technology* (4).

Qiao, Y. J. & Ma, X. R. (2015). Analysis on shear stress design index of long-life asphalt pavement under complex traffic conditions. *Applied Mechanics and Materials*, 743.

Roque R., Birgisson B., Sangpetngam B. & Zhang Z. (2002). Hot mix asphalt fracture mechanics: a fundamental crack growth law for asphalt mixtures. *Asphalt Paving Technology: Association of Asphalt Paving Technologists-Proceedings of the Technical Sessions, 71*, 816–827.

Song, S. & Yeom, C. (2021). Reduction of plastic deformation in heavy traffic intersections in urban areas. *Sustainability*, 13(7), 4002.

Tian J.J. (2013). Evaluation of fatigue properties of asphalt mixtures based on dissipative energy. *Chinese and foreign highways*.

Wang R.L. (2015). Experimental study on the characteristics of DOMIX modified asphalt mixture. *Highway Traffic Technology* (05), 10–13.

Wang Z.Q. (2014). Research on factors affecting shear performance of high-grade asphalt concrete pavement. *Highway Engineering*, 39 (5), 6.

Zhu T.Y. (2016). Calculation of Equivalent Temperature of Asphalt Pavement Rutting Based on Numerical Simulation. *Journal of Southeast University: English Edition*, 32 (3), 362–367.

Frontiers of Civil Engineering and Disaster Prevention and
Control – Yang & Rahman (Eds)
© 2023 The Author(s), ISBN: 978-1-032-31200-2

Design of steel structure roof of outpatient hall of a hospital in Huangpi

Wen Biao Wu*, Li Ming Yuan, Wen Jun Jing & Song Gao
Central-South Architectural Design Institute Co., Ltd, Wuhan, China

ABSTRACT: The steel roof of the outpatient hall of this project adopts the hybrid structure form of reticulated shell + grid + truss, and the roof spans three individual buildings and is connected to the wind-resistant truss below. The steel roof structure can meet the requirements of bearing capacity and deformation by reasonably setting the bearings. The wind-resistant truss and the grid are connected by pin and chain link to meet the requirements of coordinated deformation. This project uses the nodal multi-scale finite element model to analyze the stress state of the complex joints at the connection between the grid and the reticulated shell. The results show that: the bearing capacity and deformation of the steel structure roof meet the requirements of the code, the support design is reasonable, and the complex joints meet the requirements of strong joints and weak poles.

1 GENERAL INSTRUCTIONS

The project is located in the south of Huangpi District, Wuhan City, with a total construction area of about 259,660 m², mainly including: (1) Outpatient Medical Technology Building (2) Duty Apartment Building (3) Research Office Building (4) Inpatient Building (5) Infection Building. Among them, the outpatient medical-technical building is a concrete frame structure, 23.6m high, and the plane is divided into five areas A, B, C, D and E by three structural joints. Among them, the first floor to the fifth floor of the outpatient hall is of the same height. The steel structure roof adopts the structural form of reticulated shell + grid + truss. The steel roof is placed on the finished supports of the roofs of the three divisions A, B and C, and is jointly stressed with the concrete structure below. The roof cover is connected to the wind truss of the clinic hall by chain link, and the steel roof and the wind truss can be moved vertically and fixed horizontally.

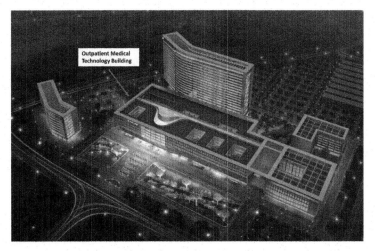

Figure 1. Overall project effect.

*Corresponding Author: 1259044966@qq.com

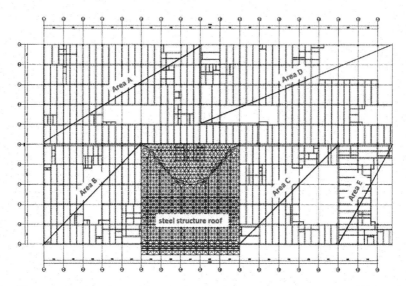

Figure 2. Diagram of zoning.

2 STRUCTURAL ANALYSIS

2.1 *Design parameters*

The service life of the steel roof of this project is 50 years, with a structural importance factor of 1.1, a basic wind pressure of 0.40kN/m² (100-year encounter) and a ground roughness category of B. The basic snow pressure is 0.6kN/m² (one hundred years), considering the temperature rise and temperature drop are 30°C; the seismic fortification intensity is 7 degrees (one degree higher for Class B buildings); the basic seismic acceleration value is 0.10g; the design seismic grouping is the first group; the site category is Class II; the damping ratio of steel structure is 0.02; the characteristic period of seismic response of the site is 0.35s; the influence coefficient of multiple encounters is 0.08.

2.2 *Computational models*

The structural calculation model established by Midas Gen (2019) is shown in Figure 3. The steel roof can be divided into three parts: the upper single-layer reticulated shell, the middle grid and the end overhanging trusses. The size of the structure is 40.5m × 40.5m. The upper single-layer reticulated shell is made of welded H400 × 160 × 12 × 16 steel beams crossed in three directions; the mesh size is 1.6m × 1.6m according to the building requirements; the central grid is in the form of orthogonal orthotropic inverted quadrangular cone with a mesh size of 2.7m × 2.5m; the height of the grid is 2.5m ~ 3.5m; the grid is made of seamless steel pipes with diameters ranging from 60mm to 325mm. Triangular trusses are used at the end to form the main keel of the building; the trusses have an overhang length of about 4.5m and are made of seamless steel tubes ranging from 60mm to 159mm in diameter.

As can be seen from Figure 2, the roof spans three individual buildings and is connected to the wind-resistant trusses at the end. The roof not only needs to consider fully releasing the relative deformation and temperature deformation generated by different individual buildings, but also needs to provide support for the lower curtain wall. After comparison, it is found that the form of fixed hinge bearing + elastic bearing can effectively solve the above problems. The bearing arrangement form is shown in Figure 3, all using finished steel bearing, ZZ1 is fixed hinge bearing, ZZ2 is elastic steel bearing, ZZ2 horizontal shear stiffness is 10kN/mm.

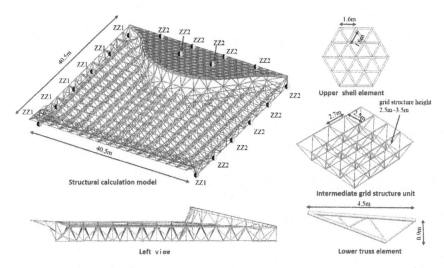

Figure 3.   Model of structural calculation.

2.3   *Static analysis of structure*

The vertical displacement of the structure under dead load + live load condition is shown in Figure 4. The vertical deformation of the structure is large in the middle and small at both ends, similar to the shape of one-way plate variation, which is consistent with the assumption of boundary constraint. The maximum downward deformation of the structure as a whole is 83mm, and the maximum deformation appears at the intersection of the grid and the reticulated shell, which is 1/481 of the span and meets the requirements of the codes (GB50017 2017, JGJ7 2010).

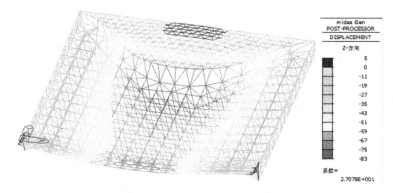

Figure 4.   Vertical deformation under dead load + live load condition.

2.4   *Structural dynamic analysis*

The first three vibration modes of the structure are shown in Table 1. From Table 1, it can be seen that the first vibration mode vertical flat motion accounts for the largest proportion, the second-order vibration has the largest proportion of translation in the Y direction, the third-order vibration has the largest proportion of vertical translation, and the first three orders of vibration are dominated by vertical vibration.

Table 1. Vibration mode parameters of structural.

Vibration models	Cycle time (s)	Direction factor of vibration pattern		
		X	Y	Z
1	0.2715	0.136	0.288	0.576
2	0.2620	0.108	0.760	0.132
3	0.2167	0.113	0.099	0.786

3 NODAL DESIGN

3.1 Support nodes

The grid and reticulated shell are connected with the lower concrete by finished steel bearing, and the connection method is shown in Figure 5. The finished bearing has the advantages of stable performance, convenient construction, controllable horizontal shear stiffness, large vertical bearing capacity, etc., and is widely used in the seismic design of buildings and bridges. The project bearing constraints are more complex, so the use of finished bearings can better meet the design requirements.

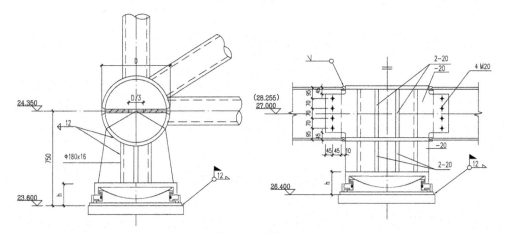

Figure 5. Detail of support node.

3.2 Connection node between wind-resistant joist and main structure

Six wind trusses are installed at the glass curtain wall of the outpatient hall as the main supporting elements of the curtain wall. The wind trusses are about 22m high and the spacing between them is about 8.1m. The left and right trusses are separated from the concrete structure, and the lower part of the trusses are connected with the concrete structure by pins, and the connection joints are shown in Figure 6. The pin joints can well meet the rotation capacity of the wind-resistant joist and avoid the bending of the joist under the wind overload, which will be transmitted to the lower concrete structure. The connection joints between the upper part of the joist and the net frame are shown in Figure 7. The joints are spaced about 2.5m apart and staggered with the wind-resistant joist, and the connection between the ball joint and the joist is realized by setting the conversion beam on the upper part of the joist. The grid and the transformation beam are connected by chain rods. The chain rods can slide vertically and provide horizontal binding force, which can better solve the deformation problem between the wind resistant joist and the net frame.

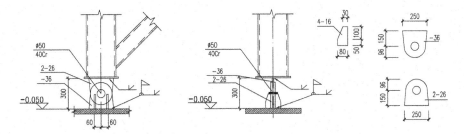

Figure 6.   Detail of the lower node of the wind-resistant truss.

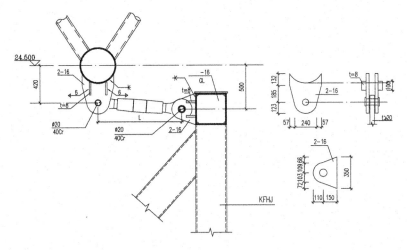

Figure 7.   Detail drawing of the connection between the upper part of the wind-resistant truss and the net frame.

3.3 *Multi-scale analysis of nodes*

The steel roof of this project is composed of three structural forms: reticulated shell, grid and truss, in which the connection between reticulated shell and grid is connected by props. The number of rods at the intersection of the joints of the braces and the grid is large and the forces are large, so further finite element analysis is required. Midas Gen (2019) is used to establish a multi-scale model as shown in Figure 8. The model is composed of a refinement model and beam-column model. Among them, the refined model is welded by the central hollow ball and 8 stell pipes, in which the size of the hollow ball is D550x20, and the other rods are P159x10, P180x14 and P76x4. In order to avoid the adverse effect of the additional stress concentration caused by the coupling region of degrees of freedom on the analysis results, the range of refinement model is 1.5D, where D is the diameter of pipe (Luo 2011; Zhang 2019). After meshing the node with HyperMesh, the grid is imported into Midas Gen. The triangular thin shell element with real thickness is established in Midas Gen. The rigid connections are used to connect the beam-column model and the refinement model in the overall model. This method can effectively simulate the boundary conditions of the detailing model and examine the stress and strain states of the nodes under all operating conditions (Wang 2014; Yin 2017).

Figure 8 shows the maximum equivalent stress distribution of the key nodes. It can be seen from the figure that the maximum stress appears on the left and right chords of the joint, and the maximum value is 213MPa, which does not exceed the design strength of steel 310MPa. Also, it can be seen from the figure that the ball node stress is less than the rod stress, which indicates that the joint can meet the design principle of strong joints weak pole.

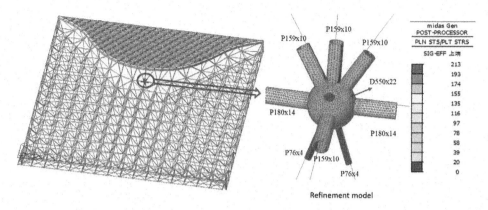

Figure 8. Finite element analysis of nodes.

4 CONCLUSION

The steel structure roof cover of this project is composed of reticulated shell, grid and truss, with moderate span, spanning three individual buildings, and connected with the wind- resistant truss below. The bearing constraints need to be reasonably designed to achieve reasonable bearing capacity under different conditions such as dead load, live load, wind load, earthquake action and temperature action.

(1) By using fixed bearing and elastic bearing, the roof can be more reasonably stressed.
(2) The deformation and bearing capacity requirements can be well met by setting the chain link between the wind- resistant truss and the roof cover.
(3) By establishing a multi-scale model of the joints at the connection between the grid and the reticulated shell, it was found that the joint conforms to the design principle of strong joints weak pole.

5 LIMITATIONS AND RECOMMENDATIONS FOR FUTURE RESEARCH

It is worth noting that the calculation of the overall deformation of the structure and the stress of joints is mainly based on the results of finite element analysis without corresponding tests. Field deformation observation and node stress test shall be carried out to verify the future finite element analysis results.

In addition, the stress level and seismic performance of the structure under large earthquake also need to be further studied in the future.

REFERENCES

GB50017. (2017). Design standards for steel structures. Peking, China. (in Chinese)
JGJ7. (2010). Technical regulations for space grid structures. Peking, China. (in Chinese)
Luo, Y.Z. Liu, H.F. Lou R. (2011) Multi-scale finite element simulation of reticulated shell considering deformation of welded hollow spherical joints. Engineering Mechanics. 28(11):190–196. (in Chinese)
Wang, F.Y., Xu, Y.L., Qu, W.L. (2014) Mixed-Dimensional Finite Element Coupling for Structural Multi-Scale Simulation. Finite Elements in Analysis and Design. 92: 12–25. (in Chinese)
Yin, X, Yu, F. (2017). Structural design and analysis of Meitan stadium canopy in Guizhou. Building Structure, 47(11): 91–95. (in Chinese)
Zhang, S. Cheng, M. Wang, J. et al. (2019). Elastic-plastic analysis of the cast steel joints of Dongjin railway station in Xiangyang based on a multi-scale model. Steel Structure. 10(34):36–42. (in Chinese)

Frontiers of Civil Engineering and Disaster Prevention and
Control – Yang & Rahman (Eds)
© 2023 The Author(s), ISBN: 978-1-032-31200-2

Energy-based optimal design of viscous dampers

Q.X. Shi* & G.Q. Lai*
Xian University of Architecture and Technology, Xian, China

ABSTRACT: Based on the concept of energy balance, the energy distribution analysis is carried out for single degree of freedom system and double degree of freedom system, and then an energy-based optimal design method is proposed, which takes the discrete degree of the ratio of hysteresis energy dissipation between storeys and viscous damper energy dissipation as the optimization objective, aiming to make each story of viscous dampers play the maximum damping efficiency. Taking the 10-storey reinforced concrete frame as an example, the UD method is used for the initial arrangement, on which the energy-based optimized design is carried out and compared with the UD method. The results show that the total hysteresis energy consumption of the optimized structure is reduced by about 17%, and the seismic damping efficiency of the viscous dampers arranged in each story is maximized, which also ensures that the maximum inter-story displacement angle meets the code requirements.

1 INTRODUCTION

In modern buildings, energy dissipation damping devices can effectively improve the seismic performance of buildings, and their mechanism of action is to dissipate the seismic input energy, thereby reducing the damage of building structural elements in earthquakes (Kasai 1998). At present, the types of energy dissipation damping devices at home and abroad include: friction dampers, viscous dampers and viscoelastic dampers (Zhou 2006). Among them, viscous dampers are good damping devices commonly used in earthquake-prone areas (Crewe 1998). Its viscous dampers are velocity dependent, produce forces with different phases from displacement, do not significantly increase the stiffness of the structure, are insensitive to temperature, have stable performance and are fatigue free (Constantinou 1994; Soong 2014).

Many researchers have conducted numerous studies to better exploit the damping effect of viscous dampers and found that the damping effect of the same number of viscous dampers with different arrangement schemes is not consistent (De Sila 1981; Gürgöze 1992; Zhou 1998). Thanks to the development of computers, more complex and refined procedures for the optimal design of viscous dampers are available (Garcia 2001; Jing 2019; Singh 2001; Takewaki 1997). These procedures use different methods, but most of them only address the amount of response generated indirectly by energy (De Domenico 2019). Analyzing the response of the structure under earthquake action from the energy perspective can visualize the absorption and dissipation of structural energy, and also accurately reflect the effect of earthquake on the structure, so the energy-based design strategy can fully control the earthquake response and also reduce the optimization procedure is effective in terms of complexity and economic cost (Liu 2013; Peng 2011).

Choi H (2006) considers the total support flexural dissipation energy as equal to the total seismic input energy, and obtains the cross-sectional area of the support required to satisfy a given target displacement by equating the hysteretic energy demand with the accumulated plastic energy dissipated by the support. Song (2014) establishes the relationship between seismic input energy and damper dissipation energy, and estimates the total number of dampers required to be installed in the damping structure using the seismic energy response spectrum of the single-degree-of-freedom system

---

*Corresponding Authors: 1157537194@qq.com

and the hysteretic energy dissipation characteristics of the dampers. Then, with the maximum total energy dissipation of dampers as the optimization objective, the relationship between the number of dampers arranged in each story of the damping structure and the vibration pattern of the structure is derived according to the equation of motion of the multi-degree-of-freedom system, so as to arrive at the optimal arrangement of dampers. Leden (2009) investigates the factors influencing the proportional relationship between the total energy dissipation of the dampers and the distribution of the damping coefficients among the floors with dampers arranged, and compares the values of the damping coefficients with the proportional values of the time-scale analysis. Neda Nabid (2020) improves the seismic performance of the structure by redistributing the slip loads of the friction dampers based on an energy-based approach to obtain a more uniform distribution of the energy dissipation capacity along the height to achieve the maximum dissipation capacity.

In this paper, based on the energy balance equation and using Matlab, the effect of increasing damping coefficient on hysteretic energy dissipation and damper energy dissipation in the single-degree-of-freedom system and the double-degree-of-freedom system is discussed, and based on this, a function with the dispersion degree of the ratio of hysteretic energy dissipation and damper energy dissipation as the optimization objective is proposed. By establishing a practical and straight-forward iterative optimization method, the dampers are arranged for a 10-story reinforced concrete frame and compared with the UD method for analysis.

2 ENERGY BALANCE THEORY

The energy balance equation for the single degree of freedom system with additional viscous dampers can be obtained by integrating the equation of motion over time t as in Equation (1) (Uang 1990).

$$\int_0^t m\ddot{x}(t)\dot{x}(t)dt + \int_0^t c\dot{x}(t)\dot{x}(t)dt + \int_0^t f_s(t)\dot{x}(t)dt + \int_0^t f_d(t)\dot{x}(t)dt = -\int_0^t m\ddot{x}_g(t)\dot{x}(t)dt \qquad (1)$$

where $m, c, \ddot{x}(t), \dot{x}(t), and \ddot{x}_g(t)$ are the mass, inherent damping coefficient, acceleration, velocity and seismic acceleration of the single-degree-of-freedom system, respectively. $f_s(t)$ is the structural elastic force, and $f_d(t)$ is the viscous damper restoring force, and the viscous damper adopts Maxwell restoring force model as Eq. (2), where C_d is the damping coefficient of the viscous damper, and α is the damping index of the viscous damper.

$$f_d(t) = C_d abs(\dot{x}(t))^\alpha \ sign(\dot{x}(t)) \qquad (2)$$

From the energy point of view, the building structure is a process of energy input and dissipation during the earthquake, while at the end of the earthquake, the kinetic and elastic strain energy of the structure is consumed by the inherent damping of the structure, the elastic-plastic recovery force and additional damping devices, so that Eq. (1) can be simplified to the following equation.

$$E_c + E_H + E_D = E_i \qquad (3)$$

where E_c, E_H, E_D and E_{in} are the inherent damping energy of the structure, the elastic-plastic energy of the structure, the viscous damper energy and the seismic input energy, respectively. When the original structural system, its own inherent damping energy is not enough to offset all the seismic input energy, the structure will occur elastic-plastic damage, that is, the generation of elastic-plastic energy dissipation. The energy dissipation and damping structural system is to dissipate the seismic energy through additional devices, thus reducing the elastic-plastic energy dissipation of the structure.

3 ENERGY-BASED OPTIMIZATION DESIGN OF NONLINEAR VISCOUS DAMPERS

The optimal design of viscous dampers involves the location, number and support conditions of the viscous dampers. In order to reduce the amount of variation in the optimal design, only the arrangement of the number of dampers along the height is studied. Further, this can be expressed in terms of the coefficient column matrix of the following dampers.

$$[C_d] = [C_{d1}, C_{d2}, C_{d3} \cdots C_{dn}] \tag{4}$$

where n represents the number of floors. The energy balance equation for the system of n degrees of freedom in terms of energy is obtained from Eq. (3), as in Eq. (5). The objective of the optimal design of the viscous dampers is to combine $[C_d]$ of Eq. (5) to find the best set of damping coefficients to achieve the required EDi for each floor.

$$\begin{cases} E_{cn} + E_{Hn} + E_{Dn} = E_{in} \\ \quad\quad\vdots \\ E_{c1} + E_{H1} + E_{D1} = E_{i1} \end{cases} \tag{5}$$

3.1 Energy analysis of single- and two-degree-of-freedom systems

To better illustrate the energy balance equation expressed by Eq (5), a single degree of freedom system with additional nonlinear viscous dampers is considered, where the structural parameters of the single degree of freedom system are m = 1000kg, c = 400N·s/mm, k = 1 × 10N/mm5, where the elasto-plasticity of the system is expressed as a trifold reinforced concrete model with cracking and yielding displacements of 4mm and 8mm, respectively, and discounted stiffness ratios 0.4 of and 0.1. The damping index in the viscous dampers is 0.3.

As represented in Figure 1(a), in the single-degree-of-freedom system, as the damping coefficient of the viscous damper increases, the structural Ec, EH and Ei all decrease accordingly, because the increase in the damping coefficient reduces the response of the structure. While E_D increases first and then decreases, the maximum dissipation point D is reached at Cd = 900. After point D, the reduction of structure E requires more viscous damping coefficients, which is not costly. Using point D as the best arrangement point for optimal design is a convenient method, which implies that the viscous dampers maximize the damping efficiency. But often the D point may not satisfy the EH requirement of the structure. In order to make an optimal choice between E_D and EH to make an optimal choice, an energy parameter R is considered, as in Eq. (6). As in Figure 1(b), when the constraint limits of the optimal design are determined, such as the total damping coefficient and the additional damping ratio, taking the smallest value of R can show the damping efficiency and the role of the viscous dampers in the structure.

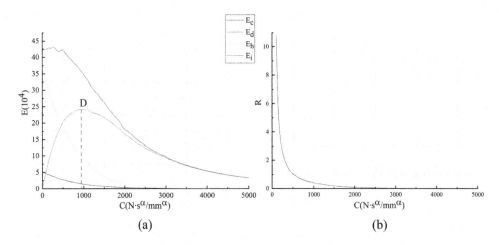

Figure 1.   (a) Energy diagram for a single degree of freedom (b) Energy ratio R.

$$R = \frac{E_H}{E_D} \tag{6}$$

To be more explicit about the variation of energy in Eq. (5), two variables C_{d1} and C_{d2} are considered from the two-degree-of-freedom system ($\alpha = 0.3$). Consider two system parameter arrangements S1 and S2, as in Table 1.

Table 1. Parameter settings for the 1two-degree-of-freedom systema.

	Floor	$m (10^3)$ kg	$k (10^5)$ N/m	C N·s/m	Discount Stiffness ratio	Cracking displacement mm	Yield displacement mm
S1	1	2	1	400	0.4/0.1	4	8
	2	1	1	400	0.4/0.1	2	5
S2	1	2	3	400	0.4/0.1	4	8
	2	2	1	400	0.4/0.1	2	5

In double degrees of freedom, R-minimum is not the only solution under the constraint limit, and R-minimum may not satisfy the seismic requirements on the interstory. If R_i is minimized from story to story, it will ensure that the seismic requirements are met at each story. However, R_i is interrelated, for example, when the total damping coefficient is used as the constraint limit, R_2 grows as R_1 decreases, which involves the choice of interstory R_i. In this paper, it is assumed that when the distribution of R_i along the height reaches uniformity, the optimal seismic effect can be ensured by arranging sufficient dampers in each story. To verify this assumption, a parameter Cov_R is established to analyze the degree of dispersion of R_i.

$$Cov_R = \left(\frac{\sqrt{Var_R}}{Ave_R} \right) \tag{7}$$

As shown in the Figure 2, like the single-degree-of-freedom system, both S1 and S2 show the E_D maximum dissipation point, but the location of its appearance is not the same. the fastest falling trend of E_H contour is also different, S1 is biased towards C_{d1} and S2 is biased towards C_{d1}. Therefore, it can be shown that different storeys of the structure do not have the same demand for viscous dampers, such as the uniform arrangement method (UD) for designing viscous dampers is not feasible. The Cov_R value is ray-like, large on both sides and small in the middle. The ray with $Cov_R < 0.1$ is close to the maximum point of E_D and also along the direction where E_H decreases fastest. If the total damping coefficient (dashed line in the Figure) is used as the constraint limit, the smaller the Cov_R value, then the larger the E_D is, the smaller the E_H is. It can be seen that the Cov_R value can be used as an optimization target for optimal design.

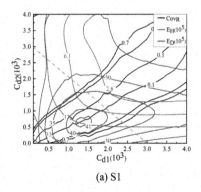

(a) S1

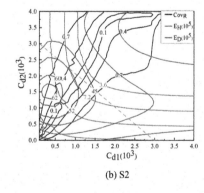

(b) S2

Figure 2. E_H, E_D and Cov_R contour maps of S1 and S2.

3.2 Energy-based optimization design method

In this paper, an energy-based optimization design method is proposed from the above-mentioned energy analysis of single degree of freedom and double own degree, as in Eq. (8).

$$\begin{cases} Subject\ to\ \sum C_d = C_{d1} + C_{d2} + C_{d3} \cdots + C_{dn} \\ \min Cov_R = \left(\frac{\sqrt{Var_R}}{Ave_R} \right) \end{cases} \tag{8}$$

In this paper, for simple and practical considerations, an iterative approach is used to solve for the optimal arrangement, and the viscous damping coefficient of each story is redistributed according to Eq. (9) when the total damping coefficient is determined.

$$(C_{di})_{j+1} = (C_{di})_j \times \left(\frac{R_i}{R_{ave}} \right)_j^{\alpha} \tag{9}$$

where j is the number of iterations. α To prevent the coefficient of dispersion during the iteration, it is generally taken 0.3. To ensure that the total damping coefficient of the structure remains unchanged, so the viscous damping coefficient is adjusted from the new, and its Eq. (9) calculated by $(C_{di})_{n+1}$ Scaling is performed as follows.

$$\left[(C_{di})_{j+1} \right]_{scaled} = (C_{di})_{j+1} \times \left(\frac{\sum (C_{di})_0}{\sum (C_{di})_{j+1}} \right) \tag{10}$$

Repeated iterations according to Eqs. (9) and (10) are effective in reducing the Cov_R value.

4 COLUMN ANALYSIS

4.1 Project overview

This paper uses a 10-story reinforced concrete frame. The design seismic grouping is set as group II, site category II, intensity8 degree 0.2 g, and structural characteristic period 0.4. The concrete grade of beam-column-slab of the structure is C30 level, the main reinforcement of beam-column is HRB400 level, and the hoop reinforcement is HRB300 level. The constant load on the floor of the structure is 6 kN/m^2 and the live load is 2 kN/m^2, the constant load on the roof is 5 kN/m^2 and the live load is 1.5 kN/m^2. The plan layout of the 10-floor frame structure is shown in the Figure 3. The viscous dampers are arranged in the middle two spans and the direction is Y direction.

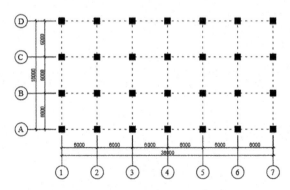

Figure 3. Ten-Storey reinforced concrete frame plan layout.

The 10-story reinforced concrete frame structure was structurally designed using YJK to meet the inter-story displacement angle limit of 1/550 under multiple encounter earthquake, and then modeled using SPA2000 and imported into Perform 3D for elastic-plastic analysis under fortified earthquake. In order to better simulate the elasto-plastic behavior of the structure, in Perform 3D, a five-fold model is used for C30, a double-fold model is used for HRB400, a fiber section model is used for the elasto-plasticity of beams and columns, and a Maxwell model is used for the viscous dampers.

The seismic excitation uses the seismic acceleration time data downloaded from the Pacific Seismic Center, which satisfies the standard response spectra under structural fortification earthquakes with the classification of RSN6, RSN18 and RSN161.

4.2 *Contrast analysis*

In order to verify the effectiveness of the optimized design method proposed in this paper, the uniform arrangement method (UD), which is commonly used in engineering, is selected for comparative analysis. Because the optimization method proposed in this paper is limited by the total damping coefficient as a constraint, in order to ensure the reasonableness of the comparison, the optimized design is carried out on the basis of the UD method arrangement to ensure that the total damping coefficients of both are the same.

As shown in the Figure 4, the Cov_R values under all three seismic excitations are decreasing with the increasing number of iterations, which indicates that the iterative method used in this paper can effectively reduce the Cov_R values. the Cov_R under the UD method arrangement is approximately equal to 1, indicating that the energy distribution of each story is different under the UD method arrangement, although the interstory C_{di} is uniformly arranged.

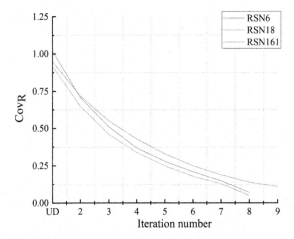

Figure 4. Cov_R variation with iteration.

To prove the validity of the Cov_R values, the analysis is carried out from C_{di}, E_{Di}, E_{hi} and the interstory displacement angle, as shown in Figures 5~7. The black curve is the UD method arrangement, and the rest of the colored curves are the curves with different iterations, and the higher iteration number represents the lower Cov_R value. It can be seen that the changes are about the same for three different seismic excitations. It shows that the earthquake will not affect the energy-based optimization design method in this paper. Under the UD method arrangement, although C_{di} is uniform, E_{Di} and E_{Hi} are not the same, such as the lowest E_{di} in the first 10 story and the lowest E_{Hi} in the first story, while the E_{Hi} in the 1 story is much larger than the first 10 story, which means that the viscous dampers arranged in the first 10 story by the UD method are wasted, while the viscous dampers arranged in the 1 story are not enough. As the number of optimization iterations increases, the C_{di} in storeys 1~2 in the Figure 4 is increasing, and the C_{di} in storeys 3~6 does not change much. 7~10 is decreasing, and the first 10 story decreases the most. While E_{Di} is increasing in 1~2 storeys and decreasing in 3~10 in Figure 5. And in the Figure 6, E_{hi} decreases in 1~5 storeys and does not change much in 5~10 storeys.

174

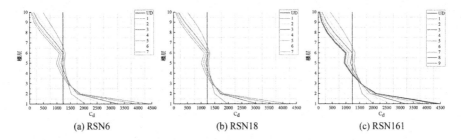

Figure 5.   10-story building UD iterative process C_{di}.

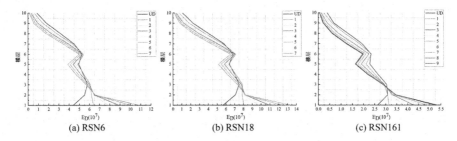

Figure 6.   10-story building UD method iterative process E_{Di}.

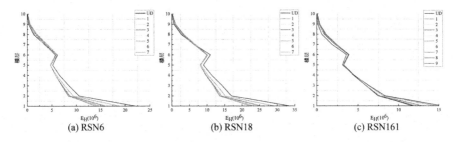

Figure 7.   10-story building UD method iterative process E_{Hi}.

For better illustration, as in Table 2, ΔC_{di}, ΔE_{Di} and ΔE_{Hi} are the differences between the optimized storeys C_{di}, E_{Di} and E_{Hi} and UD divided by their average values, which can indicate the change of each story. It can be seen that UD method is optimized by this paper, C_{di} increases in storeys 1~3, and the corresponding E_{hi} are decreased. While 4~6 storeys, C_{di} decreases, but the corresponding E_{Hi} also decreases, which is well explained in Equation (5), the increase of C_{di} in 1~3 storeys decrease the seismic energy input in 4~6 storeys. 7~10 storeys C_{di} decreases very much, but E_{Di} decreases not much, while E_{Hi} rises negligibly. All these show that UD method is unreasonable, and the optimized design method proposed in this paper can make the arrangement of viscous dampers in each story reasonable.

Table 2.   Rate of change of 10 storeys frame C_{di}, E_{Di} and E_{Hi}.

Floor	RSN6			RSN18			RSN161		
	ΔC_{di}	ΔE_{Di}	ΔE_{Hi}	ΔC_{di}	ΔE_{Di}	ΔE_{Hi}	ΔC_{di}	ΔE_{Di}	ΔE_{Hi}
1	2.58	1.55	−1.26	2.39	1.43	−1.05	2.64	1.34	−0.83
2	0.52	0.26	−0.34	0.48	0.22	−0.32	0.72	0.18	−0.20

(continued)

Table 2. Continued.

Floor	RSN6			RSN18			RSN161		
	ΔCdi	ΔEDi	ΔEHi	ΔCdi	ΔEDi	ΔEHi	ΔCdi	ΔEDi	ΔEHi
3	0.14	−0.02	−0.23	0.13	−0.03	−0.24	0.24	−0.01	−0.08
4	−0.07	−0.17	−0.13	−0.05	−0.16	−0.15	−0.04	−0.23	−0.01
5	−0.20	−0.25	−0.06	−0.18	−0.23	−0.09	−0.24	−0.32	0.04
6	−0.11	−0.06	−0.01	−0.08	−0.06	−0.07	−0.20	−0.19	0.07
7	−0.39	−0.25	0.09	−0.35	−0.26	0.03	−0.50	−0.34	0.12
8	−0.65	−0.34	0.11	−0.60	−0.33	0.08	−0.75	−0.35	0.12
9	−0.86	−0.29	0.05	−0.81	−0.28	0.05	−0.90	−0.23	0.06
10	−0.96	−0.18	0.01	−0.94	−0.17	0.01	−0.97	−0.17	0.01

As shown in Table 3, for the total energy change after UD method and optimization, the total E_D is not changed much after optimization, while the total E_H can be reduced a lot after optimization. It means that the optimization can reduce the elastic-plastic damage of the structure. the total E_D and E_H of RSN161 both decrease because, when the constrained target limit is behind the maximum point of viscous damper dissipation, as in Figure 2, the decrease of E_H, necessarily E_D will also decrease. And as 8 shown in the Figure, the maximum interlaminar displacement angle in the UD method and after optimization meet the specification requirements.

Table 3. Total E_D and E_H rate of change for 10-story frame.

	RSN6	RSN18	RSN161
ED	0.02	0.01	−0.04
EH	−0.17	−0.17	−0.06

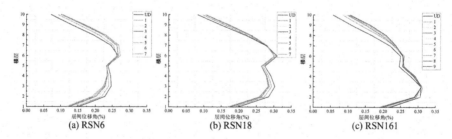

(a) RSN6   (b) RSN18   (c) RSN161

Figure 8. Inter-story displacement angle for the iterative process of UD method for 10-story building.

5 CONCLUSIONS

The energy-based optimal design method is proposed by analyzing the energy distribution in single and double degrees of freedom, and the conclusions drawn are as follows, taking a 10-storey reinforced concrete frame as an example and comparing the analysis with the UD method.

(1) UD method for arranging viscous dampers, on the story, is not reasonable, it has too many viscous dampers arranged in some storeys, thus causing waste, and also too few arranged in some storeys, causing too much structural damage in that story.
(2) In this paper, it is effective to take Cov_R value as the optimization objective. The reduction of Cov_R value can make the viscous dampers arranged in each story play the maximum damping

efficiency, and the story with the greatest structural damage can be identified, and sufficient viscous dampers can be arranged in that story.

(3) Cov_R is reduced, which can make the total hysteresis energy consumption of the structure lower and reduce the overall damage of the structure, and after optimization, the maximum inter-story displacement angle meets the structural code requirements.

6   LIMITATIONS AND FURTHER RESEARCH

(1) The structure types established in this paper are insufficient, and the effectiveness of COVR in different structure types is not analyzed
(2) The iterative method established in this paper relies on the initial arrangement of the damper, and does not consider the effects of different initial arrangements
(3) The optimization objectives established in this paper based on the energy concept should be applicable to other types of dampers, and further research can be conducted and discussed for other types of dampers.

REFERENCES

Choi, H., & Kim, J. (2006). Energy-based seismic design of buckling-restrained braced frames using hysteretic energy spectrum. Engineering Structures, 28(2), 304–311.
Constantinou, M. C. (1994). Principles of Friction, Viscoelastic., Yielding Steel and Fluid Viscous DAMPERS: Properties and Design. In Passive and active structural vibration control in civil engineering (pp. 209–240). Springer, Vienna.
Crewe, A. (1998). Passive energy dissipation systems in structural engineering. Structural Safety, 2(20), 197–198.
De Domenico, D., Ricciardi, G., & Takewaki, I. (2019). Design strategies of viscous dampers for seismic protection of building structures: a review. Soil dynamics and earthquake engineering, 118, 144–165.
De Silva, C. W. (1981). An algorithm for the optimal design of passive vibration controllers for flexible systems. Journal of sound and Vibration, 75(4), 495–502.
Garcia, D. L. (2001). A simple method for the design of optimal damper configurations in MDOF structures. Earthquake spectra, 17(3), 387–398.
Gürgöze, M., & Müller, P. C. (1992). Optimal positioning of dampers in multi-body systems. Journal of sound and vibration, 158(3), 517–530.
JinBo, Xinyi Li, ZhouWang, Liyin Tang, Zaolong Jang. (2019). Optimization analysis of damper position and number based on improved genetic algorithm. Journal of Hunan University (Natural Sciences), 46(11), 114–121.
Kasai, K., Fu, Y., & Watanabe, A. (1998). Passive control systems for seismic damage mitigation. Journal of Structural Engineering, 124(5), 501–512.
LeDeng, ZhouYun, Xunsong Deng. (2009). Study on the ratio of interlayer distribution of total energy dissipation of energy dissipators in energy-consuming vibration-damping structures. Earthquake Resistant Engineering and Retrofitting, 31(003): 26–33.
LiuLei. (2013). Viscous damping optimization analysis based on energy method. (Doctoral dissertation, Kunming University of Science and Technology).
Nabid, N., Hajirasouliha, I., Margarit, D. E., & Petkovski, M. (2020, October). Optimum energy based seismic design of friction dampers in RC structures. In Structures (Vol. 27, pp. 2550–2562). Elsevier.
PengZhu. (2011). Research on energy-based analysis and design methods for energy dissipative vibration reduction structures. (Doctoral dissertation, Central South University).
Singh, M. P., & Moreschi, L. M. (2001). Optimal seismic response control with dampers. Earthquake engineering & structural dynamics, 30(4), 553–572.
Soong, T. T., & Constantinou, M. C. (Eds.). (2014). Passive and active structural vibration control in civil engineering (Vol. 345). Springer.
Takewaki, I. (1997). Optimal damper placement for minimum transfer functions. Earthquake Engineering & Structural Dynamics, 26(11), 1113–1124.

Uang, C. M., & Bertero, V. V. (1990). Evaluation of seismic energy in structures. Earthquake Engineering & Structural Dynamics, 19(1), 77–90.

Zhigang Song, ZhangYao. (2014). Optimal arrangement of viscoelastic dampers based on energy principle. Journal of Kunming University of Science and Technology(Natural Science Edition), 39(6), 5.

Zhouyu, Zhaodong Xu, & XueSong Deng.(1998). Optimal setting of dampers in viscoelastic damping structures. World earthquake engineering, 14(3), 6.

ZhouYun. (2006). Energy-consuming seismic reinforcement technology and design methods. Science Press.

Frontiers of Civil Engineering and Disaster Prevention and
Control – Yang & Rahman (Eds)
© 2023 The Author(s), ISBN: 978-1-032-31200-2

Tunnel stability analysis considering the influence of bias voltage

Tianxue Xu*
China Road and Bridge Corporation Co, Ltd. Beijing, China

Qi Jin*
Xi'an University of Architecture and Technology, Xi'an, China

Xiaojun Peng* & Huxiang Hou*
China Road and Bridge Corporation Co, Ltd. Beijing, China

Liang Zhang*
CCCC Fourth Highway Engineering Co., Ltd. Beijing, China

ABSTRACT: Considering the problems of shallow buried bias tunnel, taking the north-south cross-ridge tunnel in Kyrgyzstan as the study object, using the improved calculation formula of surrounding rock pressure for shallow bias tunnel and numerical simulation, the interaction between the slope and the shallow buried bias tunnel is analyzed. The results show that when the main tunnel is excavated without any construction support, the safety factor is reduced to 1.16. The load of the primary support structure may be higher than the design value. Therefore, primary support and secondary lining of the tunnel should be completed in time during the construction, and the surrounding convergence in this section should be strengthened.

1 INTRODUCTION

During tunnel construction, the tunnel excavation will lead to disturbance in the lower slope and affect the stability of the shallow buried bias section at the tunnel entrance. The stability of the shallow buried bias section is a key factor in tunnel construction (Song 2006; Wang 2005). Because of the stability of shallow buried bias section at tunnel entrance during construction, scholars have done the following research: Ma (2016) regarded tunnel and landslide as a system, analyzed several typical deformation modes of the tunnel-landslide system and proposed disaster prevention measures. Zhang (1999) analyzed that the creep of the subordinate slope would change the mechanical characteristics of a tunnel support structure, and pointed out that the pressure generated by the creep of the subordinate slope was an important reason for tunnel deformation and cracking. Zhu (2006) considered the Xiaomansa River as an example to analyze the disease countermeasures of tunnel lining cracking.

Although the abovementioned studies analyzed the interaction mechanism of tunnel and slope and studied the influence of tunnel excavation on slope stability, they did not put forward how to consider the role of stratum thrust and did not provide any calculation method for surrounding rock pressure of shallow bias tunnel considering stratum thrust. Based on the imbalance thrust force method of slope stability analysis, this paper analyzes and improves the calculation formula of surrounding rock pressure of shallow-buried bias tunnel, combined with the numerical simulation method. The interaction between slope and cave in the construction of shallow buried bias section at tunnel entrance is analyzed.

---

*Corresponding Authors: xutx@crbc.com, 1378968913@qq.com, pengxj@crbc.com, houhx@crbc.com and 1132897100@qq.com

2 CALCULATION OF TUNNEL SURROUNDING ROCK PRESSURE CONSIDERING THE INFLUENCE OF BIAS VOLTAGE

The sections above the potential sliding surface of the tunnel slope are divided and numbered. A total of n, assuming that the tunnel is located in section i, as shown in Figure 1. According to the Mohr-Coulomb strength criterion, the i-1th and ith soil strips are analyzed, and the force balance equations are listed in the tangent and normal directions of the bottom surface of the soil strip, to obtain P_{i-1} and P_i.

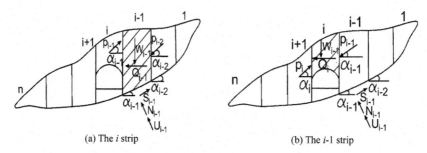

(a) The i strip   (b) The i-1 strip

Figure 1. The schematic diagram of soil strip stress.

The ground thrust ΔP_i and $\Delta P_i'$ acting on the lining due to the change of slope stability is:

$$\Delta p_i = \frac{p_{i-1}'}{H_{i-1}'} \cos \alpha_{i-1}' - \frac{p_{i-1}}{H_{i-1}} \cos \alpha_{i-1}$$
$$\Delta p_i' = \frac{p_i'}{H_i'} \cos \alpha_i' - \frac{p_i}{H_i} \cos \alpha_i \tag{1}$$

Where P' = unbalanced thrust before the original slope stability changes; P = unbalanced thrust after the stability changes; H_{i-1} and H_i = height of the soil strip. At this time, the horizontal pressure on the tunnel is the resultant force of the stratum thrust and the horizontal surrounding rock pressure is calculated by the specification.

3 STABILITY ANALYSIS OF TUNNEL ENTRANCE

3.1 Overview of tunnel engineering

Section 3-A of Kyrgyzstan north-south crossing ridge tunnel is excavated on the Paleozoic strata with sedimentary rocks. The rock is strong to medium strength, and the tunnel passes through some structural faults. The main tunnel is constructed by the upper and lower steps method, and the excavation width is 12.6 m. The long pipe shed section at the entrance of the service guide hole is constructed by the step method, and the other places are constructed by the full section method.

3.2 Tunnel stability analysis model and parameters

The K432+20 section is selected as the calculation model. Model size parameters are given in Figure 2. The surrounding rock material of the tunnel is considered according to homogeneous elastic-plasticity, and Mohr-Coulomb strength criterion is adopted. The primary support and anchor are considered based on elasticity. Beam element simulation and embedded truss element simulation are respectively for shotcrete and bolt.

In the numerical simulation, the specific excavation process is as follows: the first step is the full-face excavation of the pilot tunnel, the second step is the construction of anchor and primary support of the pilot tunnel, the third step is the excavation of the upper bench of the main tunnel, the fourth step is the construction of the anchor and primary support of the upper bench of the main tunnel, the fifth step is the excavation of the lower bench of the main tunnel, and the sixth step is the construction of anchor and primary support of the upper bench of the main tunnel. Calculation parameters are shown in Table 1.

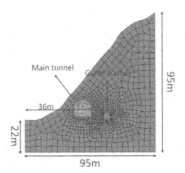

Figure 2.   Finite element model.

Table 1.   Parameters of finite element materials.

Material name	E/MPa	γ/(kN · m-3)	ν	c/kPa	ϕ/°
Wall rock	1500	20	0.3	125	25
Primary support	5000	24	0.2	–	–
Bolt	210,000	78.5	0.3	–	–

4   ANALYSIS OF SIMULATION RESULTS

4.1   *Displacement analysis*

It can be seen from Figure 3 that after excavation, due to stress release, the surrounding rock of the pilot tunnel begins to collapse along the fracture surface. The deformation of the surrounding rock of the main tunnel is characterized by the vertical displacement of the bottom plate, the horizontal displacement of the arch, and the horizontal displacement of the vault, which is slightly larger than the vertical displacement. Due to the presence of bias voltage, the displacement values of the right arch and side wall are higher than those of the left arch and side wall. This indicates that due to the influence of tunnel excavation, the deformation trend of the tunnel is no longer a simple collapse of the fracture surface of the eye, but the overall shear of the soil above the tunnel to the slope toe.

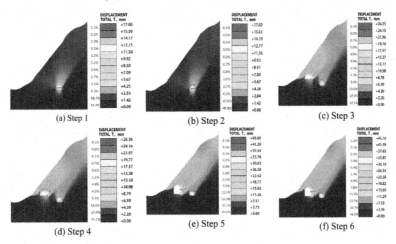

Figure 3.   Displacement vector diagram.

181

4.2 Slope stability analysis

As can be seen from Figure 4 and Table 2, when the tunnel is not excavated, the main tunnel is located on the potential sliding surface, the guide tunnel is located below the potential sliding surface, and the safety factor of the slope is 1.44. After the guide tunnel excavation, the sliding surface shows a slight change, and the safety factor of the slope is 1.31. After the completion of the main tunnel excavation, the sliding surface changes significantly, and the shear section at the foot of the slope is penetrated. At this time, the safety factor of the slope is 1.16, and the slope is in a more dangerous state. Therefore, the primary support and secondary lining of the tunnel should be completed in time during the construction, and the vault subsidence and surrounding convergence in this section should be strengthened.

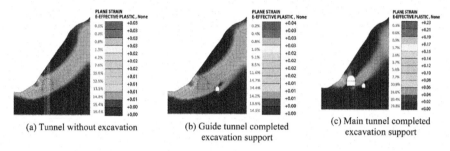

(a) Tunnel without excavation   (b) Guide tunnel completed excavation support   (c) Main tunnel completed excavation support

Figure 4. Trend diagram of sliding surface development.

Table 2. Safety factor for construction steps.

Construction steps	0	1	2	3	4	5	6
Factor of safety	1.44	1.31	1.43	1.21	1.33	1.16	1.31

5 COMPARATIVE ANALYSIS

5.1 Comparison and analysis of the surrounding rock pressure

Figure 5 shows the surrounding rock pressure figures calculated by two methods. By comparison, it can be seen that the surrounding rock pressure calculated by the specification in the improved method is smaller than that obtained by the numerical simulation, and after adding the formation thrust, it is closer to the numerical calculation results. Therefore, the surrounding rock pressure calculated by the improved method is closer to the numerical calculation results.

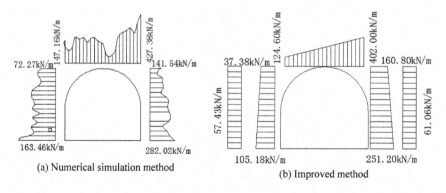

(a) Numerical simulation method   (b) Improved method

Figure 5. Pressure of surrounding rock.

5.2 Comparative analysis of the safety of the supporting structure

It can be seen from Table 3 that the overall safety factor calculated by numerical simulation is higher than that calculated by the improved method. This is because the improved method considers the formation thrust and the most unfavorable conditions. But this is more in line with the actual complex engineering situation. It has a reference value for actual engineering construction.

Table 3. Safety factor of primary support.

	Vault	Right arch waist	Right arch foot	Right wall	Right wall foot	Left arch waist	Left arch foot	Left wall	Left wall foot
Numerical simulation method	4.51	1.27	1.29	2.25	3.36	4.75	1.2	3.17	2.57
Improvement method	3.61	1.45	1.27	1.78	3.48	4.25	1.74	2.49	2.54

6 CONCLUSIONS

Based on the results and discussions, the conclusions obtained are as follows:

(1) For the calculation of surrounding rock pressure of shallow bias tunnel, the horizontal pressure should be calculated according to the resultant force of stratum thrust and horizontal surrounding rock pressure by specification, and the vertical surrounding rock pressure should be calculated according to the conventional method.
(2) In this paper, the calculation method of surrounding rock pressure considering the influence of bias was proposed. The unfavorable conditions, such as seepage and construction disturbance, arising during tunnel construction are comprehensively considered, but the influence of various unfavorable conditions on the calculation results is not specifically refined. The results are more conservative, so the safety factor is low; however, it is more in line with the complex engineering environment. It has an important reference for the safe construction of shallow buried bias tunnel engineering.
(3) In the construction, when the main tunnel is completed without construction support, the safety factor is minimized, and the slope is in a relatively dangerous state. The primary support and secondary lining of the tunnel should be completed in time, and the vault subsidence and surrounding convergence in this section should be strengthened.
(4) In the following research, a more detailed calculation formula of surrounding rock pressure are required to be made according to the possible influencing factors in construction to make the calculation results more realistic.

REFERENCES

Ma, H.M., Wu, H.G. (2016). Progress and expectation of research on tunnel-landslide System. J. Chinese Journal of Underground Space and Engineering, 02: 522–530.
Song, Z.P., Zhang, X.Q. (2006). Hypotaxis stability analysis of seem-continuum medium slope. J. Bulletin of Soil and Water Conservation, 26 (01): 16–19.
Wang, Y.Q. (2015). Analysis of comprehensive treatment and monitoring of landslide at tunnel face. J. Journal of Railway Engineering Society, 32 (10): 103–108.
Zhang, L.X., Zhou, D.P. (1999). Analysis of the causes of the peristaltic landslide and the tunnel deformation mechanism. J. Chinese Journal of Rock Mechanics and Engineering, 18 (02): 99–103.
Zhu, K.Z., Zhu, H.H.(2006). Example analysis on mechanism of interaction between landslide and tunnel. Chinese Journal of Underground Space and Engineering, 05:809–812+817.

Frontiers of Civil Engineering and Disaster Prevention and
Control – Yang & Rahman (Eds)
© 2023 The Author(s), ISBN: 978-1-032-31200-2

The structural design of a hospital in Hainan Province

Wen Jun Jing*, Song Gao*, Yong Mei You* & He Ping Cao*
Central-South Architectural Design Institute Co., Ltd, Wuhan, China

ABSTRACT: Adopting the traditional structural design concept of "hard resistance" in high seismic intensity areas will cause huge cost waste. The project adopts shock absorption technical measures to increase the structural resistance by adding buckling restrained brace (BRB) at the appropriate position of the structure, so as to achieve the purpose of energy consumption and reduce the seismic force borne by structural members, which has achieved good architectural and economic benefits.

1 GENERAL INSTRUCTIONS

This project is located in the east of Mission Hills Area in Haikou City. The effect picture is shown in Figure 1, and the structural partition diagram is shown in Figure 2. The total construction area is 103,206 m². The plane size of the hospital is 255 m*429 m. It is mainly composed of five single buildings: outpatient and emergency building, hospital street, medical technology building, inpatient building, scientific research administrative building and standby building. This project is provided with a basement (partial second floor).

Since the seismic fortification intensity of the project is 8 degrees, simply relying on structural members to resist seismic force will lead to huge structural members and need to increase a large number of reinforcements. After comparative calculation, the final choice is to carry out damping design for the structure. The specific implementation method is to add buckling restrained support in the structure to increase structural damping.

Figure 1. Project overall rendering.

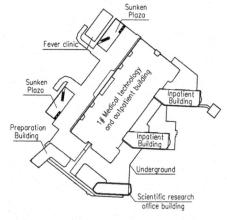

Figure 2. Schematic structure partition.

---

*Corresponding Authors: 272890091@qq.com, 374720039@qq.com , 1240094909@qq.com and 455633855@qq.com

DOI 10.1201/9781003308577-24

2 STRUCTURAL FACTORS

2.1 *Design parameters and natural conditions*

The structural design working Life of this project is 50 years. About the basic wind pressure, we adopted the 10-year recurrence wind pressure of 0.45 kN/m² for structural comfort checking calculation, and the 50-year return period wind pressure of 0.75 kN/m² is adopted for structural displacement calculation. The ground roughness is class A. For high-rise buildings with a height greater than 60m, the bearing capacity shall be designed according to 1.1 times of the basic wind pressure. This project is located in a tropical area, so snow load does not need to be considered. The most important seismic force calculation, according to Code for seismic design of buildings (GB50011-2010), the seismic fortification intensity of the project is 8 degrees, and the design basic seismic acceleration is 0.3 g. The site category is II Class, the structural elastic damping ratio is 0.05, and the characteristic period of the site seismic response is 0.45 s.

2.2 *Building classification level*

Safety grade of building structure: According to "Unified standard reliability design of building structures" (GB50068-2001), it is Grade I. Structural importance factor $\gamma o = 1.0$ (inpatient building, medical technology and outpatient building is 1.1). Foundation design level: According to the *Code for design of building foundation* (GB50007-2002), Grade A. Seismic fortification of buildings: According to" Standard for Classification of Seismic Protection of Building Constructions" (GB50223-2008), it is a key type of fortification. Structural earthquake-resistant grades: According to "Code for Seismic Design of Buildings" (GB50011-2010) and "Technical specification for concrete structures of tall building" (JGJ3-2002), Frame Level II and Shear Wall Level I.

Table 1. Overview and seismic grade of single building.

	Medical technology and outpatient building	Scientific research office building	Preparation building	Inpatient building
Safety classes of building structures	Grade one	Grade one	Grade one	Grade one
Seismic fortification category for structures	Key fortification	Standard fortification	Standard fortification	Key fortification
Seismic grade of structure	Grade one	Grade one	Grade one	Grade one
Structural height	18.0 m	23.4 m	16.2 m	55.000 m
Type of building structure	Reinforced concrete frame+Buckling restrained brace	Reinforced concrete frame+Buckling restrained brace	Reinforced concrete frame+Buckling restrained brace	Steel frame+Buckling restrained brace

3 STRUCTURAL DESIGN

According to the application of vibration reduction and isolation technology in housing construction projects issued by the Department of housing and urban rural development of Hainan Province, when schools, kindergartens, hospitals, and other densely populated public buildings with more than 3 floors (including 3 floors) are newly built in the 8-degree area, vibration reduction and isolation technology shall be strictly preferred for design. In combination with the actual situation, this project plans to make shock absorption design, and set energy dissipation and shock absorption devices in the structures of medical technology and outpatient building and inpatient building.

3.1 *Design of buckling restrained brace*

Buckling restrained brace (short for BRB) is a new type of shock absorption product that combines support and energy dissipation damper. The sleeve is used to provide constraints to the support core to restrain the compression buckling of the support. Ordinary braces will buckle under compression, and their hysteretic performance is poor under repeated load. The buckling restrained brace only connects the core plate with other members, and all the loads are borne by the core plate. The outer sleeve and filling materials only restrict the compression buckling of the core plate, so that the core plate can enter yield under tension and compression, and the hysteretic performance is excellent. On the one hand, buckling restrained braces can avoid the defects of significant difference in tensile and compressive bearing capacity of ordinary braces. On the other hand, they have the energy dissipation capacity of metal dampers, which makes the main structure basically in the elastic range.

Since the buckling restrained brace is in an elastic state under small earthquake, and its stress is characterized by axial stressed members hinged at both ends, in the small earthquake design, the stiffness of the buckling restrained brace to the structure is simulated through inclined bars in YJK program, and the model of buckling restrained brace is selected by extracting the maximum load combination internal force of the brace. Then ensure that the design bearing capacity of the buckling restrained brace is greater than the maximum load combination internal force, that is, keep the buckling restrained brace in an elastic state in small earthquakes.

The basic composition of BRB is shown in Figure 3 below, and the common layout mode is V-type. Single inclined rod type. The upper V-shaped layout is used in most positions of the project, and the detailed layout is shown in Figure 4.

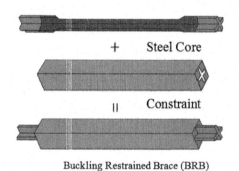

Figure 3. Detail of BRB.

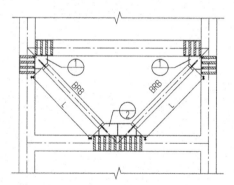

Figure 4. Elevation Layout of BRB.

In this project, buckling restrained brace has the following functions: (1) Increase the torsional stiffness of the structure; in this project, the torsional effect of the structure without buckling restrained brace is obvious, and the structural failure caused by torsion is a common failure mode of building structures under earthquake. In order to solve the torsion problem, buckling restrained braces are set at the two ends of the structure along the longitudinal direction, and the torsional stiffness of the structure is increased by buckling restrained braces, so as to limit the torsional deformation of the structure (Hao 2010).

(2) Increase the energy dissipation capacity of the structure; according to a series of tests on the seismic performance of BRB concrete frames (Gowda 2010; Gu 2011; Guo 2013; JI 2011).

186

The BRB can increase the energy dissipation capacity of the structure. In this project, the buckling restrained brace is set at the end of the structure, and the end of the structure is the position with the largest deformation. The buckling restrained brace is a displacement damper, which can give full play to the energy dissipation effect of the buckling restrained brace.

(3) Buckling restrained braces enhance the collapse resistance of the structure; in this project, the buckling restrained brace, as the main energy dissipation member, will first yield and consume energy under earthquake. After the seismic energy is absorbed by the buckling restrained brace, the main load-bearing members of the structure-beams and columns will be protected, so that the anti-collapse ability of the structure under earthquake will be strengthened.

3.2 *Design of viscous fluid damper*

The traditional anti vibration of structure is to resist natural disasters such as earthquake, wind, snow and tsunami by enhancing the anti-vibration performance (strength, stiffness and ductility) of the structure itself (Meng 2014; Zhou 2005). Due to the uncertainty of the action intensity and characteristics of natural disasters, the structure designed by the traditional anti earthquake method does not have the ability of self-regulation. Therefore, when the earthquake comes, it will often cause heavy economic losses and casualties.

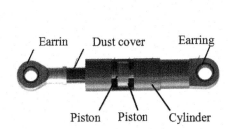

Figure 5. Detail of Viscous Fluid Damper.

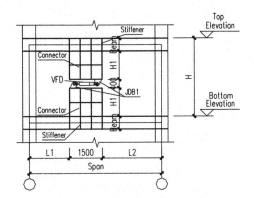

Figure 6. Elevation Layout of VFD.

The development and application of viscous energy dissipation damper is equivalent to installing "airbag" on buildings or bridges. When the earthquake comes, the damper absorbs and consumes the impact energy of the earthquake on the building structure to the greatest extent, and greatly alleviates the impact and damage of the earthquake on the building structure.

Viscous fluid damper (short for VFD) is generally composed of front earrings, dust cover, piston rod, cylinder barrel, rear earrings and so on. The damper is filled with damping medium. When the piston of the viscous damper moves back and forth, the damping medium generates damping force through the damping channel on the piston.

3.3 *Structural schemes comparison of inpatient building*

The inpatient complex building is the tallest building in the project. The structural plane is flat, and it is an offshore site. The basic wind pressure and seismic force are large. The greater the self-weight of the structure, the greater the seismic force. At the beginning of structural selection, we tried to calculate with reinforced concrete frame structure, and found that many beam column sections are too large, which seriously affects the use function of the building. After selecting the layout scheme of steel frame+Buckling restrained brace, the self-weight of the structure is lighter, the member section is smaller, and the foundation design is more economical.

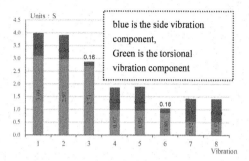

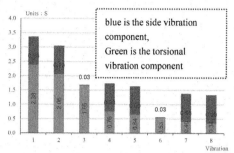

Figure 7. The vibration mode cycle diagram of inpatient building Without BRB&VFD.

Figure 8. The vibration mode cycle diagram of inpatient building With BRB&VFD.

Figures 7 and 8 are the vibration mode cycle diagram of inpatient building, the First translation period of inpatient building with BRB&VFD is 2.38s, and the model which without BRB&VFD is 3.09 s. Cycle ratio of this building with BRB&VFD is 0.69, and the model without BRB&VFD is 0.88 (Close to specification limit 0.9). Through the comparison of the results, we can see that when BRB and VFD are set in the monomer, the structural period can be significantly reduced, the overall stiffness of the structure is improved, and the ratio of the overall torsional period to the first translational period is more reasonable.

The layout of damping components needs to be fully combined with the medical process flow line and building functions. Considering that the BRB setting may have an impact on the functional channel, VFD shall be selected at the position where there is a door opening or channel, and BRB shall be selected for the whole partition wall or area without functional impact. Figures 10 and 11 are the layout of BRB and VFD of the project.

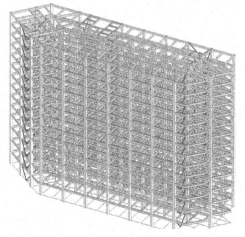

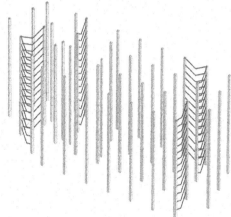

Figure 9. Seismic isolation structure model.

Figure 10. Layout of vertical components.

3.4 *Structural schemes comparison of medical technology and outpatient building*

The business work of the medical technology and outpatient building is oriented to the whole hospital. The technical level and work quality of the medical technology department directly affect the effect of the medical treatment, scientific research and teaching work of the whole hospital, especially the clinical work. The design of single structure should also meet the needs of medical process as much as possible. The plane size of the medical technology and outpatient building is

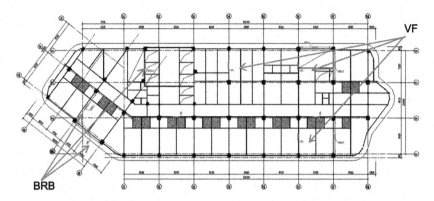

Figure 11. Seismic isolation structure model.

large, and there are irregular items such as opening and concave convex. Before the scheme selection, two groups of simple structure trial calculations are carried out, one is the pure frame structure model, and the other is the damping structure model with BRB&VFD. Through comparison, it is found that the period ratio of the pure frame structure model does not meet the requirements of the specification, and the vibration reduction and isolation model can meet the requirements of the specification. Because the damping members absorb most of the seismic forces, the section dimensions of the vertical members of the damping model are smaller than those of the pure frame model. Finally, it is confirmed that the mixed damping scheme of BRB and VFD is selected for the medical technology building.

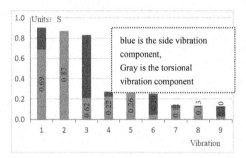

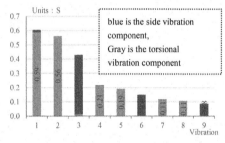

Figure 12. The vibration mode cycle diagram of medical technology and outpatient building.
First torsion cycle/first translation cycle $= 08305/0.9046 = 0.92 > 0.90$, which does not meet the specification requirements.

Figure 13. The vibration mode cycle diagram of medical technology and outpatient building.
First torsion cycle/first translation cycle $= 0.4293/0.6049 = 0.71 < 0.90$, meeting the specification requirements.

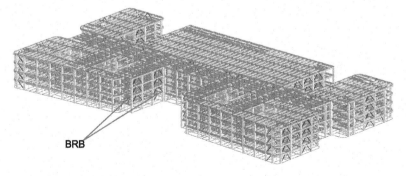

Figure 14. Seismic isolation structure model.

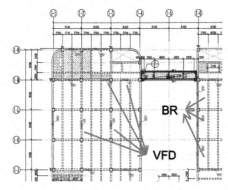

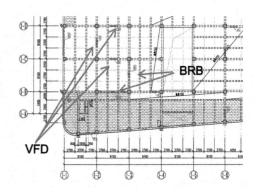

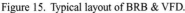

Figure 15. Typical layout of BRB & VFD.　　　　Figure 16. Typical layout of BRB & VFD.

4　SUBMISSION OF MATERIAL TO THE EDITOR

This project is a typical application example of damping measures in important buildings in high-intensity areas. Through comparison, by setting BRB at a reasonable position, the torsional stiffness of the structure can be increased and the energy dissipation capacity of the structure can be improved. we find that after the damping measures are adopted for the same building, it can bring additional structural damping to the building, and the damping components can absorb a large amount of seismic force in the earthquake, which can not only ensure the structural safety under strong earthquake, but also reduce the components section size of the structure, It can bring good building use effect, save engineering steel consumption, and achieve good social and economic benefits. Therefore, we suggest that in the structural design in the high intensity area, if the building conditions permit, vibration reduction and isolation measures should be adopted as far as possible.

REFERENCES

Gu Luzhong, Gao Xiangyu, Xu Jianwei, Hu Chuheng, Wu Na (2011). Experimental research on seismic performance of BRB concrete frames. Journal of Building Structures.

Guo Qiang Li, Peng Wang, YuShu Liu (2013). Story Drift Limit For BRB-Concrete Frame Structures and Design Suggestions. Earthquake Resistant Engineering and Retrofitting.

Hao Guiqiang, Du Yongshan, Qi Jianwei (2010). The application of buckling-restrained braces (BRB) on earthquake resistant engineering and retrofitting. Building Structures.

JI Jun-jie (2011). The Application of Buckling Restrained Braces (BRB) in the Structural Design of Irregular Concrete Structures. Progress in Steel Building Structures.

N.B. Gowda, R.K.C. Gowda, N.N.Patil (2010). Experimental and analytical studies of 3D frame structure with Buckling Restrained Bracing (BRB). International Journal of Civil Engineering & Technology.

WeiHua Meng. DaJun Wang. YuanGuang Lv (2014). Design principle and calculation analysis method of BRB energy dissipation and damping. The 14th National Symposium on modern structural engineering.

Yun Zhou (2005). Design and application of anti-buckling energy dissipation brace structure. China Architecture and Building Press.

Frontiers of Civil Engineering and Disaster Prevention and
Control – Yang & Rahman (Eds)
© 2023 The Author(s), ISBN: 978-1-032-31200-2

Research on the construction system of steel structure and timber structure modular buildings

Zhenlei Guo*, Feihua Yang, Zhijie Gao & Zhongjian Duan
Beijing General Research Institute of Building Materials Co., Ltd., Beijing, China

ABSTRACT: Modular architecture has the advantages of high integration degree, fast and efficient assembly, and good product income. With the maturity of modular building technology, its
rapid construction mode has attracted the attention of the academic community. This paper analyzes timber and steel structures in terms of structural design, construction process, advantages and
disadvantages, and cases. The finite element method is adopted to simulate the mechanical performance and airtightness of four connection methods of timber structure nodes. The advantages and
disadvantages of each node are analyzed. Also, the finite element method is employed to analyze
and compare the modular steel structure and the traditional steel frame of a 4-story building. The
five major structural indicators and their steel consumption are compared, and the advantages and
disadvantages of the modular steel structure are summarized.

1   INTRODUCTION

Modular building is a new way of building construction. Each room is produced as a unit in a
factory, transported to the construction site, and connected in a reliable way to form a unified
whole.

At present, modular buildings are vigorously promoted abroad. Most of them are manufactured in
factories, and only simple assembly is carried out on the construction site, which is the development
model of industrial buildings.

Although modular architecture has many advantages over traditional architecture, how to
promote high-quality modular architecture is a problem faced by various enterprises.

This paper focuses on the finite element analysis and comparison of four connection modes of
wood structure and two construction modes of steel structure, providing design reference for the
application of modular construction.

2   TIMBER STRUCTURE MODULAR BUILDING

Timber structure modular building refers to using timber as the main load-bearing member to
bear the building load, assembling them in the factory assembly line according to the design
requirements, and installing with a maintenance system, electromechanical system, or interior
system to form a modular room. Finally, they are transported to the construction site and connected
as a whole through timber nodes.

2.1   *Modular design of timber structure*

Timber structure modular buildings are commonly used in buildings of three stories and below. For
buildings with more than three stories, it is required to install reinforcing elements inside, which
will increase the cost (Smith 2010).

---

*Corresponding Author: 424285621qq.com

Shang 2016 proposed four connection methods: embedded connection, overlapped connection, tongue and tenon connection, and eccentric connection, as shown in Figure 1.

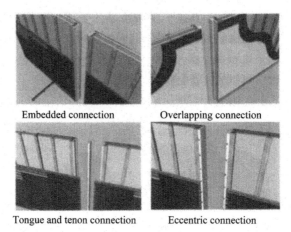

Embedded connection   Overlapping connection

Tongue and tenon connection   Eccentric connection

Figure 1.   Four modular connection methods of timber structure.

The finite element software Solidworks is used to verify the mechanical performance of the four methods. The standard material SPF is used, the material strength is 20 MPa, the Poisson's ratio is 0.29, the wind load is 50 N, and the floor live load is 2 kN/m^3.

The finite element analysis results are as follows:

Figure 2.   Stress distribution of embedded connection model stress distribution of overlapping connection model stress distribution.

Figure 3.   Stress distribution of tongue and tenon connection model stress distribution of eccentric connection model.

The study found that the embedded and eccentric connectors are more evenly stressed, but there is a concentrated stress area outside the eccentric connection, which is easy to damage. The embedded and tongue-and-tenon connections are more stable, but the tongue-and-tenon connection is more consumable and less practical. Through the airtightness analysis of the nodes, it is found that the embedded connection has a long air exchange path, which can reduce the heat loss in the house. Overall, the embedded connection is the best choice for performance among the four connection methods.

Shu 2021 advocated that structural modularity, skin modularity, and spatial modularity should be considered in the modular design of timber structures, and modular design monolithic and group combination diagrams should be considered from the perspective of architectural aesthetics.

Figure 4.   Modular design monolithic combination.

Figure 5.   Modular design group combination.

Gao 2011 analyzed the combination of design, green, and construction process of the modular house of Temple University in the United States and concluded that the modular building should start from the design stage and use the characteristics of modular components to design excellent works. This house has been certified by the U.S. Green Building Council, indicating that modular buildings effectively promote green building. The construction mode of factory production and site assembly can effectively reduce environmental pollution and improve construction efficiency and quality.

2.2   *Modular construction process of timber structure*

The construction process of timber structure modular building is shown in Figure 6:

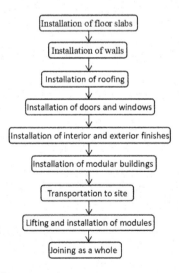

Figure 6.   Timber structure modular construction process.

The modularization of the timber structure adopts the assembly line method, which can effectively improve production efficiency and manufacturing accuracy and reduce labor costs. However, some manufacturers still directly install the water and electricity system after assembling the wall part and then install the peripheral protection system and decoration system. Instead of installing the roof part, they transport the floor and wall to the site for installation (Jian 2018). The advantage of this solution is that the components are placed in a container for overall transportation and are not easily damaged. Still, the disadvantage is also obvious: the on-site assembly is complicated, the workload is large, and there are management risks.

2.3 *Advantages and disadvantages of modular timber structure*

As a renewable resource, timber belongs to green building materials. The timber itself has low thermal conductivity and good thermal insulation performance. It naturally fits the concept of "green building" and has advantages that other building materials do not have (Zhang 2021, Fan 2009, Ding 2014, Zhu 20013, Shan 2013). The sewage discharge, waste residue, and waste gas in the production process are far lower than in other structures. Timber structures have performed well in repeated earthquakes, demonstrating reliable mechanical and seismic performance.

The disadvantages of timber structures are also obvious, such as poor fire resistance and flammability; if not handled properly, the durability of the structure will be reduced.

2.4 *Timber structure modular building case*

American Iron and Steel built the 3,500-square-foot (about 325 square meters) custom timber structure modular building in 22 weeks, with 14 weeks for in-factory production and eight weeks for on-site installation, as shown in Figure 7.

Figure 7.   Lake Tahoe timber structure modular homes.

3   STEEL STRUCTURE MODULAR BUILDING

Steel structure modular building refers to single and multi-story modular steel structure buildings that use steel as the main force-bearing component, are produced, assembled, and connected in the factory. Common types are light steel structures and box steel structures.

The Ministry of Housing and Urban-Rural Development issued the 14th Five-Year Plan for the Development of Construction Industry and proposed to vigorously disseminate the application of prefabricated buildings, actively promote high-quality steel structure residential buildings, and encourage schools, hospitals, and other public buildings to give priority to steel structures. The active development of steel structure modularization is in line with the 14th Five-Year policy orientation and practical needs.

3.1 *Modular design of steel structure*

According to the lateral force resistance system, Wang 2021 divided steel structure modular buildings into pure steel modular structures, modules, and other hybrid structures.

The horizontal lateral force of the pure modular structure system is borne by the modular units, such as Wuhan Huoshenshan and Leishenshan Hospital, Xiongan Citizen Service Center, and Silodam Apartment in the Netherlands.

When the building is tall, or the height and width are relatively large, it is recommended to use modules and other hybrid structures, that is, the horizontal lateral force is jointly borne by the module unit and other systems, mainly including module-frame, module-steel frame support, module-cylinder/shear force walls, and other mixed structural forms. For instance, the comprehensive experimental building of China Construction (Shenshan) Green Industrial Park, the Brook single apartment in New York, the dormitory building in Bristol, the public rental housing in Zhenjiang Gangnan Road, the Bank of China Cabin Building in Tokyo, Japan, etc. applied this approach.

Wen Xiao 2021 suggested improving the construction quality and shortening the project cycle by means of steel structure modular building integrated design and integrated industrial manufacturing.

There are several combinations of box-type modular buildings, as shown in the table of combination methods of box-type modular buildings.

Table 1. Combination of box-type modular buildings.

Combination	3D representation
Parallel	
Criss-cross	
Concave and convex facade	
Crossed and occlusive	

A 4-storey steel structure frame is proposed to be built, and the model is constructed according to the modular building and the traditional steel frame. The finite element software Midas is used to compare four aspects respectively from the period and vibration pattern, structural vertex displacement, inter-story displacement angle, displacement ratio, stress ratio, and steel consumption.

Via research comparison, we can see in Table 2:

Table 2. Comparison of modular and traditional steel structures.

Index \ Construction method	Modular steel structure	Traditional steel frame
Stiffness	Small	Big
Cycle ratio	0.62	0.93
Structural vertex displacement	7.518 mm	4.987 mm
Displacement ratio	1.0	1.0
Stress ratio	<0.6	<0.6
Steel consumption	172.5t	133.9t

The study found that the frequency of modular buildings is lower than that of frame buildings, but the steel consumption is greater than that of frame buildings, and further structural optimization is needed to reduce the steel consumption.

It is recommended to adopt high-strength bolts for the node connection of steel structure modular buildings. Common supporting connections include extended steel column end plates, channel steel beams, corner pieces + positioning pins, etc., as shown in Figures 8–9.

Figure 8.   Extended steel column end plate connection.

Figure 9.   Channel steel beam connection.

3.2   *Modular construction of steel structure*

The construction process of steel structure modular building is shown in Figure 10:

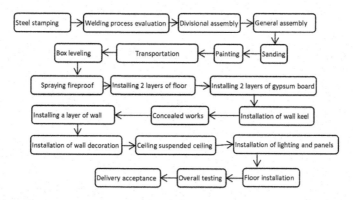

Figure 10.   Steel structure module manufacturing process.

3.3   *Modular assembly process of steel structure*

After the steel structure module is transported to the construction site, it should be installed according to the following process, as shown in Figure 11.

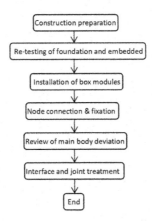

Figure 11.   Steel structure module assembly process.

3.4 *Steel structure modular cases*

Leishenshan Hospital in Wuhan has a construction area of 80,000 square meters, provided with 1,600 beds, and 2,700 experts, medical staff, and officers for work and life. The construction started on January 27, 2020, and was completed on February 6, taking 12 days.

Figure 12.   Leishenshan hospital in Wuhan.

4 CONCLUSION

(1) Through the finite element analysis of four connection modes of wood structure modularization, the analysis shows that the embedded connection performance is better.
(2) Because China mainly relies on imported wood, the application scope of wood structure modular building is limited and can only be used in a specific environment.
(3) Through the comparison between modular steel structure and traditional steel structure, it is found that modular building has a lower frequency than frame building, but its steel consumption is higher than frame building, so further structural optimization is needed to reduce steel consumption.
(4) It is recommended to use high strength bolts for joint connection of steel modular buildings.

REFERENCES

Ding M, Li P. Advantages and development of wood structure buildings [J]. Forest Products Industry, 2014, 41(4): 41–43.
Fan C M, Wang Y W, Pan J L. Wood Structure [M]. Beijing: Higher Education Press, 2009: 171.

Gao S. Temple University Modular Residence - Modular Design and Construction [J]. Residential District, 2011(6): 92–97.

Jian W T, Shen D E. The development trend of wood structures in my country under the background of construction industrialization [J]. Chongqing Architecture, 2018, 17(8): 10–11.

Ryan E. Smith. Prefab Architecture: A Guide to Modular Design and Construction[M]. John Wiley & Sons, Inc. 2010: 154.

Shan B, Gao L, Xiao Y, et al. Experiment and application of prefabricated round bamboo structure house [J]. Journal of Hunan University (Natural Science Edition), 2013, 40(3): 7–14.

Shang P, Sun Y F, Gao H R, et al. Research on Design and Manufacture of Modular Timber Structure Building and Node Connection [J]. Forest Products Industry, 2016, 43(10): 37–41.

Shu X. Modular Design—Introduction to Small-Scale Timber Structure Buildings [J]. World Architecture, 2021(9):74–78.

Standard of China Engineering Construction Standardization Association. Technical regulations for box-type steel structure integrated modular building: T/CECS 641-2019 [S]. Beijing: China Planning Press, 2019.

Wang Wenjing, Li Zhiwu, Yu Chunyi, et al. Research progress of modular steel structure building structural system [J]. Construction Technology, 2020, 49(11): 24–30, 36.

Wen X Y, Li C T, Chen Y, et al. Key technologies for integrated construction of modular steel structure buildings [J]. Construction Technology, 2021, 50(18): 65–70.

Zhang G Y, Zeng H, Cao J L. The development status and prospects of Chinese wood structure buildings [J]. Chongqing Architecture, 2021(11): 22.

Zhu Q. Talking about the advantages of wood structure and its rational application in modern architecture [J]. Urban Construction Theory Research (Electronic Edition), 2013(5).

*Frontiers of Civil Engineering and Disaster Prevention and
Control – Yang & Rahman (Eds)*
© 2023 The Author(s), ISBN: 978-1-032-31200-2

Comparative study on design response spectra in seismic design specifications of China

Songshan Niu
State Key Laboratory of Bridge Engineering Structural Dynamics, Chongqing, China
CMCT Research & Design Institute Co., LTD, Chongqing, China

Haoyu Xie*
State Key Laboratory of Bridge Engineering Structural Dynamics, Chongqing, China
CMCT Research & Design Institute Co., LTD, Chongqing, China
School of Civil Engineering, Chongqing University, Chongqing, China

ABSTRACT: Artificial ground motion simulation is of great significance for the study of structural seismic mitigation. For the major cases of simulating artificial ground motions, people often utilize design response spectrum fitting to generate artificial ground motions. In order to figure out the differences among each structural seismic design specifications in China, in total 5 major Chinese specifications are selected to operate the comparative study on them, including *Specifications for Seismic Design of Highway Bridges, Code for Seismic Design of Urban Bridges, Code for Seismic Design of Buildings, Standard for Seismic Design of Nuclear Power Plant* and *Code for Seismic Design of Electrical Installations*. The result of the comparative study of the paper is considered the reliable basis for structural seismic design.

1 INTRODUCTION

China is located in the southeastern part of the Eurasian Plate, and it is affected by Pacific Rim Seismic Belt and Eurasian Seismic Belt. It is a country prone to earthquakes. It can be seen from the current 'China Earthquake Intensity Zone Division Map' that nearly 40% of Chinese land is located in areas with a basic earthquake intensity of 7 degrees or above, and a large area is located in areas with 6 degrees and above. In the past 20 years, China has suffered many strong earthquakes, which have caused great losses to people's lives and property. In an earthquake disaster, the seismic performance of the building structure directly determines the loss of life and property caused by the earthquake disaster, and if roads and bridges can be maintained in good working conditions, the post-disaster rescue work can be greatly accelerated, and disaster losses can also be minimized (Lu et al. 2019, Qiu & Du 2021). Therefore, structural seismic research has become the focus of attention of earthquake workers in China.

For the dynamic time history analysis and model shaking table tests in structural seismic research, the dynamic response of multi-degree-of-freedom nonlinear structures under different ground motions is very complex and highly uncertain, so the choice of input ground motions is crucially important. Historical time history records of natural earthquakes are widely utilized as input ground motions. However, the collection of high-quality natural seismic wave time history data has only been realized in the 1940s, and the number is very limited (Honda & Ahmed 2011). The performance-based seismic design conception that requires a large amount of ground motion time history data, such as structural seismic fragility analysis (Deierlein et al. 2003, Lu et al. 2006), is

*Corresponding Author: xiehaoyu@cmhk.com

DOI 10.1201/9781003308577-26

difficult to meet the requirements for time history, so artificial fitting of ground motion acceleration time history becomes indispensable means to increase the number of input ground motions. In various artificial ground motion simulation theories (Boore 2003, Iyama & Kuwamura 1999, Liu et al. 2018), numerical simulations and shaking table tests that aim at structural seismic design usually require the artificial ground motion acceleration time histories fitted to design response spectra as the input ground motions.

In order to study the input ground motion simulation, a comparative study on design response spectra in seismic specifications is needed. In the paper, five specifications are selected as the research targets, including 'Specifications for Seismic Design of Highway Bridges' (Ministry of Transport of the People's Republic of China 2020), 'Code for Seismic Design of Urban Bridges' (Ministry of Housing and Urban-Rural Development of People's Republic of China 2011), 'Code for Seismic Design of Buildings' (Ministry of Housing and Urban-Rural Development of People's Republic of China 2016), 'Standard for Seismic Design of Nuclear Power Plant' (Ministry of Housing and Urban-Rural Development of People's Republic of China, 2019) and 'Code for Seismic Design of Electrical Installations' (Ministry of Housing and Urban-Rural Development of People's Republic of China 2013).

2 CHARACTERISTICS OF EACH SPECIFICATION

2.1 Specifications for seismic design of highway bridges

In specifications for seismic design of highway bridges, the design response spectrum is determined by the following equation:

$$S(T) = \begin{cases} (0.6T/T_0 + 0.4)\,S_{\max} & T \le T_0 \\ S_{\max} & T_0 < T \le T_g \\ (T_g/T)\,S_{\max} & T_g < T \le 10\text{s} \end{cases} \tag{1}$$

where $S(T)$ denotes the design response spectrum, T denotes the period, T_0 denotes the maximum period of the linear rising section of the response spectrum, T_g characteristic period, and $S_{\max}$ denotes the maximum design acceleration response spectrum.

The maximum design acceleration is determined by the following equation:

$$S_{\max} = 2.5 C_i C_s C_d A \tag{2}$$

where C_i denotes the seismic importance factor, C_s denotes the site factor, C_d denotes the damping factor, and A denotes peak acceleration of horizontal ground motion.

According to the specification, the characteristic period is determined by three coefficients, including the direction of artificial ground motion, characteristic period in the intensity zone division map, and site category. The seismic importance factor is determined by two coefficients, including seismic design category and seismic design level. The site factor is determined by three coefficients, including the direction of artificial ground motion, site category, and seismic fortification intensity. The damping factor is decided by the structural damping ratio.

2.2 Code for seismic design of urban bridges

In code for seismic design of urban bridges, the design response spectrum is determined by the following equation:

$$S(T) = \begin{cases} [0.45 + (10\eta_2 - 4.5)\,T]S_{\max} & T \le 0.1\text{s} \\ \eta_2 S_{\max} & 0.1\text{s} < T \le T_g \\ \left(\frac{T_g}{T}\right)^{\gamma} \eta_2 S_{\max} & T_g < T \le 5T_g \\ [0.2^{\gamma}\eta_2 - \eta_1\,(T - 5T_g)]S_{\max} & 5T_g < T \le 6\text{s} \end{cases} \tag{3}$$

where $S(T)$ denotes the design response spectrum, T denotes the period, T_g characteristic period, S_{max} denotes the maximum design acceleration response spectrum, η_1 denotes the falling slope adjustment factor of response spectrum function, η_2 denotes the damping adjustment factor, and γ denotes falling slope decay index of response spectrum function.

The maximum design acceleration is determined by the following equation:

$$S_{max} = 2.25A \tag{4}$$

where A denotes the peak acceleration of horizontal ground motion under earthquake action E1 or E2.

According to the specification, the characteristic period is determined by two coefficients, including site category and characteristic period division. The peak acceleration of horizontal ground motion is determined by three coefficients: seismic intensity, seismic design level, and seismic fortification category. The shape factors, including the falling slope adjustment factor of response spectrum function, the damping adjustment factor, and the falling slope decay index of response spectrum function, are determined by the structural damping ratio. In addition, the vertical design acceleration response spectrum is obtained by multiplying the horizontal design acceleration response spectrum by 0.65.

2.3 Code for seismic design of buildings

In code for seismic design of buildings, the design response spectrum is determined by the following equation:

$$\alpha\,(T) = \begin{cases} [0.45 + (10\eta_2 - 4.5)\,T]\alpha_{max} & T \leq 0.1s \\ \eta_2\alpha_{max} & 0.1s < T \leq T_g \\ \left(\frac{T_g}{T}\right)^\gamma \eta_2\alpha_{max} & T_g < T \leq 5T_g \\ [0.2^\gamma\eta_2 - \eta_1\,(T - 5T_g)]\alpha_{max} & 5T_g < T \leq 6s \end{cases} \tag{5}$$

where $\alpha(T)$ denotes the design response spectrum, T denotes the period, T_g characteristic period, α_{max} denotes maximum horizontal seismic influence coefficient, η_1 denotes the falling slope adjustment factor of response spectrum function, η_2 denotes the damping adjustment factor, and γ denotes falling slope decay index of response spectrum function.

In the specification, the maximum horizontal seismic influence coefficient is determined by two coefficients, including seismic fortification intensity and seismic design level. The characteristic period is determined by two coefficients, including the site category and design ground motion division. The shape factors, including the falling slope adjustment factor of response spectrum function, the damping adjustment factor, and the falling slope decay index of response spectrum function, are determined by the structural damping ratio. In addition, the vertical design acceleration response spectrum is obtained by multiplying the horizontal design acceleration response spectrum by 0.65.

2.4 Standard for seismic design of nuclear power plant

In standard for seismic design of nuclear power plant, the design response spectrum is referred to as the response spectrum in the guidance of RG 1.60 of Nuclear Regulatory Commission in the United States, in which the coefficients of response spectrum are in Tables 1 and 2.

In the specification, the basic seismic intensity at 33 hertz, the structural damping ratio, and the ground motion direction are supposed to be specified.

Table 1. Amplification factors for horizontal response spectrum control points in RG 1.60.

Damping ratio (%)	Amplification factors for control points in the spectrum				
	Acceleration				Displacement
	A(33Hz)	B(9Hz)	C(2.5Hz)	D(0.25Hz)	D(0.25Hz)
0.5	1.0	4.96	5.95	0.74	3.20
2.0	1.0	3.54	4.25	0.57	2.50
5.0	1.0	2.61	3.13	0.47	2.05
7.0	1.0	2.27	2.72	0.43	1.88
10.	1.0	1.90	2.28	0.39	1.70

Table 2. Amplification factors for vertical response spectrum control points in RG 1.60.

Damping ratio (%)	Amplification factors for control points in the spectrum				
	Acceleration				Displacement
	A(33Hz)	B(9Hz)	C(2.5Hz)	D(0.25Hz)	D(0.25Hz)
0.5	1.0	4.96	5.67	0.49	2.13
2.0	1.0	3.54	4.05	0.38	1.67
5.0	1.0	2.61	2.98	0.32	1.37
7.0	1.0	2.27	2.59	0.29	1.25
10.	1.0	1.90	2.17	0.26	1.13

2.5 Code for seismic design of electrical installations

In code for seismic design of electrical installations, the design response spectrum for sites of II class is determined by the following equation:

$$
\alpha\,(T) = \begin{cases} [0.4 + (10\eta_2 - 4)\,T]\alpha_{\max} & T \leq 0.1\mathrm{s} \\ \eta_2\alpha_{\max} & 0.1\mathrm{s} < T \leq T_g \\ \left(\frac{T_g}{T}\right)^{\gamma} \eta_2\alpha_{\max} & T_g < T \leq 5T_g \\ [0.2^{\gamma}\eta_2 - \eta_1\,(T - 5T_g)]\alpha_{\max} & 5T_g < T \leq 6\mathrm{s} \end{cases} \tag{6}
$$

where $\alpha\,(T)$ denotes the design response spectrum, T denotes the period, T_g characteristic period, $\alpha_{\max}$ denotes maximum horizontal seismic influence coefficient, η_1 denotes the falling slope adjustment factor of response spectrum function, η_2 denotes the damping adjustment factor, and γ denotes falling slope decay index of response spectrum function.

For sites of non-II class, the spectrum is modified by the following equation:

$$
\alpha_S = \eta_3\alpha \tag{7}
$$

where α_s denotes the seismic influence factor of different site classes, and η_3 denotes the modification coefficient for the seismic influence factor.

In the specification, the maximum horizontal seismic influence coefficient is determined by seismic fortification intensity. The characteristic period is determined by three coefficients, including the site category, seismic design level, and design ground motion division. The shape factors, including the falling slope adjustment factor of response spectrum function, the damping adjustment factor, and the falling slope decay index of response spectrum function, are determined by the structural damping ratio. The site modification coefficient is determined by seismic fortification intensity and site category. In addition, the vertical peak ground motion acceleration is 65% of the peak horizontal acceleration.

3 COMPARISON OF EACH SPECIFICATION

The differences between seismic design response spectrum of highway and urban bridges include site factor, the shape of response spectrum function, design spectrum maximum calculation constant, and characteristic period. For site factor, the code for seismic design of urban bridges does not consider the direct influence of site effects on peak acceleration of response spectrum. For the shape of the response spectrum function, the highway bridge specification stipulates it to be a 3-stage form, while the urban bridge code stipulates it as a 4-stage form. For design spectrum, maximum calculation constant, highway bridge specification is considered more cautious as it modifies from 2.25 in the 2008 version to 2.5 in the 2020 version. For the characteristic period, highway bridge specification has a different period for vertical and horizontal ground motions, while urban bridge code considers them the same characteristic period.

The major differences between the seismic design response spectrum of bridges and buildings include the seismic fortification category and seismic design level. For the seismic fortification category, the building's seismic category is considered from the structure and layer shear amplification, but not when calculating the seismic force, so that the design response spectrum is not affected by the seismic fortification category. For the seismic design level, the bridge specifications follow the 2-level design criterion, while building codes usually follow the 3-level design criterion.

4 CONCLUSION

The paper introduces five major structural seismic design specifications in China, and a comparative study is conducted. The result of a comparative study of the paper is considered the reliable basis for structural seismic design and artificial ground motion simulation.

ACKNOWLEDGMENTS

The work was financially supported by grant No. 12032008 of the National Natural Science Foundation of China.

REFERENCES

Boore, D.M. (2003) Simulation of ground motion using the Stochastic Method. Pure and Applied Geophysics, 160(1): 635–676.

Deierlein, G.G., Krawinkler, H., Cornell, C.A. (2003) A framework for performance-based earthquake engineering. In: 2003 Pacific Conference on Earthquake Engineering. New Zealand. pp. 140–147.

Honda, R., Ahmed, T. (2011) Design input motion synthesis considering the effect of uncertainty in structural and seismic parameters by feature indexes. Journal of Structural Engineering, 137(1): 391–400.

Iyama, J., Kuwamura, H. (1999) Application of wavelets to analysis and simulation of earthquake motions. Earthquake Engineering and Structural Dynamics, 28(1999): 255–272.

Liu, S., Pan, C., Zhou, Z. (2018) Discussions on the response spectral solution and fitting of spectrum-compatible artificial seismic waves. Acta Seismologica Sinica, 40(4): 519–530.

Lu, D., Li, X., Wang, G. (2006) Global seismic fragility analysis of structures based on reliability and performance. Journal of Natural Disasters, 15(2): 107–114.

Lu, X., Wu, D., Zhou, Y. (2019) State-of-the-art of earthquake-resilient structure. Journal of Building Structures, 40(2): 1–15.

Ministry of Housing and Urban-Rural Development of the People's Republic of China. (2011) Code for seismic design of urban bridges. China Architecture & Building Press, Beijing.

Ministry of Housing and Urban-Rural Development of the People's Republic of China. (2013) Code for seismic design of electrical installations. China Architecture & Building Press, Beijing.

Ministry of Housing and Urban-Rural Development of the People's Republic of China. (2016) Code for seismic design of buildings. China Architecture & Building Press, Beijing.

Ministry of Housing and Urban-Rural Development of the People's Republic of China. (2019) Standard for seismic design of nuclear power plant. China Planning Press, Beijing.

Ministry of Transport of the People's Republic of China. (2020) Specifications for seismic design of highway bridges. China Communications Publishing, Beijing.

Qiu, C., Du, X. (2021) A state-of-the-art review on the research and application of self-centering structures. China Civil Engineering Journal, 54(11): 11–26.

Frontiers of Civil Engineering and Disaster Prevention and
Control – Yang & Rahman (Eds)
© 2023 The Author(s), ISBN: 978-1-032-31200-2

Research on monitoring technology of mud-water shield crossing the Weihe River based on semi-flexible inclined settlement benchmarks

Wenhao Liu
China Eleventh Engineering Bureau of Water Resources and Hydropower Co. LTD, Qinghai, China

Xiaohong Du*
Department of Human Settlements and Civil Engineering, Xi'an Jiaotong University, Xi'an, China

RengShan Li
China Eleventh Engineering Bureau of Water Resources and Hydropower Co. LTD, Qinghai, China

Dong Luo
Department of Human Settlements and Civil Engineering, Xi'an Jiaotong University, Xi'an, China

Cuigang Kong
Xi'an Rail Transit Group Co. LTD, Xi'an, China

Jiangxun Ma
Department of Human Settlements and Civil Engineering, Xi'an Jiaotong University, Xi'an, China

Gang Yang
Xi'an Diheng Engineering Supervision Co. LTD, Xi'an, China

Tian Lu, Jianhui Guo & Siyal Insaf Imdad
Department of Human Settlements and Civil Engineering, Xi'an Jiaotong University, Xi'an, China

ABSTRACT: With the development of China's transportation rail network, there are more and more cross-river tunnels. To solve the monitoring problem that the settlement benchmarks used for river bed deformation are prone to incline and result in large measurement errors during the construction of the shield tunnel under the river, this paper takes Xi'an Metro Line 1 under the double shield tunnel in the Weihe river engineering as an example and introduces the total station combined with semi-flexible inclined settlement benchmarks static observation methods in the application of the riverbed deformation monitoring for a long time. The observation results show that during the crossing of the leading line, the riverbed settlement above the left and right lines was always controlled within a reasonable range (within 6 mm), and the total settlement of the measuring points on the left line was slightly larger than that on the right line.

1 INTRODUCTION

During the construction of the shield tunnel, the tunnel is affected by subjective and objective factors and will produce certain deformation. If the amount of deformation exceeds the prescribed limit, it will affect the normal use of surrounding buildings, even endanger the safety of buildings, and bring huge losses to society and people's life (Qing 2017). In recent years, shield tunneling technology has been widely used in highway tunnels, urban subways, and other municipal

---

*Corresponding Author: 1960004587@qq.com

DOI 10.1201/9781003308577-27

public facilities. Accordingly, the problem of ground deformation caused by shield construction is gradually highlighted, such as in Ningbo Rail Transit Line 3 (Xu 2020), Shijiazhuang Subway Line 1 (Guo et al. 2020), Dalian Metro Line 5 (Jia et al. 2021), etc.

For settlement monitoring of shield tunnels, settlement monitoring technologies at present mainly include: (1) Leveling monitoring technology. It has been widely used in shield tunnels. For example, in the settlement monitoring of the second phase of the northern section of Shanghai Line 11[5], the leveling monitoring technology has high precision, but the measuring process is tedious, the measuring work efficiency is low, and the intensity is high. (2) Total station. For example, Chi Ma uses a total station to monitor the overall settlement of the ground surface without a prism, the measurement accuracy is guaranteed, and the measurement time is greatly shortened (Ma 2016). (3) Static leveling instrument. For example, Shaohong Bai used a static leveling instrument to monitor the deformation of the Beijing Metro. However, the system has long-term performance problems such as head loss. (4) 3D laser scanning technology. For example, Yongbi Li obtained the deformation inside the tunnel quantitatively by analyzing the change of centroid position (Li et al. 2021). The method is highly automated, efficient, and fast, but lacks precision. (5) Electric level monitoring technology. For example, Xiangqi Xu used this technology to monitor the operation of the Shanghai subway settlement (Xu 2009). This method has high resolution and good stability, but it is expensive and difficult to deploy on a large scale. (6) Distributed optical fiber sensing technology. For example, Jian Wu used distributed optical fiber sensing technology to carry out indoor model tests (Wu et al. 2021). This method has a large sensing distance and good stability, but it can only judge the tunnel settlement qualitatively but not quantitatively. Existing technology is difficult to combine low-cost, large-scale layout and accurate measurement. The monitoring technology combining total station and semi-flexible inclined settlement benchmark proposed in this paper has the advantages of good stability, strong anti-interference ability, and low cost, and can be arranged in a large area, so it has a broad prospect in the channel monitoring of shield tunnel. It can meet the accuracy requirements of river subsidence monitoring and effectively guide the project of Xi'an Metro Line 1 crossing the Weihe River.

The section from Zhonghua West Road Station to Angu Road Station (phase iii project of the Xi'an Metro Line 1) is the section (YCK4+159.00 to YCK3+494.00) crossing the Weihe River. The site is relatively open, with few ground buildings. There is foundation pit excavation in the site under construction of the Yuhong Block, which has great fluctuation and a great change of buried depth. Figure 1(a) Shows the field survey.

2 THE TOTAL STATION MONITORING SCHEME COMBINED WITH SEMI-FLEXIBLE INCLINED SETTLEMENT BE-NCHMARK

2.1 *Monitoring principle and data processing of total station combined with semi-flexible inclined settlement benchmark*

The project adopts the monitoring method combining the traditional total station and semi-flexible inclined settlement benchmark as the main monitoring method. Among them, the new settlement scale is the main measuring instrument. The design of semi-flexible inclined settlement benchmarks is mainly due to the following reasons: (1) Settlement benchmarks take different length dimensions mainly due to the different measuring points in different depths. (2) The reflecting prism is used to receive the light signal sent by the total station and reflect it back, and then the distance between the total station and the reflecting prism is obtained by the principle of phase displacement. (3) The combination of dial and pointer is used to measure the tilting deformation of the river bed. (4) The concrete blocks are placed in the base part for a counterweight to fix the settlement benchmarks and prevent them from being washed away by water. (5) In addition, the concrete blocks are pre-cast so that they can be easily transported and assembled on site. The marking intention of the semi-flexible inclined settlement benchmarks is shown in Figure 1(b).

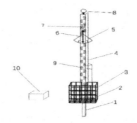

(a) The field survey	(b) The semi-flexible inclined settlement benchmarks

Figure 1.   Site and installation photos.

2.2   *Layout of monitoring points*

The total monitoring length of the shield crossing section in this project is 400 m, as shown in Figure 3. The first 140 m is flood area, the next 160 m is deep water area, and the last 100 m is shallow water area. Figure 3 shows the layout of measuring points through the Weihe River.

This monitoring scheme is mainly formulated according to Shaanxi Engineering Construction Standard-Xi'an Urban rail transit engineering monitoring technical specification. The monitoring scheme of the project is designed as follows: (1) The vertical deformation monitoring points of the river bed are mainly arranged along the axis of the pipeline (1# —12#). It needs to set up the settlement benchmarks of different length scales to achieve lower costs and more comprehensive and accurate monitoring. Specific arrangements of measuring points are shown in Figure 2. (2). The settlement benchmarks on the left and right lines are arranged with uniformly symmetrical measuring points in this scheme.

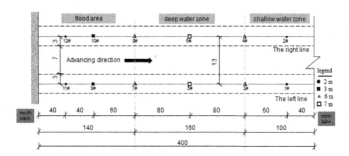

Figure 2.   Layout plan of measuring point.

3   MONITORING DATA PROCESSING

3.1   *Stability analysis of monitoring data*

The semi-flexible inclined settlement benchmarks are arranged on the bed of the Weihe River, and it is necessary to consider the influence of many factors on the slope of the settlement benchmarks and carry out a theoretical analysis. The influence of water flow on the tilt angle of the inclined settlement benchmarks can be calculated by the following formula, and the derivation schematic diagrams are shown in Figure 3.

In the construction process of the shield machine crossing the Weihe River, the elevation change of the measuring rod was measured by the total station instrument (The settlement measured on-site is H_1). The reason for the tilt of the settlement benchmarks is the action of water flow and the uneven support of mud in the river bottom; the measured settlement is greater than the actual amount of settlement by reading the pointer on the dial indicator scale value. The amount of settlement caused by the tilt of the settlement benchmarks is ΔH; thus, the amount of settlement (H_2) at the bottom of the mark caused by shield construction is calculated. According to the settlement amount (H_2), the settlement deformation of the river bed in shield construction is obtained. The specific conversion process is shown in Figure 3(a).

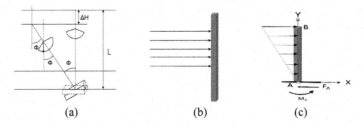

| (a) | (b) | (c) |

Figure 3. (a) Schematic diagram of conversion process (b) Formula derivation one (c) Formula derivation two.

Settlement value measured in the field (H_1) is:

$$H_1 = H_2 + \Delta H \tag{1}$$

Where, H_2 is the actual settlement value of the tilting scale, and ΔH is the settlement value caused by the tilting of the tilting scale.

In addition, the influence of water flow on the tilt angle of the inclined settlement benchmarks can be deduced by the following formulas, and the schematic diagrams are shown in Figures 5 and 6.

$$F = \rho v^2 S \tag{2}$$

Where, F represents the force of water flow acting on the rod per unit of time, ρ is the density of water, v is the velocity of water, and S represents the contact area between water and the rod.

$$q(x) = \rho \left(\frac{v}{h}y\right)^2 D \tag{3}$$

Where, h represents the depth of the water, y represents the distance from end A, and D represents the rod width.

The bending moment equation is:

$$M(x) = F_A y - M_A - \frac{\rho V^2 D y^4}{12h^2} = \frac{\rho V^2 Dh}{3}y - \frac{\rho V^2 Dh^2}{4} - \frac{\rho V^2 D y^4}{12h^2} \tag{4}$$

Assume that A is the fixed end, so

$$\Theta_B = \frac{\rho V^2 Dh^3}{10EI}\text{(clockwise)} \tag{5}$$

$$w_B = \frac{13\rho V^2 Dh^4}{180EI} \tag{6}$$

According to the above formula, a square steel tube with an elastic modulus of E = 200 Gpa, a cross-sectional area of 30 mm* 30 mm, and a thickness of 5 mm is taken. The water rate was 0.067 m/s, 0.05 m/s and 0.033 m/s respectively. The tilt angle (θ) and disturbance (w) of the settlement mark were calculated when the water depth was 2 m, 4 m, and 6 m. We considered the actual situation that segment A is not a fixed segment and the influence of other factors. Combined with the actual measured data, the influences of the flow rate and water depth on the tilt angle (θ) and disturbance (w) of the settlement benchmarks were given in Table 1.

Table 1. Table of the actual influence of flow velocity and water depth.

	h=2 m	h=4 m	h=6 m
$v = 0.067$ m/s	$\theta = 0.859°$ w $= 2.16*10^{-2}$m	$\theta = 0.988°$ w $= 3.46*10^{-2}$m	$\theta = 2.3°$ w $= 4.29*10^{-2}$m
$v = 0.05$ m/s	$\theta = 0.478°$ w $= 1.21*10^{-2}$m	$\theta = 0.583°$ w $= 1.93*10^{-2}$m	$\theta = 1.3°$ w $= 2.76*10^{-2}$m
$v = 0.033$ m/s	$\theta = 0.209°$ w $= 5.25*10^{-3}$m	$\theta = 0.317°$ w $= 8.4*10^{-3}$m	$\theta = 0.561°$ w $= 1.42*10^{-2}$m

According to the above table, it can be seen that the factors of flow velocity and water depth of the Weihe River in this project have a certain influence on the tilt of the settlement benchmarks, which should be taken into consideration when processing monitoring data. From the point of theoretical analysis, the settlement mark has good stability.

3.2 Analysis of river bed deformation

The Xianyang independent coordinate system is rotated 50° counterclockwise in the horizontal plane. We can establish the shield advancing direction as the longitudinal direction, and the vertical tunnel axis direction is taken as the horizontal plane coordinate system (positive from the right line to the left line). This study mainly analyzes the vertical settlement of the upper riverbed of the shield tunnel.

3.2.1 Settlement of riverbed
According to the data measured at the settlement monitoring point from July 5th to July 26th, the influence of the right line shield on the settlement of the riverbed above the leading line and the trailing line in the process of the right line shield crossing the Weihe River is shown in Figures 4 and 5. According to the data measured at the settlement monitoring point from July 29th to August 23rd, the influence of the left line shield on the bed settlement of the leading line and the trailing line during the right line shield crossing the Weihe River is shown in Figures 6 and 7.

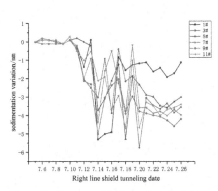
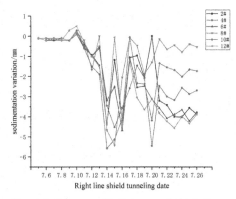

Figure 4. Diagram of influence on the left line.   Figure 5. Diagram of influence on the right line.

It can be seen from Figures 5 and 6 that before the right line shield reaches the south bank of the Weihe River, there is a certain deviation between the total thrust of the jack and the positive compressive stress of the soil, resulting in slight uplift of the river bed near the south bank. When the shield tunneling is directly below the monitoring point, the soil under the river bed is subjected to the shear action of the shell of the shield tunneling machine, and the stress state of the original soil changes, resulting in the settlement or uplift of the river bed in the process of crossing the river under the shield tunneling. The settlement deformation is controlled within 6 mm. When the shield tail emerges from the riverbed, the surface settlement of the riverbed is reduced by synchronous grouting technology, and the riverbed settlement gradually tends to be stable. The final settlement values of the left and right lines are controlled within 5 mm. In general, after crossing the leading line, the total settlement of the measuring points on the left line is slightly larger than that on the right line.

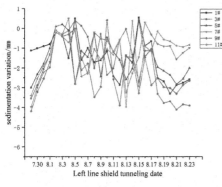

Figure 6. Influence diagram of the left line.　　　Figure 7. Influence diagram of the right line.

It can be seen from Figures 7 and 8 that during the whole tunneling process of the backline, the settlement deformation of the riverbed above the left line is basically controlled within 4.5 mm, and that above the right line is basically controlled within 4 mm. After crossing the backline, the settlement deformation of the riverbed above the left line decreases by 0.5 mm, and that above the right line decreases by 1 mm. This indicates that the backline will have a certain influence on the left line and the right line, which will reduce the settlement value of the riverbed above the double lines. Moreover, the influence of the trailing line on the right line is greater than that of the left line.

4 CONCLUSION

This study uses a traditional total station and a new semi-flexible inclined settlement benchmarks monitoring method. During Xi'an Metro Line 1 crossing the Weihe River, it can be implemented in monitoring the shield over double river bed settlement and lateral horizontal displacement of the high precision monitoring. The specific conclusions are as below:

(1) During the crossing of the leading line, the riverbed settlement above the left and right lines was always controlled within a reasonable range (within 6 mm), and the total settlement of the measuring points on the left line was slightly larger than that on the right line.

(2) During the crossing of the trailing line, it will have a certain influence on the left line and the completed right line, which will reduce the settlement value of the riverbed above the double line. Moreover, the influence of the trailing line on the right line is greater than that of the left line.

The shortcoming of this paper lies in that the reading is too dependent on manual reading. The next research plan is to add a data recorder based on the inclined settlement mark to realize semi-automation.

ACKNOWLEDGMENTS

This work was supported by the Sinohydro Bureau 11 Co., Ltd 20201225, National Natural Science Foundation of China (52078418), National Postdoctoral Science Foundation

(2019M653645), Project of Strategic Planning Department of the Ministry of Science and Technology (HXJC2019FG/072HZ), Science and technology project of Yulin City (CXY-2020-046), Key R & D plan of Shaanxi Province (2021GY-273).

REFERENCES

Baoxin Jia, Zongxian Gao, Pengfei Hui. (2021) Research on surface settlement caused by shield construction of upper soft and lower hard strata tunnel. J. Journal of Safety and Environment, 21(03):1083–1088.

Changjie Qing. (2017) Research on settlement monitoring method of long-distance tunnel across the Yangtze River. J. Surveying and mapping, 40(06): 263–266.

Chi Ma. (2016) Construction monitoring and data simulation analysis of Sijiazi Tunnel. J. highway, 61(07): 318–322.

Jian Wu, Tao Wang, Bo Li, Qiangdong Zeng, Yandong Yao. (2021) Research on settlement monitoring method of shield tunnel based on distributed optical fiber sensor. J. Henan Science, 39(06): 943–947.

Jianfeng Shi, Zhanjun Li, Bo Zhou. (2014) Application of leveling survey in settlement monitoring of subway tunnel. J. Surveying and mapping equipment, 16(04):61–64.

Shaohong Bai. (2003) Application of static level in deformation monitoring of Beijing city Railway. J. Chinese instrument, (11): 34–36.

Wei Xu. (2020) Application analysis of leveling technology in Ningbo Rail Transit Line3 Project. J. Beijing surveying and mapping, 34(01): 126–130.

Wen Guo, Pengfei Li, Yan Bao, Mingju Zhang, Yang Gao, Guoquan Wang, Rui Li, Xianfeng Duan. (2020) Monitoring channel bed deformation under shield tunnel based on high precision GPS. J. Journal of Beijing University of Technology, 46(3): 490–499.

Xiangqi Xu. (2009) Application of electric level settlement automatic telemetry system in subway monitoring. J. Geotechnical engineering, 12(02): 71–74.

Yongbing Li, Chengming Gao, Yingying Ma, Jie Wang, Yin Zhou. (2021) Application of 3D laser scanning technology in tunnel deformation monitoring and detection. J. Science Technology and Engineering, 21(12): 5111–5117.

*Frontiers of Civil Engineering and Disaster Prevention and
Control – Yang & Rahman (Eds)
© 2023 The Author(s), ISBN: 978-1-032-31200-2*

Finite element analysis of U-shaped concrete column under vertical load

J.Y. Li*, J.P. Guo*, X.Y.Wang* & G.C. Hu*
Changchun Institute of Technology, Changchun, China

ABSTRACT: In order to explore the influence of different design parameters on the bearing capacity of U-shaped concrete columns under vertical load, the finite element software ABAQUS is used to simulate two groups of six U-shaped concrete columns under vertical load. And the influence of different parameters (wall thickness, height) on the bearing performance of the specimen is compared and analyzed. The results show that under the same other conditions, with the increase in wall thickness, the bearing capacity of the U-shaped concrete column has been significantly improved; the increase in height significantly reduces the bearing capacity and ductility of the specimen. The research in this paper provides a theoretical basis for the experimental study and engineering application of U-shaped special-shaped concrete columns.

1 INTRODUCTION

With the rapid development of the social economy, many high-rise buildings have been widely used in housing construction. When the building adopts square and rectangular columns, the size of the column is often larger than the wall thickness, so the column highlights the wall surface and corner (Wang 2021). These protruding parts are usually difficult to deal with, which brings great trouble for decoration and also affects the spatial availability of the house in daily use. To better solve this problem, special-shaped columns were created. Special-shaped column refers to frame columns with L-shaped, T-shaped, and cross-shaped sections, whose sections can be determined according to the design position of the wall. Special-shaped column frame structure is formed when part or all of the structure adopts a special-shaped column. Special-shaped column frame structure is a special form of the frame structure. Its mechanical and deformation performance is similar to the ordinary frame, but it can reduce the weight of the structure, improve the natural vibration period of the structure, and effectively reduce the earthquake. When an earthquake occurs, it has good deformation resistance and seismic recovery performance (Liu 2016). In addition, due to the large cross-sectional stiffness of special-shaped columns, it is particularly suitable for building structures with large storey heights. Tianjin started the research and preliminary application of reinforced concrete special-shaped columns in the 1970s. The concept of a special-shaped column frame structure system was first proposed by the Ministry of Housing and Urban-Rural Development in 1998. Over the years, experts and scholars at home and abroad have studied special-shaped columns. Chen Zongping (Chen 2006) carried out axial compression and eccentric compression tests on four truss steel T-shaped concrete special-shaped columns. The results show that the bearing capacity of truss steel T-shaped concrete special-shaped columns under axial compression is significantly improved compared with that of ordinary reinforced concrete special-shaped columns. Under eccentric action, its mechanical characteristics are similar to those of ordinary reinforced concrete special-shaped columns. Zhang Jianshan (Zhang 2019) and other scholars used the fiber model method to analyze the influence of concrete strength, steel strength, wall width-thickness ratio, loading direction, aspect ratio, and slenderness ratio on the stability bearing capacity of T-shaped

---

*Corresponding Authors: 429838648@qq.com, gjpgjp1002@163.com, 1241957603@qq.com and 2048839443@qq.com

DOI 10.1201/9781003308577-28

concrete-filled steel tubular axial compression column. They proposed a simplified calculation method for the stability bearing capacity of T-shaped concrete-filled steel compression. BEYER et al. (K, A, N 2008) completed quasi-static cyclic tests on two U-shaped reinforced concrete walls, which provided a theoretical basis for studying the seismic performance of U-shaped walls and formulating appropriate design guidelines. The bottom of a local subway operation depot is an operation workshop. The upper part of the building is a 16-story residential building, and the bottom operating workshop is 10.2 m high. The upper structure has a large load. In order to solve the safety problems of components with high height and large load, the super-large section U-shaped reinforced concrete column composed of multiple shear walls is adopted at the bottom of the project. In order to study the safety of special-shaped columns at the bottom of the project, the research group carried out an in-depth study of U-shaped concrete columns. In this paper, a 1:5 scale model is established by finite element software to analyze the stress and deformation performance of U-shaped concrete columns under vertical load, which provides a basis for the subsequent research and engineering application of the subject.

2   INTRODUCTION OF ABAQUS SOFTWARE

Finite element analysis is a commonly used method to simulate engineering applications, which plays an important role in different industries and fields. The most obvious advantage of the software is that it has high computational efficiency in linear analysis. In the calculation process, the software can automatically select the load increment suitable for the model by analyzing the characteristics of different models and adopting reasonable convergence criteria. In addition, the software can also automatically adjust the parameters in a timely manner, so that the solution results can meet various accuracy requirements and greatly improve the analysis efficiency. The analysis module of ABAQUS is divided into two parts: ABAQUS / Standard suitable for static analysis and ABAQUS / Explicit for dynamic analysis. This paper uses ABAQUS/Standard for static analysis.

3   FINITE ELEMENT ANALYSIS PROCESS

3.1   *Test conditions*

In order to investigate the influence of thickness and specimen height of U - shaped column cantilever on bearing capacity, the research group designed six U-shaped concrete column specimens according to the 1:5 scale model for finite element simulation. The web height of the U-shaped column is 800mm, the thickness is 150mm, and the width of the cantilever is 875 mm. In the first group of specimens, the thickness of the U-shaped section column is 150 mm, 200 mm, and 250 mm, respectively. In the second group of specimens, the thickness of the U-shaped cross-section column cantilever flange was 150 mm, and the heights of specimens were 1900 mm, 2200 mm, and 2500 mm, respectively. The specific parameters of the two groups of specimens are shown in Table 1.

Table 1.   U-shaped column finite element condition table.

Working condition	Test specimen number	Reinforcement ratio	Stirrup reinforcement ratio	Wall thickness/mm	Height/ mm
	1	0.29%	0.58%	150	2500
Group 1	2	0.29%	0.58%	200	2500
	3	0.29%	0.58%	250	2500
	4	0.29%	0.58%	150	1900
Group 2	5	0.29%	0.58%	150	2200
	6	0.29%	0.58%	150	2500

3.2 Properties and constitutive relations of concrete materials

The determination of material constitutive relation is an important issue in finite element analysis. The constitutive relation models of concrete summarized by domestic scholars are as follows: the linear elastic and inelastic constitutive models based on elastic theory; the elastic-plastic and elastoplastic hardening constitutive models based on classical plasticity theory; the internal time theoretical model obtained from the constitutive equation of viscoplastic materials; the damage theory model and the model obtained by combining different theories (Guo 2017). ABAQUS / Standard module provides elastic-plastic fracture model and plastic damage model of concrete. This analysis uses the plastic damage model. When setting the plastic properties of concrete, the stress-strain relationship data of concrete under uniaxial tension and compression should be input. The stress-strain curve of concrete under uniaxial tension and compression is shown in Figure 1.

C50 is used for the concrete of the U-shaped concrete column in this simulation. According to the 'Code for design of concrete structures' GB50010-2010, the mechanical properties of C50 concrete material parameters are determined, as shown in Table 2.

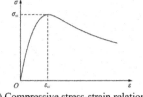
(a) Compressive stress-strain relationship

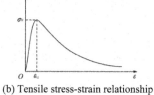
(b) Tensile stress-strain relationship

Figure 1.   Stress-strain curve of concrete

Table 2.   Mechanical properties of C50 concrete.

Concrete grade	Elastic modulus $E_c/N/mm^2$	Poisson ratio λ	Standard value of tensile strength $f_t/N/mm^2$	Standard value of compression strength $f_t /N/ mm^2$	Compression stiffness recovery factor	Recovery factor of tensile stiffness
C50	2.69×10^4	0.2	1.89	23.1	0.9	

3.3 Properties and constitutive relation of steel bar

The steel used in construction engineering can be regarded as an ideal homogeneous material. The mechanical properties under tension and compression show little difference. Especially in small deformation, the properties of steel are approximately isotropic. The U-shaped concrete columns used in this experiment are arranged with long horizontal bars, stirrups, tie bars, and vertical longitudinal bars. The ideal elastic-plastic model is adopted in the constitutive relation of steel bars in finite element analysis. It belongs to the ideal elastic state before yielding, yielding strength is 360 Mpa. The material is completely plastic after reaching the yield strength. Its mechanical properties are shown in Table 3.

Table 3.   Mechanical properties of reinforcement.

Steel grade	Design tensile strength $f_y/N/mm^2$	Elongation $f_c/N/mm^2$	Elastic modulus $E_c/N/mm^2$	Poisson ratio λ
HRB400	360	7.5	2.00×10^5	0.25

3.4 *Modeling process*

Using finite element analysis software ABAQUS to create components for a 1:5 scale U-shaped concrete column. According to the characteristics of the reinforced concrete structure itself and referring to the experience in the finite element analysis results (Guo 2017), the displacement coordination modeling method is adopted in this analysis. Each component is assembled orderly by the assembly module. Then assembly positioning is carried out to form the assembly body. Finally, through the interaction module, the long bars, stirrups, tie bars, and longitudinal bars are bound as a whole and embedded into the whole concrete component. The assembly of finite element model components is shown in Figure 2.

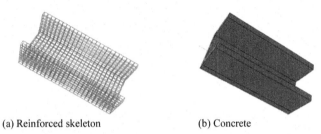

(a) Reinforced skeleton   (b) Concrete

Figure 2.   Finite element components.

3.5 *Unit type selection*

In the modeling process, the parameters of the component are selected according to the physical properties and geometric properties of the material. The solid element is the most widely used in ABAQUS. The solid element can be used in any shape and any load numerical model, and it is more convenient to connect with the surface of other elements. Assuming that concrete columns are homogeneous, three-dimensional solid elements (C3D8R) are selected, the unit is six node entity units, and three-dimensional linear truss elements (T3D2) are used in all kinds of steel bars.

3.6 *Element grid division*

In finite element analysis, the division of element mesh directly affects the accuracy of the calculation results and also has an important influence on the calculation efficiency. If the mesh division of the model unit is not detailed enough, it will deviate from the actual situation because the error is too large and even directly lead to errors. If the division of the model cell grid is too detailed, it will greatly prolong the calculation time of the model and cause a waste of resources for the computer. In the process of model generation, the network partition density is finally determined through multiple grid experiments. The specific meshing results are shown in Figure 3.

Figure 3.   Meshing of finite element simulation.

3.7 *Boundary conditions and loading methods*

In the process of finite element analysis, the simplification of the boundary conditions of the model is an essential step. In the finite element simulation analysis, under the premise of ensuring the convergence of the calculation results, the boundary conditions of the model are as far as possible in line with the real situation. Combined with the subsequent experiments on this subject, the boundary conditions of finite element analysis are shown in Figure 4. The load control method was adopted in this analysis. The vertical downward axial pressure is applied on the top of the column to act on the centroid of the specimen. The initial load increment step is 0.001, and the maximum is 0.01 (Zhou, Xiao 2020).

Figure 4. Simplified boundary diagram of the reinforced concrete column model.

4 ANALYSIS OF FINITE ELEMENT RESULTS

The mechanical properties of U-shaped concrete columns under vertical load were obtained by finite element software analysis. The deformation is shown in Figures 5 and 6. According to the comparative analysis of the factors that affect the deformation of U-shaped columns under Tables 1 and 2, the influence of wall thickness and height of U-shaped columns on the bearing capacity of U-shaped reinforced concrete is obtained.

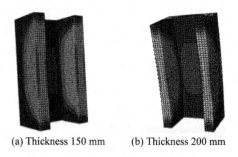

(a) Thickness 150 mm   (b) Thickness 200 mm

Figure 5. Deformation diagram of group 1.

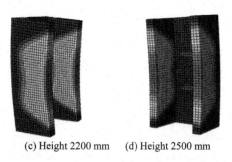

(c) Height 2200 mm   (d) Height 2500 mm

Figure 6. Deformation diagram of group 2.

4.1 *Effect of wall thickness and height variation on U-shaped reinforced concrete columns*

4.1.1 *Bearing capacity*

Figure 7 is the load-displacement curve of the two groups of specimens. It can be seen from the figure that: At the initial stage of loading, the vertical displacement of the specimen at the stage of dry projectile increases with the increase of load, and the corresponding curve is approximately linear. By comparing the ultimate load in the two graphs, it can be concluded that: (1) Appropriately increasing the thickness of the cantilever can significantly improve the bearing capacity of the specimen. (2) Increasing the height of the specimen leads to a significant decrease in the bearing capacity of the specimen. After reaching the ultimate load, the curve decreased, the first group of specimens decreased gently, and the second group of specimens decreased steeply.

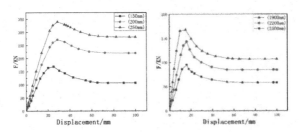

Figure 7. Load-displacement curves of two groups of specimens.

4.1.2 *Forms of destruction*

In this analysis, the deformation and failure characteristics of specimens under axial compression are basically the same under different wall thicknesses. The deformation and failure characteristics of specimens were analyzed by the failure factor of concrete and the equivalent plastic strain of reinforcement, as shown in Figure 8.

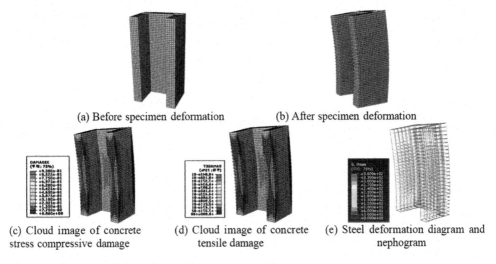

(a) Before specimen deformation   (b) After specimen deformation

(c) Cloud image of concrete stress compressive damage   (d) Cloud image of concrete tensile damage   (e) Steel deformation diagram and nephogram

Figure 8. Component deformation and damage, stress nephogram.

From Figure 8(a) and (b), it can be seen that the central deformation of U-shaped concrete column under axial compression is large, and obvious buckling failure occurs. There is no obvious deformation at the upper and lower ends, which conforms to the deformation characteristics of hinged components at both ends.

217

Figure 8(c), (d) shows that U-shaped concrete column under axial load, the cantilever part is weak, compressive failure occurs. The concrete is crushed and destroyed uniformly, and the damage factor is as high as 0.93. The tensile stress at the connection between the suspension wing and the web is large, so concentrated tensile failure occurs. The damage factor reached 0.88.

It can be seen from Figure 8 (e) that the longitudinal reinforcement and horizontal stirrup in the middle of the column show a red area, indicating that the stress value is large, reaching the yield strength of 360 MPa. Under the action of axial load, the stress of longitudinal reinforcement is all red, indicating that the stress of longitudinal reinforcement is uniform to reach the yield stress. The horizontal stirrup stress of the upper and lower ends of the specimen is small, and the middle stirrup stress is large. It can be seen from the deformation diagram that the specimen bulged outward in the middle of the column, and the horizontal stirrup yielded.

5 CONCLUSIONS

(1) In this paper, the axial compression of the U-shaped reinforced concrete column is simulated and analyzed by finite element analysis software ABAQUS, and the load-deformation curve of the compression specimen is obtained. The mechanical properties and deformation characteristics of two groups of U-shaped concrete columns were compared. The results show that under the same other conditions, the bearing capacity of U-shaped concrete columns has been significantly improved with the increase in wall thickness. However, the increase in height leads to a decrease in the bearing capacity and ductility of concrete columns.
(2) The failure mode of U-shaped reinforced concrete column under axial load is that concrete is crushed, and the longitudinal reinforcement between stirrups is laterally convex. The failure mode of the whole column is that the middle convex is sinusoidal. After the upper and lower ends are stressed, the force is effectively transferred from the upper and lower ends to the middle part, and the load is concentrated in the middle of the column. So, the upper and lower parts of the damage are small, the middle part of the sinusoidal damage.
(3) Through the research, the U-shaped section reinforced concrete column is suitable for the specific condition of high layer height and large load, and the safety performance also meets the requirements. It can be widely used in similar projects in the future. In addition, when the load is large, the thickness of the suspended wing can be appropriately increased to meet the requirements of improving the bearing capacity.

6 OUTLOOK

ABAQUS software was used to simulate the axial compression of the specimen, and some research results were obtained for the U-shaped concrete special-shaped column proposed in this paper. However, due to the complex environmental conditions in the practical application of components, this study still has many deficiencies. For example, special-shaped columns in practical engineering work will not have absolute axial force, and out-of-plane deformation cannot be completely avoided. This should be reasonably considered in subsequent studies. In the following work, the seismic performance of the component should also be analyzed.

REFERENCES

Beyer K & Dazio A & Priestley M J.N (2008). Quasi-static cyclic tests of two U-shaped reinforced concrete walls[J]. Journal of earthquake engineering, 12(7): 1023–1053.
Chen Z.P (2006). Experimental study on normal section bearing capacity of trussshaped steelreinforced concrete columns [J]. Journal of Xi'an University of Architecture and Technology (Natural Science Edition), 109–114.

GB 50010-2010 'Code for design of concrete structures' [S].

Guo Y.F (2017). The seismic performance of T-section concrete-filled steel tubular special-shaped columns based on ABAQUS [D]. North China University of Science and Technology.

Liu J.J (2016). Experimental study on axial compression and eccentric compression bearing capacity of L-shaped corroded reinforced concrete columns [D]. Harbin Institute of Technology.

Wang W.P (2021). Study on compressive mechanical properties of L-shaped reinforced concrete-filled steel tubular columns with unequal legs [D]. East China Jiaotong University.

Zhang J.S (2019). Analysis of Bearing Capacity of T-shaped Concrete-filled Steel Tubular Column under Axial Compression [D]. Southwest Jiaotong University.

Zhou Y.Y & Xiao Y.M (2020). Finite Element Analysis of Rear Steel Tube Reinforced Concrete Column Based on ABAQUS[J]. Anhui Architecture, 27(09).

Frontiers of Civil Engineering and Disaster Prevention and
Control – Yang & Rahman (Eds)
© 2023 The Author(s), ISBN: 978-1-032-31200-2

Study on influence of compaction degree on subgrade deformation

Pingsheng Dong*, Xiang Wang & Hu Wang
Jinan Urban Construction Group Co. Ltd., Jinan, Shandong, China

ABSTRACT: The degree of compaction has a significant impact on the deformation and stability of the roadbed. In this paper, a numerical simulation model is established based on the actual situation of a highway project, and the filling rate of the horizontally layered roadbed is analyzed under the condition of 2m/month, the loose paving thickness of 25cm, and filling height of 8 meters. The influence of compaction degree on subgrade deformation is studied. The step-by-step loading analysis using ABAQUS software shows that the horizontal displacement of the subgrade mainly occurs in the soil layer below the toe of the slope; the maximum horizontal displacement to the outside of the embankment is 19.7 mm for expanding to the surrounding area; the maximum settlement of the graded filling occurs in the subgrade filling.

1   INTRODUCTION

The degree of compaction has a significant impact on pavement deformation and roadbed stability, so it is necessary to carry out relevant impact research on high-grade highway projects (Cao et al. 2009; Thakurpk et al. 2013; Wichtmann et al. 2005). For this paper, a finite element calculation model is established based on the actual situation of a highway project, the deformation of the subgrade height is 8 m, and the subgrade slope ratio is 1:1.5. Table 1 shows the thickness of the main subgrade soil compaction distribution. Since the subgrade cross-section is a symmetrical structure, half of the subgrade structure is taken to establish a finite element model in the calculation. Considering the influence of boundary effects, the calculated width of the subgrade is 120 m, and the depth is 40 m. The distribution of foundation soil layers in the analysis model is set with reference to typical strata conditions, and the specific thickness of each soil layer is shown in Table 1.

2   BASIC ASSUMPTIONS OF THE MODEL

The embankment is analyzed by unit thickness according to the plane strain problem. The element type is CPE4, which is a plane four-node type. Each node has three degrees of freedom: the displacements in the x and y directions. The mesh element of the foundation part is CEP4P, which is a four-node plane quadrilateral element, bilinear deformation, and bilinear hole pressure.

The materials of each layer are homogeneous, continuous, and isotropic, and the Mohr-Coulomb model is used for the soil constitutive relationship.

Pavement load and traffic load equivalent to a uniformly distributed load of 20kPa act on the subgrade.

The boundary constraints of the model are: the left and right sides of the model constrain the horizontal displacement; the bottom of the model is a fixed constraint; the groundwater level line is located 14.5 m below the foundation surface. Considering the foundation drainage consolidation, the groundwater level remains unchanged, the foundation surface is the permeable boundary, and other boundaries are impervious boundaries.

---

*Corresponding Author: 376338859@qq.com

                                                  DOI 10.1201/9781003308577-29

The calculation starts from clearing the road surface, clearing the displacement field inside the foundation, and retaining the stress field as the initial stress state for the calculation of filling the roadbed to ensure that the filling of the roadbed generates the settlement value of the calculation result. According to the specification, the calculation period of the roadbed is 15 years after construction. The filling rate of the horizontally layered subgrade is 2m/month. In the subgrade loading process, each layer is filled with 25cm, and the filling height is 8 meters. The layered filling process of the expressway is realized by the transient analysis step of ABAQUS. The load is applied at the beginning of the analysis step and remains unchanged.

Table 1. Subgrade compaction degree.

Cut and fill type		Depth below the bottom of the pavement (cm)	Compactness (%)
Fill roadbed	Upper roadbed	0~30	>96
	Lower roadbed	30~120	≥96
	Upper embankment	120~190	>94
	Lower embankment	Below 190	>93

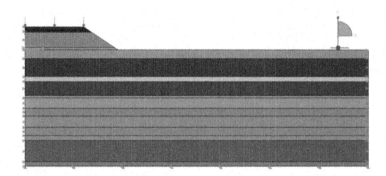

Figure 1. Finite element model of 8-meter-high subgrade.

3 MECHANICAL PARAMETERS OF FOUNDATION AND SUBGRADE

The calculation parameters of the foundation of this project are shown in Table 2, and the measured mechanical parameters corresponding to different compaction degrees are shown in Table 3, where Es is the compressive modulus; μ is the Poisson's ratio; c is the cohesion Force; φ is the angle of internal friction; k_v, k_h are the vertical and horizontal permeability coefficients.

Table 2. Distribution of foundation soil layers.

Strata	Thickness (m)	Es (Mpa)	Internal friction angle (°)	Cohesion (kN/m^2)	k_v (*10^{-6}cm/s)	k_h (*10^{-6}cm/s)
① Layer of silty clay	3	8.52	21.4	28	70	70
② Layer of silty clay	6.2	9.07	23.1	29.7	82	63
② Layer of silt	1.7	10.27	26.1	28.1	73	50
② Layer of fine sand	4.7	18	0	32	910	770
③ Layer of fine sand	4.4	22	0	35	880	780

(*continued*)

Table 2. Continued.

Strata	Thickness (m)	Es (Mpa)	Internal friction angle (°)	Cohesion (kN/m²)	k_v (*10⁻⁶cm/s)	k_h (*10⁻⁶cm/s)
④ Layer of silt	2.7	14.3	29.3	29	90	60
④ Layer of silt	4	13.1	21.1	37.3	82	61
④ Layer of silt	2.8	14.3	29.3	29	90	60
④ Layer of silt	1.7	13.1	21.1	37.3	82	61
⑤ Fine sand	7.3	26	0	37	860	770
⑥ Layer of silty clay	1.5	14.5	29.1	29.9	89	67

Table 3. Calculation parameters of subgrade soil layer.

Soil layer	Es (Mpa)	μ	c (kN/m²)	φ
Area 93	27	0.25	29.1	36.7
Area 94	33	0.25	25.6	39.8
Area 96	46	0.25	33.6	36.2

4 CALCULATION RESULTS

Figures 2 and 3 show the results of horizontal displacement and vertical displacement 15 years after the completion of subgrade filling. It can be seen from the cloud map of horizontal displacement that the horizontal displacement mainly occurs in the soil layer below the toe of the slope, and the maximum horizontal displacement to the outside of the embankment is 19.7 mm, expanding to the surrounding area. Therefore, in order to prevent the stress concentration at the toe of the slope and cause the horizontal slip of the subgrade during construction, the subgrade slope should not be too large. For high-fill roadbeds, to ensure the roadbed's stability, it is necessary to slow down the slope or support the side slope to prevent the slope from collapsing and sliding.

It can be seen from the displacement cloud of the vertical displacement map that the maximum settlement of the graded filling does not occur on the top surface of the fill, but in the middle and lower parts of the fill. This is because when the lower subgrade is filled, the upper subgrade has not yet been filled. No compression occurs, but when the upper subgrade is filled, the load on the lower subgrade increases gradually with the increase of the filling height. The settlement of the expressway is composed of foundation settlement and subgrade settlement. Reducing subgrade settlement can be achieved in two ways: one is to strengthen the foundation to reduce the settlement of the

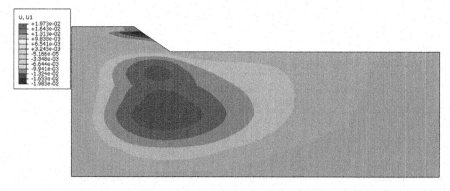

Figure 2. Horizontal displacement cloud map of 15 years after the completion of roadbed filling.

foundation, such as replacement filling method, drainage consolidation method, reinforcement, deep compaction method, heap load preloading method, etc.; the other is to control the amount of compression of the subgrade itself, such as improving the subgrade compaction, reinforcement treatment, etc.

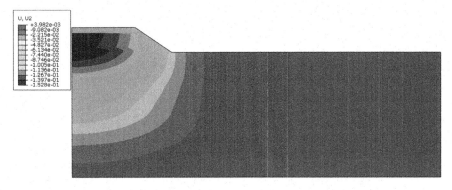

Figure 3.   Vertical displacement cloud map of 15 years after the completion of roadbed filling.

5   CONCLUSIONS

During rolling construction, the compaction effect decreases with the increase of depth. After the compaction construction equipment is determined, in order to ensure that the compaction degree of the soil under the paving layer meets the requirements, the loose paving thickness should be controlled not to exceed the maximum value of the road roller. If you need to consider controlling the construction period, the effective compaction depth is recommended to use an impact vibratory roller for compaction to increase the loose thickness and shorten the construction period.

REFERENCES

Cao Zhigang, Cai Yuanqiang, Xu Changjie. Dynamic response of pavement subjected to moving traffic load[J]. Journal of Zhejiang University: Engineering Science, 2009, 23(4):777–781. (In Chinese))
Thakurpk, Vinod J S, Indraratna B. Effect of confining pressure and frequency on the deformation of ballast[J]. Geotechnique, 2013, 63(9): 786–790.
Wichtmann T, Niemunis A, Triantafyllidis T. Strain accumulation in sand due to cyclic loading: drained triaxial tests[J]. Soil Dynamics and Earthquake Engineering, 2005, 25(12): 967–979.

*Frontiers of Civil Engineering and Disaster Prevention and
Control – Yang & Rahman (Eds)
© 2023 The Author(s), ISBN: 978-1-032-31200-2*

Influence of ground settlement on stress and deformation of CRTS slab ballastless track

Liu-Hua & Li-Bin
School of Civil Engineering, Lanzhou Jiaotong University, Lanzhou, China

Han-Tilei
Gansu Road and Bridge Construction Group Co., Ltd., Lanzhou, China

Fan-Xianghui & Zhang-Yongfu
School of Civil Engineering, Lanzhou Jiaotong University, Lanzhou, China

ABSTRACT: In order to study the influence of uneven ground settlement on the stress and deformation of CRTs III slab ballastless track, a three-dimensional finite element model of CRTs III slab ballastless track subgrade ground is established in this paper. The stress and deformation of CRTs III slab ballastless track under the condition of the uneven ground settlement are analyzed by using this model. The results show that when the uneven ground settlement occurs, the ballastless track structure will follow the settlement, the bonding between the layers of the ballastless track structure is good, and there is a gap between the base plate, the subgrade bed, and the ground. At the beginning and end of the ground settlement, the upper surface of the track structure is in a tension state, and at the center of the ground settlement, the upper surface of the track structure is in a compression state. When the amplitude of ground settlement increases, the settlement displacement and longitudinal stress of ballastless track structure increase linearly, and the void between ballastless track and subgrade bed, subgrade bed, and ground intensifies. When the land settlement wavelength increases, the settlement displacement of the ballastless track structure first increases and then tends to be stable, and the longitudinal stress of the track structure first increases and then decreases. When the land settlement wavelength is 20m, the longitudinal stress of the track structure reaches its peak.

CLC number: U211; U213   Document identification code: A

1 INTRODUCTION

The ballastless track structure form has become the main track structure form of high-speed railways and has been widely used worldwide due to its advantages of high smoothness and high stability (He 2005). However, the ballastless track is very sensitive to the settlement deformation of the lower foundation. When the high-speed railway foundation settles and deforms, it will affect the safety, stability, and comfort of the train operation, aggravate the vibration of various parts of the locomotive and also increase the stress and deformation of the track structure and the destruction (Guo 2018; Jia 2015). There have been discussions at home and abroad around the foundation settlement of high-speed railways, and it is required to limit the settlement to a certain range (Chen et al. 2008; Cheng 2015; Lu et al. 2014; Wang & Jiang 2013; Yang 2013). Hunt (1997) studied the uneven settlement in the transition area between the subgrade and the abutment and considered the changes in the stiffness and geometry of the foundation in the transition area of the road and bridge. Namura and Suzuki (2007) analyzed the transition from slab track to ballast track by combining field measurement and numerical simulation and evaluated the measures to prevent uneven settlement. Cai Xiaopei (Cai et al. 2014) established a ballastless track-subgrade-ground finite element model and studied the influence of the settlement amplitude and settlement range of uneven ground

DOI 10.1201/9781003308577-30

settlement on the smoothness of the double-block ballastless track; Xu Qingyuan (Xu et al. 2012; 2013; 2012) et al. based on the train-track coupling dynamics theory, under the effect of the uneven settlement of the subgrade, a comparative study of the CRTS-I type slab ballastless track and the CRTS-II type slab ballastless track on the subgrade. The dynamic characteristics of the train-ballastless track coupling system on the subgrade and the high-speed running on the double-block ballastless track and suggestions on the settlement limit of the subgrade were put forward. Zhang Keping et al. (Chen 2020; Zhang et al. 2020; Zhang et al. 2020; Zhong et al. 2021) established the Based on the three-dimensional space model of the overall track bed track-subgrade of urban rail transit, the track deformation and the dynamic characteristics of the vehicle-track system under the condition of uneven settlement of the subgrade are studied.

Most of the existing studies focus on the stress and deformation of the ballastless track under the uneven settlement of the subgrade, driving safety, settlement limit, etc. Still, the impact of the uneven ground settlement on the stress and deformation of the CRTS III slab ballastless track is still lacking research. Therefore, this paper establishes a spatial finite element beam-body model of CRTS III slab ballastless track, subgrade, and ground considering the self-weight load of the track and analyzes the influence of different ground subsidence conditions on the force and deformation of CRTS III slab ballastless track. The maintenance and repair of the ballastless track line provide some reference.

2 THE ESTABLISHMENT OF THE MODEL

2.1 *CRTS III slab ballastless track-subgrade model*

This paper uses the CRTS III slab ballastless track as the research object, and Abaqus is used to establish the CRTS III slab ballastless track-subgrade-ground three-dimensional finite element model, as shown in Figure 1. CRTS III slab ballastless track comprises C60 steel rail, WJ-8 type fastener, track plate, self-compacting concrete, and base plate from top to bottom. In the finite element model, the rail is regarded as an elastic body and is simulated with Eular beam elements; the fastener system is simulated with Cartesian elements as a solid unit.

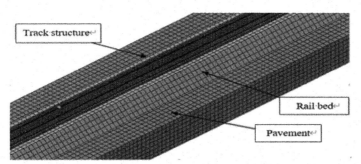

Figure 1.   Ballastless track subgrade ground finite element model.

The contact relationship between the track slab, the self-compacting concrete, and the base slab is set as a binding constraint. Considering the possible gaps and voids between the base slab and the subgrade surface and between the soil layers, the base slab, and the subgrade surface. The surface-surface contact method is used to simulate the surface-to-surface contact between the soil layers, and the tangential friction coefficient is 0.3 (Chen 2018).

The boundary conditions of the model are set as follows: the longitudinal displacement and lateral displacement of the upper track structure of the model are constrained, and free deformation is allowed in the vertical direction. Therefore, no constraint measures are taken in the vertical direction. The bottom of the foundation is completely fixed. The ground settlement imposed by this is achieved by setting the forced displacement boundary conditions on the soil surface.

Table 1. Track model parameters.

Structure	Thickness /m	Elastic Modulus /MPa	Poisson's ratio	Density /(kg/m^3)
Track plate	0.2	3.6×104	0.2	2500
Self-compacting concrete	0.1	3.25×104	0.2	2500
Base plate	0.3	3.25×104	0.2	2500
Base bed surface	0.4	180	0.25	2300
Base of bed	2.3	130	0.25	1800
Bottom fill	3.3	50	0.25	1700

2.2 *Model of uneven ground settlement*

At present, for the land subsidence curve under the influence of various factors, the cosine subsidence curve is mostly used in my country (Li et al. 2008), and the subsidence curve is shown in Figure 2. The expression of the cosine type uneven settlement model is:

$$y = \frac{f_0}{2}\left[1 - \cos\left(\frac{2\pi(z - z_0)}{l_0}\right)\right]$$

Where, f_0 is the amplitude of cosine subgrade settlement; Z is the location where uneven settlement occurs; is the initial position of settlement; is the wavelength of cosine subgrade settlement; is the amplitude of uneven subgrade settlement.

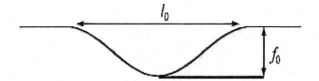

Figure 2. A model of angular differential settlement.

3 THE INFLUENCE OF LAND SUBSIDENCE AMPLITUDE

In order to study the influence of land subsidence amplitude on track force and deformation, assuming that the land subsidence wavelength is constant at 20m, according to the different land subsidence amplitudes, six working conditions (amplitude/wavelength) are selected for calculation, as shown in Table 2.

Table 2. Working condition of land subsidence amplitude (mm / m).

Working condition 1	Working condition 2	Working condition 3	Working condition 4	Working condition 5	Working condition 6
5/20	10/20	15/20	20/20	25/20	30/20

3.1 *Influence of land subsidence amplitude on track structure deformation*

Figure 3 shows the vertical deformation of the superstructure under the uneven settlement of the 15mm/20m ground. It can be seen from Figure 3 that when the ground subsides, the rails, track slabs, self-compacting concrete, base slab, subgrade surface, and subgrade bottom are all deformed, and the deformation curves basically show a "like cosine" law due to the overall stiffness of the track structure. The difference in the settlement between each structural layer of the ballastless track is not large, and good bonding is still maintained. Still, the settlement difference between the base plate and the surface layer of the base bed, the bottom layer of the base bed and the ground is large, that is, the base plate and the base bed. There are gaps between the surface layer, the bottom layer of the base bed, and the ground. At the starting and ending positions of uneven ground subsidence, the track structure also has the phenomenon of reverse bending and arching.

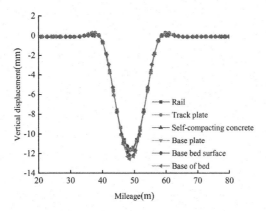

Figure 3.   Settlement curve of superstructure under subgrade settlement of 15mm / 20m.

Figure 4 shows the variation law of track slab settlement under the action of different ground settlement amplitudes. It can be seen from Figure 4 that when the wavelength of land subsidence is constant, as the amplitude of land subsidence increases from 5 mm to 30 mm, the subsidence displacement of the track slab increases linearly, from 4.57 mm to 21.17 mm, and the maximum subsidence occurs at the time of land subsidence. In the center position, with the increase of the amplitude of ground subsidence, the degree of back-bending and arching of the track slab at the edge of the subsidence range also increases.

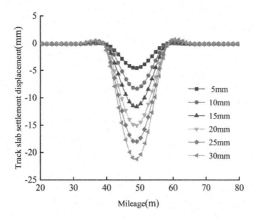

Figure 4.   Settlement of track slab under different settlement amplitude.

When the land subsidence wavelength is 20m, the variation law of the maximum subsidence displacement of each structural layer of the CRTS III slab ballastless track with the land subsidence amplitude is shown in Figure 5. It can be seen from Figure 5 that when the wavelength of land subsidence is 20m, with the increase of the amplitude of land subsidence, the maximum subsidence of the rail, track slab, self-compacting concrete, base plate, subgrade surface, and subgrade bottom also increases. There is a linear relationship between structural displacement and ground subsidence amplitude. With the increase of the amplitude of the ground subsidence, the separation value between the base plate, the surface layer of the subgrade, the bottom layer of the subgrade and the ground gradually increases; that is, the track structure layer and the subgrade, the subgrade and the ground subgrade appear to be separated. At this time, the track structure will appear periodic "beating" phenomenon under the action of the trainload, resulting in greater damage and damage to the track structure. Therefore, a large-scale settlement should be avoided in railway maintenance and repair.

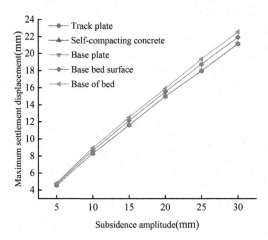

Figure 5.   Maximum deformation of track structure under different settlement amplitude.

3.2   Influence of ground subsidence amplitude on track structure stress

The maximum longitudinal stress of the superstructure under the action of different land subsidence amplitudes is shown in Table 3. It can be seen from Table 3 that when the land subsidence wavelength is 20m, with the increase of the land subsidence amplitude, the longitudinal stress of each structural layer also increases accordingly, basically a linear increase trend. The longitudinal stress of the surface layer and the bottom layer of the foundation bed is very small, far less than the longitudinal stress of the track structure.

Table 3.   Maximum stress of superstructure under different settlement amplitude.

Working condition /mm/m	Track plate /MPa	Self-compacting concrete /MPa	Base plate /MPa	Foundation bed surface layer /MPa	Foundation bed bottom layer /MPa
5/20	0.234	0.151	0.127	0.001	0.007
10/20	0.565	0.401	0.293	0.002	0.013
15/20	0.895	0.652	0.464	0.004	0.019
20/20	1.227	0.898	0.629	0.007	0.026
25/20	1.559	1.147	0.797	0.009	0.030
30/20	1.889	1.394	0.966	0.011	0.036

Figure 6 shows the variation law of the longitudinal stress of the track slab along the longitudinal direction of the track under the action of different ground subsidence amplitudes. It can be seen from Figure 6 that the longitudinal stress of the track slab increases linearly with the increase of the ground settlement amplitude. At the settlement center, the surface of the track slab is in a state of compression, resulting in large compressive stress. At the edge of the settlement range, the track slab is in a state of compression. The surface of the slab is in tension, resulting in large tensile stress; outside the settlement range, the track slab stress is almost unaffected.

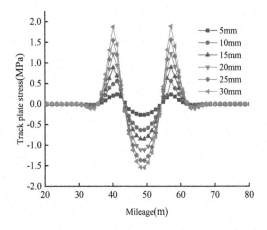

Figure 6.   Tress of track slab under different settlement amplitude.

4   INFLUENCE OF THE WAVELENGTH OF LAND SUBSIDENCE

4.1   *Influence of subgrade settlement wavelength*

In order to study the influence of land subsidence wavelength on track force and deformation, assuming that the land subsidence amplitude is constant at 15 mm, according to the different land subsidence wavelengths, seven working conditions (amplitude/wavelength) are selected for calculation, as shown in Table 4.

Table 4.   Working conditions of land subsidence wavelength (mm/m).

Working condition 7	Working condition 8	Working condition 9	Working condition 10	Working condition 11	Working condition 12	Working condition 13
15/10	15/15	15/20	15/25	15/30	15/35	15/40

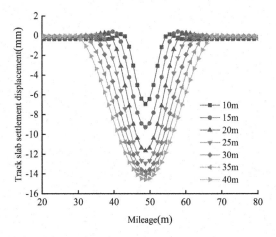

Figure 7.   Settlement of track slab under different settlement wavelengths.

Figure 7 shows the subsidence of the track slab under the action of different ground subsidence wavelengths. It can be seen from Figure 7 that the amplitude of land subsidence is constant, and the maximum subsidence displacement of the track slab increases gradually with the increase of the wavelength of land subsidence, and finally tends to be stable. When the wavelength of land subsidence is 10 m~20 m, the maximum subsidence of the track slab is 6.95 mm, 9.28 mm, and 11.64 mm, respectively, which is much smaller than the amplitude of land subsidence. When the wavelength of land subsidence continues to increase, the subsidence of the track slab increases slowly. The settlement displacement between the track structure and the subgrade surface and between the subgrade surface and the subgrade bottom layer is getting closer and closer, and the structural layers are more fit. At the same time, with the increase of the ground subsidence wavelength, the degree of back-bending and upward arching of the track slab gradually decreases at the starting and ending positions of the ground subsidence.

When the land subsidence amplitude is 15mm, the variation law of the maximum subsidence displacement of each structural layer of the CRTS III slab ballastless track with the land subsidence wavelength is shown in Figure 8. It can be seen from Figure 8 that when the ground settlement wavelength increases from 10m to 40m, the maximum settlement displacement of the rail, track slab, self-compacting concrete, and base slab is basically the same, while the displacement difference between them gradually decreases; that is, the separation value decreases.

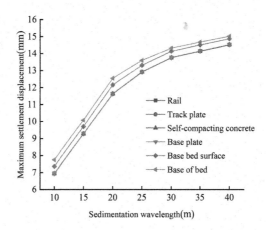

Figure 8. Maximum settlement of track structure under different settlement wavelengths.

4.2 *Influence of land subsidence wavelength on track structure stress*

The maximum longitudinal stress of the superstructure under the action of different land subsidence wavelengths is shown in Table 5. It can be seen from Table 5 that when the land subsidence wavelength is 20 m, with the increase of the land subsidence wavelength, the maximum longitudinal stress of the track structure first increases and then decreases. When the settlement wavelength reaches 20 m, the maximum longitudinal stress of the track structure reaches the maximum, and the variation law of the longitudinal stress of the subgrade surface and the subgrade bottom is not obvious.

Table 5. Maximum stress of superstructure under different settlement wavelengths.

Working condition /mm/m	Track plate /MPa	Self-compacting concrete /MPa	Base plate /MPa	Foundation bed surface layer /MPa	Foundation bed bottom layer /MPa
15/10	0.351	0.237	0.183	0.003	0.031
15/15	0.621	0.443	0.326	0.003	0.024
15/20	0.895	0.652	0.464	0.004	0.019

(continued)

230

Table 5. Continued.

Working condition /mm/m	Track plate /MPa	Self-compacting concrete /MPa	Base plate /MPa	Foundation bed surface layer /MPa	Foundation bed bottom layer /MPa
15/25	0.814	0.586	0.422	0.004	0.017
15/30	0.576	0.407	0.298	0.004	0.017
15/35	0.445	0.308	0.230	0.004	0.017
15/40	0.332	0.219	0.169	0.004	0.017

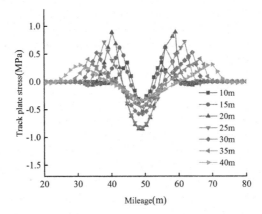

Figure 9. Stress of track slab under different settlement wavelengths.

Figure 9 shows the variation law of the longitudinal stress of the track slab along the longitudinal direction of the track under the action of different ground subsidence wavelengths. It can be seen from Figure 9 that the longitudinal stress of the track slab increases first and then decreases with the increase of the settlement wavelength. When the subgrade settlement wavelength is within the range of 5~20 m, the tensile stress and compressive stress of the track slab both increase. When the subgrade settlement wavelength is in the range of 20~40 m, both the tensile stress and the compressive stress of the track slab decrease. It can be seen from this that when the wavelength of ground subsidence is 20 m, the tensile stress and compressive stress of the track slab reach the peak value, so it is suggested that the limit of the wavelength of uneven ground subsidence is 20 m.

Figure 9 shows the variation law of the longitudinal stress of the track slab along the longitudinal direction of the track under the action of different ground subsidence wavelengths. It can be seen from Figure 9 that the longitudinal stress of the track slab increases first and then decreases with the increase of the settlement wavelength. When the subgrade settlement wavelength is within the range of 5~20 m, the tensile stress and compressive stress of the track slab both increase. When the subgrade settlement wavelength is in the range of 20~40 m, both the tensile stress and the compressive stress of the track slab decrease. It can be seen from this that when the wavelength of ground subsidence is 20 m, the tensile stress and compressive stress of the track slab reach the peak value, so it is suggested that the limit of the wavelength of uneven ground subsidence is 20 m.

5 CONCLUSION

(1) When uneven settlement occurs on the ground, the ballastless track structure will follow the settlement to a certain extent under the action of its weight. There is a gap between the ground and the ground. At the edge of the ground subsidence, the ballastless track structure has a

phenomenon of reverse bending and arching; the longitudinal stress of the ballastless track structure is affected by the ground subsidence. The maximum compressive stress occurs on the upper surface, and the maximum tensile stress occurs on the upper surface of the ballastless track at the starting and ending positions of settlement.

(2) When the wavelength of land subsidence is constant, with the increase of the amplitude of land subsidence, the settlement displacement and the maximum longitudinal stress of the ballastless track structure increase linearly. Voids appear between the track structure layer and the subgrade and between the subgrade and the ground subgrade, and the voiding intensifies with the increase of the settlement amplitude.

(3) When the amplitude of land subsidence is constant, with the increase of the wavelength of land subsidence, the settlement displacement of the ballastless track structure first increases and then tends to be stable. The void gradually decreases; the longitudinal stress of the ballastless track structure first increases and then decreases with the increase of the land subsidence wavelength. When the land subsidence wavelength is 20m, the maximum longitudinal stress of the track structure reaches its peak value.

REFERENCES

Cai Xiaopei, Liu Wei, Wang Pu, Ning Xing. Influence of ground settlement on the smoothness of double block ballastless track on Subgrade [J] Engineering mechanics, 2014, 31 (09): 160–165.

Chen Jinhao. Study on the influence of uneven settlement of subgrade on rail surface deformation of urban rail transit [J] Subgrade Engineering, 2020 (04): 16–20 DOI: 10.13379/j.issn. 1003-8825.201 907016.

Chen Peng, Gao Liang, Ma Mingnan. Limited value of subgrade settlement and its influence on mechanical characteristics of ballastless track in high-speed railway[J]. Construction & Design for Project, 2008(5): 63–66. (In Chinese)

Chen Qian. Study on track dynamic characteristics of transition section between ballasted track on Heavy Haul Railway Subgrade and ballastless track in tunnel [D]. Lanzhou: Lanzhou Jiaotong University, 2018

Cheng Qunqun. Study on dynamic influence and settlement limit of subgrade differential settlement on ballastless track structure [D] Southeast University, 2015

GUO Yu. Effects of Differential Subgrade Settlement and its Evolution in High-speed Railways on Mechanical Performance of Vehicle-Track Coupled System [D]. Chengdu: Southwest Jiaotong University, 2018.

HE Hua-wu. Ballastless Track Technology [M]. Beijing: China Railway Press, 2005.

Hunt H E M. Settlement of railway track near bridge abutments[J]. Transport, 1997, 123(1): 68–73.

JIA Jun-wang. Research of treatment technologies for subgrade settlement based on the Shanghai-Nanjing Intercity High-speed Railway(K235) [J]. Shanxi architecture, 2015, 41 (02): 130–131.

Li Guohe, Sun Shuli, Xu Zailiang. Influence of land subsidence on the bridge of high-speed railway and its engineering countermeasures [J]. Journal of Railway Engineering Society, 2008(4): 37–41.

Lu Wei, Tang Guanghui, Ji Wenli, Deng Shuai, Zhang Xiongfeng. Discussion on differential settlement limit of pile slab structure of high-speed railway foundation [J] Railway standard design, 2014, 58 (11): 57–61 DOI: 10.13238/j.issn. 1004-2954.2014.11.014.

Namura A, Suzuki T. Evaluation of Countermeasures against Differential Settlement at Track Transitions[J]. Quarterly Report of Rtri, 2007, 48(48): 176–182.

Wang Liwei, Jiang Guanlu. Discussion on limit standard of convergence settlement of subgrade behind abutment [J] Sino foreign highway, 2013, 33 (02): 38–40.

XU Qing-yuan, et al. Limited value for uneven settlement of subgrade under CRTS-II type slab track [J]. Journal of Central South University (Science and Technology), 2013, 44 (12): 5038–5044.

XU Qing-yuan, LI Bin, FAN Hao. Influence of uneven settlement of subgrade on dynamic characteristic of train-ballastless track on subgrade coupling system [J]. Journal of Railway Science and Engineering, 2012, 9 (03): 13–19.

XU Qing-yuan, LI Bin, ZHOU Zhi-hui. Study on the Limited Value for the Uneven Settlement of Subgrade under CRTS-I Type Slab Track [J]. China Railway Science, 2012, 33 (02): 1–6.

Yang Weiliang. Numerical simulation analysis of subgrade settlement of Expressway [J] China building metal structure, 2013 (08): 70–71.

Zhang Keping, He Zhenxing, Shi Guangtian, Jia Xiaohong, Zhai Zhihao. Mapping relationship between urban rail transit subgrade settlement and ballastless track surface irregularity [J] Railway standard design, 2020, 64 (02): 82–87.

Zhang Keping, Shi Guangtian, He Zhenxing. Study on the influence of uneven subgrade settlement on the dynamic characteristics of Metro type a vehicles Vibration and shock, 2020 39(17): 165–170.

Zhong Yanglong, Ma Chaozhi, Gao Liang, Cai Xiaopei, Zhao Wenqiang. Theoretical study on evaluation index of uneven settlement of ballastless track subgrade based on vehicle response [J] Engineering mechanics, 2021, 38 (12): 147–157.

*Frontiers of Civil Engineering and Disaster Prevention and
Control – Yang & Rahman (Eds)
© 2023 The Author(s), ISBN: 978-1-032-31200-2*

Research and application of non-removal expanded metal mesh technology

Guo Cheng
Xianning Vocational Technical College, Xianning, China

Jun Xie*
School of Art and Design, Hunan First Normal University, Changsha, China

Zhengtai Bao
China Railway Urban Construction Group Co., Ltd., Changsha, China

Zheng Liu
Zhuyou Zhizao Construction Technology Group Co., Ltd., Guangdong branch, Guangzhou, China

Peng Liu
The Oxbridge College, Kunming University of Science and Technology, Kunming, China

ABSTRACT: With the advantages of simple production, low cost, and convenient construction, the non-removal expanded metal mesh can be used as permanent formwork. Combining the characteristics of the non-removal expanded metal mesh, it can be applied in the process of beam-column joint splicing, composite beam cast-in-place part, shear wall cast-in-place part, floor slab seam processing, and hole formation, and the production of the rough surface of the sunk key of various prefabricated components in the prefabricated concrete structure. Therefore, the research of non-removal expanded metal mesh technology has great economic value in the construction of prefabricated concrete structures.

1 INTRODUCTION

To achieve the principle of an equivalent cast-in-place prefabricated concrete structure system, cast-in-place joints are often used between prefabricated shear wall panels and frame beam-column joints. Generally, formwork is set up on-site, which will increase the construction process and delay the construction progress. The formwork technology directly affects the quality, cost, and benefit of the project construction. Some permanent formworks have been developed both in China and abroad in the prefabricated concrete structure to replace temporary formworks, such as steel wire concrete sheets, extruded polyethylene benzene board, FRP panel, permanent fiberglass formwork, and ICF insulating formwork (Li et al. 2008; Mi 1999; Sun 1999; Yu 2009).

Non-removal expanded metal mesh formwork is a jointless expanded metal mesh made of thin hot-dipped steel plate. It is processed by a special professional machine into a steel mesh formwork composed of continuous convex V-shaped mesh bones and unidirectional three-dimensional meshes. The formwork is processed by factory machines (Ai 2008). It can be pre-pressed and formed at one time, with simple processing procedures and high production efficiency. The thickness of the thin steel sheet for processing the expanded metal mesh is generally 0.2~0.8mm, the weight is relatively light, and the length direction can be controlled according to the engineering requirements. The formwork can be cut on-site. With small lateral stiffness, it is easy to bend and install (Ai 2008). Combined with its characteristics, the non-removal expanded metal mesh

---

*Corresponding Author: 56865080@qq.com

DOI 10.1201/9781003308577-31

is applied to the fabricated structure, forming a prefabricated component non-removal technology (Lu 2014).

The non-removal expanded metal mesh is embedded in the part that needs to be spliced or joint processed when the prefabricated component is formed. It forms an integral part of the fabricated components and is used for the edge retaining formwork for post-cast. First of all, the prefabricated component non-removal expanded metal mesh technology is simple to operate. It eliminates the temporary support formwork of cast-in-place concrete at the seams of the prefabricated components and the joint connections, significantly improving the efficiency of installation joints and seams processing of fabricated components. Second, the prefabricated component non-removal expanded metal mesh has a large longitudinal stiffness. As a permanent formwork of the structure, it does not require other temporary formwork materials, formwork removal operations, and formwork removal equipment. Thereby, the installation speed of the fabricated component is improved, and the engineering cost is saved. Therefore, it can be seen that the non-removal expanded metal mesh technology in prefabricated buildings has greater economic value (Rizkallash et al. 1990).

2 ADVANTAGES OF NON-REMOVAL EXPANDED METAL MESH

2.1 *Mechanical characteristics of non-removal expanded metal mesh*

The non-removal expanded metal mesh has small lateral stiffness and is easy to bend into shape in the lateral direction. Therefore, it facilitates the installation and assembly of prefabricated components into various forms of formworks. At the same time, it has a large longitudinal stiffness, which reduces the bending deformation caused by the excessive vertical pouring concrete pressure when the prefabricated components are spliced. In some prefabricated components, such as prefabricated shear walls, due to the small thickness of the thermal insulation layer and the outer decorative layer, excessive pressure on the side of the cast-in-place concrete results in squeezing damage to the wall panels (Xie & Xie 2020). After such damage, it is difficult to maintain, which brings unnecessary trouble and loss to the project. However, the non-removal expanded metal mesh technology can effectively reduce the lateral pressure caused by the cast-in-place concrete, thereby protecting the thermal insulation layer and the outer decorative layer. Meanwhile, the non-removal expanded metal mesh can achieve good bonding with thermal insulation and decorative materials due to its misaligned three-dimensional meshes and improve the integrity of the thermal insulation layer, the decorative layer, and the wall. Additionally, with convenient processing and small thickness, the non-removal expanded metal mesh is easy to cut into meshes of different sizes to meet the requirements of construction and structure (Zhuang et al. 2017).

2.2 *Porosity of non-removal expanded metal mesh*

The pores of the non-removal expanded metal mesh are formed by the dislocation of the plate pressure gaps. It has a higher porosity, and its mesh eyelets can disperse the water pressure generated when the concrete is poured. The eyelets are smaller, which can reduce the loss of cement mortar. The concave-convex surface formed by the mesh holes allows the mortar to penetrate the interface through the mesh holes during pouring to form a coarse-grained interface with ideal shear resistance (Xie & Wu 2017).

2.3 *Bonding performance of non-removal expanded metal mesh and concrete*

Due to the misaligned hole pattern, the non-removal expanded metal mesh is connected with concrete to form a rough corrugated surface. It shows good bonding performance with concrete, improving the bonding degree of secondary pouring concrete and enhancing the integrity. The splitting tensile strength, axial tensile strength, body compressive strength, friction coefficient, and bonding strength of the joint surface between secondary pouring concrete and primary pouring concrete are better than the roughened concrete in terms of performance indicators. At the same time, due to the enhancement of concrete bonding ability, the process of digging outside the surface of the prefabricated component is eliminated, and the prefabricated component is protected (Ju et al. 2016).

The prefabricated component non-removal expanded metal mesh technology fundamentally improves the joint splicing and formwork erection of the prefabricated component during the installation process, eliminating the on-site formwork erection and removal construction process, speeding up the construction operation, and saving labor and material costs. In addition, the mechanization of the production has greatly improved the quality of the project. Therefore, the unique characteristics of the prefabricated component non-removal technology dramatically promote the development of construction industrialization (Xie et al. 2018).

3 APPLICATION OF NON-REMOVAL EXPANDED METAL MESH

3.1 *Application of non-removal expanded metal mesh in frame structure*

In the frame structure system, the non-removal expanded metal mesh is pre-embedded in the prefabricated component in the factory. It acts as an edge retaining formwork when it is cast in place. Figure 1 shows a schematic diagram of the non-removal expanded metal mesh pre-embedded in the prefabricated composite beam. The non-removal expanded metal mesh is pre-embedded on the outside of the composite beam and serves as the outer formwork for the composite cast-in-place. As shown in Figure 2, the steel mesh can also form the U-shaped notch at the beam end to replace the notch mold and reduce the difficulty of manufacturing (Jiang et al. 2018). Moreover, the formed surface is relatively rough, increasing the bonding performance with the post-pouring concrete. Figures 3, 4, and 5 show that the non-removal expanded metal mesh can be pre-embedded for the primary and secondary beam lap joints and the beam-column cast-in-place joints. At the laps of the primary and secondary beams and the beam-column joints, the steel mesh can be woven into many shapes, such as I-shaped, L-shaped, and U-shaped(Jun 2021). Then it is tied together with the steel bars at the joints to form an edge retaining formwork to prevent the deformation caused by the lateral pressure of the concrete during pouring. As shown in Figure 6, the non-removal expanded metal mesh can be applied to the cast-in-place joints in the middle of the precast column. This non-removal formwork serves as an outer formwork when the precast column needs to be disconnected and connected to the beam in the middle. Figure 7 shows that the non-removal expanded metal mesh can be used for the cast-in-place joints between the floor slabs, replacing the suspended formwork there. The non-removal expanded metal mesh application in the fabricated frame structure eliminates the high-altitude formwork installation and formwork removal process and the additional scaffolding process, which strengthens construction safety, saves formwork materials, and improves assembly efficiency (Xie et al. 2018).

Figure 1.   Non-removal expanded metal mesh applied to the beam side as a formwork of the composite cast-in-place side.

Figure 2.   Non-removal expanded metal mesh applied to the U-shaped notch at the beam end.

236

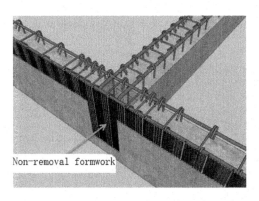

Non-removal formwork

Figure 3. Non-removal expanded metal mesh applied to the primary and secondary beams.

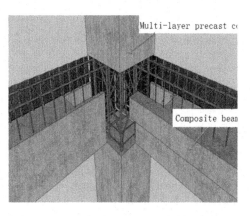

Multi-layer precast c

Composite bean

Figure 4. Non-removal expanded metal mesh applied to beam-column joints.

Figure 5. Non-removal expanded metal mesh applied to the top of the column.

Figure 6. Non-removal expanded metal mesh applied to the cast-in-place joints in the middle of the precast column.

Figure 7. Non-removal expanded metal mesh applied to the joints between the floor slabs.

3.2 *Application of non-removal expanded metal mesh in prefabricated shear wall*

In the fabricated shear wall structure system, the cast-in-place part is inevitable at the wall splicing. Due to the thinness of the thermal insulation layer and the outer decorative layer of the prefabricated external shear wall and the small stiffness, excessive pressure on the side of the cast-in-place concrete results in squeezing damage to the thermal insulation layer and the outer decorative layer. After such damage, it is difficult to maintain, which brings unnecessary trouble and loss to the project. However, the non-removal expanded metal mesh can effectively reduce the lateral pressure caused by the cast-in-place concrete. The maximum lateral pressure is 60% of general formwork, protecting the thermal insulation and the outer decorative layers (Xie et al. 2018). Meanwhile, the non-removal expanded metal mesh can bond well with thermal insulation and decorative materials due to its misaligned three-dimensional meshes. Thus, it improves the integrity of the thermal insulation layer and decorative layer with the wall and shows sound effects in applying shear walls. Figures 8~13 show the application of non-removal expanded metal mesh in shear walls. The application parts include edge members, wall-beam connections, beam-column

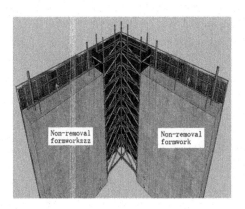

Figure 8.   Non-removal expanded metal mesh applied to constrain the cast-in-place joints of edge members.

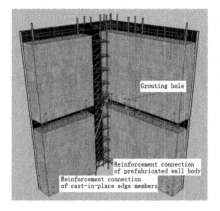

Figure 9.   Non-removal expanded metal mesh applied to construct cast-in-place joints of edge members.

Figure 10.   Non-removal expanded metal mesh applied to shear wall edge members.

Figure 11.   Non-removal expanded metal mesh applied to shear wall edge members.

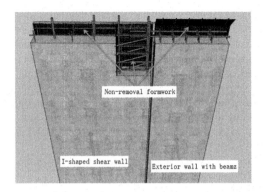 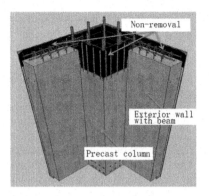

Figure 12. Non-removal expanded metal mesh applied to the connection joints of beam and shear wall.

Figure 13. Non-removal expanded metal mesh applied to the connection joints of beam and precast columns.

connections, primary and secondary beams, and floor slab seams. From the perspective of the shear wall structure system, the non-removal expanded metal mesh can completely replace temporary formwork, eliminate the materials and construction procedures for formwork erection and removal, reduce construction costs and improve assembly efficiency (Xie et al. 2019).

4 CONCLUSION

Based on the results and discussions presented above, the conclusions are obtained as below:

(1) As a permanent formwork combined with concrete, the metal structure of the non-removal expanded metal mesh formwork can strengthen the surface strength of concrete, prevent cracks on the concrete surface due to dry shrinkage and temperature difference, and improve the internal quality of concrete.

(2) The non-removal expanded metal mesh formwork technology has almost no outflow waste on the construction site and reduces the discharge of construction waste. On the contrary, it can effectively protect the dwindling natural resources, adapt to environmental protection requirements, improve the increasingly severe environmental problems, accelerate the development of green buildings, and promote the progress of science and technology.

(3) The prefabricated component non-removal expanded metal mesh technology fundamentally improves the joint splicing and formwork erection of the prefabricated component during the installation process, eliminating the on-site formwork erection and formwork removal construction process, speeding up the construction operation, and saving labor and material costs.

(4) In addition, the mechanization of the production and the feasibility of the quality have greatly improved the quality of the project. Therefore, the unique characteristics of the prefabricated component non-removal technology dramatically promote the development of construction industrialization.

5 PROSPECT

Combined with the development of the prefabricated building, the non-removal technology can be studied from the following aspects.

(1) The first is to develop from traditional on-site formwork erection to industrialized, prefabricated formwork. Thus, the non-removal formwork can be pre-made and installed in the factory to avoid removing the on-site installation process and improve construction efficiency.

(2) The second aspect is to develop from a single-function formwork to a multi-function formwork. That is to overcome the single function of the traditional formwork and develop into a multi-functional non-removal formwork, including heat preservation, heat insulation, sound insulation, moisture resistance, corrosion resistance, fire resistance, and non-combustibility.

(3) The third aspect is to develop from the complex construction process to a simple, lightweight, convenient, and feasible direction.

(4) The fourth aspect is to introduce and develop new materials and develop high-strength, lightweight, and high-performance non-removal formwork, to improve the strength, stiffness, and corrosion resistance of the non-removal expanded metal mesh and meet the usability requirements of the building structure.

(5) The fifth is to implement the corresponding formwork technical regulations and specifications to standardize the prefabricated component non-removal technology.

(6) The final aspect is to implement the comprehensive application of a variety of non-removal formwork technology. In actual projects, the different building types, the use requirements of components, and the variability of the construction environment require the construction personnel to flexibly apply the non-removal technology, thereby increasing the construction speed, saving costs, and improving engineering quality.

ACKNOWLEDGMENTS

This work was financially supported by the State Key R&D Program "Research and demonstration of flexible manufacturing technology for complex-shaped concrete building parts and decorative components" (2017YFC0703705).

REFERENCES

Ai Yichun. (2008) The Performance of Non-removal Formwork Mesh and Its Application in Hydropower Projects. Hunan Hydro & Power,01.

Jiang Difei, Xie Jun, Zhuang Wei. (2018) Technical and Economic Analysis of An Assembled Monolithic Shear Wall Residence. Building Structure, 2: 37–39.

Ju Xiaoqi, Zhuang Wei, Xie Jun. (2016) Pocket Manual for Structural Engineers. In: China Building Industry Press. Beijing.

Jun Xie, Difei Jiang, Zhengtai BAO, and Qiguo LI. (2018) Discussion about the Analysis and Design of Over-Height High-Rise Structure.The Paper of 2018 International Conference on Construction and Real Estate Management(ICCREM), 5: 56–61.

Jun Xie, Difei Jiang, Zhentai BAO, and Pin ZHOU. (2018) BIM Application Research of Assembly Building Design: Take ALLPLAN as an Example.The Paper of 2018 International Conference on Construction and Real Estate Management(ICCREM), 5: 36–39.

Jun Xie, Difei Jiang, Zhentai BAO. (2019) Study on Elastic-plastic Performance Analysis of A Prefabricated Low Multi-story Villa. The Paper of The 5th International Conference on Civil Engineering, 5: 36–39.

Jun Xie. (2021) Deviation detection design for prefabricated building Exterior Wall. In: China Building Industry Press. Beijing.

Li Zhu, Su Dongyuan, Liu Yuanzhen. (2008) Formwork Design of Composite Shear Wall System of Non-removal Thermal Insulation Formwork. Wall Materials Innovation & Energy Saving in Buildings, 1: 53–55.

Lu Wenliang. (2014) Fast-EZ High-ribbed Permanent Concrete Formwork. Construction Technology.

Mi Jiaping. (1999) Trend of Development of New-type Formworks in China. Architecture Technology, 8: 546–549.

Rizkallash, Serretterl, Heuvel J Setal. (1990) Multiple Shear Key Connections for Precast Concrete-framed Structures. ACI, Structure Journal, January, 1: 53–59.

Sun Nanping. (1999) Cast-on-site Wall Technology with Foamed Polystyrene Mould Board without Dismount. New Building Materials, 10: 41–42.

Xie Jun, Jiang Difei, Zhou Ping. (2018) Research on Precast Rate and Cost of Prefabricated Shear Wall Structure. Building Structure, 2: 33–36.

Xie Jun, Wu Xinshao. (2017) Design and Construction of Fabricated Shear Wall Structure.In: China Building Industry Press. Beijing. pp. 33–55.

Xie Lunjie, Xie Jun. (2020) Practice of Prefabricated Buildings in High-intensity Seismic Fortified Areas. In: China Building Industry Press. Beijing. pp. 50–68.

Yu Qingyuan. (2009) Experimental Research on Glass Fiber Concrete Non-removal Formwork. North China University of Technology.

Zhuang Wei, Li Hengtong, Xie Jun. (2017) Examples of Underground and Foundation Engineering Software Operation (Including PKPM and Rationality). In: China Building Industry Press. Beijing. pp. 86–93.

Frontiers of Civil Engineering and Disaster Prevention and
Control – Yang & Rahman (Eds)
© 2023 The Author(s), ISBN: 978-1-032-31200-2

Underground foundation structure design of Kunshan Xintiandi project

Xiaomeng Zhang* & Wenting Liu
China Architecture Design & Research Group, Beijing, China

Yao Feng
Beijing Guobiao Building Technology Co., LTD, Beijing, China

Xiao Yang
China Architecture Design & Research Group, Beijing, China

ABSTRACT: The project is located in Kunshan city, Jiangsu Province, including one office high-rise, four residential high-rises, and one commercial high-rise. Each unit shares the basement chassis, and the basement has a total of three floors. This paper analyzes the selection of the basement structure, and the corresponding foundation selects different pile foundations for different buildings, and studies the facts that require attention in the design process.

1 PROJECT OVERVIEW

The construction site of the project is located in Kunshan city, Jiangsu Province. The proposed site has West Qianjin Road on the north, Louchuang Road on the south, Sichang Road on the east, and Zu Chong Road on the West. The project includes one office high-rise, four residential high-rises, and one commercial high-rise. Each unit shares a large basement chassis with a total of three floors. The absolute elevation of this project is 3 m plus or minus zero elevation. There are three floors underground, one to three floors above ground for business, and four floors above ground for office. The height of the building is 142 m. There are 30 floors above ground. The foundation adopts pile foundation + raft board.

Figure 1. Architectural renderings.

---

*Corresponding Author: 155203255@qq.com

DOI 10.1201/9781003308577-32

The engineering design service life is 50 years, the building structure safety grade is two, the building structure fire prevention grade is one, and the foundation design grade is A grade. According to the Code for Seismic Design of Buildings (GB50011-2010) (2016 edition) (Beijing China building industry press 2010), the site in the project area lies in the seismic group I, the site category is CLASS IV, the seismic level is 7 degrees 0.1 g (Beijing China building industry press 2012, Beijing China building industry press 2011), and the office area is general fortification class (CLASS C). The foundation is grade A (Beijing China building industry press 2011, Beijing China building industry press 2008).

2 BASEMENT SELECTION AND LAYOUT

2.1 *Topography and landform*

Kunshan city lies southeast of the Yangtze River Delta and the middle of Taihu Lake water network plain. According to regional geological data, the crustal movement since the Quaternary is mainly subsidence, widely accepted accumulation, forming a broad and single accumulation plain, which belongs to the delta alluvial plain landform. The Quaternary strata are widely distributed and thick.

The west side of the proposed site is Zuchong Road, and the north side is Qianjin Road. During investigation, the maximum ground elevation of the proposed site is 3.57 m, the minimum is 1.51 m, the relative height difference of the surface is 2.06 m, and the terrain is slightly undulating. Ponds are distributed on the site. According to the investigation, ponds are formed from soil pits. The water level of the site reservoir was about 1.94 m after the rainstorm during the first survey approach.

2.2 *Site form and foundation selection*

The proposed building load is large, and there is no good shallow foundation bearing layer at the shallow part of the site. Therefore, a pile foundation scheme is recommended.

According to the nature of the proposed building and combined with the geological conditions of the site, the available holding layers of the site are as follows:

The 9th layer of silty sand has medium to dense, medium to low compressibility. The average QC of cone tip resistance is 9.605 MPa, and the average N of standard penetration is 32.0. The characteristic value of foundation bearing capacity is 200 kPa, which is widely distributed on the site. The single pile with high bearing capacity can be used as the pile foundation bearing layer of the commercial and underground garage.

The 10th layer of silty sand has dense, medium, and low compressibility. The average value of cone tip resistance QC is 14.321 MPa, the average value of standard penetration strike number N is 48.4 strikes, and the characteristic value of foundation bearing capacity is 220 kPa, which is widely distributed in the site. The single pile with high bearing capacity can be used as the pile foundation bearing layer of residential and office buildings.

The 11th layer of coarse sand has dense, medium-low compressibility, standard penetration hit number N average of 68.0 hits, foundation bearing capacity characteristic value of 250 kPa, the whole field distribution.

When the strength of the underlying layer of the pile foundation is relatively weak, it is suggested that the thickness of the hard bearing layer under the pile tip should not be less than 3d.

2.3 *Proposed base form*

This project plans to adopt the scheme of pile foundation plus raft board, with the 10th layer of silt as the foundation bearing layer. The main office building core tube adopts a pile with a diameter of 1000 mm and a length of 65 m, with an ultimate bearing capacity of 9600 kN, and a pile distance of

3 m. The raft is 2.5 m thick. Four pile caps are used under the frame columns. Residential buildings adopt an 800 mm diameter pile, pile length of 38 m, and ultimate bearing capacity of 4400 kN.

The groundwater level of the site is high. According to the geological survey report, the anti-floating fortification water level should be 0.50 m below the outdoor surface design elevation and not lower than 2.36 m above the Yellow Sea.

Because of its large deadweight, the anti-floating problem with the main tower is absent. A permanent anti-floating problem must be considered for the pure basement part, and corresponding anti-floating measures must be taken to adopt anti-pulling piles. The anti-floating design water level is recommended to take the high value between 0.50 m below the outdoor design floor and 2.36 m (plus or minus 3.0 m) of the highest diving level in Kunshan city. The tensile pile is 25 m long, and the ultimate bearing capacity is 2000 kN.

Table 1. Bearing capacity results of pile foundation

Construction	Bearing course at pile end	Pile top elevation (m)	Pile length (m)	Pile type
4# residence	⑩	−14	38	Φ800 mm bored pile
5# office building	⑩	−14	65	Φ1000 mm bored pile
The underground garage (Pull-out pile)	⑨/⑨T	−14	25	Φ600 mm bored pile

3 MATTERS NEEDING ATTENTION IN PILE FOUNDATION DESIGN

3.1 The construction, pile cutting, engineering pile inspection, and foundation groove excavation of pile foundation should strictly comply with the relevant codes, regulations, and standard requirements and the depth of pile end entering bearing layer and effective pile length as required in the drawings shall be ensured. The deviation of the hole-forming verticality of the cast-in-place pile is not more than 1%, and the deviation of the hole-forming aperture is not more than 50 mm. The sediment at the bottom of the hole must be removed after the pile hole is formed, and the thickness of the sediment at the bottom of the hole should not be more than 50 mm before the concrete is poured. Concrete should be poured immediately after the quality of the hole is qualified.

3.2 Main materials used are C35, C40, C50underwater concrete, steel bar HRB400, the concrete protection layer with a thickness grade of 1 pile reinforcement 50 mm, and the pile concrete environment is class B. Groundwater and shallow soil in the proposed site are slightly corrosive to concrete structures and steel bars in reinforced concrete structures. Concrete alkali content should be less than 3 kg/m, and chloride ion content should be less than 0.2%. The maximum water-cement ratio should be 0.55, and the minimum cement dosage should be greater than 350 kg/m^3.

3.3 Mechanical connection or welding connection should be adopted for the longitudinal reinforcement of pile body, and the percentage of the sectional area of longitudinal reinforcement with joints in the same connection section to the total sectional area of longitudinal reinforcement should be less than 50%.

3.4 *Requirements for post-grouting construction*

3.4.1 This test pile adopts the post-grouting method of the cast-in-place pile (PPG method for short) to reinforce the soil in a certain range at the bottom and side of the pile. PPG construction method shall be carried out by professional construction teams or construction units with experience in this construction method. The construction shall be carried out following the Code for Design of Building Foundation GB50007-2011, Technical Code for Building Pile Foundation JGJ 94-2008, or by referring to Technical Code for grouting after construction of filling pile >Q/JY14-1999. The construction operation and quality standard of post-grouting shall comply with QB/SY-2003-01 construction and inspection standard. Compound grouting is adopted in construction; that is, post-grouting pipe valves are set at the bottom and side of the pile, and two pipes and pipe valves should be set symmetrically along the circumference of the pile reinforcement cage.

3.4.2 Compound grouting shall be adopted in construction; which means post grouting pipe valves shall be set at the bottom and side of the pile, and two pipes and pipe valves shall be set symmetrically along the circumference of the reinforcement cage of the pile. The port (not including the pile end grouting valve) is about 400 mm away from the bottom end of the reinforcement cage, close to the next reinforcing hoop of the reinforcement, and the grouting pipe is tied to the main reinforcement along the whole length.

3.4.3 The inner pipe of post-grouting at the bottom of the pile should be welded with a steel pipe with a nominal diameter of 25 mm and a wall thickness of steel pipe which is not less than 3.0 mm. When it is needed to be used for ultrasonic testing of pile structural integrity, it shall be appropriately enlarged according to testing requirements. The inner pipe of post grouting on the pile side shall be welded steel pipe with a nominal diameter of 20 mm, and the thickness of the steel pipe wall is 2.75 mm. The inner pipe of post grouting on the pile bottom and pile side shall be welded or bound with the reinforcement cage stiffeners, and the upper end shall be temporarily sealed with a plug.

3.4.4 Grouting should be started 2d after pile forming, and not later than 30d after pile forming. Post-grouting should be done first at the pile side and then at the pile end, and the grouting interval at the pile end should not be less than 2 h. 42.5 ordinary Portland cement should be used for post-grouting, the water-cement ratio of grout should be 0.45–0.65, and the grouting pressure of pile end grouting should be determined according to the property of the soil layer and the depth of the grouting point. Pile side grouting soil layer, grouting quantity, and termination pressure can be carried out according to relevant technological parameters accumulated during the pile test. Meanwhile, grouting pressure should not be less than 1.2 MPa, and grouting flow should not exceed 75 L/min. According to the preliminary estimation of geological prospecting data and Article 6.7.4 jGJ94-2008 technical Code for Building Pile Foundation, the total grouting volume of the 1000 mm pile is 6T. The total grouting amount of the 800 mm diameter pile is 3.2T. The total mm grouting volume of 600 diameter piles is 2.3T. At the same time, it is suggested that the grouting should be carried out according to the more accurate experience of the test pile before the test pile unit, and the total grouting amount of 1000 mm pile is 6T. The total grouting amount of 800 mm diameter pile is 4T. The total grouting amount of 600 mm diameter pile is 3T. The actual grouting amount shall be used as construction reference and shall be determined in combination with the pile forming process. Moreover, the grouting amount of each engineering pile shall meet the termination conditions in Section 6.7 of Technical Code for Building Pile Foundation JGJ 94-2008.

4   PILE FOUNDATION FORM

Piles are arranged according to different areas. Piles are arranged under the core tube, as shown in Figure 2, and piles are arranged in the waterproof plate area, as shown in Figure 3.

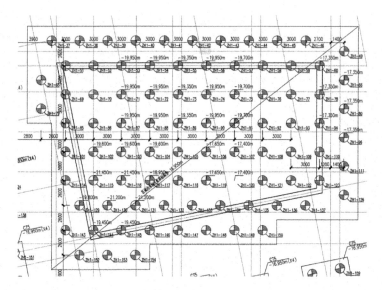

Figure 2.   Layout of core tube pile foundation.

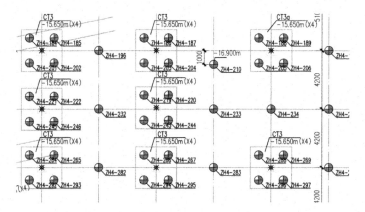

Figure 3.   Basement pile foundation layout.

5   CONCLUSION

Through analyzing the geological prospects and geological conditions, the reasonable structure and foundation form was selected, the foundation layout was analyzed, the cast-in-place pile was selected, and the corresponding suggestions were put forward for the design and construction of the cast-in-place pile.

REFERENCES

Code for seismic design of buildings GB 50011-2010 [s]. Beijing China Building Industry Press.
Load code for the design of building structures GB 50092012[s]. Beijing China Building Industry Press.
Code for design of concrete structures GB50010-2010: GB 50010—2011[s]. Beijing China Building Industry Press, 2011.
Code for design of building foundation GB50007-2011[s]. Beijing China Building Industry Press.
Technical code for building plie foundations JGJ94-2008[s]. Beijing China Building Industry Press.

Frontiers of Civil Engineering and Disaster Prevention and
Control – Yang & Rahman (Eds)
© 2023 The Author(s), ISBN: 978-1-032-31200-2

Mechanical behavior analysis of swinging column structure

Xiaomeng Zhang*, Wenting Liu*, Qingying Ren*, Xiao Yang* & Ying Wang*
China Architecture Design & Research Group, Beijing, China

ABSTRACT: In this paper, an experimental study is carried out on a swinging column structure system. The mechanical properties and mechanical properties are verified through the experimental study on a swinging column, which lays a foundation for the subsequent experimental and theoretical research.

1 PROJECT OVERVIEW

For the new swinging column structure system, a large number of scientists have carried out various experimental studies. Han Qiang (Han et al. 2021) et al. systematically combed and summarized the research process and progress of the swinging bridge structure. This paper first introduces the origin of swing structure and its application in bridge engineering and then discusses the research and experimental study of the swing bridge structure, pseudo-static behavior analysis method, dynamic behavior analysis model, and seismic design method, respectively. Deng (Zhang et al. 2018) expounds on the importance of self-resetting structure seismic performance research, and the reset component and classification of the structural system are introduced to relax the pedestal between the base and the constraint of the reset, loosen constraints between beam transverse swing structure frame structure since the reset support and energy consumption since the reset support frame of the seismic performance of the latest research progress. The results show that the seismic performance of the structure can be improved effectively by applying the rocking technology to the structure and setting the energy dissipation element at the swinging (opening and closing) interface, the damage can be controlled to the replaceable energy dissipation element, and the residual deformation of the main structure can be reduced. Lu Westwood (Zhou & Lu 2011) reviewed the swing and the restoration of the structure of the development history, this paper briefly introduces the swing and the restoration of the structure of the basic principle, swing, swing, and piers are reviewed since the restoration of the reinforced concrete frame structure, swing, and since the restoration of steel frame structure and the reset between, swing frame shear wall structure of the different structural system, such as development present situation, summarizes the swing and the development trend of self-reset structure is pointed out, and the combined application of post-tensioning prestress and energy dissipation shock absorption technology is one of the future development directions of self-reset structure. Palermo et al. (Palermo et al. 2007) proposed to combine unbonded prestress and the energy dissipating device and apply it to swing bridge pier structure, and carried out a quasi-static test. Its energy dissipation device for the built-in energy dissipation steel bar. The experimental results show that the system has good self-reset ability and energy dissipation ability, and the hysteresis curves of the components are in a typical banner shape. Studies have shown that the isolation device must have the following characteristics:

(1) It has a large vertical bearing capacity. In the use of the building, it can safely support all the weight of the superstructure and the use of load, with a large vertical bearing capacity safety factor.
(2) It can make the structure move flexibly on the foundation, so that the natural vibration period of the structure system is greatly increased, and the ground vibration can be separated and effectively reduce the acceleration response of the structure.

---

*Corresponding Authors: 155203255@qq.com

(3) The system has enough elastic stiffness to meet normal use requirements under wind load or mild earthquakes with sufficient initial stiffness. When a strong earthquake occurs, the device moves flexibly, and the system enters the isolation state.

(4) It can provide appropriate damping and quick decay resonance when a larger displacement occurs.

(5) It has the ability to eliminate excessive residual displacement, have a restorative force of rapid reset, or limit the displacement of the structure within the service area.

According to the characteristics of the isolator, it is concluded that the isolator system should be composed of an isolator, damper, seismic micro-vibration, and wind response control device.

The function of the isolator is to support all the weight of the building and change the dynamic characteristics of the structural system, mainly to prolong the natural vibration period of the building to reduce the seismic response of the building effectively. The isolators must be able to withstand the expected relative horizontal displacement between the foundation and the superstructure without losing their carrying capacity.

Isolators include rubber pads, rolling balls (or rollers), sliding materials, suspension isolators, Sleeved-Pile, etc.

2 THE EXPERIMENTAL SCHEME

Swing column structure system proposed in this paper, cylinder and the longitudinal reinforcement foundation and pile caps is disconnected, only on a steel bar connection among them, and in the adjacent position setting rubber cushion layer, the effect of rubber mat is full weight bearing structures on the one hand, on the other hand, is to change the dynamic characteristics of the structure system, main is to extend the natural vibration period of buildings, so as to effectively reduce the seismic response of the building. The new swinging column structure can not only bear the expected relative horizontal displacement between the foundation and the superstructure, but also not lose its vertical bearing capacity. According to the special structure of the swinging column, the seismic behavior of a full-scale single column member is studied, and its mechanical behavior is analyzed.

2.1 Element design

A specimen was designed and made for the test. The geometric size and section reinforcement structure of the specimen is shown in Figure 1. Specimen 1 comprises a loading block, a swinging

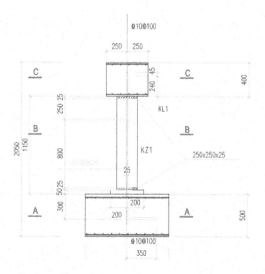

Figure 1. Detail drawing of swing column design.

column, and a foundation beam. The size of the loading block is 400 mm × 400 mm × 400 mm, the swinging column is 250 mm×250 mm×1050 mm, and the base beam ruler is 500 mm×600 mm×1000 mm.

2.2 *Material property test*

The concrete in the test was made according to three procedures, and the strength design grade of all specimens was C40. Six standard concrete cube blocks with a side length of 150 mm were reserved for all batches of concrete when the specimens were made, and they were naturally cured under the same conditions as the specimens. On the day of the test, the compressive strength of the test block was tested according to the national standard GB50081-2002, and the average value of the measured compressive strength of the concrete cube was 43.7 MPa. According to the standard tensile test method, the material properties of the steel bar are tested, and the mechanical properties of the steel bar measured are shown in Table 1.

Table 1. Wall yield strength.

Materials	Yield strength (MPa)	Ultimate strength (MPa)	Yield ratio
Q345	389	601	0.65
	415	599	0.69
	404	600	0.67

2.3 *Loading device*

The specimen floor beam is fixed to the rigid ground by a steel compression beam and anchor bolt. In the test, a predetermined vertical axial pressure was first applied on the top of the column and kept constant during the loading process, and then monotonically increasing horizontal load was applied on the top of the specimen through a 100-ton horizontal jack. According to GB/T50152-2012 "concrete structure test method standard" regulation, test load solution for formal preload, before loading to check the instrument is working correctly, to ensure good formal test parts contact load calculation using hierarchical load, load incremental until 2.5 mm specimen damage per level, should not continue to load, the load shall be terminated.

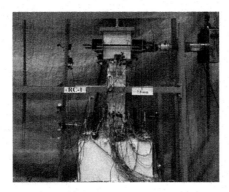

Figure 2. Loading device diagram.

3 TEST RESULTS

Test process and failure characteristics.

As there is a rubber cushion between the RC-1 column pier, top beam, and foundation, the load-displacement curve of the specimen develops linearly in the initial loading stage (Figure 3).

Figure 3. The overall diagram of the force on specimen RC-1 when the displacement loading is 2.5 mm.

When the specimen RC-1 was loaded to 7.5mm, the pressure side of the rubber pad at the bottom of the swing column was slightly bulging, and there was no gap between the rubber pad at the tension side of the swing column and the foundation. When the specimen RC-1 was loaded to 10mm, a gap appeared between the upper rubber pad on the tensile side of the swing column and the top beam, and a gap appeared between the lower rubber pad on the tensile side of the swing column and the foundation. With the increase of the horizontal displacement load of the top beam, the gap between the rubber pad on the tension side of the upper part of the swing column and the top beam does not change significantly. Still, the rubber pad on the compression side of the lower part of the swing column pops out obviously. The gap between the rubber pad on the tension side of the bottom of the swing column and the foundation gradually increases. When the horizontal load is loaded to 60mm, the maximum gap is 10mm. When the specimen was loaded to 55mm, the gap between the rubber pad on the tension side of the lower part of the swing column and the foundation basically remained stable. The rubber pad on the compression side of the lower part of the swing column popped out, and the rubber pad and the swing column were empty at the position of the steel rod. At this time, the load-displacement curve began to rebound, indicating that the steel rod began to be subjected to tension deformation (Figure 4).

Figure 4. Overall stress diagram of SPECIMEN RC-1 and compression condition of lower rubber pad (55mm).

As the horizontal displacement load continues to increase, the gap between the rubber pad on the tension side of the lower part of the swing column and the foundation continues to increase, and the compression deformation of the rubber pad on the compression side of the lower part of the swing column does not change significantly. When the specimen RC-1 is loaded to 120 mm, the compression side of the swing column decreases by 10mm, and the tension side is 15 mm. When specimen RC-1 was loaded to 130 mm, the load continued to rise. As the section size of the specimen RC-2 swing column was 250 mm×250 mm and the specimen was loaded to 130 mm, half of the column section size had been exceeded, and the loading was stopped (Figure 5). The rubber pad at the bottom of SPECIMEN RC-1 was twisted after the test stopped.

Figure 5. Overall stress diagram of SPECIMEN RC-1 and compression condition of lower rubber pad (130mm).

4   CONCLUSION

The mechanical properties and failure mechanism of the swinging column structure system were verified by the pseudo-static test of the structure, the failure mode of the swinging column structure was obtained, and the mechanical properties of the new swinging column system were verified.

REFERENCES

Han Qiang, Jia Zhen-lei, Zhou Yu-long, Du Xiu-li. Review of Seismic Resilient Bridge Structures: Rocking Bridges[J]. China Journal of Highway and Transport, 2021, 34(2): 118–133.
Palermo A. Pampanin S Marriott D. Design, Modeling, and Experimental Response of Seismic Resistant Bridge Piers with Posttensioned Dissipating Connections [J]. Journal of Structural Engineering, 2007, 133 (11), 1648–1661.
Zhang, G., Zhao, Z., & Sun, Z. (2018). Review of the seismic behavior of self-centering structure. *Building Structure*.
Zhou, Y., & Lu, X. (2011). State-of-the-art on rocking and self-centering structures. Jianzhu Jiegou Xuebao/Journal of Building Structures, *32*(9), 1–10.

Frontiers of Civil Engineering and Disaster Prevention and Control – Yang & Rahman (Eds)
© 2023 The Authors(s), ISBN: 978-1-032-31200-2

Application study on multi-stage drilling technology in super large diameter pile foundation construction

Tao Ye*
China Communications Construction Second Harbor Engineering Company Ltd., Wuhan, China
Department of Bridge Engineering, Tongji University, Shanghai, China

Hongchun Qu, Bindian Zhu & Yongxin Zhong
China Communications Construction Second Harbor Engineering Company Ltd., Wuhan, China

Yucai Li
Key Laboratory of Transport Industry for Long-span Bridge Construction Technology, Wuhan, China

ABSTRACT: Based on the newly-built Maanshan Yangtze River Bridge for dual-purpose functions of highway and railway, that is, the three-tower steel truss cable-stayed bridge with a main span of 1120 m × 2 m, this paper studies the construction of a large-diameter rotary excavation pile foundation and puts forward the construction technology of super large-diameter pile foundation based on multi-stage drill grading hole forming technology, which effectively solves the problem of insufficient one-time hole forming torque of rotary drilling rig. The general drilling mode for overburden is studied. For the rock strata, the method of pressure grading drilling replacement in rock entry mode is adopted. In addition, the parameter requirements of pressurized medium speed drilling under rock strata are studied. The results show that through the positioning casing control, the analysis of the drilling process, the optimization of drilling parameters and mud indexes, the drilling efficiency has been greatly improved, the pile quality has been improved, and the construction period has been reduced, which can provide a reference for similar projects.

1 PROJECT OVERVIEW

The Yangtze River Bridge of the Ma'anshan Highway and Railway is the largest span multi-tower cable-stayed bridge for highway and railway globally. It is also the longest highway and railway cable-stayed bridge in the world (3248 m long steel beam). It is a newly-built cable-stayed bridge with (112 + 392 + 1120 × 2 + 392 + 112 = 3248) m three-tower steel truss supporting. Among them, 36 bored piles with a diameter of 4.0m are used for the pile foundation of No. Z5# pier of the main branch channel bridge, and the length of the pile ranges from 67 m to 80 m. It is required that the weakly weathered tuff breccia at the bottom of the pile shall be not less than 4 m. The actual drilling depth is up to 90 m, and the depth into the rock is up to 20 m. The pile foundation entering the rock needs to adopt the step-by-step reaming drilling method to form the hole step by step. The overburden is about 68.4∼72.5 m thick, of which the silty sand layer is about 35∼46 m thick. There is a high risk of diameter shrinkage and hole collapse when drilling through the silty sand layer. The hole forming quality needs to be guaranteed through the analysis and research of hole wall stability and mud performance (Zhou & Miao, 2019). The underlying bedrock is a tuffaceous breccia, and the measured ultimate compressive strength is 64∼110 MPa.

---

*Corresponding Author: yetao1982827@163.com

DOI 10.1201/9781003308577-34

Table 1. Comparison and analysis of construction schemes.

Serial number	Comparison items	Whirling drilling rig	Rotary drilling rig
One	Quality	Less disturbance to the hole wall; Small risk of hole collapse	Poor stability of hole wall; High requirements for mud quality; more accurate geological verification
Two	Security	Large lifting equipment is required for displacement, and the lifting risk is high	Self-shifting can be realized
Three	Effect	Low efficiency of overburden drilling	Fast overburden drilling efficiency: the efficiency of hole formation can be accelerated by cutting the rock layer in stages
Four	Construction period	Hole forming experience of similar projects: 20d/piece, eight months	Hole forming experience in similar projects: 10d/piece, four months
Five	Environment protection	It can make slurry independently, with a low utilization rate of slurry and large treatment capacity of waste slurry; low noise	The mud can be recycled, and the mud volume is only 1/3 of that of a whirling drilling rig

To complete the construction of all bored piles before the flood season, based on analysis and the successful experience of a domestic rotary drilling rig in the construction of rock socketed piles, the high torque rotary drilling rig combined with a multi-bit graded hole forming scheme for the construction of 4.0m large-diameter rock socketed piles is selected.

At the construction site, XR550D, XR600E, XR700E, and XR800E large rotary drilling rigs produced by Xuzhou Construction Machinery Group Co., Ltd. are used for drilling (Hou, 2016). Among them, the XR800E drilling rig adopts the world's largest tonnage rotary drilling rig, with the maximum output torque of powerhead of 792kN·m, a maximum drilling diameter of 4.6m, and a maximum drilling depth of 150m.

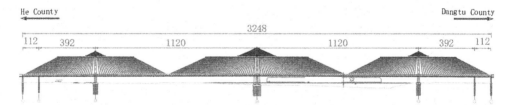

Figure 1. Vertical layout of bridge type of Ma'anshan highway railway Yangtze river bridge.

2 CONSTRUCTION PREPARATION

According to the hydrological characteristics of the river, the island-building method is used to build the drilling platform for the foundation of the No. Z5# pier pile in the main bridge. In this method, the steel casing construction is carried out based on the drilling platform. Then, the advanced XR550D, XR600E, XR700E, and XR800E drilling rig series hydraulic power head drilling rig construction is adopted for φ 4m bored pile, the large rotary drilling rig combined with the multi-stage drill bit is used to drill the hole in stages, and the monitoring of pile foundation is strengthened. The construction process of a large-diameter rotary drilling pile is as follows:

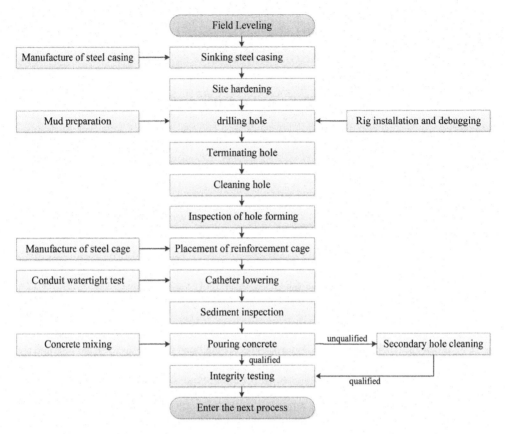

Figure 2.   Construction process flow of large diameter rotary bored pile.

2.1 *Construction of island-building and drilling platform*

The drilling platform of No. Z5# pier is constructed by the island-building backfilling method, and the plane size is 118 m × 56 m, the top elevation after backfilling is + 8.0 m (original ground + 7.0 m), and the bearing capacity of the foundation after leveling and compaction is not less than 80 kPa. Annular construction channels are set around the drilling area for the passage of mechanical equipment. The circular channel is hardened by a concrete precast slab combined with joint filling concrete. The drilling area is hardened with C25 reinforced concrete with a thickness of 25 cm. One slurry pit (divided into the slurry tank, slurry storage tank, and sedimentation tank) is set at the side of the annular channel near the embankment, with a plane size of 53 m×20 m. The layout of the drilling platform is shown in Figure 3.

2.2 *Steel casing construction*

The bored pile of No. Z5# pier is constructed with a 14 m steel casing with an inner diameter of 4.2m, a wall thickness of 24 mm, and a weight of 37.7t. To prevent the steel casing from deformation during transportation, the upper and lower parts of the steel casing shall be installed. Four [16a channel steels are set at the lower opening and welded into a "" 米 "" shaped support. After the steel

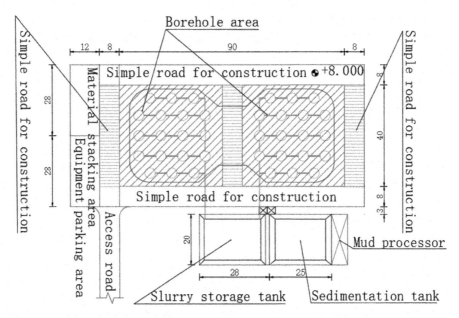

Figure 3. Layout plan of Z5# pier bored pile construction.

casing is manufactured in the professional processing plant, it is transported to the site wharf by 600t deep cabin barge, and the steel casing of No. Z5# pier is sunk with the cooperation of 320t crawler crane + Yong'an YZ-400L vibrating hammer and guide frame. The sinking sequence of the casing is to advance one by one from both sides to the middle (i.e., A1 → A2 → B1 → B2 → C1). During the construction, after the sinking of 4 pile casings on both sides is completed, the sinking of the last pile casing in the middle can be carried out (Wu, Zhou, 2018).

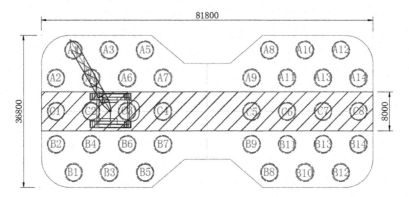

Figure 4. Construction sequence of steel casing.

3 DRILLING CONSTRUCTION

3.1 *Graded drilling construction*

According to the hydrological characteristics of the construction site, the static mud wall protection bucket process is used for soil borrowing by the No. Z5# pier drilling. During the drilling process, it is necessary to ensure that all mud indexes are qualified, and the drilling rig shall be able to rotate well locally. When drilling for soil, the soft soil layer is mainly cut by the weight of the combined drill pipe and the drill barrel with different diameters. The soil is circumscribed by the oblique drill teeth during the rotation of the drill barrel, and the soil is collected by the cavity in the drill barrel. When entering the hard formation, if the weight of the drilling gear is too small to cut into the formation, the self-contained oil cylinder can be used to pressurize, increase the force on the drill pipe, and cut the drilling gear into the corresponding formation by relying on the external pressure (Long, 2013). After the cavity in the drill barrel is filled with soil, the bucket door at the end of the drill barrel is rotated and closed counterclockwise, then, the drill pipe and the drill barrel are lifted out of the hole, and the bucket door is rotated clockwise to let the soil fall out of the cavity under the action of gravity. The drill pipe is repeatedly in place, and the bucket door is rotated and closed counterclockwise again. Repeat the above steps for subsequent soil collection. For weakly weathered argillaceous sandstone and other formations with high strength, the drill bit form can be adjusted, and the short screw type can be adopted.

In view of the disadvantages of the traditional large-diameter bored pile with a whirling drilling rig processes, such as great difficulty, low efficiency, and serious mud pollution, this paper puts forward a new method of multi-stage drilling with a large rotary drilling rig. This method is characterized by the early use of an XR550D drilling rig with a small torque value φ 1.8m and φ 2.8m drill bit for staged combination drilling and hole forming. At the later stage, the torque value of the drilling rig was gradually increased according to the change in geological conditions, and XR600E and XR800E drilling rigs were selected with φ 3.5m and φ 4.0m drill bits for staged combination drilling and hole forming; through the flexible switching of three types of drill bits (see Figure 5) of coring barrel drilling bit, sand fishing drilling bit and expanding "primary-secondary" drilling bit with different diameters (Liu, Zhou, Dai, 2018), the drilling slag is fished out and φ 4.0m bored pile is effectively drilled in the project.

(a) Coring barrel drilling bit   (b) Sand fishing drilling bit   (c) Expanding "primary-secondary"
                                            with pick   drilling bit with different diameters

Figure 5.   Three types of drill bits for hole forming.

(1) Pore forming sequence

Due to the close distance between the holes of the bored pile and the silty fine sand layer below the bottom opening of the pile casing, the construction of the bored pile shall be arranged in strict accordance with the principle of hole spacing construction in order to prevent hole string during drilling and concrete pouring (Zhong, 2016). Four drilling rigs (two XR550D are pilot drilling rigs, one XR600E and one XR800E are final drilling rigs) are drilled in 18 batches with the numbers of A and B (Wang, 2021).

(2) Pore forming process

This project uses four rotary drilling rig combinations of XR550D, XR600E, XR700E, and XR800E at φ 4.0m bored pile for staged reaming construction. The overburden is divided into

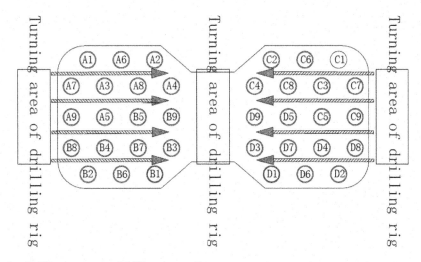

Figure 6. Drilling sequence and drilling rig traveling route.

three stages of 2.8m, 3.5m, and 4.0m, and the rock strata are divided into four stages of 1.8m, 2.8m, 3.5m, and 4.0m, as shown in Figure 7 below.

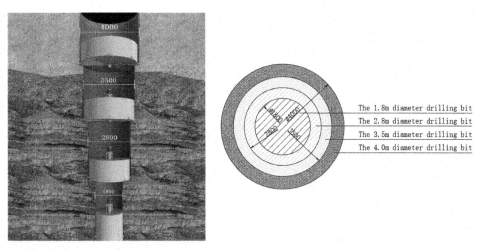

(a) Diagram of graded hole forming of 4m super large diameter bored pile

(b) Diagram of drilling sequence of rotary drilling rig

Figure 7. Construction diagram of multi-stage bit grading hole forming.

The technical innovation belongs to a new method of hole forming of bored piles. It is suitable for a hole forming of a large diameter rock socketed pile. The specific drilling sequence is as follows. The corresponding drilling steps are shown in Figure 8 below.

(1) Firstly, an XR550D rotary drilling rig with φ 2.8m sand fishing drilling bit is used to drill into the overburden to the top of the rock strata;

(2) The XR550D rotary drilling rig equipped with φ 2.8m-φ 1.8m "primary-secondary" drilling bit is used for fixed core drilling of about 2m;

(3) φ 1.8 m coring barrel drilling bit and φ 1.8m sand fishing drilling bit are used alternately to drill to the hole bottom elevation;

(4) φ 2.8m coring barrel drilling bit and φ 2.8m sand fishing drilling bit are used alternately to drill to the hole bottom elevation;

(5) XR600E rotary drilling rig with φ 3.5m sand fishing drilling bit is used to drill into the overburden layer to the rock strata;

(6) XR800E rotary drilling rig with φ 4.0m sand fishing drilling bit is used to drill into the overburden layer until the rock strata cannot be drilled (12.2m for weakly weathered layer, 9m for drilling further);

(7) XR800E rotary drilling rig with φ 3.5m coring barrel drilling bit and φ 3.5m sand fishing drilling bit are used alternately to drill until they reach the elevation of the hole bottom.

(8) XR800E rotary drilling rig with φ 4.0m coring barrel drilling bit and φ 4.0m sand fishing drilling bit are used alternately to drill until they reach the elevation of the hole bottom.

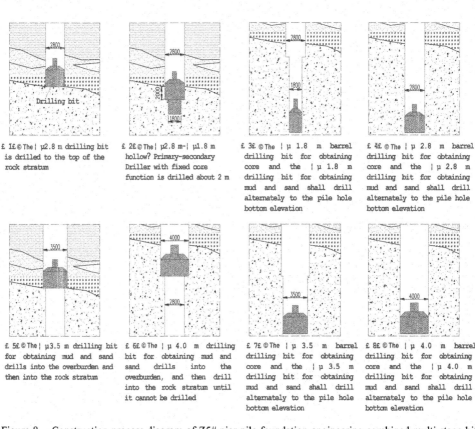

Figure 8.   Construction process diagram of Z5# pier pile foundation engineering combined multi-stage bit drilling in stages.

(3) Selection of the drill bit in general mode (non-rock overburden)

In non-rock strata such as muddy clay and silty sand layer, the general process is adopted. In this case, large-diameter rotary sand bailing coring barrel drilling bit with the same diameter pile is selected to drill directly to the end. During the construction process, the allocation of mud and the maintenance of the water head should pay attention to, and the proportion of mud should be adjusted according to the changes in strata (Liu, 2012). In addition, the electronic control system instrument panel equipped with the rotary drilling rig itself monitors the verticality of the borehole in real-time and controls the verticality of the hole in non-rock strata through the vertical joint measurement and inspection of the total station (Wen, Wu, 2017).

(4) Selection of the drill bit in rock entry mode

When the construction is completed with the general model, the bit will enter the rock strata. In this case, the strata density increases, and the sand fishing rotary coring barrel drilling bit cannot meet the requirements of drilling capacity. According to the grading plan, the 1.8m small-diameter short spiral cone coring barrel drilling bit or cone pick combined coring barrel drilling bit is selected firstly, and the cone or cone pick is staggered inward and outward, so that the free ends can be formed between the bit teeth (Ji, fan, Li, 2017).

When the drilling process with a 1.8 m coring barrel drilling bit is completed, the pilot drill bit with 1.8 m in front and 2.8 m in rear shall be used. Among them, the 1.8 m front end acts as the guide head of the 2.8m tail end for positioning and guiding, which controls the rock cutting positioning of a large-diameter drill bit and is beneficial to ensuring the verticality of the pile foundation (JIA, MENG, 2018). The same as above, and the drilling process is carried out with a 2.8 m diameter pilot drill bit in advance, followed by a 3.5m diameter pilot drill bit to continue drilling until the design hole depth is reached, and then replaced with the 4 m diameter pick coring barrel drilling bit to sweep the hole to ensure the hole depth and hole diameter. During construction, it should be paid attention to control the footage and rotation speed of different strata(Wang, Xia, 2017). At the same time, the drill bit should be checked timely to replace and adjust the drill teeth in time according to the judgment of the rock strata according to the pressure value fed back by the drill bit to the instrument and slag sample.

3.2 *Drilling efficiency analysis*

According to the statistics of drilling records, the fastest drilling of No. Z5# main pier bored pile takes seven days (68m pile length), and the slowest drilling takes 20 days (80m pile length). Without the influence of special reasons, the normal drilling time of each pile is 8-9 days. According to the normal drilling speed on-site, the footage shall be measured every certain time in graded drilling, records shall be made, and the drilling record form shall be filled in. According to the actual drilling situation of 36 piles of No. Z5# pier pile foundation, the drilling speed of bits at all levels in each layer is summarized as follows:

Table 2.   Efficiency analysis of 2.8m sand fishing drilling bit drilling overburden.

Geological category	Soil layer conditions	Drilling rig	Drilling rate	Remarks	Drilling conditions
Planting soil	Loose		5.0-6.0m/h		
Silty clay	Soft plastic		3.0-4.5m/h		
Muddy silty clay	Flow plastic		4.0-5.0m/h		
Silt	Loose		4.5-6.5m/h		
Silt	Medium density		1.5-3.0m/h		
Coarse sand	Dense	XR550E	1.5-2.5m/h		
Fine round gravel soil	Dense		0.5-1.0m/h	Replace with 1.8m coring barrel drilling bit	

Table 3. Analysis of subsequent drilling efficiency at all levels.

Drilling rig	Drilling bit	Geological category	Soil layer conditions	Drilling rate	Drilling conditions
XR550E	1.8m coring barrel drilling bit	Weakly weathered tuff breccia	Soft rock~Hard rock	0.5-1.0m/h	
		Slightly weathered tuff breccia	Relatively hard rock	0.3-0.5m/h	
	2.8m coring barrel drilling bit	Weakly weathered tuff breccia	Soft rock~Hard rock	0.5-1.5m/h	
		Slightly weathered tuff breccia	Relatively hard rock	0.1-0.5m/h	
XR600E	3.5 reaming drill	Overburden	Soil-layer	7.0-9.0m/h	
	3.5m coring barrel drilling bit	Weakly weathered tuff breccia	Soft rock~Hard rock	0.5-1.0m/h	
		Slightly weathered tuff breccia	Relatively hard rock	0.1-0.3m/h	
XR800E	4.0m reaming drill	Overburden	Soil-layer	10.0-12.5m/h	
	4.0m coring barrel drilling bit	Weakly weathered tuff breccia	Soft rock~Hard rock	0.5-0.8m/h	
		Slightly weathered tuff breccia	Relatively hard rock	0.2-0.4m/h	

According to the analysis of drilling efficiency completed (excluding special circumstances): the 2.8m sand fishing drilling bit has high drilling efficiency in the overburden. When it reaches the fine round gravel soil, the drilling efficiency begins to decline. It is necessary to immediately replace the 1.8m coring barrel drilling bit to improve the drilling efficiency. Due to the graded drilling, the borehole diameter increases at all levels, and the drilling capacity is also higher and higher. According to the field equipment, the configuration combination of the drilling rig and the drill bit is solidified, and the drilling efficiency is greatly improved.

3.3 *Drilling index parameter control*

The drilling speed can be appropriately increased after drilling to 3m from the bottom of the casing. While drilling, the verticality of the drill pipe needs to be monitored by a specially assigned person, and the drilling position can be understood by operating the hole depth digital display instrument number on the instrument panel. If the drilling is stopped, the water head in the hole shall always be more than 2 m higher than the water level outside the hole (GUO, CHEN, WANG, et al., 2021). During drilling, various indexes of mud should be measured (Li, Zhong, 2000; YANG, HU, 2020), and mud performance indexes should be adjusted timely for different strata to ensure hole forming quality, geological slag samples should be collected in time, and drilling speed and footage should be adjusted according to different strata (Liu, Yu, Bai, et al.2009). See Tables 4 and 5 for details.

Table 4.　Drilling reference index parameters of rotary drilling rig after statistical analysis.

Soil-layer	Mud weight / $(g\cdot m^{-3})$	Bit pressure	Rotation rate	Lifting and lowering bit	Drifting footage / （m/h）
Muddy silty clay	1.08~1.12	Light, not pressurized	Slow	Turn and lift 、slowly	4.0~5.0m/h
Silty clay	1.05~1.12	Light, not pressurized	Slow	Turn and lift 、slowly	3.0~4.5m/h
Medium sand	1.15~1.18	Light, not pressurized	Slow	Turn and lift 、slowly	1.5~3.0m/h
Round gravel sand	1.15~1.23	Light, not pressurized	Slow	Turn and lift 、slowly	0.5~1.0m/h
Weakly/strongly weathered tuff breccia	1.15~1.25	Apply pressure	Medium speed	Medium speed	0.1~0.5m/h

Table 5.　Construction progress indicators of bored cast-in-place pile with rotary drilling rig (diameter 4m, drilling depth 69m).

Subdivisional works	Rig preparation /h	Drilling /h	First hole cleaning and hole forming inspection /h	Hoisting reinforcement cage /h	Install the guide tube /h	Second hole cleaning /h	Pouring underwater concrete /h	Total /h
Single pile operation time	3.0	192.0	36.0	10	2.0	12.0	10	265
The comprehensive pile forming capacity is 8~11d/1 pile. Compared with other similar projects, the time of a single pile is saved by more than half								

4 CONCLUSIONS

In this paper, the construction technology of super large diameter pile foundation based on multi-stage drill staged hole forming technology is proposed, which effectively solves the problem of insufficient torque of one-time hole forming of the rotary drilling rig. The general drilling mode is adopted for a rotary drilling rig in non-rock strata, and the pressure grading drilling method is adopted for rock strata construction. In addition, the parameter requirements of pressurized medium speed drilling under rock strata are studied. This process can significantly improve the drilling efficiency of bored piles, save construction period, reduce pollution, and have obvious advantages of energy conservation and environmental protection. It fills the gap in using a large rotary drilling rig and provides a reference for the construction of a large bridge pile foundation under complex conditions.

(1) Basis for switching between two-stage drilling: the switching time of the graded drilling bit can be judged according to the torque of the drilling rig torque display screen and the fluctuation of the drill pipe.

a. The switching between different types of bits with the same diameter is determined by the functions of different bits.

For example, after the φ 2.8m coring barrel drilling bit is used for operation and the separated rock sample of the cylinder or annular body is formed, it is generally replaced by a φ 2.8m sand fishing drill bit to pick up broken slag.

b. The switching between bits of the same type and different diameters is determined by different requirements for drilling torque.

For example, when the φ 2.8m coring barrel drilling bit fails to move, it is generally replaced with a smaller one φ 1.8m same type of coring barrel drilling bit operation to reduce damping torque.

c. In the process of graded hole forming, in order to prevent the pile position eccentricity of graded hole forming, the expanded "primary-secondary" drilling bit is adopted.

(2) If the drilling footage is obviously slow or stagnant during on-site drilling, it is needed to switch to a smaller one-stage bit to continue drilling. A total of seven kinds of drill bits with diameters of φ1.5m, 1.8m, 2.2m, 2.8m, 3.2m, 3.5m, and 4.0m can be combined with multiple selections and switched flexibly. To strengthen the quality of mud wall protection in the upper strata prone to collapse, the footage rate of silty fine sand strata shall be appropriately slowed down, and the WOB shall be appropriately increased to speed up the footage rate after entering the rock is stable.

(3) The secondary hole cleaning mud has a high sand rate and takes a long time. In order to reduce the fine silty sand in the formation dissolved in the mud during the hole forming process, the hole forming diameter of the overburden is adjusted from 2.8m to 3.5m, reducing the content of fine silty sand brought in by the subsequent 4.0m bit drilling. At the same time, to improve the sand filtering effect, a 400-mesh screen (0.425mm) sand filter is added. After passing through the 400-mesh screen sand filter, the sand content of the mud is reduced, the sand rate is between 4% - 8%, the specific gravity of the mud is between 1.08-1.15, and the viscosity is 18-22s, which not only shortens the sand filtering time, but also takes into account the sand rate and viscosity index of the mud. This will also greatly reduce the pile forming period of a single pile and reduce the risk of hole collapse.

(4) The function of the first cleaning is to stabilize sand; the second cleaning is used for sand removal. In the sand removal stage, the higher the colloid rate, the higher the sand removal efficiency.

Data availability

The data used to support the findings of this study are included in the article.

Conflicts of interest

The authors declare that there are no conflicts of interest regarding the publication of this paper.

ACKNOWLEDGMENTS

The work described in this study was supported by the Major Scientific Research Project of China Railway Shanghai Group Company Ltd. in 2021: "Study on the full set of rotary drilling technology for 4m super large diameter bored pile" (Project No.: 2021226). The field construction/test data of bored piles in this study come from this project, which guarantees the completion of this study. Any opinions and concluding remarks presented here are entirely those of the authors.

REFERENCES

Guo Hang, Chen Kai-qiao, Wang Tong-min, et al. Key Construction Techniques for North Navigation Channel Bridge of Bianyuzhou Yangtze River Bridge[J] World bridges, 2021, 49 (6): 14–20. in Chinese.

Hou Fuzhan. Research on construction technology of bored cast-in-place pile for modern bridge[J]. Construction engineering technology and design, 2016 (34).

Ji Zunzhong, fan Lilong, Li Zhihui. Study on comprehensive technology of foundation construction of sea-crossing bridges in typhoon area[J]. Railway construction technology, 2017 (8) 1–4, 36.

Jia Wei-zhong, Meng Chao. Key Construction Techniques for Foundation of Main Pier of Dongting Lake Bridge on Mengxi-Huazhong Railway[J] Bridge Construction, 2018, 48(5):1–5. in Chinese.

Li Ming. Zhong Xiang. Deepwater foundation construction technology for main pier of Hanjiang River Super Large Bridge[J]. Railway construction technology, 2000 (6): 10–13.

Liu Weifang, Zhou Changdong, Dai Mingjing Application of high-pressure grouting method in the treatment of thick sediment layer of large diameter cast-in-place pile[J] World bridge, 2018, 46 (4): 46–49.

Liu Yang. Discussion on construction technology of bridge pile foundation[J]. Urban architecture, 2012 (15): 148.

Liu Yaodong, Yu Tianqing, Bai Yinghua, et al. Research on key technology of deep water super long and large diameter bored pile construction[J]. Journal of civil engineering and management, 2009, 26 (2): 57–60.

Long Bo. Key points of bridge bored pile construction control[J]. Traffic world, 2013 (7): 236–237.

Wang Qingjun, Xia Wenchao. Research on construction technology of bridge pile foundation[J]. Engineering construction and design, 2017 (17): 155–156.

Wang Tongmin Construction technology of super-long pile for N15 pier of Beicha channel bridge of bream Chau Yangtze River Bridge[J] Bridge construction, 2021, 51 (6): 118–125.

Wen Pei, Wu Dongdong Preparation and application of high-quality PHP Slurry for large diameter and super long bored pile in silt, sand, and gravel comprehensive geological layer[J] World bridge, 2017, 45 (5): 17–21.

Wu Xianfeng, Zhou Dong, Shi De Fast hole forming construction technology of hard rock bored pile by rotary drilling, ring cutting, and coring[J] China new technology and new products, 2018 (13): 90–92.

Yang Xue-xiang, HU Hao. Key Pile Hole Forming Techniques for Pile Foundation of Main Pier No. 15 of Wuxue Yangtze River Highway Bridge[J] Bridge construction, 2020, 50 (1): 111–115. in Chinese.

Zhong Tianbin. Discussion on pile foundation construction of road and bridge[J]. Architectural development orientation, 2016, 14 (6): 222–223.

Zhou wai-nan, MIAO Yu-lu. Using Rotary Drilling Rigs to Make Holes for Bored Piles in Hard Rock Strata[J]. World Bridges, 2019, 47(4):27–31. in Chinese.

Frontiers of Civil Engineering and Disaster Prevention and
Control – Yang & Rahman (Eds)
© 2023 The Author(s), ISBN: 978-1-032-31200-2

Seismic performance analysis of a super high-rise office building

Xiaomeng Zhang*, Wenting Liu, Xiao Yang & Ziao Liu
China Architecture Design & Research Group, Beijing, China

Yao Feng
Beijing Guobiao Building Technology Co., Ltd, Beijing, China

ABSTRACT: A super high-rise building, 238-m high, four floors underground, 50 floors above the ground, the structural system is the frame-core tube, super B height building, for the ultrastructure. After conducting elastic time history analysis of small earthquakes and critical component analysis of medium earthquakes, the load-bearing capacity and mechanical performance are analyzed, and the safety of structure design is determined to ensure that the requirements of seismic performance of structure are met.

1   PROJECT OVERVIEW

The project is a 31-floor office building standing and a height of 149 m. It is composed of concrete-filled steel tube frame and concrete-filled steel tube core. As the main anti-lateral force system, the shear wall provides the ability to resist wind load and horizontal earthquakes and bears the additional torsional effect caused by the non-coincidence of mass center and stiffness center. The outer frame mainly bears the vertical load and the role of the second seismic defense line. Due to the high floor, it belongs to the over-limit structure. The key indexes of seismic resistance are analyzed by elastic time-history analysis of small earthquakes and elastic-plastic analysis of medium earthquakes.

The design life of the project is 50 years, the safety level of the building structure is grade 2, and the fire level of the building structure is grade 1, which refers to the Code for Design of Concrete Structures GB50010-2010: (Beijing China building industry press 2010), and the design level of the foundation is Grade A. According to the "Code for Seismic Design of Buildings" (GB50011-2010) (2016 edition) (Beijing China building industry press 2010), the aseismic group 1 in the area where the project is located, the site category is category IV, the aseismic grade is 7 degrees 0.1g, and the office area is general fortification category (category c). The load refers to the Load Code for the Design of Building Structures (Beijing China building industry press 2012).

2   STRUCTURE SELECTION AND CLASSIFICATION

The elastic time history analysis of towers subjected to multiple earthquakes is carried out using YJK as a supplement to the calculation of the response spectrum of multiple earthquakes. Five natural waves and two artificial waves were used in the analysis. The acceleration time history curves of seismic waves are shown in Figure 2. When calculating the seismic response, the natural wave is input in three directions, and the artificial wave is input in two directions. The peak acceleration in the main direction and the second direction is adjusted in a proportion of 1:0.85. The calculation results are as follows:

---

*Corresponding Author: 155203255@qq.com

DOI 10.1201/9781003308577-35

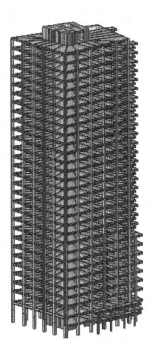

Figure 1. Structural calculation model.

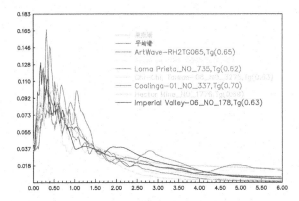

Figure 2. Comparison of gauge spectrum and seismic wave mean spectrum.

Floor shear under seismic wave action.

According to article 5.1.2 of the seismic code, in the elastic time-history analysis, the base shear calculated by each time-history curve should not be less than 65% of that calculated by using response spectrum method, and the average base shear calculated by multiple time-history curves should not be less than 80% of that calculated using response spectrum method.

The base shear under the action of 7 groups of time-history curves is shown in the table below. The ratio of the base shear under each time-history wave to the response spectrum results is 70%–123%, meeting the requirements of 65%–135% in the code. Its average value in two directions is 93% and 95%, respectively, which meets the requirements of specification 80%–120%.

The comparison of shear force between time history analysis and reaction spectrum analysis in the X and Y directions is given below. Each layer shear after the seismic time history analysis and

Table 1. Base shear forces under different seismic waves

Name of seismic wave	X-direction		Y-direction		Specification limits
	Base shear MN	Base shear/Response spectrum base shear	Base shear MN	Base shear/Response spectrum base shear	
ArtWave-RH2 Artificial wave	11.2	0.76	17.9	0.98	≥ 65%, ≤ 135%
ArtWave-RH1 Artificial wave	12.6	0.86	17.5	0.96	≥ 65%, ≤ 135%
Loma Prieta_NO_735 Natural wave	18.8	1.28	24.1	1.33	≥ 65%, ≤ 135%
Chi-Chi, Taiwan-06_NO_3275 Natural wave	12.3	0.84	14.9	0.81	≥ 65%, ≤ 135%
Coalinga-01_NO_337 Natural wave	13.7	0.93	15.6	0.86	≥ 65%, ≤ 135%
Hector Mine_NO_1776 Natural wave	13.3	0.91	12.3	0.67	≥ 65%, ≤ 135%
Imperial Valley -06_NO_178 Natural wave	13.2	0.90	22.5	1.24	≥ 65%, ≤ 135%
Means	13.6	0.92	17.9	0.98	≥ 80%, ≤ 120%
Response spectrum values	14.6		18.1		–

its average envelope value compared with the response spectrum analysis of the shear layer, two directions response spectrum analysis of the main shear layer are slightly lower than the average time history analysis of the shear layer, the layer average layer shear response spectrum analysis and time history analysis are in good agreement. In the design of components, the average value of time-lapse analysis shear force calculated by the response spectrum can be amplified to obtain greater structural safety.

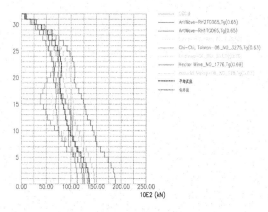

Figure 3. Comparison of shear force between time history analysis and reaction spectrum analysis (X direction).

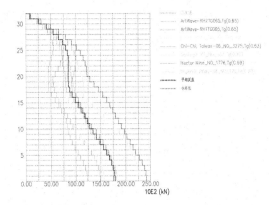

Figure 4. Comparison of shear force between time history analysis and reaction spectrum analysis (Y direction).

3 CALCULATION AND ANALYSIS UNDER MEDIUM EARTHQUAKE

In areas of high intensity, shear wall limbs, especially those on the periphery of the structure, are prone to large tensile stress. This section helps provide the ratio between the tensile stress of the wall limbs at the bottom of the tower core tube (i.e., the first floor of the building) and the standard value of concrete tensile strength. It can be seen that only the wall at the corner of the outer ring shows tensile stress, and the value is small. The remaining wall without numerical values is free from tensile stress. The results show that the average nominal tensile stress generated by the axial force of the whole section of the wall limb under the action of a bi-directional horizontal earthquake does not exceed the standard value of the tensile strength of concrete, which is in line with the code and the relevant requirements of over-limit engineering.

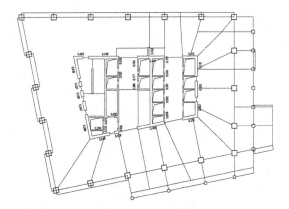

Figure 5. First floor wall limb tension stress.

Wall reinforcement checking

According to the structural performance target, the shear capacity of the frame column and core tube under moderate earthquake is designed according to elasticity, and the flexural capacity is designed according to the unyielding. During the design process, the seismic grade of the wall and frame column in the bottom strengthening area is one level.

The reinforcement results of the shear wall and frame column on the bottom of the building, the body intake, and the main floor are shown in Figure 6. It can be seen that the reinforcement of the normal section of the wall on the other floors is small, except for the reinforcement at the corner of the bottom, which meets the performance target of flexural resistance and non-yield in a medium earthquake.

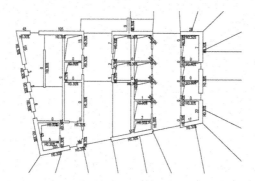

Figure 6.   Reinforcing bars for concrete members on the first floor.

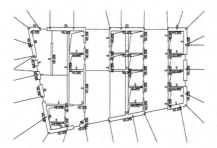

Figure 7.   Reinforcing bars for concrete members on the second floor.

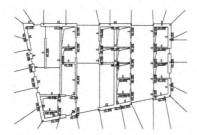

Figure 8.   Reinforcing bars for concrete members on the third floor.

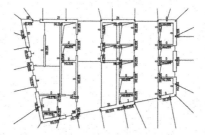

Figure 9.   Reinforcing bars for concrete members on the fourth floor.

The horizontal reinforcement arrangement of the wall, except for some walls, according to the structure of the reinforcement, all meet the shear non-yield performance target.

REFERENCES

Code for seismic design of buildings GB 50011-2010 [s] Beijing China building industry press.
Code for design of concrete structures GB50010-2010: GB 50010—2011 [S] Beijing China building industry press, 2011.
Load code for the design of building structures GB 50092012[s] Beijing China building industry press.
Technical specification for concrete structures of tall building JGJ 3-2010[s] Beijing China building industry press.

Frontiers of Civil Engineering and Disaster Prevention and
Control – Yang & Rahman (Eds)
© 2023 The Author(s), ISBN: 978-1-032-31200-2

Analysis of bearing characteristics of soft soil foundation strengthened by bamboo net covered with sand

Shouxin Han
CCCCRoad & Bridge Special Engineering Co., Ltd., Wuhan, Hubei, China

Jihui Ding*, Shengkang Yan, Zenghui Yu & Hanchen Wang
College of Civil Engineering and Architecture, Hebei University, Baoding, China

ABSTRACT: In soft soil foundation treatment, the bamboo net method can quickly and effectively solve the problem that machinery cannot enter the site, but its bearing capacity calculation lacks a theoretical basis. Based on the theoretical analysis of the double-layer foundation, the calculation formula of the bearing capacity of the soft soil foundation strengthened with bamboo nets is given. The composition and law of bearing capacity of the foundation are analyzed through an indoor load test. Under the pressure of overlying sand and load plate, the strain of the bamboo rod at the bottom of the bamboo net mainly occurs below the load plate. The strain of the outermost bamboo pole of the load plate is 8.3% ∼ 18.57% of the strain of the bamboo pole under the load plate. The corresponding stress law of the bamboo net is similar to the strain law. It provides a basis for the design of strengthening soft soil foundation with the bamboo net method.

1 INTRODUCTION

For sites with thick silt or soft soil layers, high water content, and low bearing capacity, construction machinery cannot operate. The use of reinforced cushions can improve the bearing capacity of silt and soft soil foundations. Currently, commonly used reinforced materials, such as geogrids, have a certain degree of pollution to the environment. However, bamboo has a short growth period, high tensile strength, and flexural rigidity. Using bamboo nets as a reinforcement to reinforce silt or soft soil has the advantages of stable structure, low price, convenient construction, energy-saving, and environmental protection.

In the 1990s, Malaysia, Myanmar, South Korea, and other countries with a large number of soft soils carried out "bamboo support + geotextile" reinforcement for soft foundation treatment, and the bearing capacity of the foundation after treatment was significantly improved. Chen Chunan, Zhang Jichao, et al. (2007) (Chen & Xu 2007). In the land reclamation project, bamboo nets + geotextile cushions were used to treat soft soils with thin mud on the mud surface and water content greater than 100%. Huang Songtao (2012) (Zhang et al. 2012) used bamboo frames to be laid in vertical and horizontal directions, and woven fabrics were laid on top of it to treat the soft soil foundation of the Dachan Bay Phase II Project. Mokhammad Farid Ma'ruf (2012) (Mokhammad Farid Ma'ruf, 2012) used large box-type direct shear test results to show the peak shear strength of bamboo-root-reinforced soil increased with the increase in soil-root volume ratio. Aminaton Marto, Bakhtiar Affandy Othman (A Marto 2005; Aazokhi 2014), etc. through field comparative tests show that compared with the high-strength geogrid system, bamboo mesh reinforced peat soil can save about 46.4% of the cost per square meter, and reduce about 28 % Of consolidation settlement. Hegde and Sitharam (2015) (Hegde & Sitharam 2015) indoor tests show that the tensile strength of bamboo is about nine times higher than that of commercial geocells. The bearing capacity of the foundation is increased by 4 to 5 times after the bamboo cells are used for reinforcement. Asaduzzaman, Islam

---

*Corresponding Author: dingjihui@126.com

DOI 10.1201/9781003308577-36

(2018) (Amarnath & Sitharam 2015) obtained through laboratory experiments, when the bamboo reinforcement is placed within the failure envelope depth, the bearing capacity of the single-layer reinforced soil and the multi-layer reinforced soil system are respectively increased by 1.77 than that of the unreinforced soil. Times and 2.02 times. Aazokhi Waruwu (Md & Muhammad, 2014) experimentally found that bamboo grid reinforcement can increase the subgrade modulus and soil shear modulus. Engineering practice has proved that in the treatment of soft soil foundation, the bamboo net method can quickly and effectively solve the problem that personnel and machinery cannot enter the site. It is difficult, but the calculation of the bearing capacity of the bamboo net method is mainly judged based on the experience of the engineer and lacks a theoretical basis. Therefore, this paper adopts the double-layer foundation theory to analyze the bearing capacity of soft soil foundation reinforced by bamboo nets and verifies it through an indoor load test, which provides a basis for similar soft soil foundation treatment.

2 CALCULATION OF BEARING CAPACITY OF BAMBOO NET REINFORCED SOFT SOIL SURFACE FOUNDATION

The bamboo net method to reinforce the super-soft soil surface foundation by hydraulic filling consists of four parts, namely the upper cover (usually medium-coarse sand, as a drainage cushion for subsequent treatment), geotextile (located between the bamboo net and the upper cover), bamboo net, and the super-soft soil on the surface of the hydraulic fill foundation. The bamboo net-geotextile-soil interaction and coordination form an inseparable whole. The bearing capacity of the super-soft soil foundation reinforced by the bamboo net method is composed of two parts: the bearing capacity of the unreinforced material foundation + the bearing capacity increment produced by the reinforcement.

2.1 *Calculation of bearing capacity without the bamboo net*

In the overlying sandy soil layer on the soft soil layer, the thickness H of the hard sandy soil layer is smaller than the grounding width B of the construction machinery and equipment. Hanna A.M.'s research results show that punching shear failure will occur in the upper soil layer, and overall shear failure will occur in the lower soil layer. The foundation bearing capacity of the upper hard and lower soft soil layer is composed of the upper and lower layers, namely:

$$q_u = q_b + 2\frac{c_a H}{B} + \gamma_1 H^2 (1 + \frac{D}{H})\frac{k_s \tan \varphi_1}{B} - \gamma_1 H \tag{1}$$

The ultimate bearing capacity of the underlying soil layer is:

$$q_b = c_2 N_{c2} + \gamma_1 (D + H) N_{q2} + \frac{1}{2}\gamma_2 B N_{\gamma 2} \tag{2}$$

In the formula, q_u is the ultimate bearing capacity of the upper hard and lower soft soil layer; D is the buried depth of the foundation; H is the thickness of the upper hard soil; q_b is the ultimate bearing capacity of the lower layer; c_a is the attachment on the upper punching and shearing surface focus; k_s is the impact coefficient; γ_1 and γ_2 are the weights of the upper and lower soil layers respectively; N_{c2}, N_{q2}, $N_{\gamma 2}$ is a function of the lower soil layer φ_2, k_s is the Meyerhof bearing capacity coefficient; c_2 is the cohesive force of the underlying layer.

When $c_1 = 0$, $\varphi_2 = 0$, $D = 0$, substituting formula (2) into formula (1), then the formulas can be simplified to:

$$q_u = 5.14c_2 + \gamma_1 H^2 \frac{k_s \tan \varphi_1}{B} \tag{3}$$

$$q_b = 5.14c_2 + \gamma_1 H \tag{4}$$

Considering the influence of the basic shape, the formulas (3) and (4) are revised as follows:

$$q_u = \left(1 + \frac{B}{L}\right) \gamma_1 H^2 \frac{k_s \tan \varphi_1}{B} + 5.14 c_2 \left(1 + 0.2\frac{B}{L}\right) \tag{5}$$

$$q_b = 5.14 c_2 \left(1 + 0.2\frac{B}{L}\right) + \gamma_1 H \tag{6}$$

In the formula, B and L are the base width and length. The impact coefficient k_s is a function of $\frac{q_2}{q_1}$ and φ_1, and the expression of $\frac{q_2}{q_1}$ is:

$$\frac{q_2}{q_1} = \frac{10.28 c_2}{\gamma_1 B [\exp(\pi \tan \varphi_1) \tan^2 \left(45° + \frac{\varphi_1}{2}\right) - 1] \tan 1.4\varphi_1} \tag{7}$$

In the formula, q_1 and q_2 are the ultimate bearing capacity when the foundation is assumed to be located in the upper and lower soil layers. The relationship curve between k_s, $\frac{q_2}{q_1}$, and φ_1 (Qian & Yin 2012) can be obtained.

2.2 Ultimate bearing capacity of soft soil layer with the bamboo net on the surface

The bamboo net acts on the surface of the soft soil layer, which is covered with a sandy soil cushion. It can be considered that the bamboo net increases the bearing capacity of the soft soil layer, and the increase of Δq_b is calculated according to the Takeuchi formula.

$$q_{un} = \left(1 + \frac{B}{L}\right) \gamma_1 H^2 \frac{k_s \tan \varphi_1}{B} + 5.14 c_2 \left(1 + 0.2\frac{B}{L}\right) + \frac{2}{B} \sum \left(1 + \frac{H_i}{B}\right) T_{ai} \sin \theta_i \tag{8}$$

$$q_{bn} = 5.14 c_2 \left(1 + 0.2\frac{B}{L}\right) + \gamma_1 H + \frac{2}{B} \sum \left(1 + \frac{H_i}{B}\right) T_{ai} \sin \theta_i \tag{9}$$

In the formula, H_i is the thickness of the layer i bamboo net covering layer; B is the grounding width of the equipment; T_{ai} is the allowable tensile force of the reinforcement material; θ_i is the angle formed by the reinforcement material and the horizontal plane; σ_b is the tensile strength of the bamboo; R is the outer surface of the bamboo Diameter; r is the inner diameter of the bamboo; s_i is the bamboo spacing of the i-layer bamboo net; σ_{gc} is the tensile strength of the geotextile.

3 TEST ON BEARING CHARACTERISTICS OF BAMBOO NETS FOR STRENGTHENING SOFT SOIL SURFACE FOUNDATION

3.1 Bamboo

The bamboo net in the experiment uses Jiangsu Moso bamboo with a diameter of 1.0cm~1.2cm, the moisture content is 11.0%, the length of the bamboo joint is 17.5cm~23.3cm, and the average is 22.4cm. A layer of the bamboo net is 15cm×15cm apart. The flexural test of the single rod and bamboo net used in the experiment showed that the ratio limit of the single rod was 56.6MPa~107.2MPa, and the ultimate stress was 71.9MPa~117.8MPa, and the modulus of elasticity was 11.6GPa~19.6GPa.

3.2 Geotextile

The geotextile weighs 150N/m³, and its longitudinal tear strength is 5.1kPa. The transverse tear strength is 5.2kPa, the longitudinal elongation is 65%, the transverse elongation is 66%, the burst strength is 1kN, and the longitudinal and transverse tearing strength is 0.15kN.

3.3 *Test soil*

The test soft soil layer is taken from the soft soil belonging to Wenzhou Economic and Technological Development Zone. The site soil layer has 27.20m∼33.60m thick fluid plastic, high compression, high sensitivity soil, and the soft soil layer has a weight of 17.4kN/m^3 ∼18.3kN /m^3, the water content is between 47.31%∼61.43%, the liquidity index is greater than 1.50, and the plasticity index is greater than 18. The on-site soft soil is evenly saturated with water, and the water content is 60%. The gravity is 16.5kN/m3, and the void ratio is 1.55. As can be seen from Figure 2, $q_{bcr} = 6.06$kPa, $q_{bu} = 9.09$ kPa, $q_{bu}/q_{bcr} = 1.5$. According to formula (6), $c_2 = 1.47$ kPa can be obtained.

The sand used in the test passes through a 5-mm sieve. The content of particles larger than 2mm is 6.53%, the content of particles larger than 0.5mm is 48.68%, the content of particles larger than 0.25mm is 90.16%, and the content of particles smaller than 0.075 mm is 1.98%. The unevenness coefficient of the soil sample $C_u = 2.36$ and the curvature coefficient $C_c = 0.83$, which belong to poorly graded sand. The sandy soil has a gravity of 16.5 kN/m^3 and a moisture content of 0.25%. The friction angle is $\varphi = 30.5°$.

3.4 *Loading test of the bamboo net to strengthen the surface of soft soil*

As shown in Figure 1, the experimental device is mainly composed of a model box and a hydraulic jack. The length, width, and height of the model box are 100cm×100cm×100 cm. The thickness of the soft soil layer in the model box is 55cm, and a 15cm×15cm bamboo net is placed on the bamboo net. A layer of geotextile is arranged on the bamboo net, and the sand of different thicknesses is spread flat on it, and then the load test is carried out. The ratio of the thickness of the sand-covered sand on the bamboo net to the width of the load plate is called the thickness-to-width ratio (H/D=0.0∼0.78). The size of the square load plate is 256mm×256mm. Figure 1(b) shows the bamboo net with the sand cushion removed after the load test. Five strain gauges (SN-1∼SN-5) are arranged on the bamboo net before the test. The bamboo net is bent under the action of the overlying sand and the load, and the bamboo net is in close contact with the foundation. After the test, the bamboo net deforms and starts to rebound. After the load test, the bamboo net is taken out from the foundation, and the bamboo net is bent. Almost all the deformation is restored.

(a) Load test

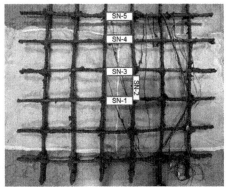
(b) Bamboo net after the test

Figure 1.   Load test of bamboo net reinforcement of soft soil surface treatment.

Figure 2 shows the $p \sim s$ curve when H/B=0.0∼0.78. When the H/B value is small, the non-linear section of the $p \sim s$ curve is not obvious. The survey system of the Ministry of Metallurgy (China Construction Industry Press, 2018) conducted a comparative test of the size of the bearing plate in Taiyuan and showed that with the change in the size of the bearing plate, the value of s/b is inversely proportional to b, and the nonlinearity of s/b increases with the increase of b. According to the building foundation design code (China Construction Industry Press, 2012), the ultimate bearing capacity of the foundation is taken as the corresponding static pressure when $s/b=0.06$. It can be seen from Figure 3 that when $s/b=0.06$, due to the size effect, the foundation soil layer has not reached the limit state. Based on the characteristics of this test and the size of the small load plate, the temporary plastic load of this test is taken as the pressure corresponding to $s/b=0.07$, and the ultimate load is taken as the pressure corresponding to $s/b=0.14$. Table 2 shows the analysis of load test results and calculation results when $H/B=0.0\sim0.78$.

Table 1. Analysis of bearing capacity.

H/B	P_{cr}(kPa)	P_u(kPa)	P_u/P_{cr}	q_b(kPa)	q_u(kPa)	Δq_b(kPa)	q_{bn}(kPa)	q_{bn}/q_{b0}
0	9.09	15.87	1.75	9.07	9.07	6.80	15.87	1.75
0.19	8.52	17.11	2.01	9.87	9.95	7.16	17.05	1.88
0.39	10.29	21.60	2.10	10.72	12.61	8.99	19.71	2.17
0.58	19.83	32.09	1.62	11.54	17.03	15.06	26.60	2.93
0.78	25.82	43.12	1.67	12.37	23.32	19.89	32.26	3.55

It can be seen from Table 1 that after the bamboo net + sand soil, the ratio of the surface ultimate bearing capacity to the characteristic value of the bearing capacity is between 1.75 and 2.10. The ultimate bearing capacity q_{bn} of the soft soil + bamboo net and the ultimate bearing capacity of the soft soil. The force q_{b0} ratio q_{bn}/q_{b0} is related to the H/B ratio. As H/B increases, the q_{bn}/q_{b0} ratio increases from 1.75 to 3.55.

Table 2. Strain value of measuring point on the bamboo net.

Number	$\mu\varepsilon(P_{cr})$			$\mu\varepsilon(P_u)$		
	$H/B=0.19$	$H/B=0.39$	$H/B=0.59$	$H/B=0.19$	$H/B=0.39$	$H/B=0.59$
SN-1	1511.8	1204.2	1706.1	3589	3209.9	2938.3
SN-3	1912.1	1048.9	1959.3	4492.7	3450.6	3577.9
SN-4	689.1	401.1	956.9	1705.1	1485.5	1661.1
SN-5	93.8	32.2	256.1	419.3	286.4	550.9

It can be seen from Figure 3 and Table 2 that the bamboo net is mainly bent under the pressure of the overlying sand and the load plate, and the strain value increases with the increase of the load plate pressure, and the relationship is basically linear. The tensile strain is greatest directly under the load plate, and as the distance from the center of the load plate increases, the strain decreases. When the pressure of the load plate is the characteristic value of the bearing capacity, the strain of the outermost layer is 2.57%~15.01% of the strain under the load plate; when the pressure of the load plate is the limit value of the bearing capacity, the strain of the outermost layer is the strain under the load plate 8.3% to 18.57% of strain.

The elastic modulus of bamboo is 15.6 GPa, and the bending tensile stress of the measuring point is shown in Table 3. From Table 4, it can be seen that the bending tensile stress law of the measuring point is consistent with the tensile strain law. When the load plate pressure is the characteristic value of the bearing capacity, the maximum tensile stress value under the load plate is between 16.36MPa~30.57MPa. When the load plate pressure is the limit value of the bearing capacity, the tensile stress value is between 45.84MPa~70.09MPa, which is less than the bending strength value. Therefore, the pile net is not damaged when the bearing capacity reaches the limit value. After the test, the bamboo net was taken out, and the bending deformation was basically restored.

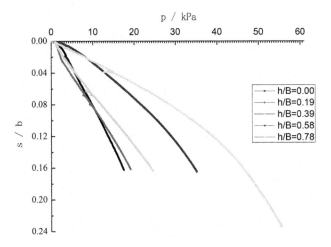

Figure 2. $p \sim s$ curve.

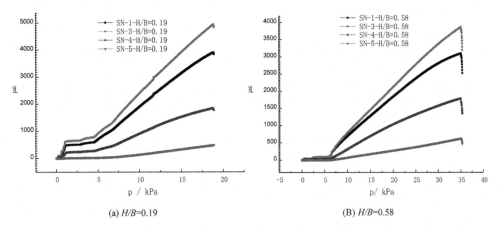

(a) *H/B*=0.19   (B) *H/B*=0.58

Figure 3.   The relationship curve between bamboo mesh strain and load plate pressure.

Table 3.   Stress values of measuring points on the bamboo net.

number	$\sigma_T(P_{cr})$ (MPa)			$\sigma_T(P_u)$ (MPa)		
	H/B=0.19	*H/B*=0.39	*H/B*=0.59	*H/B*=0.19	*H/B*=0.39	*H/B*=0.59
SN-1	23.58	18.79	26.62	55.99	50.07	45.84
SN-3	29.83	16.36	30.57	70.09	53.83	55.82
SN-4	10.75	6.26	14.93	26.60	23.17	25.91
SN-5	1.46	0.50	4.00	6.54	4.47	8.59

4   CONCLUSION

(1) The bearing capacity of the overlying sand + bamboo net to strengthen the soft soil is regarded as the upper hard and lower soft double-layer foundation. The foundation bearing capacity comprises three parts, the overlying sand, soft soil, and bamboo net bearing capacity. When

the thickness of the overlying sand and the grounding width of the equipment is small, the calculation of the bearing capacity of each part is given.

(2) The indoor bearing capacity test shows that the ratio of the ultimate bearing capacity of sand + bamboo net + soft soil to the characteristic value is between 1.75~2.10. The bearing capacity of bamboo net + soft soil increases significantly with the increase of H/B. When H/B=0.0~0.78, the ultimate bearing capacity of bamboo net + soft soil is 1.75~3.5 times the bearing capacity of soft soil.

(3) Under the pressure of the overlying sand and the load plate, the strain value of the bamboo net increases with the increase of the load plate pressure. The strain below the load plate is much greater than the strain of the bamboo rod outside the load plate when the load plate pressure is the bearing capacity. At the characteristic value, the strain of the outermost layer is 2.57%~15.01% of the strain under the load plate. When the pressure of the load plate is the limit value of the bearing capacity, the strain of the outermost layer is 8.3%~18.57% of the strain under the load plate.

(4) The bending tensile strain law of the bamboo net is similar to the strain law. Under the ultimate load, such as the load plate pressure, the corresponding bending normal stress under the load plate is 45.84MPa~70.09MPa, which is less than the ultimate bending strength. Therefore, when the foundation is damaged, the bending normal stress of the bamboo net has not yet reached the ultimate strength.

(5) Bamboo has high tensile strength and flexural rigidity. The use of the bamboo net as a reinforcement to reinforce soft or super soft soil has the advantages of stable structure, low price, convenient construction, energy-saving, environmental protection, and a wide range of application prospects. The formula for calculating the bearing capacity of the overlying sand + bamboo net to strengthen the soft soil and the deformation and stress law of the pile net provided in this paper provide a theoretical basis for the bamboo net to strengthen the soft soil foundation.

REFERENCES

A Marto. Bearing capacity of soil overlaid by sand with interfacial bamboo-geotextile composite reinforcement[J]. Advances in Environmental Biology, 2005, 5(7): 1751–1755.

A. Hegde, T. G. Sitharam. Use of Bamboo in Soft-Ground Engineering and Its Performance Comparison with Geosynthetics: Experimental Studies[J]. Journal of Materials in Civil Engineering. 2015, 27(9).

Aazokhi W. Bamboo Reinforcement in Shallow Foundation on the Peat Soil[J]. Journal of Civil Engineering Research. 2014, 4(3A): 96–102.

Amarnath H, Sitharam T G. Experimental and Analytical Studies on Soft Clay Beds Reinforced with Bamboo Cells and Geocells[J]. International Journal of Geosynthetics and Ground Engineering. 2015, 1(2): 1–13.

Building foundation design specification (GB 50007-2012) [C]. China Construction Industry Press, 2012, Beijing.

Chunan Chen, Jinming Xu. Application of bamboo in the treatment of silt surface in super-soft foundation engineering[J]. Water transportation project, 2007(10): 75–77.

Engineering Geology Manual (Fifth Edition) [M]. China Construction Industry Press, 2018.

Jiahuan Qian, Zongze Yin. Geotechnical principles and calculations[M]. China Water Resources and Hydropower Press (2nd Edition), 2012, Beijing.

Jichao Zhang, Tiemao Liang, Yongkang Yang. Research on the key technology of dynamic drainage consolidation method for soft soil foundation treatment[J]. Construction Technology, 2012, 41(379): 42–45.

MD A, MUHAMMAD I I. Soil Improvement Using Bamboo Reinforcement[J]. American Journal of Engineering Research. 2014, 362–368.

Mokhammad Farid Ma'ruf. Shear strength of Apus bamboo root reinforced soil[J]. Ecological Engineering, 41(2012): 84–86.

*Frontiers of Civil Engineering and Disaster Prevention and
Control – Yang & Rahman (Eds)*
© 2023 The Author(s), ISBN: 978-1-032-31200-2

Design of large-scale studio in Hengdian film and television industrial architecture

Haibo Ren*
China Architecture Design & Research Group, Beijing, China

ABSTRACT: This project is the 1# studio of Hengdian Film and Television Industrial Park (Phase I). This paper introduces the design idea, structural system, structural calculation method, and some design measures of this project in detail and introduces the design of the double-height column and the design of the steel truss roof in detail. This paper has a strong referable value for the structural design of the large space studio.

1 INTRODUCTION

This project is the 1# studio of Hengdian Film and Television Industrial Park (Phase I). It is located in the west of Film and Television Avenue and north of Huaxia Avenue in Hengdian Town, Dongyang, Zhejiang Province, with a total construction area of 58518.89m². The construction area of 1# shed is 7480m², the building height is 27.140m, and the main function is a 3400m² studio and auxiliary rooms. The studio renderings and structural model diagrams are shown in Figures 1 and 2.

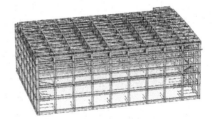

Figure 1.   Studio renderings.   Figure 2.   Structure model of the studio.

   The main structure is a reinforced concrete frame structure, and the large space roof is made of steel trusses. The structural design life is 50 years. The seismic fortification intensity is 6 degrees, and the design basic seismic acceleration is 0.05g; the design earthquakes are grouped into the first group. The building site category is Class II, and the characteristic period of the site is 0.35s. The seismic fortification of the project is classified as the standard fortification category, and the structural safety level is Class II. The seismic grade of the frame is grade 3, the seismic grade of the seismic measures of the surrounding columns of the studio is grade 3, and the seismic grade of the seismic structural measures is grade 2. The basic wind pressure at the location is 0.35kN/m², and the ground roughness category is Class B. The foundation adopts bored piles with enlarged heads.

   At present, the national film and television culture industry is further developed. In the future, we will continue to increase investment in film and television culture. The demand for studios will increase, and many studios will be built. The research results of this paper can be used as a reference for the construction of similar projects. This is the purpose and important value of this study.

*Corresponding Author: renhaibo@tju.edu.cn

2 STRUCTURAL CALCULATION

2.1 *Structural system*

The entire structural system is a lower reinforced concrete frame and an upper steel truss roof. In the overall calculation, the interaction needs to be considered[1], the YJK program is used for the overall calculation, the SATWE is used for the overall calculation check, and the PK module is used to check the frame column. The layout of the typical structure plan is shown in Figure 3. The studio part is five-storey high, and the opening area of each floor is greater than 60%. The layout of the steel truss roof is shown in Figure 4.

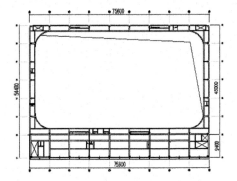

Figure 3.   Standard layer structure plan.   Figure 4.   Structure layout of steel truss roof.

2.2 *Structural calculation*

The overall calculation adopts YJK and SATWE, both three-dimensional space analysis software. The cross-layer columns around the studio were checked with PK software.

2.2.1 *Overall calculation result*

Using the spatial structure layer of YJK, the steel truss layer is modeled into the overall structure, the relationship between the steel truss and the concrete column is defined using two-point constraints, and the first 15 mode shapes are taken for the calculation. The calculation results are shown in Table 1. It can be seen from the data that the calculation results can meet the requirements of the specification.

Table 1.   Calculation results under frequent earthquakes.

Category		Content
T	T1(X+Y)	1.4348 (0.00+1.00)
	T2 (X+Y)	1.3821 (1.00+0.00)
	T3 (X+Y)	1.2259 (0.05+0.00)
	T3/ T1	0.85
	Participation factor	X: 91.20%; Y: 91.58%
	Earthquake displacement	X: 1/1755; Y: 1/1671
	Wind load displacement	X: 1/3086; Y: 1/2022
Rigid weight ratio	X direction	23.530 (Floor 2)
	Y direction	22.403 (Floor 2)
Shear weight ratio	X direction	1.560% (Floor 1)
	Y direction	1.506% (Floor 1)

2.2.2 Calculated length of one-way spanning column

The YJK software can automatically judge the one-way spanning column and calculate the reinforcement according to the appropriate calculation length coefficient and column length. However, it should be noted that different from PKPM software, YJK displays and calculates the length coefficient according to a whole section of a column for a spanning column, and the column length is also given according to the whole spanning column (Shu & Yan 2020). (Column length can be found in "Component Information" of "Design Results"). The PKPM software provides the column length and the converted calculated length coefficient by layer for the cross-layer column.

For the beam on the column, the software judges it as the constraint of the frame column and judges it as the floor, which is unreasonable, and the calculation length coefficient should be manually modified to ensure the accuracy of the calculation.

2.2.3 Single-story double-height columns around the studio

The single-storey full-height column is an important component of this project. The design and analysis are based on three considerations, and the envelope is used for reinforcement (Zong & Wu 2018). (1) For the calculation under the frequent earthquakes in the overall model, the seismic level of the seismic structural measures is increased by one level, and it is in accordance with the second level. (2) Take a single model, input the load effect, optimize the calculation of the length coefficient, and use the PK software to check. (3) In the overall model, the mid-seismic elasticity is used to check the column reinforcement.

2.2.4 Structural irregularity judgment

According to "Code for Seismic Design of Buildings (GB50011-2010)" (2016 edition) 3.4.3, (1) the first is the torsional irregularity judgment. Under the action of the specified horizontal force with accidental eccentricity, the displacement ratio is less than 1.2. (2) The irregularity of the unevenness is satisfied. (3) the local discontinuity of the floor slab is judged. In each structural layer, the opening area is about 80% of the floor area of the layer; it does not meet the requirements. (4) The lateral stiffness is irregularly judged. The 2nd, 3rd, and 4th layers do not meet the requirement of stiffness ratio. (5) The last is the judgment of sudden change of bearing capacity. The fourth layer belongs to the sudden change of bearing capacity. To sum up, this project also has three of the six main irregular types listed in Table 3.4.3 of the code, which are special seismically irregular structures. However, this project does not belong to high-rise buildings, so it does not belong to earthquake-resistant overrun structures.

2.2.5 Weaknesses in floor slabs and strengthening measures

The studio part is five-storey high, and the opening area of each floor is greater than 80%. In the structural analysis, the floor slab is designated as an elastic membrane (Wang et al. 2020), the floor slabs are all double-layered with steel bars, and the thickness of the floor slabs is appropriately thickened.

2.2.6 Beams and armpits around the studio

The size of the frame column around the large space is 700x1100, and the width of the frame beam is 350. Due to the requirements of the building wall layout, the frame beam is attached to the outer skin of the column, and the eccentricity of the beam and column does not meet the requirements of Article 6.1.7 of "Technical Regulations for Concrete Structures of High-rise Buildings," and needs to be calculated by adding haunches. The axillary is added horizontally, as shown in Figure 6.

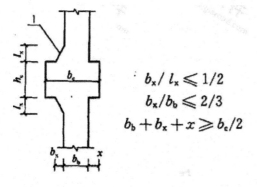

$$b_x/ l_x \leqslant 1/2$$
$$b_x/b_b \leqslant 2/3$$
$$b_b + b_x + x \geqslant b_c/2$$

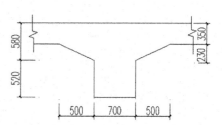

Figure 5. "High Regulations" beam level and haunch requirements.

Figure 6. Beam level with haunches.

3 STEEL TRUSS ROOF DESIGN

The large-span steel truss adopts a top-supported flat steel truss, and the support is fixed, hinged steel support. The truss is provided with horizontal supports, tie rods, and vertical supports. The roof is made of a metal plate roof, and the surface layer is a colored steel plate. The upper chord layer of the truss has horizontal supports and purlins, and the lower chord layer has tie rods and horse track beams (Kong et al. 2007). And has vertical support. The top chord of the main truss is sloped at 5%, and the splicing of the steel members adopts the rigid connection method. The main truss is considered 1/500 of the arch, and measures should be taken to ensure the cross-sectional size of the components.

4 CONCLUSION

The studio is a single-story large space structure, the lower part is a concrete frame structure, and the upper part is generally a steel truss or steel grid structure. In structural calculation and analysis, various models need to be used for envelope design. The frame columns around the studio are the key and important components of the structure. The calculated length coefficients need to be accurately determined. The cantilever beams cannot be used as constraints for the frame columns. The calculated length coefficients need to be adjusted manually, and the seismic grade of the seismic structural measures needs to be improved according to the specifications.

REFERENCES

Ganping Shu and Xin Yan 2020 Comparison and optimization of steel structure roof design scheme of a university gymnasium *Building Structure* vol 9 pp37–43.
Hai Zong and Qi Wu 2018 Structural Design of Qingdao Oriental Movie Metropolis Theater. *Engineering Construction and Design* vol 1 pp35–38.
Hongchen Wang, Tao Zhang and Ling Chu 2020 Structural Design of Yan'an Grand Theater *Building Structure* vol 11 pp27–32.
Jianghong Kong, Mingquan Liu and Jianli Zhao 2007 Structural Design of Fujian Grand Theater *Architectural Technology* vol 2 pp123–126.

Frontiers of Civil Engineering and Disaster Prevention and Control – Yang & Rahman (Eds)
© 2023 The Author(s), ISBN: 978-1-032-31200-2

Hierarchical multi-step accumulative jacking construction technology of large-span special-shaped roof grid

Yantao Zhang*, Fajiang Luo, Zhen Wang, Zhixiong Guo & Shengxi Wang
China Construction Third Engineering Group Co., Ltd., Chengxi District, Xining City, Qinghai Province, China

ABSTRACT: Combined with Haidong Sports Center Project in Qinghai Province, this paper introduces the multi-step accumulative jacking construction technology of a large-span special-shaped roof grid. The lifting installation of assembled floors in limited sites and different elevations breaks through the limitation of the site and structural elevation difference. In the process of construction simulation calculation, a sensor is set at the maximum stress and strain in the middle of the span for stress and strain monitoring and connected to the sensor data acquisition instrument to collect the stress and strain in the process of hierarchical multi-step cumulative jacking. The graph is used to display the measurement value of each monitoring interval of the built structure and the actual displacement and deformation of the structure at each measurement point to achieve the purpose of process structure monitoring. At the same time, it solves the problem of high-altitude construction difficulty, and provides valuable construction reference for large-span complex space steel structure construction.

1 INTRODUCTION

With the development of society, the form of buildings is becoming more and more complex, the number of buildings with novel shapes is increasing, and the functional requirements of buildings pose a greater challenge to the design and selection of structural forms. In the structural design of stadiums, in order to meet the needs of the building function, the roof often needs to increase the large-span steel structure to ensure structural safety. Due to the need for large lifting equipment, large-span steel structures in the construction process have a complex installation process and poor economic and technical indicators.

Combined with the construction practice of the roof steel grid of Haidong Sports Center in Qinghai Province, a set of hierarchical multi-step cumulative jacking construction technology of long-span special-shaped roof grid is studied. This construction technology solves the safety problem of long-span steel structure hoisting structures. The construction process can effectively solve the problems of the limited lifting capacity of existing tower cranes, and the site does not have large lifting equipment. In addition, it can address the inconvenience of supporting the tire frame installation, the height difference of the structure, and the height difference of the site caused by the assembly inconvenience, etc., to ensure the quality and safety of the large-span network frame construction and reduce its construction costs.

2 OVERVIEW OF THE PROJECT

Qinghai Haidong Sports Center project is located in Vocational Education City, Ledu District, Haidong City, covering an area of about 232453 m². The sports center consists of a stadium and a sports natatorium, with a total construction area of 63345 m², including 61238 m² above ground

---

*Corresponding Author: 284411875@qq.com

and 2107 m^2 underground, as shown in Figure 1. There are three floors in the stadium. The average height of roof elevation is ≤ 30m. The sports natatorium is divided into gymnasium and natatorium, and the average height of roof elevation is ≤ 24m. The main body adopts a reinforced concrete frame structure, the stadium roof adopts a steel truss structure system, and the sports natatorium adopts a roof truss and large-span steel grid structure system, as shown in Figure 2.

Figure 1.   Renderings of Haidong Sports Center.

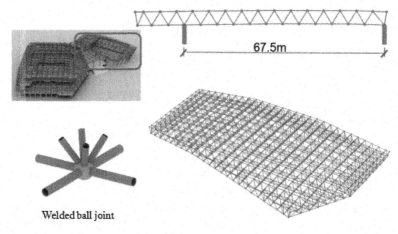

Welded ball joint

Figure 2.   Layout of roof grid of natatorium.

3   ENGINEERING DIFFICULTIES

The space truss span of the natatorium is about 67.5m. There are three structural, functional areas of swimming pool, play pool, and platform on the assembled floor of the jacking space truss, corresponding to three different elevations of -2.2m, - 1.2m, and -0.4m. There are four different slopes of 3%, 4%, 6%, and 12% in the jacking part of the grid. Due to the limitation of the ground assembly site, it is impossible to complete the grid installation by the overall jacking or lifting process. After the project research, the hierarchical multi-step cumulative jacking construction technology is adopted, and the research methods are mainly jacking system and hierarchical multi-step cumulative jacking[5].

The jacking system is mainly composed of a jack, support frame, and stable support. At the same time, when multiple jacking systems operate, the hydraulic cylinder synchronous control system is added to realize the synchronization of multiple jacking points. In fact, it is to jointly realize the vertical movement of heavy objects[3], as shown in Figure 3.

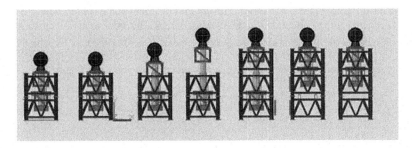

Figure 3. Process flow chart of hydraulic jacking.

Hierarchical multi-step cumulative jacking is based on the hydraulic jacking principle. The civil structures in the jacking assembly area are assembled in different sections at different elevations and floors on the same floor. The space frame with a long span and a large slope is assembled in blocks. After assembly, the lowest assembled space frame partition (block) is lifted first. Then the adjacent lowest assembled space frame partition (block) is lifted. The reinforcement is welded to form a whole, increases the jacking point, expands the jacking range, jacks up to the adjacent lowest assembled grid partition (block), and then welds the complementary rod to form a whole. The jacking point is increased, and the jacking range is expanded. According to this, we accumulate the jacking in multiple steps, increase the jacking weight by stages, gradually expand the jacking range, form all the grid frames in the jacking area as a whole, and finally realize the overall jacking in place[4].

4 CONSTRUCTION PROCESS FLOW

4.1 *Construction process*

Figure 4. Process flow chart.

4.2 Key points of each process operation

Construction process simulation

There are six different planes in the roof grid and four different planes in the jacking area, which leads to a horizontal force component in the jacking process. In order to avoid the excessive load combination value in the construction stage, which will affect the safety of the jacking process, the finite element analysis software Midas/Gen is used to calculate the structure of the swimming pool during the three jacking processes before construction[2], as shown in Figure 4.

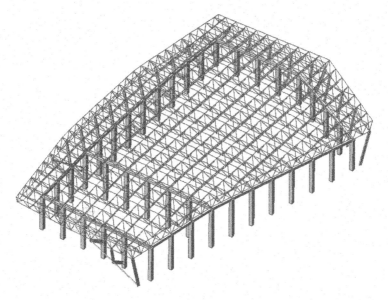

Figure 5.   Finite element analysis model.

Through simulation analysis, the maximum vertical displacement during the construction of the jacking grid is - 3mm, and the maximum horizontal displacement is 0.8mm before the unloading of the jacking grid. The maximum combined stress is 16.34 n/mm², and the design value of material strength of Q345 steel is 295 N/mm², which meets the requirements of strength design.

Through simulation analysis, in Figures 5-12, the maximum vertical displacement at the completion of structural construction is -22.57mm (the maximum displacement occurs in the middle,

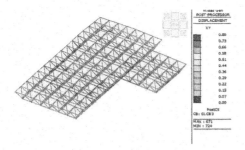

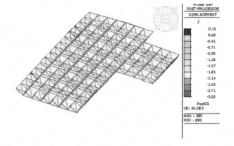

Figure 6.   Cloud diagram of XY displacement in jacking construction (mm).

Figure 7.   Cloud diagram of Z-direction displacement in jacking construction (mm).

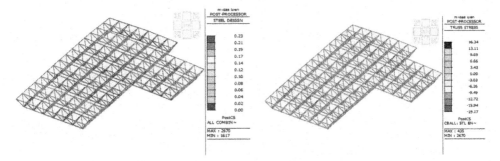

Figure 8.   Cloud diagram of von Mises stress ratio of jacking.

Figure 9.   Cloud diagram of von Mises stress during jacking (Mpa).

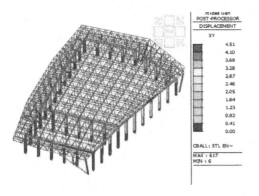

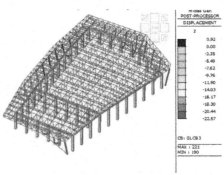

Figure 10.   Cloud diagram of XY displacement after overall construction (mm).

Figure 11.   Cloud diagram of Z-direction displacement after overall construction (mm).

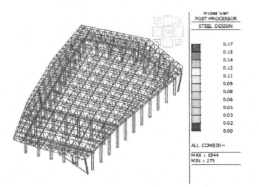

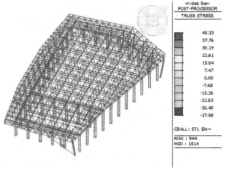

Figure 12.   Cloud diagram of stress ratio after overall construction.

Figure 13.   Stress nephogram after overall construction (Mpa).

the middle span is about 56m, and the structural displacement is less than 1/1600 of the span, meeting the design requirements); The maximum horizontal displacement is -4.51mm, the maximum combined stress is 45.33 N/mm^2, and the material strength design value of Q345 steel is 295 N/mm^2, which meets the requirements of strength design.

4.3   Hierarchical multi-step cumulative jacking construction process

4.3.1   Allocation principle of hydraulic equipment

1) The hydraulic equipment shall meet the requirements of hydraulic jacking force of grid steel structure, and try to make each hydraulic equipment load evenly;
2) Try to ensure that the number of hydraulic equipment driven by each hydraulic pump station is equal in order to improve the utilization rate of the hydraulic pump station;
3) During the general layout, the safety and reliability of the system shall be carefully considered to reduce the project risk.

4.3.2   Allocation principle of hydraulic equipment

The maximum dead weight of the jacking part of the grid is about 130T, calculated by 150T. A total of 14 jacking platforms bear the load, and the load of each jacking platform is about 11t. Each jacking point is equipped with a DS750-30-50 hydraulic jack. The rated jacking force of the DS550-40-50 hydraulic jack is 50t. The safety factor of the lifter is n = 50 / 11 = 4.5 (considering the horizontal force generated by the jacking of a special-shaped grid), and the safety factor of the lifter is not less than 2. The above configuration can meet the requirements[6].

4.3.3   Jacking equipment

The maximum design pressure of the DS550-40-50 series oil cylinder is 65T (the explosion-proof valve cannot be damaged under 65T), and the design working top lift is 50T. When the 50T series works, when the pressure at the jacking point exceeds 50T, the oil pump stops supplying oil to the oil cylinder. When the pressure of the oil cylinder is less than 65T, the oil cylinder will not unload.

a. The oil cylinder and hydraulic oil supply system is produced by professional manufacturers. The jacking force of a single oil cylinder is 50 tons. The maximum oil supply pressure of the hydraulic oil supply system is 31.5 MP. After many successful linkage tests in the factory, the maximum stroke of the hydraulic jack is 770mm.
b. The jacking frame is designed by the manufacturer's technology and construction personnel according to the actual needs of the construction site and produced and processed by the processing plant. The maximum jacking force is 50t, the maximum jacking height is 16m, the main material is Q235B, the pipe diameter is φ 146 * 10, the plane projection size is 1.0m * 1.0m (center distance), and the standard node height is 770mm. The maximum height of the grid frame of the project is less than 20m, and the jacking height of the frame of this specification is 20m, which meets the actual requirements.

4.3.4   Technical measures for jacking points

Set the jacking frame according to the location of the jacking point layout, and lay a 20 * 1500 * 1500 steel plate on the ground where the jacking frame is located to strengthen the ground. At the same time, the elevation shall be measured with a level gauge, and a lateral positioning steel wire rope (equipped with chain fall) shall be set on the jacking frame for fine adjustment of the horizontal position when necessary.

For the connection between the jacking point of the upper bracket and the lower chord ball of the steel grid, a 319 * 16 circular pipe (height 300 ~ 500mm, Q345) and four limit plates are set under the lower chord ball of the grid to resist the horizontal component together with the jacking frame. The welding ball is located just above the circular tube to limit the horizontal displacement of the ball, and the center of the circular tube is in the same line as the center of the ball.

4.3.5   Hierarchical multi-step cumulative jacking construction process

The grid span of the natatorium reaches 67.5m, the structure adopts a large-span hollow structure, and the hollow part of the grid adopts hydraulic jacking construction to reduce the risk of high-altitude operation. There are three structural, functional areas of the swimming pool, play pool, and platform on the assembled floor of the jacking grid of the swimming pool, corresponding to three different elevations of -2.2m, - 1.2m, and -0.4m. There are four different slopes of 3%, 4%, 6%, and 12% in the jacking part of the grid. According to the assembled floor structure, the jacking part of the grid is divided into two zones: assembly in the swimming pool and assembly outside the swimming pool. There are three different gradients of 3%, 4%, and 12% for the partition grid in the swimming pool, with a span of 48.75m. If the overall assembled grid is 4m above the ground, the swimming pool will be assembled in two parts to reduce the risk of high-altitude operation and facilitate assembly construction. The swimming pool is divided into two jacking assembly areas, and the first-floor platform is divided into one jacking assembly area, with three jacking assembly areas. For the first time, jack up the lowest block grid of the swimming pool against the slope direction, as shown in Figures 13-14, and connect it with another block of the swimming pool as a whole, as shown in Figure 15. Increase the jacking point and expand the jacking range, as shown in Figure 16. For the second jacking, the whole grid of the swimming pool is assembled with the grid of the first-floor platform to form a whole, as shown in Figure 17. We increase the jacking point, expand the jacking range, and jack the grid in the jacking area for the third time[1], as shown in Figures 18-20.

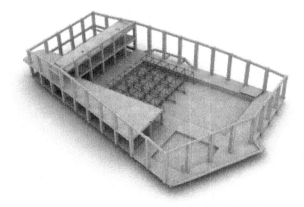

Figure 14.   Completion of the assembly of the lowest block grid in the anti-slope direction of the swimming pool.

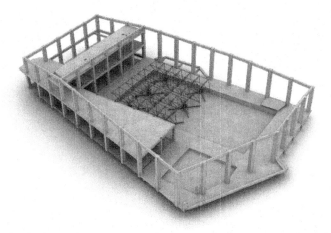

Figure 15.   Layout of jacking equipment, commissioning of hydraulic jacking system, trial jacking of the grid for 100mm, and inspection of stress deformation and hydraulic jacking working state of the grid.

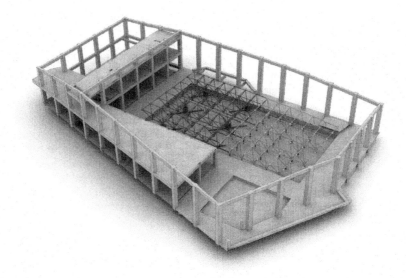

Figure 16. The first time the roof is lifted to the bottom of the swimming pool and connected with another block, and the two grid frames at the bottom of the swimming pool are connected as a whole by welding the middle reinforcing bar.

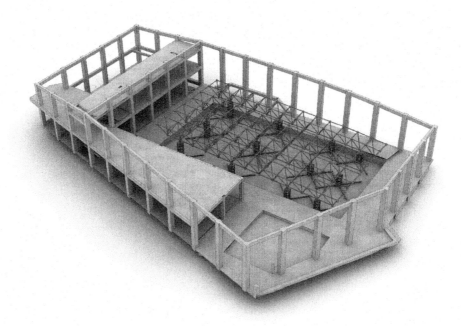

Figure 17. Add jacking points, arrange jacking equipment, expand the jacking range, and start the second jacking after the trial jacking.

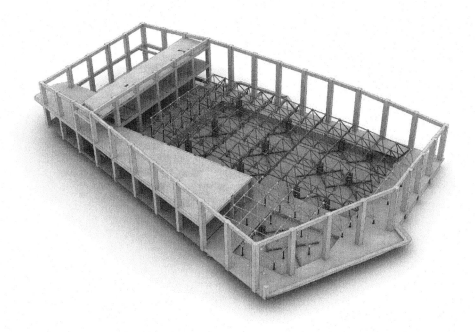

Figure 18. At the junction of the second jacking to the platform partition, the reinforcement welding connects the swimming pool bottom with the platform grid to form a whole, increases the jacking points, arranges the jacking equipment, expands the jacking range, and prepares for the third jacking.

Figure 19. Start the third jacking. After the jacking frame exceeds 6m, add a round pipe diagonal brace to enhance stability and continue jacking.

Figure 20.   The third jacking is completed, the jacking grid is in place, and the complementary rod is welded to form an integral with the peripheral grid.

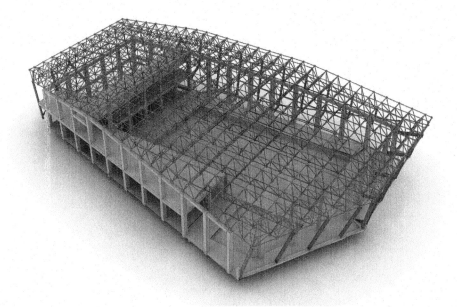

Figure 21.   If the stress monitoring is normal and the measurement deviation meets the specification requirements, the jacking frame shall be removed and handed over to the working face.

4.3.6 *Control measures for hierarchical multi-step cumulative jacking over*

1) The stress on each vertex shall be uniform.
2) In order to make the jacking speed of each hydraulic jack equal to the cumulative jacking process of the structure, the computer control system adjusts and controls the jacking process through the feedback data of the displacement sensor.
3) According to the pre-calculated jacking support reaction value under jacking conditions, the maximum jacking force of each hydraulic jack is set in the computer synchronous control system. When the jacking force exceeds the set value, the lifter will automatically stop jacking to prevent serious uneven load distribution at the jacking point, resulting in damage to structural members and jacking facilities.
4) Through the self-locking device and mechanical self-locking system set in the hydraulic circuit, the lifter can automatically lock the steel strand for a long time to ensure the safety of jacking components after the completion of jacking in the previous stage and before jacking in the next stage, when the lifter stops working or encounters power failure.
5) When each jacking frame is lifted to 6m, four round pipes (114 * 6) diagonal braces are added at 5m, and four cable wind ropes are added at 10m. Enhance the stability of the frame and resist the horizontal component in the construction process.

4.4 *Construction process monitoring*

4.4.1 *Displacement, stress, and strain monitoring during multi-step cumulative jacking*

1) In order to monitor the deformation caused by steel grid jacking, reflective stickers are pasted at the jacking points during the jacking process, a high-precision total station is used to monitor and record the displacement, and a line hammer is used to assist in monitoring the horizontal displacement, as shown in Figure 21. According to the construction simulation calculation, a sensor is set at the maximum stress and strain in the middle of the span for stress and strain monitoring, as shown in Figure 22, and connected to the sensor data acquisition instrument to collect the stress and strain data in the process of hierarchical multi-step cumulative jacking.
2) Data collection and analysis are carried out on the changes of standard section and jacking frame before and after unloading.

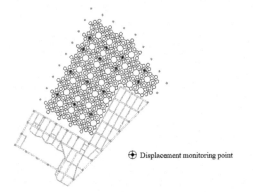

⊕ Displacement monitoring point

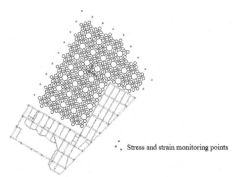

•, Stress and strain monitoring points

Figure 22.   Layout of displacement monitoring points.

Figure 23.   Layout of stress and strain monitoring points.

4.4.2 *Analysis and feedback of monitoring data*

By recording the monitoring data of all monitoring points during construction and performing necessary data processing, the measurement value of each monitoring section of the built structure and the actual displacement and deformation of the structure at each measuring point are displayed by graphics.

At the same time, the corresponding values expected in the simulation analysis of construction simulation software are compared. Visually compare the preset value with the real value. If the deviation between the stress or deformation caused by multi-step cumulative jacking and the construction simulation data is too large, the member shall be reinforced in time and reported to the design institute for analysis.

5 CONSTRUCTION EFFECT

5.1 Social benefits

Aiming at the process of constructing a large-span roof special-shaped space truss, the jacking construction does not need to set up a full scaffold and ultra-high support jig frame to avoid the entry and exit of jig frame materials and the selection of a large tower crane. The construction saves the rental cost of measuring materials and equipment and effectively saves resources. The hierarchical multi-step cumulative jacking solves the influence of the structure itself and the assembled floor height difference on the grid installation during the construction of the jacking grid and avoids a large number of high-altitude construction operations in such dangerous projects. At the same time, the hierarchical multi-step cumulative jacking installation enables the grid assembly to be carried out close to the structural plane, greatly optimizes the labor environment and reduces the construction cycle, improves the construction speed of the project, and is beneficial to quality, safety, and construction period. The successful implementation of this technology will provide experience for the construction of other complex venue projects in the future, and the application prospect is very broad. Haidong Sports Center has been interviewed and reported by Haidong TV station many times during the construction of the project. After completion, it will fill the gap of Haidong's lack of large stadiums, provide high-level and highly configured venues and facilities for the development of competitive sports in Haidong City, and play an important role in promoting the all-round progress of sports in Haidong City.

5.2 Economic benefits

The use of hierarchical multi-step cumulative jacking construction technology can save 160t of jig frame steel, and the material unit price is calculated as 3800 yuan/ton. The total material cost is 160 × 3800 = 608000 yuan. If the steel is amortized and recovered by 40%, the material cost will be saved by 60.8% × 0.6 = 364800 yuan. Save the labor cost of jig frame assembly and disassembly. The labor cost is calculated according to 630 yuan/ton. The total labor cost is: 160 × 630 = 100800 yuan = 100800 yuan. Save the production and transportation costs of the jig frame. The production and transportation costs of the jig frame are calculated according to 1660 yuan/ton, and the total production and transportation costs are saved: 160 × 1660 = 265600 yuan = 2655600 yuan. The construction period of half a month is saved, and the cost of each tower crane is 80000 yuan/set. If there are two tower cranes, the total cost of tower cranes saved is two sets × 80000 yuan/set/month × 0.5 months = 80000 yuan. The total cost saved by this technology is 811200.

6 CONCLUSION

Hierarchical multi-step cumulative jacking construction technology and floor assembly enable most manual operations to be completed on the floor. Various safety measures are easy to set, which can greatly reduce the risks of high-altitude falling casualties and high-altitude falling objects. The technical grid assembly is not limited by the height difference of the civil engineering floor. In the assembly process of a large-span special-shaped space truss, due to the special-shaped structure and slope of the structure itself, the longer the assembly span will lead to the higher the space truss from the assembly ground. In order to reduce the risk of high-altitude operation, the assembled space truss with too large a span is divided into blocks for low-level assembly on the floor. Generally, 30 ~ 40m is taken as a block assembly unit for a special-shaped space truss, and the jacking range is expanded after it is lifted to the corresponding height in the jacking process. The cumulative jacking finally realizes the overall jacking.

According to the displacement monitoring conducted by pasting reflective stickers at the jacking points during the jacking process and the simulation calculation, a sensor is set at the maximum stress and strain in the middle of the span for stress and strain monitoring, and the sensor data acquisition instrument is connected to collect the data of stress and strain in the hierarchical multi-step cumulative jacking process. Data collection and analysis are carried out on the changes of standard section and jacking frame before and after unloading to achieve the purpose of structural monitoring.

Through computer construction simulation technology, the multi-stage construction simulation of a large-span special-shaped grid hierarchical multi-step cumulative jacking system is carried out to calculate the safety and reliability of the structure in the process of static load and dynamic construction to meet the construction safety requirements. Through design analysis, this technology calculates the deformation during component installation, pre-cambers the components, analyzes the possible causes by monitoring the deformation and stress during construction and comparing it with the construction simulation data, and adjusts the construction preset value to meet the construction quality requirements.

The jacking construction speed is fast, which is more than double the efficiency of the traditional high-altitude splicing method. Moreover, the key lines and resources constructed by other units are not occupied before and after jacking, which makes a great contribution to the saving of the overall construction period.

REFERENCES

Binbin Ma, Application and analysis of overall synchronous jacking construction technology of mass steel grid, 2018, Issue 19.
Guoping Zhu, On the application of integral jacking method of space truss in engineering, Issue 30.
Julong Wu, Integral jacking construction technology of large roof, 2008, Issue 03.
Weihai Li, On the construction technology of integral jacking method of space truss, 2018, Issue 11.
Weiqing Zhang, Jiangbin Wang, Sipeng Chen, Ming Rao, Fan Zhao, Research and Application of Integral Jacking Technology of Long-span Space Truss, 2018, Issue 018.
Zhenghao Cui, Shoucheng Zhang, Hydraulic synchronous jacking 500t large steel grid, 2011, Issue 6.

Frontiers of Civil Engineering and Disaster Prevention and
Control – Yang & Rahman (Eds)
© 2023 The Author(s), ISBN: 978-1-032-31200-2

Application of dynamic monitoring system in the reinforcement of Ganquan tunnel in Gansu province

Jie Wang, Xipeng Wang, Xin Zhang & Fang An
Tianshui Highway Bureau, Tianshui City, Gansu Province, China

ABSTRACT: This paper takes the application of a dynamic monitoring system in the reinforcement of the Ganquan tunnel of the Baotian Expressway in Gansu Province as an example. Numerical simulation is used to analyze the reinforcement of tunnel surrounding rock systematically. The dynamic monitoring system of the tunnel is established to analyze and verify the tunnel reinforcement effect from the data of subsidence, convergence, and uplift. Judging from the system data in the past two years, the monitoring system has operated normally and achieved the expected purposes, providing a new method for the research on tunnel reinforcement acceptance.

1 FIRST SECTION

With the rapid development of a new generation of information technology, especially the application of concepts such as the Big Data and Cloud Computing, the birth of the dynamic monitoring system provides a new method and new idea for the study of tunnel reinforcement and acceptance (Shi et al. 2004). The tunnel dynamic monitoring system integrates the tunnel deformation data system through the analysis of the Big Data. Through a simple line-type comparison, the tunnel maintenance management personnel can clearly and accurately grasp the tunnel operation status and judge the reinforcement effect and the next maintenance measures (Wang 2009).

This paper applies the tunnel dynamic monitoring system in the Baotian Expressway Ganquan Tunnel reinforced by Ganquan Highway Bureau. By comparing and analyzing various system data, it is found that the Ganquan Tunnel has special temporal and spatial laws of deformation. By comparing the effects before and after reinforcement, the feasibility of the tunnel reinforcement scheme is studied, and the good effects of the system application are analyzed and summarized.

2 OVERVIEW AND MAINTENANCE OF THE TUNNEL

Baotian Expressway is located in Tianshui City, Gansu Province, with bridges and tunnels accounting for 65.4%. The beginning and end pile numbers of the Ganquan Tunnel downstream line are XK90+530~XK92+240, the total length is 1710 m, the longitudinal slope of the tunnel is +2.35%, and the entrances and exits adopt a bamboo-truncating portal. It was completed and opened to traffic in 2009.

In 2013, local defects such as floor heave, cracking, and water seepage occurred in the tunnel pavement, which seriously affected the stability of the tunnel structure and driving safety. In 2014, the Tianshui Highway Bureau of Gansu Province adopted the method of grouting steel pipe piles + steel concrete. A total of 257 m of inverted arch reinforcement and maintenance were carried out of the down line that was seriously subjected to tunnel defects and achieved good effects.

In October 2017, the Tianshui Highway Bureau of Gansu Province organized a working group to conduct regular defects inspections on the tunnel site. It was found that the defects in the untreated floor heave section had developed rapidly, and there was a trend of further development. To completely eliminate the hidden dangers of Ganquan Tunne, the Tianshui Highway Bureau of Gansu Province carried out maintenance and treatment in 2018 for the defects of the pavement

DOI 10.1201/9781003308577-39

floor heaves 60 m standard section, 20 m widened section, and the pavement longitudinal crack 140 m standard section.

3 TECHNICAL ROUTE

The application of the dynamic monitoring system mentioned in this paper in tunnel reinforcement is mainly to conduct in-depth research on the prevention and treatment effects of tunnel reinforcement projects and long-term safety evaluation after construction. The main technical routes adopted are as follows:

3.1 Engineering geological survey

Through on-site engineering-geological surveys, the characteristics of the tunnel body, such as the boundary, scale, structure, and support of the tunnel, are ascertained, and the geometric character-istics, causes, and failure mechanisms of the tunnel are preliminarily inferred. The targeted survey workload is arranged to verify and modify the geometric, structural, and mechanical characteristics of the tunnel body.

3.2 Analysis of dynamic geological information

The dynamic geological information obtained during the construction is analyzed to supplement and improve the characteristics of the tunnel body, providing a detailed basis for the theoretical calculation of the tunnel and the optimization of the design scheme.

3.3 Experimental and theoretical research

The on-site direct shear, indoor triaxial shear, and indoor repeated shear tests are used to obtain the shear strength parameters of the tunnel, which provide force analysis for the structural design of reinforcement projects (Wang 1997).

3.4 Dynamic real-time monitoring

The laser range finder, single-point settlement meter, temperature, humidity sensor, and remote data wireless transmission technology are applied to dynamically monitor the displacement and temperature, and humidity of the supporting structure in the tunnel reinforcement and treatment project, give real-time feedback on the displacement distribution of the supporting structure.

4 DYNAMIC MONITORING AND DATA ANALYSIS

4.1 Dynamic monitoring

The Tianshui Highway Bureau of Gansu Province adopted the dynamic monitoring system men-tioned in this paper for the reinforcement of the Ganquan Tunnel. The tunnel deformation management level (see Table 1) is determined based on the comparison between the measured displacement and the ultimate displacement.

Table 1. Deformation management level of surrounding rock.

Management level	Management displacement	Construction status
III	$U_0 < \frac{2}{3}U_n$	Normal construction is available
II	$\frac{2}{3}U_n \leq U_0 \leq U_n$	Supporting should be strengthened
I	$U_0 > U_n$	Special measures should be taken

Note: U_0 – Measured displacement value; U_0 – Ultimate displacement value.

The measured displacement value of clearance convergence should also be based on the allowable relative clearance convergence value of the tunnel in Table 2 to determine the stable state of the surrounding rock.

Table 2. Tunnel allowable relative clearance convergence value (%).

Surrounding rock level	Buried depth (m)		
	Less than 50	50~300	Greater than 300
III	0.10~0.30	0.20~0.50	0.40~1.20
IV	0.15~0.50	0.40~1.20	0.80~2.00
V	0.20~0.80	0.60~1.60	1.00~3.00

The stable state of the surrounding rock is judged according to the displacement and deformation rate. The judgment standard is shown in Table 3 for the displacement change rate.

Table 3. Displacement change rate table.

Deformation rate	Deformation stage
Greater than 3mm/d	Sharp increase in deformation
1~3mm/d	Slow growth in deformation
Less than 1 mm/d	Basically stable

4.2 Data analysis

In order to ensure the best results of actual testing and analysis of monitoring data, two monitoring sections were arranged, and four monitoring equipment were installed on the downline of the Ganquan Tunnel. The specific monitoring data results are as follows:

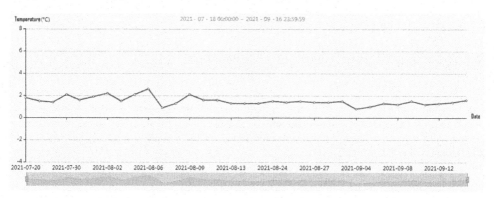

Figure 1. Vault subsidence curve of Ganquan tunnel.

As shown in Figure 1, the maximum change rate of the maximum vault subsidence in the Ganquan tunnel is normal. The surrounding rock is always in a basically stable state.

As shown in Figure 2, the maximum vault subsidence rate of the Ganquan tunnel is normal. The surrounding rock is always in a basically stable state.

Based on the above two monitoring data, the surrounding rock of the Ganquan Tunnel has always been in a basically stable state during its operation.

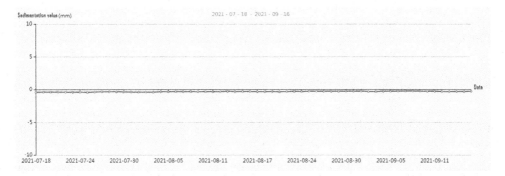

Figure 2.   Vault subsidence curve of Ganquan tunnel.

5   TECHNICAL PROBLEMS SOLVED

This paper establishes the dynamic monitoring system, a relatively accurate and easily identifiable tunnel reinforcement effect evaluation system. The main technical problems to be solved are as follows:

5.1   *Analysis of tunnel characteristics based on construction information feedback*

Based on collecting, sorting, and analyzing the survey and design data of the Ganquan Tunnel reinforcement treatment project of Baotian Expressway in Tianshui City, Gansu Province, and the geological information fed back during the construction process, combined with the tunnel dynamic monitoring data, a comparative analysis is conducted on the geometric and material structure characteristics of the tunnel and failure model.

5.2   *Research on evaluation technology of tunnel reinforcement effect*

The dynamic monitoring system mentioned in this paper establishes a reasonable, unified, and easy-to-operate tunnel reinforcement project prevention effect and long-term post-construction safety evaluation methods and technical standards, including tunnel reinforcement effect evaluation content, tunnel reinforcement effect evaluation index, tunnel reinforcement effect evaluation standard, tunnel reinforcement effect evaluation method and model, and tunnel reinforcement effect evaluation system.

5.3   *Long-term safety evaluation technology of tunnel reinforcement and treatment project based on monitoring data*

The safety of the tunnel reinforcement treatment project changes with the geological conditions and external conditions. Under certain unfavorable conditions, the treatment project structure will cause partial or overall instability. This requires effective monitoring of the treatment project. Through structural displacement, temperature and humidity monitoring data, inverting the stress state and safety factor of the tunnel support structure, the short-term and long-term prediction and evaluation of the post-construction safety of the tunnel reinforcement and treatment project are performed to ensure the long-term safety of the protected objects and prevent the recurrence of disasters and accidents (Wu 2001). The dynamic monitoring system applied in the Ganquan Tunnel uses the post-construction long-term safety evaluation model to analyze the dynamic monitoring information and achieves the goal of comprehensive analysis and evaluation of the long-term safety of the tunnel reinforcement and treatment project.

6 CONCLUSION

The dynamic monitoring system studied in this paper is applied in Gansu Province for the first time. It uses dynamic geological information acquisition and analysis, theoretical calculations, field tests, real-time monitoring, and other means to achieve dynamic real-time online monitoring for the deformation data of subsidence, convergence, and uplift after the tunnel reinforcement.

The successful application of the dynamic monitoring system in the tunnel reinforcement in Gansu provides a new management model for the tunnel maintenance and management personnel. The tunnel maintenance engineer can judge the health status of the tunnel operation through the simple line shape presented by the system, which fundamentally solves technical difficulties in the process of tunnel maintenance.

REFERENCES

Shi Bin, Xu Hongzhong, Zhang Dan, etc. Feasibility Study on the Application of BOTDR Strain Monitoring Technology in Large-scale Foundation Engineering Health Diagnosis[J]. Chinese Journal of Rock Mechanics and Engineering. 2004, 23(3): 493~499.

Wang Jianyu. Tunnel Engineering Monitoring and Information Design Principles[M]. China Railway Publishing House. 2009.

Wang Tiemeng. Engineering Structure Crack Control[M]. China Building Industry Press. 1997.

Wu Faquan. Several Major Engineering Geological and Environmental Issues in the 21st Century in China[J]. Journal of Engineering Geology, 2001.9 (2): 115~120.

Frontiers of Civil Engineering and Disaster Prevention and Control – Yang & Rahman (Eds)
© 2023 The Author(s), ISBN: 978-1-032-31200-2

Design and construction of adjustable fabricated sling for steel box girders of large offshore interchange

Linting Li & Haiqing Cao
China State Construction Bridge Corp. Ltd., Chongqing, China

Penglin Xie, Yinyong Zeng & Lei Zhao
China Construction Sixth Engineering Bureau Corp., Ltd., Tianjin, China

ABSTRACT: Taking Changbai Interchange at Section DSSG02 of Zhoushan-Daishan Cross-sea Bridge as an example, the paper designs an adjustable fabricated sling for quick lifting, in-air posture adjustment, and stable lowering in place of steel box girders, considering the diversified geometric dimensions of these steel box girders and the eccentric gravity of the special beam segments on Changbai Interchange. The finite element analysis (FEA) results reveal that this adjustable fabricated sling has reasonable overall stress and fulfills the rigidity, strength, and stability requirements.

1 BACKGROUND

According to the lifting techniques, the slings special for lifting eccentric box girders fall into manually adjustable slings and powered adjustable slings. In the construction of the Jiaozhou Bay Cross-sea Bridge in Qingdao, a new multi-functional sling was developed for the lifting of 2100 t precast box girders, which consists of braces, sling ropes, connectors, and equalizing beams. This sling, however, cannot adjust the center of gravity. In the construction of the Hong Kong-Zhuhai-Macao Bridge, for the lifting of 2900 t L132 m × W33 m steel box girders, two sling options, equalizing beam slings and truss slings, were compared, and the latter was finally used. Though both can simplify the erection of large offshore fabricated box girders, they fail to solve the eccentricity issues. For land engineering, a stepless adjustable spreader for 850 t eccentric box girder was developed considering the diversified types and the large number of 850 t box girders and the serious eccentricity and different eccentric distances of beam segments on the Yuci Section.

2 PROJECT OVERVIEW

Zhoushan-Daishan Cross-sea Bridge, located at Zhoushan Islands in Zhejiang Province, stretches over Huibie Yang Sea, connecting Zhoushan Island to the south and Daishan Island to the North. With an overall length of 16.347 km, it is a standard two-way four-lane expressway, with the offshore Changbai Interchange linking to Changbai Island. Changbai Interchange is horn-shaped, with five ramps (A, B, C, D, and E) in the interchange area and a minimum curve radius of 60 m.

The superstructure of Changbai Interchange is a monolithic structure of 62 steel box girders with cantilever diagonal webs. Such structure is fabricated and installed as a whole. Most steel box girders are placed on the curve ramps. They have special shapes with large variations in length, width, and weight, and different centers of gravity from that of standard girders. There are eight lifting points for lifting steel box girders of Ramp A and four lifting points for Ramps B~E each. The longitudinal spacing between these lifting points ranges from 8 m to 25.5 m.

Figure 1.   Layout of ramp bridges of changbai interchange.

If steel wire ropes are directly used for lifting steel box girders, operators have to change different lengths of ropes repeatedly and prepare a large number of ropes, which means a combination of difficulty and lower efficiency. Besides, the lifting of special girder segments is challenged by the difficulty in keeping balance in the air and the risk of instability. Therefore, adjustable fabricated slings are needed for lifting steel box girders. The diversified special steel box girders raise more stringent requirements for the universality of slings.

Table 1.   Parameters of steel box girders of changbai interchange.

Position	Number of Steel Box Girders	Plan of Steel Box Girder (cm)	Number of Lifting Points	Longitudinal Spacing between Lifting Points L (m)	Parameters of Steel Box Girder
Ramp A	16		8	14.0 19.5 22.0 25.5	**Length:** 24.72~36.72 m **Width:** 16.5~21.794 m **Weight:** 187.8~326.8 t
Ramps B~E	46		4	8.0 10.0 14.0 19.5 23.5 25.5	**Length:** 14.4~36.92 m **Width:** 9.0 m or 9.25 m **Weight:** 53.4~145.4 t

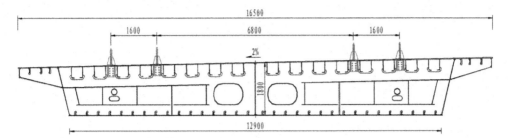

Figure 2.   Typical cross-section of steel box girders of ramp a (mm).

3   STRUCTURAL DESIGN

3.1   *Design principles*

Upon rational structure selection and calculation, a kind of sling structure was developed that meets the requirements for lifting 62 pieces of special steel box girders of Changbai Interchange.

(1) The sling structure, featuring longitudinal adjustment, caters to the variable longitudinal spacing range (8~25.5 m) of lifting points.
(2) The sling structure, featuring longitudinal adjustment, caters to the lifting mode switch between four and eight lifting points.
(3) The sling structure, featuring center of gravity adjustment, can adjust the in-air posture of steel box girders to keep them balanced during lifting.
(4) The sling structure, featuring longitudinal gradient adjustment, can adjust the longitudinal gradient of steel box girders to lower them in place stably.

3.2   *Design parameters*

(1) With eight lifting points, the maximum lifting weight is about 326.8 t, the maximum longitudinal spacing between lifting points is 25.5 m, the minimum lifting weight is about 187.8 t, and the minimum longitudinal spacing is 14 m.
(2) With four lifting points, the maximum lifting weight of the steel box girder is about 145.4 t, the maximum longitudinal spacing between lifting points is 25.5 m, the minimum lifting weight is about 53.4 t, and the minimum longitudinal spacing is 8 m.
(3) The longitudinal and transverse adjustment of the centers of gravity of steel box girders is from 0 m to 100 cm.
(4) The longitudinal gradient adjustment of steel box girders is equal to or greater than a 1.5 m/31 m span.

3.3   *Structural design*

(1) Overall Design
   The adjustable fabricated sling consists of ① longitudinal beams, ② cross beams, ③ lifting components, ④ longitudinal cylinder, ⑤ transverse cylinder, ⑥ distributive beams, ⑦ electric winch, and ⑧ manual winch, with a length of 23 m, a width of 8.4 m, a deadweight of 90 t, and a maximum design lifting load of 330 t.
   (2) Structure of Longitudinal Beam and Distributive Beam
   Longitudinal beams are variable-box-section structures welded with Q345 steel plates, with a top and bottom plate thickness of 35 mm, a web thickness of 16 mm, a width of 65 cm, a height of 90 cm on both sides, and 150 cm on the middle for increased bending strength of the section. Distributive beams are fixed-box-section structures welded with Q345 steel plates, with a top and

301

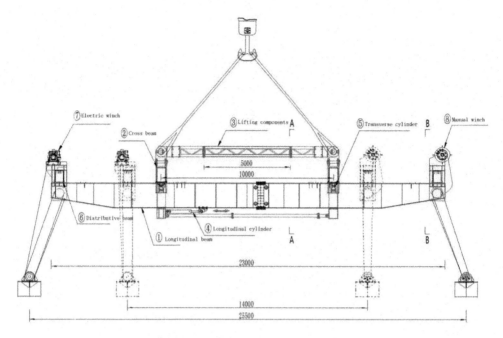

Figure 3. Elevation of adjustable fabricated sling (mm).

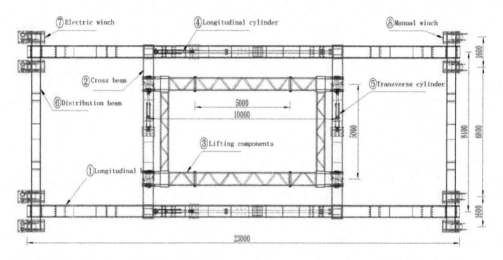

Figure 4. Plan of adjustable fabricated sling (mm).

bottom plate thickness of 16 mm, a web thickness of 12 mm, a width of 50 cm, and a height of 71 cm.

Each longitudinal beam has multiple set bolt holes. Distributive beams are bolted on longitudinal beams depending on the longitudinal spacing between lifting points. Distributive beams are designed to be longitudinally adjustable to cater to the lifting of steel box girders with different lengths.

(3) Electric/Manual Winch

Four electric winches are installed on distributive beams on one side of the sling, with variable-length steel wire ropes (4~8 m); four manual winches on the other side, with fixed-length steel

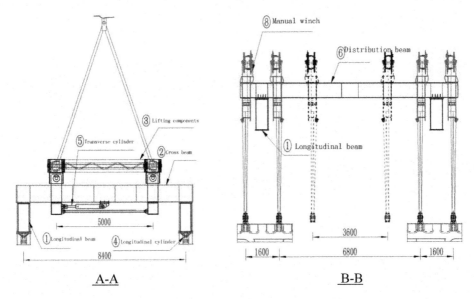

Figure 5.   Side view of adjustable fabricated sling (mm).

wire ropes (6 m). Both the electric and manual winches have a rated load of 60 t. After lifting the steel box girder to the right place, operators can start electric winches to speed up the longitudinal gradient adjustment within a 2 m/31 m span so that the steel box girder can be lowered to place stably.

Each distributive beam has multiple set bolt holes. Electric/manual winches are bolted on distributive beams depending on the number of lifting points for lifting steel box girders. The steel box girders of Ramp A are lifted by eight winches, and those of Ramps B~E are lifted by four winches on the inner side. These winches are designed to be transversely adjustable to switch between the lifting modes with 4 and 8 lifting points.

(4) Lifting System

The sling lifting system consists of cross beams, lifting components, longitudinal cylinder, and transverse cylinder, among which the cross beams and lifting components are made of Q345 steel. Crossbeams are fixed-section-box structures welded with a width of 50 cm, a height of 90 mm, a top and bottom plate thickness of 30 mm, and a web thickness of 16 mm. Lifting components are truss structures flanged with segments. If the plane dimension is 10×5 m, steel box girders are lifted at the longitudinal spacing between lifting points ranging from 14 m to 25.5 m. If the plane dimension is 5×5 m, steel box girders are lifted at the longitudinal spacing of 8 m or 10 m.

Cylinders are installed underneath longitudinal beams and cross beams for the longitudinal and transverse shift of the lifting system and eccentric adjustment of special steel box girders during their lifting. The longitudinal cylinder has the designed load of 140 t and an effective stroke of 1,000 mm; the transverse cylinder has the design load of 65 t and an effective stroke of 1,000 mm. The longitudinal and transverse cylinders enable the bidirectional adjustment of the lifting system on the plane, ensuring that steel box girders are in equilibrium during lifting.

3.4   Structure checking

The Midas finite element model is used to check the adjustable fabricated sling structure. It mainly checks the calculation of overall strength, rigidity, and stability of longitudinal beams, crossbeams, and distributive beams and enables the detailed calculations of connecting plates between cross

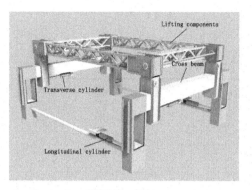

BIM simulation

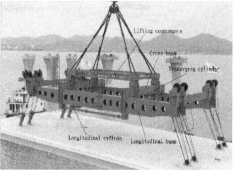

Photo

Figure 6.   Lifting system of adjustable fabricated sling.

beam and longitudinal beam, transverse/longitudinal cylinders, connecting bolts between the lon-gitudinal beam and distributive beam, and electric/manual winches. The calculations consider the impact of adverse loads, such as unbalanced loads, dynamic loads, wind loads, etc. The finite element model regards longitudinal beams as beam units to establish simply supported constraints at contact points between longitudinal beams and cross beams and imposes vertical lifting loads on both sides of longitudinal beams.

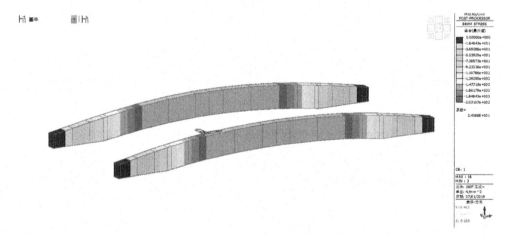

Figure 7.   Combined stress calculation of longitudinal beams (mm).

4   LOAD TEST

The project department of Section DSSG02 of Zhoushan-Daishan Cross-sea Bridge completed the load test on the adjustable fabricated sling for special steel box girders on August 28, 2019. The sling went through the static load test by loading water bags in increments, respectively 40%, 60%, 80%, 100%, and 120% of the load, and then the dynamic load test by unloading to 110%. The field inspection and measured stress reveal that the adjustable fabricated sling has its structural strength, rigidity, and stability consistent with the design and specification requirements and can be put into normal operation.

Table 2. Summary of calculation results for main components of adjustable fabricated sling.

SN	Most Adverse Conditions	Position	Calculation Parameter	Calculated Value	Allowable Value	Compliant or Not
1	1) There are eight lifting points	Longitudinal beam	Combined stress	203.1 Mpa	230 Mpa	Yes
2	2) The longitudinal spacing between lifting	Crossbeam	Combined stress	147.8 Mpa	230 Mpa	Yes
3	points is 25.5 m	Distributive beam	Combined stress	110.0 Mpa	230 Mpa	Yes
4	3) The lifting load is 330 t	Connecting plate between the cross beam and longitudinal beam	Tensile stress	78.7 Mpa	230 Mpa	Yes
5		Longitudinal cylinder	Hydraulic load	117.9 t	140 t	Yes
6		Transverse cylinder	Hydraulic load	59.8 t	65 t	Yes
7		Connecting bolt between the longitudinal beam and distributive beam	Shear stress	89.9 Mpa	250 Mpa	Yes
8		Electric/manual winch	Lifting capacity	49.9 t	60 t	Yes

Figure 8. Load test of adjustable fabricated sling.

5 FIELD APPLICATION

The lifting of steel box girders for Changbai Interchange officially commenced on August 31, 2019. The adjustable fabricated sling worked well with safe and reliable structures during operation.

(1) After the steel box girder was lifted 50 cm above the deck of the transport ship, all of the floating cranes were braked and stopped for 15 minutes. During this period, operators observed the in-air posture of the steel box girder and adjusted it to horizontal equilibrium with the longitudinal/transverse cylinders of the sling to ensure lifting safety.

(2) The floating crane lifted the steel box girder and moved by fetching the anchor to keep the suspended steel box girder right above the proposed place.

(3) After starting the electric winches on one side of the sling, operators rounded up and down the steel wire ropes to adjust the longitudinal gradient of the steel box girder and adjusted the position slightly by fetching the anchor of the floating crane while lowering the steel box girder onto the temporary supports on the top of piers. Then operators completed the fine adjustment of the steel box girder and lowered it in place with the three-dimensional jacks on the top of the piers.

Figure 9.   Construction application of adjustable fabricated sling.

6   CONCLUSION

The successful application of the adjustable fabricated sling for Changbai Interchange symbolizes the easier lifting of steel box girders for offshore interchanges and good serviceability for lifting the steel box girders of different sizes and special shapes. The adjustable fabricated sling can quickly adjust the in-air posture of steel box girders for lifting safety and the longitudinal gradient of steel box girders for stable lowering in place. The sling is easy to operate, with a safe and reliable structure. It greatly increases the efficiency of steel box girder construction, bringing favorable economic benefits to the project and providing a reference for subsequent construction for lifting steel box girders of similar interchanges.

REFERENCES

GB 50017-2020, Code for Design of Steel Structures.
GB/T3811-2008, Design Rules for Cranes.
Hu XD, Shen C. Research of Stepless Adjustable Spreader for 850t Eccentric Beam [J], Construction Machinery and Equipment, February 2017, 42–46.
Huang Y, Zheng HH, Chen JR. Design of Lifting Sling for 85m-span Combined Girder J], Journal of China & Foreign Highway, October 2013, 141–144.
Mo XL. Construction Technology Research on Non-navigable Span Bridge of Large Cross-sea Bridge (Master Dissertation) [D]. Dalian: Dalian University of Technology, 2015.
Wen ZB. Hong Kong-Zhuhai-Macao Bridge CB04 Blocks Steel Box Girder Crane Structure Design and Finite Element Analysis (Master Dissertation) [D]. Dalian: Dalian Jiaotong University, 2015.
Xu JS. Ansys-based Design of the Sling for Steel Box Beam of Hong Kong-Zhuhai-Macao Bridge [J], Hoisting and Conveying Machinery, July 2016, 24–29.

Frontiers of Civil Engineering and Disaster Prevention and
Control – Yang & Rahman (Eds)
© 2023 The Author(s), ISBN: 978-1-032-31200-2

Bearing capacity analysis of uplift piles during deep excavation and a simple test method

Yibin Wu*

College of Civil Engineering and Architecture, Xiamen University of Technology, Xiamen, China
Key Laboratory of Transport Infrastructure Health and Safety of Xiamen, China

Hongbin Ge, Guangqing Liu & Haiwang Liu
Xiamen Xiang'an airport investment and Construction Co., Ltd, Xiamen, China

ABSTRACT: The bearing capacity of uplift piles would inevitably decrease owing to the change of the ground stress field and the normal stress of the pile-soil interface under deep excavation. Unfortunately, due to the limitation of field conditions, it was almost impossible to field test under deep excavation. Therefore, it is important to develop an economic and replaceable test method. In this paper, different test methods (regular field test, steel stress test, double-pile test, and pit-bottom test) are simulated through the axis-symmetric finite element model. Through the deep excavation simulation, the load transfer mechanism and settlement of the pile loaded after deep excavation are studied. Next, a parametric study of the pile testing using a steel bar meter suggested in this paper is conducted. The results show that as the excavation depth increases, the bearing capacity of uplift piles after excavation can have a reduction of 10% to 15%, according to the in-situ test results using a steel bar meter.

1 INTRODUCTION

With the booming of cities and large-scale public facilities construction, the area of substantial deep foundation pit engineering grows larger and larger. Meanwhile, the excavation goes much deeper. As one of the main forms of the uplift foundation, the uplift pile is widely used in such kinds of projects. Hao Zhe et al. (2018) used the FLAC3D program to simulate and analyze the bearing capacity and deformation law of a single pile and pile group during the excavation of a deep foundation pit. Mu Rui et al. (2019) used Kotter limit equilibrium equation to analyze the uplift capacity of piles in composite rock strata. V. B. Deshmukh et al. (2010) used the Kotter equation to analyze the pulp-out resistance of piles in cohesive soil. However, due to the construction restrictions, the pit pile should be accomplished before the excavation of the foundation pit. Also, the accompanying large-scale unloading of overlaying soil makes it more complicated to determine the bearing state of the uplift piles.

The vertical stress of soil mass below the excavation surface after large-scale deep excavation decreases. Meanwhile, the normal stress in the pile-soil interface decreases. As a result, the stress field and displacement field of soil around the pile change. This results from the fact that the vertical unloading is in an over-consolidation state. In the unloaded state, the soil mass below the excavation level will bounce, which may have an up-pull action on the pile block to a huge extent. The above process may impact the uplift pile in three aspects:

(1) After large-scale deep excavation, the nominal stress of the pile-soil surface decreases, which may result in a decrease in uplift pile ultimate bearing capacity and vertical stiffness below the excavation interface.

---

*Corresponding Author: 2015000004@xmut.edu.cn

(2) The rebound of soil mass will drive the rebound of foundation piles below the excavation interface, and there may be relative displacement on the pile-soil face and initial side friction on the side of the pile. Furthermore, it may produce a pull force on the pile, which will further impact the uplift bearing capacity.

(3) In existing studies, the double pipe method is employed for construction. It can result in the separation of the pile and the soil mass from the ground to the bottom of the foundation pit. It can deduct the side friction of the soil layer in the excavation section, but it cannot reflect the influence of decreasing nominal stress and soil rebound caused by the unloading of excavation on the bearing capacity of the uplift pile. In addition, this method is relatively high in cost, therefore be difficult for engineering promotion.

Since the actual uplift pipe enters the bearing stage with the gradual addition of engineering load, the large-scale deep excavation will have a huge impact on the bearing performance of the uplift pile. At present, there has not been a mature computing theory considering such impacts on the design. Furthermore, due to the restrictions of construction, the test pile of the uplift pile is carried out before the excavation. The fact is that there are usually huge differences between the ground test pile results and accurate pile bottom test results. To fill in this gap in practice, it is very significant to grasp the bearing features of a large-scale deep excavation uplift pile. Also, there exists an urgent need to seek simple and practical bottom pile bearing test method(s).

Due to the limitation of engineering conditions, most of the existing domestic studies are mainly based on numerical simulation analysis and lack corresponding engineering test cases for verification. In this paper, the addition of a rebar strainmeter for testing the bearing capacity of the uplift pile in the static load test is proposed in the engineering context of the static load test of a certain plot in Shanghai. Meanwhile, the uplift single-pile analysis model after the large-scale deep excavation is established with the ABAQUS finite element software. Based on the finite element models, the regular field test, steel stress test, double-pile test, and pit-bottom test are simulated. Next, the influence on single-pile load transmission rules under different test methods is analyzed. Furthermore, the mechanism of large-scale deep-excavation of the uplift single pit is revealed, and suggestions for engineering application are also proposed.

2 PROJECT DESCRIPTION AND TEST RESULTS

The project site is located in Shanghai, and it is planned that the project covers 29820 m2, consisting of two 30-floor buildings and a 4-floor annex. The entire site has a 4-floor basement. The excavation depth of the foundation pit is 22.1m, and both the engineering pile foundation and uplift pile employ a bored pile.

In order to seek the pile bearing capacity below the pile block (22.1m), three groups of SZ-1b piles are designed with tests, about 600mm in diameter and 47.5m in length. Twenty HRB400 reinforcing steel bars of 25mm in diameter were applied. The depth of the pit is 22.1m, and the excavation width is 20m. The reinforcement meter is buried underneath (-22.1m) in the test. The test pile should smash the concrete to form a total cross-section construction joint. Furthermore, the rebar and reinforcement meter near the position should be wrapped with a foamed plastic sheath, as shown in Figure 1. The test employs the steel plate as the counter-force platform, and the jack will gradually exert pulls on the pile. Furthermore, the reinforcement meter can be read through the vibration wire reader, and the pull of the rebar can be calculated to reflect the bearing of the pit bottom for the uplift pile.

It can be seen from Table 1 that the bearing capacity of each pile reaches 3500kN, which satisfies the design demands. The uplift pile bearing capacity below the excavation interface measured by the reinforcement meter ranges between 2000kN and 2600kN (with a percentile difference of 25.7%~42.9%). It shows that the large-scale deep excavation has a significant impact on the bearing of the uplift pile.

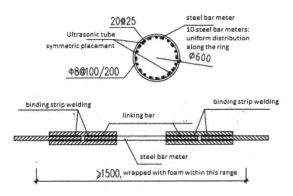

Figure 1. Diagram of steer bar meter installation.

Table 1. Result of pile load test.

NO.	Test method	Forcemax (kN)/ uplift amount (mm)	Average pull of the rebar (kN)	Side friction of pile above the excavation face (kN)
KBZ-1	Slow	3504/29.92	100.82	746~872
KBZ-2	sustaining	3504/33.99	120.42	·
KBZ-3	load method	3504/38.05	133.08	

3 NUMERICAL SIMULATATIONS

According to the test pile cases provided in this paper, the general finite element ABAQUS is employed to establish the axial symmetry model for different test piles. The pile-soil employs the entity unit of axial symmetry. The working conditions analyzed with different test pile methods are expressed as follows (Table 2).

Table 2. Load cases of finite element model.

Analysis Type	Foundation Excavation	Pile Side friction	Foundation Pit Rebound	Excavation Surface Gap
field Test	×	√	×	×
Steel Stress Test	×	√	×	√
Double Pile Test	×	×	×	×
Pit-bottom Test	√	×	√	×

The pile length in the analysis model is 47.5m, the diameter is 600mm, and both the depth and width of the calculation model are 100m. The grid size within the pile length scale is 0.5m, and the meshing size in the width direction is 0.15m. The soil mass employs the Moore Coulomb elastic-plastic model. KBZ1 soil layer distribution employs the distribution near the drill hole. Meanwhile, the finite element trial is employed for adjusting the soil layer parameter. The settings of soil layering and soil parameters adopted are shown in Table 3.

Thickness-free surface contact is adopted between the soil and pile with a penalty function algorithm. The contact constitutive relation employs limited slipping coulomb friction model simulation. After several trials and analysis, the friction coefficient is 0.28, and the slip tolerance is 0.005. The working condition boundary conditions and load analysis STEP are shown in Table 4.

It can be seen from the comparison of the regular field experiment and simulated Q-S curve (Figure 2) that the two curves fit well. The soil layer, pile-soil contact, pile material, as well as all mesh generations suggested in this paper are quite reasonable and can simulate the regular pile test working conditions well. Consequently, the model can be applied to other working conditions.

Table 3. Soil parameters.

Soil Type	h(m)	γ(kg/m3)	E0(MPa)	μ	φ	c (kPa)
Silty clay within clay	2	1800	30	0.38	20	0
Mucky clay within Silty clay	2.9	1870	32.4	0.35	21	19
Mucky clay	8.7	1810	36.4	0.315	32	0
Silty clay	17.6	1660	10.87	0.38	13.4	12
Silty clay	31.0	1800	23.1	0.346	21.2	20
Silty clay	39.0	1810	26.1	0.33	22	22
Silty sand	40.7	1940	31.8	0.33	20.5	45
Silty clay within sand	45.5	1890	63.7	0.3	31	0
Silty sand within clay	59.0	1760	23.8	0.32	18.9	21
Conglomeratic medium sand	75.0	1820	27.7	0.32	19.7	21
Silty sand	88	1930	200	0.3	32	10
Conglomeratic medium sand	100	1960	20	0.35	30.5	10

Table 4. Step cases in different models in Abaqus.

Working condition	Step1	Step2	Step3	Step4
Regular Field Test Steel Stress Test	Balanced soil stress	Excavate soil at the pile; activate the contact; exert the pile gravity	Pile block loading	– –
Double Pile Test		constrain the radical displacement of soil mass at the casing pipe	Excavate soil at the pile and double-pipe; activate the contact of pile and soil mass; exert the gravity of the pile,	Pile block loading
Pit-bottom Test		Constrain the radical displacement of the soil on the side of the foundation pit	Excavate soil at the pile and foundation pit; activate the contact of the pile and soil; exert the pile gravity.	

*Remarks: In the steel stress test, the piles near the excavation surface keep continuous and break at the concrete surface. The other is the same as a regular field test.

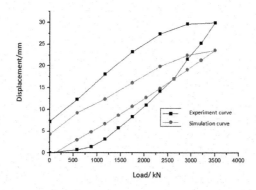

Figure 2. Comparison of Q-S curves of KBZ-1.

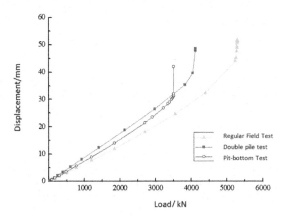

Figure 3.   Comparison of Q-S curves in three test methods.

4   ANALYSIS OF RESULTS

4.1   Q-S Curves

It can be seen from the Q-S curve (Figure 3) that after large-scale excavation of an uplift pile, the soil below the excavation interface usually experiences vertical unloading and rebound, decreasing normal stress at the pile-soil surface. As a result, it lowers the shear strength and stiffness of the interface and the ultimate bearing capacity and stiffness of the uplift pile. The load at the knee of the Q-S curve obtained with the pit-bottom test is about 3507kN. The corresponding deformation is 31.3mm, while that is obtained by the double-pile test. The regular field test method is 4038kN, and 5252kN, respectively, and the corresponding deformation is about 39.7mm and 51.1mm, respectively, as shown in Table 5. The ultimate bearing capacity obtained with the pit-bottom test is reduced by 13.1% and 33%, respectively, when compared to the double-pile test and regular field test, and the corresponding deformation of the Q-S knee increases by 21.7% and 30.9%, respectively.

When the load reaches 600kN in the steel stress test, the concrete reaches the extension strength and starts cracking since 20 reinforcing steel bars of around 25mm in diameter are employed. The cracking concrete nearly bears no pull, of which the pull is mainly undertaken by the rebar. The pull of the rebar is balanced with the frictional resistance at the side of the concrete pile. When the load reaches 3500kN, the rebar has not reached the yield strength. According to the test result of reinforcement meter at pile block (-22.10m), the total pull of the rebar is 100.82*20=2016.4kN, while in regular field test simulation result, the reinforcement stress at the bottom of the excavation pit is far from yielding, and the reinforcement stress is around 200MPa. Consequently, the total pull is 210×20×490/1000 =2058kN. The close values between these two cases suggest the reliability of test pile simulation.

Table 5.   Key point of Q-S curves in different model.

	Q (kN)	(Qi–Q3)/Qi	S (mm)	(Si –S3)/Si
Regular Field Test	Q1=3504	–	S1=29.2	–
Steel stress test	Q1=5252	33.2%	S1=51.1	30.9%
Double pile test	Q2=4038	13.1%	S2=39.7	21.7%
Pit-bottom Test	Q3=3507	–	S3=31.3	–

4.2 *Axial force and pile side friction resistance*

The distribution of axial force and side friction along the effective pile in different test methods is shown in Figure 4, in which a load of each level is the pile block load.

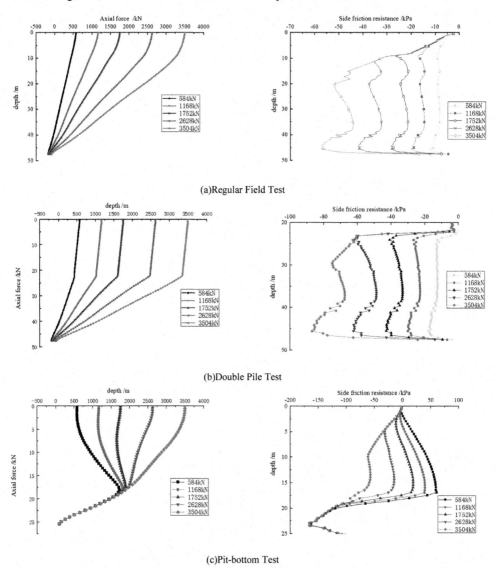

(a)Regular Field Test

(b)Double Pile Test

(c)Pit-bottom Test

Figure 4.   Comparison of axial force and side friction.

As for the regular field test, after the pile bears the up-pull load, the pile may generate upward displacement, and meanwhile, it will produce tensile deformation, and the pile side surface will stand downward friction. A pile load will be transmitted to the surrounding soil layer through the side friction, which may gradually decrease the pie load and tensile deformation. However, with the increase of pile load, the displacement of the pile increases, promoting the exertion of side resistance from top to bottom till the side friction resistance of each point reaches the limit value.

As for the pit-bottom test, after large-scale deep excavation, the positive frictional resistance of the pile is balanced with the negative frictional resistance born at the bottom, and it may produce pull force on the pile. The point with maximum pull is on the neutral surface, namely, the neutral point of side friction, and the maximum value of 1717kN is about 18m below the excavation surface. With the gradual increase of pile up-pull load, the pile surface presents the process of reverse shear, represented by the gradually decreasing positive friction, and it will turn to be negative friction and give full play to its limit. The negative friction of the bottom pile develops gradually, and the pile-soil friction whose negative friction reaches the limit will stay the same. With the increase of pile up-pull load, the axial force of the upper pile increases gradually, and the bottom pile stays the same.

The axial force of pile distribution and side friction at the pull F=3504kN of the regular field test, double-pile test, and pit-bottom test are compared. As shown in Figure 5, the pile side friction at the bottom of the pile is obviously greater than the regular field test and double-pile test method. It gives full play to the pile side friction, for the rebound of excavation soil has already exerted the side friction within a certain scope.

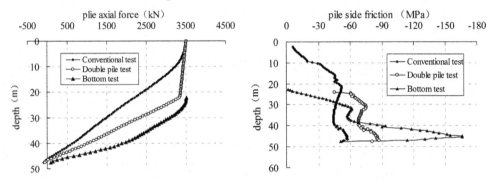

Figure 5. Comparison of axial force and side friction at F=3504kN.

It can be seen by comparing the regular field test and double-pile test that although the normal stress of the soil-pile surface under the excavation surface is the same (Figure 8), since the upper pile of the regular field test method suffers from the up-pull, the sliding deformation of soil-pile expands gradually downward, and side friction also distributes along the pile body. Since the upper pile of the double-pile test is separated from the soil, the sliding deformation of the bottom pile body is also greater than in the regular field test. Consequently, there are differences in the exertion degree of side friction.

In conclusion, there are certain differences in the method of placing the reinforcement meter at the pile block (bottom) proposed for measuring the uplift bearing capacity and the pipe test pile. The main distinctions lie in the different exertion efficiency of side friction within effective pile length, which will certainly result in differences in axial force of the pile. Consequently, it should seek appropriate engineering suggestions to analyze and judge the differences between the regular field test, steel stress test, and pit-bottom test.

4.3 *Pile-soil relative displacement*

The relative displacement in Figure 6, in which the relative displacement varies with depth, shows that the rebound displacement of soil mass is larger than that of the pile, and there may be positive frictional resistance around the pile; the negative relative displacement means that the soil rebound displacement is smaller than that of the pile. According to the shear properties of the pile-soil surface, different relative displacements will give rise to different degrees of pile side friction exertion. It can be seen from the figure that the soil rebound of the upper pile is larger than the rebound of the pile, while the soil rebound is smaller than that of the lower pile. With the neutral surface of the two as the transitional surface, the relative displacement of the soil and pile is zero.

It can be seen from the comparison of side friction distribution of different test methods that when it gives full play to the friction, the maximum displacement of the regular field test method

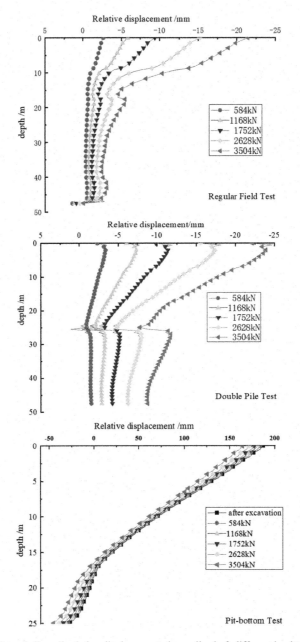

Figure 6. Distribution of pile-soil relative displacement along pile shaft different load tests.

and double-pile test is 51.1mm and 39.7mm, respectively. As for the pit-bottom test, the pile-soil relative displacement caused by the excavation should be deducted, and the maximum displacement when it gives full play to the friction is 31.3mm. In excavation conditions, the soil-pile slippage when it exerts the side friction completely is smaller than the other two test methods. The main reason is that for the pit-bottom test (as shown in Figure 7), a certain range of side friction produces enough pile-soil relative displacement due to the soil rebound, while in the bottom negative friction distribution region, the pile-soil friction is downward. When it starts loading, the negative side friction can increase rapidly till the ultimate side friction (the pile-soil relative slippage is smaller than that of regular field test and double-pile test), which results that the exertion of side friction only requires small relative pile-soil slippage.

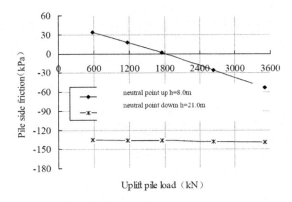

Figure 7.   Side friction under neutral point in excavation test.

4.4   *Normal stress of the pile-soil interface*

It can be seen from the comparison of nominal stress changes with the depth of pile-soil surface at F=3504kN in different test methods (Figure 8) that the normal stress of pile-soil surface below the pit bottom of regular field test and double-pile test is basically the same. But after excavation, the soil at the side of the pile unloads vertically, and the nominal stress of the pile-soil surface is lower than the normally consolidated state, which leads to a decrease in pile bearing capacity and vertical stiffness.

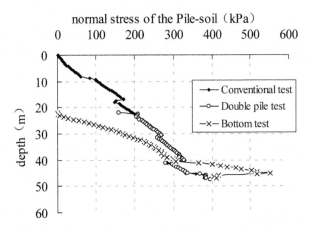

Figure 8.   The normal stress of the Pile-soil interface at F=3504kN.

5   TEST PARAMETER VALIDATION

Based on the analysis of load transferring rules of the uplift pile in different test methods, the test method proposed in this paper is sought to estimate the bottom pile bearing capacity and achieve appropriate engineering suggestions. Further parameter analysis should be carried out from the three methods.

In this chapter, the extreme pulling resistance of non-excavation, 10m, 20m, and 30m of excavation depth, is analyzed respectively by designing15m, 30m, and 50m of the effective pile length, respectively. The pile bearing capacity below the excavation surface can be obtained from the rebar pull at the excavation surface, and it is compared to the analysis result before (regular field test) and after (pit-bottom test) the excavation (Table 6).

Table 6. The loss ratio between steel bars in different load tests.

| excavation depth (m) | Type | 15m | | 30m | | 50m | |
		Qmax (kN)	Smax (mm)	Qmax (kN)	Smax (mm)	Qmax (kN)	Smax (mm)
10m	P1	1478.21	23.65	4658.20	36.59	10847.00	65.32
	P2	1189.00	–	4269.00	–	10612.00	–
	P3	1040.40	15.00	3773.70	28.60	9819.30	58.92
	loss ratio 1	29.62	36.50	18.99	21.69	10.63	9.81
	loss ratio 2	19.56	–	8.36	–	2.17	–
	loss ratio 3	12.50	–	11.60	–	7.47	–
20m	P1	2014.65	27.60	7455.80	56.59	15879.54	92.10
	P2	1421.00	–	5812.00	–	13255.00	–
	P3	1265.10	16.37	5129.34	32.97	12306.72	74.20
	loss ratio 1	37.21	40.76	31.20	41.75	22.50	19.41
	loss ratio 2	29.47	–	22.05	–	16.53	–
	loss ratio 3	10.97	–	11.75	–	7.15	–
30m	P1	4129.32	39.65	11987.32	79.65	23658.30	128.65
	P2	2281.00	–	7536.00	–	16788.00	–
	P3	2087.44	18.14	6941.14	40.23	15593.56	88.60
	loss ratio 1	49.45	54.24	42.10	49.49	34.09	31.10
	loss ratio 2	44.76	–	37.13	–	29.04	–
	loss ratio 3	8.49	–	7.89	–	7.11	–

*Remarks: P1 is Bearing capacity limit before excavation pile; P2 is Steel bar tension by using steel stress test; P3 is Bearing capacity limit after excavation pile; loss ratio 1 = (P1-P3)/P1; loss ratio 2 = (P1-P2)/P1; loss ratio 3 = (P2-P3)/ P2.

It can be seen from the contrastive analysis of loss proportion in different test conditions (Figure 9) that in the same effective pile length, with the increase of excavation depth, the loss proportion of ultimate bearing capacity of uplift pipe also increases. But the reinforcement pull at the excavation surface is close to the bearing capacity of the pile after the excavation. Consequently, compared to bearing capacity loss before excavation, the reinforcement meter increases gradually, and it is quite similar to the bearing capacity loss trend before and after the excavation; However, the reinforcement meter pull decreases with the excavation depth when compared to the bearing capacity after excavation, but the variation range is basically small, about 10% to 15%. Under the same excavation depth, the longer the effective uplift pile is, the smaller the bearing capacity loss after excavation will be, and the smaller the loss proportion of reinforcement meter will be.

To sum up, among each influencing factor of large-scale deep excavation on the uplift pile bearing capacity below the excavation surface, the pile side friction within the range of excavation takes a dominant position since the nominal stress of pile-soil surface caused by the excavation turns to be the secondary influence. In this paper, the method of testing bearing capacity with a reinforcement meter at the bottom pit fails to consider the decreasing nominal stress of the pile-soil surface, just like the double-pile test method. However, it is still basically realistic to employ the reinforcement meter to test the bottom pile bearing capacity. The bearing capacity may be determined by the ultimate bearing capacity after excavation with a reduction of 10%~15%.

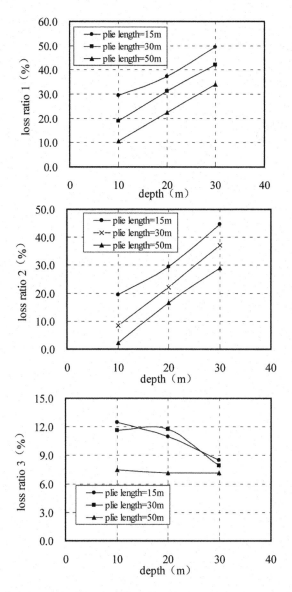

Figure 9.   The loss ratio between steel bars in different load tests.

To be sure the proposed pit location method to test the bottom bearing capacity of pile reinforcement project although like casing pile test method cannot consider normal stress of pile-soil interface is reduced, but seen from the above calculation and analysis results, still can use steel bar meter test lower bearing capacity of pile body, and by 10% to 15% reduction to consider the ultimate bearing capacity after excavation. In the follow-up work, the influence of soil consolidation, pile forming technology, different positions of piles in the foundation pit, and other factors will be taken into account to further carry out pit bottom pile test site tests to verify the conclusions of this paper.

6 CONCLUSIONS

In this chapter, according to the regular field experiment result of the project, the analysis model of large-scale deep excavation on the uplift pile bearing capacity in the different test methods is established with ABAQUS. The influence of different test methods on the load transmission rules is analyzed, resulting in the following conclusions and suggestions.

(1) Compared to the regular field test and double-pile test, the unloading effect generated by the large-scale deep excavation reduces the nominal stress, which may reduce the side friction and the ultimate bearing capacity. Meanwhile, large-scale deep excavation will also result in the unloading rebound of soil, which may drive the rebound of the foundation pile, produce a pull on the pile, and further impact the load transmission in post bearing process. Consequently, reducing the reinforcing steel bars in a pile in deep excavation conditions is not suggested, resulting in the fracture or broken piles during the construction process.

(2) Although the normal stress distribution of pile-soil surface under the excavation surface in regular field test and double-pile test is similar, the upper pile is separated from the soil. In the same uplift loading action, the slippage deformation of the bottom pile is greater than in the regular field test. As a result, the exertion of side friction is much more remarkable than that of the regular field test. However, the side friction at the end of excavation is produced due to the rebound of excavation soil. As a result, the pile side friction of the bottom pile in the pit-bottom test is obviously larger than that of the regular field test and double-pile test.

(3) In the same effective pile length, with the increase of excavation depth, the pull of the reinforcement meter at the excavation surface increases gradually when compared to the bearing capacity loss before excavation, and it is quite similar to the bearing capacity loss trend before and after the excavation. However, the variation range of pull of reinforcement meter compared to the bearing capacity loss after the excavation is 10% to 15%. At the same excavation depth, the longer the effective length is, the smaller the loss of reinforcement meter pull will be. Based on the work in this paper, we suggest measuring the bottom pile bearing capacity with a reinforcement meter in a practical project. Also, the bearing capacity should be determined by the ultimate bearing capacity after excavation with a reduction of 10% to 15%.

REFERENCES

Chen Jin-jian, Wang Jian-hua, Fan Wei, et al. Behavior of up-lift pile foundation during large-scale deep excavation[J]. Chinese Journal of Geotechnical Engineering, 2009, 31(3): 402–407.

Hao Zhe. Research on bearing capacity and deformation characteristics of uplift piles during deep foundation pit excavation[J].Journal of underground space and engineering, 2018, 14(S2):673–678.

Huang Maosong, Li Jianjun, Wang Wei dong, Chen Zheng. Loss ratio of bearing capacity of uplift piles under deep excavation[J]. Chinese Journal of Geotechnical Engineering, 2008, 30(9): 1291–1297.

Huang Mao-song, Ren Qing, Wang Wei-dong, Chen Zheng. Analysis for ultimate uplift capacity of tension pile under deep excavation[J]. Chinese Journal of Geotechnical Engineering, 2007,4(11): 1689–1695.

Mu, R.; Pu, S.-Y.; Huang, Z.-H.; Li, Y.-H.; Zheng, P.-X.; Liu, Y.; Liu, Z.; Zheng, H.-C., Determination of ultimate bearing capacity of uplift piles in combined soil and rock masses(Article)[J]. Yantu Lixue/Rock and Soil Mechanics, 2019,40(7):2825–2837.

V. B. Deshmukh and D. M. Dewaikar and Deepankar Choudhury. Computations of uplift capacity of pile anchors in cohesionless soil[J]. Acta Geotechnica, 2010, 5(2) : 87–94.

Wang Wei dong, Wu Jiangbin. Design and analysis of uplift pile under deep excavation[J]. Chinese Journal of Building Structures, 2010, 31(5): 28–35.

Zheng Gang et al. Finite element analysis on mechanism of the effect of extra-deep excavation on vertical load transfer and settlement of a single pile[J]. Chinese Journal of Geotechnical Engineering, 2009,31(6): 937–845.

Frontiers of Civil Engineering and Disaster Prevention and Control – Yang & Rahman (Eds)
© 2023 The Author(s), ISBN: 978-1-032-31200-2

Deformation mechanism and stability analysis of a high slope during excavation and unloading

Sheng Xiao, Jie Yang* & Xiaoyan Xu
Institute of Water Resources and hydro-electric Engineering, Xi'an University of Technology, Xi'an, China

Pengli Zhang & Dewei Liu
Hanjiang-to-Weihe River Vally Water Diversion Project Construction Co, Ltd., Xi'an, China

ABSTRACT: Many cracks and faults were found in the high slope of a hydraulic project during construction. The slope cracking and abrupt deformation occurred, which might lead to slope instability during excavation and unloading. According to the project's geological survey and design data, a three-dimensional numerical calculation method is adopted to simulate the slope excavation process and analyze the law of slope deformation. The results show that the deformation of the high slope is mainly manifested as the deformation of the structural plane, and its instability mode is sliding failure along fault fz39 and crack L920. After considering slope cutting and load reduction, the slope safety is significantly improved. The research provides essential technical support for the project construction, and it also enriches the study of slope stability in practical engineering.

1 INTRODUCTION

The stability of bank slope in water conservancy and hydropower engineering is one of the core problems in slope design and construction, which directly impacts the operation safety of dam and hydropower stations. In the construction of hydropower projects, it is often necessary to excavate and unload the slope, which will change the original stress state and destroy the balance condition of the slope. The excavated slope will be deformed or even unstable under the conditions of dead weight, rainfall, and construction disturbance. Therefore, it is necessary to analyze and evaluate the stability of the high slope in the process of slope excavation and take corresponding engineering measures to improve its safety when necessary.

Numerical simulation analysis is an effective means to study slope stability, and the finite element method has become the primary calculation tool. Yingren Zheng et al. (Zheng & Zhao 2004)applied the finite element strength reduction method to the jointed rock slope stability analysis, which sets a precedent for finding the sliding surface and stability safety factor of jointed rock slope. Based on the theory of random fields, Yiwei Niu et al. (Niu & Zhou, Qian 2017) analyzed the stability of rock slopes by considering the spatial variability of the rock materials. Three-dimensional stability assessment of rock slopes is analyzed to obtain the safety factor and probability of failure of rock slopes. Shuangli Cao et al. (Cao et al. 2021) analyzed the key factors affecting the slope deformation and used the three-dimensional numerical calculation method to demonstrate the reinforcement effect and check the slope stability. Yingfa Lu et al. (Lu et al. 2021) proposed five mechanical failure modes for thrust- and pull-type slopes, respectively, and five field forms of thrust-type slopes are described. The research results can be used to evaluate slope stability feasibly. Jian Huang et al. (Huang et al. 2021) developed a parallel computing program of peer-to-peer architecture for the contact stability of high rock slope, which avoids solving the nonlinear problem of complex rock

---

*Corresponding Author: yjiexaut@163.com

DOI 10.1201/9781003308577-42

mass materials and has apparent advantages in solving the high rock slope stability problem These achievements have led to the rapid development of slope stability research. In addition, the research on the slope deformation mechanism is of great significance to the slope treatment. Xuefeng Mei et al. (Mei et al. 2021) established a set of multiphase field geological surveys combined with GPS, inclinometers, and piezometer monitoring systems to analyze the deformation and failure mechanism of rock slope. Peng Lin et al. (Lin et al. 2016) proposed a new toppling mode, i.e. the so-called 'conjugate block' mode, to explain the large deformation mechanism of the slope. Xiaobing Song et al. (Song et al. 2012) simulated the deformation controlled by the slightly-inclined angle dislocation interfaces and the NW fault zone by the distinct element numerical analysis code UDEC and 3DEC after the development mode of unloading deformation was summarized. These findings have important implications for the analytical method and reinforcement design with geological settings.

Based on the geological survey and design data of the high slope in a hydraulic project, this paper analyzes the deterioration law and deformation mechanism of the high slope in the excavation process, simulates the slope excavation process by three-dimensional elastic-plastic numerical calculation method, and analyzes its stability in each construction stage.

2 PROJECT BACKGROUND

The construction task of the hydraulic project is mainly for water supply, taking into account power generation and improving water transport conditions. The stability of the high slope of the project is directly related to the safety of hydraulic structures and the long-term normal operation of the project. During the construction of the slope, however, there are some phenomena such as the cracking of the excavated slope, the settlement of the prestressed anchor cable, and the dislocation of the intercepting ditch. With the progress of slope excavation, the deformation continues to develop and may cause unstable failure. The high slope of this project is divided into area I ~ IV, area I is the slope of the pump station intake, area II is the slope of the water-retaining dam section, area III is the slope of the hydropower house, and area IV is the slope of hydropower station tailrace section. The natural slope of the dam shoulder is about 315°, and the elevation of the bank slope is 695m. The exposed stratum of the slope is an ancient intrusive rock mass with single lithology and a simple geological structure. The main structural types are minor faults and fissures. The faults are mainly NNE and NNW directions, all with medium and steep dip angles. The dominant orientation of the slope fissures in this area is strike NEE, NNW~NNE and NWW. The dip angle is mainly steep, followed by a middle dip angle, which accounts for about 90%, and the gentle dip angle is not well developed. The wedge-shaped block KT21 composed of fault fz39(fz39-1) and fissure L920 of the slope in Area I is a potentially unstable block with a large scale. Many groups of structural planes are developed in the rock mass in area II~IV. The developed faults are generally fissured faults, and the larger-scale faults are F1, f8 and f15.

3 FINITE ELEMENT NUMERICAL SIMULATION

3.1 *Basic calculation conditions*

3.1.1 *Calculation model*
The finite element software ABAQUS is used to recheck the stability of the high slope of this project, calculate and simulate the slope excavation process, and demonstrate the influence of excavation on the slope. According to the geological engineering conditions revealed in the process of slope excavation and cleaning, as well as the original topography and excavation form before destruction, the refined finite element numerical simulation model of the high slope is established, as shown in Figure 1(a). The rock strata involved in the model are mainly diorite, divided into four layers from top to bottom: strongly weathered zone, weakly weathered upper zone, weakly

weathered lower zone and micro-new rock mass. The unfavorable structural planes mainly consider F1, f15, f8, fz39-1, fz39 and L920. The slope excavation process is simulated according to the original design scheme in the calculation. The first simulation is from the excavation of EL590m to the EL470m, and the second simulation is from the excavation of EL695m to the EL470m. The models after the two excavations are shown in Figure 1(b) and 1(c) respectively.

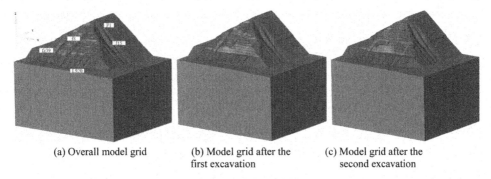

(a) Overall model grid   (b) Model grid after the   (c) Model grid after the
first excavation   second excavation

Figure 1. Numerical calculation model of the high slope.

The coordinate system of the model is shown in Figure 1(a), in which the X-axis is along the river, and it is positive to point downstream. Y-axis is the axial direction of the dam, and it is positive to point to the inner side of the mountain body. The Z-axis is vertical, and the upward is positive. The calculation range along the X and Y axes is 916m and 728m from the coordinate origin to the X and Y axes, respectively, and the calculation range along the Z-axis is from the elevation of −106m to the surface, with the maximum height difference of 801m.

3.1.2 *Mechanical parameters of rock mass*

In the calculation, the ideal elastic-plastic model with the Mohr-Coulomb criterion as the yield function is adopted in the constitutive model of rock mass and fault. According to the results of the rock laboratory test, an in-situ test of rock mass and the development of fissures and structural planes of rock mass with different weathering degrees, combined with the characteristics of excavated slope rock mass and the inversion of strongly weathered rock mass parameters, the mechanical indexes of the sloping rock mass and principal unfavorable structural planes of this project are given in Table 1 (Ma et al. 2020).

Table 1. Mechanical index of slope rock mass and unfavorable structural planes.

Rock mass and fault	Unit weight $\gamma/(kN/m^3)$	Deformation parameters		Strength parameters	
		E/GPa	μ	$f''/°$	c'/MPa
Strongly weathered zone	26.5	0.5	0.32	24	0.15
Weakly weathered upper zone	28.2	4.5	0.27	40	0.8
Weakly weathered lower zone	28.4	6.5	0.26	45	0.9
Micro-new rock mass	28.5	18.0	0.25	50	1.1
f8	27.0	0.5	0.32	15	0.24
F1	26.0	0.5	0.32	16	0.25
f15	26.0	0.5	0.28	16	0.3
fz39-1/fz39	26.0	0.5	0.32	16	0.22
L920	27.0	3.5	0.26	23	0.75

3.1.3 *Boundary conditions*

The initial in-situ stress only considers the self-weight stress in the excavation simulation and slope stability calculation. The finite element model imposes a fixed constraint on the bottom of the foundation, a normal constraint on the periphery of the foundation, and the slope surface is free. In the process of simulating bank slope excavation, the excavated slope soil units are deleted layer by layer according to the actual construction sequence.

3.2 *Analysis of calculation results*

3.2.1 *Initial in-situ stress*

The initial in-situ stress field of the high slope is obtained through in-situ stress analysis, as shown in Figure 2. Under the action of self-weight, the distribution characteristics of displacement before and after balancing are shown in Figure 3.

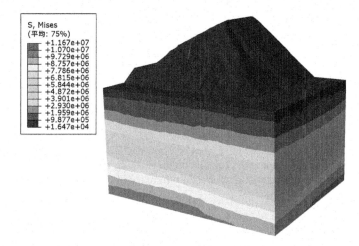

Figure 2.   Distribution of initial in-situ stress.

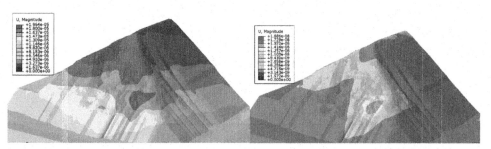

(a) Combined displacement before balance   (b) Combined displacement after balance

Figure 3.   Distribution characteristics of initial displacement.

Under the action of self-weight, the initial in-situ stress of the high slope is larger under the slope than in the valley, which conforms to the distribution characteristics of in-situ stress. The initial displacement of the valley is mainly vertical, and the area with larger displacement is mainly distributed in the ridge at the upstream side of the slope in area I and the lower part of the slope in area II ~ IV.

3.2.2 *Deformation characteristics of slope excavation*

The distribution characteristics of displacement after the first excavation are shown in Figure 4 and the surface displacement vector diagram is shown in Figure 5.

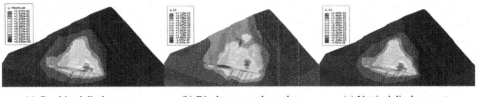

|(a) Combined displacement|(b) Displacement along slope|(c) Vertical displacement|

Figure 4.   Distribution characteristics of displacement after the first excavation.

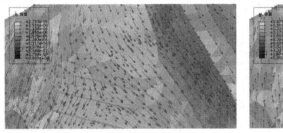 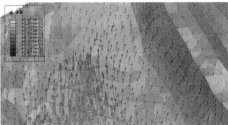

(a) Initial state of excavation   (b) Balanced state of excavation

Figure 5.   Displacement vector after the first excavation.

Due to the influence of excavation and unloading, the settlement of the slope gradually recovered after the first excavation. Compared with the state of initial in-situ stress, the displacement mainly occurred in the vertically upward direction, indicating that the displacement of the slope gradually decreased, and a rebound occurred in the excavation scope and nearby soil. In the initial stage of excavation, the deformation of the slope surface mainly develops in the downward direction, but in the later stage of excavation, the displacement mainly develops upward as a whole. When the excavation is completed, the slope deformation increment is between $3\sim160$mm, and the maximum value mainly occurs at the elevation of $465\sim480$m.

From Figure 4 (b), it can be seen that the displacement along the slope at the elevation of 615m above the upper edge of excavation is about $1.5\sim2.5$ times that at 590m. This phenomenon may be caused by the rebound deformation of the foundation soil due to excavation and unloading. However, due to the existence of fault F1, the deformation is not coordinated, and the fault becomes loose.

The displacement distribution after the second excavation is shown in Figure 6.

(a) Combined displacement   (b) Displacement along slope   (c) Vertical displacement

Figure 6.   Distribution characteristics of displacement after the second excavation.

Compared with the initial in-situ stress equilibrium state, the larger deformation area after the second excavation mainly occurs at the original maximum displacement. In the second excavation process, the main deformation area of the first excavation was further excavated. Due to further load reduction on the slope top, rebound deformation also occurred within the excavation range. The displacement increment between the first excavation and the second excavation is about 1.7 times the displacement increment between the first excavation and the initial in-situ stress. However, the displacement increment along the slope is only about 1.2 times, which indicates that under the second excavation load reduction, the overall displacement of the slope is still mainly vertical upward. That is, compared with the vertical displacement, the displacement along the slope develops less.

3.2.3 *Safety factor of slope stability*

The essence of sliding failure of the slope is strength failure caused by insufficient shear strength of rock soil or structural plane, so the strength reduction method is used to check the safety factor of slope stability (Zheng & Zhao 2004). In the numerical calculation, the instability criterion is used to judge whether the slope enters the critical state, such as the non-convergence of the calculation, the inflection point of the curve between the displacement of the key points of the slope, and the strength reduction coefficient, and the penetration of the shear strain concentration zone (Zheng et al. 2005).

The original model, the first excavation model, and the second excavation model are used to calculate the stability safety factor of the high slope. The linear increase from 0.75 to 2.5 is set in each model, and the calculation does not converge. According to the proportion of non-convergent frame time to analysis step time, the safety factor is calculated by linear interpolation between 0.75 and 2.5. The results show that the safety factor of the slope before excavation is 1.09 and then increases to 1.10 after the first slope cutting and load reduction. During the normal operation period after the second excavation, the safety factor of the slope is 1.27. Since the bolt support is not considered in the calculation, the safety factor of the slope is relatively low.

4 CONCLUSION

Based on the results and discussions presented above, the conclusions are obtained as below:

(1) For the typical structural plane-controlled slope, its excavation and unloading deformation is mainly manifested as the structural plane deformation, and the slope deformation is mostly composed of structural plane dislocation deformation. The structural plane deformation is mainly controlled by the exposed situation of the structural surface when excavating the free face and the excavation scale.

(2) Under the action of excavation and unloading, the slope partially rebounds and forms a loose structural plane at the back edge fault of the first excavation slope, which may lead to a landslide. Moreover, the block deformation of the high slope is also an important problem. Fault fz39 is the upstream cutting surface of the block, and the fracture 1920 is the lateral cutting surface of the block. The instability mode is sliding along the above two structural planes.

(3) When slope reinforcement is carried out, the density, length, and tonnage of the pre-stressed anchor cables should be increased first; secondly, it is reasonable to adopt "load reduction" measures in combination with the engineering form of block; finally, the excavation should be stopped before the slope reinforcement is completed. During the excavation process, the blasting intensity should be strictly controlled, and the dynamic adjustment should be carried out in sections and sequences.

ACKNOWLEDGMENTS

This work was supported in part by the Joint funds of natural science fundamental research program of Shaanxi province of China and the Hanjiang-to-Weihe river valley water diversion project under Grant 2019JLM-55.

REFERENCES

Cao Shuangli, Hu Zhongping, Ding Xiuli, et al. Study on Deformation Mechanism and Stability of Slope Controlled by Structural Plane of Left Bank Slope in Zone I of Huangjinxia Water Conservancy [J]. Water Resources and Power, 2021, 39(02): 114–118.

Huang Jian, Cheng Peng, Zheng Xiaohe. Parallel Computing for the Stability of High Rock Slopes and Its Application [J]. Journal of China Institute of Water Resources and Hydropower Research, 2021, 19(05): 457– 468.

Lin Peng, Liu Xiaoli, Hu Senying, et al. Large Deformation Analysis of a High Steep Slope Relating to the Laxiwa Reservoir, China [J]. Rock Mechanics and Rock Engineering, 2016, 49(6): 2253–2276.

Lu Yingfa, Liu Gan, Cui Kai, et al. Mechanism and Stability Analysis of Deformation Failure of a Slope [J]. Advances in Civil Engineering, 2021, 2021: 1–16. DOI:10.1155/2021/8949846.

Ma Chunhui, Yang Jie, Cheng Lin, et al. Adaptive Parameter Inversion Analysis Method of Rockfill Dam Based on Harmony Search Algorithm and Mixed Multi-Output Relevance Vector Machine [J]. Engineering Computations, 2020, 37(7): 2229–2249.

Mei Xuefeng, Wang Nengfeng, Ma Guotao, et al. Deformation Process and Mechanism Analyses of a Rock Slope Based on Long-Term Monitoring at the Pubugou Hydropower Station, China [J]. Geofluids, 2021, 2021: 1–17.

Niu Yiwei, Zhou Xiaoping, Qian Qihu. Three-dimensional Stability Assessment of Rock Slopes Based on Random Fields [J]. Journal of Civil, Architectural & Environmental Engineering, 2017, 39(03): 129–137.

Song Xiaobing, Shi Anchi, Zheng Weifeng, et al. Analysis of Slope Deformation Characteristics and Mechanism in Left Bank of Baihetan Hydropower Station Jinsha River [J]. Chinese Journal of Rock Mechanics and Engineering, 2012, 31(S2): 3533–3538.

Zheng Hong, Tian Bin, Liu Defu, et al. On Definitions of Safety Factor of Slope Stability Analysis with Finite Element Method [J]. Chinese Journal of Rock Mechanics and Engineering, 2005(13): 2225–2230.

Zheng Yingren, Zhao Shangyi. Application of Strength Reduction Fem in Soil and Rock Slope [J]. Chinese Journal of Rock Mechanics and Engineering, 2004(19): 3381–3388.

Frontiers of Civil Engineering and Disaster Prevention and Control – Yang & Rahman (Eds)
© 2023 The Author(s), ISBN: 978-1-032-31200-2

Research on modular hoisting construction technology of fabricated steel structure residence

Liying Wang*, Yingying Guo & Limei Lei
Chongqing Jianzhu College, School of Intelligent Construction, China

ABSTRACT: With the deepening of China's urbanization construction process, large-scale steel structure buildings are favored by people because of their advantages of fast construction, large span, and high strength. This paper studies the steel structure building, analyze the safety risks existing in its construction, puts forward targeted solutions, and puts forward a targeted, dynamic risk prediction and evaluation system combined with modeling technology, hoping to provide scientific suggestions for workers in the same industry.

1 PREFACE

In steel structure construction engineering, various modern construction materials such as steel plates are mainly used to link with each other to form a modern building component form that can bear and transmit load pressure. Compared with the traditional form of the concrete building, it has higher overall strength, lighter self-weight, more prominent seismic resistance, strong toughness, and a beautiful shape. Based on the above advantages, building developers and building designers will give priority to this building form to a certain extent. Let's take a project in a city as an example to analyze this architectural form.

2 PROJECT OVERVIEW

A large steel structure residential building is located in the mountainous section of a city. It is a fully welded steel pipe truss structure building, which has become a landmark building in the region. The safety level of the steel structure is level II. The steel is q295nhb weather-resistant steel. The site category is level II. A steel structure company is responsible for the welding and erection of all steel pipe structures.

3 MODULAR CONSTRUCTION TECHNOLOGY ANALYSIS

3.1 *Frame module assembly construction and hoisting operation*

In full combination with the actual situation of the construction site, the local steel frame construction mode is selected to complete the ground assembly and modular hoisting tasks. Taking building 1 as an example, the modular assembly area is a circular pipe column (center distance is 1.414m) and a box steel column in the staircase and elevator room (center distance is 2.750m). The corresponding standard floor plan and modular combination details are shown in Figure 1:

---

*Corresponding Author: 1079242017@qq.com

                                          DOI 10.1201/9781003308577-43

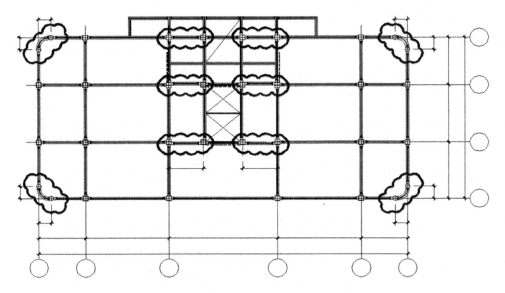

Figure 1.　schematic plan of building standard floor (cloud line part is modular combination).

Building 1 has 16 floors on the ground. The steel column on the ground is divided into six different sections. The main structure is composed of two circular pipe columns and three rectangular arc beams. The cumulative number is 24 groups. This method can reduce 96 lifting operations at different levels. Another combination method is composed of two rectangular columns and three H-shaped steel girders, which can reduce 144 different lifting times. For the nine buildings, the lifting times are about 17000, respectively. The modularization of the first nine buildings is statistically analyzed. After modular assembly, the number of lifting times of the tower crane at high altitude can be further reduced to 2384. It can be seen that this method can significantly improve the construction efficiency on site. At the same time, it can also provide a reliable guarantee for the construction process of steel structures.

3.2　Modular combined hoisting analysis

Taking building 1 as an example, after the modular assembly of its steel frame is completed, the actual lifting weight of components can be reanalyzed in combination with tower crane equipment. At this time, the combined one mode weighs 4.3t, and the weight is within the 44m rotation range of the tower crane (model stt293, boom length 50m). At this time, the lifting capacity is 5.62t > 4.30t, which can fully meet the use requirements of hoisting. The actual weight of combination 2 is 5.8t, and the quantities are within the 38m rotation range of the tower crane (model stt293, boom length 50m). At this time, the lifting capacity is 6.73t > 5.80t, which can fully meet the use requirements of hoisting construction.

3.3　On-site assembly during modular lifting operation

The modular hoisting unit shall be completed in the way of flat assembly, and the high-altitude operation shall be reduced as far as possible, so as to more directly control the assembly accuracy and further improve the welding quality. When all the assembled jig frames in the building are located at the top plate of the first floor underground, the corresponding thickness of the top plate is 450mm. 16mm thick embedded steel plate materials can be reserved in the top plate in advance. Hw200x200 standard steel materials are required for the jig frame column and beam structure (Yao et al. 2019).

1) While the ground is setting out an operation on the construction site, it is necessary to use the total station to measure the centerline of the circular pipe column in the top plate position of all underground floors 1, and determine the final control line of the bracket position of the circular pipe column and the final jig frame position control line.

2) For modular assembly operation, it is necessary to use total station equipment to place the centerline of two round tubular columns and the position of the I-shaped bracket on the jig frame, and then use a 25t truck crane to lift the round tubular column according to the ratio of the center control line of the round tubular column to the control line of I-shaped bracket to make it located on the jig frame. Then a tape measure is used to measure the size of the center position of the two circular pipe columns. At the same time, it's important to ensure that the actual size of the upper and lower openings of the box bracket and the corresponding linear size of the I-shaped bracket meet the expected setting standards, as shown in Figure 2:

Figure 2.   Schematic diagram of modular unit assembly.

3) After all calibration and assembly operations are completed, the final accuracy of all dimensions needs to be rechecked with the help of a level, plumb bob, steel tape, and other equipment. When deviation problems are found, they need to be adjusted immediately until all data are accurate (Ca anguish 2020).

3.4   *Standard for accuracy control of cylindrical tube woodblock assembly*

According to the relevant acceptance standards of gb50205-2020 acceptance standard for construction quality of steel structures, the final acceptance standard for modular unit assembly is formulated. The details are shown in Table 1, which can fully meet the quality control requirements of steel structure installation.

Table 1.   Allowable deviation table of modular cell eye assembly.

Project	Weld gap (with backing plate)	Inspection methods
Axis positioning	1.0	Total station check
Deviation of a reference elevation	±2.0	Level inspection
Center distance of steel column	±3.0	Steel rule check
Length of the steel column	±3.0	Steel rule check
Docking hi-lo	t/10.& ≤2.0	Weld gauge inspection
Weld gap (with backing plate)	−2.0∼3.0	Weld gauge inspection

4   PRECAUTIONS DURING CONSTRUCTION

4.1   *Risk analysis of original structure demolition*

The steel truss is a long-span structure, and the demolition process is actually the process of internal force unloading. Therefore, structural stability should be ensured during the whole demolition process (Lin & Ja 2021).

4.2 Analysis of falling accident

High altitude falling accidents are divided into personnel falling and hoisting component falling, including the following hoisting accidents and the probability value of base events. The principle is that the probability of base events is very high, and the consequences are particularly serious is 0.9; the probability that the base event is more likely to occur and the consequent impact is very serious is 0.7; the probability that the base event may occur and the consequent impact is serious is 0.5; the probability of possible but infrequent base events is 0.3; the probability of almost impossible occurrence is 0.05.

4.3 Analysis of fire during construction

Combustibles and oxidants are the necessary conditions for fire. To fundamentally eliminate fire, it is necessary to manage the combustibles on the construction site strictly.

4.4 Using 4D technology to control dangerous construction projects

Using the 4D virtual simulation technology of BIM, combined with the WBS of the project and its corresponding schedule, it can demonstrate the whole process of steel structure construction and the construction process of each process, including the travel path of hoisting machinery, complex node installation, complex construction process, etc., to intuitively display the changes of the construction site at each stage and truly reflect the risk situation, It makes the identification of hazard sources comprehensive, systematic and dynamic. Based on this information, we can accurately complete the hazard source evaluation and response and improve the efficiency of risk management (Duo Benny 2021).

The installation of a large steel structure is to install the steel members assembled in the factory to the design position and connect them to form the whole structure gradually. The key process is to hoist the transported large steel members to the designated position. When the members reach the designated position, they are connected (including welding and bolt connection) and continue to be assembled in turn until the assembly is completed to form a stable structure.

According to BIM virtual simulation technology, visually simulate the construction path and behavior of construction machinery, accurately identify the hazard sources in the process of travel, and dynamically and visually understand the area where the hazard sources are located. No construction personnel can stay or pass in this area, so as to block people's unsafe behavior and prevent accidents due to the site layout. For the generation of traveling hazard sources caused by collision, reasonably plan the construction site, make the whole construction process reasonable and orderly, and minimize the incidence of hazard sources. The hazard sources of various construction schemes are continuously identified and controlled through the continuous preconstruction simulation, and hazard source management is continuously improved [5].

5 MODULAR COMBINED HOISTING ANALYSIS

When the frame combination construction is completed, the spacing standards of frame columns at this time are 1.414m (circular pipe shape) and 2.750m (rectangular shape), respectively. In order to further ensure the final hoisting construction stability of the composite structure, it is also necessary to minimize the stress on the frame steel beam during the on-site hoisting construction, and select the steel shoulder pole to complete the corresponding hoisting task. At this time, the steel carrying pole adopts φ 180mmx8mm steel pipe, and a 20mm thick steel plate can be made of steel plate. The φ180mmx8mm steel pipe shall be grooved at the lifting point to make the 20mm thick steel plate lifting lug pass through the steel pipe, and the fillet weld can be welded with the steel pipe to ensure the connection effect. At this time, the weld height is h = 8mm.

Before hoisting, a positioning measuring piece shall be pasted on the top of the steel column. After the combined module is hoisted in place, it shall be temporarily fixed with connecting bolts, and the elevation of the steel column shall be corrected with a total station and a level gauge; Place two theodolites in the vertical and horizontal directions of the "cross" centerline of the steel column to correct the criticality of the steel column. After correction, semi-automatic CO_2 gas shielded welding shall be adopted for welding. After welding, the appearance of the weld shall be inspected, and the welding shall be carried out 24 hours after it is qualified to conduct ultrasonic nondestructive testing.

6 CONCLUSION

To sum up, there are more and more potential safety hazards during large-scale steel structure construction projects. Therefore, we need to pay attention to construction safety risk management. The traditional construction safety risk management method cannot give full play to its role, and the control of construction safety accidents cannot achieve the ideal effect. Therefore, under the background of the sustainable development of information technology, BIM Technology can be used to lay a good technical foundation for the construction of large-scale steel structure construction projects in the future and provide an all-around guarantee for construction safety management.

FUND PROJECT

Science and Technology Research Project of Chongqing Education Commission "Application Research of Prefabricated Modular Steel Structure Building in Emergency Medical Engineering." Project Number: KJQN202104302

REFERENCES

Ca anguish Installation technology of anti-buckling steel plate shear wall in fabricated steel structure residence [J] Construction Technology, 2020,49 (16): 5–6

Duo Benny Construction technology of ALC exterior wall panel of a fabricated steel structure residence [J] Shanxi Architecture, 2021,47 (15): 31–32

Lin Wei, Ja Li, Chen Anyway Research on the modular design of fabricated steel structure residence based on BIM Technology [J] Anhui Architecture, 2021,28 (11): 3–4

Yao Kaifeng, Chen Injun, Bhang Guise, et al. BIM simulates the assembly construction technology of large group fabricated steel structure residential buildings [J] Urban Housing, 2019,26 (08): 4–5

Frontiers of Civil Engineering and Disaster Prevention and
Control – Yang & Rahman (Eds)
© 2023 The Author(s), ISBN: 978-1-032-31200-2

Seismic behavior of CFST column with FRP-confined UHPC core-to-steel beam joints with external diaphragm

Zhiheng Chen* & Yi Tao*
Xi'an University of Architecture and Technology, China

ABSTRACT: A novel composite column, which has been developed from the integration of a concrete filled steel tube filled with an FRP confined UHPC core, presents the advantages of high load-bearing capacity, excellent ductility and seismic resistance. This paper introduced the SCF-UHPC composite column-steel beam joint with external reinforcing diaphragm. This joint effectively overcomes the challenge of connection configuration for solid double skin composite column, and guarantees seismic requirement of strong joint and weak members. A quasi-static test was conducted to investigate the seismic performance the joints, and the effects from the FRP thickness, UHPC core diameter and axial compression ratio were analysed. The results showed that increases of the FRP thickness had a limited improvement on the bearing capacity of the joint, but it showed a significant improvements on the ductility and energy dissipation capacity of the joints. Greater UHPC core diameter led to a significant improvement on the bearing capacity of the joint, and showed excellent inhibition effects on the shear deformation of the joint. Increasing the FRP thickness represented more effectiveness on the improvement of ductility and energy dissipation capacity of joints than increasing the UHPC core diameter. Greater axial compression ratio resulted in the decrease of the bearing capacity, but the increase of the ductility.

1 INTRODUCTION

With the development of building structures towards super high-rises and large spans, the performance requirements of the structures are further improved. Therefore, as the key part of the whole structure, the node needs to be further studied. The beam and column connections are a critical structural member as it transfers load between beams and columns. Therefore, the structural safety and reliability are mainly dominated by the mechanical performance of connections between columns and beams. Various joint configurations have been proposed in recent decades. These joints can be classified into three categories, which are joints of column penetrating through the beam, joints of beam penetrating through the column and joints of column partly penetrating through the beam. The joints using external diaphragms are more applicable for the composite column having double skins, such as SCF-UHPC columns, because of the convenient construction. Yufen Zhang (2019), Guochang Li (2014), Dongfang Zhang (2018) found that this type of node has good performance. The external diaphragms joint connection is usually classified as the through-columns joint having higher stiffness. It thus can satisfy the seismic requirements of "strong column weak beam" and has been widely applied for the connection between CFST and steel beam.

2 EXPERIMENTAL PROGRAM

The composite column is constructed by placing the GFRP tube confined UHPC core into the center of the square steel tubes and casting ordinary concrete in between. The formed H-shape steel beams were employed to prepare the beams in the test. The external diaphragms and anchorage webs were

---

*Corresponding Authors: 1293255068@qq.com and y.tao@xauat.edu.cn

DOI 10.1201/9781003308577-44

manufactured using cold formed steel plates with the same thicknesses as the steel beams. The schematic of the joint configuration and the cross-section of SCF-UHPC composite column were shown in Figure 1 and 2, respectively. Total of four connection specimens were tested in the present study.

2.1 Specimens design and preparation

The diameters of UHPC core that was encased in SCF-UHPC composite columns were selected as 100 mm and 150 mm. The number of fiber layers of GFRP tubes were taken as 8 and 10, with a winding angle of 85°. The outer square steel tube with a cross-section of 250 mm×250 mm and a thickness of 8 mm was employed in the present study. The height of column was 1946 mm. The flanges of steel beams had the same thickness of 9 mm as the external diaphragm, and the web of the steel beams has the same thickness of 6.5 mm as the anchorage webs. The overall size diagram is shown in Figure 3. The specific design parameters of specimens are given in Table 1.

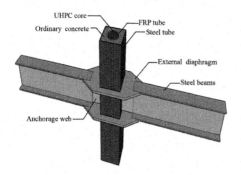

Figure 1.   Schematic diagram of joint configuration.

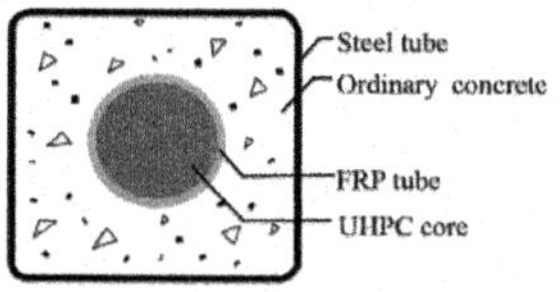

Figure 2.   Schematic diagram of the cross-section of SCF-UHPC composite column.

Table 1.   Characteristic of specimens.

Specimen ID	Diameter of UHPC core (mm)	GFRP tube thickness (mm)	Axial compression ratio n	Steel tube thickness (mm)	Column height (mm)	Steel beam length (mm)	Vertical load (kN)
JD-100-8-0.25	100	2.92	0.25	8	1946	1000	1283
JD-100-10-0.25	100	3.72	0.25	8	1946	1000	1365
JD-100-10-0.5	100	3.72	0.5	8	1946	1000	2731
JD-150-10-0.5	150	3.72	0.5	8	1946	1000	3280

2.2 Test Setup and loading protocol

Figure 4 shows the test setup which was used for the pseudo-static test on beam-to-column connections in the present study. A constant vertical compression was applied to the head of column by a hydraulic jack. The cyclic lateral load was applied at the top end of the column by an MTS actuator.

The vertical load was firstly applied according to the axial compression ratio listed in Table 1. A displacement control regime was adopted to apply the cyclic lateral load in the test. The loading protocol used in the present study is shown in Figure 5. The test was terminated when the lateral load dropped below 85% of its peak value or the failure of the joint caused potential danger.

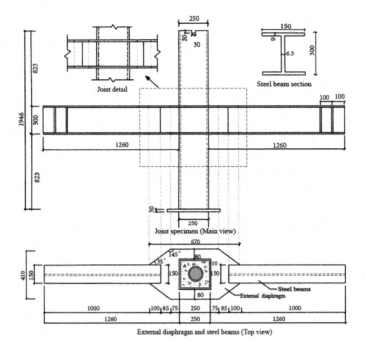

Figure 3. Schematic diagram of the cross-section.

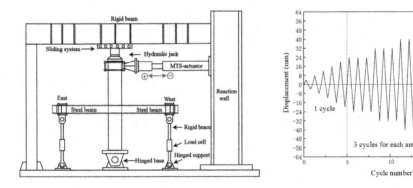

Figure 4. Test setup.

Figure 5. Loading history.

3 TEST RESULTS AND ANALYSIS

3.1 *Failure mode*

The failure of all specimen initiated from the yielding on the beam flanges. All the specimens suffered a failure in the order of the buckling of flanges at the beam end, buckling of webs at the beam end, the formation of the plastic hinge at the beam end, and cracking in the beam flanges in the plastic hinge zone. The failure sequence indicates that the joints satisfy the principles of strong column and weak beams as well as strong joint and weak member. The cracking on the flanges in the beam plastic hinge zone occurred on all specimens, indicating that there is appearance of high stress concentration phenomenon. The specimens under the higher axial compression load, JD-100-10-0.5 and JD-150-10-0.5, exhibited the buckling deformation on the steel tube in both the panel zone and plastic hinge at column end.

3.2 Hysteretic behavior

Figure 7 shows the lateral load-displacement hysteresis curves at the column top of each specimen. The hysteresis curves of all joints showed spindle shape. It indicates that the shape of hysteresis curve is full and the seismic performance of the joint is prominent. The shape of hysteresis loop was still full with the increasing lateral displacement, indicating that the energy dissipation capability of joints was excellent. Comparing specimens JD-100-8-0.25 and JD-100-10-0.25 as shown in Figure 6(a), it is found that the hysteresis loop of specimen with thicker FRP tube (JD-100-10-0.25) was more plumping. This is because the lateral stiffness of the SCF-UHPC composite column increases with the thicker FRP tube. The buckling and cracking at the beam end thus occurred earlier and dramatically developed, and led to the joint exhibiting a better energy dissipation capability. The hysteresis loop of specimen that was under larger axial compression ratio (JD-100-10-0.5) was more plumping than the specimen received the lower axial compression ratio (JD-100-10-0.25) as shown in Figure 6(b). This is because the SCF-UHPC composite column has superior load bearing capacity, the beam-to-column external diaphragm joint thus showed stable energy dissipation capability under greater axial compression. The diameter of UHPC core had a negligible effect on the hysteresis performance of the joints as shown in Figure 6(c). Because the FRP confinement to UHPC core and yield resistance of steel tube of the SCF-UHPC column were kept constant, the specimen having larger size of UHPC core (JD-150-10-0.5) still failed due to the plastic hinge at the beam end and buckling in the panel zone, and its hysteresis behaviour was thus similar as specimen JD-100-10-0.5.

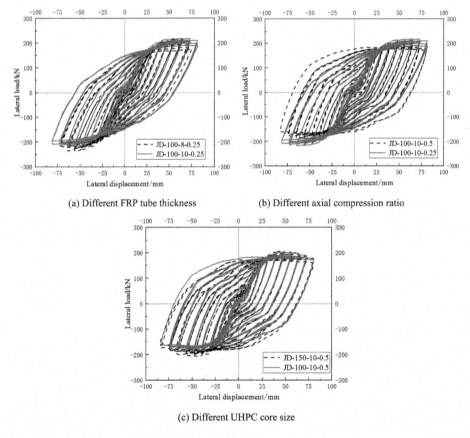

(a) Different FRP tube thickness   (b) Different axial compression ratio

(c) Different UHPC core size

Figure 6.   Hysteresis curves for load-displacement of specimens.

3.3 Skeleton curves

Figure 7 shows the lateral load-displacement skeleton curves of all specimens. It is found that specimen having thicker FRP tube (JD-100-10-0.25) showed a slower softening after the peak load. The specimen under larger axial compression ratio (JD-100-10-0.5) exhibited a smaller load bearing capacity,and the specimen having greater UHPC core (JD-150-10-0.5) led to a higher load bearing capacity.

3.4 Ductility analysis

The ductility coefficient is defined by the ratio of the horizontal displacement Δu of the column top to the yield displacement Δy when the specimen fails. The expression of the ductility coefficient μ is shown in Equation (1). In this paper, the geometric method is used to solve the yield point(Wang 2019). The coefficient calculation results are shown in Table 2. It indicated that increasing the FRP tube thickness is more effectively to enhance the ductility of joints compared with increasing the diameter of the UHPC core.

$$\mu = \Delta_u / \Delta_y \tag{1}$$

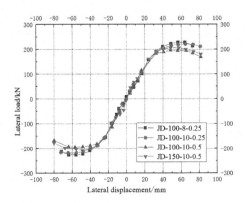

Figure 7. Skeleton curves for load-displacement of specimens.

3.5 Stiffness degradation

The joint stiffness gradually decreased as the specimen suffered the damage accumulation during cyclic loading. In the present study, the loop stiffness K_j was used to describe the stiffness variation as follows (Nie 2012):

$$K_j = \frac{\sum_{i=1}^{n} P_j^i}{\sum_{i=1}^{n} \Delta_j^i} \tag{2}$$

where K_j = denotes loop stiffness; P_j^i = peak load of the ith loading step at displacement coefficient j ($\Delta/\Delta_y = j$); Δ_j^i = corresponding displacement of P_j; and n denotes the total cycle times at each loading level. Figure 8 depicts the relationship of stiffness degradation of $K_j/K_{j,yield}$ and $\Delta/\Delta_{y,yield}$. It can be found that the specimen with a high axial compression ratio has a larger initial stiffness, but the stiffness degrades faster; under the same axial compression ratio, the specimen with a larger diameter of the core column has a larger initial stiffness. FRP thickness has less effect on stiffness degradation.

Table 2. Test Results.

Specimen number	Loading direction	Yield load P_y/KN		Yield displacement Δ_y/mm		Peak load P_{max}/KN		Failure displacement Δ_u/mm		Displacement ductility coefficient μ	Mean value μ
JD-100-8-0.25	Positive direction	194.8		41.4		215.2		71.9		1.74	
			197.2		43.0		225.6		71.3		1.67
	Negative direction	199.6		44.6		236.0		70.6		1.59	
JD-100-10-0.25	Positive direction	191.6		38.4		217.8		81.2		2.11	
			196.3		38.5		217.7		81.7		2.12
	Negative direction	200.9		38.6		217.6		82.1		2.13	
JD-100-10-0.5	Positive direction	172.5		31.2		198.6		78.3		2.51	
			172.0		31.8		197.2		78.6		2.47
	Negative direction	171.4		32.5		195.7		79.0		2.43	
JD-150-10-0.5	Positive direction	167.8		31.3		206.6		80.0		2.55	
			169.2		32.5		204.4		81.7		2.51
	Negative direction	170.5		33.8		202.1		83.4		2.46	

3.6 *Energy dissipation*

The energy dissipation capability of the joints can be evaluated by the cumulative energy dissipation that is defined as the sum of the areas enclosed by each force-displacement hysteresis loop. As shown in Figure 9, with the increase in FRP tube thickness, the cumulative energy consumption increases significantly, while the change of axial compression ratios has less effect on the energy consumption of the joint. The increase in the diameter of the core column has a constant improvement in the hysteretic performance of the specimen under each load.

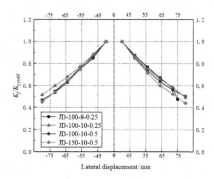

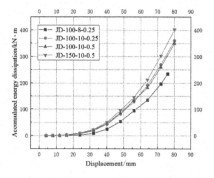

Figure 8. Stiffness degradation curves of specimens.

Figure 9. Cumulative energy consumption curves.

4 CONCLUSIONS AND PROSPECT

Through the research on the seismic performance of four CFST column with FRP-confined UHPC core-to-steel beam joints with external diaphragm, the following conclusions are drawn:

This type of node failure mode is a hybrid failure mode under the combined action of beam end buckling failure, node domain shear failure and composite column buckling. The load-displacement hysteresis curve of each specimen is a plump fusiform, and the skeleton curve is divided into elastic section, yield section and softening stage. The average ductility coefficient of each specimen is above 2, and the ring stiffness degradation curve decreases steadily. Increasing the axial compression ratio and FRP thickness will reduce the bearing capacity of the specimen, while increasing the diameter of the core column will increase the bearing capacity of the specimen; increasing the thickness of FRP, the axial compression ratio, and the diameter of the core column will improve the displacement ductility of the specimen; Increasing the thickness of the FRP and the diameter of the core column will greatly increase the cumulative energy consumption, and the change of the axial pressure ratio has almost no effect on the node energy consumption.

In this paper, the design node is the center column node, and the side column node is not studied for this node form, and its applicability to the side column needs to be further studied. The floor slab is not added in the test of this joint, and the improvement of the lateral stiffness of the steel beam by the floor slab cannot be reflected.

REFERENCES

Li Guochang, Li Donghui, Liu Yu. 2014. Static performance of square steel tube high-strength concrete column-steel beam joints with built-in CFRP circular tubes. *Journal of Shenyang Jianzhu University* (*Natural Science Edition*) 30(03): 442–449.

Nie Jianguo, Wang Yuhang, Tao Muxuan. 2012. Experimental study on seismic behavior of laminated steel tube column-concrete beam joint with outer stiffening ring. *Journal of Building Structures* 33(7): 88–97.

Wang Ying, Bi Lingyun. 2019. Research on seismic performance of new composite concrete-filled steel tube beam-column joints. *Building Structure* 49(14): 42–47.

Zhang Yufen, Jiang Zonghao, Zhu Ge. 2019. Experimental study on seismic performance of CFST externally reinforced ring-slab joints. *Industrial Architecture* 49(06):167–175.

Zhang Dongfang, Zhao Junhai, He Shuanhai, et al. 2018. Study on stress distribution and force transfer mechanism of CFST column-steel beam joints. *Building Structure* 48(15): 37–43.

*Building material properties and
highway bridge construction*

*Frontiers of Civil Engineering and Disaster Prevention and
Control – Yang & Rahman (Eds)
© 2023 The Author(s), ISBN: 978-1-032-31200-2*

A brief review of self-healing concrete: typical mechanisms and approaches

Z.Q. Gu, X.L. Ji* & H.H. Zhang
*School of Civil Engineering and Architecture, Nanchang Hangkong University, Nanchang, Jiangxi,
P.R. China*

ABSTRACT: Due to the automatic detection and repair ability, self-healing (SH) concrete with various healing mechanisms has received tremendous attention in recent years. Traditional approaches (such as experimental, theoretical, and numerical tools) and modern machine learning methods, for instance, the artificial neural network (ANN) tools, are frequently adopted in the studies of SH concrete. In this paper, the well-known healing mechanisms of SH concrete are introduced for the first time. Then, traditional methods to reveal the healing properties of SH concrete are briefly reviewed. The following is concisely introduced the status of the healing performance study with the ANN. Finally, conclusions and some future works are presented.

1 INTRODUCTION

Concrete is one of the most important materials in the civil engineering field due to its good durability, affordability, high compressive strength, and flexible shaping property (Muhammad et al. 2016; Ramin et al. 2016). Suffering from limited tensile strength, asymmetrical shrinkage, and temperature gradient, concrete structures are prone to cracks or micro-craters during their service life. The occurrence of cracks leads to the degradation of bearing capacity, applicability and durability of concrete structures, and may even cause more serious consequences such as collapse (Li & Herbert 2012).

To improve the performance of the conventional concrete, especially its crack resistance, many efforts have been put into the development of SH concrete. There exists a lot of work on the healing performance of SH concrete by traditional methods, experimentally, theoretically or numerically. Using three-point bending test, Liu et al. (2021) studied the bending resistance of the samples with preset cracks in SH concrete. By Poiseuille's law, Lee et al. (2021) calculated the water flow through cracked SH concrete. Through SH permeability model, Yuan et al. (2019) predicted the healing rate of SH concrete with mineral additives. Wang et al. (2018) predicted the healing performance of BK7 glass material using finite element method. However, traditional methods encounter insufficient experiment observation, lacking theoretical model or constrained numerical simulation (Mauludin & Oucif 2019).

Fortunately, originally used to simulate the function and structure of human brain, a data-driven technology, i.e., ANN (Feng et al. 2019), provides a new path for the prediction of the healing performance of SH concrete by a multi-nonlinear fitting process. One of the representatives of the ANN is the back propagation neural network (BP) (Wen & Ru 2019), which can be regarded as an implicit nonlinear fitting tool with memory function, automatic optimization function, intelligence and high accuracy. Adesanya et al. (2021) demonstrated that BP neural network not only

---

*Corresponding Author: xiaoleiji@nchu.edu.cn

can quickly and accurately predict healing performance, but also can further study the effects of various parameters on healing performance. Guo et al. (2018) predicted the compressive strength of recycled thermal insulation concrete through the BP neural network. Considering the drawbacks of BP neural network, i.e., slow convergence speed, easy to fall into local minima and weak network generalization, some optimizing algorithm is proposed to construct hybrid ANN methods, e.g., the PSO-BP network (Wang 2015).

In this paper, a brief review of typical healing mechanisms and analysis approaches of SH concrete is provided. Firstly, typical healing mechanisms for SH concrete are presented. Subsequently, traditional approaches and advanced ANN tools for healing study are discussed. Finally, some future perspectives on the research of SH concrete are highlighted.

2 HEALING MECHANISMS OF SH CONCRETE

Based on physical, chemical and biological principles (Jiang 2018), as shown in Figure 1, the healing mechanism of SH concrete can be roughly divided into three kinds, i.e. shape memory alloy (SMA) healing mechanism, capsule healing mechanism and biochemical reaction healing mechanism. For the SMA mechanism, Chen et al. (Chen et al. 2021) proposed a composite SMA ECC and studied its crack-healing performance; It is found that this composite can reduce damage, heal cracks, improve structural durability, and reduce maintenance cost. By capsule mechanism, Shen et al. (2020) tested the healing performance of SH concrete containing epoxy resin microcapsules with a universal tensile machine; Litina and Abir (2020) developed a polymer capsule containing liquid sodium silicate that can repair cracks. With biochemical mechanism. Feng et al. (2021) used Bacillus subtilis M9 as healing agent and added polyvinyl alcohol fiber to prepare SH concrete beam specimens, and it was found that micro cracks were filled and healed by calcium carbonate precipitation after 28 days of curing.

 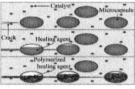 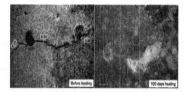

| (a) SMA mechanism (Kuang & Ou, 2008) | (b) capsule mechanism (White, Sottos, Geubelle, et al., 2001) | (c) biochemical mechanism (Joshi, Goyal, Mukherjee, et al., 2017) |

Figure 1.   The schematic of three types of healing mechanism (Jiang, 2018).

3 TRADITIONAL APPROACHS FOR SH CONCRETE

In this section, the status of traditional methods, i.e., experimental, theoretical and numerical methods, for SH concrete are introduced. Using an experimental approach, De Nardi et al. (Nardi et al. 2021) studied the effect of the mixing ratio of modified cyanoacrylate adhesive on healing performance. Based on the Hopkinson pressure bar device. Huang et al. (2021) conducted the test on SH concrete. Based on ion diffusion theory and thermodynamic theory. Huang et al. (2012) found that the additional water can promote the further hydration of un-hydrated cement particles in concrete. With the basic principles of probability theory and geometry. Zemskov et al. (2011) established the mathematical model of crack impact capsule probability under two modes, which provided a theoretical basis for selecting the best capsule size and quantity of SH concrete. In the numerical aspect, Algaifi et al. (2018) built a numerical model composed of first-order ordinary differential equations and second-order partial differential equations to predict the healing process and verified the predicted crack-healing results by a scanning electron microscope and energy-dispersive X-ray analysis.

4 ANN FOR SH CONCRETE

4.1 *A brief of the ANN*

ANN was first proposed by McCulloch and Pitts (McCulloch & Pitts 1943) in 1943. ANNs are highly interconnected through a large number of nodes (or neurons), fully parallel work, and finally realize the operation of highly nonlinear mapping, as shown in Figure 2.

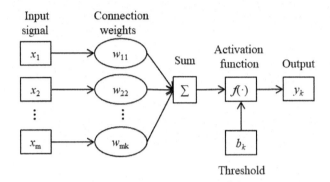

Figure 2.   ANN model (Qiu 2020).

ANN modeling has been applied in almost all engineering fields. For civil engineering in particular, ANN modeling has been employed to solve complex problems in the areas of structural, construction, geotechnical, environmental, and management engineering (Adeli 2001). For instance, Wang et al. (2020) used the ANN model to predict the airport road surface response in the surface deformation of the airport road and showed that the prediction accuracy of the ANN model was better than that of the traditional inverse algorithm. Moradi (2021) used the ANN model to predict the compressive strength of metakaolin concrete.

4.2 *Application of ANN for SH concrete*

As a typical approach to ANN, the BP neural network is a very effective multilayer neural network. With the BP neural network, a lot of work has been carried out. Luo et al. (2021) predicted the healing and mechanical properties of SH epoxy resin system through BP neural network. Chaitanya et al. (2020) predicted the SH properties of superabsorbent polymer added to M40 grade concrete. However, there still exist defects in the BP neural network, for example, easy convergence to local minimum points, and the contradiction between network learning ability and generalization ability (Cheng 2011). To improve the shortcomings of the standard BP neural network, many collaborative methods are proposed, among which the genetic algorithm-optimized BP neural network is representative. Considering that the genetic algorithm (GA) has global random search ability and strong robustness, the GA-BP neural network can prevent the standard BP neural network from falling into local optimum. Presently, GA-BP neural network has been extensively studied for SH concrete. For instance, Suleiman and Nehdi. Suleiman and Nehdi (2017) predicted the SH crack ability in concrete through GA-BP neural network. Huang et al. (2021) predicted the healing performance of SH concrete by GA-BP neural network. Suleiman and Nehdi (2017) developed a GA-BP neural network model to predict the change in airflow permeability in SH concrete. Based on the standard BP and also the GA-BP neural network, Jiang (2018) predicted the life of the SH concrete after the healing of the cracks.

5 CONCLUSIONS AND FUTURE PERSPECTIVES

The demand for high-performance SH concrete in civil engineering is increasing. In this work, three types of healing mechanism are briefly introduced, i.e., the SMA mechanism, the capsule mechanism and the biochemical mechanism. Hereafter, the applications of traditional approaches including experimental, theoretical and numerical methods, in the analysis of healing performance of SH concrete are concisely presented. Following, modern machine learning tools, especially ANN methods (e.g., the BP and GA-BP neural networks), are succinctly reported to predict the SH process. Upon this brief review, and concerning the mechanism, analysis, improvement and design of the SH concrete, some future work is suggested as follows.

(1) Further study of the existing healing mechanism and, at the same time, development of new healing mechanisms with higher healing efficiency, lower cost, and wider applicability.
(2) Development of advanced experimental, theoretical and numerical methods or their combinations, for accurate and efficient study of healing performance of SH concrete.
(3) Development of advanced machine learning tools to analyze the healing property and optimize the design of new SH concrete.
(4) Combinations of the traditional solutions and modern computer tools (e.g., combined numerical-machine learning-experimental tools) for the analysis, improvement and design of SH concrete.

ACKNOWLEDGMENTS

This work was supported by the National Natural Science Foundation of China (Grant No. 12062015), the Jiangxi Provincial Natural Science Foundation of China (Grant No. 20212BAB211016) and the Scientific Research Foundation (Grant No. EA202011154).

REFERENCES

Adeli, H. (2001) Neural Networks in Civil Engineering. Computer-Aided. Civ. Inf., 16: 126–142.
Adesanya, E., Aladejare, A., Adediran, A., et al. (2021) Predicting shrinkage of alkali-activated blast furnace-fly ash mortars using artificial neural network (ANN). Cement. Concrete. Comp., 124: 104265.
Algaifi, H.A., Bakar, S.A., Sam, A.R.M., et al. (2018) Numerical modeling for crack self-healing concrete by microbial calcium carbonate. Constr. Build. Mater., 189: 816–824.
Chaitanya, M., Manikandan, P., Prem Kumar, V. et al. (2020) Prediction of self-healing characteristics of GGBS-admixed concrete using Artificial Neural Network. Journal of Physics: Conference Series. 1716(1): 012019.
Chen, W.H., Feng, K., Wang, Y., et al. (2021) Evaluation of self-healing performance of a smart composite material (SMA-ECC). Constr. Build. Mater., 290: 123216.
Cheng, Y. (2011) Study on improved algorithm and application of BP neural network. Chongqing University.
Feng, J., Chen, B.C., Sun, W.W., et al. (2021) Microbial induced calcium carbonate precipitation study using Bacillus subtilis with application to self-healing concrete preparation and characterization. Constr. Build. Mater., 280: 122460.
Feng, Z.J., Li, C.L., Ren, Z.Z. (2019) Application of artificial neural network model in concrete field. Southern Agricultural Machinery. 50(09): 206–207.
Guo, Y.D., Liu, Y.Z., Wang, W.J., et al. (2018) Prediction of compressive strength of regenerated thermal insulation concrete based on BP neural network. Concrete. (10): 33–35+39.
Huang, H.L., Ye, G. (2012) Simulation of self-healing by further hydration in cementitious materials. Cement. Concrete. Comp., 34(4): 460–467.
Huang, X., Wasouf, M., Sresakoolchai, J., et al. (2021) Prediction of healing performance of autogenous healing concrete using machine learning. Materials (Basel). 14(15): 4068.
Huang, Y.J., Wang, X.F., Sheng, M., et al. (2021) Dynamic behavior of microcapsule-based self-healing concrete subjected to impact loading. Constr. Build. Mater., 301: 124322.
Jiang, S.Y. (2018) Experimental study and life prediction on the self-healing concrete with sodium silicate healing agent. South China University of Technology.

Joshi, S., Goyal, S., Mukherjee, A., et al. (2017) Microbial healing of cracks in concrete: a review. J Ind Microbiol Biotechnol. 44(11): 1511–1525.

Kuang, Y.C., Ou, J.P. (2008) Passive smart self-repairing concrete beams by using shape memory alloy wires and fibers containing adhesives. J. Cent. South Univ. T., (03): 411–417.

Lee, K.M., Kim, H.S., Lee, D.K., et al. (2021) Self-healing performance evaluation of concrete incorporating inorganic materials based on a water permeability test. Materials (Basel). 14(12): 3202.

Li, V.C., Herbert, E. (2012) Robust self-healing concrete for sustainable infrastructure. J. Adv. Concr. Technol., 10(6): 207–218.

Litina, C., Abir, A. (2020) First generation microcapsule-based self-healing cementitious construction repair materials. Construction and Building Materials. 255: 119389.

Liu, C., Zhang, R.F., Liu, H.W., et al. (2021) Experimental and analytical study on the flexural rigidity of microbial self-healing concrete based on recycled coarse aggregate (RCA). Constr. Build. Mater., 285: 122941.

Luo, H., Jin, K., Tao, J., et al. (2021) Properties prediction and design of self-healing epoxy resin combining molecular dynamics simulation and back propagation neural network. Mater. Res. Express., 8(4): 045308.

Mauludin, L.M., Oucif, C. (2019) Modeling of self-healing concrete: a review. J. Appl. Comput. Mech., 5(3): 526–539.

McCulloch, W.S., Pitts, W. (1943) A logical calculus of the ideas immanent in nervous activity. Bulletin of Mathematical Biophysics. 5: 115–133.

Moradi, M.J., Khaleghi, M., Salimi, J., et al. (2021) Predicting the compressive strength of concrete containing metakaolin with different properties using ANN. Measurement. 183: 109790.

Muhammad, N.Z., Shafaghat, A., Keyvanfar, A., et al. (2016) Tests and methods of evaluating the self-healing efficiency of concrete: A review. Constr. Build. Mater., 112: 1123–1132.

Nardi, C.D., Gardner, D., Cazzador, G., et al. (2021) Experimental investigation of a novel formulation of a Cyanoacrylate-based adhesive for self-healing concrete technologies. Frontiers in Built Environment. 7: 660562.

Qiu, X.P. (2020) Neural Networks and Deep Learning. China Machine Press

Ramin, A., Abd Majid, M.Z., Hussin, M.W., et al. (2016) Optimum concentration of Bacillus megaterium for strengthening structural concrete. Constr. Build. Mater. 118: 180–193.

Shen, J., Kan, L.H., Zheng, T.X. (2020) Study on properties of self-healing epoxy resin microcapsule concrete. China Building Materials Technology. 29(02): 28–29.

Suleiman, A.R., Nehdi, M.L. (2017) Modeling self-healing of concrete using hybrid genetic algorithm-artificial neural network. Materials (Basel). 10(2): 135.

Suleiman, A.R., Nehdi, M.L. (2017) Predicting change of concrete air-flow permeability due to self-healing. Conf. Leadership in Sustainable Infrastructure.

Wang, C., Wang, H.X., Shen, L., et al. (2018) Numerical simulation and experimental study on crack self-healing in BK7 glass. Ceram. Int., 44(2): 1850–1858.

Wang, G.M. (2015) Based on PSO-BP network learning method research. Anhui University, Hefei.

Wang, H., Xie, P.Y., Ji, R., et al. (2020) Prediction of airfield pavement responses from surface deflections: comparison between the traditional back calculation approach and the ANN model. Road. Mater. Pavement., 22(9): 1930–1945.

Wen, C.B., Ru, F. (2019) Theory and application of artificial neural network. Xidian University

White, S.R., Sottos, N.R., Geubelle, P.H., et al. (2001) Autonomic healing of polymer composites. Nature: International Weekly Journal of science. 409(6822): 794–797.

Yuan, Z.C., Jiang, Z.W., Chen, Q. (2019) Permeability modeling of self-healing due to calcium carbonate precipitation in cement-based materials with mineral additives. J. Cent. South. Univ., 26(3): 567–576.

Zemskov, S.V., Jonkers, H.M., Vermolen, F.J. (2011) Two analytical models for the probability characteristics of a crack hitting encapsulated particles: Application to self-healing materials. Comp. Mater. Sci., 50(12): 3323–3333.

*Frontiers of Civil Engineering and Disaster Prevention and
Control – Yang & Rahman (Eds)*
© 2023 The Author(s), ISBN: 978-1-032-31200-2

Study on the foam process of different types of warm-mix asphalt

Rong Chang
Research Institute of Highway, Ministry of Transport, Beijing, China

ZeWen Tan*, ZhouShuai Wei, XinYe Cao & DeNing Cai
Guangxi Xinfazhan Communication Group Co., Ltd, Nanning, Guangxi, China

ABSTRACT: The mechanical foaming warm-mix asphalt technology uses foaming equipment
to form foamed asphalt, which effectively improves the construction and workability of the asphalt
mixture and achieves road performance equivalent to that of the hot-mix asphalt mixture under
the condition of cooling construction. This paper uses different foaming equipment and different
foaming methods to conduct foaming tests on base asphalt, styrene butadiene styrene (SBS) modified asphalt, and rubber-modified asphalt with different oil sources and provides data support and
reference for determining the control standards of asphalt foaming parameters. The results show
that within a certain range of temperature and water consumption, as the foaming temperature
and water consumption increase, the expansion rate of different types of asphalt increases and the
half-life decreases. Different foaming equipment and foaming methods have a certain impact on
the foaming effect of asphalt. The expansion rate of the 3 kinds of asphalt in the aeration condition
of the foaming equipment increased, and the half-life decreased. The foaming effect of bitumen
with the same label will be different due to different oil sources. Therefore, the selection of suitable
foaming asphalt, foaming temperature, and water consumption are the key factors that determine
the foaming effect of asphalt.

1 INTRODUCTION

The mechanical foaming type warm-mix asphalt pavement technology uses asphalt and trace water
through mechanical foaming equipment to form foamed asphalt, which increases the specific surface area of the asphalt and reduces the viscosity of the asphalt binder so that it can be combined
with coarse and fine aggregates in the mixing station. Minerals, powders, etc. are mixed at a lower
temperature (Hailesilassie et al. 2014; You et al. 2018). Warm-mix asphalt mixture can effectively
reduce the mixing temperature by 10–40°C, improve the coating property and construction workability of the mixture, and the road performance is equivalent to that of the hot-mix asphalt mixture
(He et al. 2015; Ma et al. 2015; Zhang et al. 2012). Compared with other warm mixing technologies, this technology has a low input cost and only needs foaming equipment. Compared with
hot-mix technology, it can effectively reduce the construction temperature. This can bring a series
of benefits, such as reducing asphalt aging, improving pavement durability, saving fuel consumption, reducing asphalt smoke and other toxic gas emissions, etc., which can bring considerable
economic and social benefits (Akisettyckk 2008; Chen, 2014; Hajj et al. 2011).

At present, the mechanical foaming and warm mixing technologies have been studied in depth,
and they have been promoted and applied in many provinces, but there is no standardized evaluation
standard for the foaming effect of asphalt (Guo et al. 2014; Mo et al. 2012Wrustje & Putmanbj
2013). This study uses different foaming equipment and different foaming methods to conduct

---

*Corresponding Author: roadzz2022@163.com

                                    DOI 10.1201/9781003308577-46

systematic foaming tests on different oil sources of base asphalt, SBS modified asphalt, and rubber-modified asphalt, in order to provide a reference for determining the control standards of asphalt foaming parameters.

2 MATERIALS AND PROGRAMS

2.1 *Asphalt*

Asphalt was used for foaming tests of matrix asphalt, SBS-modified asphalt, and rubber-modified asphalt. Select five different oil sources for asphalt and modified asphalt, respectively. Asphalt is marked as A, B, C, D, E; SBS modified asphalt is marked as H, I, J, K, and L. Three types of rubber-modified asphalt with different rubber powder numbers are marked as O, P, and Q, respectively. The basic technical indexes of asphalt, SBS modified asphalt, and rubber-modified asphalt are shown in Tables 1–3.

Table 1. Technical indicators of asphalt.

Index	A	B	C	D	E
Penetration/0.1 mm (25°C)	70	80	62	65	85
Soft point/°C	51.5	49.0	48.0	48.0	46.5
Ductility/cm (15°C)	>100	>100	>100	>100	>100
(RTFOT) Residue					
Quality loss/%	−0.17	−0.24	−0.10	−0.14	−0.21
Penetration/0.1 mm (25°C)	46	51	42	42	57
Penetration ratio (25°C)/%	66	64	68	64	67
Ductility/cm (15°C)	27	66	21	23	57

Table 2. Technical indexes of SBS modified asphalt.

Index	H	I	J	K	L
Penetration/0.1 mm (25°C)	57	62	51	65	56
Soft point/°C	61.5	60.4	84.0	58.5	62.0
Ductility/cm (5°C)	24	45	31	34	28
Elastic recovery/%	89	85	95	80	92
(RTFOT) Residue					
Quality loss/%	0.4	−0.2	−0.16	−0.36	0.21
Penetration/0.1 mm (25°C)	37	42	37	40	38
Penetration ratio (25°C)/%	65	68	73	62	67
Ductility/cm (5°C)	11	19	18	22	15

Table 3. Technical Indexes of rubber-modified asphalt.

Index	O	P	Q
Penetration 100 g, 5 s, 25°C/0.1 mm	37	42	45
Ductility 5°C, 1 cm/min/cm	9	11	15
Soft point $T_{R\&B}$/°C	76	75	75
Rotational viscosity 180°C/Pa·s	4.075	3.864	3.912
Elastic recovery 25°C/%	91	92	94

2.2 *Design of the experiment scheme*

Choose one or more of the 3 types of asphalt for different foaming tests. According to experience, the heating temperature of each asphalt used for foaming is selected: 140, 150, 160°C, and the water consumption is 1%, 2%, 3%; the heating temperature of SBS modified asphalt is 150, 160, 170°C, and the water consumption is 2%, 3%, 4%. The heating temperatures of rubber-modified asphalt are 160, 170, and 180 degrees Celsius, respectively, and the water consumption is 1%, 2%, and 3%.In order to analyze the impact of different foaming equipment and different foaming methods on the foaming effect of asphalt, 2 types of foaming equipment are used in this article: imported foaming equipment (W-type) and domestic foaming equipment (N-type). Adopt 2 foaming methods: aerated or non-aerated.

3 DATA ANALYSIS

3.1 *The impact of foaming equipment*

Different types of foaming equipment directly affect the foaming result of asphalt. Using W-type and N-type foaming equipment, the foaming test is carried out with and without aeration, respectively. Choose Asphalt E, SBS Modified Asphalt L, and rubber-modified asphalt P as foamed asphalt to compare the effects of different types of foaming equipment on the foaming effect. The test results are shown in Figures 1–5.

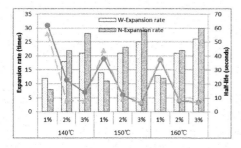

Figure 1.   Asphalt E (aerated).

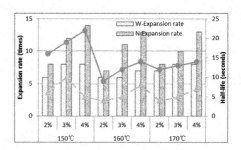

Figure 2.   SBS modified asphalt L (aerated).

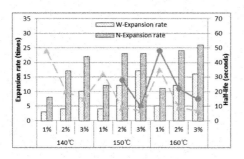

Figure 3.   Asphalt E (no air added).

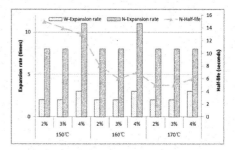

Figure 4.   SBS modified asphalt L (no air added).

In the non-aerated state, the half-life of the asphalt after foaming with W-type foaming equipment exceeds 10 minutes, so we do not make a comparison here. It can be seen from the comparison of the above figures:

(1) In the aerated state, the two-foaming equipment have little effect on the foaming effect of asphalt; in terms of the foaming effect of SBS modified asphalt, the expansion ratio results

348

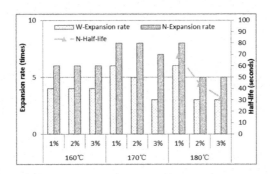

Figure 5.   Rubber modified asphalt P (aerated).

in W-type foaming equipment being better than N-type foaming equipment, and the half-life result is the opposite.

(2) In the unaerated state, as for the expansion rate results, the foaming effect of the N-type foaming equipment on the asphalt is better than that of the W-type foaming equipment. The half-life after foaming through W-type foaming equipment is generally relatively long, but the expansion rate is relatively small. Rubber-modified asphalt cannot obtain a good foaming effect when the foaming equipment is not aerated.

(3) Comparing the differences between the two-foaming equipment under different foaming methods, it can be seen that different types of foaming equipment have different effects on the foaming effect of asphalt. Therefore, it is very important to choose the right foaming equipment.

3.2   *Influence of foaming method*

Different asphalts and different foaming methods are analyzed to determine the impact. When foaming, it is divided into 2 cases: aerated and non-aerated. C is used for asphalt, L is used for SBS-modified asphalt, and O is used for rubber-modified asphalt. The test results are shown in Figures 6–8. Those with a longer half-life are not marked in these figures.

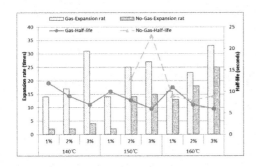

Figure 6.   Asphalt C.

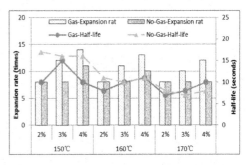

Figure 7.   SBS modified asphalt L.

3.3   *Different asphalt foaming effects*

At present, most of the domestic on-site foaming equipment is mainly non-aerated. Therefore, the N-type foaming equipment is used to separate five types of base asphalt, five types of SBS-modified asphalt, and three types of rubber-modified asphalt in an unaerated state. Carry out a foaming test to compare the foaming effects of different oil sources on asphalt. Due to the long half-life of some asphalts, only the expansion rate is analyzed, and the test results are shown in Figures 9–11.

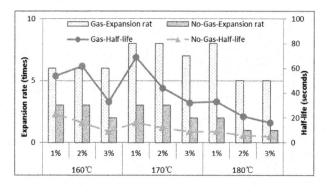

Figure 8.   Rubber-modified asphalt O.

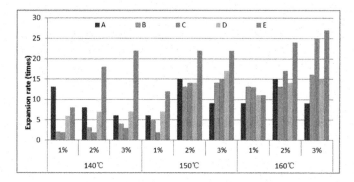

Figure 9.   Test result of asphalt expansion rate.

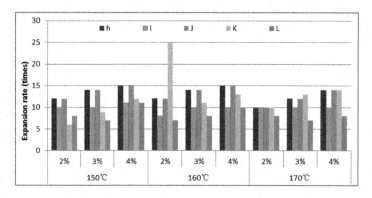

Figure 10.   SBS modified asphalt expansion rate test results.

From the results of the expansion rate test, it can be seen that the final foaming effect is different between different oil sources with the same label and under the same foaming conditions. The difference in foaming effects of asphalt with different oil sources is more obvious, while SBS modified asphalt is relatively stable. With the increase of the rubber powder mesh, the rubber-modified asphalt becomes more delicate and stable, the expansion rate of the rubber foam asphalt increases, and the half-life increases. Therefore, before the construction of a mechanically foamed warm-mix asphalt mixture, it is necessary to conduct a foaming test on the existing asphalt and determine the best foaming parameters through its foaming effect.

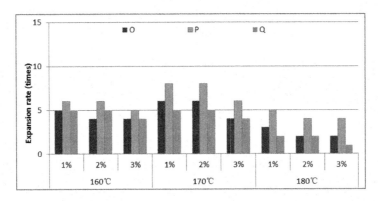

Figure 11.   Test results of expansion rate of rubber-modified asphalt.

4   CONCLUSIONS

According to the existing foaming indicators (expansion rate and half-life), the three types of asphalt (asphalt, SBS-modified asphalt, and rubber-modified asphalt) are tested with different foaming methods and different foaming equipment under different temperatures and different water consumption conditions. The main conclusions of the research are as follows:

(1) On the whole, no matter what kind of asphalt, as the foaming temperature increases and the water consumption increases, the expansion rate increases and the half-life decreases; on the contrary, the expansion rate decreases and the half-life increases.
(2) Foaming equipment has a certain influence on the foaming effect of asphalt, so it is recommended that the foaming device in the laboratory should be matched with the foaming device on the construction site.
(3) The different foaming methods have a significant impact on the foaming effect of the asphalt. Under aerated conditions, the expansion rate of the asphalt is increased and the half-life is reduced.
(4) Due to different oil sources, the foaming effect of bitumen with the same label will be different. Choosing the right asphalt, foaming temperature, and water consumption are the key factors that determine the foaming effect of asphalt.

ACKNOWLEDGMENTS

This work was financially supported by the Basic Scientific Research of Central Institute (Grant No. 2020-9049).

REFERENCES

AKISETTYCKK. (2008) Evaluation of warm asphalt additives on performance properties of CRM binders and mixtures [M]. Clemson: Clemson University.
Chen Wei-hao. (2014) Experimental study on application performance of warm mix rubber asphalt mixture [J]. Petroleum Asphalt, 28(6): 41–45. (in Chinese)
Guo Nai-sheng, YOU Zhan-ping, ZHAO Ying-hua, et al. (2014) Durability of warm mix asphalt containing recycled asphalt mixtures[J]. China Journal of Highway and Transport,27(8): 17–22. (in Chinese)
Hailesilassie B, Schuetz P, Jerjen I, et al. (2014) Evolution of bubble size distribution during foam bitumen formation and decay [M]. Asphalt Pavements, 33–40.
Hajj E Y, Sebaaly P E, Hitti E, et al. (2011) Performance evaluation of terminal blend tire rubber HMA and WMA mixtures-case studies[J]. Journal of the Association of Asphalt Paving Technologists, 80(4): 665–696.

He Liang, Ling Tian-Qing, Ma Yu, et al. (2015) Rheological properties of warm mix asphalt rubber in wide range of pavement temperature[J]. Journal of Traffic and Transportation Engineering, 15(1): 1–9. (in Chinese)

MA Yu, HE Zhao-yi, HE Liang, et al. (2015) Analysis on aging characteristics and infrared spectroscopy of warm mix asphalt-rubbe [J]. Journal of Highway and Transportation Research and Development, 32(1): 13–18. (in Chinese)

Mo Lian-Tong, Li Xun, Fang Xing, et al. (2012) Laboratory investigation of compaction characteristics and performance of warm mix asphalt containing chemical additives[J]. Construction and Building Materials, 237: 239–247.

Wrustje, Putmanbj. (2013) Laboratory evaluation of warm-mix open graded friction course mixtures[J]. Journal of Materials in Civil Engineering, 25(3): 403–409.

You L, You Z, Yang X, et al. (2018) Laboratory Testing of Rheological Behavior of Water-Foamed Bitumen [J]. Journal of Materials in Civil Engineering, 30(8): 63–69.

Zhang Ya-Tao, Wu Qi-Feng, Zhang Zheng-qi, et al. (2012) Evaluation on performance of Sasobit warm mix asphalt rubber [J]. Journal of China & Foreign, 32(3): 295–300. (in Chinese)

Frontiers of Civil Engineering and Disaster Prevention and Control – Yang & Rahman (Eds)
© 2023 The Author(s), ISBN: 978-1-032-31200-2

Laboratory evaluation of pavement performance for permeable asphalt concrete using in seasonal frozen region of China

Yu Baoyang & Sun Zongguang
Transportation Engineering College, Dalian Maritime University, Dalian, China

Qi Lin
Department of Civil Engineering, Shenyang Urban Construction Institute, Shenyang, China

ABSTRACT: This paper analyzes the disadvantages of the original calculation method when determining the asphalt content from the leakage test results. A new calculation formula was used to determine the optimum amount of asphalt. The limit value of the leakage experiment was corrected. The road performance test of the permeable asphalt mixture is carried out, including high temperature performance, low-temperature performance, and water stability performance. Dynamic stability was evaluated for high temperature stability. The test conditions for freeze-thaw splitting were improved. At the same time, the low-temperature bending test was carried out. Although the oil-aggregate ratio of the asphalt is slightly less stable at high temperatures, the low-temperature cracking resistance and water resistance can be substantially improved, and it is therefore better suited to the climate of the northeast seasonal frozen region.

1 INTRODUCTION

Permeable pavement has good drainage and water purification performance to prevent waterlogging in the city and purify the water. It has good anti-sliding performance and can improve the driving safety factor. It can also reduce noise pollution and relieve urban heat islands by weakening the air pump effect of vehicle tires and road surfaces (Tong 2015). The region in northern China is 787,300 square kilometers. The temperature is about 30°C in the summer, which is close to 40°C in extreme weather. The average temperature in winter is −16.2°C, and the temperature in extreme weather is close to −37°C. The cold period lasts about 5 months and the freezing period is long. The climate characteristics of large thermal differences and low temperatures lead to more problems in the paving and operation process of the permeable asphalt test road.

At present, many paved permeable pavement roads have been laid in southern China, and there are few pavements in the seasonal frozen region. Permeable pavements will face more problems, such as low-temperature cracking resistance, high temperature performance, and water stability. Therefore, it is necessary to enhance the permeable asphalt mixture according to various climate conditions of the northern seasonal frozen region and to test the road's performance.

2 MIX DESIGN OF PERMEABLE ASPHALT MIXTURE

According to the analysis of the annual average precipitation of typical cities in the northern seasonal frozen region, the target void ratio of the proposed asphalt mixture is 20%. Wu Jinhang et al. selected the appropriate gradation accurately and reasonably to avoid redundant test work. The factors affecting the void ratio were analyzed by the orthogonal test and back-off variable screening technique (Wu 2012), and the target porosity and key mesh were established. The regression equation for the pass rate was established as follows:

$$y = 28.724 + 0.04P_{13.2} - 0.677P_{2.36} - 0.878P_{0.075}(R^2 = 0.879) \tag{1}$$

$$y = 32.470 - 0.677P_{2.36} - 0.878P_{0.075}(R^2 = 0.876) \tag{2}$$

$$y = 27.644 - 0.677P_{2.36}(R^2 = 0.874) \tag{3}$$

Where: y-mixture void ratio (%).
$P_{13.2}$ − 13.2 – mm mesh pass rate (%).
$P_{2.36}$ − 2.36 – mm mesh pass rate (%).
$P_{0.075}$ − 0.075 – mm mesh pass rate (%). $y_1 = 19\%$, $P_{2.36} = 12.8\%$; $y_2 = 21\%$, $P_{2.36} = 9.8\%$.

When the target void ratio is 20%, the pass ratio of 2.36-mm mesh is 9.8%-12.8%. The requirements and the specific conditions of the passing rate of the aggregate are quickly determined by the grading (Putman & Lyons 2014). The selection of gradation is shown in Table 1.

Table 1. Selected grading range.

Grading	Pass rate through the following sieve holes (%)									
	16.0	13.2	9.5	4.75	2.36	1.18	0.6	0.3	0.15	0.075
Lower limit	100.0	90.0	60.0	12.0	10.0	6.0	4.0	3.0	3.0	2.0
Upper limit	100.0	100.0	80.0	30.0	22.0	18.0	15.0	12.0	8.0	6.0
Proposed grading	100.0	95.0	65.0	19.4	11.9	8.4	7.1	5.9	5.4	4.8

2.1 Determination of the optimum asphalt dosage

Many scholars have different methods for testing data and determining the optimum amount of asphalt, and the results are highly biased. Therefore, a new method is used for the calculation. The stability and flow index are used to further verify the optimum asphalt dosage. The formula is as follows:

$$OAC = OAC_{\min} + 0.75 \, (OAC_{\max} - OAC_{\min}) \tag{4}$$

Where: OAC – the optimum amount of asphalt;
OAC_{Max} – the minimum amount of asphalt determined based on the loss of scattering;
OAC_{Min} – the maximum amount of asphalt determined by the leakage loss.

The test pieces were prepared for the permeable mixture of 4.4%, 5.0%, 5.3% and 5.6% of the oil-aggregate ratio. The results show that the Marshall stability is greater than 3.5 kN when the oil-aggregate ratio is greater than 4.4%. The specific test results are shown in Table 2.

Table 2. Marshall stability and flow test data.

Oil-aggregate ratio	Void ratio (%)	Marshall stability (kN)	Flow value (mm)
4.4%	22.35	3.32	1.86
5.0%	20.91	3.85	2.44
5.3%	20.23	4.00	2.97
5.6%	19.54	3.67	3.08

The CT and the SBDT are carried out for the oil-aggregate ratios of 4.4%, 4.7%, 5.0%, 5.3% and 5.6%. The specific test results are shown in Figures 1 and 2.

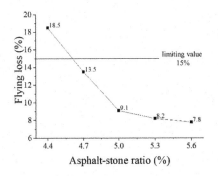

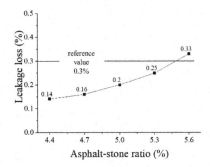

Figure 1.   Dispersion loss with oil-aggregate ratio.   Figure 2.   Leakage loss with oil-aggregate ratio.

Figure 1 shows that the scattering loss of the oil-aggregate ratio in the range of 4.6%–5.6% meets the requirements. As the oil-aggregate ratio increases, the scattering loss of the mixture decreases rapidly. When the oil-aggregate ratio is less than 4.6%, the scattering loss value of the mixture does not meet the specification requirements.

Figure 2 shows that the leakage loss of the oil-aggregate ratio in the range of 4.4%–5.5% meets the requirement. There is no obvious inflection point, and it cannot be used as the best asphalt dosage, it exposes the disadvantages of the mapping method. Using the new method, the optimal asphalt dosage is 5.3%.

When permeable asphalt mixture is used in the northeast seasonal frozen region with large thermal differences and long-term winter temperatures, it has higher requirements for low-temperature cracking resistance. Guo Yong paved the experimental road with a leakage loss limit of 0.8%, and the road performance was good (Guo 2006). Therefore, based on the optimum asphalt dosage of 5.3%, the production of test pieces was done for the large oil-aggregate ratio of 5.6% and 5.8%.

3   STUDY ON ROAD PERFORMANCE OF PERMEABLE ASPHALT MIXTURE

3.1   *Evaluation of high temperature stability*

For the mixture of 4.4%, 5.0% and 5.3% of the oil-aggregate ratio and 5.6% and 5.8% of the large oil-aggregate ratio, the calculation of the dynamic stability was carried out after the data was obtained. The test results are shown in Figure 3.

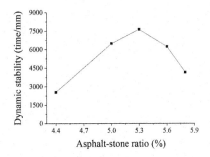

Figure 3.   Trend of dynamic stability with oil-aggregate ratio.

Figure 3 shows that when the oil-aggregate ratio increases from 4.4% to 5.3% (the optimum amount of asphalt), the dynamic stability of the mixture increases linearly. When the oil-aggregate ratio increases from 5.3% to 5.8%, the dynamic stability of the mixture decreases parabolically. The free asphalt gradually increases, which has an adverse effect on the high temperature ability.

When the oil-aggregate ratio is 5.8%, the dynamic stability is 4155 times/mm, which meets the requirements of the specification.

3.2 Evaluation of water ability

The water stability of the asphalt mixture was evaluated by a freeze-thaw splitting test. The specimens are saturated with water for 15 minutes according to the standard method; then the specimens are immersed in normal temperature water for 24 hours (the specification requires 0.5 hours of water) to ensure that the water can fully enter the internal space of the mixture structure and is fully saturated (Yi 2012). The water state was frozen at −18°C ± 2°C for 16 hours, and then taken out and placed in a 60°C water bath for 24 hours; the first set of test pieces was placed at room temperature. Finally, the two sets of test pieces were placed in 25°C water for two hours. The test results are shown in Figure 4.

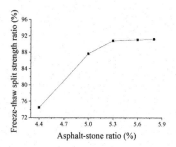

Figure 4.   Freeze-thaw splitting strength ratio with oil-aggregate ratio.

Figure 4 shows that as the oil-aggregate ratio increases, the ratio of freeze-thaw splitting strength also increases. When the oil-aggregate ratio reaches 5.3% (the optimum amount of asphalt), the ratio of freeze-thaw splitting strength has reached 90.86%. As the oil-aggregate ratio continues to increase, the ratio of freeze-thaw splitting strength rises slowly. The ratio of freeze-thaw splitting strength increased by 0.42% after increasing the oil-aggregate ratio from 5.3% to 5.8%, indicating that when the thickness of the asphalt film exceeds a suitable value, the effect of resisting water damage is small.

3.3 Evaluation of low-temperature cracking resistance

The daily thermal difference in the northeast seasonal frozen region can reach 30°C, and the annual thermal difference can reach 70°C. The temperature shrinkage of asphalt under large thermal differences leads to obvious low-temperature cracking of pavement, which restricts the application of permeable asphalt concrete in the northeast seasonal frozen region. The trabecular low-temperature bending test was carried out for the permeable asphalt mixture with a different oil-aggregate ratio (large oil-aggregate ratio). The test procedures and results are shown in Figures 5–7.

Figures 5 and 6 show that as the oil-aggregate ratio increases, the tensile strength and maximum tensile strain of the mixture increase continuously, and the low-temperature cracking resistance becomes better. When the oil-aggregate ratio increases from 5.3% to 5.8%, the maximum bending strain increases rapidly. Asphalt plays the most important role in improving low-temperature cracking resistance. When using a permeable asphalt mixture in the northeast seasonal frozen region, the amount of asphalt should not be limited to the requirement of selecting the maximum amount of asphalt in the leakage test.

Figure 7 shows the void ratio values of the mixture with different oil-aggregate ratios. The larger the void is, the worse the low-temperature cracking resistance of the mixture will be. Additionally, water and frost can heave in the gap, causing the asphalt to crack and damage. This series of

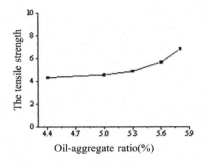
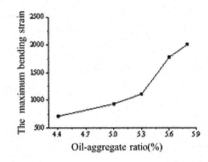

Figure 5. Tensile strength with oil-aggregate ratio.

Figure 6. Bending strain with oil-aggregate ratio.

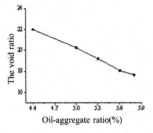

Figure 7. Void ratio with oil-aggregate ratio.

chain reactions seriously reduces the effect of the adhesion between asphalt and aggregate at low temperatures (Tan et al. 2017). It is therefore essential to determine the optimum asphalt in the northeast seasonal frozen region as the relationship between the amount of integrated asphalt and the size of the void ratio is mutually restrictive.

3.4 Determination of water permeability coefficient

The water permeability coefficient was measured. The results showed that the water permeability coefficient was good and all met the requirements of the specification. The results are shown in Figure 8.

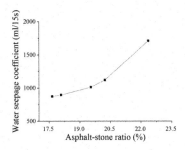

Figure 8. Relationship between water permeability coefficient and porosity.

4 CONCLUSIONS

(1) The void ratio can be measured by using a formula so that the target porosity can be determined quickly and accurately. A new calculation formula was used to determine the optimal oil-aggregate

ratio. (2) In the experimental study of road performance, the high temperature performance under the large oil-aggregate ratio also meets the specification requirements. The ability to absorb water increases with the oil-aggregate ratio. The low-temperature performance increases greatly, and when the oil-aggregate ratio is 5.8%, the maximum bending strain is 80%, greatly improving the low-temperature cracking resistance of the mixture. The permeable asphalt mixture maintains good road performance in the northeast seasonal frozen region.

REFERENCES

Guo Y. (2006) Application research of highway drainage asphalt mixture. Southeast University.

Putman B J, Lyons K R. (2014) Laboratory evaluation of long-term drain down of porous asphalt mixtures. Journal of Materials in Civil Engineering, 27(10): 04015009.

Tan Y Q, Xing C, Ren J D, et al. (2017) Study on Microstructure Characteristics of asphalt mixture based on particle stacking theory. China Journal of Highway, 30(7): 1–8.

Tong W G. (2015) Structural design and performance study of ecological permeable asphalt pavement. Chang'an University.

Wu J H. (2012) Research on key performance of drainage asphalt mixture. Chongqing Jiaotong University.

Yi J Y. (2012) Research on freeze-thaw damage characteristics of porous asphalt mixture based on interface behavior. Harbin Institute of Technology.

*Frontiers of Civil Engineering and Disaster Prevention and
Control – Yang & Rahman (Eds)*
© *2023 The Author(s), ISBN: 978-1-032-31200-2*

Experimental study on the influence of interface on concrete properties

Bin Yang*
Qinghai College of Architectural Technology, Xining, Qinghai, China

ABSTRACT: Taking fly ash as the admixture and the cement mortar aggregate interface as the research content, the breaking probability of aggregate is used to qualitatively characterize the interface strength of cement mortar coarse aggregate, and then the enrichment phenomenon on the cement mortar aggregate interface is observed. An experimental study is carried out on the influence of the cement mortar coarse aggregate interface on the performance of concrete. The results show that bubbles and particles are enriched in the interface area within 0.2mm away from the coarse aggregate. The greater the fracture probability of the aggregate, the higher the interface strength, and vice versa.

1   INTRODUCTION

Concrete is a multi-heterogeneous complex system on macro, sub micro and micro scales. It is a common method to analyze and study concrete from different scales. On the macro scale, the constitutive relationship and fracture performance of concrete can be studied. On the sub micro scale, the pore structure and pore characteristics, interface relationship and interface characteristics of concrete can be studied. On the micro scale, the characteristics and structure of hydration products such as aft can be studied (Guo 2004; Tang 2008; Wang 2000; Zhang 2015). As a multiphase composite, the interface of concrete is the key area affecting the mechanical properties of concrete. Scholars at home and abroad have done a lot of research work on the formation mechanism, performance characteristics and improvement methods of interface transition zone (Chen & Sun 2004a, 2005, 2004b, He Xiaofang et al. 2009), but most of the research is carried out from the micro level on the influencing factors and deterioration mechanism of interface transition zone, because there is still a long gap between micro research and macro performance of concrete, it is difficult to establish a direct influence relationship. In the research on the mechanical properties of concrete, in order to simplify the mechanical analysis and more clearly reveal the influence of various components of concrete on the strength of concrete, researchers regard concrete as a binary model composed of cement mortar, coarse aggregate and interface transition zone. As the weakest link in concrete, the interface area affects the performance of concrete to a great extent. Concrete is divided into cement mortar and coarse aggregate, and the characteristics of interface transition area are deeply studied. The research results have a good correlation with the macro performance of concrete, which is an important method for the study of modern concrete mechanical behavior.

The interface area between cement mortar and coarse aggregate has the characteristics of loose structure and high porosity. Thomas T C et al. Found that there are a large number of cracks in the concrete before it is stressed. These cracks mainly exist in the interface between cement mortar and coarse aggregate. It is this characteristic of the interface that makes it easier for the cracks to expand and grow at the interface when the concrete is stressed, and finally form through cracks, resulting in the failure of the concrete component structure. Jiang Lu et al. Calculated the volume fraction of the interface transition zone, which laid a foundation for the quantitative study of the influence

*Corresponding Author: yangbin@qhavtc.edu.cn

DOI 10.1201/9781003308577-48

of the interface on the performance of concrete. Minedss et al. (Minedss, 1996) determined the quantitative relationship between cement slurry strength, interfacial bond strength and concrete strength through regression analysis of test data. The publicity shows that the influence of cement slurry strength on concrete strength is about twice that of interfacial bond strength. Darwin et al. (970) found that when the interfacial bonding strength between cement mortar and coarse aggregate decreased significantly, the concrete strength decreased slightly. Integrating the previous research results, Darwin et al. Pointed out that the macro strength of concrete is controlled by the strength of each component of concrete materials, not the interface alone.

Based on the research results of scholars at home and abroad, concrete is regarded as a composite material composed of cement mortar, coarse aggregate, and interface. The study of the properties of three-phase materials has become the main problem affecting the mechanical properties of concrete. Based on the submicroscopic scale, the effects of the composition and properties of the interface transition zone on the mechanical properties of concrete are experimentally studied in this paper.

2 RESEARCH AND TEST ON INTERFACIAL PARTICLE ENRICHMENT

Particles with different particle sizes will form stratification under stirring and gravity, that is, small particles will move downward to fill the pores between large particles. There are a large number of particles with different particle sizes in concrete. Under the action of mixing, the particles with smaller particle sizes will occupy the gap between large particles. In order to observe the enrichment of finer particles on the aggregate surface under the optical microscope, fly ash is used as the admixture in this test. The specific particle size distribution of fly ash is shown in Table 1.

Table 1. Particle size distribution of fly ash.

Particle size (um)	0–1.06	1.06–4.88	4.88–10.48	10.48–22.49	22.49–76.32	>76.32
Proportion (%)	9.94	38.95	19.46	16.06	14.80	0.79

2.1 Test method

A cement paste, with water binder ratio of 0.35, is prepared, and the substitution rate of fly ash is 0–50%. After mixing according to GB17671 standard, a small amount of coarse aggregate is added, and 70.7mm is selected for manual mixing $\times$ 70.7 mm $\times$ 70.7 mm cube specimen shall be formed and demolded after standard curing for 24 hrs. Take about 10g of cement paste at the interface between cement mortar and coarse aggregate on the test piece. The samples were added to absolute ethanol to stop hydration, and the samples were analyzed after seven days. Put the sample whose hydration is terminated into absolute ethanol for 2–3 min grinding, in order to fully crush the large fast particles, dry the sample, configure the sample according to the mass ratio of sample: absolute ethanol = 1:100, put it into the ultrasonic disperser for 5min ultrasonic dispersion, so as to completely disperse the fly ash particles and cement particles, absorb the sample with a dropper and place it on the slide, observe and analyze after alcohol volatilization.

2.2 Result analysis

Figure 1 shows the content relationship of fly ash in hydration products in different areas of the same test block. It can be seen from the figure that the content of fly ash in the slurry coarse aggregate interface is high, the content of fly ash in the *large interface* between the formwork and the test block is second, and the content of fly ash in the base phase is the lowest. The hydration process of

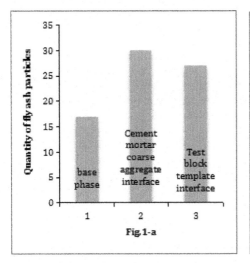
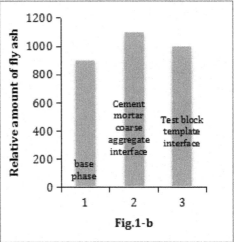

Figure 1. Relationship between the content of fly ash in hydration products in different areas of the same test block.

fly ash is slow and will reduce the strength of concrete. The enrichment of fly ash exacerbates the reduction of concrete strength.

Using the method of equivalent interface, the distribution relationship between interface pores and base phase pores of different specimens is measured. The results are shown in Figure 2. It can be seen from the figure that the values of interface pores/base phase pores are greater than 100%, indicating that the content of interface pores is higher than that of base phase pores, and there is pore enrichment in the interface area. With the increase of air entraining agent content, the interfacial pores/base phase pores gradually increase, indicating that the enrichment of pores on the interface is aggravated with the increase of air entraining agent. When the output of air entraining agent is 2/10000, the interface pore is 50% higher than the base phase. The pore on the interface is bound to weaken the bond strength in the interface area and reduce the strength of concrete.

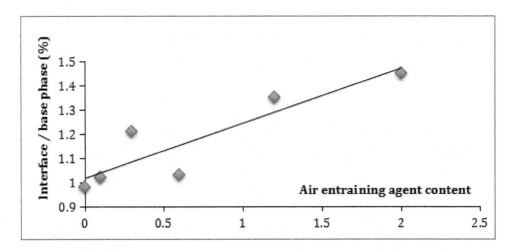

Figure 2. Test results of equivalent interface method.

3 RESEARCH AND TEST OF INTERFACE MECHANICAL PROPERTIES

The mechanical properties of the interface can directly reflect the state of the interface. The study of the mechanical properties of the interface is of great significance to the exploration of the failure mechanism of concrete. In concrete, coarse aggregate and cement mortar are connected together through the interface. Generally speaking, the higher the interface strength, the greater the probability of aggregate breaking when concrete breaks. On the contrary, the smaller the probability of aggregate breaking. Because there are still insurmountable difficulties in studying the mechanical properties of concrete interface alone, the indirect method is selected to qualitatively characterize the interfacial strength of cement mortar coarse aggregate by using the tensile fracture probability of aggregate.

3.1 Test method

The concrete test shall be carried out according to the required test mix proportion, and the forming size of the test block is 100 mm × 100 mm × 100 mm. After formwork removal, it shall be cured in the standard curing room for 28 d. After curing to a certain age, the concrete splitting test shall be carried out, and the section size obtained is 100 mm × 100 mm.

3.2 Result analysis

Figure 3 shows the test results of interface mechanical properties analyzed by splitting method. The change conditions of a, b, c, and d, are water binder ratio, air entraining agent, fly ash and silica fume in turn. For the water binder ratio, the greater its value, the lower the strength of the concrete, the greater the thickness of the interface transition zone and the more serious the deterioration.

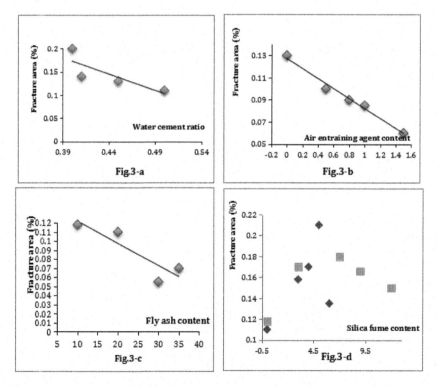

Figure 3.   Analysis results of interface mechanical properties.

The mechanical properties of the interface will be reduced accordingly, resulting in the reduction of the proportion of coarse aggregate broken. It should be pointed out that when the water binder ratio decreases, the change of interfacial mechanical properties on the water binder ratio is more obvious; as for air entraining agent, it can be seen from Figure 3-b that with the increase of air entraining agent content, the proportion of coarse aggregate broken shows an obvious downward trend, which is consistent with the research results of pore enrichment, that is, the greater the air entraining agent content, the more serious the pore enrichment on the interface, the worse the mechanical performance of the interface and the lower the fracture proportion of aggregate; for fly ash, generally speaking, the impact of fly ash on concrete performance is very complex, but it generally has a negative impact on strength.

As shown in Figure 3-c, before the content is 20%, the influence of fly ash on the interface is small. After exceeding 20%, the influence of fly ash on the interface is more significant, the interface deterioration is aggravated, and the fracture proportion of coarse aggregate is reduced; Silica fume has a dual effect on the strength of concrete. It can be seen from Figure 3-d that when silica fume is added into the base phase, the proportion of aggregate broken first increases and then decreases with the content of silica fume, because a certain amount of base phase silica fume can improve the interface area between cement hydration conditions and cement stone and coarse aggregate, and fill the pores between cement stone at the same time, it plays a certain compaction role and is conducive to the increase of concrete strength. When silica fume is over mixed, the workability of concrete decreases, the hydration of cement becomes worse and the strength decreases. When silica fume is added to the interface, the proportion of aggregate broken can be seen to increase, indicating that the addition of silica fume at the interface can improve the interface between cement mortar and coarse aggregate and improve the mechanical properties of the interface.

4   CONCLUSIONS

Cement mortar aggregate interface is the key area to analyze and solve the mechanical properties of concrete from the sub micro level, which can clearly reveal the origin and development of concrete mechanical properties. Due to the complexity of concrete raw materials, there are certain particle material enrichment and bubble enrichment in the interface area, which are the key characteristics affecting the performance of concrete.

(1) The interface between cement mortar and coarse aggregate has the enrichment characteristics of bubbles and particles;
(2) At the interface thickness of 0.2 mm, it shows that the boundary decreases with the increase of fly ash content Flour and coal ash particles increased first and then tended to be stable.
(3) It has a certain significance to express the interfacial bonding performance by the tensile fracture probability of aggregate on the fracture surface representativeness. The greater the breaking probability of aggregate, the higher the interfacial strength, and vice versa; The enrichment of interfacial bubbles and inert particles will lead to the deterioration of the interface and the decline of interfacial mechanical properties, but the enrichment of active particles can strengthen the interface.

REFERENCES

Chen Huisu, Sun Wei, Stroeven Piet. A review of the research on the interface between aggregate and slurry of cement-based composites (2): formation, deterioration mechanism and influencing factors of interface microstructure [J]. Journal of silicate. 2004a, 01: 70–79.
Chen Huisu, Sun Wei, Stroeven Piet. Effect of interface on macro properties of cement-based composites [J]. Journal of building materials. 2005, 8(1): 51–62.
Chen Huisu, Sun Wei, Stroeven Piet. Summary of research on the interface between aggregate and slurry of cement-based composites (I): experimental technology [J]. Journal of silicate. 2004b, 01: 63–69.

Darwin D, Slate F O. EFFECT OF PASTE AGGREGATE BOND STRENGTH ON BEHAVIOR OF CONCRETE [J]. Journal of Materials. 1970, 03: 96–98.

Guo Jianfei. Theoretical study on the relationship between concrete pore structure and strength [D]. Zhejiang University, 2004.

He Xiaofang, Miao Changwen, Zhang Yunsheng. Review on analysis methods of structure and properties of transition zone of cement-based composites [J]. Concrete, 2009, 10: 19–23.

MINEDSS S. Tests to determine the mechanical properties of the interfacial zone [J]. RILEM report. 1996, 47–63.

Tang Xinwei. Research on damage behavior of concrete based on macro meso mechanics [D]. Beijing: Tsinghua University, 2008.

Wang Zongmin, Zhu Mingxia, Zhao Xiaoxi. Hierarchical method for the study of concrete fracture [J]. Journal of Zhengzhou University of technology, 2000, (04): 16–22.

Zhang Xiong, Huang Tinghao, Zhang Yongjuan, Gao Hui, Jiang man. Image Pro Plus image analysis method for concrete pore structure [J]. Journal of building materials, 2015, (01): 177–182.

Frontiers of Civil Engineering and Disaster Prevention and Control – Yang & Rahman (Eds)
© 2023 The Author(s), ISBN: 978-1-032-31200-2

Research and application of integrated water pollution prevention and control technology in expressway construction

Qun-long Mao
ZCCC Road and Bridge Construction Co., Ltd, Zhejiang, China

Chen Lu
Zhejiang Expressway Construction Management Co., Ltd, Zhejiang, China

Dong Zhang*
Research Institute of Highway Ministry of Transport, Zhejiang, China

Meng-lin Yang
Zhejiang Expressway Construction Management Co., Ltd, Zhejiang, China

ABSTRACT: In view of water environmental protection requirements during construction of the Lin'an-Jiande Section project of the Zhejiang Lin'an-Jinhua Expressway, field investigation and test analysis are adopted to determine the pollution link in the project construction process, put forward the special measures for construction wastewater treatment of roadbed, bridge, tunnel, and precast mixing station, form an integrated prevention and control technology system for water environmental pollution in expressway construction, and effectively guarantee the water environmental safety of project construction.

China has continued to make efforts in pollution prevention and control in recent years, with the goal of achieving clear water, blue sky, and pure land, as well as improving overall ecological and environmental protection. A wide range of technology research and application is carried out for water pollution prevention and control during the construction of the Zhejiang Expressway construction project. By establishing a sound system for prevention and control measures (JTJ 1997), effective technologies for water pollution prevention and control are implemented to improve environmental protection in the expressway area. This paper focuses on the Lin'an-Jiande Section project of the Zhejiang Lin'an-Jinhua Expressway. It conducts the investigation and analysis of water pollution sources in expressway construction, puts forward the water pollution prevention and control measures against each pollution link, sets up an integrated prevention and control system, and adopts various physical and chemical treatment processes for purification and disposal of the polluted water, thus effectively ensuring the water pollution prevention and control in expressway construction.

1 SYSTEM ESTABLISHMENT OF INTEGRATED WATER POLLUTION PREVENTION AND CONTROL TECHNOLOGY IN CONSTRUCTION

1.1 *Water pollution prevention and control in roadbed construction*

During the construction, practical measures should be taken to prevent rain washing on the roadbed, which may result in sediment scouring and deposition of farmland, blocking of ditches, etc. The following measures can be taken to avoid rain wash: e.g., using geotextile cloth or plastic film to

---

*Corresponding Author: 421312606@163.com

DOI 10.1201/9781003308577-49

cover the slope; using plastic film to build a temporary collecting ditch; building a silt arrest dam or cutting ditches at the foot of the slope; setting up a sedimentation tank, etc.

The maintenance of the construction machinery should be strengthened to prevent the mechanical oil from leaking into groundwater or municipal sewers. For operation water flushing aggregates or containing sediments, treatment through filtration, sedimentation tanks, or other ways should be adopted, so that the sediment does not exceed the amount of sediment discharged into rivers and lakes together with water before construction.

Oily wastewater from cleaning construction machinery and equipment, sewage from truck wash platforms, wash water for concrete curing, wash water for sand and gravel, and drainage water from earthwork excavation are all examples of production wastewater. The wastewater containing oil and the wastewater containing sand and gravel should be subjected to treatment, respectively. The oil separation tank should be used to de-oil wastewater containing oil, while the sedimentation tank should be used to remove solid materials from wastewater containing sand and gravel. Arbitrary discharge is strictly prohibited. Oil-water separation tank should be built at each construction wastewater discharge point. The construction wastewater after stage filtration can be used for sprinkling and dust suppression. The collected oil-immersed wastes should be packed and sealed and then disposed of together with other hazardous solid wastes by the qualified units.

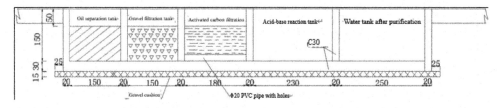

Figure 1. Schematic diagram of oil-water separation tank.

It is illegal to discharge polluted construction oil on the job site. In the event of overflow or leakage, the polluted oil should be immediately isolated and adsorbed. It is strictly prohibited to discharge water containing pollutants or visible suspended solids into rivers.

1.2 Water pollution prevention and control in bridge construction

During the construction of bridges near rivers (water), attention should be paid to the protection of water resources to avoid water pollution caused by boring wall protection slurry and construction garbage. The specific prevention and control measures are as follows:

(1) The earthwork and waste slurry generated from the bored pile construction should be carried away by the tank truck and then transported to the spoil area for centralized burial. (2) Cement, bentonite, and other admixtures should be piled up safely and covered properly, and they should not be dumped into rivers. (3) The wastewater from the concrete mixing station should be collected into the sedimentation tank and discharged after purification treatment. The capacity of the tank should be larger than the calculated amount of slurry to prevent its overflow, which may result in pollution of the surrounding environment. (4) "Environment-friendly toilets" should be set up on the construction platform, where the feces should be regularly collected and transported to the septic tank of the onshore living area for unified treatment. (5) It is forbidden to use disposable plastic tableware, in order to prevent white pollution. The waste oil and lubricating oil of the construction machinery must be collected and transported to the designated spoil area for deep burial. (6) The oil used in production must be tightly controlled to avoid leakage or pollution of river channels. (7) The connection between the water inlet/outlet of the culvert and the side ditch/natural ditch should be smooth. (8) To accomplish environmental protection and civilized construction, the boring waste (waste slurry) should be placed in a

designated place and should not be arbitrarily piled up at the construction site or directly discharged into ponds and rivers so as to avoid environmental pollution. (9) During bridge construction, measures should be taken to prevent petroleum pollutants from being discharged into water bodies (Liu 2011; Wang 2002). The slurry generated from the pile foundation boring construction should be removed from the sediment through precipitation and separation, and then the sediment should be transported to and disposed of in a place designated by the local environmental protection department, while the wastewater can be reused or used for dust suppression and greening of the sites and roads. A sewage sedimentation tank should be set up for domestic sewage generated from field construction, and the domestic sewage after precipitation treatment can be used for construction dust suppression or greening. The waste machines and tools, accessories, packages, and various solid oil-immersed wastes generated from the construction should be collected and sealed in a centralized way, and then transported to the spoil area for disposal or recycling.

Before the bored pile construction begins, a construction plan should be formulated and submitted to the supervisor for approval. The construction plan should meet the relevant requirements on environmental protection, and the construction can only be carried out after approval. During the bored pile construction, the slurry and boring waste should not be dumped into rivers, otherwise it may cause pollution of water sources. The waste from the bored pile should be transported to the designated spoil area according to the design of the construction organization. The spoil area should be preferably located in a low-lying place, and should occupy the arable land as little as possible, but should also occupy the abandoned ditches and ponds as much as possible, where the dumped soil can be used to fill them flat so that these ditches and ponds become arable land, creating benefits for the local people. If there is no low-lying place nearby, a retaining wall should be set up first in the spoil area, and then disposal should be allowed, but the surface runoff should not be blocked. A slurry tank and sedimentation tank should be set up for the bored pile to prevent water loss and soil erosion. After the water in the slurry tank is fully permeated, the slurry tank should be leveled to restore the surface level. Maintenance of the boring machinery needs to be enhanced to prevent leakage of lubricating oil, which can cause pollution of the soil and water.

Table 1. Pollution sources and control measures in bridge construction.

S/N	Pollution sources	Control measures
1	Boring waste	The slurry tank is set up for the boring construction. The boring waste is discharged into the slurry tank, and the boring waste in the slurry tank is dug out by an excavator, loaded onto a dump truck, and then transported to the designated place.
2	Slurry	A slurry tank is set up for the bored pile construction to prevent the outflow of slurry, and the slurry should not be arbitrarily dumped to avoid water pollution of rivers and the surrounding environment, but should be uniformly discharged into the tank truck through the pipeline, and then dumped from the tank truck in a designated discharge place.
3	Domestic garbage	Garbage bins are set up in the construction area and cleaned regularly by the special personnel; the garbage is piled into the garbage bins and should not be discarded carelessly.
4	Grease	Whenever grease drips onto the ground, use a shovel to clean it up and dispose of it with the slurry.
5	Waste oil generated from mechanical maintenance	The waste oil generated by the maintenance and servicing machinery is uniformly collected and stored at a fixed point, and should not be arbitrarily discharged to cause pollution of farmland and construction sites.

1.3 Water pollution prevention and control for tunnel construction

Through field investigation and analysis, the main water pollution links arising from tunnel construction are tunnel water gushing and construction site sewage and wastewater. The analysis of the pollution link is shown in Table 2.

Table 2. Analysis on water pollution in tunnel construction.

S/N	Site	Work content	Environmental impact	Impact level
1	Tunnel	Concrete pouring	Wastewater	Slight
	construction	Concrete curing	Wastewater	Slight
		Waste concrete	Solid waste	Slight
		Vehicle and site flushing	Wastewater	Medium
		Repair and maintenance of mechanical equipment	Oil and other hazardous wastes	Severe
2	Others	Domestic sewage	Wastewater	Severe
		Domestic garbage	Solid waste	Medium
		Rainwater runoff	Wastewater	Severe

The tunnel structure should be equipped with a complete water-proof and drainage system, and the original groundwater system should be maintained as much as possible. Once damaged, engineering measures (e.g., diversion tunnels, etc.) should be taken for repair (Zhou 2010). In accordance with the slope condition of the line at both ends, a larger longitudinal slope is adopted, and in combination with its internal contour, a larger size of ditch is adopted, so as to avoid flooding of the ballast bed and improve the engineering operation quality of the tunnel. For karst tunnels with groundwater development, the principle of "drainage first, with a combination of drainage and plugging" may cause massive loss of groundwater and environmental damage, thus the design principle of "plugging first, with limited drainage" should be adopted.

The following are emergency measures for water gushing accidents:

(1) In the event of a large area of water leakage, it is necessary to order the workers to stop work at once and evacuate to a safe place, and then report to the construction and design unit immediately; at the same time, observe and record the leaking part, water volume, change rules and water turbidity, take necessary protective measures, and report to the supervisor.
(2) If, during the excavation of deep tunnels, a severe water gushing accent occurs suddenly outside the established safe distance, the high-power pump installed in the tunnel should be used to drain the section of water gushing, and the workers must evacuate immediately using the established escape route.
(3) In case of severe water gushing during the excavation, the personnel at the excavated surface should evacuate immediately out of the tunnel along the escape route; at the same time, start the alarm system to send the alarm signal, quickly cut off the power supply and start the emergency lighting.
(4) When there is a water gushing accident, it is important to inform relevant department in time, and arrange for first aid for those in danger or injured.
(5) After the gushing water is kept stable, the high-power pumping equipment should be used for drainage. After the gushing water volume and the water pressure have been reduced, first aid for mechanical equipment should be carried out.

1.4 Water pollution prevention and control at temporary sites

The wastewater and waste materials generated during the construction period of the temporary sites (e.g., construction, production and living areas, reinforcement processing plant, etc.) may lead to water pollution. The following protective measures should be implemented in order to save water, mitigate water loss, and soil erosion, as well as reduce or avoid environmental pollution caused by construction sewage:

(1) All discharge standards should be strictly observed with respect to the discharge of wastewater. When the wastewater is discharged into natural water bodies, the suspended solids (SS) should strictly comply with the secondary standard of 150 mg/L in the Integrated Wastewater Discharge Standard (GB8978, 1996).

(2) Construction and production waste should not be directly discharged into water bodies. The construction of the site drainage and wastewater treatment facilities should be completed before the commencement, a sewage treatment system should be set up in the living area, and the temporary domestic sewage treatment facilities (see Figure 2 for treatment process) should be provided to realize cyclic utilization and prevent sewage from being directly discharged into streams, ponds or irrigation/drainage systems; the effectiveness of the site drainage and wastewater treatment facilities (see Figure 3 for treatment process) should be ensured throughout the whole construction process, so that there is no water accumulation at the site, no overflow during drainage, no blockage, and the water quality is up to standard.

(3) The management of transportation, stacking, and use of the construction and building materials should be strengthened. A guide ditch should be dug along the edge of the temporary yard, and covers should be placed on the yard. Lime and cement should not be piled up in the open-air during transportation and storage, and proper arrangement of materials should be made to reduce the storage time of the building materials. During the construction of a bridge or section near rivers, the yard should be away from the river channel as far as possible. In particular, building materials containing harmful substances (e.g., asphalt) should not be piled up near water bodies and should be covered to avoid entering the water bodies along with rain wash.

(4) Waste oil and other solid wastes created during construction should be stored as far away from water bodies as feasible, and should be removed as soon as possible to avoid infiltrating the water bodies with rains.

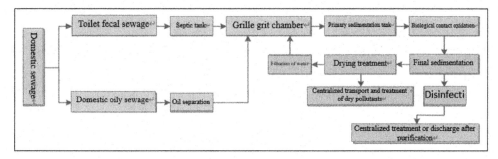

Figure 2. Treatment process of domestic sewage.

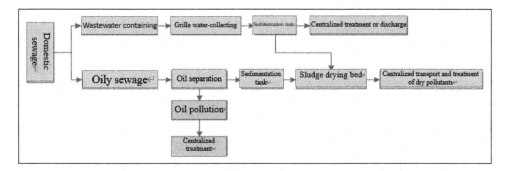

Figure 3. Treatment process of production sewage.

2 INTEGRATED PREVENTION AND CONTROL EFFECTS

The water environmental safety protection level is enhanced in an all-round way by developing an integrated water pollution prevention and control plan for the project construction, setting up various water pollution treatment facilities in roadbeds, bridges, tunnels, and temporary stations (see Figure 4 for site measures), strictly implementing all prevention and control measures, and forming a complete environmental protection system.

| Cleaning of construction trucks | Dust suppression by fog gun |
| Sewage treatment facility | Station runoff three-stage sedimentation tank |

Figure 4.   Site photos of water pollution prevention and control measures in construction.

3 CONCLUSION

The level of water pollution prevention and control technology in expressway construction can be effectively improved by establishing a perfect system for water pollution prevention and control measures in expressway construction, vigorously promoting the technique and technology featured in low-carbon energy-saving and environmental protection construction, and using highly efficient sewage treatment facilities. Moreover, by strengthening the process management and putting the fine construction and strict control into practice, the expressway construction's green environmental protection will be greatly promoted, accomplishing the goal of green development of the transportation infrastructure.

REFERENCES

China Ministry of Communications. JTJ 018-97 Specification for Drainage Design of Highway[S]. Beijing: China Communications Press, 1997–08–04.

Liu Yang-hua, Ao Hong-guang, Feng Yu-jie, et al. Progress of Environmental Risk Assessment Research[J]. Environmental Science and Management, 2011, 36(8): 159–163.

Wang Hua-cheng, Zhu Xin-chun, Wang Zi-fu, Liu Qiang, Liu Hai-sheng. Research on Online Monitoring Technology System of Green Expressway During Operation Period, Transport Energy Conservation & Environmental Protection, 2022, 18: 139–144.

Wang Xiao-feng, Wang Xiao-yan. Analysis on Foreign Surface Runoff's Pollution Process and Control Management[J]. Journal of Capital Normal University (Natural Science Edition), 2002(1).

Zhou Hai-yan. Analysis on Highway Bridge Runoff Treatment Method[J]. Journal of Highway and Transportation Research and Development, 2010(10): 488–490.

Frontiers of Civil Engineering and Disaster Prevention and Control – Yang & Rahman (Eds)
© 2023 The Author(s), ISBN: 978-1-032-31200-2

Axial compressive behavior of grout-filled double-skin tubular (GFDST) columns with stainless-steel outer tubes: Experimental investigation

Guiwu Li & Yue Wang
CCCC Fourth Highway Engineering CO., LTD., Beijing, China

Huanze Zheng & Ruilin Ding*
College of Architecture and Civil Engineering, Beijing University of Technology, Beijing, China

ABSTRACT: In this paper, the axial compression performance of GFDST columns with stainless-steel outer tubes was investigated. A total of 24 column specimens were fabricated and tested. The test specimens' parameters include the thickness of the stainless-steel tube, the width of the carbon steel tube, and the grouting, if any, between the stainless-steel tube and the carbon steel tube. Failure mode, load-displacement curve, and ultimate strength were investigated. Results show that the local buckling of the stainless-steel tube occurs, while the carbon steel tube shows no obvious deformation when there is no grout interlayer. The local buckling of the carbon steel tube occurs when there is a grout interlayer. Before reaching the peak load, the N-Δ curves first shows linearity, then a transition to plasticity. The ultimate strength increases by 31.56%–78.02% with increase in the thickness of stainless-steel tube. When there is no grout between the stainless-steel tube and the carbon steel tube, increasing the width of the carbon steel tube can increase the ultimate strength by 97.03%. The ultimate strength of the specimens filled with grout can be improved by 66.73%.

1 INTRODUCTION

Concrete-filled steel tube (CFST) components exhibit full advantages of steel and concrete and have excellent mechanical properties and fire performance. Hence, they are been widely used in practical engineering (Abed et al. 2013; Alhatmey et al. 2020; Han et al. 2014; Han 2007; Jayaprakash & Manvi 2020; Zhang et al. 2021; Zhu et al. 2021). However, the corrosion of CFST components used in underground and marine structures is a serious problem, which affects the normal usage of the components leading to high maintenance costs (Dewanbabe & Das 2013; Han et al. 2014; Zhang et al. 2020). In recent years, concrete-filled stainless-steel tube (CFSST) components are being widely studied due to excellent corrosion resistance properties. Uy et al. (2011) carried out a series of experimental studies on CFSST long and short columns, confirming that CFSST members have a wide applicative prospect in engineering practice. Liao et al. (2017) conducted a study on the seismic performance of CFSST columns. CFSST columns have good seismic performance. Compared with CFST columns, CFSST columns exhibit better ductility, deformation capacity, and seismic energy absorption capacity. In addition, the utilization of stainless-steel pipes instead of carbon steel pipes can lower the price of the whole life cycle of members. Nevertheless, the initial cost of stainless-steel tubes is significantly higher than that of carbon steel tubes, which restricts the promotion and application of CFSST members in practical engineering (He et al. 2019). To promote the application of stainless-steel tubes in practical engineering and reduce the

---

*Corresponding Author: Birdie2017@163.com

DOI 10.1201/9781003308577-50

self-weight of the components to reduce its response to seismic effects, a GFDST column with stainless-steel outer tubes (see Figure 1) is proposed. It is composed of two concentric steel pipes of equal-height with stainless-steel pipes inside and carbon steel pipes outside. Grout is filled between them. The insertion of carbon steel tubes can reduce the cost of using stainless-steel tubes for the component. Replacing concrete with grout and emptying the internal space can reduce the weight of the component and reduce the response of the component under seismic action.

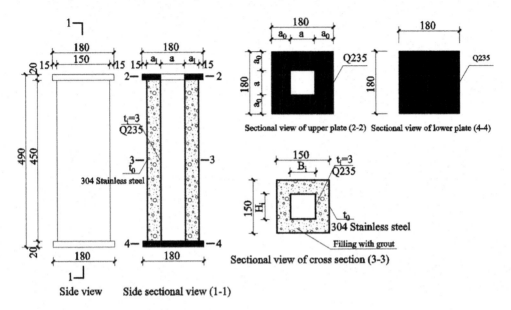

Figure 1. Basic dimensions of the test specimen.

2 TEST OVERVIEW

2.1 *General*

In this paper, 24 specimens (12 groups) were fabricated and tested under axial compression, including 12 double-skin tubular short columns with stainless-steel outer tube and no grout-filled in between the 12 GFDST short columns with stainless-steel outer tube. The stainless-steel tubes and the carbon steel tubes are both square sections. The height of each test specimen is 450 mm, the thickness of the base plates is 20 mm, and the extension distance of the base plates is 15 mm. The parameter variables of the test include: (1) the thickness of the stainless-steel tube: 2 mm, 3 mm, and 4 mm. (2) Carbon steel tube has a width of 50 mm or 100 mm. (3) If or not grout is used between the stainless-steel tube and the carbon steel tube. The design parameters of the specimens are shown in Table 1 and Figure 1.

2.2 *Material properties*

For determining the mechanical properties (GB/T 228.1-2010, 2010), a few tests were conducted. The yield ratio of stainless-steel is lower than that of carbon steel. Stainless-steel has better ductility and greater elongation. The yield strength, ultimate strength, and elastic modulus of stainless-steel tubes and carbon steel tubes are listed in Table 2. The cube compressive strength of grout is 36.67 MPa.

Table 1. Design parameters of each specimen.

Specimen	B_o/ mm	t_o/ mm	A_o/ mm^2	B_i/ mm	t_i/ mm	Specimen	B_o/ mm	t_o/ mm	A_o/ mm^2	B_i/ mm	t_i/ mm
S3S3-50-1a/b	150	3	1,764	50	3	S3S3-100-1a/b	150	3	1,764	100	3
S3S3-50-2a/b	150	3	1,764	50	3	S3S3-100-2a/b	150	3	1,764	100	3
S2S3-50-1a/b	150	2	1,184	50	3	S2S3-100-1a/b	150	2	1,184	100	3
S2S3-50-2a/b	150	2	1,184	50	3	S2S3-100-2a/b	150	2	1,184	100	3
S4S3-50-1a/b	150	4	2,336	50	3	S4S3-100-1a/b	150	4	2,336	100	3
S4S3-50-2a/b	150	4	2,336	50	3	S4S3-100-2a/b	150	4	2,336	100	3

Table 2. Mechanical properties of stainless-steel tubes and steel tubes.

Steel type	Yield strength f_y (MPa)	Ultimate strength f_u (MPa)	Elastic modulus E (MPa)
2-mm stainless-steel tube	335.00	690.67	174,131
3-mm stainless-steel tube	397.33	696.00	167,155
4-mm stainless-steel tube	406.00	696.67	171,711
50-mm carbon steel tube	376.33	457.00	186,286
100-mm carbon steel tube	367.33	441.00	187,876

2.3 Instrument and test setup

Four LVDTs were provided at the upper and lower ends of the plate to measure the axial deformation. Two sets of strain gauges were attached to the inner sides of adjacent faces, and four sets of strain gauges were attached to the outer sides of each face. The test loading is controlled by both load and displacement. When the load is lower than 70% of the estimated bearing capacity, the load control is adopted. Each increment is 10% of the estimated bearing capacity. The load is held for 2 minutes after reaching the specified value. When the load reaches 70% of the estimated bearing capacity, the displacement control is used. The loading rate is 0.5 mm/min, and the loading is completed when the steel tube experiences severe local buckling or the axial compression rate of the specimen reaches 5%.

3 TEST RESULTS AND PARAMETER ANALYSIS

3.1 Failure mode

The typical final failure mode of specimens is shown in Figure 2. The failure mode of the specimen without interlayer grout is that the stainless-steel tube displays an obvious concave-convex deformation in the middle, overlapping on the four sides, while the carbon steel tube shows no obvious deformation. The final failure mode of the specimen with interlayer grout is local buckling or corner tearing. Carbon steel tubes also exhibit local buckling of the same height as the stainless-steel tubes.

3.2 Axial load versus displacement relationship

Load (N)–displacement (Δ) curves of all the specimens are displayed in Figure 3. The N-Δ curves first show linearity, followed by a transition to plasticity. As the thickness of the stainless-steel tube increases, the ultimate strength, stiffness, and deformation capacity of the specimens are increased, and the specimens show good ductility. As the width of the carbon steel tube increases, the ultimate

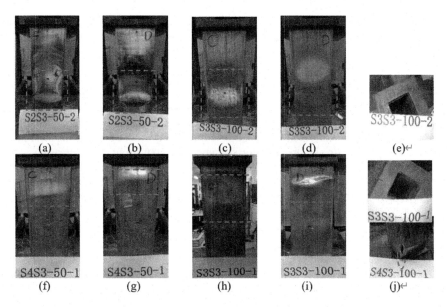

Figure 2.   Failure modes.

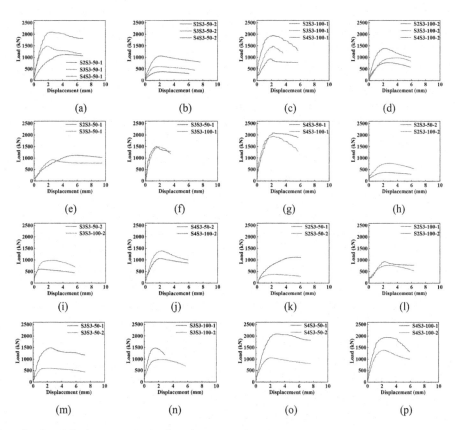

Figure 3.   Load-displacement curves of specimens.

strength and deformation capacity of the specimen are increased. In addition, the ultimate strength of the specimen filled with grout is stronger than that of the specimen without grout, but the load decreases faster after reaching the peak value.

3.3 *Ultimate strength*

The ultimate strength of all specimens is shown in Table 3. Increasing the thickness of the stainless-steel tube can increase the ultimate strength of specimens, and the increase in range is from 31.56% to 78.02%. For the specimens filled with grout, the increasing width of the carbon steel tube has little influence on the ultimate strength; for specimens without grout, the increasing width of the carbon steel tube has a significant impact on the ultimate strength. Furthermore, increasing the width of the carbon steel tube can ameliorate the ultimate strength of specimens with thinner stainless-steel tubes more significantly, up to 97.03%. The ultimate strength of the specimens filled with grout is higher than that of the specimens without grout.

Table 3.　Ultimate strength of all specimens.

No.	Ultimate strength P_u/kN	No.	Ultimate strength P_u/kN	No.	Ultimate strength P_u/kN
S2S3-50-1	1144.69	S3S3-50-1	1514.85	S4S3-50-1	2149.76
S2S3-50-2	380.79	S3S3-50-2	606.12	S4S3-50-2	1079.01
S2S3-100-1	973.55	S3S3-100-1	1503.22	S4S3-100-1	2077.71
S2S3-100-2	750.27	S3S3-100-2	987.04	S4S3-100-2	1399.10

4　CONCLUSIONS

(1) The failure mode is an obvious four-sided overlapping concave-convex deformation that occurs in the specimen without grout, while local buckling occurs in the specimen with grout.

(2) Before reaching the peak load, the N-Δ curves initially exhibit linearity before transitioning to plasticity. With increase in the thickness of the stainless-steel tube, the stiffness, ultimate strength, and deformation capacity of the specimens increases, and the specimens show good ductility.

(3) Increasing the width of the carbon steel tube can significantly improve the ultimate strength of specimens with thinner stainless-steel tube.

According to the characteristics of high material strength, large bending stiffness, fire resistance, corrosion resistance, and certain internal space, GFDST columns with stainless-steel outer tubes can be applied to power transmission, small marine buildings, and bridge towers.

ACKNOWLEDGMENTS

We appreciate the National Key Research and Development Program of China (Grant No. 2019YFD1101003) for funding the study.

REFERENCES

Abed F, AlHamaydeh M, Abdalla S. Experimental and numerical investigations of the compressive behavior of concrete filled steel tubes (CFSTs). *J. Constr. Steel Res.* 2013; 80: 429–39.

Alhatmey Ihssan A and Ekmekyapar Talha and Ayoob Nadheer S. (2020). Post-fire resistance of concrete filled steel tube columns. IOP Conference Series: Materials Science and Engineering, 988(1), pp. 012–036.

Dewanbabee H, Das S. Structural behavior of concrete steel pipes subject to axial compression and internal pressure: experimental study. *J. Struct. Eng.* 2013; 139 (1): 57–65.

Zhang FJ et al. (2020). Degradation of axial ultimate load-bearing capacity of circular thin-walled concrete-filled steel tubular stub columns after corrosion. Materials, 13(3), pp. 795–795.

GB/T 228.1-2010. Metallic Materials-Tensile Testing, Part 1: Method of Test at Room Temperature. Standards Press of China, Beijing China, 2010 (In Chinese).

Han LH, Hou CC, Wang QL. Behavior of circular CFST stub columns under sustained load and chloride corrosion. *J. Constr. Steel Res.* 2014; 103: 23–36.

Han LH, Li W, Bjorhovde R. Development and advanced application of concrete-filled steel tubular (CFST) structures: Members. *J. Constr. Steel Res.* 2014; 100: 211–28.

Han LH. *Concrete Filled Steel Tube Structure—Theory and Practice*. Beijing: Science Press, 2007.

He A, Wang FY, Zhao O. Experimental and numerical studies of concrete-filled high-chromium stainless steel tube (CFHSST) stub columns. Thin-Walled Struct. 2019; 144: 106273.

Jayaprakash Narayana S. and Manvi Rao Kolan. (2020). A Study on Effect of Steel Tube in Axial Behavior of CFST Stubs. IOP Conference Series: Materials Science and Engineering, 1006(1), pp. 012–019.

Liao FY, Han LH, Tao Z, et al. Experimental Behavior of Concrete-Filled Stainless Steel Tubular Columns under Cyclic Lateral Loading. *J. Struct. Eng.* 2017; 143 (4): 04016219.

Uy B, Tao Z, Han LH. Behavior of short and slender concrete-filled stainless steel tubular columns. *J. Constr. Steel Res.* 2011; 67: 360–78.

Zhang J et al. (2021). Experimental study of axial compression behavior of circular concrete-filled steel tubes after being loaded at an early age. Construction and Building Materials, 300.

Zhu JY and Chen JB and Chan TM. (2021). Analytical model for circular high strength concrete filled steel tubes under compression. Engineering Structures, 244.

Frontiers of Civil Engineering and Disaster Prevention and Control – Yang & Rahman (Eds)
© 2023 The Author(s), ISBN: 978-1-032-31200-2

Research on distribution characteristics of extraction radius in different areas of the overlying coal seam of floor roadway

Luanluan Sun
Chongqing Energy Industry Technician College, Chongqing, China

Zhonghua Wang*
National Key Laboratory of Gas Disaster Detecting, Preventing and Emergency Controlling, Chongqing, China
China Coal Technology Engineering Group Chongqing Research Institute, Chongqing, China

ABSTRACT: In order to explore the distribution characteristics of the drainage radius in different areas of the overlying coal seam of the floor roadway, the mine gas occurrence characteristics are analyzed, and the field investigation shows that the drainage radius of the overlying coal seam of the floor roadway is directly above the floor roadway and 7.5 m below the bottom. The upper bank is 7.5 m, the lower bank is 15 m, and the upper bank is 15 m away. This enables the drainage boreholes to be oriented in a targeted manner, avoiding the disadvantage of arranging them all along the same drainage radius, which could waste significant drilling time and provide a basis for arranging the drainage boreholes along the coal road.

1 INTRODUCTION

1.1 *The basic situation of the mine*

Shangzhuang Minefield is a part of the Wusheli mining area in Fengcheng Hexi Coalfield, located in Fengcheng City in the middle and lower reaches of the Ganjiang River. The mine is about 5.1 km long, about 2.65 km wide, and has an area of about 13.288 km^2. The approved production capacity in 2005 was 350 kt/a, and the annual output has stabilized at 350,000 to 400,000 tons in recent years. Geological factors affecting the zoning of Shangzhuang Coal Mine gas include coal seam burial depth, coal seam thickness, coal seam gas storage conditions, and regional geological structure.

1.2 *The gas content varies greatly between east and west*

The thickness of the B4 coal seam in the east wing is more than 3m, and the roof is layered with sandy rock with poor air permeability. The B4 coal seam in the west wing is bifurcated into B4b and B4a, and the coal thickness becomes thinner. The gas content is 24.58 m^3/t in the east wing mining area and 18.81 m^3/t in the central and west wing mining areas.

1.3 *Gas zoning characteristics*

Above -110 m is a weathered zone; -110 m to -220 m is a gas zone; -220 m to -330 m is a high gas zone; -330 m to -650 m is a gas outburst zone. At the current level of mine production, it is divided into the east wing mining area as a serious outburst danger area and the central mining area and the west wing mining area as general outburst danger areas. Based on the geological data, it is subdivided into two zones—the severe gas outburst zone and the general outburst zone—and a hierarchical management system is in place.

---

*Corresponding Author: boaidajia2007@126.com

DOI 10.1201/9781003308577-51

1.4 Gas outburst

As the burial depth of the coal seam increases, the gas pressure gradually increases, the number of coal and gas outbursts increases, and the outburst strength also increases. From the analysis of the areas that have been mined at the third and fourth levels, the gas pressure, gas content, and outburst risk of the fourth level B4 coal seam are greater than those of the third level. All mining areas at the third level are dangerous areas for gas outbursts. Although stricter comprehensive anti-outburst measures were taken in the mining areas during the third-level production, coal and gas outbursts still occur from time to time, with a total of 32 occurrences at the third level since the fourth level was put into operation. It has happened 29 times.

The first gas outburst was on November 24, 1977. The outburst elevation was −304 m, and the outburst position was 3,001 m. Up to now, a total of 61 outbursts of coal and gas have occurred, of which the largest outburst intensity was on the 3652-coal road, with an elevation of 587 m on October 4, 2004. The amount of outburst coal was 895 t and the amount of gas ejected was 51,500 m³.

2 TEST PLAN

2.1 Drainage drilling layout

The test site is the 710-floor lane of Shangzhuang Coal Mine. The 15 m range on both sides of the coal roadway strip in the overlying coal seam of the floor road is very important for coal road outburst prevention. Therefore, five rows of boreholes are arranged within 15 m on both sides of the test coal road strip, with five boreholes in each row. Each row of drilling holes is arranged at 15 m on the lower side, 7.5 m on the lower side, directly above, 7.5 m on the upper side, and 15 m on the upper side.

2.2 Calculation method of drainage radius

First, determine the gas drainage rate according to the actual conditions of the drainage coal seam (Zou 2020), and calculate the total amount of gas when the drainage area is effective (Zhou 2017). Then the effective influence radius of the drainage borehole is calculated using the formula according to the statistics, and the drainage attenuation negative exponential curve is fitted according to the statistical results of the average single-hole daily gas drainage. Finally, the effective radius of the hole calculated by the statistics is compared with the effective radius of the hole calculated by the negative exponential curve integral of the drainage attenuation, and the effective radius of the hole is finally determined (Xu 2018). According to the measurement results, construct residual gas content test boreholes at the expected effective influence radius, measure the residual gas content of the boreholes, calculate the gas drainage rate of the gas content test boreholes, and compare, analyze, and evaluate the drainage effect.

The number of drilled holes is calculated as follows (Liu 2020):

$$N = Q_{Total} / Q_{sheet} \tag{1}$$

Where Q_{Total} is the total amount of single-hole draining counted for a certain period of draining time;

Q_{sheet} is the statistical calculation.

One is to directly accumulate based on the test data; the other is to fit the negative exponential curve of drainage attenuation based on the test data and then integrate to find the total amount of single-hole drainage in different time periods.

By distributing the N boreholes evenly in the control area of the drainage boreholes, it is possible to calculate the effective influence radius for borehole drainage under the conditions of a given drainage time.

The effective influence radius of the borehole drainage can be calculated (Liu 2016), as shown in Formula (2):

$$r = \frac{L}{2\sqrt{N}} \tag{2}$$

Where r refers to the effective radius of influence of the borehole;

 N refers to the number of boreholes in the drainage area;

 L refers to the equivalent drainage length of the drainage area.

3 RESEARCH ON THE CHARACTERISTICS OF THE DRAINAGE RADIUS DISTRIBUTION

In order to achieve targeted outburst prevention in the mine, it is important to accurately determine the drainage radius of the coal seam. Therefore, in the 710-floor lane of Shangzhuang Coal Mine, the distribution characteristics of the extraction radius in different areas of the overlying coal seam were studied to provide a basis for the layout of the extraction radius in the strip area of the coal roadway.

3.1 Overview of the test area

The bottom extraction lane of 702 wind lane is located at the fourth level (−650 m), belonging to the east III uphill mining area. The ground elevation is +34.6 m to +38.2 m. The underground elevation is 527.6 m to 575.3 m. The design section of this lane is a semi-circular arch, with a clear width of 4,200 mm, a clear height of 3,300 mm, and a design tunneling area of 11.92 m². The roadway is 12 m to 26 m away from the B4 coal seam floor.

3.2 Measurement results and analysis

3.2.1 Determination of drainage rate

According to the actual measured maximum gas pressure of coal seam B4 of 710-floor lane is 6 MPa, the extraction rate when the residual gas pressure is reduced to 0.74 Mpa, $\eta^3 = 53.4\%$.

According to the measured gas content of B4 coal seam of 710-floor lane, which is 13.03 m³/t, which is higher than the 8 m³/t required by the "Regulations on Prevention and Control of Coal and Gas Outburst", the extraction rate when the residual gas content is reduced to 8 m³/t, $\eta^1 = 38.6\%$.

Therefore, the coal seam gas drainage rate is greater than or equal to 53.4% as the basis for determining the effective influence radius of drainage.

3.2.2 Analysis of measurement results

The effective influence radius of coal seam extraction in the B4 coal seam of the 710-floor lane is investigated. The amount of gas that needs to be extracted in the borehole control area is:

$$Q_{Total} = L1 \times L2 \times h \times \gamma \times W \times \eta = 20 \times 4 \times 3.2 \times 1.4 \times 13.03 \times 53.4\% = 4670 \text{ m}^3$$

According to statistics calculations and integral calculations, the effective influence radius of drainage of 710-floor lane is shown in Table 1.

Table 1. Calculated results of effective influence radius of drilling and extraction in coal seam B4 of 710-floor lane.

Draw time (d)	category	Lower cut 15m	Lower cut 7.5m	Right above	Upper cut 7.5m	Upper cut 15m
15	integral	1.16	1.28	1.31	1.24	1.13
	statistics	1.15	1.26	1.30	1.25	1.15
30	integral	1.52	1.65	1.71	1.60	1.49
	statistics	1.53	1.66	1.70	1.65	1.54
60	integral	1.88	2.00	2.08	1.91	1.85
	statistics	1.86	1.90	2.00	1.94	1.86
90	integral	2.05	2.14	2.24	2.03	2.02
	statistics	2.06	2.10	2.20	2.05	2.00

It can be seen from Table 1 that the effective influence radius of the borehole drainage calculated by the statistical calculation and the integral calculation is basically the same, which mutually verifies the accuracy of the effective influence radius of the drainage. Taking the average value of the two sets of data for the effective influence radius of the extraction at different locations, it can be obtained that the effective influence radius of the drilling extraction with a diameter of Φ75 mm in the B4 coal seam of the 710-floor lane of Shangzhuang Coal Mine under the condition of 13kPa extraction negative pressure. The same method is used to obtain the effective influence radius of drainage every day, and then draw the curve of the effective influence radius of drainage over time as shown in Figure 1.

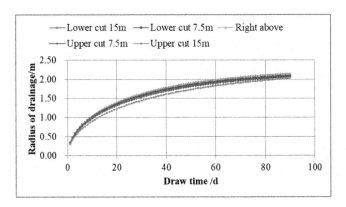

Figure 1. Changes in the effective influence radius of drainage.

4 CONCLUSION

Field tests show that the extraction radius of different areas within 15 m of the upper and lower sides of the overlying coal seam of the floor roadway is directly above the floor roadway, 7.5 m of the lower side, 7.5 m of the upper side, 15 m of the lower side, and 15 m of the upper side.

According to the results of the inspection of the extraction radius, different locations are arranged with different extraction effective influence radii, which avoids the disadvantage of using the same extraction radius to arrange the extraction holes and wastes a large amount of drilling construction.

FUND PROJECT

This work was financially supported by the Tiandi Technology Co., Ltd. Special Project of Science and Technology Innovation and Entrepreneurship Fund (Grant No. 2021-2-TD-ZD008), and Chongqing Natural Science Foundation General Project (Grant No. cstc2021jcyj-msxmX1149).

REFERENCES

Liu Dianping, Ma Wenwei. Liu Dianping, Ma Wenwei. Research on the method of measuring the effective drainage, radius of gas drainage boreholes[J]. Industry and Mine Automation, 2020, 46(11): 59–64.

Liu Jun, Shan Wenjuan, Liu Guanpeng. Analysis on the difference of the drainage radius between the through-bed borehole and the bedding borehole[J]. Coal Technology, 2016, 35(08): 148–150.

Xu Qingwei, Wang Zhaofeng, Wang Liguo. Research on the relationship between the effective extraction radius and the survey area shape and hole spacing[J]. Coal Mine Safety, 2018, 49(04): 144–147.

Zou Shichao, Xin Song. Research on the effective drainage radius of coal seam gas boreholes[J]. Chinese Safety Science Journal, 2020, 30(04): 53–59.

Zhou Liang, Dai Guanglong, Qin Ruxiang. Research on the effective drainage radius measurement technology of boreholes in deep mining of high pressure coal seams[J]. Coal Technology, 2017, 36(04): 27–29.

*Frontiers of Civil Engineering and Disaster Prevention and
Control – Yang & Rahman (Eds)
© 2023 The Author(s), ISBN: 978-1-032-31200-2*

Research on failure characteristics of overlying rock strata in floor roadway

Zhonghua Wang*

*National Key Laboratory of Gas Disaster Detecting Preventing and Emergency Controlling,
Chongqing, China*
China Coal Technology Engineering Group Chongqing Research Institute, Chongqing, China

ABSTRACT: With the increase in coal demand and mining intensity, there are obvious differences between the deep floor roadway and the shallow part. In order to study the failure characteristics of the overlying rock in the floor roadway, the YTJ20 rock formation detection recorder was used to detect the damage of the surrounding rock, and it was concluded that the surrounding rock within 15 m of the upper and lower sides of the floor rock roadway has zonal rupture, that is, the fracture zone and the intact zone. It appears that intervals occur, the monitoring section currently has 4 to 5 rupture areas, and the rupture areas show a trend of decreasing from the outside into the interior. The average initial rupture position and size of the rupture zone of the roadway section increased with the increase of the roadway excavation time. The pressure relief effect of the roadway surrounding rock is more sufficient and significant. According to the zoning fracture characteristics of the surrounding rock, the optimized sealing length is greater than the pressure relief range; that is, the sealing length is not less than 9 m.

1 INTRODUCTION

With the increase in coal demand and mining intensity, domestic and foreign coal mines have successively entered deep mining. According to incomplete statistics, in 2015, the average mining depth of large and medium-sized coal mines in my country was close to 600 m, and it extended to the deep at a speed of nearly 10–20 m/a. Coal mines with a mining depth greater than 600 m have accounted for 30% of the total. There are more than 50 coal mines within a kilometer. Scholars at home and abroad have done a lot of research on the deformation and failure of roadways surrounding rock (Wang et al. 2018; Xu & Yuan 2015; Yuan et al. 2014). As the mining depth increases, the coal seam mining conditions become more complex. In particular, the displacement failure law of the deep roadway is different from that of the shallow part, which cannot be explained by the traditional continuum mechanics theory, such as the phenomenon of partitioned fracture of the surrounding rock (Chen et al. 2011; Li et al. 2008). This paper studies the displacement and failure characteristics of surrounding rock in deep mines through field tests in order to provide a reference for similar mines.

2 OVERVIEW OF THE TEST MINE

2.1 *Mine coal seams*

In the Shangzhuang Minefield, the B4 coal seam in the Lower Laoshan section of the Leping Coal System of the Upper Permian was mined. Coal measure strata coal-bearing seams include the Guanshan, Laoshan, and Wangpanli sections. The Guanshan section, from the bottom to the

---

*Corresponding Author: boaidajia2007@126.com

DOI 10.1201/9781003308577-52

upper section, is called the A coal group, the Laoshan section is called the B coal group, and the Wangpanli section is called the C coal group. Among them, the A coal group contains two layers of coal, which are not mineable; the B coal group contains four layers of coal, and only the B4 and B5 coals are mineable; and the C coal group contains 21 layers of coal, which are basically unmineable.

The main coal seam B4 is a medium-thick coal seam, with a coal thickness of 2.0 m to 3.0 m, an average of 2.2 m, and an inclination angle of 10 to 23°. The coal seam structure is complex. The coal seam to the west of the shaft is bifurcated into B4a and B4b. The B4b coal seam has a complex structure, is stable, and can be mined in its entirety. The B4a coal seam has a simple structure, an unstable occurrence, and is partially mined.

The lithology of the pseudo roof of the coal seam is carbonaceous siltstone or carbonaceous shale, generally 0 m to 0.4 m thick, with an average thickness of 0.2 m; the direct roof is gray-black siltstone with a thickness of 2m; the old roof is light gray quartz fine sandstone and coarse powder The interbedded rocks have well-developed fissures with a thickness of 7.0 m.

The false bottom of the coal seam is scaly gray-black carbonaceous mudstone, with a thickness of 0 m to 0.3 m, with an average of 0.1 m; the direct bottom is dark gray and light gray mudstone, which is rich in plant root fossils, which is easy to swell in contact with water, and is generally 2.0 m thick; Gray-black argillaceous sandy siltstone, containing more coal line and pyrite lineage, clear horizontal bedding, containing a lot of plant leaf fossils, 4 m to 8 m thick.

2.2 *Mine geological structure*

The Shangzhuang mining area is located in the middle section of the PingxiangLeping anti-sag belt in the South China Sea Basin and at the northwestern edge of the WugongshanHuanyushan uplift belt. The former occurred in the depression on the south side of the Jiangnan geology and generally deposited paleozoic shallow seas. The coal measures of limestone and sea-land alternating facies have stable deposits and well-developed coal seams; in the Mesozoic, due to the influence of the Yanshan Movement, the original sediments and their structures were transformed and the structural traces (folds and fractures) were large. The part has a north-east direction, and the structure is basically finalized. At the same time, a faulted basin is formed near the depression, and it accepts the formation of deposits such as the Anyuan coal measure. It is characterized by a small coal-bearing basin, and the coal measure basement is the Xiangxi of the Pre-Sinian system. The group is surrounded by geological structure, and the thickness of the coal measures varies greatly.

3 TEST PLAN FOR FAILURE OF OVERLYING ROCK STRATA IN FLOOR ROCK ROADWAY

3.1 *Observations*

The YTJ20 rock formation detection recorder can clearly distinguish the internal damage of different rock formations. According to the different degrees of damage, it can be divided into complete, broken, fissured, and cracked. The measured image is shown in Figure 1.

3.2 *Analysis of observation results*

The borehole diameter of the surrounding rock deformation survey is Φ32 mm, and the YTJ20 rock formation detection recorder is used to detect the damage of the surrounding rock. After the drilling construction, five inspection holes are arranged for each inspection location, that is, five inspection locations, which are 15 m on the lower side, 7.5 m on the lower side, directly above, 7.5 m on the upper side, and 15 m on the upper side. The location of the investigation is shown in Table 1.

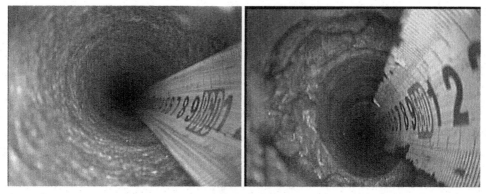

(a) Complete (b) Broken

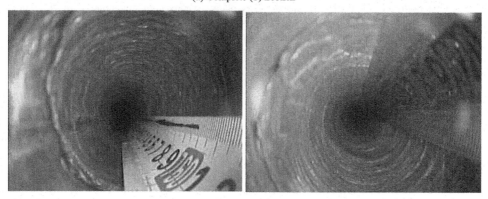

(c) Fissured (d) Cracked

Figure 1.   Varying degrees of damage during borehole drilling.

Table 1.   Inspection plan for decompression effect of floor rock roadway.

Roadway	Inspection location (from the head of the tunneling: m)		
	The first group	The second group	The third group
710-Floor Lane	335	230	135
505-Floor Lane	260	120	/

4   EXPERIMENTAL STUDY ON FAILURE OF SURROUNDING ROCK OF FLOOR ROCK ROADWAY

The regional fracture distribution of surrounding rock within 15 m on both sides of the roof of the 710 and 505 floor lanes in Shangzhuang coal mine is shown in Tables 2 and 3 respectively.

It can be seen from Tables 2 and 3 that there are generally 3 to 5 fracture zones in the surrounding rock of the 710 and 505-floor lanes. The damage of surrounding rock in the range of 0.84 m to 1.17 m from the roadway surface is the most serious. Mainly broken, this area can be considered as a loose circle of the surrounding rock of the roadway in the traditional sense. Outside the surface failure zone is the intact zone, and then into the second rupture zone, with an average depth from 1.61 m to 3.11 m. The damage to the surrounding rock in this rupture zone is also relatively large,

Table 2. The rupture of surrounding rock in the monitoring section of 710 floor roadway.

Type of data	Number of groups	Rupture area (m)				
		I	II	III	IV	V
Initial rupture zone Location	710-1	0.00	3.11	5.45	/	/
(from the roadway wall)	710-2	0.00	1.84	4.16	6.97	8.47
	710-3	0.00	2.10	3.63	4.98	7.06
Rupture zone size	710-1	0.93	0.36	0.26	/	/
	710-2	1.13	0.25	0.39	0.19	/
	710-3	1.17	0.15	0.42	0.25	0.28

Table 3. Fracture situation of surrounding rock in the monitoring section of 505 floor roadway.

Type of data	Number of groups	Rupture area (m)				
		I	II	III	IV	V
Initial rupture zone Location	505-1	0.00	1.61	2.39	3.87	8.57
(from the roadway wall)	505-2	0.00	2.13	3.80	6.10	/
Rupture zone size	505-1	1.04	0.34	0.47	0.51	0.05
	505-2	0.84	0.29	0.44	0.31	/

mainly broken and fissures. Then there is the intact zone, and then the third fracture zone, with an average depth from 2.39 m to 5.45 m. The damage to the surrounding rock in this fracture zone is relatively small, mainly cracks. The interval is complete and enters the fourth fracture zone with an average depth from 3.87 m to 6.97 m. The surrounding rock fractures in this zone are mainly fractures. It is separated from the complete zone and enters the fifth fracture zone with an average depth of 8.2 m to 9.2 m. The surrounding rock fractures in this zone are dominated by fissures.

The average initial rupture position of the rupture zone of the roadway section basically increases linearly, and the average width of the rupture zone of the roadway section basically shows a trend of gradually decreasing with the increase of the depth of the rupture zone, which shows that due to the time rheological effect, the more severe the rock damage, the greater the pressure relief of the floor roadway.

5 CONCLUSIONS

The surrounding rock within 15 m of the upper and lower sides of the pressure relief floor rock roadway appears to have a zonal rupture; that is, the rupture zone and the complete zone appear at intervals. There are 4 to 5 rupture zones in the monitoring section, and the rupture zone gradually decreases from the outside to the inside. The size of the rupture zone ranges from 0.25 m to 1.17 m, and the farthest position of the rupture zone from the orifice is 8.62 m.

The average initial fracture location and size of the fracture zone of the roadway section increase linearly with the increase of the roadway excavation time, indicating that the time-space effect of the roadway excavation makes the pressure relief effect of the roadway's surrounding rock more sufficient and significant.

According to the zoning fracture characteristics of the surrounding rock in Shangzhuang coal mine, the maximum pressure relief of the floor roadway is 8.62 m, and the optimized sealing length is greater than the pressure relief range; that is, the sealing length is not less than 9 m.

FUND PROJECT

This work was financially supported by the Tiandi Technology Co., Ltd. Special Project of Science and Technology Innovation and Entrepreneurship Fund (Grant No. 2021-2-TD-ZD008), and Chongqing Natural Science Foundation General Project (Grant No. cstc2021jcyj-msxmX1149).

REFERENCES

Chen X.G., Zhang Q.Y., Yang W.D. et al. Comparative analysis of test and field monitoring on the zonal failure of surrounding rock in deep roadway[J]. Chinese Journal of Geotechnical Engineering, 2011, 33(01): 70–76.
Li S.C., Wang H.P., Qian Q.H. et al. Field monitoring research on zonal fracture of surrounding rock in deep roadway[J]. Chinese Journal of Rock Mechanics and Engineering, 2008(08): 1545–1553.
Wang M.Y., Chen H.X., Li J. et al. Comparative study on calculation theory and actual measurement of zonal fracture of deep roadway[J]. Chinese Journal of Rock Mechanics and Engineering, 2018, 37(10): 2209–2218.
Xu Y., Yuan P. Model test study on zonal failure of deep surrounding rock under explosive load[J]. Chinese Journal of Rock Mechanics and Engineering, 2015, 34(S2): 3844–3851.
Yuan L. Gu J.C., Xue J.H. et al. Model test study on zonal fracture of deep surrounding rock[J]. Journal of China Coal Society, 2014, 39(06): 987–993.

Frontiers of Civil Engineering and Disaster Prevention and
Control – Yang & Rahman (Eds)
© 2023 The Author(s), ISBN: 978-1-032-31200-2

Effect of siliceous parent rock mechanism sand fineness modulus and sand ratio on concrete performance

Xiandong Cai*
HuiTong Construction Group Corporation Co., Ltd., Hebei Gaobeidian, China

Xingang Wang
CCCC Tianjin Harbor Engineering Research Institute Co., Ltd., Tianjin, China

ABSTRACT: The current specifications for the use of machine-made sand do not consider the influence of lithology, grain type, and sand ratio on concrete performance. The actual production of machine-made sand concrete performance is not satisfactory. In this paper, the effect of each grain size on the fineness modulus of machine-made sand is studied to guide the production of machine-made sand. The bleeding of granite machine-made sand concrete can be improved by reducing the grading range of machine-made sand required by the specification. At the same time, a sand rate correction method considering fineness modulus, stone powder content, and water-cement ratio is proposed, which is a new technology for the use of granite machine-made sand concrete and provides a reference for the future use of granite machine-made sand concrete preparation projects.

1 INTRODUCTION

At home and abroad, the fine aggregate grain type, grading, and sand ratio on concrete performance are many research results. The influence of sand on concrete performance has many research results, but for mechanism, related technical requirements in the specification still have some limitations and do not consider such as lithology, grain shape, and other effects on the grain size distribution, which also did not give specific indicators. And particle level pair concrete workability, construction convenience, and final strength will have an impact, but only in accordance with the requirements of the specification to prepare. Concrete performance may be difficult to meet the needs of the project.

Based on the granite machine-made sand used in a certain physical engineering project, according to the requirements on the grading of machine-made sand in the literature (Celik & Marar 1993), the range of its value is reduced through experiments to determine a reasonable grading interval for granite machine-made sand. At the same time, a method to modify the existing mix ratio is proposed to consider the influence of fineness modulus (Guimaraes et al. 2007) and stone powder content on the mix ratio of granite machine-made sand concrete, and the correction relationship is clear. The above research and indoor tests have demonstrated that machine-made sand concrete can perform poorly in the actual production process, so that improved working performance of granite machine-made sand concrete may be achieved, as well as accumulating experience for engineering promotion of granite machine-made sand concrete (Huang 2009, Huang & Lv 2015, He et al. 2007, Mostofinejad & Reisi 2012, Shilstone 1990, Wang 2005, Wallevik & Wallevik 2011, Xie 2016, Zeng 2010, Zhao et al. 2012).

---

*Corresponding Author: 15933902269@139.com

DOI 10.1201/9781003308577-53

2 CHARACTERISTICS OF RAW MATERIALS

Cement: PO 42.5 cement of Linjiang Fengcheng Cement Co., Ltd.

Coarse aggregate: 5 mm to 16 mm (mud content 0.4%), 16 mm to 31.5 mm (mud content 0.3%), gravel from Shiling Quarry in Lianjiang City. The mixing ratio is 50:50.

KTPCA polycarboxylic acid high performance water reducing agent, produced by Shanxi Kantel Fine Chemical Co., Ltd., has a water reducing rate of 27%.

Fine aggregate: mechanism sand from Shiling quarry in Lianjiang City, fineness modulus 3.18 to 2.58, crushing value 22.1%, MB value 0.6. The river sand of Qingping sand field in Lianjiang city has a fineness modulus of 2.22 to 2.86 and mud content of 2.0%.

3 MODULUS OF FINENESS OF FINE AGGREGATE

3.1 *Influence of fine aggregate gradation*

Screening tests were carried out on 30 groups of mechanistic sand and 15 groups of river sand, respectively. The fineness modulus of mechanistic sand ranged from 2.58 to 3.18, mainly concentrated around 2.80, and the fineness modulus of river sand ranged from 2.22 to 2.86, mainly concentrated around 2.40. According to reference 2, river sand and mechanistic sand were classified as medium sand (3.0 to 2.3).

As shown in Figure 1, in order to determine if the gradation of mechanized sand and river sand is good or not, an analysis of the gradation is carried out by comparing several groups of representative mechanized sand and river sand groups and analyzing the median value of the specification.

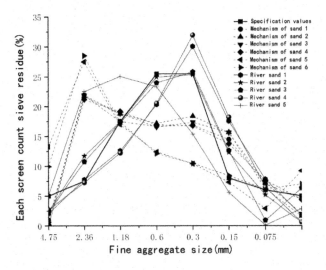

Figure 1. The curve of percentage retained.

Comparing the remaining percentage of sieve of 6 groups of mechanized sand and 5 groups of river sand in Figure 1 with the standard value, it can be seen that the machine-made sand has an obvious common problem of being "large at both ends and small in the middle". The grain content of machine-made sand with a grade above 1.18 mm is higher, and the particle content of 0.075 mm to 0.60 mm is less. Most river sands are graded near the median value of the specification, with low particle content of 0.075 mm to 0.30 mm grade, and a few river sands have similar gradation with the same mechanism, with high particle content of above 1.18 mm and low particle content of 0.075 mm to 0.60 mm grade. After tracking, it was found that these river sands generally have high mud content. The water washing method was adopted to reduce the mud content. In the process

388

of water washing, the particle content of 0.075 mm to 0.30 mm was reduced and the gradation became bad.

According to the regulations, the fineness modulus of mechanism sand is determined to be medium sand, and the fineness modulus is generally around 2.8. According to the previous experience in using river sand, concrete is generally preferred when the fineness modulus of river sand is around 2.8. However, the use of machine-made sand is not as optimistic as imagined. The reasons why this may occur are explained by analyzing the relationship between the sieve residue and the fineness modulus of each sieve pore meter below.

Regression analysis of each stage and the fineness modulus is carried out for the sieve residue of each stage to calculate the fineness modulus. The regression formula is as follows:

$$M_x = C_1 \cdot \alpha + C_0 \tag{1}$$

Where

M_x — Mechanism sand fineness modulus;

α — All levels of screening (%);

C_1, C_0 — Regression constant.

The regression analysis was conducted on the fineness modulus of the sample and the residue of each sieve size meter, and the correlation coefficient was 0.80 or more. Among them, the residue of 2.36 mm and 1.18 mm particles had a positive correlation, and the other two, 0.30 mm and 0.15 mm particles, had a negative correlation, as shown in Table 1.

Table 1. Regression statistics of percentage retained of sieve holes at all levels and fineness modulus.

Regression parameters	Screen diameter of each stage (mm)					
	4.75	2.36	1.18	0.60	0.30	0.15
C_1	0.0127	0.0419	0.0849	−0.0508	−0.0550	−0.0547
C_0	3.2856	1.9234	1.3110	3.5638	3.7005	3.511
R	0.1175	0.9382	0.8308	0.5136	0.9145	0.8215

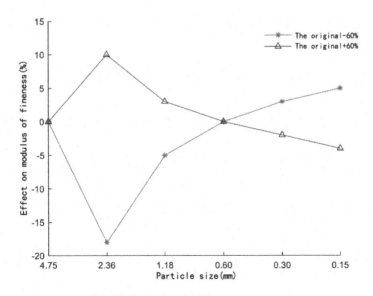

Figure 2. Temperature test model of hydration heat inhibitor.

By adjusting the remaining screen mass of one gear to increase 60% and decrease 60%, while the remaining screen mass of the other gear remains unchanged, the change of the fineness modulus is studied, as shown in Figure 2. It is found that 4.75 gear and 0.60 mm gear have little influence on the fineness modulus, while the remaining screen mass of 2.36 mm and 1.18 mm gear decreases the fineness modulus, and the increase in fineness modulus decreases the fineness modulus. In contrast to the 0.30 mm and 0.15 mm files, the fineness modulus increases when the remaining screen mass decreases, and the fineness modulus decreases when the remaining screen mass increases. This is consistent with the results of the regression analysis done above. The variation range of 0.30 mm is smaller than 0.15 mm, and 2.36 is larger than 1.18, showing an antisymmetric butterfly type. It can also be found from the figure above that 2.36 is more sensitive to a decrease than an increase. The control of fineness modulus in machine-made sand production can be guided by adjusting the sieve residual percentage of 2.36 mm, 1.18 mm, 0.30 mm, and 0.15 mm.

3.2 *Influence of stone powder*

Stone powder is a general term for stone powder, which is defined in this paper as the composition of machine-made sand with a particle size of less than 0.075mm. Taking the median value of the specification as the benchmark, it can be found that the fineness modulus decreases by 0.3 when the stone powder content increases by 10% by adjusting the proportion of stone powder in machine-made sand under the condition that the proportion of stone powder particle size above 0.075mm remains unchanged.

In the process of machine-made sand production, machine-made sand is gradually broken from large particles and ground into smaller particles. By controlling the residual percentage of sieve and stone powder content of 2.36mm, 1.18mm, 0.30mm and 0.15mm, the fineness modulus of finished sand will not have great deviation.

4 CONCRETE STATE

According to the calculation of the median value of GB/T 14684-2011 machine-made sand specification, the particles above 1.18mm account for 30% of the total mass, that is, the ratio of thickness to fineness is 3:7. According to the design test of different ratio of thickness to fineness, the granite machine-made sand concrete can have better working performance by reducing the grading interval of granite machine-made sand. See Table 2 for specific test design.

Table 2. Mechanism sand gradation adjusts mix ratio (kg/m³).

No.	Cement	Water	Admixtures	Coarse aggregate	Fine aggregate Species	Dosage	Stone powder (%)
B-1	427.5	171	5.13	998	River sand (3:7)	756	0
B-2	427.5	171	5.13	998	River sand (3:7)	756	8
B-3	427.5	171	5.13	998	Mechanism sand (3:7)	756	8
B-4	427.5	171	5.13	998	Mechanism sand (4:6)	756	8
B-5	427.5	171	5.13	998	Mechanism sand (5:5)	756	8
B-6	427.5	171	5.13	998	Mechanism sand (6:4)	756	8
B-7	427.5	171	5.13	998	Mechanism sand (7:3)	756	8

By Table 3 can be found, river sand preparation mechanism of concrete strength is about 5 Mpa sand concrete strength small, at the same time, the air content to 1%–2% higher than mechanism sand and river sand adding 8% powder can increase the strength, at the same time reduce exudation,

Table 3. Test data.

No.	Slump (mm)	Extension degree (mm)	Air content (%)	Workability	Strength of 28 d (MPa)
B-1	200	51	3.10	4	39.3
B-2	215	51	/	5	44.3
B-3	220	50	1.80	4	51
B-4	180	48	2.00	4	49.9
B-5	170	53	/	3	56
B-6	143	47	1.10	1	/
B-7	152	48	1.20	2	45.9

can slightly improve the slump may be 8% stone powder increases the flow of the slurry ability to have the lubrication effect. Judging by the performance indexes of concrete such as water leakage, the concrete with the ratio of fine aggregate thickness between 3:7 and 5:5 is in better condition. When the ratio of thickness to fineness is greater than 5:5, the bleeding water of concrete increases, the slump and expansion decrease obviously, the workability is poor, and the strength decreases.

5 CORRECT SAND RATE

In the mix ratio design of this system, only the sand ratio is modified, and the rest, such as water-cement ratio, water consumption and cement consumption, are carried out according to JGJ 55-2011.

In the process of determining the sand ratio of mechanized sand concrete, the consideration of mechanized sandstone property is added. When the variation of mechanized sandstone property is relatively large, the mix ratio should be readjusted. The adjustment direction is given according to the fineness modulus of mechanism sand and the content of stone powder. According to JGJ 55-2011, according to coarse aggregate variety and particle size, water-cement ratio and slump requirement, sand yield β S is determined by conventional method, and then corrected by water-cement ratio, fineness modulus and stone powder content correction coefficient.

For machine-made sand with different lithology, correction should be made according to the test. The empirical parameter θ W of water-cement ratio correction for granite is given here, as shown in Table 4.

Table 4. Correction parameters of water-cement ratio.

Water cement ratio	0.4	0.5	0.6	0.7
$\theta_{w,c}$	1%	3%	5%	7%

This parameter is determined based on experience. On the premise of ensuring similar working performance, the sand rate of mechanism sand is several percentage points higher than that of river sand.

For the correction of stone powder content, the correction coefficient θ_{rd} of stone powder content is multiplied by subtracting the stone powder content of the current mechanism sand from a reference stone powder content (determined according to lithology), and the correction empirical coefficient of granite is 12.

The correction of fineness modulus is as follows: the fineness modulus of machine-made sand produced is divided by the fineness modulus of machine-made sand calculated by the median value of machine-made sand specification 2.54, plus the modification value of stone powder, and then multiplied by the lithology correction coefficient θ_{mx}. The correction experience coefficient of granite is 2.5. The sand ratio of granite machine-made sand concrete is:

$$\beta_{s,new} = \beta_s + \theta_{w,c} + \theta_{mx} \times \left(\frac{M_x}{2.54} + (0.12 - r_d) \times \theta_{rd} \right) \tag{2}$$

In the formula:

$\beta_{s,new}$ — Modified sand ratio of mechanized sand concrete, %;

β_s — Reference sand ratio, %, determined according to (JGJ 55-2011);

$\theta_{w,c}, \theta_{mx}, \theta_{rd}$ — The modified parameters of water-cement ratio, fineness modulus and stone powder content are determined according to lithology.

M_x — Fineness modulus of machine-made sand;

r_d — Stone powder content;

2.54—The fineness modulus calculated in the specification (0.075mm sub-screen residual is set as 6%, followed by 8%);

0.12—The content of boundary stone powder determined according to the test.

Above formula of fixed sand ratio, and through the water cement ratio, stone powder content, fineness modulus of amended, revised relationship clear, fixed sand ratio is proportional to the water-cement ratio, fineness modulus, is inversely proportional to the stone powder content, through the six groups of different water cement ratio, stone powder content and the fineness modulus (listed in Table 5) fixed sand ratio, good use effect. At the same time, the correction of fineness modulus may include the correction of lithology, which has not been further studied here. Later, the application scope of the formula can be extended to different lithologies through further research to consider the influence of different lithologies on the mix ratio.

Table 5. Parameter modification for laboratory test of mechanical sand yield.

No.	1	2	3	4	5	6
Water cement ratio	0.4	0.4	0.5	0.5	0.6	0.6
Stone powder content (%)	6	16	9	16	6	9
Fineness modulus	3.24	2.50	2.50	2.80	3.24	2.80
Fixed sand ratio (%)	45	41	48	47	55	54

Table 6. Mechanical sand index used in engineering site.

Fineness modulus	Stone powder content (%)	Water cement ratio	Methylene blue (MB) value	Fixed sand ratio	slump	Intensity of 3d
2.78	6.6	0.44	0.6	46%	180mm	22.6MPa

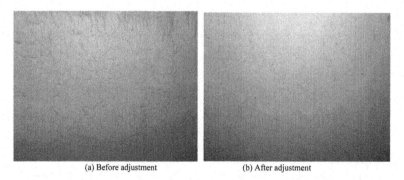

(a) Before adjustment   (b) After adjustment

Figure 3. Comparison of appearance of mechanical sand before and after adjustment.

The component shown in Figure 3 is the physical engineering guardrail of this project. It can be seen from the figure that the use of the modified sand rate has significantly improved the situation of mechanical sand bleeding. Although there are some bubbles in the adjusted appearance, the overall quality has been greatly improved compared with that before, which verifies the applicability of the modified sand rate in field use. See Table 6 for the on-site use of mechanical sand parameters.

6 CONCLUSIONS

In this paper, by reducing the value range of machine-made sand gradation required by the specification, the bleeding situation of granite machine-made sand is improved. Meanwhile, a sand rate correction method is proposed, and the conclusions are as follows:

(1) Regression analysis was conducted on the fineness modulus of the sample and the residue of each sieve hole score. The positive correlation was 2.36 mm and 1.18 mm particles, and the negative correlation was 0.30 mm and 0.15 mm particles. The correlation coefficient was above 0.80, which was consistent with the conclusion obtained by adjusting the quality of a certain gear. It is also found that 2.36 mm is more sensitive to decrease than increase.

(2) Based on the median value of the specification, it is found that the fineness modulus decreases by 0.3 when the stone powder content increases by 10% when the proportion of stone powder particle size above 0.075 mm remains unchanged. In the process of machine-made sand production, the fineness modulus of finished sand can be well controlled by controlling the residual percentage of sieve and the content of stone powder at 2.36 mm, 1.18 mm, 0.30 mm, and 0.15 mm.

(3) Under the same conditions, the strength of concrete prepared by river sand is about 5MPa less than that of mechanized sand, and the gas content is 1%-2% higher. Adding 8% stone powder to river sand can increase strength and work performance. Concrete with coarse aggregate ratio greater than the lower limit of specification and less than 5:5 is in better condition. When the ratio of thickness to fineness is greater than 5:5, the bleeding water of concrete increases, the slump and expansion decrease obviously, the workability is poor, and the strength decreases.

(4) A new sand rate correction method is proposed, which considers the influence of water cement ratio, stone powder content and fineness modulus on sand rate. The correction relationship is clear and the effect is good.

REFERENCES

Celik T, Marar K. (1993) Effect of crushed stone dust on some properties of concrete. Cement and Concrete Research, 26(7): 1124–1130.

Guimaraes M.S., Valdes J.R., Palomino A.M. (2007) Aggregate production: Fines generation during rock crushing. International Journal of Mineral Processing, 81(4): 237–247.

Huang, Q. (2009) Relationship between gradation type and segregation degree for asphalt mixture. Journal of Traffic and Transportation Engineering, 9(2): 1–6.

Huang J., Lv Y. (2015) Study on the Performance of Concrete with Different Fineness Modulus Artificial Sand. Beton Chinese Edition - Ready-mixed Concrete,05: 51–53.

He, T.S., Zhou, M.K., Li, B.X. (2007) Influence of micro- fines on bleeding rate of fresh concrete of manufactured- sand. Concrete,02: 58–60.

Mostofinejad D, Reisi M.A. (2012) New DEM-based method to predict packing density of coarse aggregates considering their grading and shapes. Construction and Building Materials, 35: 414–420.

Shilstone J S M. (1990) Concrete mixture optimization. concrete international, 12(6): 33–39.

Wang, S.Y. (2005) Effects of grain size distribution and structure on mechanical behavior of silty sands. Rock and Soil Mechanics, 26(7): 1029–1032.

Wallevik O H, Wallevik J E. (2011) Rheology as a tool in concrete science: The use of rheographs and workability boxes. Cement and Concrete Research., 41(12): 1279–1288.

Xie, H.B. (2016) The Characteristics of Shape and Grading of Manufactured Sand and theirs Effects on the Properties of Concrete. South China University of Technology, Guang Zhou.

Zeng, D. Q. (2010) Study on the Effect of Granular Grade on the Performance of Slag Cement. China Concrete and Cement Products, 6: 1–4.

Zhao H, Sun W, Wu X. (2012) The effect of coarse aggregate gradation on the properties of self-compacting concrete. Materials & Design., 40: 109–116.

Frontiers of Civil Engineering and Disaster Prevention and Control – Yang & Rahman (Eds)
© 2023 The Author(s), ISBN: 978-1-032-31200-2

Study on the dynamic response of separated steel-concrete continuous curved composite box-section girder Under the action of vehicles

Jianqing Bu*
School of Traffic and Transportation, Shijiazhuang Tiedao University, Shijiazhuang, China

Zhiqiang Pang* & Lingpei Meng*
School of Civil Engineering, Shijiazhuang Tiedao University, Shijiazhuang, China

Jingchuan Xun*
China Construction Road and Bridge Group Co. Ltd, Shijiazhuang, China

ABSTRACT: The separated curve composite box girder bridge is widely used in urban bridges due to its reasonable section and smooth linear. Under the action of eccentric vehicle, the inner and outer box girders of separated curve composite box girder influence each other, and the bending and torsion effect is more complex. In order to study the partial load effect under vehicle loading, the dynamic deflection distribution of separated curve composite box girder under multi-vehicle-bridge coupling was analyzed by ANSYS based on the separation iteration method, and the effects of curvature radius, transverse connection number, bridge deck roughness and driving speed on the dynamic deflection of separated curve composite box girder were explored. The research shows that the dynamic deflection of the separated curve composite box girder is the largest when the vehicle runs along the lateral box girder. The mid-span dynamic deflection of curved composite box girder decreases with the increase of the longitudinal spacing of the fleet. The smaller the radius of curvature, the more obvious the influence on the dynamic deflection of separated curve composite box girder. With the increase of bridge deck roughness, the influence of vehicle speed on the mechanical response of separated curve composite box girder gradually increases. The bridge deck should be maintained regularly during the actual operation of the small radius bridge, and the driving speed should not exceed 50 km/h.

1 INSTRUCTIONS

Steel-concrete composite beams are widely used in small and medium-span urban curved girder bridges due to their light weight and short construction period (Huang 2004; Zhou 2018). With the increasing traffic demand, some new girder bridges adopt wide multi-box structure. Multiple box girders are connected by multiple separated box girders through flat plates to form a whole (V 1988). The bending-torsion effect of separated curved box girder is more complicated under eccentric load (Zhuang 2017). When the vehicle passes the bridge, the vehicle will cause the bridge vibration, and the bridge vibration will react on the vehicle (Deng et al. 2018). Under the action of vehicle-bridge coupling, the bending-torsion effect of separated steel-concrete curve composite box girder bridge is intensified, and it is easy to cause fatigue and damage in the weak parts of the force.

In recent years, Zhang et al. (2015) has carried out systematic research on steel-concrete curved composite box girder. Eccentric load aggravated the bending-torsion effect of curved composite girder. And curvature radius and diaphragm setting have obvious influence on the bending-torsion effect of curved composite girder. Huang (2010) used ANSYS to explore the influence of vehicle

*Corresponding Authors: bujq2004@163.com, 1259602220@qq.com, menglingpei@126.com and 76201366@qq.com

DOI 10.1201/9781003308577-54

eccentricity on the vibration response of curved beam. The vehicle traveled along the outer side of the curved beam, and the displacement of the outer web was the largest. Chen (2020) studied the influence of the number of longitudinal loading vehicles on the dynamic response of the main beam. The number of longitudinal loading vehicles has a great influence on the deflection impact coefficient of the main beam. Scholars have studied more on the bending-torsion effect of curved composite box girder under static action. The dynamic response of curved composite girder under vehicle-bridge coupling is insufficient. The eccentric vehicle makes the spatial stress of the separated curved composite box girder complex. The dynamic deflection is the most real response of the bridge stiffness. Therefore, to study the variation law of the dynamic deflection of the separated curved composite box girder under the action of the vehicle is necessary.

ANSYS is used to establish the finite element model of the separated curved composite box girder and the vehicle. By coordinating the degree of freedom of the vehicle-bridge contact point and considering the influence of the bridge deck flatness, the Newmark method is used to solve the vehicle-bridge coupling problem. The dynamic deflection of the separated curved composite box girder under multi-vehicle-bridge coupling is analyzed. The influence of curvature radius, transverse connection number and vehicle speed on the dynamic deflection of the curved composite box girder is explored.

2 AXLE COUPLING SYSTEM

2.1 *Vehicle model*

At present, vehicle models commonly used in vehicle-bridge coupling problems include moving mass model, quarter vehicle model and quarter vehicle model. Compared with the quarter car model, the half car model considers the bumpy vibration of the vehicle and the degree of freedom of the car body. And it better simulates the state of the car body along the longitudinal direction of the bridge. The vehicle model is shown in Figure 1.

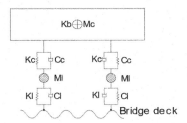

Figure 1.   Vehicle model.

The vehicle motion equation can be expressed by

$$M_v \ddot{Y}_v + C_v \dot{Y}_v + K_v Y_v = P_v \tag{1}$$

where M_v, C_v and K_v are the mass matrix, damping matrix and stiffness matrix of the vehicle; P_v is the external load vector of the vehicle; Y_v is the node displacement matrix of the vehicle; $\dot{Y}_v$ is the node velocity matrix of the two vehicles; $\ddot{Y}_v$ is the node acceleration matrix of the vehicle.

2.2 *Bridge model*

The bridge is a multi-degree-of-freedom system. And the finite element discrete method is used to analyze it. The motion equation of the bridge can be expressed by

$$M_b \ddot{Y}_b + C_b \dot{Y}_b + K_b Y_b = P_b \tag{2}$$

where M_b, C_b and K_b are the mass matrix, damping matrix and stiffness matrix of the vehicle, respectively; P_b is the external load vector of the vehicle; Y_b is the node displacement matrix of the vehicle; $\dot{Y}_b$ is the node velocity matrix of the two vehicles; $\ddot{Y}_b$ is the node acceleration matrix of the vehicle.

The curve composite box girder adopts Rayleigh damping, the calculated mass damping coefficient is 1.3105, and the stiffness damping coefficient is 0.87×10^{-3}.

2.3 Simulation of bridge deck irregularity

The flatness coefficient of bridge deck is obtained by Chinese standard GB7031 spectrum (Zhou 2016). And the unevenness of bridge deck is simulated by trigonometric series method. After Fourier transform, the vertical excitation r_x can be expressed by

$$r_x = \sum_{k=1}^{N} \sqrt{2G_q(n_k)\Delta n} \cos(2\pi n_k x + \theta_k) \tag{3}$$

where; $n_k = n_1 + (k - 0.5)\Delta n$; n_1, n_2 represent the lower limit and upper limit of the spatial frequency of the road power spectrum, as suggested in are $n_1 = 0.011$, $n_2 = 2.83$; Δn represents the interval of the spatial frequency, $\Delta n = (n_2 - n_1)/N$; θ_k represents the random phase angle; x represents the longitudinal length of the bridge deck; n represents the number of samples of the road surface roughness. $G_d(n_k)$ is the displacement power spectral density.

The bridge deck roughness is simulated by MATLAB platform. The bridge deck length is 90 m, the sample interval is 0.375 m, and the number of samples is 240. The random samples of bridge deck roughness under the deck grades of A, B, C and D with the deterioration of bridge deck conditions are obtained.

2.4 Vehicle-bridge coupled model

During the driving process of the vehicle, it is assumed that the vehicle tire keeps contact with the bridge deck all the time. Considering the influence of the bridge deck roughness, the interaction force F_{vi} between the wheel and the bridge can be obtained from the force and displacement coordination relationship of the vehicle-bridge contact point (Huang 2016).

$$F_{vi} = c_{ti}\left[\dot{Y}_{vi} - \dot{Y}_{bi} - \dot{r}_{xi}\right] + k_{ti}[Y_{vi} - Y_{bi} - r_{xi}] \tag{4}$$

where $i = 1, 2$; c_{ti}, k_{ti} are equivalent stiffness and damping coefficient of the first wheel; $\dot{r}_{xi}$ is the first derivative of the vertical excitation of bridge deck.

Firstly, the finite element model of the steel-concrete curved composite beam and the fixed half car model are established by using ANSYS. Then, the vehicle node position is changed by transforming the node coordinate system (Yang 2019). The vehicle and the curved composite beam are connected by the coordination relationship between the geometric compatibility equation of the contact point and the dynamic interaction force. Each element is iteratively solved, and the sampling time interval is 0.034 s. Considering the influence of vehicle speed and bridge deck roughness, Newmark method is used to solve the vehicle-bridge coupling problem.

3 CASE ANALYSIS

3.1 Introduction of project

Referring to a separated steel-concrete continuous curved composite girder bridge in Zhanjiang City. The curvature radius of the separated curved composite box girder is 120 m, the width of the main beam is 7.6 m, the height is 1.6 m, the thickness of the bottom plate is 24 mm, and the thickness of the web plate is 16 mm. The transverse connection of the curved composite box girder includes the diaphragm in the box and the connection beam between the boxes. The number of transverse connections is five. And the thickness of transverse connection is 22 mm. The section size of the separated curved composite box girder is shown in Figure 2.

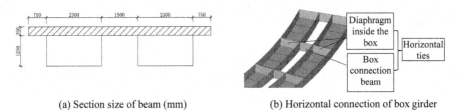

(a) Section size of beam (mm)　　　　　　(b) Horizontal connection of box girder

Figure 2.　Separated Curved Composite Box Girder.

3.2　Finite element model

3.2.1　Bridge finite element model

The separated steel-concrete continuous curve composite box girder is established by ANSYS. The calculation span is 3×30 m, and the section size is shown in Figure 2. The separated curved composite box girder adopts torsional bearing. The torsional bearing is arranged on the bottom plates of inner box girder and outer box girder respectively. The concrete roof of separated steel-concrete curve composite beam is simulated by Solid45 element. The steel beam is simulated by Shell63 element. The Targe element and Contact element elements are used to simulate shear connectors between concrete roof and steel box girder. The material parameters in the modeling process are taken from the values of C50 concrete and Q345C steel in the specification. The elastic modulus of C50 concrete is 0.35×10^5 MPa, and the Poisson ratio is 0.2. The elastic modulus of Q345C steel is 2.06×10^5 MPa, and the Poisson ratio is 0.28. The finite element model of separated steel-concrete curved composite box girder is shown in Figure 3.

Figure 3.　Separated steel-concrete curved composite box girder.

3.2.2　Finite element model of vehicle

The vehicle adopts half-car model, the vehicle specific parameters reference 12, as shown in Table 1. The vehicle adopts a biaxial vehicle with a distance of 2.4 m between the front and rear axles. The body and tire are simulated by Mass21. The spring damping of suspension system and wheel is simulated by Combine14. The connection between the body and suspension system is simulated by MPC184.

3.3　Dynamic response analysis

3.3.1　Influence of vehicle lateral position

In order to study the influence of vehicle lateral position on the dynamic deflection of separated steel-concrete composite curved beam, the factors of vehicle lateral position were simplified. The

Table 1. Vehicle model parameters (Zhong 2020).

Parametric classification	Mass/kg		Stiffness coefficient/Nm^{-1}			Damp coefficient /kgs^{-1}	
	M_c	M_l	K_b	K_c	K_l	C_c	C_l
Vehicle parameters	3.8×10^4	4.3×10^3	2.4×10^4	4.2×10^6	2.5×10^6	9.8×10^4	1.9×10^5

single vehicle runs unilaterally on the *A* level bridge deck at 40 km/h. The lateral position of the vehicle is shown in Figure 4. In Figure 4, the right side is the outside of the box girder, and the left side is the inside of the box girder.

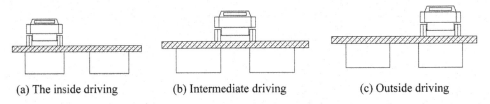

(a) The inside driving   (b) Intermediate driving   (c) Outside driving

Figure 4. Vehicle lateral position.

Table 2. Influence of vehicle lateral position on dynamic deflection of mid-span box girder.

Position of the vehicle	Side span cross section			Mid-span mid-section		
	Inner box girder w_1/mm	External box girder w_2/mm	w_2/w_1	Inner box girder w_3/mm	External box girder w_4/mm	w_4/w_3
Self-weight effect	−8.02	−8.75	1.09	−2.71	−2.88	1.06
The inside driving	−12.47	−12.39	0.99	−6.50	−5.65	0.87
Intermediate driving	−12.04	−13.15	1.09	−5.97	−6.40	1.07
Outside drive	−11.48	−13.79	1.20	−5.49	−7.21	1.31

Table 2 shows the influence of vehicle lateral position on the dynamic deflection of curved composite box girder across the middle and outer box girder. The lateral driving position of the vehicle has an obvious influence on the dynamic deflection of the separated curve composite box girder. The dynamic deflection caused by the vehicle driving along the outside is the largest, which occurs in the side span of the box girder is 13.79 mm, and the deflection span ratio is 1/2178. The value is less than the limit 1/600 of deflection-span ratio specified in the Code for Design of Steel-concrete Composite Bridges. When the vehicle runs along the inner box girder, the dynamic deflection of the inner box girder is greater than that of the outer box girder. The partial load effect of the inner and outer box girder is small. When the vehicle runs along the middle of the double box girder, the dynamic deflection ratio of the inner and outer box girder is close to the ratio under the action of self-weight. When the vehicle runs along the outer box girder, the dynamic deflection ratio of the inner and outer box girders is the largest. The partial load effect is the most serious. When the vehicle is running on the separated curved girder bridge, it is recommended to avoid running along the outer box girder, which is conducive to alleviating the eccentric load effect of the inner and outer box girders.

3.3.2 *Influence of vehicle longitudinal position*
In order to study the influence of vehicle longitudinal position on the dynamic response of separated steel-concrete composite curved beam, two vehicles were selected to drive at a constant speed of

40 km/h on the A-level bridge deck. The distance between the two vehicles was 2.5 m, 5 m, 15 m, and 30 m, respectively. The vehicle was driven along the middle position of the outer box girder.

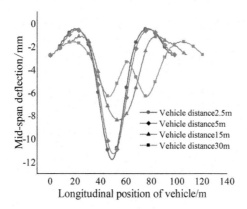

Figure 5.   Effect of vehicle longitudinal position on midspan deflection.

Figure 5 shows the influence of vehicle longitudinal position on dynamic deflection of mid-span section. The longitudinal vehicle spacing has a great influence on the mid-span dynamic deflection. When the longitudinal distance is small, the midspan deflection is about 1.45 times of the single vehicle load. With the increase of longitudinal vehicle spacing and less than the length of single span beam, the maximum dynamic deflection of midspan decreases. And the vehicle gradually moves away from midspan when the maximum dynamic deflection occurs. When the longitudinal vehicle spacing is greater than the single-span beam length, the dynamic deflection of mid-section of the mid-span decreases. When the maximum dynamic deflection occurs, the specific location of the vehicle is related to the longitudinal vehicle spacing.

Table 3.   Maximum dynamic deflection of midspan (mm).

Box girder	Single vehicle	Distance between two vehicles			
		2.5 m	5 m	15 m	30 m
Inner box girder	−4.34	−5.72	−5.58	−4.61	−3.72
External box girder	−8.14	−12.35	−11.84	−8.57	−7.19

Table 3 shows the influence of vehicle longitudinal position on the mid-span maximum dynamic deflection. When the distance between two cars is small, the maximum dynamic deflection of the outer box girder is 12.35 mm. When the distance between two vehicles is 15 m, the mid-span dynamic deflection of the double vehicle loading is equal to that of the single vehicle loading. When the distance between two cars is 30 m, the maximum dynamic deflection of the outer box girder of the two cars is reduced to 0.58 times that of 2.5 m.

4   ANALYSIS OF INFLUENCING FACTORS

In order to comprehensively study the dynamic deflection distribution law of separated steel-concrete curved composite box girder, the influence of curvature radius, transverse connection number, bridge deck flatness and vehicle speed on the dynamic deflection of curved composite box girder is explored.

4.1 Influence of curvature radius

To study the influence of curvature radius on the dynamic deflection of separated steel-concrete curved composite box girder, a single vehicle is selected to travel at a constant speed of 40 km/h on the *A* level bridge deck. And the curvature radius of the bridge is 60 m, 90 m, and 120 m. The vehicle travels along the middle position of the outer box girder. The influence of curvature radius on the transverse dynamic deflection of the middle span along the steel beam bottom plate is shown in Table 4.

Table 4. Dynamic deflection of middle span floor (mm).

Radius of curvature/m	Inner box girder			External box girder		
	Inner web	Intermediate position	Lateral web	Inner web	Intermediate position	Lateral web
60	−6.24	−7.13	−7.69	−8.78	−9.78	−10.29
90	−5.10	−5.77	−6.15	−6.96	−7.72	−8.03
120	−4.88	−5.49	−5.80	−6.52	−7.21	−7.46

The curvature radius has a great influence on the dynamic deflection of the separated curved composite box girder. The mid-span dynamic deflection increases with the decrease of curvature radius. When the curvature radius is greater than 90 m, the decrease of curvature radius has little effect on the dynamic deflection of curved composite box girder. When the curvature radius is less than 90 m, the decrease of curvature radius has a significant influence on the increase of dynamic deflection of curved composite box girder. With the decrease of curvature radius, the dynamic deflection increment of the outermost web of curved composite box girder is twice that of the innermost web.

4.2 Number of horizontal contacts

For studying the influence of the number of transverse connections on the dynamic deflection of the separated steel-concrete curved composite box girder, the single vehicle was selected to travel on the A level bridge deck at a constant speed of 40 km/h. The number of transverse connections of each span was 2, 4, 6 and 9. The vehicle traveled along the middle position of the outer box girder. The influence of the number of transverse connections on the dynamic deflection of the middle span along the steel beam bottom plate is shown in Table 5.

Table 5. Dynamic deflection of midspan floor (mm).

The number of transverse connections	Inner box girder			External box girder		
	Inner web	Intermediate position	Lateral web	Inner web	Intermediate position	Lateral web
2	−2.11	−5.11	−4.85	−6.84	−9.47	−9.02
4	−4.82	−5.46	−5.77	−6.51	−7.20	−7.43
6	−4.88	−5.49	−5.80	−6.52	−7.21	−7.46
9	−4.94	−5.54	−5.84	−6.53	−7.21	−7.21

From Table 5, it can be seen that with the increase of transverse connection number, the dynamic deflection of inner box girder increases and the dynamic deflection of outer box girder decreases. It indicates that setting transverse connection in separated curve composite box girder is beneficial to increase the transverse stiffness of double box girder and strengthen the mutual connection between inner and outer box girder. When the transverse connection is arranged at the middle span, the dynamic deflection of the middle span floor is greatly affected, and the eccentric load effect of the inner and outer box girder is the smallest. Therefore, in order to reduce the eccentric load effect of box girder, 6 to 9 transverse connections in per span are recommended for small and medium-span steel-concrete curve composite box girders.

4.3 *Influence of bridge deck flatness and vehicle speed*

In order to study the influence of vehicle speed and bridge deck flatness on the dynamic deflection of the separated curved composite box girder bridge, the single vehicle was selected to drive along the lateral box girder under eccentric load. The radius of curvature of the bridge was 120 m, and the speed of 30 km/h, 40 km/h, 50 km/h and 60 km/h were used to drive on the bridge deck with grade *A*, *B*, *C* and *D*. The dynamic deflection of the middle span section under vehicle speed and pavement flatness was obtained as shown in Figure 6.

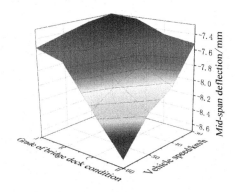

Figure 6.   Effect of vehicle speed and bridge deck irregularity on midspan deflection.

Figure 6 shows that the influence of vehicle speed and bridge deck roughness on the dynamic deflection of curved composite box girder is obvious. With the deterioration of bridge deck condition, the influence of vehicle speed on dynamic deflection of curved composite box girder increases. When the vehicle speed is less than 50 km/h, the dynamic deflection of curved composite box girder changes little with the deterioration of bridge deck grade. Compared with the speed of 50 km/h and the bridge deck grade *A*, the mid-span dynamic deflection of the curved composite box girder increases by 19.4% when the speed is 60 km/h and the bridge deck grade is *D*.

5   CONCLUSIONS

This paper analyzes the vehicle-bridge coupling vibration response of separated steel-concrete continuous curve composite box girder, and studies the influence of driving position, transverse connection, and vehicle speed on the eccentric load effect of separated curve composite box girder, which provides reference for the design of related box girder bridges.

(1) Under the action of vehicles, the influence of vehicle position on the mechanical response of separated curve composite box girder is complex. When the vehicle runs along the outer box girder, the dynamic deflection of the outer box girder is the largest, and the eccentric load effect of the curved composite box girder is obvious. With the increase of longitudinal spacing, the maximum dynamic deflection of the mid-span box girder gradually decreases. The vehicle should avoid driving along the outside of the curved bridge as far as possible when driving on the separated curved composite box girder bridge.

(2) The curvature radius of box girder is the smaller, the dynamic deflection of mid-span section is the larger and the increase is the more obvious. The lateral connection improves the eccentric load effect of the inner and outer box girders and strengthens the connection between the inner and outer box girders. The transverse connection arrangement improves the eccentric load effect of the inner and outer box girders and strengthens the connection between the inner and outer box girders.

(3) With the deterioration of bridge deck condition, the influence of vehicle speed on dynamic deflection of separated curve composite box girder is obvious. When the bridge deck condition is the worst, the mid-span deflection caused by vehicle driving at 60 km/h increases by nearly 20 % compared with that at 30 km/h. During the operation of small radius bridge, it is necessary to carry out regular inspection and maintenance of bridge deck and speed limit of driving vehicles.

(4) This paper only studies the dynamic response of separated curved composite box girder under vehicle action. In future research, it is suggested to explore the influence and optimization of the form and number of connections between box and box on the eccentric load effect.

ACKNOWLEDGMENT

This research was financially supported by the Key R&D Program of Hebei Province (Grant Number:19275405D).

REFERENCES

Chen Shuisheng & Luo Hao & Gui Shuirong. (2020). Dynamic response of continuous curved girder bridge under multi-vehicle load J. Journal of Jiangsu University (Natural Science Edition), 41 (01): 118-124.

Deng Lu & He Wei & Yu Yang, et al. (2018). Research progress on theory and application of vehicle-bridge coupled vibration of highway J. Journal of China Highway, 31 (07): 38-54.

Huang Qiao. Bridge (2004). steel-concrete composite structure design principle M. China Communications Press.

Huang Xinyi & Zhuo Weidong & Sheng Hongfei. (2010). Effect of vehicle eccentric driving on vehicle-bridge coupling vibration of curved girder bridge J. Journal of Wuhan University of Technology, 32 (09): 318-322.

Huang Xinyi. (2008). Research on dynamic response of concrete continuous curved girder bridge under vehicle load D. Harbin University of Technology.

V. Christian. (1988). Box girder theory M. China Communications Press: 161-193.

Yang Xiaotian. (2019). Indirect identification of continuous girder bridge frequency under uneven bridge deck D. Shandong Jianzhu University.

Zhang Yanling, et al. (2015). The flexural-torsional performance of diaphragm steel-concrete curve composite beams J. Journal of Jilin University (Engineering Edition), 45 (04): 1107-1114.

Zhong Chengxing. (2020). Vehicle-bridge coupled vibration analysis of continuous rigid frame box girder bridge with corrugated steel webs under earthquake D. Chang'an University.

Zhou Changcun. (2016). Study on vehicle-bridge coupled vibration of externally prestressed steel-concrete composite continuous box girder D. Fuzhou University.

Zhou Liang & Liu Zhiwen & Yan Xingfei, et al. (2018). Curved girder bridge structure design theory and analysis method M. Beijing: China Communications Press.

Zhuang Shusen. (2017). Experimental and theoretical study on bending-torsional coupling of curved double box girders D. China University of Mining and Technology.

Frontiers of Civil Engineering and Disaster Prevention and Control – Yang & Rahman (Eds)
© 2023 The Author(s), ISBN: 978-1-032-31200-2

The effect of corner radius on the local buckling of cold-formed steel tubes

Zhi Xiao Wu*
Central-South Architectural Design Institute Co., Ltd, Wuhan, China

Gong Wen Li*
Central-South Architectural Design Institute Co., Ltd, Wuhan, China
School of Urban Construction, Wuhan University of Science and Technology, Wuhan, China

Wen Jun Jing* & Han Ji*
Central-South Architectural Design Institute Co., Ltd, Wuhan, China

ABSTRACT: Nowadays, with the improvement of manufacturing technique and facilities, it is no longer a problem to manufacture cold-formed sections with wall-thickness greater than 6 mm. With the increase of wall thickness, the corner radius became larger and thus might affect the local buckling behavior of the steel plate. Thus, the effect of corner radius on the local buckling behavior of rectangular steel tubes was studied and the limitation of width-to-thickness ratio was proposed. The proposed formulas were compared with the stipulation in the design codes. It was concluded that the definition of the width of the flat plate in AISI S100 and AS/NZS 4600 might predict lower results, the outline dimension B was recommended for the calculation of width-to-thickness ratio, the limitation of width-to-thickness ratio in the Chinse and Eurocode code might lead to unsafe design results when $\lambda_n < 0.55$.

1 INTRODUCTION

Nowadays, with the improvement of manufacturing technique and facilities, it is no longer a problem to manufacture cold-formed sections with wall-thickness greater than 6 mm. As we know, cold-forming process has a significant effect on the material properties and would induce residual stress in cold-formed steel sections, and thus influences the bearing capacity of cold-formed steel members. For this reason, many researches have been carried out on the material properties, residual stress and bearing capacity of cold-formed thick-walled steel members for the past several decades.

Hou (2011) conducted a series of tests on the material properties and stub columns of 6 cold-formed square steel tubes. The strength grade was of 235MPa or 345MPa, and the plate thickness ranged from 8 to 16 mm. The test results showed that the yield stress of the flat parts and the corner parts increased by 14% and 38% respectively compared with the virgin material. On the basis of test results, Hou (2011) proposed the material model for the flat and corner coupons. Guo (2007), Hu (2010), Wang (2011), Li (2015), Shen (2016) and Wen (2016) also conducted a series of tests on the material properties and stub columns of cold-formed square and rectangular steel tubes. The strength grade was of 235MPa or 345MPa. The plate thickness of the specimens ranged from 4 to 16 mm. The test results showed that there was no yield plateau for most of the coupons especially for the corner coupons and the yield stress of the corner coupons increased by 12%–54% compared with the flat coupons. A distribution model of yield and tensile strength along the cross section was also proposed by Shen (2016).

*Corresponding Authors: 66146794@qq.com, lgw19890806@163.com, 272890091@qq.com and csadi_jih@zonaland.cn

Except for the material properties, theoretical and experimental studies on the residual stress of cold-formed steel sections were conducted by several scholars. Distribution models of longitudinal residual stress for "direct forming" and "round to square forming" square hollow sections were proposed by Tong (2012). It was shown that the residual stress could be as large as 66% of the yield strength for flat plates. Li (2019) also proposed the distribution models of bending residual stress and membrane residual stress based on the experimental investigation on the cold-formed steel tubes using the sectioning method.

For the bearing capacity of cold-formed steel members with wall thickness larger than 6 mm, Li (2018) conducted the tests on the overall stability behavior of axially compressed cold-formed steel columns with square and rectangular hollow sections. The calculation formula of overall stability coefficient was proposed and the column curves were compared with that of common steel members and cold-formed thin-walled steel members. It was shown that the column curves of cold-formed thick-walled steel members were much different.

The effective width method and direct strength method were the most commonly used methods for the design of cold-formed steel members in the design codes (AISI-S100 2016, AS/NZS- 4600 2005, GB-50018 2002). The width-to-thickness ratio was an important parameter in the design methods. However, with the increase of wall thickness, the corner radius became larger and thus might affect the local buckling behavior of the steel plate. Therefore, the effect of corner radius on the local buckling behavior of rectangular steel tubes will be studied and the limitation of width-to-thickness ratio will be proposed on this basis.

2 CRITICAL STRESS OF STEEL PLATES

2.1 Theoretical derivation

For a simple supported thin plate subjected to uniform pressure load N_x at both ends (Figure 1), assume that the plate buckles when the load reaches to the critical buckling load and the deflection in the middle is w.According to the elastic theory, the buckling equilibrium differential Formula of the plate can be given as

$$D\left(\frac{\partial^4 w}{\partial x^4} + 2\frac{\partial^4 w}{\partial x^2 \partial y^2} + \frac{\partial^4 w}{\partial y^4}\right) + N_x\frac{\partial^4 w}{\partial x^2} = 0 \tag{1}$$

where D is the bending stiffness per unit width of the plate and it is given as

$$D = \frac{Et^3}{12(1 - v^2)} \tag{2}$$

where E is the elastic modulus, t is the thickness of the plate and v is the Poisson's ratio which is taken as 0.3 for steel.

For simple supported plate, the solution of formula (1) can be given as

$$w = \sum_{m=1}^{\infty}\sum_{n=1}^{\infty} A_{mn} \sin\frac{m\pi x}{a} \sin\frac{n\pi y}{b} \tag{3}$$

where m is the number of half-waves along x direction, n is the number of half-waves along y direction, a is the length along x direction, b is the length along y direction. Substitute formula (3) into formula (1), the critical value of N_x can be got as

$$N_{xcr} = \frac{\pi^2 D}{b^2}\left(\frac{mb}{a} + \frac{n^2 a}{mb}\right)^2 \tag{4}$$

N_{xcr} comes to the minimum value when n = 1. Then the critical value of N_x comes to

$$N_{xcr} = \frac{\pi^2 D}{b^2}\left(\frac{mb}{a} + \frac{a}{mb}\right)^2 \tag{5}$$

Define the stability coefficient k as

$$k = \left(\frac{mb}{a} + \frac{a}{mb}\right)^2 \tag{6}$$

Then

$$N_{\mathrm{xcr}} = k\frac{\pi^2 D}{b^2} \tag{7}$$

Substitute Formula (2) into Formula (7), then the critical stress can be given as

$$\sigma_{\mathrm{xcr}} = k\frac{\pi^2 E}{12(1 - v^2)}\left(\frac{t}{b}\right)^2 \tag{8}$$

$k = 4$ for thin plate with 4 simple supported edges.
    For elasto-plastic buckling

$$\sigma_{\mathrm{xcr}} = \sqrt{\eta}\,k\frac{\pi^2 E}{12(1 - v^2)}\left(\frac{t}{b}\right)^2 \tag{9}$$

where η is the reduction factor of elastic modulus, $\eta = E_t/E$, E_t is the tangent modulus.

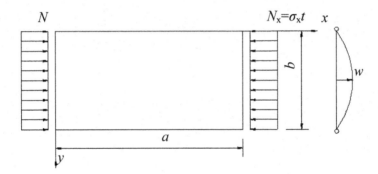

Figure 1.   Definition of symbols.

2.2   FEA verification

Finite element analysis (FEA) model was built to verify the theoretical calculation formula and to check the computational accuracy of FEA. A square steel plate with $a = 1.0$ m, $b = 1.0$ m and $t = 0.01$ m was chosen as the analytical object. The ANSYS software was used to conduct the FEA. The elastic material model was used and the elastic modulus E was set as 2.06×10^5 MPa, v was set as 0.3. Shell 181 element was used to mesh the area and the element division along the four sides was 20. The boundary condition was that the displacement along Z direction at the 4 edges was set as zero, the displacement along X, Y direction at the center of the plate was set as zero. The uniform pressure load was applied to the two vertical edges. The FEA model was shown in Figure 2. The elastic buckling analysis was conducted and the Z direction deformation of the steel plate for the first buckling mode was shown in Figure 3 (cloud map). It was shown that the deformation at the center was the largest which corresponded to the peak point of half-wave. The critical load N_{xcr} calculated by Formula (7) was 744.7 kN. It could be seen from Figure 3 that the critical load of FEA (N_{xcre}) was 746.3 kN. $N_{\mathrm{xcre}}/N_{\mathrm{xcr}}=1.002$, which indicated that the result of FEA was very close to the theoretical solution.

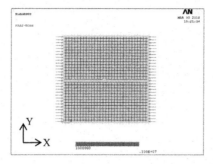

Figure 2.   FEA model of the steel plate.

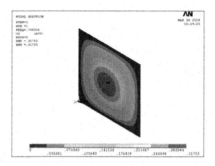

Figure 3.   Deformation of the steel plate for the first buckling mode.

3   EFFECT OF CORNER RADIUS ON THE LOCAL BUCKLING

The typical section of cold-formed thick-walled square and rectangular steel tubes was shown in Figure 4 where B was the depth of the section, W was the width of the section, b was the length of flat plate, R was the outer radius of the corner, t was the wall thickness. The calculation formula of the nominal outer radius provided by Chinese code (JG/T-178, 2005) was given as:

$$R = 2.0t \quad (t \leq 6) \tag{10.1}$$

$$R = 2.5t \quad (6 < t \leq 10) \tag{10.2}$$

$$R = 3.0t \quad (t > 10) \tag{10.3}$$

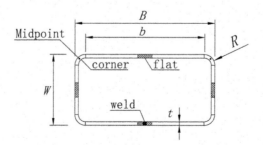

Figure 4.   Typical section.

For sections with thickness of 16 mm, R could be 48 mm. Thus, the effect of corner radius on the local buckling could not be ignored. In AISI-S100 (2016) and AS/NZS-4600 (2005), the width of the flat plate was taken as b for the calculation of width-to-thickness ratio. In BS EN1993-1-3 (2006), the width of the flat plate was measured from the midpoints of the adjacent corners as indicated in Figure 4. There was no unified conclusion about the effect of corner radius on the local buckling of cold-formed steel sections. As was introduced above, the FEA method had high accuracy for the calculation of local buckling. FEA model of square steel tubes with wall thickness of 4 mm, 6 mm, and 16 mm were built in the ANSYS software. The parameter b/t ranged from 12.5–200 and the length of the tube L was taken as b. According to formulas (10)–(12), R was taken as 8 mm, 20 mm, and 48 mm respectively. The element division of flat plates was 25 and element division was 5 for the corners. The displacement along Z direction and the displacement vertical to the surface was restricted at the bottom of the tube. The displacement vertical to the surface was also restricted at the top of the tube. Uniform pressure load along Z direction was applied to the edges at the top. The typical FEA model was shown in Figure 5 and the corresponding FEA result was shown in Figure 6. According to Formula (7), the equivalent width b_e could be derived as

$$b_{\mathrm{e}} = \sqrt{\frac{k\pi^2 D}{N_{\mathrm{xcre}}}} \tag{11}$$

where N_{xcre} was the critical load of FEA. k was taken as 4 for plate with 4 simple supported edges. The calculation result of b_e for each FEA model was listed in Table 1. It was shown in Table 1 that b_e was larger than b for all the specimens especially when b/t was small. Therefore, it was not proper to take the width of the flat plate as b for the calculation of width-to-thickness ratio as it might predict lower results. Table 1 also listed the calculation result of b_e/B. The relationship between b_e/B and b/t was shown in Figure 7. It could be seen form Table 1 and Figure 7 that b_e was smaller than B when $b/t < 25$ and was slightly larger than B when $b/t > 75$. On the whole, b_e was very close to B. Thus, it was more reasonable to take the width of the flat plate as B for the calculation of width-to-thickness ratio.

Table 1. Calculation result of be.

N_{xcre} (kN)	b (mm)	b_e (mm)	b_e/B	N_{xcre} (kN)	b (mm)	b_e (mm)	b_e/B	N_{xcre} (kN)	b (mm)	b_e (mm)	b_e/B
	t=4 mm				t=8 mm				t=16 mm		
11365	50	64.8	0.981	20941	100	134.9	0.964	39297	200	278.6	0.941
3564	100	115.6	0.997	6734	200	238.0	0.991	12869	400	486.9	0.982
1727	150	166.1	1.001	3307	300	339.6	0.999	6378	600	691.6	0.994
1017	200	216.5	1.002	1962	400	440.8	1.002	3808	800	895.1	0.999
669	250	266.9	1.003	1299	500	541.7	1.003	2548	1000	1094.1	0.998
474	300	317.1	1.004	929	600	640.7	1.001	1814	1200	1296.7	1.001
355	350	366.5	1.001	693	700	741.6	1.002	1358	1400	1498.9	1.002
274	400	416.9	1.002	537	800	842.4	1.003	1054	1600	1700.8	1.003
218	450	467.2	1.003	429	900	943.2	1.003	843	1800	1902.5	1.003
178	500	517.5	1.003	350	1000	1043.8	1.004	689	2000	2104.0	1.004
148	550	567.8	1.003	291	1100	1144.4	1.004	574	2200	2305.4	1.004
125	600	618.0	1.003	246	1200	1245.0	1.004	485	2400	2506.6	1.004
107	650	668.3	1.003	211	1300	1345.5	1.004	416	2600	2707.8	1.004
92	700	718.5	1.004	182	1400	1446.0	1.004	361	2800	2908.8	1.004
81	750	768.7	1.004	159	1500	1546.5	1.004	315	3000	3109.8	1.004
71	800	819.0	1.004	141	1600	1647.0	1.004	278	3200	3310.8	1.004

Note: $B = b + 2R$.

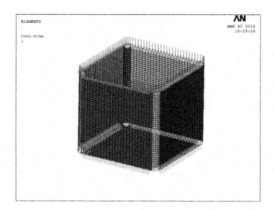

Figure 5. Typical FEA model.

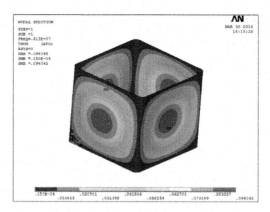

Figure 6. Typical FEA result.

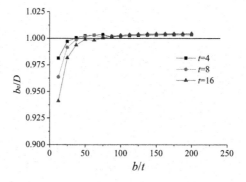

Figure 7. Relationship between b_e/B and b/t.

4 LIMITATION FOR WIDTH-TO-THICKNESS RATIO

For axial compression members that local buckling was not allowed, the design criteria were that the critical local buckling stress should not be smaller than the critical overall stability stress, that is

$$\sqrt{\eta}k\frac{\pi^2 E}{12(1-\nu^2)}(\frac{t}{b})^2 \geq \phi f_y \tag{12}$$

where φ was the overall stability coefficient, f_y was the yield strength of the steel plate. Then the width-to-thickness ratio could be derived as

$$\frac{b}{t} \leq \left[\frac{\sqrt{\eta} k \pi^2 E}{12(1-\nu^2)\phi f_y} \right]^{\frac{1}{2}} \tag{13}$$

According to the tangent modulus theory (Gerard 1962), the critical overall stability stress could be given as

$$\sigma_{cr} = \frac{E_t}{E} \frac{\pi^2 E}{\lambda^2} = \frac{\eta \pi^2 E}{\lambda^2} = \frac{\eta f_y}{\lambda_n^2} \tag{14}$$

where λ_n was given as

$$\lambda_n = \frac{\lambda}{\pi} \sqrt{f_y/E} \tag{15}$$

then η could be derived as

$$\eta = \frac{\sigma_{cr}}{f_y} \lambda_n^2 = \phi \lambda_n^2 \tag{16}$$

The authors have studied the overall stability behavior of cold-formed thick-walled square and rectangular steel tubes in reference (Li 2018), and have proposed the calculation formula of φ which is given as

$$\text{For } \lambda_n \leq 0.215 \qquad \phi = 1 - \alpha_1 \lambda_n^k \tag{17}$$

$$\text{For } \lambda_n > 0.215 \qquad \phi = \frac{1}{2\lambda_n^2} [(\alpha_2 + \alpha_3 \lambda_n + \lambda_n^2) - \sqrt{(\alpha_2 + \alpha_3 \lambda_n + \lambda_n^2)^2 - 4\lambda_n^2}] \tag{18}$$

Parameter α_1, α_2, α_3 and k were shown in Table 2.

Table 2. Values of α_1, α_2, α_3 and k.

Strength grade	α_1	α_2	α_3	k
Q235	0.2322	1.0215	0.2545	0.7504
Q345	0.2093	1.0333	0.2299	0.6294

The limitation of width-to-thickness ratio could be got by substituting Formula (16) into Formula (13). The correlation between b/t and $\lambda_n \sqrt{235/f_y}$ was shown in Figure 8. It was shown in Figure 8 that b/t was positively proportional to $\lambda_n \sqrt{235/f_y}$. A linear fitting of the curves was conducted and the fitted line was shown in Figure 8. Then the simplified calculation formula for the limitation of b/t could be given as

$$b/t \leq (10.65 + 52.35\lambda_n)\sqrt{235/f_y} \tag{19}$$

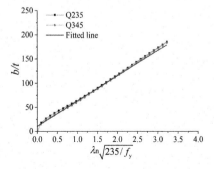

Figure 8. Correlation between b/t and $\lambda_n \sqrt{235/f_y}$ for axial compression members.

5 CONCLUSION

The effect of corner radius on the local buckling behavior of rectangular steel tubes was studied and the limitation of width-to-thickness ratio was proposed. The main conclusions were drawn as follows:

(1) The corner radius had a significant effect on the local buckling of flat plates. The definition of the width of the flat plate in AISI S100 and AS/NZS 4600 might predict lower results.
(2) The outline dimension B was recommended for the calculation of width-to-thickness ratio for cold-formed rectangular steel tubes.

6 LIMITATIONS AND RECOMMENDATIONS FOR FUTURE RESEARCH

It is worth noting that the conclusion about the effect of corner radius here is mainly based on the FEA results while there are no corresponding tests. Tests on stub columns of cold-formed sections should be conducted to validate the FEA results in the future.

Also, the residual stress and initial geometric imperfection may have an effect on the local buckling behavior which also needs further investigation in the future.

7 ACKNOWLEDGEMENTS

The research was supported by the Educational Commission of Hubei Province of China (No. Q20211105). Any opinions, findings, and conclusions or recommendations expressed in this article are those of the authors and do not necessarily reflect the views of the sponsors.

REFERENCES

AISI S100. (2016). North American specification for the design of cold-formed steel structural members. Washington, D.C.

AS/NZS 4600. (2005). Australian/New Zealand Standard (AS/NZS). Cold-formed steel structures. Sydney, Australia.

BS EN1993-1-3. (2006). Eurocode 3–Design of Steel Structures–Part 1-3: General rules–Supplementary rules for cold-formed members and sheeting. Brussels, Belgium: European Committee for Standardization.

GB 50018. (2002). Technical code of cold-formed thin-walled steel structures. Peking, China. (in Chinese)

Gerard. (1962). Introduction to Structural Stability Theory. McGraw-Hill, New York.

Guo, Y.J., Zhu, A.Z., Yong, L.P. et al. (2007). Experimental study on compressive strengths of thick-walled cold-formed sections. Journal of Constructional Steel Research. 63(2007): 718–723.

Hou, G. (2011). Experimental research and numerical analysis on axial compression performance of cold-formed non-thin-walled square hollow sections (Doctoral thesis). Tongji University, Shanghai, China. (in Chinese)

Hu, S.D., Li, L.X., Zhou, J.L. et al. (2010). Comparative analysis on strain hardening of thick-wall cold formed steel tube with square and rectangular hollow section. Journal of Materials Science & Engineering. 28(1): 76–80. (in Chinese)

JG/T 178. (2005). Cold-formed steel hollow sections for building structures. Peking, China. (in Chinese)

Li, G.W., Li, Y.Q., Xu, J. et al. (2019). Experimental investigation on the longitudinal residual stress of cold-formed thick-walled SHS and RHS steel tubes. Thin-Walled Structures. 138(2019): 473–484.

Li, G.W., Li, Y.Q. (2018). Overall stability behavior of axially compressed cold-formed thick-walled steel tubes. Thin-Walled Structures. 125(2018): 234–244.

Li, Y.Q., Li, G.W., Shen, Z.Y. et al. (2015). Modification method for yield strength of cold-formed thick-walled steel sections considering cold-forming effect. Journal of Building Structure. 36(5):1–7. (in Chinese)

Shen, Z.Y., Wen, D.H. Li, Y.Q. et al. (2016). Distribution patterns of material properties for cross-section of cold-formed thick-walled steel rectangular tubes. Journal of Tongji University (Natural Science). 44(7): 981–990. (in Chinese)

Tong, L.W., Hou, G., Chen, Y.Y. et al. (2012). Experimental investigation on longitudinal residual stresses for cold-formed thick-walled square hollow sections. Journal of Constructional Steel Research. 73(2012): 105–116.

Wang, L.P. (2011). Experimental investigation on cold-forming effect of thick-walled steel sections (Master degree thesis). Tongji University, Shanghai, China. (in Chinese)

Wen, D.H., Shen, Z.Y., Li, Y.Q. et al. (2016). Experimental research on cold-formed thick-walled steel box tubes and comparison of results with related codes. Journal of Tongji University (Natural Science). 44(8): 1190–1198. (in Chinese)

Frontiers of Civil Engineering and Disaster Prevention and Control – Yang & Rahman (Eds)
© 2023 The Author(s), ISBN: 978-1-032-31200-2

Experimental study on the mechanical properties of lead rubber bearings in acid rain environment

JiYu Xia*

College of Civil Engineering, Xi'an University of Architecture and Technology, Xi'an, China

ABSTRACT: In order to study the law of the mechanical properties of lead rubber bearing (LRB) after being corroded by acid rain, the artificial acid rain corrosion solution with pH of 4.5 and 3.0 was used to conduct corrosion tests on 9 LRBs with a diameter of 300 mm for a total duration of 92 days. During the test period, the basic properties of LRB were tested regularly, and the test results before and after the experiment were compared. Through the analysis of the results, the law of the mechanical properties of the LRB with the increase of acid rain corrosion time was obtained. The test results show that, the effect of acid rain corrosion on the vertical compression stiffness of LRB is about 10%. The effect of acid rain corrosion on the compressive performance of LRB is not obvious. The yield force (Q_d), and horizontal equivalent stiffness (K_h), the equivalent damping ratio (h_{eq}) of the LRB, will decrease at first and then increase with acid rain corrosion time. When the shear strain is 100%, the maximum decrease of Q_d are 15.483%, 21.827%, and 14.388%, respectively. When the acid rain is strong, the degradation of K_h is more obvious. The Q_d K_h h_{eq} of the unprotected support decreases faster, indicating that the shear performance of LRB is more sensitive to the corrosive effect of acid rain. Actually, we found the degree of shear performance degradation is strongly influenced by the acidity, the presence of the protective layer.

1   INTRODUCTION

As the lead rubber bearing (LBR) has the advantages of simple manufacture, convenient replacement and excellent seismic isolation effect, it is widely used in key parts of houses and bridges, to connect the upper structure and the lower structure (Li 2021, 2007; Liu 2021). In actual use, the lead rubber bearing is under the combined action of load and environmental factors for a long time, which may lead to the deterioration or even failure of the bearing performance, which seriously reduce the reliability of the seismic isolation system. There are many environmental factors that cause the failure of lead rubber bearing, among which the corrosion of lead rubber bearing by acid rain in the atmosphere has gradually become a new issue. In recent years, with the acceleration of industrialization and urbanization, industrial activities and automobile exhaust produce a large amount of CO_2, SO_2, H_2S and other polluting gases. After these polluting gases are discharged into the atmosphere, chemical reactions would generate sulfuric acid, nitric acid and other secondary pollutants, and finally they are transferred to the ground through rain, snow, fog and other forms, forming the acid rain is formed (Niu 2008; Yu 2021; Zhang 2010). The long-term contact between acid rain and lead rubber bearing will lead to corrosion of the metal part of the bearing (Zheng 2021; Zu 2020). At the same time, acid rain will gradually penetrate through the protective layer and then react with the natural rubber and steel plate inside the bearing, causing the deterioration of the bearing performance.

At present, some scholars at have conducted studies on the degradation of rubber material properties under acid rain environment. The rubber material will be corroded by factors such as ultraviolet rays, ozone and acid rain at room temperature, which will lead to the degradation of its physical properties. Itoh Y et al (2005) and Gu et al (2010) found that the rubber on the bearing

---

*Corresponding Author: xiajiyu123@gmail.com

DOI 10.1201/9781003308577-56

surface gradually hardens and the tensile strength slowly decreases with the increase of acid rain spraying time. And on this basis, a method to predict the performance of rubber vibration isolation bearings after acid rain corrosion was determined and the relevant verification is carried out. Liu S J et al (2014) carried out a study on the performance of natural rubber bearings in various erosive environments such as sulfuric acid, nitric acid, damp heat, freeze-thaw, etc., obtained the change law of bearing compressive performance and shear performance, and established decay models of bearing elastic modulus considering the effect of acid rain erosion. Zhang Y N et al (2014) adopted the method of complete immersion in acid rain solution with pH=4.5 to accelerate the corrosion of the chloroprene rubber bearing, and gave the change of the bearing's ultimate strength after immersion for 100 days. Si Y H et al (2013) evaluated the change of compressive properties of neoprene bearing by acid rain accelerated corrosion test and established the decay model of compressive elastic modulus of neoprene bearing. It is found that the objects of the current researches on the durability of rubber isolation bearings in acid rain environment are mostly laminated rubber bearings.

Through summarizing the existing research results, it is easy to find that the research objects of the durability of rubber isolation bearings under acid rain environment at domestic and overseas are mostly laminated rubber bearings, while the research on the most widely used in engineering, the lead rubber bearing's durability under acid rain environment is still very limited. In this paper, 9 comparative acid rain solution corrosion tests of lead rubber bearings are designed to simulate different corrosion of LBRs by acid rain in the actual engineering. By testing the compressive properties and shear properties of the lead rubber bearings before and after the test, the change law of the mechanical properties of the lead rubber bearings under acid rain corrosion was obtained.

2 TEST OVERVIEW

2.1 The specimen designs

In order to study the law of performance deterioration of lead rubber bearing in acid rain environment, 9 lead rubber bearings with an effective diameter of 300mm were designed. Considering that the protective layer of the bearing may be damaged and fail in extreme environments, the specimens were divided into 3 groups, the first 2 groups with protective layer and the last 1 group without protective layer, and 3 bearings in each group. The construction parameters of the 9 lead rubber bearings are exactly the same except whether they have a protective layer or not, and you can find the parameters in Table 1.The bearings are processed by Shaanxi Yongan Damping Technology Co, and the mechanical property tests of the bearings are conducted at Xauat Engineering Technology Co.

2.2 Test method and procedure

The main anions in atmospheric precipitation are SO_4^{2-}, NO^- and Cl^-, etc., while the cations, on the other hand, are NH_4^+, Ca^{2+}, Na^+, H^+ etc., in the acid rain according to the long-term observations of atmospheric precipitation in Shanghai, Jiangsu, Chongqing and Guangzhou by the Chinese meteorological departments. In which, except for a few areas, the ratio of $[SO_4^{2-}]/[NO^-]$ in the acid rain decreased in the past few years, but in general, the acid rain in China is still sulfuric acid rain. The acidity of the acid rain is mainly determined by the concentration of free H^+ and SO_4^{2-} in the atmospheric precipitation.

According to the relevant reports on the actual situation of acid rain in the Yangtze River Delta region, the annual acid rain rainfall in such region is estimated. In the past 5 years, the region has been experiencing about 22 days of heavier acid rain on average per year. The type of acid rain is generally sulfuric acid, and the pH of acid rain is roughly ranging from 4.5 to 5.0, in some areas the pH value has reached as low as 2.85 (Cai 2018; Du 2015; Xuan 2021). In this paper, sulfuric acid, sodium chloride, sodium nitrate and distilled water were used to configure the acid rain corrosion solution, and the pH of the solution is 4.5 and 3.0. According to the study of Zhang Y N et al (2013), it can be considered that the results of 22 days of acid rain solution corrosion at room temperature are equivalent to the acid rain corrosion of lead rubber bearings in actual engineering within one year. After the test commenced, the pH of the solution was tested once a day and the pH of the acid rain solution was adjusted using sulfuric acid with a concentration of 30% to keep its acidity was stable; the acid rain solution was replaced every 3 days to ensure that the solution always had a high transparency to facilitate observation of the change in the appearance of the support.

Table 1. Parameters of lead rubber bearing.

Bearings Type	Number	Thickness of protective layer/mm	Thickness of rubber layer (layers)/mm	Thickness of steel plate (layers)/mm	Diameter of lead/mm	modulus of elasticity/ $N \cdot mm^{-1}$
With protective	1#, 2#	10	2.875(16)	2(15)	60	0.388
Without protective	3#	0				

In this study, nine lead rubber bearings were subjected to acid rain corrosion tests for up to 92 days, and the test placement pattern is shown in Figure 1. Before the test started, the basic performance of the lead rubber bearings was tested first, and then the bearings were soaked in the artificial acid rain test chamber for 22 days, and after 22 days, the bearings were taken out and rinsed with water, and then they were dried in the constant temperature room for 1 day, and this was taken as a cycle, and the cycle was repeated 4 times until the test was completed at 92 days. The ambient temperature is kept constant at 20°C throughout the test. The experiment conditions are shown in Table 2.

Table 2. Test parameters and content of the corrosion of acid rain for lead rubber bearings.

Number	pH of acid rain	Total test time/day	The corrosion cycle/day	Test content
1#	4.5	92	23	Test of fundamental performance
2#	3.0			
3#	4.5			

Figure 1. A photograph of LRB soaking in acid rain.

Since the sulfuric acid in the acid rain solution will react with the sealing steel plate and the inner steel plate of the lead-core rubber support as shown in formula (1), the pH of the solution will rise rapidly. After the test starts, the pH of the solution is tested once a day. In the test, dilute sulfuric acid with a concentration of 30% was used to adjust the pH, so that the acidity of the acid rain was always kept in a relatively stable state; the solution was replaced every 3 days to ensure that the solution always had a high transparency, to keep it convenient to observe the appearance of the bearings.

$$\begin{cases} Fe_2O_3 + 3H_2SO_4 \rightarrow Fe_2(SO_4)_3 + 3H_2O \\ Fe + H_2SO_4 \rightarrow FeSO_4 + H_4 \uparrow \end{cases} \tag{1}$$

2.3 Basic performance tests

During the acid rain corrosion test, the basic mechanical properties of the nine lead rubber bearings were tested according to the sampling time nodes, and the loading system was based on the national specification GB/T 20668.1-2007, as shown in Table 3.

Table 3. Basic mechanic properties of lead rubber bearings.

Parameters	Vertical load/kN	Compressive stress/MPa	Horizontal displacement/mm	Shear strain/%	Frequency/Hz
Testing of Compressive properties	707	10	0	0	0.03
Testing of shear properties	707	10	±46	±100	0.05

3 ANALYSIS OF THE TEST RESULTS

3.1 Change in the appearance of the bearings

The appearance of the 3 groups of lead rubber bearings after 92 days of acid rain corrosion is shown in Figure 2. It can be seen from the Figure 2 that after being corroded by acid rain, the exposed metal part of the lead rubber bearings got rusted in a short time. As the corrosion time increase, the protective layer rubber gradually turned white from the original classic rubber black; while the surface became sticky and less smooth. Purely from the appearance, the rubber of the nine bearings did not appear cracking, or delamination.

In addition, after a period of acid rain corrosion, the surface of the lead core gradually becomes white, and the original metallic bright silver color of the lead core surface becomes completely white in about 46 days. Then the color of the lead core no longer changes with the time of acid rain corrosion, and the time taken of the lead core becoming white under the two pH acid rain solutions is about the same. A small amount of white powder was scraped on the surface of the lead core with a scraper and tested, he main component of the white powder was lead sulfate. The chemical reaction of formula (2) occurred on the surface of the lead core under the corrosion of acid rain.

$$Pb + H_2SO_4 \rightarrow PbSO_4 + H_2 \uparrow \tag{2}$$

3.2 Compression properties

Considering the manufacturing error of the bearing itself, the vertical compression stiffness at moment t is noted as K_{vt}, the vertical compression stiffness at the initial moment noted as K_{v0}. The variation law of K_{vt}/K_{v0} with time increasing is shown in Figure 3.

As can be seen from Figure 3, with the time increase of acid rain corrosion, the compression stiffness of lead rubber bearing rises first and then falls; among them, the maximum values of compression stiffness ratio of bearing with and without protective layer are 1.187, 1.437 and 1.06, and the minimum values are 1.07, 1.034 and 1.025 respectively; within 46 days after the beginning of the test, the compression stiffness of two groups of bearing with protective layer is in the rising The maximum increase was 18.698% to 43.718%, indicating that the influence of acid rain corrosion

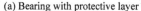

(a) Bearing with protective layer

(b) Bearing without protective layer

Figure 2.   Appearance of lead rubber bearing after acid rain corrosion test.

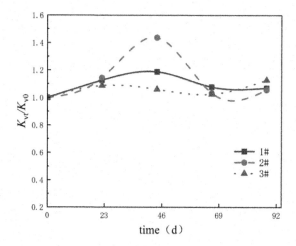

Figure 3.   Curve of K_{vt}/K_{v0}.

on the compression performance of the lead rubber bearing was greater in the short term, and the influence was more obvious when the acidity of acid rain was greater; while the compression stiffness of the bearing without protective layer increased only in the first corrosion cycle, and the increase was significantly lower than that of the bearing with protective layer, which was only 8.774%; the compression stiffness of the two groups of bearings After rising to the maximum value will begin to decline, but this decline is not infinite, after four cycles of acid rain corrosion test, the compression stiffness of the specimen is basically stable. Generally speaking, after a total duration of 92 days (about equivalent to 4 years of acid rain encountered in the actual environment of the region) of acid rain corrosion, the compression stiffness of the lead rubber bearing has a certain increase compared with the initial value, and the increase is about 10%, indicating that the long-term acid rain has little effect on the compression performance of the lead rubber bearing.

Liu S J et al (2014) found that the compression stiffness of natural rubber bearings decreased exponentially after acid rain corrosion. In this study, the compressive stiffness of the lead rubber bearing first increased after the acid rain corrosion test and then began to decrease. This phenomenon may be due to the fact that during the corrosion test, the acid rain solution penetrated into the gap between the bearing and the lead core and contacted with the lead core, and the chemical reaction generated $PbSO_4$, which caused the increase of the compressive stiffness of the bearing; as the chemical reaction proceeded, the generated $PbSO_4$ gradually covered the surface of the lead core, forming a protective layer and preventing the further chemical reaction, and the color change of the surface of the lead core with the increase of the test time can support this speculation; on the other hand, due to the existence of the protective layer of rubber, the solution penetrated into the interior of the bearing at a slower rate, and when the solution entered the interior of the bearing and reacted with the internal rubber and steel plate, and when the rate of decrease in compression stiffness caused by the reaction between acid rain and the interior of the bearing was greater than the rate of increase caused by the reaction with the lead core, the compression stiffness of the lead core rubber bearing showed a decreasing trend; and for the specimen without the protective layer, the acid rain solution could directly corrode the rubber and steel plate inside the bearing, so the decreasing stage of its compression stiffness appeared earlier than that of the specimen with the protective layer.

3.3 *Shear properties*

The yield force (Q_d), the post-yield stiffness (K_d), the horizontal equivalent stiffness (K_h) and the equivalent damping ratio (h_{eq}) of the lead rubber bearing changing with the corrosion time are shown in Figure 4.

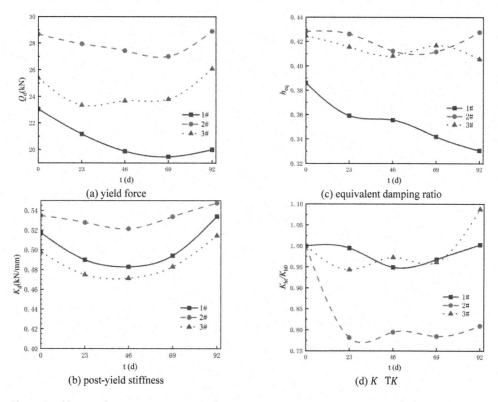

Figure 4.   Shear performance change curve of lead core rubber bearing.

From Figure 4 (a), it can be seen that the yield force of lead rubber bearing decreases and then increases with the increase of acid rain corrosion time, in which the yield force of two groups of bearing with protective layer has a similar change pattern, both dropping to the lowest value when the length of immersion is about 69 days, 19.467 kN and 27.019 kN respectively, with a decrease of 15.483% and 5.760% respectively, and the decrease is greater when the acid rain is less acidic. When the acid rain acidity is low, the magnitude of the reduction is greater; the yield force of the bearing without the protective layer drops to the lowest value in 23 days, the magnitude of the reduction is 8.066%, the pH of the acid rain is the same, the yield force of the bearing with the protective layer deteriorates more; after 92 days of acid rain corrosion test, the yield force of 1# bearing is only 86.806% of the initial value respectively, indicating that the influence of acid rain on the yield force of the lead rubber bearing cannot be ignored.

From Figure 4 (b), the post-yield stiffness of the lead rubber bearing shows a trend of first decreasing and then increasing with the increase of acid rain corrosion time, and the change curve is approximately the power of two. The most decline, 2# specimen is the least decline; With the corrosion time continues to grow, the support after yielding stiffness gradually rebound, in four cycles of corrosion test after the end of the three support after yielding stiffness are restored to or even beyond the initial state; In the early acid rain corrosion test, due to corrosion of the lead core, the support after yielding stiffness gradually decline, with the chemical reaction generated by $PbSO_4$ gradually coated As the $PbSO_4$ generated by the chemical reaction gradually covers the surface of the lead core and prevents the corrosion from continuing, the post-yield stiffness of the bearing does not decrease; the rubber gradually becomes larger with the growth of the sulfuric acid solution immersion time, and the post-yield stiffness of the bearing gradually increases.

The damping effect of the bridge superstructure is affected by the damping of the bearing, and the damping effect increases with the increase of damping. From Figure 4(c), it can be seen that when the pH=4.5 of acid rain, the equivalent damping ratio of lead rubber bearing shows a general trend of decreasing with the increase of acid rain corrosion time, and after 92 days of corrosion test, the equivalent damping ratios of 1# and 3# bearings are reduced to 85.612% and 95.505% of the initial value, respectively, and the equivalent damping ratio of bearings with protective layer is more obviously affected by acid rain corrosion. When pH=3.0, the equivalent damping ratio of the bearing is more stable and decreases to the minimum value after 69 days, but finally still has more than 96% of the initial state; in summary, the equivalent damping ratio of the lead rubber bearing is more sensitive to the corrosion of acid rain, and the influence of acid rain cannot be ignored when designing and using.

Due to the manufacturing error, there is a relatively obvious gap between the initial horizontal equivalent stiffness of the bearing, so take the acid rain corrosion time as t, the horizontal equivalent stiffness of the lead rubber bearing at any moment as K_{ht}, and the horizontal equivalent stiffness at the initial moment as K_{h0}. The horizontal equivalent stiffness ratio K_{ht}/K_{h0} is analyzed with time, and the results are shown in Figure 4(d). As can be seen from the figure, with the increase of sulfuric acid solution soaking time, the horizontal equivalent stiffness ratio of the three groups of bearings have experienced the process of first decreasing and then increasing; when the pH of acid rain is 4.5, with the increase of soaking time, the horizontal equivalent stiffness of the bearings decreased slightly, the presence or absence of the protective layer basically does not affect this result, the smallest K_h still has more than 94% of the initial state; with the acid rain corrosion time When pH=3.0, the influence of acid rain corrosion on the K_h of lead rubber bearing is very obvious, in the late test period, the horizontal equivalent stiffness ratio of 2# bearing only has a very small rebound, after 92 days, the horizontal equivalent stiffness of this group of bearings is only about 80% of the initial state, indicating that the lead rubber bearing is not sensitive to the acidic lower acid rain. Acid rain of low acidity is not sensitive, but when encountering acid rain of very strong acidity, the deterioration effect of K_h should be considered.

With the acid rain corrosion time t as the independent variable, the test result value of horizontal equivalent stiffness ratio was fitted, and the horizontal equivalent stiffness K_h of lead rubber bearing with protective layer was obtained when the pH=4.5 of acid rain changed with the acid rain corrosion time as shown in equation (3).

$$K_h = 1.363 - 0.539 - 0.00220486t + 1.833 \times 10^{-3}t \qquad (3)$$

The mean value of the ratio of the pre-experimental measured results by equation (3) is 1.0040, and the standard deviation is 0.788, indicating that the fitted and tested values are in good agreement, so equation (3) can be used to approximately describe the deterioration of the horizontal equivalent stiffness of the lead rubber bearing under the effect of acid rain corrosion.

4 CONCLUSIONS

In order to learn the actual engineering seismic isolation device in the Yangtze River Delta region encountered acid rain, to explore the influence of acid rain corrosion on the mechanical properties of lead rubber bearings, this paper carried out a total of 92 days of periodic acid rain corrosion tests on 3 groups of lead rubber bearings, and the basic mechanical properties of the bearings were tested before the start of the test and after each treatment cycle. The main conclusions were drawn as follows.

(1) When the lead rubber bearing is corroded by acid rain environment, its vertical compression stiffness firstly increases and then decreases a little bit finally stabilizes with the corrosion time. The overall trend is increase. after 92 days of acid rain corrosion, the maximum increase of vertical compression stiffness of 9 bearings is about 10%, which means that the influence of acid rain on the vertical compression stiffness of lead rubber bearing is not significant, and there is no need to consider the influence of acid rain on the vertical bearing capacity of lead rubber bearing in actual engineering.
(2) With the growth of acid rain corrosion time, the shear performance of lead rubber bearing first decreases and then increases, the maximum reduction in yield force is 15.483%, the horizontal equivalent stiffness is 21.827%, and the equivalent damping ratio is 14.388%, indicating that the shear performance of lead rubber bearing is sensitive to acid rain corrosion; and when the acid rain is more acidic, the deterioration of bearing shear performance is more obvious.
(3) In the acid rain corrosion test to about 46 days, the deterioration of the shear capacity of the lead rubber bearing is generally the most serious, indicating that the short-term acid rain corrosion on the shear properties of the lead rubber bearing influence is obvious, bearing protective layer exists on the corrosion of acid rain has a certain degree of mitigation effect, in the acid rain area building seismic design need to consider the corrosion effect of acid rain on the bearing, the building in use after the need to regularly check the integrity of the bearing protective layer and replace the seismic isolation bearing.

In this paper, the basic mechanical performance test of the lead-core rubber bearing under acid rain corrosion is carried out, and a relatively reliable conclusion is obtained, which provides a basis for further research on the impact of acid rain environment on seismically isolated buildings. Of course, due to the limitations of the test conditions, the effect of acid rain on the bearing can be further studied. For example, in the actual project, the bearing will be washed by rainwater, while in this paper, the accelerated corrosion test is carried out by the method of complete immersion. The results obtained may contain small errors. The test can be carried out by spraying acid solution with extended testing time to update the test. Secondly, this test only studies the corrosion law of acid rain on LRB under the conditions of 20°C without additional load, etc. While bearing in the real environment often experience huge temperature changes and are often in different load-bearing states. These factors may also affect acid rain's corrosion effect on the LRB.

REFERENCES

Cai P CH. (2018). Analysis of the distribution status and causes of acid rain in China. Science & Technology Information, 16(15): 127–128.
Du J F, Cheng T T, Ma J L. (2015). The climatic characteristics and formation of acid rain in Shanghai. Transactions of Atmospheric Sciences, 38(01): 137–143.
GB/T 20668.1-2007, Rubber bearings Part 1: Seismic-protection isolation test methods.
Gu H S, Itoh Y. (2010). Ageing behavior of natural rubber and high damping rubber materials used in bridge rubber bearings. Advances in Structural Engineering, 13(6): 1105–1113.

Itoh, Y. (2005). Study on environmental deterioration of rubber material for bridge bearings. Journal of Structure Mechanics and Earthquake Engineering, 794: 253–266.

Li Ch, Gan Y W, Qin S Q. (2021). Research on shock absorption effect of lead rubber bearings on beam-arch composite structure. Acta Scientiarum Naturalium Universitatis Sunyatseni, 60 (03): 138–146.

Li J A. (2007). Study on seismic absorption and isolation of railway bridge by lead-rubber bearing. Beijing Jiaotong University.

Liu S J. (2014). The study on mechanical properties of plate natural rubber bearings under the conditions of erosion environments. Shenyang Jianzhu University.

Liu W G, Ren Y, He W F, Feng D M. (2012). Aging and creep properties of LRB isolators used in building. World Earthquake Engineering, 28(04): 131–136.

Niu J G, Niu D T, Zhou H SH. (2008). Review on the harm and control of acid rain. Journal of Catastrophology, (4): 110–116.

Xuan C, Xiao ran S, Zhao ji S, et al. (2021). Analysis of the Spatio-Temporal Changes in Acid Rain and Their Causes in China (1998–2018). Journal of Resources and Ecology, 12(5): 593–599.

Yu Q, Duan L, Hao J M. (2021). Acid deposition in China: Sources, effects and control. Acta Scientiae Circumstantiae, 41(03): 731–746.

Zhang X M, Chai F H, Wang S L, Sun X Z, Han M. (2010). Research Progress of Acid Precipitation in China. Research of Environmental Sciences, 23(05): 527–532.

Zhang Y N, Ma L, Liu N, Zheng Y, Shen X J, Gao F. (2014). Experimental study on plate neoprene bearing compression based on acid corrosion. Journal of Wuhan University of Technology (Transportation Science & Engineering), 38(01): 55–58.

Zhang Y N, Si Y H, Zhang J, Liu S J, Zheng Y. (2013). Compression Tests of Plain Chloroprene Rubber Bearings of Highway Bridge under Acid Corrosion Condition. Journal of Shenyang Jianzhu University Natural Science, 29(04): 621–627.

Zheng S S, Chen J Ch, Zheng H, Zhang Ch, Shang Zh G. (2021). Experimental research on seismic behaviors of corroded reinforced concrete column failed in bending subjected to acid rain exposure. Journal of Central South University (Science and Technology), 52(10): 3680–3688.

Zu W F, Li Z L, Fan W L, Liu F Q, Yang X Y. (2020). Experimental study on corrosion fatigue behavior of Q420B angle in simulated acid rain atmospheric environment. Journal of Building Structures, 41(08): 105–115.

Frontiers of Civil Engineering and Disaster Prevention and
Control – Yang & Rahman (Eds)
© 2023 The Author(s), ISBN: 978-1-032-31200-2

Experimental study on seismic performance of new assembled joints of RC frame with bolted connection

Yan Cao
School of Arts Design, Wuchang University of Technology, Wuhan, China

Zhao Yang*
School of Urban Construction, Wuhan University of Science and Technology, Wuhan, China

ABSTRACT: In order to simplify the joint form and improve the construction efficiency, this paper proposes a new type of prefabricated concrete frame joint based on bolt connection. Through quasi-static test, the seismic performance of these new assembled joints is studied and compared with that of the assembled integral RC frame joint. The failure mode, hysteretic curve, skeleton curve, as well as the influence of axial compressive ratio on the seismic behavior of these new joints are analyzed. The results of this experimental study show that, compared with the assembled integral joint the carrying capacity and ductility of the new type joint are reduced, and the maximum ranges are 19% and 16.7%, respectively. While the failure forms of the new type joint all pose as beam-end bending failure. The research results can provide reference for assembly joints with bolted connections.

1 GENERAL INSTRUCTIONS

The prefabricated concrete frame structure has been widely studied and applied for its mass production in factory, excellent quality, short construction period, light environmental pollution and so on(Wang et al. 2018). In a frame structure, the beam-column joints are responsible for transmitting and distributing the internal force and ensuring the integrity of the structure, while the joints of a prefabricated structure are easy to become the weakest parts of the structure under the earthquake and to be destroyed first. The post-earthquake investigation (Zhang et al. 2010) shows that under most circumstances, the collapse of the frame structure is due to the loss of the original function of the joints. Therefore, the research of joints is the focus of seismic research of prefabricated concrete frame structures.

There are many studies on the connection mode of prefabricated concrete frame joints. Parastesh et al. (2014) proposed that the hollow U-shaped section and the inclined stirrups supported by longitudinal bars should be used in the connection area of the precast beam end. Guan et al. (2016) used prestressed steel strands at the core of the joints to avoid steel bar blocking. Ha et al. (2014) creatively put forward a kind of precast concrete beam-column joint which can transmit force through U-shaped steel strand. Shariatmadar and Beydokhti (2018) tested the connection performance of U-shaped steel bar protruding from the precast beam. A new type of joint proposed by Zhang et al. (2020) combines straight thread sleeve and bolt connection. To conclude, the prefabricated concrete frame joints are mostly connected by grouting sleeve, prestressed reinforcement, and section steel and so on, but the construction process is complex and uneconomical. However, the research on the way of bolt connection is not enough. It is in this case that we propose a new type of RC frame

---

*Corresponding Author: yzwh77@163.com

assembled joint with bolted connection, which has the advantages of obvious force transmission, simple construction and less wet operation in construction site.

2 EXPERIMENTAL PROGRAM

2.1 *Specimen design*

Considering the application in practical engineering, three joint specimens of RC frame (Table 1) were designed according to 1:2 scale to carry out quasi-static test. RCJ-0 is an assembled integral RC frame joint with an axial compressive ratio of 0.2. Its purpose was to compare with RCJ-1 and RCJ-2. RCJ-1 and RCJ-2 are assembled RC frame joints with bolted connection. The axial compressive ratio of RCJ-1 and RCJ-2 are 0.2 and 0.3 respectively.

Table 1. Scheme of joint specimens.

Specimen number	Joint type	Beam-column connection mode	Beam section size/mm	Column section size/mm	Axial compressive ratio
RCJ-0	assembled integral RC frame joint	Longitudinal bar welding, post-pouring of core area	150×250	250×250	0.2
RCJ-1	assembled RC frame joint with bolted connection	Connection between embedded bolt and steel plate	150×250	250×250	0.2
RCJ-2	assembled RC frame joint with bolted connection	Connection between embedded bolt and steel plate	150×250	250×250	0.3

The beam section size of all joint specimens is 150 mm×250 mm and the column section size are 250 mm×250 mm. The strength grade of concrete is C30. The average value of the measured concrete cube compressive strength is 33.2 MPa and the elastic modulus is $3×10^4$ MPa. The strength grade of longitudinal bars is HRB400 with a diameter of 12 mm. The strength grade of short anchor bars is HRB400, with a diameter of 16 mm and a length of 300 mm. The yielding strength of these steel bars is 423 MPa. The strength grade of stirrups is HPB300, with a diameter of 6 mm. The yielding strength of these steel bars is 471 MPa. The diameter of the bolt is 30 mm and the length is 500 mm. Steel plate size is 250 mm×500 mm, thickness 20 mm, steel grade is Q235 with the yielding strength 282 MPa. The elastic modulus of these steel materials is $2.05×10^5$ MPa.

The details of the specimen RCJ-0 are shown in Figure 1. In the RCJ-0 specimen, the upper and lower columns and the left and right beams are prefabricated in advance, and the shear keyway is reserved at the end of the beam. The longitudinal bars are welded in the core area of the joint and later poured in the core area. Figure 2 shows the details of specimens RCJ-1 and RCJ-2. In RCJ-1 and RCJ-2 specimens, the upper and lower columns are cast, and the bolts are embedded in the predetermined position of the column. The longitudinal reinforcements of the beams and the additional short anchor bars are welded on the end plates at the end of the beams. The welding design is based on the design principle of embedded parts and connectors in the Chinese Code: Code for Design of concrete structures GB/50010-2010 (2011). Figure 3 presents the Details of welding between beam end steel bar and end plate.

2.2 *Test setup*

In this test, the pseudo-static test method was used to study the seismic performance of joint specimens. Figure 4 presents the Illustration and picture of Test loading device. In Figure 4, the end of the column is connected to the horizontal actuator, which provides low cycle reciprocating

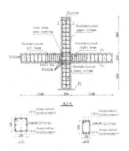

Figure 1.   Details of specimen RCJ-0 (unit: mm).

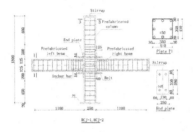

Figure 2.   Details of specimens RCJ-1 and RCJ-2 (unit: mm).

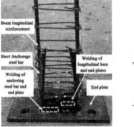
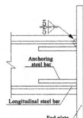
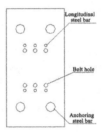

Figure 3.   Details of welding between beam end steel bar and end plate.

horizontal load, by the steel plate and the connecting rod. The top of the column exerts a constant axial compression through a hydraulic Jack, which can be moved horizontally with the sliding trolley, so as to acquire the axial compressive ratio required. And the bottom of the column is connected to the steel base by the ball hinge at the end of the column.

During the test, the vertical load was applied to the column by hydraulic Jack to achieve the required axial compressive ratio. Then the horizontal low cyclic load was applied to the end of the column by the horizontal actuator with the stable axial compressive ratio. The method of displacement-controlled loading was adopted in the test. Figure 5 shows the diagram of the loading process. The loading step is 3 mm before the longitudinal bars yield, and each load was cycled once. When the longitudinal bar had been yielded, the displacement was recorded as the yield displacement y. After that, it was loaded step by step according to the multiple of yield displacement with a 3-time-load cycle on each stage, until the bearing capacity of the specimen was reduced to less than 85% of the ultimate bearing capacity or the deformation of the member was too large to continue loading.

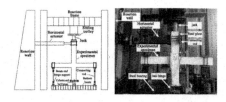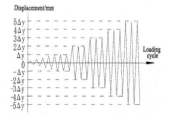

Figure 4. Illustration and picture of Test loading device.

Figure 5. Diagram of the loading process.

3 EXPERIMENTAL PHENOMENA

3.1 *Specimen RCJ-0*

The failure characteristics of specimen RCJ-0 is shown in Figure 6. At the initial stage of loading when the members are in the linear elastic stage, the specimen RCJ-0 had no obvious changes. When the displacement was loaded to 9 mm, the first bending oblique crack appeared in the lower part of the beam end. When the loading displacement was 72 mm, the through cracks were formed at the junction between the beam end and the cylinder, with the concrete at the beam end crushed and peeled off, and two cross oblique cracks appeared in the core area of the joint. When the load was reduced to less than 85% of the ultimate bearing capacity, the test was terminated. Through the test, it was found that most of the RCJ-0 failure was concentrated at the beam end. While the concrete at the beam end was crushed and peeled off, the core area of the joint was basically intact. In a word, the failure form of the member was the bending failure at the end of the beam, which belongs to ductile failure.

Figure 6. The failure characteristics of specimen RCJ-0.

3.2 *Specimen RCJ-1*

The failure characteristics of specimen RCJ-1 is shown in Figure 7. When the loading displacement was 12 mm, the first transverse crack appeared in the lower part of the beam end. When the loading displacement was 29 mm, there were several oblique cracks near the transverse cracks, and there was a gap between the concrete at the end of the beam and the end plate. When the loading displacement was 39 mm, the concrete at the bottom of the beam end was completely cracked, with the anchoring steel bar exposed, and the crack at the end of the original beam developed upward from the bottom. Due to the load was reduced to less than 85% of the ultimate bearing capacity, the test was terminated. After inspection, it was found that the bolt at the end of the beam was loose and there was no obvious damage in the core area of the member joint. The failure was concentrated at the end of the beam, and the failure form was bending failure at the end of the beam, which belongs to ductile failure.

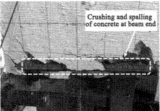

Figure 7.　The failure characteristics of specimen RCJ-1.

3.3　*Specimen RCJ-2*

The failure characteristics of specimen RCJ-2 is shown in Figure 8. At the initial stage of loading when the members are in the linear elastic stage, the specimen RCJ-2 had no obvious changes. When the loading displacement was 10 mm, the bending oblique crack appeared at the end of the beam. When the loading displacement was 24 mm, the vertical crack at the bottom of the beam end became wider, with the concrete at the end of the beam and the end plate cracked, and the gap width increased with the increase of the loading displacement. When loaded into 45 mm, cracks appear in the upper part of the beam and elongated to the end of the beam. Due to the load was reduced to less than 85% of the ultimate bearing capacity, the test was terminated. After inspection, it was found that the bolt at the end of the beam was loose and there was no obvious damage in the core area of the member joint. The failure was concentrated at the end of the beam, and the failure form was bending failure at the end of the beam, which belongs to ductile failure.

Figure 8.　The failure characteristics of specimen RCJ-2.

4　EXPERIMENTAL RESULTS AND DISCUSSION

4.1　*Hysteretic curve*

The hysteretic curves of the specimens are shown in Figure 9. It is observed from the hysteretic curve that the loading and unloading curve of the specimen at the initial stage of RCJ-0 loading is almost linear, and the residual deformation of the member is small. After the beam end cracked, the slope of the loading curve and unloading curve gradually decreased, showing a certain degree of kneading phenomenon, and the stiffness of the specimen gradually degraded. With a continuous loading, the slip between steel bar and concrete increases, and the shape of hysteretic curve changes from "shuttle" to "bow". Compared with the specimen RCJ-0, the specimen RCJ-1 shows its superiority in terms of the phenomenon of "kneading", the hysteresis loop and energy dissipation capacity. However, the horizontal displacement of RCJ-1 specimen is smaller than RCJ-0, and the ductility of RCJ-1 is reduced. In addition, the ultimate bearing capacity of RCJ-1 is reduced to 81% of that of RCJ-0. The difference between specimens RCJ-2 and RCJ-1 is that the axial force of specimen RCJ-2 is relatively large. In contrast, the "pinch" phenomenon of the specimen

RCJ-2 is aggravated, and the size of the hysteresis loop area is slightly reduced. At the same time, its maximum displacement is smaller than that of RCJ-1, and the ultimate bearing capacity of members increases.

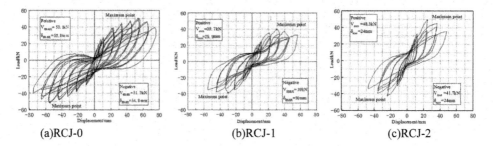

(a)RCJ-0   (b)RCJ-1   (c)RCJ-2

Figure 9.   Hysteresis curves of the specimens.

4.2   Skeleton curve

Figure 10 shows the skeleton curves of the specimens. In the Figure 10, the initial stiffness of RCJ-1 is smaller than that of RCJ-0. After the yield of the member, the yield step of RCJ-1 is not as obvious as RCJ-0, its ultimate bearing capacity is 81% of that of RCJ-0, and the displacement of RCJ-1 is smaller. This shows that the stiffness and ductility of the new joint RCJ-1 are lower than that of the assembled integral joint RCJ-0, but its reduced bearing capacity is still no less than 80% of the RCJ-0.

The initial stiffness of RCJ-2 is larger than that of RCJ-1. After the members yield, the yield step of RCJ-2 is not more obvious than that of RCJ-1, but the ultimate bearing capacity of RCJ-2 is larger than that of RCJ-1. After reaching the ultimate load, the bearing capacity of RCJ-2 decreases rapidly, and its displacement decreases relatively. This shows that with the increase of axial compressive ratio, the initial stiffness and bearing capacity of the new joints increase, but the ductility of the joints decreases.

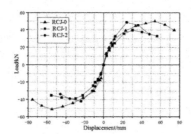

Figure 10.   Skeleton curves of the specimens.

5   CONCLUSIONS

This paper proposes a new type of RC frame assembled joint with bolt connection. Through the quasi-static test, the seismic performance of the new joint specimen was studied and compared with the prefabricated joint specimen. The conclusions can be drawn as follows.

1) Compared with the assembled integral RC frame joint, the bearing capacity and ductility of the new assembled joint are reduced. In this test, compared with the assembled integral joint

426

the bearing capacity and ductility of the new type joint are reduced, and the maximum ranges are 19% and 16.7%; the degradation of stiffness and bearing capacity is accelerated. However, due to the extrusion between bolts and steel plates, the new assembled joint has better energy dissipation capacity.

2) The failure of the new assembled joint is concentrated at the end of the beam, and there is no damage in the core area and the end of the column, which poses as the bending failure at the end of the beam. This is in line with the design principle of "strong column and weak beam, strong shear and weak bending", so the bolt connection method proposed in this paper is feasible.

3) The influencing factors such as structural measures, reinforcement ratio, concrete strength grade and bolt retightening force should be studied deeply in the future.

ACKNOWLEDGEMENTS

This research is funded by the Science and Technology Research Project of Education Department of Hubei Province (Project code: B2019289) and Philosophy and Social Science Research Project of Education Department of Hubei Province (Project code: 21G149). Their support is gratefully acknowledged.

REFERENCES

Concrete beam to column connections: experimental study. Asian J. Civil Eng,15(1), 41–59.

GB50010-2010. (2011). Code for Design of concrete structures, China Architecture & Building Press, 137–138. (in Chinese)

Guan, D. Z., Guo, Z. X., Xiao, Q. D. & Zheng, Y. F. (2016). Experimental study of a new beam-to-column connection for precast concrete frames under reversal cyclic loading. Advances in Structural Engineering, 19(3), 529–545.

Ha, S. S., Kim, S. H., Lee, M. S. & Moon, J. H. (2014). Performance evaluation of semi-precast concrete beam-column connections with U-shaped strands. Advances in Structural Engineering, 17(11), 1585–1600.

Parastesh, H., Hajirasouliha, I. & Ramezani, R. (2014). A new ductile moment-resisting connection for precast concrete frames in seismic regions: an experimental investigation. Engineering Structures, 70, 144–157.

Shariatmadar, H. & Beydokhti, E. Z. (2018). An investigation of seismic response of precast.

Wang, H. S., Marino, E. M., Pan, P., Liu, H. & Nie, X. (2018). Experimental study of a novel precast prestressed reinforced concrete beam-to-column joint. Engineering Structures, 156, 78–81.

Zhang, P., Lu, Z. H., & Dan, H. (2010). The survey analysis of prefabricated structure in Wenchuan earthquake, 36(3), 129–133. (in Chinese)

Zhang, X. Z., Hao, J. S., Duan, D. C., Xu, S. B., Zhang, S. H. & Yu, H. X. (2020). Experimental study on bolted and anchored beam-to-column joints of prefabricated concrete frames. Advances in Structural Engineering, 23(2), 374–387.

*Frontiers of Civil Engineering and Disaster Prevention and
Control – Yang & Rahman (Eds)*
© 2023 The Author(s), ISBN: 978-1-032-31200-2

Research on seismic displacement response of simply-supported girder bridges with different bearing types

W.T. Yin & K.H. Wang*
Research Institute of Highway, Ministry of Transport, Beijing, China

B.Z. Zhang & W.Z. Guo
School of Transportation, Southeast University, Nanjing, Jiangsu, China

ABSTRACT: The type of bearing is a key factor affecting the seismic performance of the simply-supported girder bridge. In this paper, a 20m span simply-supported girder bridge was selected to establish a finite element model with the input of the ground motion for analysis, and compare the effects of laminated rubber bearings and high damping rubber bearings on the main beam displacement and bearing deformation, and the following conclusions are obtained: the deformation of the laminated rubber bearings is greater than that of the high damping rubber bearings because of the obvious sliding of the laminated rubber bearings which greatly reduces the force transmitted to the piers. Under the rare earthquake, the high damping rubber bearings and laminated rubber bearings achieve slight damage and medium damage, respectively. According to the analysis results, the laminated rubber bearings are recommended to be used for simply-supported girder bridges as fuse elements.

1 INTRODUCTION

Under the action of earthquakes, the degree of damage to bridges will greatly affect traffic recovery and rescue, thus it is important to effectively reduce the impact of earthquakes on bridges (Kawashima 2011). The form of medium-and small-span girder bridges are generally used among the highway bridges, and varied support forms are the key factors affecting the seismic performance of bridges. LRBs (laminated rubber bearings) and HDRBs (high damping rubber bearings) are widely used in the seismic resistance of medium-and small-span bridge structures with the advantages of economy, simple structure, simple construction.

The advantages of HDRBs were stable performance, strong energy consumption and ductility, which could effectively control the seismic response of the structure (Zhuang 2006). The displacement could be controlled within a certain range with less influence on the substructure when the dampers were set at the bearing (Xiang 2019), but Mahmoud (2012) found that the base response peak of the nonlinear HDRBs was bigger than that of the system using linear LRBs. And the aims of HDRBs and dampers were to reduce the seismic response of the superstructure, but the force transmitted to the substructure increased. The typical transverse seismic damage forms of LRBs girder bridge were the seismic damage of the falling beam due to excessive displacement of the pier and beam, the damage of the bearing and the retainer (Tang 2016). Steelmand (2013) proposed a sliding mechanical model based on rubber bearing tests and believed that the slip performance of the LRBs in seismic resistance of the bridge could not be ignored which increased the risk of falling beams. Although the displacement response of the superstructure with LRBs was large, it could be limited by other limit devices such as retainer. And through the investigation and summary of

---

*Corresponding Author: kh.wang@rioh.cn

                                                    DOI 10.1201/9781003308577-58

the damage to beam bridges caused by the Wenchuan earthquake, it had been discovered that most girder bridges with LRBs have suffered light damage of substructure, and the bearings could be used as fuse elements for priority destruction (Wang 2012). From previous studies, there were two main ideas for the application of LRBs and HDRBs in the earthquake resistance of medium-and small-span bridges, one was to reduce the seismic force transmitted to the substructure and control the displacement response of the superstructure, and the other was to reduce the displacement response of the superstructure and control the seismic force transmitted to the substructure, thus it was necessary to compare their impact on the overall seismic performance of the bridge.

In this paper, a finite element model of a 20m span simply-supported girder bridge was established by Sap2000 to analyze and compare the effects of LRBs and HDRBs on the seismic response of the bridge. Then the damage state of the bearing under the seismic level E1 and E2 was determined, and finally appropriate proposals for the use of the bearings in the simply-supported girder bridge were made.

2 BACKGROUND AND MODELLING

2.1 Engineering background

A typical 20m simply-supported girder bridge with C50 four-piece single-box and single-chamber small box girders was used as an example in this research. The substructure is composed of a 5m high double column constructed of C35 concrete with a circular cross-section of 1.2m in diameter and the HRB 300 reinforcement rate is 1.2%. The girder is supported by two bearings at the bent top and the distance is 0.5 m, where the bearing type is GYZ D350×96 and HDR(I)-D370×177-G1.0. Each span is set up expansion joint on the both sides and the gap is 8cm. The site of the bridge is Class III with a designed seismic peak acceleration of 0.2g.

2.2 Bridge finite element modeling

A three-dimensional finite element dynamic calculation model is established for seismic performance analysis basing on the SAP200, and the longitudinal direction, the transverse direction and the vertical direction are the x-axis, y-axis, and z-axis, respectively. The superstructure and cover beam adopt linear elastic beam elements, and the substructure adopts fiber beam elements which are given the frame cross-sectional properties, including steel fibers, confined concrete fibers, and unconfined concrete fibers. The main beams are modeled according to grillage simulation method, and the cross beams are simulated by the virtual beam, and the longitudinal beams form an orthogonal grid with the virtual beams. The models of the Multi-Linear Plastic and Plastic (Wen) are used to simulate LRBs and HDRBs, respectively, and the parameters of LRBs model are calculated basing on the Guidelines for seismic design of highway bridges (JTG 2020) and HDRBs refer to the code of HDRBs (JTG 2012). The gap element is used for simulation in order to consider the collision effect at the expansion joint where the initial value of the gap was 8cm, and the axial stiffness of the main beam in this research is used for the collision stiffness (Muthukumar 2006), and all the details are shown in Figure 1.

2.3 Earthquake input

According to the seismic fortification standard and specification requirements of the simply-supported girder bridge (JTG 2020), the designed ground motion response spectra of E1 and E2 are calculated and three artificial seismic waves are synthesized that match the response spectrum respectively, as shown in Figure 2, the input direction of seismic wave is longitudinal bridge plus transverse direction, and the calculation results are the maximum value of the three artificial seismic waves.

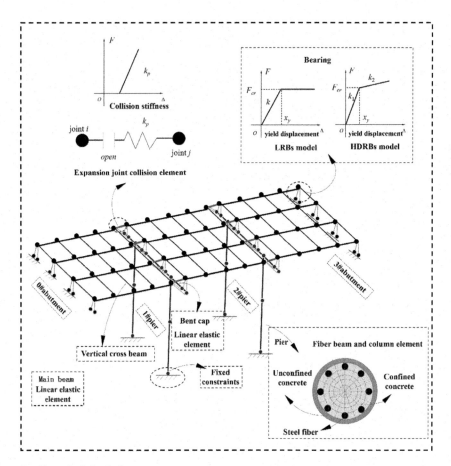

Figure 1. Numerical simulation.

3 EARTHQUAKE RESPONSE ANALYSIS

The bearing is a weak link in the seismic performance of the bridge structure system (Tang 2016; Wang 2012). The relative displacement of the bearing can consume the energy of the earthquake and reduce the seismic response of the bridge, but the influence on the displacement of the girder should be taken into consideration. Especially, in a simply-supported girder bridge, the excessive deformation of the bearing will cause the main girder to be seriously displaced, or even fall off the girder. This paper focuses on the comparative study of the deformation of LRBs and HDRBs, and the displacement of the main beam of the simply-supported girder bridge under the action of earthquake.

3.1 *Internal force of the pier*

Figure 3 shows the internal force envelope values of the pier 2, including bending moment and shear. From the Figure 3, the maximum bending moment and shear of HDRBs are greater than that of the LRBs under the earthquake action E1 and E2, and Compared with LRBs, the bottom bending moment of HDRBs increased by 48% and the shear increased by 60.0% and 77.7%, respectively.

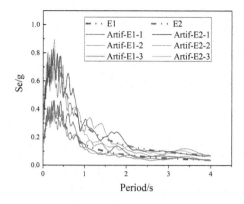

Figure 2. Acceleration spectra of ground motions.

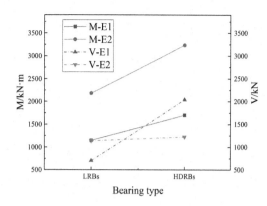

Figure 3. Internal force envelope value of the piers.

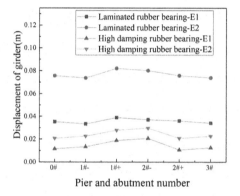

Figure 4. Beam displacement envelope value.

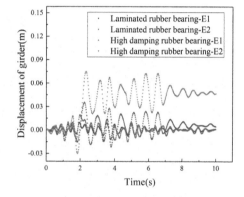

Figure 5. Displacement results of the beam.

3.2 Girder displacement

Figure 4 shows the longitudinal displacement response envelope value of all main beams under the earthquake action E1 and E2, including LRBs and HDRBs. From this figure, the displacement of the main beam with LRBs is larger than HDRBs, and the maximum displacement is about 8.2cm. However, the location of the maximum displacement is different, the largest position of the main girder displacement of LRBs is on the right side of the pier 1 and on the left side of the pier 2 of HDRBs. These different trends above may be due to the influence of the horizontal partition between the main beams, and the maximum value of the main girder displacement occurs on the pier rather than on the abutment. Hence, it's necessary to set the longitudinal lap length at the cover beam to prevent the falling beam behavior.

Figure 5 shows the time-history analysis results of the displacement of the main girders selected on the 0-abutment. As shown in Figure 4, there are significant differences of these two kinds of bearings in the displacement of the main beam. In the rare earthquake, the maximum displacement of the main girder set with HDRBs is small, and the residual displacement is close to 0, which will cause more seismic force to be transmitted to the pier and will greatly affect the seismic performance of the structure. The displacement of LRBs is suddenly increased at about 2s and then reaches the top as the result of LRBs sliding, and the residual displacement reaches 4cm which is far less than the lap length. Therefore, LRBs protect the bridge pier effectively and serve as fuse components which are beneficial to the overall bridge seismic resistance.

3.3 Bearing deformation and damage

Table 1. Damage assessments and criteria for bearing.

Symbols	Value of damage limit state
B_0	Bearing yield displacement
B_1	75% Bearing shear deformation
B_2	150% Bearing shear deformation
B_3	250% Bearing shear deformation
B_4	The minimum longitudinal lap length of the pier and cap

This paper is based on the evaluation index of the seismic performance of the bearing established by Wu (2018), and Table 1 lists the damage state and judgment criterion of the bearing longitudinal bridge to all levels, including elastic status, minor damage, moderate damage, severe damage and falling beam. Among them, the index value of the falling beam state is calculated using the provisions of the literature on the minimum longitudinal lap length of the simply-supported beam (JTG 2020). In this paper, Table 2 shows the index value of the bearing according to the actual situation.

Table 2. Damage index values for bearing.

Bearing type	B_0	B_1	B_2	B_3	B_4
LRBs	24.5	72	144	240	660
HDRBs	7.4	132.8	265.5	442.5	660

Figure 6 summarizes the maximum deformation of the bearing and the damage results at each pier and abutment. From Figure 6, under the earthquake action E1, the deformation of the LRBs on the 3-abutment is the largest and the maximum deformation is 3.8cm, but on the 1-pier the maximum deformation of HDRBs is only 1.8cm on the right side, and the difference between the right and left bearings can reach 61.1%. Under the earthquake action E2, the deformation of the LRBs on the 2-pier is the largest and the maximum deformation is 8.1cm, however, on the 1-pier, the maximum deformation of HDRBs is 2.6cm and the difference between the right and left bearings can reach 42.3%. With the increase of seismic intensity, the maximum deformation of LRBs increase by 113.2%, while HDRBs increased by 44.4%. And under the earthquake action E1 and E2, no bearing can reach 250% of its shear deformation that is the state of severe damage, and it is far less than the minimum lap length. Even in the rare earthquake, the maximum damage state of the LRBs is only moderate damage and HDRBs is only slightly damaged, but the piers were subjected to greater seismic forces, aggravating the overall damage to the bridge.

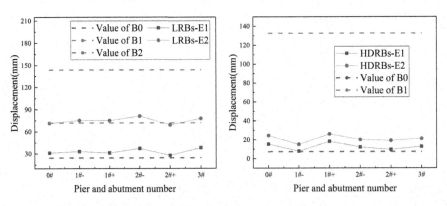

Figure 6. Bearing deformation envelope value and damage.

432

4 CONCLUSIONS

In this paper, the finite element model of a typical simple girder bridge is established to compare the influence of the seismic response between LRBs and HDRBs, and the damage state of the bearing was determined. The major findings and recommendations are summarized as follows.

(1) The displacement of the main beam is highly correlated with the deformation of bearings. Under the rare earthquake, the residual displacement of the main beam with HDRBs is close to 0 and that of LRBs can reach 4cm that is much smaller than gap length.
(2) With the increase of seismic intensity, the maximum deformation is greater than that of HDRBs, and due to the obvious slip of the LRBs, the force transmitted to the pier is greatly reduced. In the rare earthquake, the damage state of LRBs is only moderate damage, and the beam will not fall with a reasonable lap length, but the damage of HDRBs is only slight and the piers are subjected to greater seismic forces, aggravating the overall damage to the bridge.
(3) Owing to the obvious bearing slip, LRBs protect the bridge pier well and serve as fuse elements which are beneficial to the overall seismic resistance of the bridge. The LRBs are recommended for usage in the simply-supported girder bridge.
(4) The simply-supported girder bridge with LRBs has suffered a large displacement response of the superstructure under the rare earthquake action, the reasonable setting of retainer and other limit devices should be strengthened in the following research to improve the seismic performance of the bridge.

REFERENCES

JT/T 842-2012. (2012). High damping seismic isolation rubber bearings for highway bridges. China: Department of Transportation.
JTG/T 2231-01-2020. (2020). Guidelines for seismic design of highway bridges. China: Department of Transportation.
Kawashima, K., Unjoh, S., Hoshikuma, J. I., Kosa, K (2011). Damage of bridges due to the 2010 Maule, Chile, earthquake. Journal of Earthquake Engineering, 15(7), 1036–1068.
Mahmoud, S., Austrell, P. E., Jankowski, R (2012). Simulation of the response of base-isolated buildings under earthquake excitations considering soil flexibility. Earthquake Engineering and Engineering Vibration, 11(3), 359–374.
Muthukumar, S., & DesRoches, R (2006). A Hertz contact model with non-linear damping for pounding simulation. Earthquake engineering & structural dynamics, 35(7), 811–828.
Steelman, J. S., Fahnestock, L. A., Filipov, E. T., LaFave, J. M., Hajjar, J. F., Foutch, D. A (2013). Shear and friction response of nonseismic laminated elastomeric bridge bearings subject to seismic demands. Journal of Bridge Engineering, 18(7), 612–623.
Tang, H., Li, J.Z., Shao, C.Y (2016). Seismic performance of small and medium span girder bridges with plate type elastomeric pad bearings in the transverse direction. China Journal of Highway and Transport, 29(3), 55.
Wang, K.H (2015). Bridge Seismic Research (2nd ed). Beijing: China Railway Publishing House.
Wang, K.H., Li, C., Li, Y (2013). Problems in Chinese Highway Bridge Seismic Specifications and Suggestions for Improvement. Journal of Architecture and Civil Engineering. 30(2): 95–103.
Wang, K.H., Wei, H., Li, Q., Li, Y (2012). Philosophies on seismic design of highway bridges of small or medium spans[J]. China Civil Engineering Journal, 45(9): 115–121.
Wu, G (2018). Study on reasonable constraint system of curved girder bridge under seismic action (Doctoral dissertation, Southeast University).
Xiang, N., Alam, M. S., Li, J (2019). Yielding steel dampers as restraining devices to control seismic sliding of laminated rubber bearings for highway bridges: analytical and experimental study. *Journal of Bridge Engineering*, 24(11), 04019103.
Zhuang, X.Z., Shen, C.Y., Jin, J.M (2006). Earthquake Engineering and Engineering Vibration, 26(05): 208–212.

Frontiers of Civil Engineering and Disaster Prevention and
Control – Yang & Rahman (Eds)
© 2023 The Author(s), ISBN: 978-1-032-31200-2

Yield stress of Chengdu clay slurry

XianJun Ji*
School of Civil Engineering, Nanyang Institute of Technology, Singapore

Ying Liang
School of Mathematics and Physics, Nanyang Institute of Technology, Singapore

ABSTRACT: The yield stress of mud is the key to the starting analysis of debris flow. Taking the clay in Longquan District of Chengdu as the experimental object, slurries with different solid volume concentrations were prepared. Using the blade rotor system of MCR301 rheometer, the continuous increasing shear rheological test of mud is carried out, and the dynamic change process of shear stress with the increase of shear rate is recorded. According to the results, the following conclusions are drawn: Chengdu clay mud is a non-Newtonian fluid with yield stress. When the solid volume concentration exceeds 35%, the shear rate is less than 1s-1, and the shear stress increases rapidly with the increase in shear rate. When shear rate is higher than 1s-1, the shear stress decreases with the increase in shear rate, and finally tends to be stable. There is a maximum shear stress near 1s-1. When the solid volume concentration is less than 35%, the shear rate is less than 0.1s-1, and the shear stress increases with increase in shear rate. When the shear rate is higher than 0.1s-1, the shear stress changes little with the increase in shear rate. An exponential relationship between mud (dynamic and static) yield stress and solid volume concentration can be seen.

1 INTRODUCTION

Mud is a kind of non-Newtonian fluid with yield stress (Barnes 2007; Ferraris 1999; Moller et al 2009). Metzner (1985) and Utracki (1988) carried out relevant experiments on the rheological properties of mud, in which yield stress was one of the research focuses. For such materials, although the concept of yield stress is generally accepted, when the applied stress exceeds the yield stress, the material begins to flow. When applied stress is less than yield stress, the material will undergo plastic deformation and will not flow. However, there has always been a dispute about yield stress. N. Roussel (2005), N. Pashias, et al (1996) believe that the yield stress depends largely on the test conditions and techniques used. In addition, many non-Newtonian fluids have thixotropy, and the structure changes with time during the shear process (Pierre et al. 2013; Mechtcherine et al. 2014). E C. Bingham (1922) believes that this time-varying rheological property makes it difficult to find a reliable method to determine yield stress of slurry. Scott Blair (1933) proposed a practical definition of yield stress, that is, "any critical stress below which flow cannot be observed under experimental conditions." This subject is still not clearly defined in the analysis and prevention of landslide and debris flow disasters.

Debris flows often occur all over the world due to heavy rainfall. Mud (mixture of fine soil and water) is the main component of debris flow. Mud moves with particles, which gives it the potential to have a strong impact and great destruction (Ochiai, et al 2004). The flow behavior of debris flow is controlled by the rheological properties of mud (Forterre & Pouliquen 2008). The

---

*Corresponding Author: jifeng988@163.com

DOI 10.1201/9781003308577-59

rheological properties of slurry, including rheological model, yield stress, viscosity, flow index, etc, is very important. It is the key to analyze the movement of debris flow, including movement speed, impact force, and movement distance, and is helpful to evaluate the risk of debris flow. For debris flow comprising viscous slurry and inviscid particles, if the change in friction angle caused by coarse particle breakage in the movement process is ignored, the yield stress of mud becomes key to analyzing the initiation and deposition of landslide debris flow.

The yield stress of slurry is caused by the microstructure (Ye et al 2018), which is related to stress history (Barnes 1997; Wallevik 2009). According to the flow state, the yield stress was divided into static and dynamic yield stresses (Cheng 1986; Scotto di Santoloa et al 2012). The dynamic yield stress is the minimum shear stress of steady flow. The static yield stress is the peak stress and represents the maximum shear stress that must be overcome when the mud begins to flow. The processes of initiation, migration, and accumulation of debris flow show that the mud is subjected to different stress histories. Therefore, the determination of mud static and dynamic yield stress is the basis of debris flow movement analysis.

According to the actual change in solid volume concentration during the movement of debris flow (solid matter is added, mud concentration increases, water is supplemented, mud concentration decreases), Chengdu clay slurries with different concentrations were prepared. Based on the preliminary analysis of the influence of experimental methods and equipment on the results, the stress sweep rheological experiment is carried out by using the blade rotor system of MCR301 rheometer. The effects of concentration on dynamic and static yield stress are analyzed and determined. In addition, the filling mud influence to yield stress is not considered.

2 EXPERIMENTAL MATERIALS AND SAMPLE PREPARATION

Chengdu clay is brownish yellow. The liquid limit moisture content is 66.5% (w_L), and the plastic limit moisture content is 24.5% (w_P). The plasticity index (I_P) is 42 > 17 ($I_P = w_L - w_P$). The composition (see Figure 1) of Chengdu clay is analyzed by the Malvin Mastersizer 2000, UK. The particle density determined is 2.7 g/cm^3 (ds).

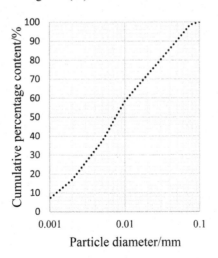

Figure 1.   Composition of Chengdu clay.

Take some Chengdu clay and soak it in a container for more than 48 hours. Determine the initial water content of slurry by drying method (w_0: Percentage of water mass and dry soil mass per unit volume). Take a certain amount of standby mud into 10 cups. Add different amounts of water to each cup, and calculate the final water content of the slurry (w_1), according to the amount of water added and w_0, calculate mud solid volume concentration (C_s):

$$C_s = \frac{V_s}{V} \times 100\% = \frac{V_s}{V_s + V_v} \times 100\%$$

V_s is the solid volume, V is the total volume ($V = V_s + V_v$), V_v is the water volume.
According to w_1 and d_s, calculate slurry solid volume concentration of cup (see Table 1).

Table 1. Moisture content and solid concentration of slurry.

Sample No	1	2	3	4	5	6	7	8	9	10
Moisture content /%	50.0	57.0	67.0	72.0	80.0	101.0	131.0	182.0	302.0	400.0
solid volume concentration /%	43.0	40.0	36.0	35.0	32.0	28.0	23.0	17.0	11.0	9.0

3 EXPERIMENTAL INSTRUMENT AND EXPERIMENT METHOD

3.1 Experimental instrument

The blade viscometer is designed by Keentok (1982) and Nguyen & Boger (1985a), which is often used in the rheological experiments conducted on mud. The viscometer blade consists of four blades around a small cylindrical axis. After the blade was slowly immersed in the fluid, torque and shear stress were recorded during testing. The rheometer can avoid wall slip effect (Barnes 1995; Barnes & Nguyen 2001). In addition, the blade has the least interference with the test sample (Nguyen & Boger 1985b; James et al 1987). Keating & Hannant (1989) compared this method with other mature experimental methods and confirmed the accuracy of it.

The blade rotor system of MCR301 rotary rheometer in Antonpa, Austria is used in this study. It consists of four blades; with radius 13 mm (R1), and height 48 mm (L); cup radius R2 = 18.5 mm. Ignoring the normal stress difference and inertia effect, the shear stress (τ) and shear rate ($\dot{\gamma}$) are calculated as follows:

$$\tau = \frac{T}{2\pi R_1^2 L}; \quad \dot{\gamma} = \frac{\omega R_1}{R_2 - R_1}$$

Where T represents the torque and ω stands for angular velocity of the blade.

3.2 Testing method

In the past, different methods were used to measure slurry yield stress, for example, stress relaxation, creep, stress growth, stress scanning, oscillatory shear, and so on. Nguyen & Boger (1992) summarized the advantages and disadvantages of each method. Although some methods are widely used, none is considered as a standard procedure for determining yield stress. Many debris flow disasters are unexpected. When the shear stress is more than yield stress of debris flow slurry, the debris flow starts and shear rate of debris flow mud increases rapidly in a short time. According to the field observation of debris flow (Lorenzini & Mazza 2004; Ochiai et al 2004), the average shear rate of debris fluid is usually between 0.1 and 20 s^{-1} (Parsons et al 2001). Therefore, continuous shear test was adopted. At the beginning of the test, take some mud into the fixed outer cylinder of the rheometer, put blade into mud and immerse it below the mud surface in the cylinder. During the test, the rotation rate exhibits continuous logarithmic growth from 0.01s^{-1} to 30s^{-1}, and the shear stress was recorded. Throughout the test, the temperature is 20°C through water circulation system.

4 THE RESULTS AND DISCUSSION

According to the above experimental method, the blade rotor system of Antonpa MCR301 rheometer is used to measure the change process of shear stress about slurry with different solid volume concentration (see Figure 2).

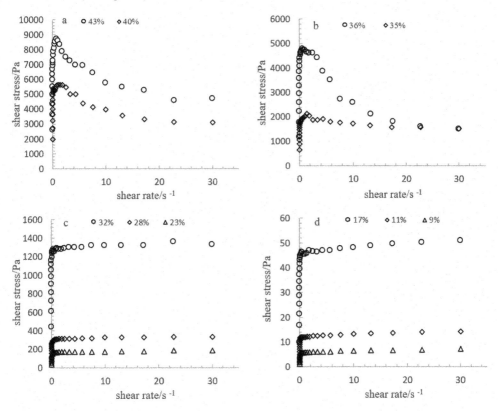

Figure 2.   Shear stress vs shear rate for various solid concentration (natural coordinate system).

4.1 *Rheological process analysis*

The variation process of shear stress (τ) with shear rate ($\dot{\gamma}$) is shown in the natural coordinate system (Figure 2a, b, c, and d). Figure 2 shows that Chengdu clay mud is a non-Newtonian fluid, and with shear thinning characteristics (Besq et al 2003; Luck-ham & Rossi 1999). At the same shear rate, shear stress decreases with decrease in solid volume concentration. However, the variation process about shear stress with shear rate is quite different for different solid volume concentrations slurries. When concentration of slurry is more than 35%, in the range of 0.01-$1s^{-1}$, the shear stress increases rapidly with shear rate, but then decreases rapidly, and finally tends to be stable in the range of 1s-1-30s-1. The shear stress has stress overshoot near the shear rate of $1s^{-1}$. For example, for slurry (solid volume concentration is 43%), when the shear rate increases from 0.01 to 1s-1, the shear stress increases rapidly from the initial value of 2000 pa to nearly 9000 pa. When the shear rate increases from 1s-1 to 30s-1, the shear stress decreases from 9000 pa to less than 8000 pa. (Figure 2a). When concentration of slurry is less than 35%, shear rate is in the range of 0.01-0.1s-1, the shear stress increases rapidly, but then the shear stress changes little when shear rate is in the range of 0.1s-1-30s-1, and the shear stress overshoot disappears. For example, for the slurry with solid volume concentration of 28%, When the shear rate is 0.01s-1, the shear stress is 300 pa, and when the shear rate is 0.1s-1, the shear stress only increases to 310 pa (Figure 2c). As Jeong et al (2010) stated, "The shear rate and particle concentration determine the flow behavior of bentonite and illite natural clay slurry." This result is consistent with the results obtained from similar experiments (Major & Pearson 1992; Schatzmann et al 2009; Sueng-won Jeong 2019). This fully proves that Chengdu clay slurry is a non-Newtonian fluid with yield stress.

4.2 Yield stress of mud

At present, there is no clear standard for definition of yield stress. When the scheme is selected to obtain yield stress, the expected application and time scale need to be considered. This paper mainly focuses on the starting process of debris flow. Under gravity action, debris flow transits from starting to stable motion in a short time. Accordingly, the shear rate of mud increases from zero to a large value in a short time. Therefore, this paper adopts the continuous shear experimental scheme to obtain yield stress of slurry according to variation process of shear stress with shear rate. According to analysis in Section 4.1, the rheological process of mud with different concentrations is not consistent. When slurry concentration is more than 35%, shear stress first increases rapidly and then decreases gradually, and finally tends to be stable. This change process is more clearly shown in Figure 3a and 3b. Cheng C H (1986) described this change process: in thixotropic fluid nonlinear viscoelastic fluid, there will be stress growth first, then stress attenuation, and then the stress change tends to be stable (equilibrium flow). Yield stress is divided into static or dynamic yield stress (Bonnecaze & Brady 1992; Pham et al. 2008). For continuous shear test, the shear stress peaked after the rapid increase in the initial stage and before reaching the steady-state flow. In this case, the peak stress is static yield stress of mud. After the peak stress, mud flow reaches a stable state. The minimum stress value at this stage helps maintain minimum stress and stable flow of mud.

According to variation process of shear stress, the peak shear stress is defined as static yield stress (τ_s) when solid volume concentration is more than 35%. The maximum shear stress occurs near the shear rate of $1s^{-1}$. The shear stress decreases with the increase in shear rate when shear rate is more than 1s-1. The reason is local liquefaction or local particle migration caused by shear. However, finally, the shear stress tends to be stable. The minimum shear stress at this stage is dynamic yield stress (τ_d). The static yield stress (τ_s) and dynamic yield stress (τ_d) of mud are highlighted in Figure 3a and b, and are listed in Table 2. Leonardo Schippa (2020) conducted rheological experiments on natural mud with 30%–42% solid concentration recombined with fine pyroclastic particles, and obtained similar results.

When concentration is less than 35%, shear stress increases rapidly during rate 0.01-$0.1s^{-1}$, then increases gradually with increase in shear rate, and stress overshoot disappears. Flow curve in semi-logarithmic coordinate system (Figure 3c and d) and natural coordinate system (Figure 2c and d) clearly show this change process. This indicates that the mud enters a stable flow state immediately after the flow starts, and static and dynamic yield stress is equal, which occurs near the shear rate of $0.1s^{-1}$ (Figure 3c and d). Wildemuth & Williams (1985), Magnin & Piau (1990) emphasized that the data reliability at low shear rate must be checked in yield stress fluid and particle systems. To avoid over prediction of yield stress, the critical shear rate is 0.1s-1, and corresponding shear stress is taken as the dynamic and static yield stress of mud with concentration less than 35% (see Table 2).

Table 2. Static and dynamic yield stress.

Mud sample	1	2	3	4	5	6	7	8	9	10
Solid volume concentration /%	43.0	40.0	36.0	35.0	32.0	28.0	23.0	17.0	11.0	9.0
Static Yield Stresses/Pa	8711	5638	4769	2115	1190	275	149	41	11	5
Dynamic Yield Stresses/Pa	4561	3119	1484	1523	1190	275	149	41	11	5

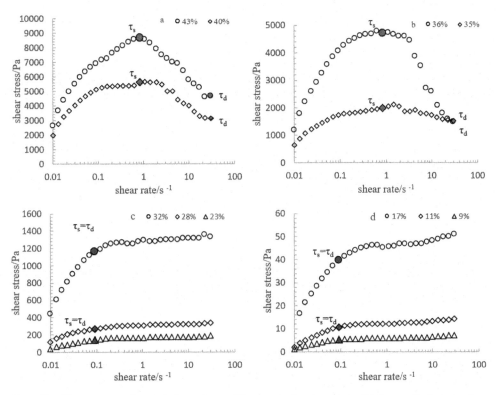

Figure 3.   Shear stress vs shear rate for various solid concentration (semi logarithmic coordinate system).

Table 2 confirms that solid volume concentration has a strong effect on the yield stress. For slurries that concentration is more than 35%, dynamic yield stress is much less than the static yield stress (J.J. Assaad et al 2006), and this difference decreases with the decrease in concentration. The yield stress has an exponential relationship with particle concentration, and has a good correlation (Figure 4).

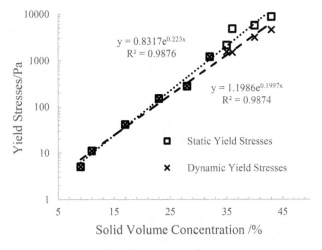

Figure 4.   Plots of yield stress versus solid volume concentration.

5 CONCLUSIONS

Taking Chengdu clay slurry as the experimental object, the continuous increasing shear experiment is carried out using MCR301 rheometer to analyze the variation process of shear stress with shear rate, and determine the static and dynamic yield stress of slurry with different solid volume concentration. According to the results, the following conclusions are obtained:

(1) Chengdu clay slurry is a non-Newtonian fluid with yield stress.
(2) When mud concentration is more than 35%, shear stress increases rapidly, then decreases rapidly, and finally tends to be stable. The maximum shear stress is recorded near the shear rate of $1s^{-1}$, which is defined as static yield stress. The minimum shear stress of stable flow was taken as the dynamic yield stress.
(3) When the mud concentration is less than 35%, mud shear stress remains basically stable after rapidly increasing, and static and dynamic yield stress is equal.
(4) The static and dynamic yield stress of slurry increases exponentially with particle concentration.

This paper only considers the influence of mud solid volume concentration on the change process of shear stress, but there are many factors affecting the rheological process of mud, such as mud composition, experimental scheme, initial shear rate, and so on. Therefore, in the next step, based on the existing research and considering relevant factors, different experimental schemes will be formulated to analyze rheological properties of debris flow slurry.

ACKNOWLEDGMENT

This work was supported by National Natural Science Foundation (Grant No.41672357); Henan Provincial Department of Science and Technology Research Project (Grant No. 182102310779).

REFERENCES

Barnes, H. A. (1995). A review of the slip (wall depletion) of polymer solutions, emulsions and particle suspensions in viscometers: Its cause, character and cure. Journal of Non-Newtonian Fluid Mechanics, 56, 221–251.

Barnes, H. A., & Nguyen, Q. D. (2001). Rotating vane rheometry – a review. Journal of Non-Newtonian Fluid Mechanics, 98, 1–14.

Besq, A., Malfoy, C., Pantet, A., Monnet, P., and Righi, D. 2003. Physicochemical characterisation and flow properties of some bentonite muds. Applied Clay Science, 23(5–6): 275–286.doi:10.1016/S0169-1317(03)00127-3.

C.F. Ferraris, Measurement of the rheological properties of high-performance concrete: state of the art report, J. Research-National Inst. Stand. Technol. 104 (1999) 461–478.

Cheng C H. Yield stress: A time-dependent property and how to measure it[J]. Rheologica Acta, 1986, 25(5):542–554.

Cheng, D. C.-II. 1984. Further observations on the rheological behavior of dense suspensions. Powder Technol. 37: 255–73.

E.C. Bingham, Fluidity and plasticity, McGraw-Hill Book Compny, Incorporated, 1922.

Gorislavets, V. M., Dunets, A. A. 1975. Rheological properties of concentrated suspensions in the presence of wall effect. Inz-Fiz Zh. 29: 273–82

H.A. Barnes, The yield stress Myth? 'Papere21 Years on, Appl. Rheol. 17 (2007), 43110–44250.

H.A. Barnes, Thixotropyda review, J. Newt. Fluid Mech. 70 (1997) 1–33.

J. Assaad, K.H. Khayat, Assessment of thixotropy of self-consolidating concrete and concrete-equivalent-mortar- effect of binder composition and content, ACI Mater. J. 101 (2004).

J.Assaad, K.H. Khayat, H. Mesbah, Assessment of thixotropy of flowable and self-consolidating concrete, ACI Mater. J. 100 (2003).

J.E. Wallevik, Rheological properties of cement paste: thixotropic behavior and structural breakdown, Cem. Concr. Res. 39 (2009) 14–29.

J.E. Wallevik, Thixotropic investigation on cement paste: experimental and numerical approach, J. Newt. fluid Mech. 132 (2005) 86–99.

J.J. Assaad, K.H. Khayat, Effect of viscosity-enhancing admixtures on formwork pressure and thixotropy of self-consolidating concrete, ACI Mater. J. 103 (2006).

James, A. E., Williams, D. l. A., Williams, P. R. 1987. Direct measurement of static yield properties of cohesive suspensions. Rheol. Acta 26: 437–46.

Jeong S W. Shear Rate-Dependent Rheological Properties of Mine Tailings: Determination of Dynamic and Static Yield Stresses[J]. Applied Sciences, 2019, 9(22):4744.

Keating, J., Hannant, D. J. 1989. The effect of rotation rate on gel strength and dynamic yield strength of thixotropic oil well cements measured using a shear vane. J. Rheol. 33: 1 0 1 1–20.

Keentok, M. 1982. The measurement of the yield stress of liquids. Rheol. Acta 2 1: 325–32.

Lorenzini, G., and Mazza, N. 2004. Debris flow: Phenomenology and rheological modeling. WIT Press, Southampton, UK.

Luckham, P.F., and Rossi, S. 1999. The colloidal and rheological properties of bentonite suspensions. Advances in Colloid and Interface Science, 82(1–3): 43–92. doi:10.1016/S0001-8686(99) 00005-6

Magnin, A, Piau, J. M. 1987. Shear rheometry of fluids with a yield stress. J. NonNewtonian Fluid Mech. 23: 9 1–106.

Magnin, A, Piau, J. M. 1990. Cone-andplate rheometry of yield stress fluids.Study of an aqueous gel. J. Non-Newtonian Fluid Mech. 36: 85–108.

Major J J , Pierson T C . Debris flow rheology: Experimental analysis of fine-grained slurries[J]. Water Resources Research, 1992, 28(3):841–857.

Metzner, A. B. 1985. Rheology of suspensions in polymeric liquids. J. Rheol. 29: 739–75.

N. Pashias, D. Boger, J. Summers, D. Glenister, A fifty cent rheometer for yield stress measurement, J. Rheology (1978-present) 40 (1996) 1179–1189.

N. Roussel, P. Coussot, "Fifty-cent rheometer" for yield stress measurements: from slump to spreading flow, J. Rheology (1978-present) 49 (2005) 705–718.

Nguyen Q. Measuring the Flow Properties of Yield Stress Fluids[J]. Annual Review of Fluid Mechanics, 1992, 24(1):47–88.

Nguyen, Q. D., Boger, D. V. 1985a. Direct yield stress measurement with the vane method. J. Rheol. 29: 335–47.

Nguyen, Q. D., Boger, D. V. 1985b. Thixotropic behaviour of concentrated bauxite residue suspensions. Rheol. Acta 24: 427–37.

Ochiai, H., Okada, Y., Furuya, G., Okura, Y., Matsui, T., Sammori,T., Terajima, T., and Sassa, K. 2004. A fluidized landslide on a natural slope by artificial rainfall. Landslides, 1(3): 211–219. doi:10.1007/s10346-004-0030-4.

P. Moller, A. Fall, V. Chikkadi, D. Derks, D. Bonn, an attempt to categorize yield stress fluid behaviour, Philosophical Transactions of the Royal Society of London A: Mathematical, Phys. Eng. Sci. 367 (2009) 5139–5155.

Pierre A, Lanos C, Estelle P. Extension of spread-slump formulae for yield stress evaluation[J]. Applied Rheology, 2013, 23(6).

R. Lapasin, V. Longo, S. Rajgelj, Thixotropic behaviour of cement pastes, Cem. Concr. Res. 9 (1979) 309–318.

Schatzmann, M., Bezzola, G. R., Minor, H. E., Windhab, E. J., & Fischer, P. (2009). Rheometry for large particulated fluids: Analysis of the ball measuring system and comparison to debris flow rheometry. Rheologica Acta, 48, 715–733.

Schippa L. Modeling the effect of sediment concentration on the flow-like behavior of natural debris flow[J]. International Journal of Sediment Research, 2020, 35(4): 315–327.

Scotto di Santolo A, Pellegrino A M, Evangelista A, et al. Rheological behaviour of reconstituted pyroclastic debris flow[J]. Geotechnique, 2012, 62(1): 19–27.

Uhlherr, P. H. T., Park, K. H., Tiu, c., Andrews, 1. R. G. 1984. Yield stress from fluid behaviour on an inclined plane. In Advances in Rheology, ed. B. Mena, A. Garcia-Rejon, C. Rangel-Nagaile, 2: 1 83–90. Mexico City: Univ. Nac. Auton. Mex.

Utracki, L. A. 1988. The rheology of two-phase flows. In Rheological Measurement, ed. A. A. ColJyer, D. W. Clegg, pp. 479–594. London: Elsevier.

V. Mechtcherine, A. Gram, K. Krenzer, J.-H. Schwabe, C. Bellmann, S. Shyshko, Simulation of fresh concrete flow using Discrete Element Method (DEM), in: Simulation of Fresh Concrete Flow, Springer, 2014, pp. 65–98.

Wildemuth, C. R., Williams, M. C. 1985. A new interpretation of viscosity and yield stress in dense slurries: coal and other irregular particles. Rheol. Acta 24: 75–91.

Ye, Qian, Shiho, & Kawashima. (2018). Distinguishing dynamic and static yield stress of fresh cement mortars through thixotropy. Cement & Concrete Composites.

Frontiers of Civil Engineering and Disaster Prevention and Control – Yang & Rahman (Eds)
© 2023 The Author(s), ISBN: 978-1-032-31200-2

Flood control water level design for the tunnel project under the estuarine area of Karnaphuli river, Chittagong, Bangladesh

Chen Xi, Dai Minglong*, Li Yanqing, Xu Gaohong & Wang Qingjing
Bureau of Hydrology, Changjiang Water Resources Commission, Wuhan, China

ABSTRACT: To design the flood control water level for multi-lane road tunnel under Karnaphuli River, Chittagong, Bangladesh, a systematic study on flood control design water level, which includes the analysis of return period of historical high tide level, the frequency analysis of design high tide level has been conducted. According to the relative position relationship of tidal station close to the proposed tunnel works, Khal No. 18 and Khal No. 10 stations are used as references for the analysis of design high tide level in the location of the proposed tunnel works. The results show that: 1) To enhance the reliability of research, it is necessary to carry out further survey close to the tide stations, measure the historical high tide levels, evaluate the reliability of hourly tide level data and the maximum tide level over the past years. 2) The return period of historical high tide level should be verified through the death toll resulting from historical storm tides and the survey data of historical flood level and high tide level. 3) According to the results of design high tide level of Khal No. 18 and Khal No. 10 stations, the design high tide levels at tunnel location for different frequencies are calculated as 7.11 m. The study provides a very helpful process and typical case in a perspective of coastal flood control design.

1 INTRODUCTION

Port cities are vital components of national and global economies in both developed and developing countries, playing an increasingly important role for the future (Krugman 1999). At the same time, these cities are exposed to significant consequences because of extreme coastal water level arising, created by considerable surge in storms. A global overview of flood exposure in world coastal cities shows that Dhaka and Chittagong have the highest rate of increase in population exposed out of the top 50 cities most exposed to present-day extreme sea level (Hanson et al. 2011). The flood risk is still increasing, owing to rapid growing populations and the sea level rising due to the changing climate, and subsidence (Hinkel et al. 2014; Nicholls & Cazenave 2010).

Chittagong, located at the southeast coast of Bangladesh, is the main sea port and largest commercial city of the country. The Karnaphuli River divides the Chittagong port area into two parts, the western part is bounded by the urban area and the port, set up on the closed protection zone of seaward flood prevention embankment and the riverward flood prevention embankment even though the dike does not meet the flood control requirements. According to current situation of the riverside and seaside embankments of Chittagong, a proposed tunnel project, the approach segment and open-cut segment, on the west bank of the Karnaphuli River are in the closed protective range formed by existing flood control embankments with the building criterion being slightly lower than the planned criterion. This is while the approach segment and open-cut segment on the east bank are in a protective range of future flood control embankments.

The proposed tunnel project is based on dual-two lane design, hence it will be considered in accordance with the flood control standard for first-class highways with Grade I protection and

---

*Corresponding Author: 12181343@qq.com

DOI 10.1201/9781003308577-60

flood control standard of 100-year recurrence period (Standard for Flood Control (GB50201-2014)). Therefore, the flood control design water levels for the east and west bank segments of the project need to be analyzed with all the factors in consideration. This research mainly includes the following contents: 1) the presentation of the topography, hydrology, and loss in tropical cyclones; 2) the analysis of return period of historical high tide level and the correction of historical high tide level; 3) the frequency analysis of design high tide level based on available tidal station measurements. The study provides a process of flood control water level design, which could be very helpful in a perspective of coastal flood control design in Chittagong.

2 OVERVIEW OF THE STUDY AREA

2.1 Topography and river system

The Bay of Bengal is located to the north of the Indian Ocean and the west of India. Bangladesh is on the top of the bay, and Myanmar and Malaysia are on the east side of the bay. The shape of the Bay of Bengal is flared, and the mouth of the bay in the south is huge, but the top of the bay in the north is narrow. The semi-closed and flared shape is highly susceptible to storm surges, causing a rise in water levels at the top of the bay accompanied by disasters. If the storm surges northward in the northern part of the Bay of Bengal, it will be deflected eastward by the effect of the Coriolis force, resulting in an increase in storm surge along the east coast. It is particularly vulnerable to cyclonic storm surge floods due to its location on the path of tropical cyclones.

The main river that enters the sea in the Bay of Bengal is the Meghna River, which is mainly formed by the upstream Ganges and Brahmaputra. At Goelundo, the Ganges meets with the Brahmaputra. The Padma, which is 105 km from the estuary, meets with the Meghna River from the left bank at Chandpur, and the merged river is also called the Meghna River. Then, the merged river flows into the Bay of Bengal through three tributaries. It is about 250 km long from Goalondo to the Bay of Bengal. The total drainage area of three rivers is about 1.75 million km^2. Among them, the Ganges is 1.05 million km^2, the Brahmaputra 622,000 km^2, and the Meghna River about 80,000 km^2. India accounts for 814,400 km^2, China accounts for 52,000 km^2, Nepal accounts for 147,000 km^2, and Bangladesh accounts for 38,000 km^2 in the 1.05 million km^2 drainage area of the Ganges. The average flow of the Ganges is 17,400 m^3/s, the average flow of the Brahmaputra is 19,600 m^3/s, the Meghna River has an average flow of 4500 m^3/s, and the total runoff of rivers into the sea is $41,500/m^3$. The Karnaphuli with the width of approximately 670 m is a mountain stream, which is located in the southeastern basin of Bangladesh. The river originates from the Lushai mountainous region of Mizoram in India, flows through 270 km and enters the Bay of Bengal via the southwest of Chittagong Hill and the Chittagong Port. The annual average flow of the Karnaphuli is 469 m^3/s.

2.2 Tidal height and human life loss in history tropical cyclone

The tidal range of the Indian border on the west side of the Bay of Bengal is generally around 3 m, and the tidal range near the Sandwip island at the estuary of the Meghna River is up to approximately 5 m. The tidal range is about 3.5 m to 4 m along the southeastern part of the Bay of Bengal. There are three tidal stations near the estuary of Karnaphuli, they are Khal No. 18, Khal No. 10, and Sadarghat. Except for Khal No. 18 tidal station, which has more complete hourly tide data from 1985 to 2016, the other tidal stations only have the highest daily tidal level and lowest tidal level characteristic value. It can be seen that the highest water level is 4.34 m (October 1984) at Khal No. 18 tidal station over the years, which appears in July at the earliest and in October (2010) at the latest; July and August accounts for 71.5%.

The coastal areas of Bangladesh comprise low-lying and poorly protected land, which supports a large population. All the ingredients for a major cyclone disaster are present and such disasters have occurred several times in the past and claimed 500,000 and 138,000 lives, notably in 1970 and 1991, respectively (Haque 1997). A brief summary of historic cyclones occurred in 1960–2015 that land along the coast of Bangladesh and caused catastrophe in terms of coastal flooding and human casualties (Chittagong Development Authority Chittagong City Outer Ring Road Project (SMEC 2012); Chittagong Storm Water Drainage and Flood Control Master Plan (CPA 1994)). It mainly includes characteristics like historical typical storm surge landing location, maximum wind speed, wind direction, maximum surge height, and central pressure. There were 4 typhoons that landed in and around Chittagong and had a maximum storm surge of more than 6.10 m, which respectively occurred in October 1960, November 1966, November 1970, and April 1991, as shown in Table 1. The maximum storm surge of the typhoon in 1970 was 10.1 m, but at the time of occurrence of the typhoon in 1970, there was no record on measured maximum water level at Khal No. 18 tide station. While from the perspective of the death toll in Bangladesh caused by storm surges shown in Table 2, the storm surges that resulted in top three death tolls in Bangladesh since 1822 occurred in 1970, 1897, and 1991, respectively.

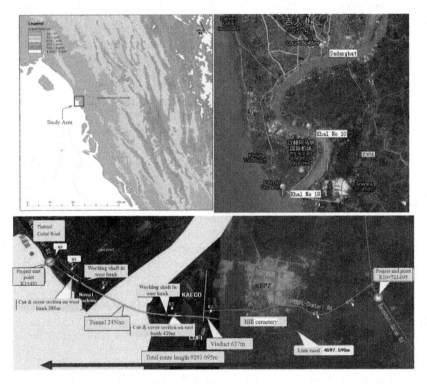

Figure 1. Study area with the location of project and hydrological station in the Karnaphuli River, Chittagong, Bangladesh.

Table 1. Maximum water level recorded during significant cyclone occurring in Chittagong area, 1960–2015. Datum: MSL

Date	Tidal Surge Height (m)	Khal No. 18 (m)	Khal No. 10 (m)	Sadarghat (m)
10.31.1960	6.1	7.129	4.819	3.575
11.01.1966	6.7	5.849	4.575	4.261
11.12.1970	10.1	–	–	–
11.28.1974	5.2	6.695	5.92	5.41
05.24.1985	4.6	7.197	5.86	4.64
04.29.1991	6.7	7.875	5.28	4.627

Table 2. Loss of human life by tropical cyclone disaster in Bangladesh.

Year	Deaths (data source a)	Deaths (data source b)
1970	500,000	Officially 200,000, unofficially 500,000
1897	175,000	/
1991	150,000	138,882
1876	100,000	/
1822	40,000	/
1965	19,279	17,279
1963	11,520	11,520
1961	11,468	11,488
1960	5149	5149

3 FREQUENCY ANALYSIS OF DESIGN HIGH TIDE LEVEL

3.1 Rechecking of tide level data

According to the relative position relationship of tidal station close to the proposed tunnel works, Khal No. 18 and Khal No. 10 stations are used as reference stations for the analysis of design high tide levels in the location of the proposed tunnel works. ISLWL is used as the height datum for observing and recording the water level data of the two stations, and its datum conversion relationship with the mean sea level height datum MSL is found.

Khal No. 18 station is located about 500 m from the downstream of the proposed tunnel works, Khal No. 10 station is located about 6,800 m from the upstream of the proposed tunnel works, it can be seen that Khal No. 18 station is very important for the engineering design. Khal No. 18 station tide level test facility, within the management scope of the Naval Academy, was visited. Tide staff is used as the tide level observation equipment at this station. According to the description of management personnel, before 2013, tide staff was used to observe the tide level once every 30 min, and high tide level was tested once every 5 min when storm tide occurs. To eliminate the effect of wave when observing the tide level, the well-barrel tide gauge was used for simultaneous observation in this station since 2013, and was connected to the online data transmission system for transmitting tide level data once every 1 min.

In this research, the measured hourly water level data of 1985–2015 at Khal No. 18 station was collected from the Chittagong Port Authority (CPA), and the data quality was analyzed. Height datum of the three data sources were converted to MSL. Moreover, according to the collected typhoon death toll of 1970, which was ranked first as mentioned in multiple research reports, surge height during the typhoon period was 10–33 ft (about 3.05 m–10.1 m), but there is no maximum typhoon level observation record in the Khal No. 18 station. According to the analysis and research, typhoon information of 1970 was incomplete, and there was a great dispute about the death toll. There is no maximum water level record of Khal No. 18 station; thus, it is inappropriate to directly use the maximum typhoon surge height of 10.1 m of 1970 for frequency analysis of Khal No. 18 station. No measured hourly water level data at Khal No. 10 station is collected from the CPA, the annual maximum high tide levels of Khal No. 10 station adopted in this research is sourced from the feasibility study for multi-lane road tunnel under River Karnaphuli, Chittagong, Bangladesh. Correlation analysis is made about high tide levels of Khal No. 18 and Khal No. 10 stations. In Figure 3, it can be seen that the high tide level correlation between the two stations is good, correlation coefficient can be up to 0.88, it can be evaluated that external factors, which affect the high tide level of the two tidal stations, are the same and stable. To enhance the reliability of research results, it is necessary to carry out further survey close to Khal No. 18 and Khal No. 10 stations, measure the historical high tide levels before establishment of the station, and evaluate the reliability of hourly tide level data since the establishment of the station, and the maximum tide level over the past years.

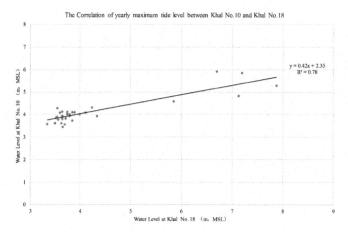

Figure 2.   Correlation of yearly maximum tide level between khal No. 18 and khal No. 10 stations.

3.2   *Analysis of return period of historical high tide level*

Survey data of historical flood level and high tide level of the Karnaphuli River estuary is incomplete; the return period of historical high tide level is verified through the death toll resulting from historical storm tides in Chittagong.

According to the Chittagong flood control plan (Chittagong Storm Water Drainage and Flood Control Master Plan (CPA 1994)), the top three highest deaths due to storm tides in Bangladesh ever since 1822 happened in 1970, 1897, and 1991 (see Table 2). The deaths during the storm tide period in 1970 were the most since 1822 and the typhoon information of 1970 was incomplete. There is dispute about the death toll, and there is no maximum high tide level record of Khal No. 18 station during the storm tide surge period in 1970. Furthermore, there is no survey result of the historical high tide level. However, with social and economic development, the flood control and disaster mitigation facilities continue to be improved, CPA (Chittagong Storm Water Drainage and Flood Control Master Plan (CPA 1994)) infers that the maximum water level of Khal No. 18 station during the storm tide period in 1970 was lower than that in 1991. However, typhoon information of 1991 is relatively complete, there is the maximum tide level record of Khal No. 18 station during the storm tide surge period, i.e., 7.875 m (MSL, the same below), death toll is 150 thousand persons, slightly less than 175 thousand deaths caused by typhoon in 1897. It can be considered that storm tide surge in 1991 is the maximum surge since 1897, and return period is 121 years; or it is in the second place since 1822, return period is 98 years. Return period for the storm tide surge of 1991 estimated by the two methods is less different. In this research, a return period of 121 years is used for the storm tide surge of Khal No. 18 in 1991.

The maximum water level of Khal No. 18 station due to storm tide in 1985 and 1960, respectively, is 7.197 m and 7.129 m, the two maximum water levels are almost the same and are tied for the second place, and the maximum water levels 6.695 m and 5.849 m of Khal No. 18 station due to storm tide in 1974 and 1966 are, respectively, ranked 4th and 5th.

Similarly, the yearly maximum high tide levels of Khal No. 10 station in 1991, 1985, 1960, 1974, and 1966 are considered exceptional values, wherein the maximum high tide level of 1974 can be considered as the highest level of Khal No. 10 station since 1960, ranked first, the return period is 59 years, and the values of 1985, 1991, 1960, and 1966, respectively, are ranked from the 2nd place to the 5th place.

3.3   *Correction of historical high tide level*

Through field survey and confirmation, there are flood marks for high tide level of 1991 marked on a tree close to the cemetery of martyrs in the Naval Academy. This tree is located at 150 m from

446

the downstream of Khal No.18 station. According to the field survey conducted by CCCC, flood mark for 1991 is 7.01 m (MSL), about 0.86 m lower than the value 7.875 m of Khal No.18 station included in the survey record (see Figure 3).

Considering that the high tide level flood marks found in survey are difficult to eliminate the wave effect, the surveyed high tide level data may be higher. Therefore, the historical high tide level values surveyed from 5 surveys are considered to be corrected together.

As the time of recording flood mark during typhoon is hard to determine, it is considered in a conservative manner that at this time, the correction is done by subtracting average wave height caused by maximum wind speed measured in local weather stations during typhoon from high tide levels obtained from Khal No. 18 and Khal No. 10 stations.

Based on investigation record of high tide level during various typical strong typhoons, topography in project area and measured data of maximum wind speed at Chittagong Weather Station, the Putian Test Station Formula (Code for Design of Levee Project (GB50286-2013)) is used to calculate the average wave heights, which is shown in Table 3.

Table 3. Correction results of high tide level survey record for khal No.18 and khal No.10 stations.

Occurrence time of strong typhoon	Khal No. 18 (m, MSL)			Khal No. 10 (m, MSL)		
	Survey record	Calculate average wave height	After correction	Survey record	Calculate average wave height	After correction
1960.10.31	7.129	0.555	6.574	4.819	0.256	4.563
1966.11.01	5.849	0.392	5.457	4.575	0.210	4.365
1974.11.28	6.695	0.493	6.202	5.920	0.401	5.519
1985.05.24	7.197	0.562	6.635	5.860	0.404	5.456
1991.04.29	7.875	0.577	7.298	5.280	0.297	4.983

Figure 3. High tide level record location (1991) to Khal No.18 Station and the measurement.

3.4 *Recheck of design high tide level*

Through analysis, the corrected historical high tide levels of Khal No. 18 station in 1991, 1985, 1960, 1974, and 1966 are considered as exceptional values, which compose a discontinuous series together with the maximum tide levels measured in1980–2015. The P-III frequency curve is used for fitting, shown in Figure 4 (Left). Long measured data series is available for the calculation of design high tide level of Khal No. 18 station, and verification of historical exceptional value has been carried out, thus it is well representative.

The corrected historical high tide levels of Khal No. 10 station and the measured yearly maximum tide levels of 1980–2010 compose a discontinuous series, the P-III frequency curve is used for fitting, less difference between the maximum high tide level in 1985 and that in 1974 is considered in fitting, and the point of 1985 is properly moved leftward (Figure 4, Right).

According to the results of design high tide level of Khal No. 18 station and Khal No. 10 station for different frequencies, the relative position of tunnel location to Khal No. 18 station and Khal No. 10 stations, the design high tide levels at tunnel location for different frequencies are calculated. For the design high tide levels at the design reference station and the proposed tunnel location for different frequencies, refer to Table 4.

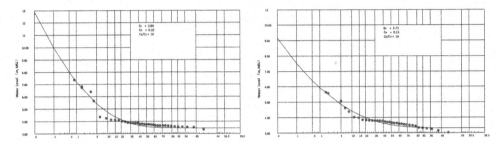

Figure 4. Frequency Curve of Yearly Maximum Tide Level of Khal No. 18 Station (Left) and Khal No. 10 Station (Right).

Table 4. Design high tide level at the design reference station and the proposed tunnel location.

Location	EX	Cv	Cs/Cv	P=0.33%	P=1%	P=2%	P=5%	P=10%
Khal No. 18	3.89	0.2	18	8.52	7.20	6.41	5.42	4.73
Khal No. 10	3.72	0.15	18	6.68	5.91	5.45	4.84	4.40
Tunnel location	/	/	/	8.39	7.11	6.34	5.38	4.71

Notes: height datum: MSL, unit: m.

4 CONCLUSION AND DISCUSSION

4.1 Conclusion

The proposed tunnel project is located at the Karnaphuli River estuary. The design reference stations Khal No. 18 and Khal No. 10 adopted for engineering design are located on the tidal river reach, where the water level is affected by upstream, ocean tide, and storm tide. According to the results of design high tide level of Khal No. 18 station and Khal No. 10 station, the design high tide levels at tunnel location for different frequencies are calculated as 7.11 m. The study provides a very helpful process and typical case in a perspective of coastal flood control design.

4.2 Discussion

As Khal No. 18 station is close to the location of the proposed tunnel works and the complete data are available, in previous research results, there are many results about the design high tide level of the station for different frequencies, and rationality analysis is conducted for design high tide level of Khal No. 18 station.

In this research, based on review of previous research results, the yearly maximum tide levels of Khal No. 18 station are extended to 2015, exceptional values of years 1991, 1985, 1960, 1974, and 1966 are added, the exceptional values of 1991 are verified for return period, the measured maximum high tide level of 1991 is identified as the highest level since 1897, and return period is 121. This is rational and the results are highly reliable. Since the data series used in the research is short, and no historical exceptional value is used in frequency calculation, the frequency design result is low. SMEC did not verify the return period of the high tide level of Khal No. 18 station caused due to several typical storm tides in 1991, etc., and directly used the storm tide maximum surge (unknown location) in 1974 as the maximum tide level of Khal No. 18 station of this year, and the calculated 50-year design result is relatively higher. In conclusion, the results of design high tide level for different frequencies of Khal No. 18 station, recommended for this research are rational and highly reliable.

ACKNOWLEDGMENTS

We would like to thank the Chittagong Port Authority and the Naval Academy of Bangladesh for providing convenience in the processes of site survey and historical high tide level investigation. We would like to thank the kind-hearted and friendly people of Bangladesh along the project route for their enthusiastic cooperation with the project team members in the flood water level survey – providing reliable historical data support for the project design for the project team members to conduct reliability verification of the design results.

REFERENCES

Chittagong Development Authority Chittagong City Outer Ring Road Project (SMEC, 2012).
Chittagong Storm Water Drainage and Flood Control Master Plan (CPA, 1994).
Code for Design of Levee Project (GB50286-2013).
Hanson, S. et al. (2011) A global ranking of port cities with high exposure to climate extremes. Climatic Change 104, 89–111.
Haque C E. (1997) Atmospheric hazards preparedness in Bangladesh: a study of warning, adjustments and recovery from the April 1991 cyclone[M]//Earthquake and Atmospheric Hazards. Springer, Dordrecht, 181–202.
Hinkel J, Lincke D, Vafeidis A T, et al. (2014) Coastal flood damage and adaptation costs under 21st century sea-level rise. Proceedings of the National Academy of Sciences, 111(9): 3292–3297.
Krugman P. (1999) The role of geography in development. International regional science review, 22(2): 142–161.
Nicholls R J, Cazenave A. (2010) Sea-level rise and its impact on coastal zones. science, 328(5985): 1517–1520.
Standard for Flood Control (GB50201-2014)

Frontiers of Civil Engineering and Disaster Prevention and
Control – Yang & Rahman (Eds)
© 2023 The Author(s), ISBN: 978-1-032-31200-2

Determination and optimization of cycle advance for the Kyrgyzstan North-South ridge crossing tunnel

J. Li*
China Road and Bridge Engineering Co., LTD, Beijing, China

Z.Z. Xia*
Civil Engineering College, Xi'an University of Architecture and Technology, Xi'an, China
Shaanxi Provincial Key Laboratory of Geotechnical and Underground Space Engineering, Xi'an, China

Z.P. Zhou*
CCCC No. 4 Highway Engineering Bureau Co. LTD, Beijing, China

ABSTRACT: Selecting a reasonable cycle advance is an important measure to effectively prevent tunnel vault collapse and heading face failure, meanwhile, the construction period and schedule are also under the influence of cycle advance. Thus, the cycle advance is an important factor that keep balance of safety and economical demand in tunnel construction. In this paper, we conclude some commonly used methods for cycle advance's determination, and choose the model of considering vault and heading face stabilities given to practical application scenarios by comparison and analysis, and according to the theoretical calculation formula to obtain the theoretical value of the construction cycle advance of the Kyrgyzstan north-south over-ridge tunnel. Then the vault settlement and horizontal convergence are selected as the control targets, the deformation characteristics of surrounding rock mass in different cycle advance parameters are compared and analyzed by numerical simulation, the optimized result of cycle advance is 1.5 m.

1 CASE OVERVIEW

The total length of the Kyrgyzstan North-South Ridge Crossing Tunnel is 3750 m. The tunnel is divided into two sections, A and B. In section 3-A, the entrance's pile number is 43+190. The length of the integrated facility at the entrance is 38.5m. The service tunnel is located in the tunnel section at 18.25 m from the centerline on the upslope side (east side). The length of the service tunnel is 3745 m without considering the auxiliary buildings and structures located at the entrance location. The main tunnel is excavated by the two-bench excavation method and excavation span is 12.6 m while the service tunnel is excavated by the full-face excavation method and excavation span is 5.2 m. The buried depth of entrance of main tunnel is 12.3 m. Primary lining consists of steel arch support, system anchor bolt and shotcrete layer with 30 cm thickness. Since the tunnels were designed by Russian engineers according to Russian specifications, but the supervisor was Kirgizia, which led to inconsistencies in construction concepts and specifications between the parties in tunneling, causing many problems for Chinese construction companies during the construction of the tunnel. So much so that later in the process of developing the construction plan and reporting the construction plan to the supervisor, the supervisor required the theoretical basis for the construction cycle advance to be given and the comparative analysis of the surrounding rock deformation under different cycle advance to ensure the construction safety. Therefore, based on the above engineering background, this paper analyzes and determines the construction cycle advance from the theoretical and numerical simulation aspects respectively.

---

*Corresponding Authors: 715276838@qq.com, kashinsyou@163.com and 3247580769@qq.com

                                            DOI 10.1201/9781003308577-61

2 THEORETICAL STUDY OF CYCLE ADVANCE

Certain scholars explored and discussed the determination of tunnel with the cycle advance. Wang and Gong (2010) deduced and proposed the calculation formula of the cycle advance of the shallow buried underground tunnel. Based on the Janssen silos theory, Shi et al. (2015) proposed the calculation formula for the cycle advance for the shallow buried section of the rock mass tunnel entrance, and analyzing the reasonable values of cycle advance under different mechanical processes and parameters' conditions. Zhang et al. (2020) proposed an improved prism-wedge model to quantify and evaluate the stability factor of a tunnel heading face in some cohesive frictional soil. Liang et al. (2017) proposed a calculation formula considering vertical force loading of vaults, which is based on the logarithmic spiral failure mode, and deduced the formula to calculate the round length in advanced core soil on the tunnel face.

At present, the main methods for determining the cycle advance are theoretical calculation method, numerical simulation method, engineering analogy method and other methods. Among them, the theoretical method mainly includes the calculation method of cycle advance based on arch effect, Terzaghi loose body theory, Janssen silo theory and full column theory.

2.1 Cycle advance determination method based on arch theory

Hui and Wang (1995) consider that there is arching effect existing in unsupported rock mass. The tunnel vault will form a pressure arch along the longitudinal direction, further, based on the pressure arch effect, a theoretical calculation formula for the cycle advance is established. The calculation model is shown in Figure 1. The calculation formula is shown in Equation 1 below.

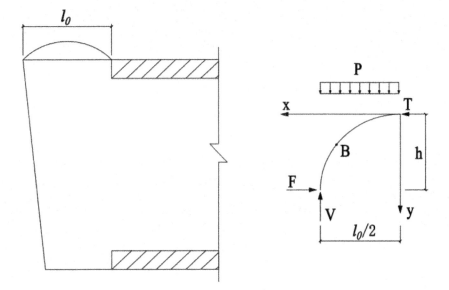

Figure 1.   Calculation model.

$$l_0 = \frac{6Cf_k}{K\gamma} \tag{1}$$

2.2 Cycle advance determination method based on Terzaghi loose body theory

Wang and Gong (2010) consider that surrounding rock affected by working still have cohesion. The collapse of the surrounding rock at the vault depends mainly on whether the frictional force on the undisturbed soil mass is greater than the self-weight of the rock at the vault. Based on Platts theory of ground pressure and Terzaghi loose body theory, the cycle advance calculation theory with consideration of vault's stability is proposed. The calculation model is shown in Figure 2, and the calculation formula is shown in Equation 2 below.

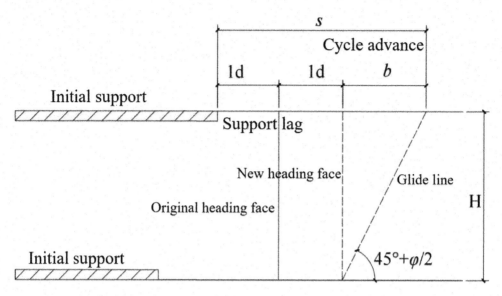

Figure 2. Calculation model.

$$\frac{qs - 2\gamma l_1 h}{2l_1 + s} \leq \left(\frac{\gamma l_1 - c}{k \tan \varphi} - q\right)\left(e^{-k \tan \varphi \frac{h}{l_1}} - 1\right) \tag{2}$$

2.3 Cycle advance determination method based on silo theory

Shi et al. (2015) consider that when it comes to the vault of tunnels in rock debris, the contribution of arch effect to stability is slight, and face stability would be affected obviously by the cycle advance's parameter. The cycle advance's calculation formulas are proposed with consideration of face stability and the formulas are as Equations 3–6 shown with the explanation of Figure 3.

$$\frac{G + V}{\tan (45° + \varphi/2)} \leq \frac{1}{K}\left[\frac{T_s + \frac{cbh}{\cos(45° - \varphi/2)}}{\sin (45° - \varphi/2) + \cos (45° - \varphi/2) \tan \varphi} + S\right] \tag{3}$$

$$G = \frac{1}{2}bh^2 \tan (45° - \varphi/2) \gamma \tag{4}$$

$$V = A\sigma_v \tag{5}$$

$$T_s = h^2 \tan (45° - \varphi/2)\left(c + k_a \tan \varphi \frac{rh + 3\sigma_v}{3}\right) \tag{6}$$

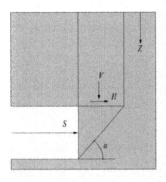
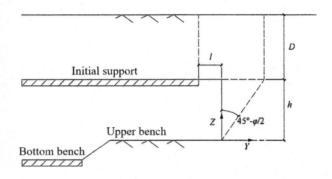

Figure 3. Calculation model.

2.4 Cycle advance determination method based on surrounding rock mass pressure theory

Li et al. (2018) think that cycle advance would affect both the face stability and vault stability, and the formulas with consideration of vault and face stability are also proposed as follows:

$$
s = \frac{k_3}{\left(dH\gamma - \gamma H^2 \lambda \tan\theta\right)\left(k_2 - \frac{2h^2}{3bd}K_a k_1\right)}p
$$
$$
+ \frac{(b+d)H\lambda \tan\theta - bd}{d - H\lambda \tan\theta} + \frac{Ch(\frac{d}{\sin\beta} + \frac{h}{\tan\beta}) + \frac{1}{3}\gamma h^3 K_a k_1 - k_2 W_3}{(dH\gamma - \gamma H^2 \lambda \tan\theta)(k_2 - \frac{2h^2}{3bd}K_a k_1)} \tag{7}
$$

$$
s = \frac{dH\lambda \tan\theta}{d - H\lambda \tan\theta} \tag{8}
$$

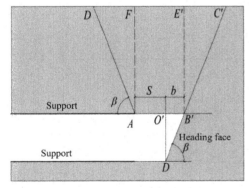
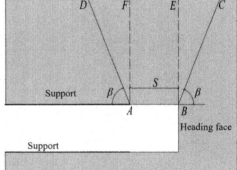

Figure 4. Calculation model.

3 DETERMINATION OF KYRGYZSTAN TUNNEL'S CYCLE ADVANCE

Through the above summary of typical cycle advance theoretical calculation methods found. Scholars have carried out theoretical calculations of cycle advance based on single arch effect, palm face stability, and vault stability, respectively. Meanwhile, this paper considers that the calculation formula based on the simultaneous stability of palm face and vault proposed by Li et al. (2018) is more reasonable. Because two phenomena, excavation face and arch collapse, often occur during tunnel construction, rather than a single situation. Therefore, the calculation method of cycle advance proposed by Li et al. (2018) is closer to the actual project. Therefore, in this paper, we adopt the method of calculating the cycle advance that can guarantee the stability of both vault and excavation face to determine the cycle advance for the construction of the bench excavation method of the North-South Crossing Tunnel in Kyrgyzstan, and use the numerical simulation method to optimize the calculated cycle advance.

3.1 Theoretical determination of tunnel advance

The tunnel surrounding rocks are mainly schist and siltstone, and the surrounding rock grade is V. In the calculation, considering the difference of rock parameters, two sets of data corresponding to higher and lower parameters of the surrounding rock grade were selected for calculation, and the calculated parameters are shown in Tables 1, 2 and 3. Then, according to equations (7) and (8), the cycle advance of the tunnel under the two sets of parameters is obtained as $s_1 = 1.5m, s_2 = 0.7m$, respectively.

Table 1. θ and φ_0 under different rock grades.

Grades	I	II	III	IV	V	VI
θ	$0.9\varphi_0$	$0.9\varphi_0$	$0.9\varphi_0$	$(0.7 - 0.9)\,\varphi_0$	$(0.5 - 0.7)\,\varphi_0$	$(0.3 - 0.5)\,\varphi_0$
φ_0	>78	70~78	60~70	50~60	40~50	30~40

Table 2. Calculation parameters.

Weights $\gamma(kN/m^3)$	Cohesion $c(kPa)$	Frictional angel φ_0	Lateral pressure coefficient K_a
20	200	27	0.57
17	50	20	0.38

Table 3. Materials' parameters of simulation.

Materials	Weight $\gamma(kN/m^3)$	Cohesion $c(kPa)$	Frictional angle φ_0	Elastic modulus E (GPa)	Poisson ratio
Rock ①	20	200	27	2	0.35
Rock ②	20	125	25	1.5	0.35
Shotcrete	24	–	–	5	0.2
Anchor bolt	78.5	–	–	210	0.3

3.2 Optimization of cycle advance

Theoretically, the cycle advance of the tunnel is 0.7 m - 1.5 m. In order to obtain the cycle advance that can meet the safety requirements and speed up the construction progress. In this paper, three cycle advances of 1.0m, 1.2m and 1.5m are selected for numerical simulation, and the optimal excavation length is determined by comparing the deformation of the surrounding rock under each cycle advance. Numerical calculation model by Rhinoceros is imported into FLAC software for calculation, constraints to the displacement surfaces in the X and Y directions of the model boundary and constraints to the displacement surfaces in the Z direction of the bottom boundary are activated, in the initial support, shotcrete is simulated by 2D unit; anchor is simulated by 1D implantable truss unit; rock body is simulated by 3D solid unit. What is more, the surrounding rock is simulated by the Mohr-Coulomb constitutional model and the support is simulated by the elastic constitutional model to realize the mechanical process. The material parameters of the numerical simulation are shown in Table 3, and the finite element calculation model is shown in Figure 5.

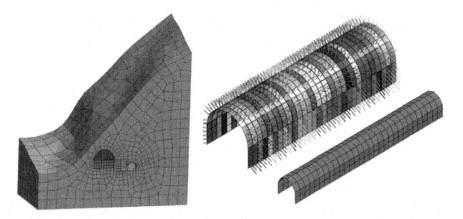

Figure 5.   FEM calculation model: Mesh model and shotcrete and anchor bolts system.

3.2.1 *Vault settlement*

The settlement pattern of the vault under different advance is shown in Figure 6. Under the two rock conditions of ① and ②, the change pattern of tunnel vault settlement under the three working conditions is basically the same, with the vault settlement gradually increasing and then stabilizing. It means that the deformation of the tunnel can be converged under all three cycle advances. At the same time, it can be found that the vault settlement increases with the increase of the cycle advance. The final settlement of the vault is 3.3 mm, 3.4 mm and 3.5 mm under the condition of the surrounding rock , while the final settlement of the vault is 5.2 mm, 5.3 mm and 5.5 mm

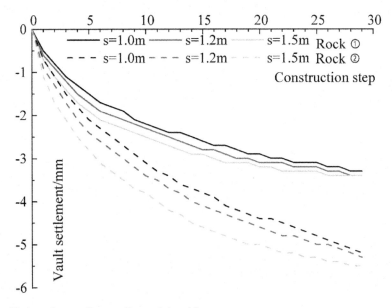

Figure 6.   Vault settlement: Surrounding rock 1 and 2.

under the condition of the surrounding rock . Although the settlement of the vault increases with the increase of the excavation, the increase is only 0.1 mm, which is almost negligible. When the excavation depth is 1.5 m, the settlement of the vault under the two rock conditions is 3.5 mm and 5.5 mm respectively, which does not exceed the specification of 20 mm and meets the requirement of settlement deformation.

3.3 *Horizontal convergence*

The deformation pattern of horizontal convergence of the surrounding rock under different excavation steps is shown in Figure 7. Under the condition of ①, the horizontal convergence tends to expand outside the tunnel before the palm face reaches the monitoring section, and the tunnel profile starts to shrink inside the tunnel after the palm face passes the monitoring section. Under the condition of ②, the horizontal convergence tends to shrink into the tunnel before the palm face reaches the monitoring section, and the tunnel profile starts to shrink into the tunnel after the palm face passes the monitoring section. Under the conditions of ① and ②, the change pattern of horizontal convergence under the three working conditions is basically the same, with horizontal convergence increasing first and then stabilizing. It means that the deformation of the tunnel can be converged under all three cycle advances. At the same time, it can be found that the horizontal convergence increases with the increase of the cycle advances. The horizontal convergence is 4.6 mm, 4.8 mm, and 4.9 mm for the three cycle advances in the surrounding rock ①, and 7.4 mm, 15.7 mm, and 14.8 mm for the three cycle advances in the surrounding rock ②. Although the horizontal convergence increases with the increase of the cycle advance, when the excavation advance is 1.5 m, the horizontal convergence in the two surrounding rock conditions is 4.9 mm, respectively. The horizontal convergence is 4.9 mm and 14.8 mm respectively when the excavation depth is 1.5 m, which do not exceed the specification of 20 mm and meet the requirement of settlement deformation.

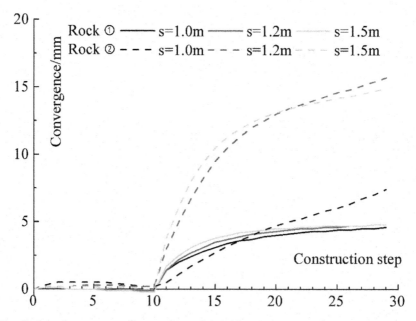

Figure 7. Horizontal convergence: Surrounding rock 1 and 2.

According to the above comparative analysis of vault settlement and horizontal convergence, it can be found: Although both horizontal convergence and vault settlement increased with the increase of the cycle advances. However, under both rock conditions, the deformation of the tunnel did not exceed the standard's limit of 20 mm at 1.5 m. Therefore, after considering the construction period and safety factors, the recirculation level of the tunnel was set to 1.5 m. This excavation level also met the requirement of 0.7 m steel arch spacing. Then, the theory and calculating equations in 2.4 are selected to verify the cycle advance value, similar result can be performed for the stability of both vault and heading face, which has slightly different with the numerical simulation in exact values. What is more, the construction project OPA indicates such length in cycle advance can fit with most work conditions in other places near this study case's location.

4 CONCLUSIONS

(1) The theoretical calculation methods of several typical tunnel cycle advances are summarized and analyzed, and based on the comparative analysis, it is concluded that the calculation formula of tunnel cycle advances given by considering the simultaneous stability of palm face and vault is closer to the engineering reality.
(2) Adopting the method of calculating the cycle advance that can guarantee the stability of both vault and palm face, we determine the cycle advance for the construction of the bench method in the north-south cross-ridge tunnel in Kyrgyzstan, and use numerical simulation to compare and analyze the settlement under different cycle advances, which shows that the vault settlement and horizontal convergence of the tunnel increase with the increase of the cycle advances.
(3) Under both rock conditions, the deformation of the tunnel did not exceed the 20 mm standard specified when the cycle advance is 1.5 m. After considering the construction period and safety factors, the tunnel cycle advance is set to 1.5 m. This excavation level also met the requirement of 0.7 m steel arch spacing.

REFERENCES

Hui Liping & Wang Liang. 1995.Theoretical discussion on the excavation progress of shallow buried soil tunnels. *Railway Standard Design*, 11: 25–27.
Li Hui,Tian Xiaoxu,Song Zhanping,Wang Juanjuan,Zhou Guannan.2018. Research on the calculation method of shallow buried tunnel excavation progress based on Xie Jiajie's formula. *Journal of Xi'an University of Architecture and Technology (Natural Science Edition)*, 50(05): 662–667.
Liang Q, Yang X.L., Chen X.2017. "Limit equilibrium analysis of round length in tunnel excavation," *Journal of South China University of Technology (Natural Science)*, 45(05) :113–119.
Shi Xianhuo, Dai Yuanquan, Guo Jianqiang. 2010.Calculation and analysis of the excavation progress of shallow buried section of rock pile tunnel opening–Take Zhaojiaya Tunnel of Ma Zhao Expressway in Yunnan as an example. *Tunnel Construction*, 35(08): 787–791.
Wang Zhida & Gong Xiaonan. 2010.Calculation method of excavation footage for shallowly buried concealed pedestrian tunnels. *Geotechnics*, 31(08): 2637–2642.
Zhang X, Wang M.N., Li J.W., Wang Z.L., Tong J.J., Liu D.G.2020. "Safety factor analysis of a tunnel face with an unsupported span in cohesive-frictional soils," *Computers and Geotechnics*, 117: Article ID 103221.

*Frontiers of Civil Engineering and Disaster Prevention and
Control – Yang & Rahman (Eds)*
© 2023 The Author(s), ISBN: 978-1-032-31200-2

Research on microscopic properties of alkali-activated cementing materials mixed with iron ore tailings powder and ground granulated blast furnace slag at different curing ages

Song Xia, Ming Zhou* & Aimin Qin
School of Civil Engineering and Architecture, West Anhui University, Lu'an, Anhui, China

Chuanming Chen
Anhui Gaudi Circular Economy Industrial Park Co. Ltd., Lu'an, Anhui, China

ABSTRACT: To study the influencing factors of activity of the materials, an experiment on alkali-activated cementing materials which are prepared with Iron Ore Tailings (IOT) micro-power and Ground Granulated Blast Furnace Slag (GGBFS) under different curing ages was carried out. In the experiment, sodium silicate and solid sodium hydroxide were used as the alkali activators. Moreover, the micro-structural properties were tested through microscopic experiments. According to SEM results, the quantity and size of SiO_2 crystals gradually increased with the increase of curing age. The flocculent-loose structure was changed to a denser platy structure, with the generation of uniform and dense products. According to EDS results, the hydration products should be silicate aluminates, and the inert RO phase also participates in the hydration process as the curing age extends. FTIR results demonstrate that the degree of anion polymerization of silicate increased, and hydration degree also increased with the increase of curing age.

1 INTRODUCTION

GGBFS and IOT are solid wastes from industrial production. Since most of mineral processing enterprises choose simple ore mining technologies, it has considerable remaining tailings and metal resources account for a high proportion in tailings (Huang et al 2012). According to statistics, the comprehensive utilization rate of tailing in China was only 26.5 % in 2016 (Behera et al. 2018), accompanied with a considerable volume of IOT deposits, accounting for nearly 1/3 of all types of tailing. However, the utilization of slags could reach more than 85% (Ministry of Ecology and Environment of the People's Republic of China 2018). The iron ores are grinded thinner with the improvement of mining technologies thus resulting in the generation of thinning tailings. Therefore, using IOT powder as the building materials (Mendes et al. 2019; Tian et al. 2016; Wang & Wang et al 2016) and increasing the utilization rate of IOT have been an important research direction.

Cheng et al. (2020) discusses the durability of concrete incorporated with a mineral admixture of IOT that were activated mechanochemically, Huang (2013) et al. demonstrated that using IOT as the cement substitute to produce more environmental-friendly cement composite (ECC). Han (2017) studied the early hydration characteristics after adding IOT into cement-based materials and found that mortar containing fine IOT powder released more hydration heats, and thus resulting in the higher early strength. Cheng et al. (2016) found that IOT were rich of non-metallic minerals and the total contents of Si and Al elements exceeded 70 wt. The global CO2 emissions from cement production accounts for 5% of total emissions (Shen et al. 2020). Compared with the ordinary concrete, concrete with alkali-activated slags can be used as the binder for concrete production under alkaline conditions, which has high strength and strong resistance to chemical erosion (Wang 1995).

---

*Corresponding Author: wxxyzm123@163.com

DOI 10.1201/9781003308577-62

In this study, GGBFS and IOT micro-powder were used as binding materials, and the influences of curing age was discussed. The micro-structures of samples were analyzed by XRD, SEM and FTIR. This lay a foundation for further practical engineering applications.

2 RAW MATERIALS AND MIX PROPORTION

The S95-grade GGBFS and IOT produced by local supplier company were chosen. The specific surface area of GGBFS and IOT were 382m2/kg and 416m2/kg, and their chemical composition was listed in Table 1.

Table 1. Chemical compositions of GGBFS and IOT micro-powder (wt %).

	SiO_2	Fe_2O_3	CaO	Al_2O_3	MgO	Na_2O	Loss on ignition
Grade slags	32.11	0.53	38.52	18.56	9.62	0.66	<1.0%
IOT micro-powder	49.01	30.02	7.94	6.67	5.73	0.63	<1.0%

The sodium silicate used as the alkali-activated agent was produced by the Wu xi Yatai United Chemical Co. Ltd. Physical and its chemical properties were shown in Table 2. The used sodium hydroxide was provided by Aladdin Company, with an analytical pure of 97.5%. Dosages of various materials were estimated, as shown in Table 2.

Table 2. Physical and chemical properties of sodium silicate.

$w(SiO_2)/\%$	$w(Na_2O)/\%$	$w(water)/\%$	Modulus	Baume/° Bé
12.8	29.20	53	2.30–2.50 (2.4)	50.0–51.0

Table 3. Mix proportions.

M	$\frac{m(Na_2O)}{m_B}/\%$	m_W/m_B	Temperature/°C	Sodium silicate/g	NaOH/g	Iron tailing powder/g	Grade slags/g	Water/g
1.35	9	0.35	35	2115.0	249.0	2160.0	3240.0	580.9

3 TEST PROCEDURE

3.1 *XRD of Raw material, energy spectrum test*

XRD and energy spectrum tests were carried out on raw materials slag powder and IOT powder. The particle size distribution of GGBFS and IOT powder was measured by laser particle size analyzer.

3.2 *Preparation of test specimen*

The preparation of test specimen will mix the GGBFS and IOT powder weighed according to the proportioning weight in the slurry mixing pot; The it adds alkali activator (prepared by water glass, water and NaOH according to the required water glass modulus m, pay attention to the temperature of alkali activator), fully dissolves the superplasticizer in the water, then pours it into the slurry mixing pot, fully stirs (slow stirring for 20s, then fast stirring for 10 s), adds hydrogen peroxide for 10s, and finally injects the slurry with the size of 40mm × 40mm × 160 mm mold. After 1d, the mould will be removed, and test specimen will be put it into the standard curing box for standard curing. The standard curing system is 20 ± 2° and relative humidity RH ≥ 95%.

3.3 *Micro experimental method of test specimen samples*

X-ray diffraction (XRD) test: the sample is broken after standard curing to the specified age, the hydration is immediately terminated with absolute ethanol, baked to constant weight at 60°C, sieved through 0.08 mm sieve, powdered, sealed and sent for testing. Use the X-ray powder diffractometer produced by Bruker D8 advance in Germany, test range: 10–80 2θ, The length of test step is 0.02°/ 0.2 s;

Scanning electron microscope test (SEM): after the sample is cured to the specified age, it is broken, the hydration is immediately terminated with absolute ethanol, dried at 60°C to constant weight, and sprayed with gold on the section for standby. TESCAN VEGA2 variable vacuum scanning electron microscope and INCA Energy 350X were used for elemental analysis. FTIR test parameters: the instrument is Thermo Fisher IS50; Transmission mode, test range: 400-4000 cm-1, Resolution: 4cm-1, Number of scans: 32.

4 TEST RESULTS AND DISCUSSIONS

4.1 *XRD and Laser particle size analysis of slag powder and iron tailing powder*

The XRD results of GGBFS and IOT was shown in Figure 1.

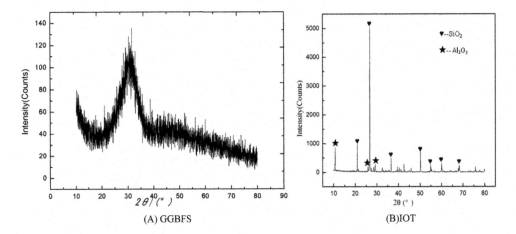

Figure 1. XRD of GGBFS and IOT.

The GGBFS used in the experiment were from Special steels of Liugang. The main chemical composition of GGBFS was shown in Table 1. The performance can meet the requirements of GB/T18046-2017. It can be seen from Tables 1 and Figure 1 that the GGBFS contained a lot of vitreous bodes but no evident diffraction peak.

The IOT used in the experiment were from Special steels of Liugang. It can be seen from Figure 1 that SiO2 has sharp peaks at 21°, 27°, 37° and 51°. This reflects that SiO2 has the high degree of crystallization, and it shows the high content of SiO2 indirectly. When the diffraction angles are 10°, 26° and 29°, there are diffraction peaks of Al_2O_3. Al_2O_3 is still a major component even though its content in iron tailing powder is not as high as SiO2 content. SiO2 and Al2O3 provide raw materials to generate calcium silicate hydrate (C-S-H) and tricalcium aluminate. Moreover, the IOT micro-powder serves as the crystal nucleus in the formation of Ca(OH)2 and C-S-H in early hydration stage, which accelerates the hydration rate of tricalcium silicate (3CaO·SiO2) from early strength.

4.2 Effects of curing time on performances of net alkali-activated IOT-slag slurry

An XRD test of the sample was carried out to further discuss the changes in the crystal alkali-activated process of samples (Figure 2). Through atlas analysis, the primary crystal phase in samples is silicon dioxide and cummingtonite (Al0.008Ca0.076Fe3.471Mg3.471O24Si8). The peaks at 20.9°, 26.7° and 50.23° are (100), (011) and (112) crystal faces of silicon dioxide (PDF 87-2096). Peaks at 10.6°, 29.1° and 32.6° are ($\bar{1}\bar{1}0$), ($\bar{3}\bar{1}0$) and ($\bar{3}\bar{3}0$) of cummingtonite (COD 9001910). According to comparison of XRD spectra, the diffraction peaks of silicon dioxide at 7d, 14d and 28d become sharper and sharper with the increase of curing age. This reflects that crystallinity of SiO2 is increasing gradually and the compressive strength of slurry specimens is increasing accordingly. It can be found from calculation of diffraction peak of SiO2 based on the Scherrer formula, the crystal sizes of SiO2 are 59.14nm, 126.77nm and 143.0nm as curing age increases, indicating that the size is increasing gradually. In a word, the growth of strength is related with the crystal expansion of SiO2.

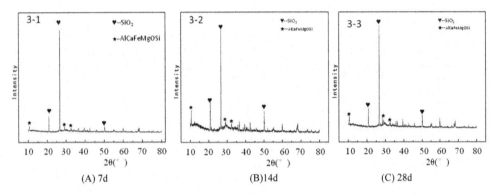

Figure 2. XRD spectra of the sample at 7d, 14d and 28d.

4.3 SEM and EDS analysis

SEM images and EDS spectra of the sample at standard curing age of 7d, 14d and 28d were shown in Figures 3, 4 and 5, respectively. Figure 3 shows that the micro-structure of sample was relatively loose at 7d. The steel slag particles were wrapped by flocculate hydration products and embedded among hydration products, with a lot of cracks. EDS spectra show that silicate aluminate products were primary products.

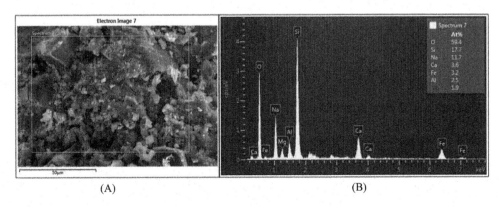

Figure 3. SEM images and EDS at 7d.

461

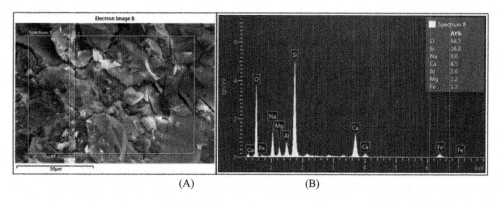

(A)   (B)

Figure 4.   SEM images and EDS at 14d.

It can be seen from Figures 4 and 5 that micro-structures of samples at 14d and 28d were basically consistent with those in Figure 3. The flocculation-loose structure was changed to a denser platy structure and products were relatively uniform and dense. There were lamellar products embedding into the structure and binding with the base tightly into an integrity, in which slag particles were rare. This means that abundant slag particles participated in the hydration. It can be seen from EDS that lamellar hydration products were similarly aluminosilicates and there were metal ions (e.g., Fe) bound into the hydration products. In other words, inert RO phase also participated in the hydration process as the curing age prolongs.

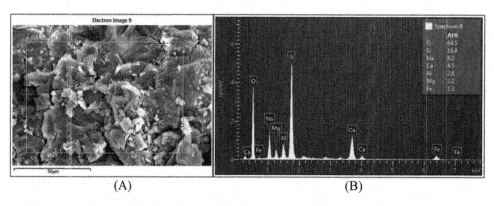

(A)   (B)

Figure 5.   SEM images and EDS at 28d.

Generally speaking, the material has an integral structure and low porosity due to the alkali-activated reaction, in which gel clusters and dense alkali-activated products are observed. Among them, the hardening structure of materials with 1.35 of sodium silicate modules shows the highest density. Various alkali-activated products interweave to form tight particle connections and a compact section, showing a high integrity. For materials with low sodium silicate modules, micro-structure of alkali-activated products generally presents joints, with a lot of sections, high porosity and relatively loose connection of products. There are many loosen-structured particular substances in the material. According to SEM results, micro-structural characteristics of materials are closely related with macroscopic mechanical properties. High-strength materials have dense and integral micro-structures, and products are connected tightly. Low-strength materials have loose and porous micro-structures and are characteristic of multiple sections.

4.4 *FTIR*

FTIR spectra of the sample at 7d, 14d and 28d were shown in Figure 6, which were basically similar after hydration.

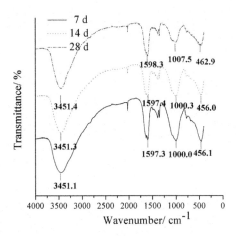

Figure 6.   FTIR spectra of samples at 7d, 14d and 28d.

The FTIR absorption peaks near 1000cm-1 and 456 cm-1 were ascribed to the asymmetric stretching vibration and bending vibration peak of the Si-O-T (T is Si or Fe) bond. It is important to note that the asymmetric stretching vibration peak (moving from 1000.0 cm^{-1} to 1007.5 cm^{-1}) and the bending vibration peak (moving 456.1cm^{-1} to 462.9 cm^{-1}) of Si-O-T in the sample at 28d both presented evident high-wavenumber drifts compared to samples at 7d and 14d. This reflects that degree of silicate anion polymerization in hydration products and the hydration degree increased after curing for 28d.

5   CONCLUSIONS

Groups of mixing ratios were designed by using the uniform design method with considerations of the effects of curing age on strength of alkali-activated slurry. Based on further analysis of microstructural experiments, some conclusions could be drawn:

Through the analysis of micro morphology, the compressive strength increases with the increase of curing age, and the diffraction peak of silica gradually becomes sharp with the curing age, indicating that the crystallinity and size of SiO2 are gradually improving. However, the flexural strength is low and changes little with age. It is also necessary to use some plasticizers or admixtures to increase toughness in practical engineering.

SEM results show that the micro morphology characteristics of materials are closely related to their macro mechanical properties. The corresponding micro morphology of high-strength materials is dense and complete, and the products are closely connected, while the micro morphology of low-strength materials is loose, porous and multi-section. The FTIR results also show that with the increase of curing time, the polymerization degree of silicate anions in hydration products.

ACKNOWLEDGMENTS

The authors are grateful to the nancial support from the Natural Science Foundation of Anhui Education Department (Grant No. KJ2021A0952, KJ2019A0622), the Initial Scientific Research Fund of High-level Faculty in West Anhui University (Grant No. WGKQ201702009), Anhui demonstration grassroots teaching and Research Office Fund (Grant No. 2020SJSFJXZZ398), Anhui ideological and political demonstration course Fund (Grant No.2020szsfkc0956), and Anhui quality engineering project Fund (Grant No.2020jyxm2144,2020jyxm2153).

REFERENCES

B.C. Mendes, L.G. Pedroti, M.P.F. Fontes, et al. (2019), Technical and environmental assessment of the incorporation of IOT in construction clay bricks. Construction and Building Materials, 227 :116669–116681, 2019

F. Han, L. Li, S. Song, et al. (2017). Early-age hydration characteristics of composite binder containing iron tailing powder. Powder Technology, 315, 322–331

K. Behera, B. P. Rose, M. K. Mondal (2018), Production of construction bricks using IOT and clay. Waste Management and Resource Efficiency, 2018(4): 583–596.

Ministry of Ecolagy and Environment of the People's Republic of China (2018). 2017 Annual report on prevention and control of solid waste in China's large and medium-sized cities. Environmental Protection (in Chinese), 46(S1). 90–106.

S. D. Wang and K. L (1995). Scrivener. Hydration products of Alkali Activated slag cement. Cement & Concrete Research, 25(3), 561-571.

Shen D, Jiao Y, Kang J, et al. (2020). Influence of ground granulated blast furnace slag on early-age cracking potential of internally cured high-performance concrete. Construction and Building Materials, 233, 117083–117093.

X. Huang, R. Ranade, V. C. Li (2013). Feasibility Study of Developing Green ECC Using IOT Powder as Cement Replacement. Journal of Materials in Civil Engineering, 25(7), 923–931, 2013.

Y. Cheng, F. Huang, S. Qi, et al. (2020). Durability of concrete incorporated with siliceous IOT. Construction and Building Materials, 242118147–118155.

Y. Cheng, F. Huang, W. Li, et al. (2016). Test research on the effects of mechanochemically activated IOT on the compressive strength of concrete. Construction and Building Materials, 118, 164–170.

Y. Huang, G. P. Xu, H. G. Cheng, et al. (2012). An Overview of Utilization of Steel Slag. Procedia Environmental Sciences, vol.16, pp.791–801.

Z.J. Wang, C. Xu, S. Wang, et al. (2016), Utilization of magnetite tailings as aggregates in asphalt mixtures, Construction and Building Materials, 114: 392–399, 2016.

Z.X. Tian, Z.H. Zhao, C.Q. Dai, et al. (2016), Experimental study on the properties of concrete mixed with IOT. Advances in Materials Science and Engineering, 2016: 1–9.

Frontiers of Civil Engineering and Disaster Prevention and
Control – Yang & Rahman (Eds)
© 2023 The Author(s), ISBN: 978-1-032-31200-2

Performance of permeable concrete based on the blocking principle

Yanqin Dai*
Yunnan Water Resources and Hydropower Vocational College, Yunnan Kunming, China

Xin Chen
Yunnan Institute of Water& Hydropower Engineering Investigation, Design and Research, Yunnan Kunming, China

Feng Pan, Jing Yuan & Dan Wu
Yunnan Water Resources and Hydropower Vocational College, Yunnan Kunming, China

ABSTRACT: Pervious concrete has good ecological and economic benefits because of its porous structure, but it is also easy to be blocked because of its porous structure and the introduction of recycled aggregate. At present, there is a lack of analysis on the change of meso-structure before and after the blocking of pervious concrete. Based on the study of clogging mechanism of pervious concrete, according to the change of permeability coefficient of pervious concrete with the increase of clogging content and the recovery of permeability coefficient after clogging removal, the meso-structural characteristics of perous concrete in each stage are extracted by analyzing the images of unclogged, clogged and cleaned pervious concrete specimens. The effects of aggregate size and cement paste content on these meso-structural characteristics are analyzed. Combining the macroscopic test results with the meso-structural characteristics, the clogging mechanism of pervious concrete and the reason for the difference of clogging removal effect are explained by the change of meso-structural characteristics. By establishing the prediction model of permeability coefficient, the deterioration of permeability performance of pervious concrete in the process of clogging can be predicted according to the meso-structure characteristic value of pervious concrete. This study can minimize the impact of man-made facilities on the surrounding environment, and has great research value.

1   INTRODUCTION

In the process of urbanization, a large number of concretes, asphalt and other impervious pavements have been built, which hinders the water cycle in the urban ecosystem and causes many ecological problems. Impervious pavement has strong heat absorption capacity, small specific heat capacity, excessive density and airtightness, all of which lead to its high heat storage but difficult to dissipate heat (Winston 2016). The extensive use of impervious pavement will cause the temperature of the city to be significantly higher than that of the suburbs, resulting in the "heat island effect" of the city. In addition, the sound absorption ability of dense pavement is poor, and the noise generated by vehicles seriously affects the work and rest of residents near the road. On the dense road surface, the surface water is difficult to penetrate into the ground, and the urban drainage system cannot effectively play the soil's own seepage and water storage function (Zhang 2018).

The "Sponge city" concept came into being. The "Sponge city" will be the construction of artificial drainage pipe network and the local natural ecosystem organic combination. In the season of high precipitation, natural waterways such as rivers and lakes are used for drainage. Natural sponges, such as gardens and meadows are used for water absorption and storage. In the season of low precipitation, the stored water is used for cleaning roads and vehicles, irrigation of farmland,

---

*Corresponding Author: dyq8183@163.com

etc. This cannot only reduce the occurrence of extreme situations such as drought and flood, but also improve the utilization efficiency of water resources (Zhang 2018).

With the progress of science and technology and the enhancement of human awareness of environmental protection, the environmentally friendly performance of pervious concrete is increasingly reflected. In 1995, Balades confirmed that the permeable pavement could effectively adsorb pollutants through in-situ tests. Through tests and γ-ray scanning, he found that the blockage usually existed within 2-3cm from the surface of the permeable pavement. The reason for the blockage was that particles smaller than sand accumulated on the surface of the permeable pavement, rather than moving to the interior of the porous pavement (Gersson 2020). In 2018, Zeng Fangui et al studied the rapid clogging process of pervious concrete by using variable head test, and found that the larger the design aperture, the smaller the clogging particles, and the more difficult it is to clog. The water-cement ratio has a certain influence on the clogging recovery effect of pervious concrete specimens. The larger the water-cement ratio is, the smaller the recovery degree after clogging is (Wu 2016).

In this paper, the characteristic value of the internal meso-structure of pervious concrete is extracted. Combined with the macro test, the influence of the change of meso-structure on the clogging and clogging removal effect of pervious concrete are explored. Air entraining agent can improve the anti - blocking performance of fine and microscopic mechanism of reclaimed aggregate pervious concrete.

2 FEATURE EXTRACTION OF AGGREGATE PARTICLE SIZE AND PORE SIZE DISTRIBUTION

The extraction methods of aggregate particle size and pore size distribution characteristics are the same. Taking the pore size as an example, the Mimincs software is used to process CT images, and the operation steps are shown in Figure 1.

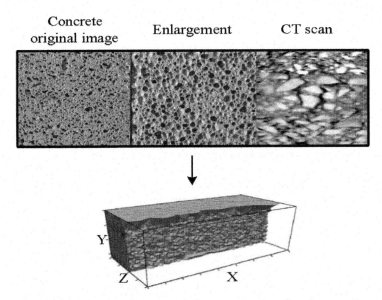

Figure 1. Aperture distribution cloud image.

Through CT scanning, the section images of the pores of the permeable concrete can be obtained. Continuously open the binary mask of the aperture until there is no pixel in the mask, record the number of pixels after each opening operation, and calculate the probability distribution density p_i of the aperture in the interval L_i by using formulas (1)–(3) (Lund 2020). It should be noted that since each opening operation is performed on both sides of the aperture, the aperture in formula (3) is multiplied by 2. p_i is the pixel value used in the ith opening operation. The number of pixels is taken. p_i is an integer, and $P_i > P_{i-1}$. l is the actual length represented by a pixel in the CT image. Since the size of a pixel in each CT window section is fixed, the smaller the area of the specimen in the window is, the smaller the l is, and the higher the accuracy is. Therefore, it is necessary to select a suitable scanning area for the specimen during CT scanning.

$$p_i = \frac{m_{1-1} - m_i}{m_0 \times L_i} \tag{1}$$

$$L_i = d_{ai} - d_{a_{i-l}} \tag{2}$$

$$dai = 2 \cdot p_i \cdot l \tag{3}$$

Where: p_i is the probability density of the aperture in the interval d_{ai-l} and d_{ai}. m_0 is the number of pixels of the mask when the aperture is not opened; m_i is the number of aperture mask pixel points after the i-th open operation; the interval length (true value) between the i-1 open operation and the i-th open operation; L_i is the aperture of the hole corresponding to the ith opening operation; l is the actual length represented by a pixel in the CT image (Tabatabaeian 2019).

3 BLOCKING EQUATION

Since aggregate particle size and porosity are the two most critical factors affecting the clogging process of pervious concrete, the fitting parameters of the clogging content change formula are shown in Table 1.

Table 1. Fitting parameters of the formula for the change of clog content.

Aggregate size (mm)	Design porosity	A	Ko	R2
5—8	0.18	−0.146	5.443	0.98
8—10	0.18	−0.085	6.818	0.99
5—10	0.24	−0.121	8.139	0.99
5—10	0.18	−0.110	6.521	0.98
5—10	0.12	−0.109	4.518	0.90

Regression analysis was adopted to fit parameters A and k_0 by means of expected value AE of aggregate particle size and designed porosity P, as shown in Equations (4) and (5).

$$A = 0.07426 \times AE - 0.09725 \times P - 0.23173 (R^2 = 0.82) \tag{4}$$

$$K_o = 1.81509 \times AE + 30.16867 \times P - 2.43773 (R^2 = 0.99) \tag{5}$$

Under the condition that the expected value AE of aggregate particle size and the design porosity P are known, the parameters A and k_0 can be estimated by formula (4) and (5), and substituted into formula (6) to predict the permeability coefficient of pervious concrete specimens with different contents of obstructions, which provides a reference for the maintenance of pervious concrete (Adil 2020).

$$K(m) = A \times m + K_0 \tag{6}$$

In Formula 6, $K(m)$ is the permeability coefficient of pervious concrete when m grams of clogging are added; M is the mass of clogging; A is a parameter, reflecting the decrease speed of permeability coefficient; k_0 is a parameter, reflecting the initial permeability coefficient.

4 EXPERIMENTAL ANALYSIS

4.1 *Experimental material ratio*

In order to analyze the influence of air-entraining agent content on the performance of the specimens, the freeze-thaw test specimens were designed according to the bond-aggregate ratio. The bond-aggregate ratio was 0. 18, 0. 24 and 0. 30. When the bond-aggregates ratio was 0. 24 and there was no air-entraining agent, the porosity was 0. 18. The aggregate size is 5–10 mm, and the amount of air-entraining agent is expressed by the percentage of the mass of air-entraining agent in the mass of cementitious material. Due to the strong water absorption of recycled aggregate, in order to prevent it from absorbing the water in the cement paste and ensure that the design water-cement ratio remains unchanged, it is necessary to treat the recycled aggregate into a saturated surface dry state. The additional water consumption is the ratio of the additional water mass to the recycled aggregate mass.

4.2 *Experimental process*

The statistics in this paper are the pores larger than 7.4ym in the cement matrix. When the ratio of glue to aggregate is 0.24, the porosity of the internal pores in the cement matrix of the recycled aggregate pervious concrete specimen under different amounts of air entraining agent is shown in Figure 2.

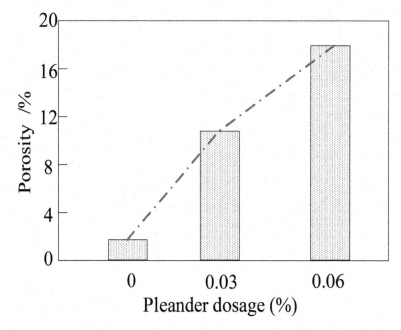

Figure 2.   The pore porosity of cement-based specimen with different amount of entrainment agent.

With the increase of the amount of entraining agent, the porosity increases significantly.

When the cement-bone ratio is 0.24, the pore size distribution of the cement-based internal pores is shown in Figure 3.

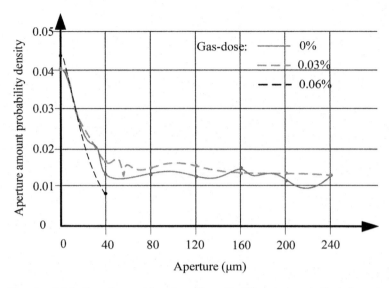

Figure 3. Pore size distribution of cement-based samples with different amount of entrainment agent.

It conforms to the two-parameter Weibull distribution, and the fitting parameters are shown in Table 2.

Table 2. Fitting parameters of cement-based inner pore size with two-parameter Weibull distribution.

Test group	λ	k
RT1	8.150	1.324
RT2	24.753	1.122
RT3	23.011	1.270
RT4	24.059	1.222
RT5	26.481	1.416

The pore size is less than 20ym without entrainer, while the pore size is mostly in the range of 20-200ym with entrainer, and there is little difference in the pore size, which indicates that the stability of air bubbles introduced by air entrainer is similar with different quality of air entraining agent. Because the pore size of these pores is more than 200 nm, which are harmful pores, the strength of cement paste decreases after adding air-entraining agent. For the pervious concrete specimens, add appropriate air-entraining agent can increase the area of cement-based overlapping area, which is conducive to enhancing the strength and frost resistance of pervious concrete. The minimum pore spacing distribution in the cement matrix is in accordance with the lognormal distribution.

With the increase of the amount of air-entraining agent, the pore spacing becomes smaller, and the distribution range of pore spacing also becomes smaller. When the air-entraining agent is not used, the pore itself is very small, and the pore spacing is very large. When the pore spacing is appropriate, the expansion stress caused by the freeze-thaw cycles can be released on the basis of not damaging the cement-based strength too much. When the pore spacing is too small and the pores are too dense, the cement-based strength will be reduced too much, even if the frost resistance is improved, its compressive strength will be reduced.

4.3 *Experimental data fitting analysis*

For pores with a design porosity of 0.18 and different aggregate sizes, the pore size follows a lognormal distribution, as shown in Figure 4.

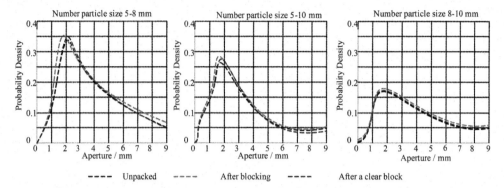

Figure 4. Fitting curves of different particle diameters.

When there is no clogging, the larger the aggregate size is, the greater the probability of large pore size is, and the larger the pore size corresponding to the peak of probability density is, which indicates that the greater the aggregate size is, the larger pores there are. After complete the blockage, the aperture corresponding to the peak of probability density becomes smaller, and the fitting curve becomes thinner. However, the fitting curves corresponding to different aggregate sizes are quite different. The smaller the aggregate size is, the greater the probability of small pore size is. After cleaning, the pore size of pervious concrete is between unclogged and clogged, the probability of small pores decreases, and the probability of large pores increases.

5 CONCLUSION

In this paper, permeable concrete with different meso-structures is designed by changing the aggregate size and the amount of cement paste, and the permeability coefficient is measured by adding different amounts of blockage and after cleaning. The change law of meso-structure of porous concrete without blocking, after blocking and clearing, is quantified. The blocking mechanism of porous concrete and the reason for the difference of clearing effect after blocking are explained by the change of meso-structure characteristics.

The blockage part does not study the influence of the ratio of the particle size of the blockage to the pore size of pervious concrete on the degree of blockage. In future work, we can take the blockage with a single particle size, count its ratio to the pore size, and explore the ratio range that is most likely to cause blockage.

ACKNOWLEDGEMENTS

Foundation for Scientific Research of Yunnan Education Department "Research on Permeable Concrete Based on Sponge City Construction" (No. 2019 J0865).

REFERENCES

Adil G, Kevern J T, Mann D. (2020). Influence of Silica Fume on Mechanical and Durability of Pervious Concrete. Construction and Building Materials, 247:118453.

GERSSON F B,ISAAC G. (2020). Hydraulic Behavior Variation of Pervious Concrete due to Clogging. Case Studies in Construction Materials, 13:1–11.

Lund M S M, Hansen K K, Brincker R. (2018). Evaluation of Freeze-Thaw Durability of Pervious Concrete by Use of Operational Modal Analysis. Cement and Concrete Research, 106: 57–64.

Tabatabaeian M, Khaloo A, Khaloo H. (2019). An Innovative High Performance Pervious Concrete with Polyester and Epoxy Resins. Construction and Building Materials, 228:116820.

Winston R J, Al-Rubaei A M, Blecken G T. (2016). Maintenance Measures for Preservation and Recovery of Permeable Pavement Surface Infiltration Rate - The Effects of Street Sweeping, Vacuum Cleaning, High Pressure Washing, and Milling. Journal of Environmental Management, 169: 132–144.

Wu H, Liu Z, Sun B B. (2016). Experimental Investigation on Freeze–Thaw Durability of Portland Cement Pervious Concrete (PCPC). Construction and Building Materials, 17:63–71.

Zhang J, Ma G D, Dai Z X. (2018). Numerical Study on Pore Clogging Mechanism in Pervious Pavements. Journal of Hydrology, 565: 589–598.

Zhang J, She R, Dai Z X. (2018). Experimental Simulation Study on Pore Clogging Mechanism of Porous Pavement. Construction and Building Materials, 187: 803–818.

Frontiers of Civil Engineering and Disaster Prevention and
Control – Yang & Rahman (Eds)
© 2023 The Author(s), ISBN: 978-1-032-31200-2

Design wind and wave study on proposed tunnel project in the Chittagong Port, Bangladesh

Chen Xi, Dai Minglong*, Zhang Dongdong, Xu Gaohong & Xiong Feng
Bureau of Hydrology, Changjiang Water Resources Commission, Wuhan, China

ABSTRACT: A continuous research of design wind speed and wave elements is carried out on the basis of flood control water level for the tunnel project in this paper. Considering a flood control standard of 100-year recurrence period, the wind and waves for this project are mainly analyzed and recommended in accordance with the flood control standard. The results show that: 1) Wind speed and direction are important elements in wave design; 2) the wind direction is determined to be S (SW, SE) after merging the direction of 45° on both left and right sides, then the design wind speeds at 100-year return period on the design high-tide level is about 28.97m; 3) an average wave height of 0.5 m and a wave run-up of about 0.6 m are recommended under a combined condition of the 100-year return period design high-tide level and the 100-year return period wind speed. The study provides effective supplement for the design of flood control water level in coastal area.

1 INTRODUCTION

Chittagong is the second largest city and the largest port city in Bangladesh. Most of the export and import activities of the country are being carried out through Chittagong Port situated in the mouth of the Karnaphuli River. The Karnaphuli River divides the Chittagong port area into two parts, the western part is bounded by the urban area and the port, set up on the closed protection zone of seaward flood prevention embankment and the riverward flood prevention embankment even though the dike does not meet the flood control requirements. The eastern part is the heavy industrial area still remained in the natural status without embankment protection. Currently there are two bridges which are more than 20km away from the estuary to cross the Karnaphuli River.

The proposed tunnel project consists of approach road segments and open segments on both the west and east sides and also a tunnel segment on the west side, of which the east approach road segment includes a viaduct over the KAFCO Fertilizer Plant. The tunnel project is located at the estuary of the Karnaphuli River where a consideration of the strong wind speed (Hussain 1986) and significant wave heights (Deb 2018) are needed, shortage of embankment on the east side exposed directly to the sea and frequent typhoons. Waves, a major impetus for the development of areas along seas and river mouths, are important factors that cannot be neglected in the design of bridges and subgrades. In addition, the wave run-up is another factor to be considered in the design of subgrades.

On the basis of the previous study on flood control water level by using the measured tidal water level data (Chen 2022). This paper is continuous research to provide the analysis of design wind speed and wave elements for the tunnel project design. The proposed tunnel project is considered flood control standard of 100-year recurrence period (GB50201-2014), the wind and waves for this project are mainly analyzed and recommended in accordance with the flood control standard in this study (GB/T 51015-2014).

---

*Corresponding Author: 12181343@qq.com

                                                    DOI 10.1201/9781003308577-64

2 MATERIALS AND METHODS

2.1 Data collection

Daily measured wind speed of both Ambagan and Chittagong weather stations have been collected since their establishment till December 2017. However, parts of data of the two stations are missing: The miss rate of the daily wind speed data at Ambagan station is 14.5% and that of Chittagong station is 11.7% (Years between 2002 and 2007 are excluded). The relative positions of the two stations and the proposed tunnel are shown in Figure 1. The data series and basic information are shown in Table 1.

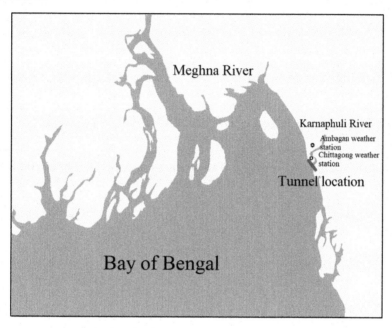

Figure 1.   The relative positions of Ambagan and Chittagong weather stations and the Tunnel.

Table 1.   Basic information of ambagan and chittagong weather stations.

Site	Longitude	Latitude	Data series	Distance from coasts	Distance from the tunnel site	Altitude (m, MSL)
Ambagan	91° 49′	22° 21′	1999~2017	About 6 km	About 13 km	33.2
Chittagong	91° 49′	22° 16′	1968~2002; 2008~2017	About 4.5 km	About 4 km	5.5

2.2 Methods

The wind speed of the design basis station at the tunnel site shall be calculated after correcting the height and land-sea wind speed in wind speed data (GB50201-2014). The design wind speed at different recurrence periods on the basis of design tidal levels shall be calculated by applying the P-III curve to analyze the frequency of wind speed at the tunnel site.

The site for wind measurement of Chittagong Station has an elevation of 5.5m (MSL). The height of basic wind speed data of the Chittagong Station shall be first corrected to the standard wind

473

speed at the height of 10 m above the design high-tide level before calculation of the design wind speed. The study area covering 40 km landward and seaward respectively is a belt where wind speeds change rapidly and there is small change of wind speed in areas beyond this scope. Therefore, whether the correction of land-sea wind speed shall be performed is based on relative locations of weather stations and the proposed project. The main correction method is to conduct correlation analysis with synchronously measured data, converting long-term wind speed data of weather stations into offshore wind speed at the proposed project location.

3 DESIGN WIND

3.1 *Wind direction analysis*

According to the daily measurement of wind speed and direction at the Chittagong Station and Ambagan Station, the frequency of wind directions during the whole year and the typhoon season from April to August is shown in Figure 2. It can be concluded that the annual prevailing wind direction at Chittagong station is SE with a frequency of 16.4%, which is followed by S, SW and SSE directions with frequencies of 16.0%, 8.4% and 7.5% respectively; the prevailing wind direction from April to August is S with a frequency of 25.9%, which is followed by SE, SSE, and SW with frequencies of 23.1%, 11.4%, and 11% respectively. The annual prevailing wind direction at Ambagan is S with a frequency of 32.4%, which is followed by SW, SE directions with frequencies of 7.9% and 6.9% respectively; the prevailing wind direction from April to August is S with a frequency of 51.3%, which is followed by SW and SE with frequencies of 10.3% and 8.4% respectively.

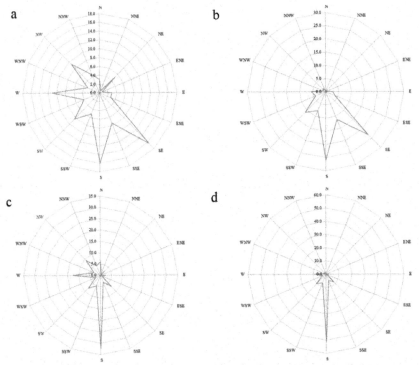

a. Annual wind frequency statistics at Chittagong Weather Station from 1968 to 2017
b. Wind frequency statistics from April to August at Chittagong Weather Station from 1968 to 2017
c. Annual wind frequency statistics at Ambagan Weather Station from 1999 to 2017
d. Wind frequency statistics from April to August at Ambagan Weather Station from 1999 to 2017

Figure 2.   Statistical chart of wind frequencies (%).

3.2 *Wind speed analysis*

Since the data series of daily wind speed at Chittagong Station covers 45 years with two time intervals, i.e. from 1968 to 2002 and from 2008 to 2017, in which each typhoon from 1970 to 1991 has been registered with good representativeness, and that the station is under the influence of typhoons with a greater annual maximum wind speed than that of Ambagan due to its nearer location to the coast, Chittagong Station is finally selected as a design basis for the analysis and calculation of the wind speed. According to relevant analysis, the above-mentioned typhoons in 1985 and 1991 are registered in data series of wind speed observation at Chittagong Station, but wind speed data for the typhoon in April 1991 are missing. The maximum measured wind speed (wind direction of NW) at Chittagong Station from 1968 to 2017 is 24.5 m/s on April 27th, 1985, which was followed by a maximum measured wind speed (wind direction of SW) of 23.3m/s on August 14th, 1979.

According to the geographical location of Chittagong and an analysis by the formation principle of storm surges, there are two main wind directions causing storm surges: Firstly, when waves pushed by fierce onshore wind towards the coast cause an abrupt rise of water level and the radius of maximum wind speed coincides with the landing point with the shoreline being vertical with the wind direction, it is extremely likely for the occurrence of wind surges, during which strong southwesterly winds are helpful for such an effect and the spatial scale for strong winds at the radius of maximum wind speed of typhoon is small for the Coriolis force taking effect. Secondly, on the east coast of the Bay of Bengal, strong winds blowing along the left side of the coast transport waves to the Ekman coast and cause the rise of water level on the coast to create storm surges, during which strong southeasterly winds are helpful for such an effect and the Coriolis force plays an important part in the water transportation under wind stress due to large spatial scale for strong winds.

Therefore, according to statistical results of daily measured wind directions of such two weather stations and the formation principles of storm surges, the prevailing wind direction for the proposed project is determined to be S after taking consideration of the unfavorable effects of the wind direction vertical to subgrades and bridges of the proposed tunnel project to engineering safety. During the calculation of design wind speed, the final wind direction for the calculation of final wind speed is determined to be S (SW, SE) after merging the direction of 45° on both left and right sides.

3.3 *Design wind speed*

Since the Chittagong Station as the design basis station of wind speed is about 4.5 km away from the coast, with the center of the viaduct on the east bank of the proposed tunnel site being about 3 km away from the coast and center of east subgrades being about 4.5 km away from the seaside, the proposed project location and the design basis station are close to the coast, which requires no correction of land-sea speed for this project.

After the height correction of maximum wind speed series at the Chittagong Weather Station, the P-III Curve shall be applied for frequency analysis. The frequency distribution curve of the design wind speed at the proposed tunnel is shown in Figure 3, and research results in Table 2.

The disparity between design wind speeds at different design high-tide levels of the tunnel site is small, which meets the standard for the control of a 100-year return period flood of the proposed tunnel project. Design wind speed results at the 100-year return period design high-tide levels of the tunnel site shall be selected for the analysis and calculation of design waves.

Table 2. Design wind speeds at different return periods on the design high-tide level. Unit: m/s

Wind direction	Frequency at the high-tide level	Design high-tide level (m)	Return periods (year)					
			300	100	50	20	10	5
S (SW, SE)	0.33%	8.39	34.09	29.26	26.23	22.19	19.09	15.95
	1%	7.11	33.75	28.97	25.96	21.96	18.90	15.79
	2%	6.34	33.54	28.79	25.80	21.83	18.78	15.69

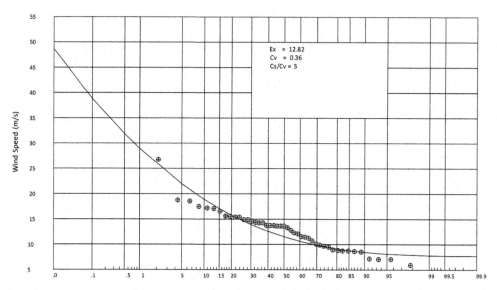

Figure 3. Wind speed frequency curve in the main wind direction at the proposed tunnel site in the condition of design high-tide level.

4 DESIGN WAVE

4.1 *Design wave calculation*

According to calculation results of water depths and design wind speeds in water areas near the proposed tunnel project, a wave height calculation formula shall be used in a reasonable manner to calculate the average wave height and average wave period at 100-year return period design high-tide levels of the tunnel site by applying a Putian Test Station Formula (GB50286-2013). According to the topographic survey data along the proposed project by CCCC and the digital elevation model (DEM) provided by Bangladesh Duct Company, the average elevation from the project location at east bank of Karnaphuli River to the coastline shall be taken as 3.7m, and the average water depth in fetch water areas shall be 3.41m (height datum is MSL) at a 100-year return period design high-tide level of 7.11 m (Chen 2022).

After the calculation of average wave heights and average periods of different frequencies at 100-year return period design high-tide levels of tunnel segments with the Putian Test Station Formula, wave heights of accumulative frequencies at the tunnel site shall be obtained with the conversion relation of wave heights between different accumulative frequencies (GB50286-2013); with the wave length ~ wave period and depth relation described in the code above, the average wave length at the tunnel site shall be calculated, and results of different accumulative frequencies and average wave lengths are shown in Table 3.

Table 3. Wind and wave elements at 100-year return period design high-tide levels of the proposed tunnel.

Design wind speed Return period (year)	Wave heights of different accumulative frequencies (m)		Average H	Average wave period T (s)	Average wave length L (m)
	$H_{20\%}$	$H_{50\%}$			
100	1.13	0.82	0.83	4.04	21.1
50	1.06	0.76	0.78	3.91	20.2
20	0.96	0.69	0.70	3.72	18.8

4.2 Wave run-ups

Wave run-ups of various design frequencies of embankments on both sides of the proposed tunnel project shall be calculated according to the structural type and segment design parameters of embankments with the consideration of factors, such as the water depth, wave-breaking forests in front of embankments below the design high-tide level at the project location, as well as conditions of embankment surface (GB/T 51015-2014). Single-slope (gradient m is 1:1.5) subgrades shall be built on two sides of the proposed tunnel. Wave run-ups shall be calculated by Appendix E.0.2 of the Code for Design of Sea Dike Engineering (GB/T 51015-2014). According to the design of the proposed project, riprap shall be designed at the bottom of the subgrade slope, with its top being covered with nets and grass. The run-up of other cumulative frequencies can be calculated through multiplying the run-up when the cumulative frequency is 1% by the conversion factor for each frequency.

When the design wind direction is S or SE, waves are dissipated by wave-breaking forests after flowing over 10km on shoals before reaching the project area, and the wave height at the project location can be neglected due to the depth-induced wave breaking (DIWB). When the design wind direction is SW, waves land on the river mouth 1km away from the project location, which shall be dissipated by two natural slopes according to site survey results on the east-coast project area, and the underlying surface is densely covered by wave breaking forests, whose efficiency of wave dissipation is between 25% and 65% according to structures and distribution of different forest forms by referring to relevant research results on wave-breaking efficiency of wave-breaking forests.

With the function of wave-breaking forests being neglected or a wave-breaking efficiency of 40% being selected below the 100-year return period design high-tide level, the 100-year return period, 50-year return period and 20-year return period wave run-up values at the proposed project location under a condition of the design wind speed shall be calculated respectively, which is shown in Table 4. An average wave height of 0.50 m and a wave run-up of 0.59 m for the wave-breaking efficiency of 40% are recommended for this project under a combined condition of the 100-year return period design high-tide level and the 100-year return period wind speed.

Table 4. Wave run-up for design wind speeds at the proposed project location.

Design conditions	Design wind speed	Wave run-ups of different frequencies (m)	
	Return period (year)	$R_{20\%}$	$R_{50\%}$
The function of wave-breaking forests is neglected	100	1.36	0.98
	50	1.35	0.97
	20	1.33	0.94
Wave-breaking efficiency of 40% upon conservative estimation	100	0.82	0.59
	50	0.81	0.58
	20	0.8	0.57

5 CONCLUSION

Under a combined condition of the 100-year return period design high-tide level and the 100-year return period design wind speed, the efficiency of energy dissipation of natural slopes and wave break of wave-breaking forests shall be recommended as 40% in this case study. The average wave height and the wave run-up for a frequency of 50% are 0.50 m and 0.59 m respectively. It is recommended that design wave height and wave run-up are not considered at the same time, i.e., in the case of vertical retaining wall at the portal, only wave height is considered, in the case of inclined retaining wall, only wave run-up is considered.

REFERENCES

Chen Xi et al. (2022) Flood control water level design for the tunnel project under the estuarine area of Karnaphuli river, Chittagong, Bangladesh. CADPC, Wuhan, accepted.
Code for Design of Levee Project (GB50286-2013).
Code for Design of Sea Dike Project (GB/T 51015-2014).
Deb M, Ferreira C M. (2018) Simulation of cyclone–induced storm surges in the low–lying delta of Bangladesh using coupled hydrodynamic and wave model (SWAN+ADCIRC). Journal of Flood Risk Management, 11: S750–S765.
Hussain M, Alam S, Reza K A, et al. (1986) A study of the wind speed and wind energy availability in Bangladesh. Energy conversion and management, 26(3–4): 321–327.
Standard for Flood Control (GB50201-2014).

Frontiers of Civil Engineering and Disaster Prevention and
Control – Yang & Rahman (Eds)
© 2023 The Author(s), ISBN: 978-1-032-31200-2

In-tunnel disassembly technology of composite earth pressure balanced shield machine

Guanjun You & Chunming Pi
CCCC Second Harbor Engineering Co., Ltd, Wuhan, China

Yangyang Chen*
School of Civil and Hydraulic Engineering, Huazhong University of Science and Technology, Wuhan, China

Wen Liu & Zhipeng Zhou
CCCC Second Harbor Engineering Co., Ltd, Wuhan, China

ABSTRACT: Given the common difficulties of poor in-tunnel disassembly conditions, complex disassembly process and high disassembly-induced risks of composite earth pressure balanced (EPB) shield machines faced by tunnel construction units in core city areas throughout China, the Phase III project of Shenzhen Urban Rail Transit Line 2 was taken for example to explore the disassembly scheme design, analyze the host system and investigate the component disassembly and transportation methods. Next, the key in-tunnel disassembly technologies of composite EPB shield machines were comprehensively analyzed and expounded, and the in-tunnel disassembly scheme was compared with the shaft hoisting scheme. The study results revealed that: An important precondition for successful disassembly lies in a thorough disassembly plan in the design phase of shield machines, and various emphases and difficulties should be fully considered, such as the disassembly and hoisting of shield machines, the transportation of shield machine components, the unloading at shield machine storage places outside the tunnel, and the repair and disposal; The host part of a shield machine should be disassembled in strict accordance with the disassembly design scheme; The pipelines for each subsequent supporting equipment must be disassembled by following the principles of "clear identification, classified storage and in situ fixation"; Compared with the shaft hoisting scheme, the in-tunnel disassembly scheme can save the direct cost and resources and shorten the construction period. The study results can serve as a reference for the follow-up disassembly of equipment of the same kind.

1 INTRODUCTION

Located in Luohu District, Shenzhen, the Phase III project of Shenzhen Urban Rail Transit Line 2 starts from the existing Xinxiu Station on Line 2 and ends at Liantang Station, with a total length of 3.82 km. A total of three stations were built, including three intervals.

The shield intervals (totaling 1,942.396 m in length) of this section include the interval from Liantang Port Station to Xianhu Road Station and that from Xianhu Road Station to Liantang Station, which was constructed using two composite earth pressure balanced (EPB) shield machines. The shield machines started separately from the left and right lines at the east end of Xianhu Road Station, received transitions at Liantang Station, and restarted from the other end (west end) of Xianhu Road Station, thus completing the excavation task from Liantang Station to Xianhu Road Station and the in-tunnel disassembly task. As for the shield machines used in the shield intervals,

---

*Corresponding Author: 824890254@qq.com

the outer diameter of the cutterhead was 6,300 mm, and the inner diameter and outer diameter of the pipe segment were 5.4 m and 6 m, respectively. There was no hoisting shaft at the tunnel border between the shield method and the mining method, so after shield tunneling was completed, transition receiving needed to be done in the tunnel, along with shell-discarding disintegration operations. After being disassembled, the shield machines returned from the tunnel to Xianhu Road Station and were then hoisted out of the ground.

As to in-tunnel shield machine disassembly technologies, tunnel design for shield machine disassembly, host disassembly technology, disassembly workflow design, and the relationship between the disassembly of subsequent supporting equipment and in-tunnel construction have been explored. Liu Chenghong et al.based on Shijiazhuang City Rail Transit Line 2 Phase I Project 4 Bid Shijiazhuang Station-Dongsanjiao Station Intersection Project, carried out practice and research on the construction technology of shield machine underground docking and in-cavity dismantling (Liu Chenghong & Chen Yubo 2020). Feng Huanhuan et al. comprehensively analyzes and introduces the key technologies of TBM in-cavity dismantling from the design of the dismantling preparation scheme to the research on the dismantling method of the host system or components, based on the characteristics of the West Qinling Tunnel (Feng Huan-huan & Chen Kui 2015). Wang Lei et al. put forward the management and maintenance measures for the construction equipment of the subway shield dismantling machine in accordance with the characteristics of the shield machine equipment (Wang Lei & Zhou Bin 2020). Xiao Haihui et al. took S301 as an example to introduce the TBM dismantling scheme in the hole, put forward the general sequence and requirements of dismantling, and accumulated TBM dismantling experience (Xiao Haihui & Guo Rongwei 2010). However, most of the above studies focus on a certain aspect or the discussion is not detailed enough, and there is a lack of systematic and in-depth research. With the characteristics of the Phase III project of Shenzhen Urban Rail Transit Line 2 combined, the shaft hoisting scheme was compared with the in-tunnel disassembly scheme from the aspects of preparations for in-tunnel disassembly and disassembly scheme design. Then, the in-tunnel disassembly technologies of composite EPB shield machines were investigated and discussed.

2 OVERVIEW OF IN-TUNNEL SHIELD MACHINE DISASSEMBLY

The in-tunnel disassembly work is divided into the disassembly and hoisting of shield machines, the transportation of shield machine components, the unloading at shield machine storage places outside the tunnel, and the repair and disposal (Li Tao-meng & Sun Hai-bobo 2015). Based on the characteristics of this project, the emphases and difficulties of in-tunnel shield machine disassembly are mainly embodied in the following aspects: (1) The construction period is tight, accompanied by heavy tasks and large disassembly quantities; (2) The disassembly space is narrow and small, which impedes the construction of large equipment. With various hoisting processes, those disassembled in the tunnel are mostly bulk parts, while large components totally rely upon manual disassembly; (3) The selection of hoisting position and method is critical in the in-tunnel decomposition of large components (main drive, screw machine, etc.); (4) The safety protection of heavy cargo transportation is very difficult during disassembly; (5) Ventilation and water drainage problems in the tunnel, and the in-tunnel working environment is harsher than the outside environment; (6) Fire safety guarantee measures are essential under the simultaneous operation of multiple working faces; (7) There is little in-tunnel shield machine disassembly experience that can be taken for reference.

After the shield machines arrived at the disassembly position in the tunnel, the shield machine disassembly operation was started (successive halt, change of transformer position and layout of power supply line for machine disassembly). To ensure the disassembly efficiency, the host disassembly group and subsequent supporting equipment disassembly group were founded to simultaneously carry out the disassembly operation in the disassembly tunnel and subsequent supporting equipment disassembly region, respectively.

3 IN-TUNNEL SHELL-DISCARDING DISASSEMBLY SCHEME OF SHIELD MACHINES AND PREPARATORY WORK

3.1 Overall scheme of in-tunnel shell-discarding disassembly

After arriving at the designated position, the shield machines were shut down to start the in-tunnel disintegration. Given the fact that the tunnel space was limited, the subsequent supporting equipment could meet the non-destructive disassembly requirements even in the tunnel, and the components of shield machines could be transported only through the tunnel exit due to the construction limitations, the shield machines were firstly disassembled, followed by subsequent supporting equipment and the host machines. The overall disintegration scheme is depicted in Figure 1.

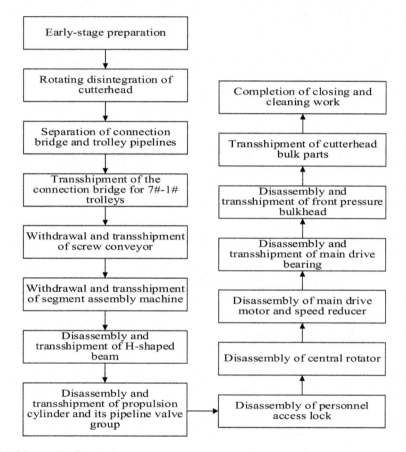

Figure 1. Disassembly flowchart.

3.2 Early-stage preparation

Before the shield machines were shut down for disassembly, the back of the pipe segment at the shield end should be grouted for the second time to form a water stop ring. Meanwhile, the 15 ring pipe segments approaching the tunnel portal should be longitudinally fixed using U-steel tensioners to form a whole and protect the gap between pipe segments from being enlarged after the recovery of the oil cylinder, since this would result in water leakage.

After the shield machines were shut down, the subsequent supporting trolley tracks were paved from the original shaft until the disassembly position, the connection bridge support frames were

481

prepared, and the hoisting points for disassembling cutterhead were arranged and installed, followed by the partitioned cutterhead positioning by marking, the welding of hoisting ears, the layout of lighting, wind power and hydroelectric equipment, and the preparation of special disassembly tools.

3.3 Fabrication of special workpiece

A portal frame was fabricated to match the hand chain hoists and cooperatively used for the hoisting of partitioned cutterhead. The self-made portal frame was welded using 200 H section steel, with a length of 7.7 m and height of 2.8 m, and reinforced welding was performed at joints using diagonal bracings. Then, the self-made portal frame was used to hoist and overturn the cutterhead components (checking calculation should be done before the fabrication of the portal frame to meet the use requirements), as shown in Figure 2.

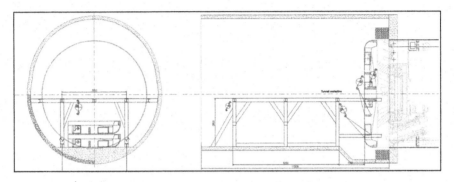

Figure 2.   Schematic diagram of cutterhead disassembly and hoisting.

3.4 Layout of hoisting points

A fixed independent hoisting point was arranged near a chord length of 2 m at the central front end of each shield machine. A total of 6 hoisting points bearing the load of 20 t were set, being hoisting points 1 and 2, 3 and 4 and 5 and 6, respectively, among which hoisting points 1 and 2 were temporary hoisting points (not serving as the main load-bearing points) of pipe segments. Hoisting ears were welded at hoisting points 1, 2, 3 and 4, on which hand chain hoists were installed. Hoisting points 1, 2, 3 and 4 were mainly used to disintegrate segment assembly machine, main drive motor and personnel access lock and place them flatwise. Hoisting points 5 and 6 were mainly used to hoist the screw conveyor and bulk parts, and segment hoisting points 1 and 2 were used to assist in hoisting the screw conveyor. The schematic diagram of the hoisting point layout inside the shield body is as shown in Figure 3.

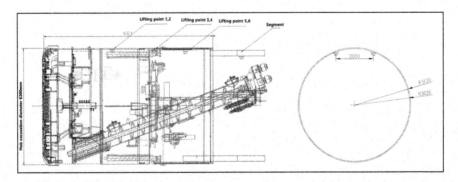

Figure 3.   Schematic diagram of hoisting point layout.

4 IN-TUNNEL DISASSEMBLY SCHEME FOR KEY SHIELD MACHINE PARTS

Based on the early-stage preparatory work, the host machine and subsequent supporting equipment were disassembled according to the overall design scheme. Next, the disassembly of key parts was described according to the sequence of construction operations.

4.1 *Cutterhead disintegration*

The cutterhead was cut into six workpieces and marked through the partitioning method, as shown in Figure 4. Each workpiece was connected to a corbel. Specifically, after being rotated to the 6-point position, the cutterhead was cut according to the sequence of 1–6, hoisting points were arranged on the cutterhead hanger beam and inside the soil cabin for hanging, and then the corbel was cut off by welding. To prevent the autorotation due to center-of-gravity shift after a piece of cutterhead was cut, the cutterhead should be temporarily fixed by welding to keep from rotation. After a single workpiece was cut, it should be placed at the rear end of the tugboat platform, and every two pieces were placed as one group and finally shipped out and hoisted to the ground surface along the already formed shield tunnel path. Comprehensive flaw detection was carried out after cleaning.

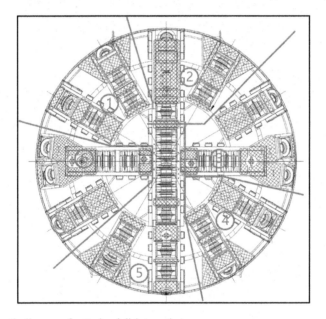

Figure 4.   Schematic diagram of cutterhead disintegration

4.2 *Disassembly of subsequent supporting equipment*

4.2.1 *Disassembly process analysis*

After the cutterhead of the shield machine entered the receiving expansion segment, 1# trolley was disassembled from the main shield body, and the subsequent trolleys were moved towards their hoisting direction using a diesel locomotive, thus providing enough space for disassembling the screw conveyor. In this case, the main shield body and subsequent trolleys could be simultaneously disassembled. The subsequent supporting equipment was disassembled according to the sequence of 7#→6#→5#→4#→3#→2#→1#→ connection bridge and transported to the original shaft via the already formed tunnel. After bulk parts like the belt frame on each trolley were disassembled off the trolley, the trolley was hoisted out through a 200 T truck-mounted crane and placed on the ground.

4.2.2 *Disassembly and identification of pipelines and routes*

Complex pipelines and routes existed in the shield machine. To facilitate the smooth connection of shield parts and the normal functional recovery in the follow-up installation and reduce the workload of the disassembly and follow-up installation, the disassembly of 1# trolley from the main shield body and the pipeline disassembly of subsequent supporting trolleys must follow the principle of "clear identification, classified storage and in situ fixation." Hydraulic lines, wind power and water pipelines, cables and belts should be marked and separated, among which hydraulic pipes and water pipes were sealed using plugs, and the disassembled cable heads must be sealed with gum.

Oil pipes were disassembled through the following steps: First, the oil return pipes and outlet pipe valves of each fuel tank were closed to prevent oil in the fuel tank from flowing out. The outlet pipes were taken apart from the control elements and marked according to the system name. To prevent the leakage of residual oil inside the system and pipelines, the oil ports of the disassembled pipelines should be plugged up rapidly. The pipelines that must be taken apart from the trolley should be stored in the ready-made wooden case with clear identification after classification.

Circuit lines were disassembled according to the following steps: First, the connecting cables of control apparatuses inside the assembly platform and shield machine were numbered and labeled. As for the connecting cables of control apparatuses inside the operating room and shield machine, after the end socket of one control cable was disassembled, this control cable must be collected, and its lock nut was fixed at the outlet of this cable on the disassembly platform. After this control cable was disassembled onto the end socket of the control apparatus, a label the same as that at the head end was pasted, and the connector was wrapped with insulated rubber tapes.

4.3 *Disassembly of screw conveyor*

The total weight and total length of the screw conveyor were about 26 t and 12.18 m, respectively. The total weight of the A sleeve and the screw shaft in this segment was 5.5 t, and the weight of the rest part of the screw conveyor was 20.5 t. The disassembly steps of the screw conveyor are as follows:

(1) The screw conveyor hoisting and supporting platform was welded and reinforced using profile steel on the assembly platform. A total of two hand chain hoists (10 t) were set as the main hoisting points to disassemble the screw conveyor.
(2) Two hoisting ears were welded at hoisting points 5 and 6 (Figure 3) at the top of the shield body.
(3) The bolts connected to the A sleeve of the screw conveyor cylinder were disassembled, and the screw shaft was cut off from the observation port, as shown in Figure 5.

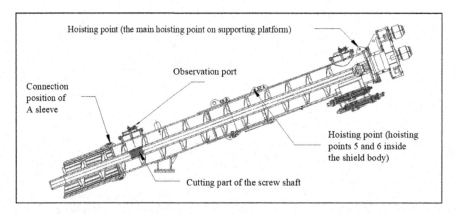

Figure 5. Schematic diagram of screw conveyor disintegration and hoisting point layout.

(4) The two main hoisting points on the assembly platform and hoisting platform and the hoisting points 5 and 6 at the top of the shield body (totally four hoisting points) were combined. A 10-t hand chain hoist was hung at each hoisting point, and the screw conveyor was slowly placed flatwise on the segment car supporting platform by regulating the hand chain hoist and then reinforced.

(5) The disassembled screw conveyor parts were transported to the original shaft by a diesel locomotive and hoisted out of the mouth of the shaft.

4.4 *Disassembly of assembly machine*

The assembly machine was firstly hoisted using hoisting points 3, 4, 5 and 6 inside the shield body, then four hoisting points of 5 t hand chain hoists were selected to install the hand chain hoists. The assembly machine was hoisted from the slideway, and the segment car was parked right below the assembly machine in advance. After the assembly machine was hoisted away from the slideway, the roller, support wheel, hydraulic motor and assembly machine cable reel were disassembled successively. In the end, structural parts like large gear rings and balancing weights were hoisted and disassembled. Afterwards, the assembly machine was turned over using a hand chain hoist relay and placed flatwise on the segment car, transported to the original shaft and hoisted out of the mouth of the shaft.

4.5 *Disassembly of parts and components inside the shield body*

The disassembly steps of parts and components (Figure 6) inside the shield body and the matters needing attention are as follows:

(1) Internal sensors, electrical control elements, electricity boxes and cables were disassembled and hoisted out after being stored in cases according to the identification;

(2) The main valves on the propulsion jack and articulated jack, valves in each subregion, connecting oil pipes and connectors were plugged up. Internal cleanness of oil pipes and valves should

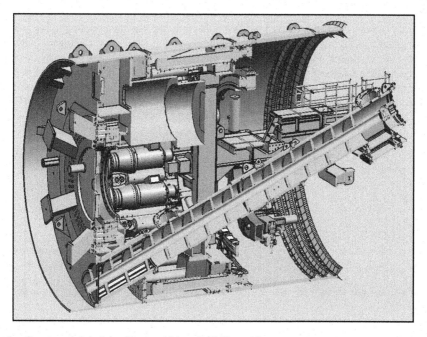

Figure 6.   Cross-sectional view of host machine of shield machine.

be ensured, and then the oil pipes and valves were hoisted out after classified encasement according to the ready-made identifications.

(3) All connecting pipelines and their connectors should be disassembled in the disassembly of the central rotator to prevent damage in the hoisting process;

(4) When the over cutter system was torn down, all the oil pipes and control lines on the hydraulic pressure station were disassembled, the pipelines and connectors were plugged up to ensure the cleanness of the oil line, and pipelines were well-organized, encased and preserved uniformly;

(5) Disassembly platform and other connecting parts were shipped out and hoisted to the ground.

4.6 *Propulsion cylinder and articulation cylinder*

When the shield machine proceeded towards the in-tunnel receiving part, special attention should be paid to the stroke of the main propulsion cylinder at the receiving ring, which should meet the optimal cylinder disassembly conditions. Before disassembly, the spacing between the first ring pipe segment behind the propulsion cylinder and the recovery state of the oil cylinder should exceed 2,000 mm as far as possible. Meanwhile, the oil inlet and return pipes in the propulsion cylinder were disassembled and plugged up, and the corresponding protective measures should be taken for the stroke sensor of the propulsion cylinder.

Through the hoisting point at the top of the shield body, a 2-t hand chain hoist and a hanger band were hung respectively before and after hoisting points 1 and 2 and 3 and 4 and fixed, and then the positioning block of the propulsion cylinder was cut off. As to the disassembly of oil cylinders, the lower oil cylinders (3 o'clock to 9 o'clock) were firstly torn down, followed by the upper oil cylinders. Next, the oil cylinders were slowly placed on the prepared segment car using the hoisting point at the top of the shield body.

There were 16 articulation cylinders in total, which could be fixed using hoisting points by reference to the disassembly method for the main propulsion cylinder. Afterwards, the positioning block of the oil cylinder was cut off, and the articulation cylinders were successively disassembled and marked.

4.7 *Disassembly of main drive*

The central main drive was the heart of the shield machine and the most important unit, containing bearings, gear rings, main drive seal and the rotator driving the cutterhead. The transportation of the central main drive, which was especially important in overall disassembly, was also the emphasis and difficulty in the disassembly. Before the disassembly of the main drive, the personnel access lock and drive motor should be firstly torn down. After the reducer and motor of the main drive were disassembled, the gear oil was drained from large gear rings, with a net weight of about 30 t (Figure 7).

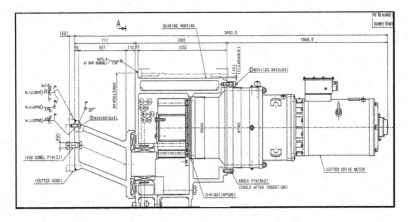

Figure 7.   Schematic diagram of main drive (motor included).

4.7.1 *Disassembly of personnel access lock and drive motor*

The personnel access lock was disassembled using hoisting points 1, 2, 3 and 4. A 5 t hand chain hoist was hung at each hoisting point, and the personnel access loc was slowly placed on the lower segment car. The six main drive motor reducers were dismounted using hoisting points 3 and 4; cover plates were installed for protection after the disassembly, connecting bolts were classified, encased and transported to the original shaft, and then hoisted out. Next, the connecting holes of the main drive and reducers were connected with guard plates to guarantee the internal cleanness of the drive. The gear ring of the motor taken apart as a whole should be cleaned and protected.

4.7.2 *Disassembly of main drive*

The disassembly steps of the main drive and matters needing attention are presented as follows:

(1) The soil bin bulkhead beneath the main drive was cut off, and the track was extended beneath the main drive to prepare for disassembling the main drive (the height should match with the baseboard height of the muck car);

(2) Diagonal bracings were fabricated to support the main drive. The baseboard of the muck car was reinforced and welded with the hoisting ears of two main hoisting points (points 11 and 1);

(3) The baseboard of the muck car was moved beneath the main drive (the overall load-bearing point), a 20 mm steel plate was paved on the baseboard of the muck car and fixed with the main drive, so the whole main drive was reinforced (see details in the reinforcement and installation drawing of flatbed truck for the main drive);

(4) After the main drive and flatbed truck were fixed, two bolted hoisting ears behind the original main drive were installed (hoisting ears in the original design) and hung by two 20 t hand chain hoists. The soil bin board welded to the main drive on the A ring was cut off, and the ribbed plates and annular plates influencing the rearward shift of the main drive were cut open once for all from the bottom up, thus not damaging the drive shell. The cutting sequence is as follows: sector regions at 12 o'clock, 3 o'clock, 9 o'clock and 6 o'clock.

(5) The fixed welding plates behind the main drive were cut off;

(6) After the main drive steadily fell on the baseboard of the muck car, the main drive was reinforced by combining the means of bracing and supporting to guarantee stable and reliable transportation;

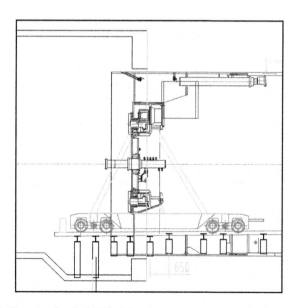

Figure 8. Schematic diagram of main drive hoisting

(7) Especially, no welding or cutting operations were allowed for the main drive shell in the supporting or reinforcement process, since these operations would damage the sealing condition and internal paint of the main drive.

4.8 Shield shell clearing, retention and disposal method

The internal part of the shield shell was cleaned and polished to meet the conditions for other construction operations in the shield shell. The objects cleared away from the shield shell were transported by a diesel locomotive to the original shaft and hoisted out of its mouth. As for the shield shell retained in the tunnel, the proprietor and supervisor should be organized to carry out the acceptance check work. After the construction conditions of the following construction procedure were met, concrete pouring was performed to form a closed loop according to the scheme given by the design organization.

5 ECONOMIC COST ANALYSIS

5.1 Shaft hoisting scheme

After the shield machine arrived at the receiving position, a concrete shaft envelop enclosure would be generally designed to ensure that the shield machine could be smoothly hoisted out after completing the excavation task. If the shaft hoisting scheme was adopted, the initial budget price of the concrete shaft envelop enclosure in this project was RMB 20,816,300, the cost of hoisting the shield machine via a shaft was RMB 930,000, and then the cost spent by the shaft hoisting scheme for the two shield machines was initially estimated as RMB 22,676,300. By reference to the shaft hoisting scheme in Liantang Station with a construction period of 11 months, the construction period of the shaft hoisting scheme was initially estimated as seven months.

5.2 In-tunnel disassembly scheme

The complexity of the in-tunnel disassembly process and the restorability of shield machines were mainly considered in the in-tunnel shield machine disassembly scheme. In the design phase of shield machines, the post-assembly restorability was specially designed, which, however, would elevate the related design cost and disassembly-induced loss cost. The accounting results showed that if the in-tunnel disassembly scheme was used, the disassembly cost would be RMB 1.6 million per unit of shield machine used in this project, the in-tunnel disassembly-induced equipment loss cost was RMB 5,935 million/unit, and the total cost spent in disassembling two shield machines was RMB 15,07 million. According to the past similar disassembly project, the construction period for the in-tunnel disassembly of one shield machine was about 40 d (Li Zhen-dong et al. 2020).

5.3 Process comparison

If the shaft hoisting scheme was taken, the shield hoisting shaft was located on the slope behind Zhongxing Garden Community between Liantang Port Station to Xianhu Road Station. In the initial design phase of this project, the shaft construction was planned after mountain truncation and slope protection. Meanwhile, the open space between Zhongxing Garden Community and the slope behind the mountain was considered as a temporary transportation corridor for shield machines to come in and out of the construction site. It was found through field investigation that a house was newly built in the open space, so the construction road could not be constructed. Besides, the roads were narrow and small, with many bends in Zhongxing Garden Community, thus failing to meet the space requirements for the transportation of shield machines.

High requirements were proposed by in-tunnel disassembly to the shield machines themselves. In the design phase of shield machines, the in-tunnel disassembly conditions should be considered, along with restorability, or otherwise, the complete scrapping of the damaged part would increase enormous additional cost to the project. Meanwhile, since the internal sectional dimension of the shield tunnel was 5.4 m, the working space was small, and the efficiency was low, the shield machines should be disassembled by personnel experienced in in-tunnel disassembly to avoid part damage and lower the construction and safety risks.

By comparing the cost of the two schemes, the in-tunnel disassembly scheme could save the cost by RMB 7,606,300 compared with the shaft hoisting scheme. Moreover, the in-tunnel disassembly scheme was finally adopted after comprehensively analyzing the safety, quality, progress and cost of the construction process. After the shield tunneling was completed by the shield machines, the in-tunnel disassembly was started at the tunnel junction between the shield method and the mining method, and the split unit was hoisted out from the original shaft at Xianhu Road Station.

6 CONCLUSION

TBM in-tunnel disassembly is characterized by a relatively narrow and small working face and high complexity of operations. During this process, contingency plans should be formulated by fully considering all kinds of disadvantages to ensure smooth disassembly. In this study, based on the characteristics of the Phase III project of Shenzhen Urban Rail Transit Line 2, the in-tunnel disassembly schemes of composite EPB shield machines were expounded in detail, expecting to provide a reference for the subsequent engineering construction. The main conclusions are drawn as follows: (1) In the design phase of shield machines, their detachability and restorability should be fully considered; (2) When the subsequent supporting equipment is detached from the main shield machine body and split, the pipelines of shield machines should be disassembled following the principle of "clear identification, classified storage and in situ fixation"; (3) The main body of shield machines should be disassembled in strict accordance with the following sequence: cutterhead, subsequent supporting equipment, assembly machine, propulsion system and main drive; (4) In comparison to the shaft hoisting scheme, the in-tunnel shield machine disassembly scheme can save the cost, take advantages of the resources of project department and equipment to the greatest extent, avoid the waste of social resources and the adverse impacts on the surrounding environment, and shorten the construction period. In this project, the in-tunnel disassembly technology of shield machines can provide a reference for the follow-up disassembly of similar equipment.

REFERENCES

Feng Huan-huan, Chen Kui. 2015. Design and study on TBM disassembled scheme in-tunnel of west Qinling tunnel. *Construction Technology* 44(23):6.

Li Tao-meng, Sun Hai-bo. (2015). Non-destructive Disassembling Technologies for Large-diameter Open-type TBMs. *Tunnel Construction* 35: 197–204.

Li Zhen-dong, Liu Tian-zheng, Sun Chang-jun. (2020). Underground Disintegration Technique for Shield Machine in Core Area of City. *Science Technology and Engineering* 20(24):10060–07.

Liu Chenghong, Chen Yubo. 2020. Study on construction technology of shield machine underground docking and dismantling in tunnel. *Construction & Design for Engineering* (18):179–180.

Wang Lei, Zhou Bin. (2020). Subway shield construction equipment management and maintenance measures. *Intelligent City* 6(8): 2.

Xiao Haihui, Guo Rongwei. (2010). Demolition scheme in TBM tunnel. *Shanxi Architecture* 36(3): 362–363.

Frontiers of Civil Engineering and Disaster Prevention and Control – Yang & Rahman (Eds)
© 2023 The Author(s), ISBN: 978-1-032-31200-2

Experimental study on flexural behavior of TRHDC-ALC composite panels

Mingke Deng* & Xian Hu*

College of Civil Engineering, Xi'an University of Architecture and Technology; Xi'an, China

ABSTRACT: Autoclaved Lightweight Concrete (ALC) panels are widely used in the internal partition wall panels of assembled buildings because of their light weight, thermal insulation, sound insulation and fire resistance properties, but when used as the perimeter protection system of high-rise buildings or as lightweight floor panels of mezzanine structures, the lower strength limits the application and promotion of ALC panels. In this paper, a TRHDC-ALC composite panel is proposed and its flexural properties are investigated. The results show that the TRHDC surface layer can significantly improve the flexural performance of ALC panels, which lays the foundation for later research and application.

1 INTRODUCTION

Autoclaved Lightweight Concrete (ALC) panels are porous concrete panels made of cement, lime and silica sand as the primary raw materials and cured by high-pressure steam (Narayanan 2000). Autoclaved aerated concrete panels are lightweight, fire-resistant, soundproof, heat insulation and other properties (Mohammed 2009, Ghazi 2015), can be used as internal partition walls, outer parapet walls, roof panels, but also as floor slabs (Liu 2013, Zhang 2019), is a new building material with superior performance. Still, ordinary autoclaved aerated concrete panels have a low load-bearing capacity and are prone to cracking and water seepage (Lv 2017), which limits their application and promotion.

In recent years, composite panels using ALC plates as cores have been studied extensively. Some scholars studied the flexural mechanical properties of FRP-ALC composite structures (Wu 2007), and the results showed that the strength and ductility of FRP-ALC composite structures were significantly increased, but brittle damage occurred in the composite panels; some scholars studied the flexural mechanical properties of paperless phosphor gypsum board-ALC composite panels (Xu 2019), but their flexural bearing capacity was increased to a smaller extent; in addition, some studies used steel wire mesh and 20 mm thickness of anti-cracking mortar combination to reinforce the ALC board (Zhang 2016), and the results showed that the load-bearing capacity of the board was significantly increased, but the thickness and self-weight of the board were increased to a larger extent.

Textile Reinforced High Ductility Concrete (TRHDC) is an optimized textile reinforced concrete (TRC) by using short polyvinyl alcohol (PVA) fibers (Dong 2020), which has high tensile strength and deformation performance. The combination of TRHDC facings and ALC panels form a TRHDC-ALC composite panel. While ensuring the functional advantages of the ALC panels themselves, TRHDC can increase their strength and flexibility and can better meet the applicable requirements. Therefore, the mechanical properties of TRHDC-ALC composite panels are of great value for research. This test investigates the effects of the type and number of layers of textile, the thickness of the TRHDC surface layer and the span-height ratio on the flexural properties of TRHDC-ALC composite panels to provide a basis for engineering applications.

---

*Corresponding Authors: 1161859746@qq.com and huxian7017@163.com

DOI 10.1201/9781003308577-66

2 BENDING TEST

2.1 *Test program*

This paper combines a TRHDC sheet and ALC plate, a total of 6 groups of specimens were designed; ALC plate size is 2000 mm*600 mm*100 mm. Test scheme are reported in Table 1. For example, A1 group is ALC plate without face layer, as a control group specimen; A5 group means 2 layers of basalt textile reinforced high ductility concrete on the tensile side, HDC face layer on the compressive side of the sandwich panel specimen, the rest are single side with TRHDC sheet on the tensile side.

Table 1. Main parameters of specimens.

Specimen number	Specimen size/mm	Facing materials	Surface thickness/mm
A1	2000*600*100	None	–
A2	2000*600*100	HDC	8
A3	2000*600*100	1-ply basalt TRHDC (B1TRHDC)	8
A4	2000*600*100	2-ply basalt TRHDC (B2TRHDC)	8
A5	2000*600*100	2-ply basalt TRHDC+HDC (B2TRHDC+HDC)	8
A6	2000*600*100	1-ply carbon TRHDC (C1TRHDC)	8

(a) ALC panels   (b) TRHDC-ALC composite panels

Figure 1.   Diagram of ALC panels before and after TRHDC topcoat application.

2.2 *Mechanical properties of the material parameters*

2.2.1 *TRHDC surface and mechanical properties*
The TRHDC surface layer combines textile and high ductility concrete. The main components of the high ductility concrete are: cement, fly ash, fine river sand, water, high-efficiency water reducing agent and 1.2% volume dose of PVA fiber; the compressive strength of the HDC was measured to be 52.0MPa using a cube standard specimen with a side length of 100mm and the tensile strength was estimated to be 3.40MPa using a dumbbell type specimen. Basalt fiber fabric was used, and the specific material properties are shown in Table 1.

Table 2. Mechanical properties of fiber meshes.

Fabric type	Modulus of elasticity/GPa	Tensile strength/MPa	Elongation/%	Density/(g/cm^3)
Basalt fiber mesh	34	860	2.5	2.65
Carbon fiber mesh	230	3600	1.5	1.74

2.2.2 *Autoclaved aerated concrete slabs*
The measured compressive strength and splitting tensile strength of ALC are shown in Table 2 according to the Test Method for Properties of Autoclaved Aerated Concrete (GB/T 11969-2020).

Table 3. Main performance parameters of ALC panels.

Grade	Mass density (kg/m³)	Compressive strength (MPa)	Splitting tensile strength (MPa)	Thermal conductivity (W/m-K)	Fire resistance (h)
B06	600∼650	5.03	0.53	0.16	6.0

2.3 *Loading options*

The test was carried out using four-point bending static loading on a 100t microcomputer-controlled electro-hydraulic servo pressure tester, with the calculated span of the plates all being 1800mm, and the basic loading device diagram shown in Figure 2.

By the provisions of the Standard for Test Methods for Concrete Structures (GB/T50152-2012), the loading procedure is as follows.

(1) Pre-load the specimen before formal loading to ensure that the loading device is in complete contact with the model without abnormalities and that the relevant instrument and data acquisition instrument are in average condition.
(2) Using displacement-controlled loading at a rate of 0.5 mm/min, the crack development pattern was observed, and the crack width was measured and labeled at the same time.
(3) When the loading reaches 85% of the peak load of the descending section, the specimen is considered to have reached a state of damage, and the loading is complete.

In order to measure the deflection of the composite plate in bending, displacement gauges were arranged in the span of the specimen and at the loading point, resulting in a load-displacement curve for the specimen.

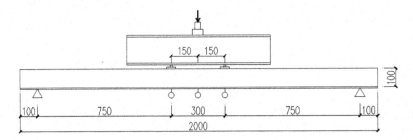

Figure 2. Loading diagram.

Figure 3. Loading photos.

3 TEST RESULTS AND ANALYSIS

3.1 *Damage pattern of the plate*

For the ALC plate, when loaded to 1.5 kN, the first bending crack appeared, accompanied by a large sound, the load dropped abruptly, continued loading, the pure bending section appeared new cracks and extended upwards, the deflection of the specimen increased while the load did not change much, the specimen bending damage occurred.

The TRHDC surface layer and the ALC slab have good ability to coordinate deformation and show good interfacial bonding performance, with no peeling damage to the reinforced surface layer before peak loads are reached.

The damage characteristics of the specimens with basalt mesh TRHDC surfacing are basically similar, the first crack appears near the bottom span of the slab, the crack width is small, as the load increases, the bending crack extends upwards and new bending cracks are produced nearby, when the mid-span crack extends to the middle of the specimen, the crack develops in an oblique direction, close to the damage load, the sound of the fibers pulling off in the TRHDC surfacing can be heard, finally the specimen reaches the ultimate tensile strain of the surfacing The specimen eventually reaches the ultimate tensile strain of the face layer and bending damage occurs.

Group A6 specimens with carbon fiber mesh TRHDC top layer, a number of vertical cracks were observed successively at the beginning of loading, when the peak load was reached, the shear diagonal crack below the loading point on the right extended through, the crack widened, the ALC plate was crushed at the loading point and the specimen suffered shear damage.

The high ductility concrete matrix in the TRHDC surface layer is mixed with PVA short fibers, after the initial cracking of the specimen, the PVA fibers across the cracks can continue to bear a small amount of tensile stress and bridge the cracks, transferring the stress to the fiber grid, so that the TRHDC-ALC specimen has an increased number of cracks compared to the comparison specimen, the crack spacing is reduced, and the bottom surface layer has a multi-crack development pattern during the bending loading process During the bending loading process, the bottom surface layer showed multiple cracks.

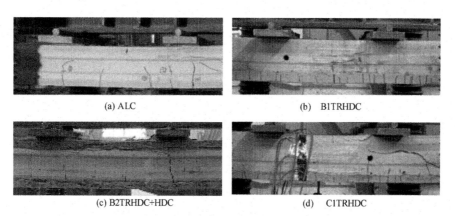

(a) ALC	(b) B1TRHDC
(c) B2TRHDC+HDC	(d) C1TRHDC

Figure 4.   Damage pattern and crack distribution of the specimen.

3.2 *Test results*

As can be seen from Table 4, the initial crack load and damage load of the TRHDC-reinforced ALC panels have increased substantially. The specimen load capacity was increased by 120% with B1TRHDC surfacing and by 229% with C1TRHDC surfacing. This is due to the higher tensile strength and modulus of elasticity of the carbon fiber mesh, which increased the flexural load capacity of the specimen to a greater extent than the basalt mesh TRHDC.

As the number of layers of basalt mesh in the TRHDC facing increases, the flexural load capacity increases accordingly. The load capacity of the specimen with two layers of basalt mesh TRHDC is 1.24 times higher than that of the specimen with one layer of basalt mesh TRHDC, but the load capacity of the specimen with one layer of basalt mesh is 1.46 times higher than that of the specimen without basalt mesh. It can be seen that the utilization of the TRHDC surface decreases as the number of layers of fiber mesh increases.

For specimens with two layers of basalt mesh TRHDC on the tensile side, a layer of HDC applied to the compressive side increased the flexural load capacity by 18% compared to the specimens without it.

Table 4. Test results.

Specimen number	Facing materials	Initial cracking load/kN	Damage load/kN	Load carrying capacity increase
A1	–	1.23	4.83	–
A2	HDC (tension zone)	3.2	7.29	51%
A3	B1TRHDC (tension zone)	3.8	10.61	120%
A4	B2TRHDC (tension zone)	5.32	13.15	172%
A5	B2TRHDC (tension zone) + HDC (compression zone)	6.08	15.57	222%
A6	C1TRHDC (tension zone)	6.77	15.90	229%

3.3 Load-medium span deflection curve

The load-span deflection curves of each specimen are shown in Figure 5. The use of TRHDC surfacing can effectively improve ALC panels' flexural strength, stiffness, and energy dissipation capacity.

At the beginning of loading, the load-span deflection curve of each specimen grows linearly and is in the elastic phase; compared with the ALC plate, the slope of the adjustable phase curve of the TRHDC-ALC model is large and the overall stiffness of the plate increases.

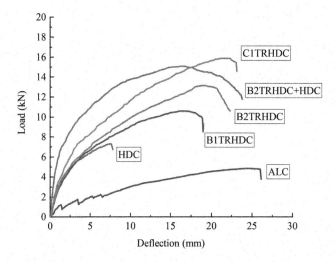

Figure 5. Load-span deflection curve.

Continuously loaded up to the crack development stage, the cracks continue to extend, the slope of the load-span deflection curve decreases and the TRHDC-ALC specimens exhibit significant plasticity.

With the increase in the number of layers of basalt fiber mesh, the flexural load capacity of the specimens increased, and the deformation capacity was also improved; for models with B2TRHDC on the tensile side, a layer of HDC was applied to the compressive side, and the flexural load capacity and deformation capacity of the specimens were improved, and the energy dissipation capacity was strong. All other conditions being equal, for the models with CTRHDC the flexural stiffness is greater than that of the specimens with BTRHDC because the modulus of elasticity of the carbon fiber mesh is somewhat more significant, but the models with CTRHDC have a linear drop in load after the peak, and the damage is more abrupt.

4 CONCLUSIONS

(1) Given the shortcomings of existing ALC panels, this paper proposes a new type of TRHDC-ALC composite panel combined with TRHDC and ALC panels. In engineering applications, TRHDC-ALC composite panel integrates lightweight, load-bearing, energy-saving, fireproof, heat insulation, and sound insulation, which can be used in the fields of building internal partition panels, high-rise structure perimeter protection systems, sandwich structure lightweight floor panels, etc. TRHDC-ALC panel can greatly reduce the self-weight of the building and shorten the construction cycle in the application of assembled buildings, which has good prospects for engineering applications.

(2) In this paper, the flexural performance of the TRHDC-ALC composite panel is studied, and the results show that the cracking load of the composite panel can be increased by 394% and the ultimate flexural load capacity can be increased by 222% after adopting TRHDC surface layer. TRHDC-ALC composite panel has very good flexural mechanical properties.

(3) The PVA short fibers in the TRHDC matrix can bridge the cracks and transfer the stress. During the loading process, the bottom surface layer of the TRHDC-ALC specimen is in the form of multiple cracks, and the number of cracks increases and the spacing of cracks decreases. Moreover, the adhesion performance of the TRHDC surface layer and ALC plate is better, the new load is borne by the fiber grid, the bearing capacity is greatly increased, and the material strength of the fiber grid is efficiently utilized.

(4) The improvement of bearing capacity and flexural stiffness of CTRHDC is greater than that of BTRHDC; for the specimen with BTRHDC, the number of layers increases, the bearing capacity and deformation capacity of the specimen increases, but the utilization rate of basalt fiber grid decreases; a layer of HDC applied on the compressed side can effectively improve the flexural bearing capacity, flexural stiffness, and energy dissipation capacity of the composite plate.

(5) This paper is an exploratory test, because a group of specimens in this test had shear damage, the tensile strength of TRHDC surface layer was not fully utilized, so the reasonable utilization rate of TRHDC surface layer in composite plate should be analyzed subsequently; secondly, the bonding performance and bonding mechanism between TRHDC matrix and ALC need to be further studied; in addition, the variable parameters considered in this paper are limited, and the follow-up can be studied in depth Different factors, such as mesh type, short fiber type and reinforcement rate of ALC slab, span, grade of autoclaved aerated concrete, etc., have effects on the mechanical properties of TRHDC-ALC composite slab; the experimental study of fire resistance of TRHDC-ALC composite slab is supplemented; finally, this paper adopts the method of manual pressure smearing to produce TRHDC-ALC slab, and the subsequent study of prefabricated TRHDC -ALC composite panels and the anchorage measures between the interfaces should be studied for engineering application and promotion.

REFERENCES

Dong Z.F.& Deng M.K. (2020). Tensile behavior of glass textile reinforced mortar (TRM) added short PVA fibers. Construction and Building Materials.

Liu Y. J.& Chen B.T. (2013). A review of aerated concrete development at home and abroad. Building Energy Conservation. 265, 30–34.

Lv C.S.& Zhang D.L. (2017). Research on the cracking of autoclaved aerated concrete (ALC) slab walls. Wall Material Innovation and Building Energy Efficiency. 8, 29–32.

Mousa, M. K.& Uddin, N (2009). Experimental and analytical study of carbon fiber-reinforced polymer (FRP)/autoclaved aerated concrete (AAC) sandwich panels. Engineering Structures. 31, 2337–2344.

Narayanan, N.& K. Ramamurthy (2000). Structure and properties of aerated concrete: A review. Cement & Concrete Composites. 22, 321–329.

Wakili, K. G.& E. Hugi (2015). Thermal behavior of autoclaved aerated concrete exposed to fire. Cement & Concrete Composites. 62, 52–58.

Wu G. & Wu Z.S. (2007).A new fiber reinforced polymer-autoclaved lightweight concrete(FRP-ALC) sandwich structure. Industrial Construction. 126–130.

Xu C.Y. & Guo Y.Y. (2019). Experimental study on flexural behavior of assembled lightweight block assembled wallboard. Journal of Shenyang Jianzhu University (Natural Science). 35(6):1004–1012.

Zhang G. W.& Chen B.S. (2016). Study on the flexural performance of autoclaved aerated concrete external wall panels. Building Structures. 46, 97–102.

Zhang J. J.& Zhang X.W. (2019). Experimental study on flexural performance of autoclaved aerated concrete floor slabs. Sichuan building science research. 45, 52–58.

*Frontiers of Civil Engineering and Disaster Prevention and
Control – Yang & Rahman (Eds)
© 2023 The Author(s), ISBN: 978-1-032-31200-2*

Scour characteristics of middle approach bridge foundations in Hangzhou Bay sea-crossing bridge

Jinquan Wang
Ningbo Hangzhou Bay Bridge Development Co., Ltd., Ningbo, China

Zhiyong Zhang*, Zuisen Li & Yuanping Yang
Zhejiang Institute of Hydraulics and Estuary, Hangzhou, China

Xiaoliang Xia
Ningbo Hangzhou Bay Bridge Development Co., Ltd., Ningbo, China

ABSTRACT: To evaluate and predict the scour status of the Hangzhou Bay Sea-crossing Bridge well, the general and local scour characteristics in a strong tidal area were analyzed based on field data of 133 middle approach bridge foundations. It is found that the scour holes are elongated upstream due to a stronger flood tide. The general scour depths are about 2-4 m, while the local scour depths are about 4-8 m, which illustrates that the local scour depth is about twice the general scour depth. In addition, the existing equations, including HEC-18 and OH equations, both overestimate the scour depth. Based on field data, an equation for local scour is established, and the relative errors are both within 20%, which can be used to predict local scour depth in conditions of strong tide and cohesive seabed.

1 INTRODUCTION

With the development of the economy and technology in China, the construction of cross-sea bridges in China is in full swing. The Hangzhou Bay Sea-crossing Bridge, Jiaozhou Bay Bridge, Hong Kong-Zhuhai-Macao Bridge, and other super-large sea-crossing bridge projects opened to traffic, marking that China has become a world power in sea-crossing bridge construction. As an important transportation facility in coastal areas, cross-sea bridges play an important role in promoting people-to-people communication and economic and social exchanges in coastal areas. However, due to the complexity of the problem of sea bridges in the long-term operation, the foundations of cross-sea bridges have been scoured to different degrees (Guo et al. 2020), such as the Donghai bridge, whose scour elevation is labeled as scouring tsunami warning (Chen et al. 2019).

On the one hand, the construction of cross-sea bridges will lead to many bridge foundations occupying a certain area of water crossing. This will narrow the water crossing section on the whole, and then cause a certain degree of narrow scouring, which is often called general scouring in China. At the same time, local scouring occurs in the local area of the bridge foundation due to the influence of the local flow around the pier. For the general scour of bridges, the scour range is mainly affected by the overall water blocking effect of the bridge, specifically the span of the bridge foundation and the water blocking scale of the bridge foundation. Melville et al. (2000) found that the general scour depth is only about 0.1 times that of the local scour depth based on laboratory tests and measured data of rivers and bridges. The general scour equations commonly used in China are mainly the non-cohesive sand 64-1 and 64-2 equations, and the general scour equation for cohesive soils in the "Specification for Hydrological Survey and Design of Highway Engineering" (JTG C30-2015) (hereinafter referred to as the Highway Hydrological Code). Li (2002) gave a general scour equation related to the narrow flow velocity within the bridge beam, the starting flow velocity

---

*Corresponding Author: 82457114zzy@163.com

of clay, and the initial water depth through the clay scour test. The general scour equation caused by engineering construction (including bridge construction) in tidal sea areas is also presented by Cao et al. (2009). The equation is related to the bridge span velocity, and initial water depth before and after the bridge is built. This equation is commonly used to analyze the scour and sedimentation effects of engineering construction.

For the local scour problem of bridge piers, many scholars have carried out extensive research on the local scour mechanism and characteristic scale. The methods used are mainly included physical model tests (Ni et al. 2020; Xiao et al. 2021; Yang et al. 2020), numerical simulation (Roulund et al. 2005; Zhao et al. 2010), and measurement analysis method (Lu et al. 2016; Yang et al. 2021). Some mathematical models (Xiong et al. 2014; Zhu et al. 2011) and calculation equations for local scour prediction of bridge piers are also obtained. Melville and Chiew (1999), Sheppard et al. (2006), and Orth et al. (2015) successively studied the local scour equilibrium depth of bridge piers with non-viscous uniform sand bed surface through physical model tests and proposed their respective pier scour equations. The US Department of Transportation recommends using the HEC-18 equation (Richardson et al. 2001) as the calculation method for the local scour of bridges. Equations 65-1 and 65-2 are commonly used for the local scour equations of bridge piers in China. However, the above equations are based on the experimental results or measured data of unidirectional flow, and the conditions of use are limited to non-viscous sand river beds. There are also some equations to predict the local scour of the clayey pier. The main difference lies in which parameters are used to reflect the sediment characteristics in the equations. For example, in the equation for local scour of the clayey pier under unidirectional flow given in the hydrologic code of highway, the sediment characteristics are reflected by the clayey liquid index. However, the scour equation of OH clay pier recommended in HC-18 abroad is related to the characteristics of the starting velocity of clay. However, most of these equations are obtained based on physical tests, and the application effect in practical engineering is rarely evaluated and compared.

Cross-sea bridge projects are often located in near-shore sea areas, and their water flow characteristics are mostly reciprocating tidal currents. In particular, the Hangzhou Bay area is the sea area with the most sea-crossing bridge projects and the strongest tidal dynamics in China. The sediment in this sea area is mostly silt and clay. For the local scour of cohesive soil seabed bridge piers under the action of reciprocating tidal currents, many scholars have used scale physical model tests to study. Some research conclusions are quite different, such as the maximum scour depth. Some scholars believe that the reciprocating flow is smaller than the constant flow (Schendel et al. 2018), while others believe that the two are similar (Escarameia et al. 1999).

As a prototype-scale scour data, the on-site observation data of bridge engineering can best reflect the real scour characteristics of bridge foundations. It effectively avoids the scale effect of physical tests and is a valuable material for evaluating the general scour and local scour equations of existing bridge piers. However, to better distinguish the general scour of bridge foundation, local scour scale, and the influence of various factors on various scour scales, more detailed hydrology, sediment, and seabed geological exploration data are needed. Based on the scour measurement data of 133 bridge foundations with the same structure in the middle approach area of the Hangzhou Bay Cross-sea Bridge, combined with relatively detailed hydrological sediment and geological exploration data, this study intends to separate the general scour and local scour scales of the bridge foundations in the middle approach bridge area. On this basis, the differences between the current general scour equation and the local scour equation are judged, and the general scour and local scour equations applicable to the muddy seabed bridge foundation in the reciprocating tidal sea area are proposed for reference to related projects.

2 PROJECT OVERVIEW

2.1 Geographical location

Hangzhou Bay Bridge is located at the mouth of the Qiantang River in Hangzhou Bay. It starts from Ping Lake in Jiaxing in the north and ends at Cixi in Ningbo in the south, with a total length of about

36 km. The construction of the bridge began in 2003 and was completed in June 2007. The bridge was officially opened to traffic in May 2008. From north to south, there is the north navigable hole area, the middle approach bridge area, the Nantong navigation area, and the south approach bridge area. This study focuses on the bridge piers in the middle approach bridge area, and the specific location and real scene are shown in Figure 1. In the area where the bridge is located, the high tide flows from the East China Sea and Hangzhou Bay to the Qiantang River estuary, while the low tide flows from the Qiantang River estuary to the Hangzhou Bay and the East China Sea.

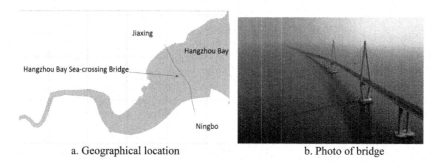

a. Geographical location	b. Photo of bridge

Figure 1.   Location of Hangzhou Bay Sea-crossing bridge.

2.2   *Structure type of bridge pier*

There are 133 piers in the middle approach bridge area, which are 3078-12000 m away from the north shore seawall. The span of all piers is 70 m, and the piers are of the same type, double-cap pile group structure. The specific structure type is shown in Figure 2. The diameter of the upper bearing cap of the bridge foundation is 10.5 m, the thickness of the bearing cap is 2.8 m, and the height of the bottom of the bearing cap is 0.2 m. The bearing cap is basically above the average mid-tide level (the moment of the rapid flow rate of fluctuation), and most of the time, there is no water, and the flow rate is generally low. Therefore, for the middle approach bridge, the erosion of the bridge foundation is mainly affected by the pile group structure at the bottom. The diameter of the single pile group at the bottom of the cap is 1.5 m, and there are nine piles under each cap with a vertical slope of 1:5–1:6. By calculating the projection along the flow direction, the width of the projection resistance of the pier below the average tide level is 7.2m.

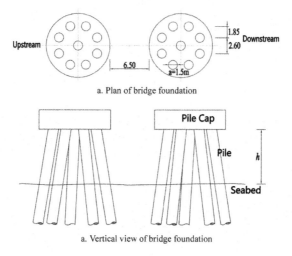

a. Plan of bridge foundation

a. Vertical view of bridge foundation

Figure 2.   Shape of middle approach bridge foundation.

2.3 Hydrological condition

The annual average wave height in the sea area where the bridge is located is 0.32 m, and the annual significant wave height above 1.5 m only accounts for 2.0%. In comparison, the tidal power is much stronger than the wave. The spring tide period at Hangzhou Bay is in autumn, and the tidal power is stronger than in other periods. According to the long-term tidal level observation data of Zhapu station on the north bank of the Bridge, the historical maximum tidal is 8.18 m, and the average monthly maximum tidal range is about 7.05 m. Therefore, the tidal power is very strong. Figure 3 shows the typical tidal current velocity process near pier 105 of the central approach bridge, and it can be seen that the sea area of the bridge is irregular semi-diurnal tide, with a tide cycle of about 12 hours. Within a tide cycle, the duration of high tide (positive flow rate) is short, while that of low tide (negative flow rate) is long, but the velocity of high tide is generally larger than that of low tide, which is characterized by the dominance of high tide. For tidal current, tidal current velocity correlates with tidal range. Under the average monthly maximum tidal range of 7.05 m, the fully verified two-dimensional fixed-bed mathematical model was applied to calculate the maximum tidal flow velocity of each pier in the sea area where the bridge is located. It was concluded that the overall trend of the velocity in front of the pier in the middle approach bridge area was greater, and the velocity varied from 2.09 to 2.63 m/s. The calculation results are shown in Figure 4.

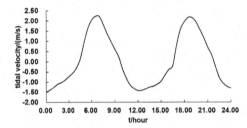

Figure 3. Typical time process of tidal current.

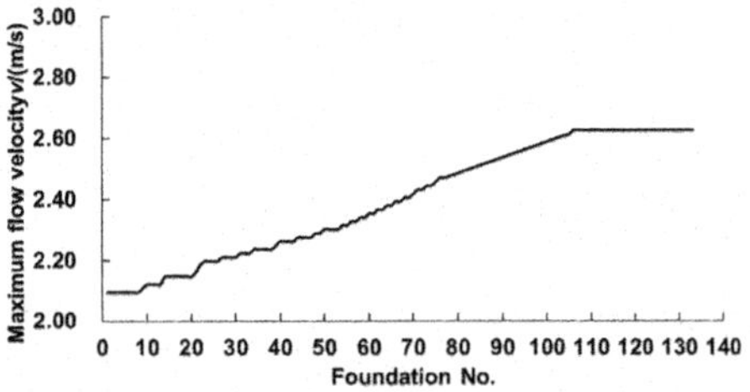

Figure 4. Maximum flood tidal velocities of each middle approach bridge foundations.

2.4 Seabed geological condition

During the construction of the bridge, geological boreholes and soil samples were analyzed at 11 locations in the middle approach bridge area, as shown in Table 1. According to the analysis, the seabed of the middle approach bridge area is mostly silt or silt clay. The median particle size of the sediment sampled by each borehole is 0.025–0.07 mm, and the average median particle size is about 0.05 mm. Zong et al. (2018) showed that the liquid index has a great influence on the initiation of sediment movement, and they also calculated the liquid index of each soil layer. Due to the difference in mechanism between general scour and local scour, the soil humoral indexes in general scour range and local scour range were averaged, respectively. The specific situation is shown in Table 1.

Table 1. Geologic bore among middle approach bridge.

Drill number	Adjacent bridge pier number	Average liquid index I_L (general scour a range of seabed)	Average liquid index I_L (local scour a range of seabed)
ZK1	11	1.25	0.99
ZK2	22	0.84	1.75
ZK3	34	0.79	0.53
ZK4	45	0.78	1.43

(continued)

Table 1. Continued.

Drill number	Adjacent bridge pier number	Average liquid index I_L (general scour a range of seabed)	Average liquid index I_L (local scour a range of seabed)
ZK5	57	1.22	1.75
ZK6	68	1.18	1.51
ZK7	80	1.11	1.29
ZK8	91	1.40	1.82
ZK9	102	1.32	1.29
ZK10	114	1.62	1.36
ZK11	125	1.32	0.97

3 MEASUREMENT AND DATA STATISTICS

After the Hangzhou Bay Cross-sea Bridge was officially opened to traffic in 2008, the bridge company commissioned a professional survey unit to conduct a multi-beam fine topographic survey on the background erosion of the bridge (Jiang et al. 2010), whose measurement range was 100 m above and downstream of the bridge. A multi-beam underwater topographic survey system is used for local topographic observation of the pier. Then we calculate the tidal level by the GPS, and data combination, total propagation error calculation, and denoising are performed. The accurate local topographic map of the pier is finally obtained. Figure 5 shows the topography of local scour pits in some pier sections of the approach bridge area in 2008. It can be seen that there are obvious scour pits in local pier sections. The scour pits of each pier are similar in shape. Influenced by the reciprocating tidal current, the scour pits extend upward and downstream. However, in general, the scour pits extend upstream for a long time because the dynamic force of flood tide is stronger than that of the ebb tide, presenting scour pits dominated by the flood tide.

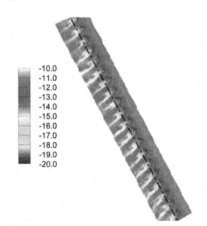

Figure 5.   Seabed bathymetry around middle approach bridge (Unit:m).

4 ANALYSIS AND DISCUSSION OF MEASURED DATA

4.1 *Discrimination of general scouring and local scouring*

After the construction of the bridge, under the influence of the bridge beam narrow flow, the overall seabed brush depth at the bridge axis position, and under the influence of the local flow around the bridge pier, the local scour pit of the bridge pier is formed. Figure 6 compares the terrain before and after the construction of the bridge by intercepting the bridge axis section between Foundation No.18 and No.25. It can be seen that before the construction in 2001, the seabed in the middle approach bridge area was relatively flat, and its seabed elevation was generally about -12m. After the bridge was built, the seabed of the axial section of the bridge was scoured as a whole and fluctuated due to the influence of local scour pits of each pier, the position of the pier is affected by local scour, and the seabed elevation is lower. The middle of the bridge span is located outside the local scour pit, the seabed elevation is higher than the local scour pit elevation of the pier, but it is also generally lower than the seabed before the bridge was built. Since the construction of Hangzhou Bay Bridge was officially started in 2005 and opened to traffic in 2008, the natural scour and silting evolution range of the seabed in the middle approach bridge area is small. The overall change of the seabed in the position of the bridge axis can be considered to be mainly influenced by the general scour caused by the bridge construction.

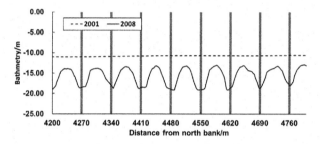

Figure 6.   Seabed elevation along bridge axis between No.18–25 before and after building the bridge.

According to the topography monitoring data of 2008, the difference between the seabed elevation outside the scour pit on the axis of the bridge and the seabed elevation before the construction of the bridge is taken as the general scour depth, while the difference between the elevation of the deepest point in the local scours pit and the elevation of the seabed after the general scour is the local scour depth. Then the general scour depth and local scour depth of each pier of the middle approach bridge are obtained. The scouring condition of each pier is shown in Figure 7. Due to the same pier type and spacing of the middle approach bridge, its water-blocking effect is relatively close, so the general scour amplitude caused by the bridge is relatively close, generally within the range of 2–4 m. The local scour amplitude of each pier in the middle approach bridge section is in the range of 4–8 m, and its fluctuation amplitude is much larger than that of the general scour fluctuation amplitude. In addition, statistically speaking, the general scour amplitude is about 0.3-0.9 times of local scour amplitude, and the average scour amplitude is about 0.53 times.

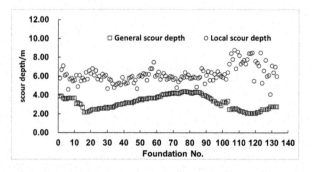

Figure 7.   General and local scour depths of each bridge foundation.

4.2 Discussion on calculation methods of local scour

The 65-1 and 65-2 equations and HEC-18 equations in the United States are commonly used to calculate the local scour of bridge piers for non-cohesive soil. The hydrologic code of highway also provides an equation for local scours of cohesive soil pier for cohesive soil. OH summarized the scour test data of cohesive soil pier by Brauid et al. (2011) and obtained the scour equation of cohesive soil pier related to the starting characteristics of silt through equation fitting. For the existing commonly used local scour equation of bridge pier, both equations 65-1 and 65-2 contain factors based on the initiation characteristics of channel sediment and local scour characteristics of bridge piers (Zhu et al. 2016), which is suitable for non-viscous river bed, but it is very poor in the clayey river bed. However, although the US HEC-18 equation is based on the results of non-viscous sediment scour, its equation does not contain the parameters of sediment characteristics. Some studies (Sonia et al. 2017) found that in some clay scour tests, the prediction results of this equation are close to the calculation results of the partial clay pier scour equation. Based on the application of existing equations, for the local scour problem of the clay bridge pier in Hangzhou Bay. In this work, the prediction effects of the HEC-18 equation and OH equation are tested by using measured data.

For the HEC-18 equation, its form is as follows:

$$\frac{h_b}{h_p} = 2.0 K_1 K_2 K_3 (B/h_p)^{0.65} Fr^{0.43} \tag{1}$$

Where h_b is the local scour depth of the pier, K_1, K_2 and K_3 are correction coefficients of pier shape (column is 1.0), flow angle of attack coefficient, and channel condition correction coefficient, respectively. B is the effective resistance scale of bridge pier, for pile group, $B = K_4 B_t$, wherein B_t is the projected width of bridge pier group pile, the middle approach bridge is 7.2 m, K_4 is pile spacing factor, $K_4 = 1 - 4/3(1 - a/B_t)[(1 - (L/a)^{-0.6}]$, where a is pile diameter, 1.5 m, and L is spacing. The average spacing of pile groups at general scouring elevation is 2.35 m. Fr is Froude number, $Fr = v/\sqrt{gh_p}$, where the v is the velocity before the pier. Here, we take the maximum flow rate of ebb and flow tide.

The HEC-18 equation was used to predict the local scour depth of 133 piers in the middle approach bridge area and compared with the measured value (Figure 8). It can be seen that the calculated value of the equation is generally 7–8 m, generally larger than the measured value (4–8m), and the maximum deviation is nearly 60%. There are two main reasons for this. First, due to the action of viscosity and inhomogeneity of clay soil, the starting velocity of clay soil is generally higher than that of non-clay sand, and the ability of clay soil to resist scour is strong, leading to the small local scour amplitude of bridge pier. Second, the effective hydrodynamic strength of the reciprocating tidal current is weaker than that of the constant current (flow velocity is equal to the maximum flow velocity), and the scour amplitude is also weaker than that of the constant flow.

Most scholars put the sediment starting velocity parameter, which reflects the anti-scour capacity of sediment, into the equation of bridge pier scour as the influencing factor, such as the local scour equation of clay pier of OH.

$$h_b = 2.2 K_1 K_2 B^{0.65} \left(\frac{2.6v - v_c}{\sqrt{g}} \right)^{0.7} \tag{2}$$

The symbolic meaning of the equation is the same as that of the HEC-18 equation. In the equation, the starting flow rate can be calculated by $v_C = 0.33(1/I_L)h_p^{0.6}$.

The OH equation was verified by applying the measured local scour depth of 11 piers. Figure 9 shows the comparison between the calculated results of the equation and the measured values. Compared with the HEC-18 equation, the deviation between the calculated results of the OH equation and the measured values is within 50%, and the deviation range of the equation is the smallest among the two equations.

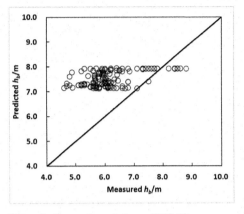

Figure 8. Comparisons between HEC-18 and field data.

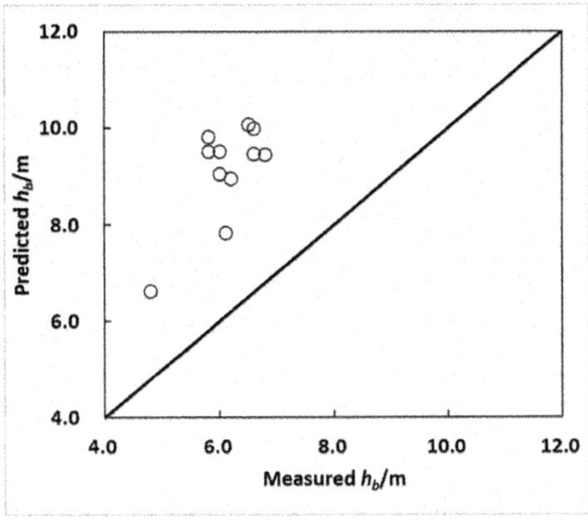

Figure 9. Comparisons between equation from OH and field data.

For the local scour of piers, the main factors are not only the scale of resistance, but also the tidal current dynamics and sediment characteristics. Here, combined with the measured data of pier scour in the middle approach area, we modified the OH equation and finally obtained an equation for predicting the local scour of clay pier under reciprocating tidal currents. Its equation is shown in Equation (3). The error between the calculated result and the measured value is within 20%, and the prediction accuracy is obviously higher than that of other equations.

$$h_b = 1.51 K_1 K_2 B^{0.65} \left(\frac{2.6v - v_c}{\sqrt{g}} \right)^{0.7} \tag{3}$$

However, in practical engineering applications, the design buried depth of bridges is generally determined from the angle of partial safety. Considering that Equation (3) has a certain probability of underestimating the local scour depth, it is suggested to enlarge the calculation results of Equation (3) by more than 1.2 times in practical engineering to ensure the safety of bridge foundation design.

5 CONCLUSION

Based on the tracking observation data of pier scour in the middle approach area of Hangzhou Bay Cross-sea Bridge, the general scour and local scour caused by the foundation of the cross-sea bridge under the condition of reciprocating tidal currents are studied. The following results are found:

(1) The pier scour pit in the middle approach area of Hangzhou Bay Bridge extends upward and downstream bidirectional, which is influenced by the reciprocating tidal current dominated by the rising tide and extends longer the upstream. The height of the deepest scour pit of each pier of the middle approach bridge is between -18 m and 24 m because of the difference in seabed bottom quality.

(2) By comparing the seabed topographic data of the middle approach bridge area before and after the construction of the bridge, it is found that the general seabed scours amplitude of the middle approach bridge area of Hangzhou Bay Bridge is about 2-4m, and the local scour amplitude is about 4-8 m, and the general scour amplitude is about 0.5 times of the local scour amplitude.

(3) For local scour, the equation of HEC-18 overestimates the local scour depth, and the OH equation also forecasts the scour rate by about 50%. With modification based on the OH equation, we obtained the method for calculating local scour depth used for local scouring of clay pier in reciprocating tidal waters, which prediction accuracy is within 20%.

ACKNOWLEDGMENTS

This study was funded by Zhejiang Transportation Science and Technology (Grant No. 2021064), the Public Welfare Research Project of Zhejiang Province (Grant No. LGF21E090003).

REFERENCES

Briaud J. L., Ting F. C. K., Chen H. C. Summary Report of the SRICOS-EFA Method[R].Texas A&M University, 2011.
ESCARAMEIA M, MAY R W P. Scour Around Structures in Tidal Flows[R]. UK: HR Wallingford, 1999.
Jian Guo, Bing Jiang. Research Progress and Key Issues of Bridge Pile Scour in Recent 30 Years[J]. China Journal of Highway and Transport, 2020, 33(07): 1–16.
JTG-C30-2015. Hydrological Specifications for Survey and Design of Highway Engineering[S].
Melville Bw, Chiew Y M . Time Scale for Local Scour at Bridge Piers[J]. Journal of Hydraulic Engineering, 1999, 125(1): 59–65.
Melville Bw, Se Coleman. Bridge Scour[M]. Water Resources Publications, Highlands Ranch, 2000.
Ming Zhao, Liang Cheng , ZhipengZhang . Experimental and numerical investigation of local scours around a submerged vertical circular cylinder in steady currents[J]. Coastal Engineering, 2010, 57(8): 709–721.
Orth, Franciska, Ettmer, et al. Live-Bed Scour at Bridge Piers in a Lightweight Polystyrene Bed[J]. Journal of hydraulic engineering, 2015, 141(9): 4015017.
Quanli Zong, Junqiang Xia, Yi Zhang, et al., Experimental Study on Scouring Characteristics of Cohesive Bank Soil in the Jingjiang Reach[J].Advances in Water Science, 2014, 25(04): 567–574.
Richardson E.V., Davis S. R. Evaluating Scour at Bridge. 4th edn. Hydraulic Engineering Circular No. 18 (HEC-18)[R]. Washington, DC: Federal Highway Administration, U.S. Department of Transportation, 2001.
Roulund A, Sumer B M, FredsE J, et al. Numerical and experimental investigation of flow and scour around a circular pile[J]. Journal of Fluid Mechanics, 2005, 534: 351–401.
Schendel A, Hildebrandt A, Goseberg N, et al. Processes and evolution of scour around a monopile induced by tidal currents[J]. Coastal Engineering, 2018, 139: 65–84.
Sheppard D M, Jr W M . Live-Bed Local Pier Scour Experiments[J]. Journal of Hydraulic Engineering, 2006, 132(7): 635–642.
Shu Chen, Study of Scouring Protection Schemes for Pier Foundation on Donghai Bridge, World Bridge, 2019, 47(04): 17–21.
Sonia D.Y., Barbhuiya A K. Bridge Pier Scour in Cohesive Soil: a Review[J]. Sādhanā, 2017, 42(10): 1803–1819.
Wen Xiong et al. CFD Simulations and Analyses for Bridge-Scour Development Using a Dynamic-Mesh Updating Technique[J]. Journal of Computing in Civil Engineering, 2014: 4014121.
Xiaojun Jiang, Nan Liu, Renyi Liu, et al., Study on the Method to validate the Pier Local Scour Model in the Strong Tide Region: A Case Study of the piers local scour of Hangzhou Bay Sea-crossing Bridge[J]. Journal of Zhejiang University (Science Edition), 2010, 37(01): 112–116.
Xuan Ni, Leiping Xue, Chao An. Experimental Investigation of Scour around Circular Arrangement Pile Groups[J]. Ocean Engineering, 2021, 219: 108096.
Xuejun Lu, Heqin Cheng, Quanping Zhou, et al., Features and mechanism of asymmetric double-kidneys scoured geomorphology of the pier in tidal estuary[J]. Haiyang Xuebao, 2016, 38(09): 118–125.
Y Li. Bridge Pier Scour and Contraction Scour in Cohesive Soils Based on Flume Tests[D]. College Station, Texas, USA: Texas A&M University, 2002.
Yaling Wang, Calculation Method for Abutment Scour of the Riverbed Consisted of Clay[J]. Journal of Xian University of Highway and Communication, 2000(01): 46–48.

Yang Xiao, Hao Jia, Guan Dawei, et al. Experimental Investigation on Scour Topography around High-rise Structure Foundations[J].International Journal of Sediment Research, 2021, 36(3): 348–361.

Yilin Yang, Meilian Qi, Xin Wang, et al. Experimental Study of Scour around Pile Groups in Steady Flows[J]. Ocean Engineering, 2020, 195: 106651.

Yuanping Yang, Zhiyong Zhang, Zuisen Li. Scour Characteristics of Sea-crossing Bridge. [J]. Hydro-Science and Engineering, 2021, 40(04):131-137.

Zhiwen Zhu, Zhenqing Liu. Three-dimensional Numerical Simulation for Local Scour Around Cylindric Bridge Pier[J]. China Journal of Highway and Transport, 2011, 24(02):42–48.

Zhiwen Zhu, Peng Yu, Comparative Study Between Chinese Code and US Code on Calculation of Local Scour Depth Around Bridge Piers[J]. China Journal of Highway and Transport, 2016, 29(01):36–43.

Zude Cao, Hui Xiao, Seabed Evolution Prediction and Application on Muddy Coast under Action of Tidal Current. Waterway and Port, 2009, 30(01):1–8.

*Frontiers of Civil Engineering and Disaster Prevention and
Control – Yang & Rahman (Eds)
© 2023 The Author(s), ISBN: 978-1-032-31200-2*

Finite element analysis of seismic performance of U-shaped reinforced concrete special-shaped columns

J.Y. Li*, X.Y. Wang*, W.B. Lu*, L. Chen*, Z.W. Wang*, J.P. Guo* & G.C. Hu*
Changchun Institute of Technology, Changchun, China

ABSTRACT: To study the seismic performance of U-shaped reinforced concrete, the finite element analysis software ABAQUS simulated the quasi-static loading test of ten URC special-shaped columns. The effects of different design parameters (concrete strength grade, stirrup spacing, tie bar spacing, slenderness ratio) on the failure mode, hysteretic characteristics, bearing capacity, elasticity, and stiffness degradation of specimens were analyzed. Results showed that each specimen's failure modes were an asymmetrical ductile bending failure. With the decrease of stirrup and tie bar spacing, the elasticity of the sample is improved. With the increase of slenderness ratio, specimen bearing capacity and ductility decrease accordingly. High-strength concrete can significantly improve the bearing capacity of anisotropic columns, but due to the brittleness of high-strength concrete, flexibility will gradually decrease. Therefore, URC special-shaped columns have good seismic performance.

1   INTRODUCTION

In recent years, with the vigorous development of China's construction industry, the form of a building structure has been diversified, and the structure's layout is more and more reasonable. The layout requirements and overall safety, and rationality have been improved to ensure beauty and the need to increase the indoor area. However, due to the different and complex stress of various cross-section forms of special-shaped columns, it is necessary to study the seismic performance of special-shaped columns.

Special-shaped column refers to the L-shaped, T-shaped, and cross-shaped columns with the height-width ratio of wall limbs not more than four and the same thickness as the wall in the structure (JGJ 149-2017). At present, Su et al. have carried out experimental studies on the seismic performance of L-shaped (4) and T-shaped (4) reinforced concrete columns and proposed the single-degree-of-freedom system restoring force model of anisotropic columns, which provides a basis for the study of special-shaped columns. A low cyclic loading test of 13 reinforced concrete columns with different parameters was carried out by Huang Chenkui and Qu Fulai. The results showed that (1) with the increase of axial compression ratio and web, the bearing capacity increases, but the deformation capacity and ductility decrease; (2) increasing stirrup ratio and strengthening and longitudinal reinforcement can improve the ductility of specimens. Diao bo et al. studied the seismic behavior of six reinforcement concrete L-section columns, and it's obtained that the loading angle has a great influence on the bearing capacity and ductility of the specimen. The finite element analysis model was established by using ABAQUS. Teng Zhengchao studied the effects of axial compression ratio and stirrup ratio on seismic performance of L-shaped columns and T-shaped columns, which laid the foundation for seismic analysis of frame systems using the kind of column. It can be seen that there is little research on the restoring force model and experimental study of U-shaped reinforced concrete (URC) special-shaped columns under different parameters in China. ABAQUS

---

*Corresponding Authors: 429838648@qq.com, 1241957603@qq.com, 1026477519@qq.com, 1063915947@qq.com, 942482763@qq.com, gjpgjp1002@163.com and 2048839443@qq.com

carried out the relevant simulation analysis of special-shaped columns, which provides a relevant reference for promoting the application of URC special-shaped column structure application.

2 FINITE ELEMENT MODEL ANALYSIS

2.1 Design of specimen

To explore the influence of different parameters on the seismic performance of URC special-shaped columns, a total of ten URC special-shaped columns were designed. Figure 1 is the detailed section size of the specimen, and Table 1 is the detailed comparison parameters of each sample.

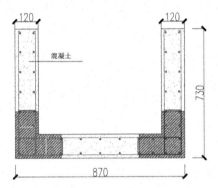

Figure 1.   Model section dimension.

Table 1.   Parameters of specimen.

Test specimen number	Concrete grades	Stirrup spacing/mm	Tie bar spacing/mm	Length of specimen/mm	Slenderness ratio λ
UG-1	C30	100	100	2050	17.08
UG-2	C40	100	100	2050	17.08
UG-3	C50	100	100	2050	17.08
UG-4	C60	100	100	2050	17.08
UG-5	C40	50	50	2050	17.08
UG-6	C40	200	200	2050	17.08
UG-7	C40	300	300	2050	17.08
UG-8	C40	100	100	1850	15.41
UG-9	C40	100	100	2450	20.42
UG-10	C40	100	100	2650	22.08

2.2 Modeling process

This paper does not use the top of the URC special-shaped column to add a sizeable rigid pad; instead, a reference point is set at a certain distance above the centroid of the URC-shaped column to impose various properties. Since the property of the concrete special-shaped section column is a homogeneous material, a three-dimensional solid element (C3D8R) is selected, which is six node solid elements. The steel frame adopts a three-dimensional linear truss element (T3D2). To consider the interaction between reinforcement and concrete, steel bars are fully embedded in concrete columns so that the stain of the support is consistent with the substantial element. But the slip between reinforcement and concrete is not considered in this simulation. The grid size is parameterized, and the optimal grid size is 50 mm. In practice, the column pier and foundation are rigid. To achieve the effect of rigid connection at the bottom of the column, the degrees of freedom in six directions at the bottom of the column pier are limited, which is conducive to making the simulation results more realistic. A constant axial load and cyclic lateral displacement are applied to the reference point for load application. The model is established in Figure 2.

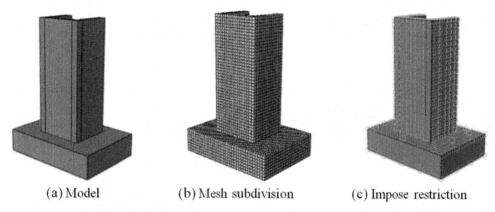

(a) Model   (b) Mesh subdivision   (c) Impose restriction

Figure 2. Modeling schematic diagram.

2.3 *Definition of material properties*

The related parameters of material properties of specimens are defined in Table 2.

Table 2. Material property definition table.

Name of the material	Elastic modulus/ $(N \cdot mm^{-2})$	Density/ $(Kg \cdot m^{-3})$	Material type	Poisson ratio
Concrete	3.00×10^5	2.50×10^3	—	0.2
Steel reinforcement	2.00×10^4	7.85×10^3	HRB400	0.3

3 VERIFICATION OF FINITE ELEMENT MODEL

By applying the constitutive model used in this model to the reference, the validity and accuracy of the finite element modeling are verified by comparing the simulation results with the experimental results, as shown in Figure 3. It can be seen from the figure that the peak load, ultimate load, and other significant parameters of the two have little difference in trend. The skeleton curve page is in good agreement, indicating that the model has certain rationality and reliability.

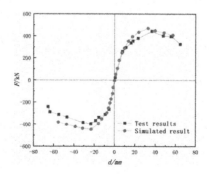

Figure 3. Comparison of load displacement curves.

4   TEST RESULTS AND ANALYSIS

4.1   *Hysteresis curve*

The P-Δcurves of selected UG-5 and UG-7 are shown in Figure 4.

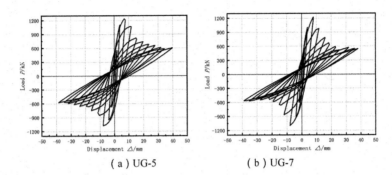

(a) UG-5   (b) UG-7

Figure 4.   Hysteresis curve.

The hysteresis curve of the specimen has the following characteristics:

1) The hysteresis curve of the specimen is generally full spindle shape, indicating that the specimen has good seismic performance. At the beginning of loading, the hysteresis curves of the specimens were linear, and the area of the curve envelope was small. With the increase of horizontal displacement, the slope of the curve gradually closes to the X-axis, and the surrounding area increases. When the peak load is reached, the slope of the hysteresis loop under each displacement cycle is gradually reduced and far less than the slope of the first hysteresis loop, indicating that the specimen has strength attenuation and stiffness degradation under reciprocating load.
2) When the ultimate displacement is reached, the plumpness of the hysteresis curves with different parameters becomes significantly worse, and the degradation of the hysteresis loop is more pronounced, which is mainly due to the bond failure between steel and concrete, and the two cannot work together.

4.2   *Skeleton curve*

The skeleton curve of the specimen refers to the envelope of the city connected by the peak point of the first cycle of load-deformation at all levels (Zhao 2018). The skeleton curves of each specimen model are shown in Figure 5 (influence of slenderness ratio, influence of spacing between stirrups and tie bars, influence of slenderness ratio).

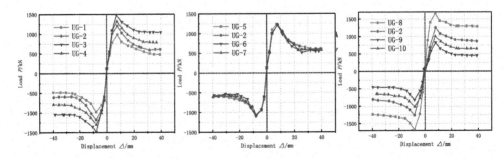

Figure 5.   Skeleton curves of each parameter.

1) The skeleton curve of the specimen is roughly divided into the elastic stage, elastic-plastic stage, and plastic stage.
2) It can be seen from Figure 5 (a) that the curves of specimens UG-1, UG-2, UG-3, and UG-4 were roughly similar, especially the curves at the initial loading stage were consistent. The maximum bearing capacity and plastic deformation resistance of URC anisotropic columns were improved at the later loading stage. The bearing capacity changes show a downward trend when the ultimate load is reached.
3) It can be seen from Figure 5 (b) that with the decrease of stirrup and tie spacing, the ultimate bearing capacity of the specimen is improved, but the influence is small. When the plastic stage is reached, the skeleton curve shows a steep slow downward trend, and the 'coupling effect' appears.
4) It can be seen from Figure 5 (c) that reducing the slenderness ratio will improve the initial stiffness and ultimate bearing capacity of the specimen. The ultimate bearing capacity of specimen UG-8 is increased by about 85.46 % compared with that of UG-10, so the slenderness ratio greatly influences the bearing capacity of URC special-shaped columns.

4.3 *Bearing capacity and ductility*

The horizontal load, displacement, and ductility coefficients of each stage of the simulated specimen are summarized in Table 3. is the cracking load; and are yield load and yield displacement, respectively. The 'Park method' may be used. is peak load, is the peak displacement; For destructive loads (taking the displacement when the horizontal force decreases to 85 % of the peak load). μ is the ductility factor. The specimens' horizontal bearing capacity and displacement ductility coefficient in Table 4 are positive values. From Table 4, it can be concluded that:

① The horizontal bearing capacity of the specimen increases significantly with the strength of concrete. When C60 concrete strength is adopted, the horizontal bearing capacity of the specimen increases by 49.75 % compared with C30, and the ductility coefficient of the specimen increases by 32.39 %.
② The spacing between stirrups and tie bars did not significantly improve the ultimate bearing capacity of specimens, but the ductility was improved.
③ With the decrease in slenderness ratio, the cracking load, yield load, and peak load increase. The slenderness ratio of specimen UG-8 was 15.41, and the ductility coefficient was 3.02 and greater than 3, indicating that the specimen had good seismic performance when the slenderness ratio was small.

Table 3. Test results.

Test specimen number	Opening point		Yield point		Peak point		Breaking point		
	P_C/kN	Δ_C/mm	P_Y/kN	Δy/mm	P_m/kN	Δm/mm	P_u/kN	Δ_u/mm	μ
UG-1	263	2.15	572.6	4.39	989.63	8.75	704.87	11.98	2.73
UG-2	305	2.21	584.4	4.59	1209.42	9.42	838.12	11.34	2.47
UG-3	394	2.37	597.0	4.67	1322.13	9.49	990.53	10.87	2.33
UG-4	405	2.40	603.8	4.65	1481.99	9.52	1209.51	10.21	2.20
UG-5	394	2.37	597.0	4.61	1232.0	9.49	1047.20	11.15	2.42
UG-6	275	2.22	574.0	4.53	1204.5	9.39	1023.83	11.06	2.44
UG-7	247	2.04	549.8	4.53	1202.3	9.36	1021.96	11.03	2.43
UG-8	415	2.42	607.6	4.69	1559.63	9.45	1076.87	14.18	3.02
UG-9	239	1.98	546.6	4.48	1035.7	9.29	880.35	10.23	2.28
UG-10	188	1.89	512.1	4.35	840.97	9.15	877.29	9.45	2.17

4.4 *Stiffness degradation*

Under cyclic loading, the specimen's stiffness decreases with the increase of the number of cycles, which is called stiffness degradation. The specimen's stiffness is taken as the ratio of the positive or negative maximum load to the corresponding displacement in each cycle. The stiffness degradation curves of specimens under various parameters are shown in Figure 6 (Influence of different concrete strength grades, influence of spacing between stirrups and tie bars, Influence of slenderness ratio).

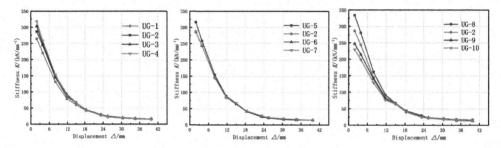

Figure 6. Stiffness degradation curve of each parameter.

It can be seen from Figure 6 that in the loading process, the secant stiffness of the specimen decreases with the increase of displacement. In the initial loading stage, the secant stiffness of the specimen decreased rapidly, and the gradual buckling of the longitudinal reinforcement in the late loading stage and the gradual widening of the cracks in the concrete led to unstable changes in the plastic bearing capacity of the component. The secant stiffness of the specimen gradually decreased and tended to be flat. Different concrete strength has a noticeable influence on the initial stiffness of the specimen curve, which indicates that the bearing capacity of concrete in the elastic stage increases rapidly. When the displacement is between 6 mm and 42 mm, the stiffness degradation curves of different stirrups and tie bars are coincident, indicating that this parameter has little effect on the stiffness degradation curve. When the slenderness ratio of the specimen decreases, the initial stiffness of the specimen increases significantly, and the horizontal load of the specimen increases gradually at each loading displacement stage, leading to the rapid degradation of the stiffness of the specimen at the later stage.

5 CONCLUSIONS

Through the numerical analysis of the quasi-static load test of URC special-shaped columns under different parameters, the following conclusions can be drawn:

(1) Based on the constitutive model of this paper, the nonlinear finite element simulation results are in good agreement with the experimental results, which verify the effectiveness and accuracy of the numerical simulation.
(2) When high-strength concrete is used, the ultimate bearing capacity of the specimen has been dramatically improved, but due to the inherent brittleness of high-strength concrete, ductility decreases with the increase of concrete strength.
(3) The influence of the change of stirrup spacing and tie bar spacing on the ultimate bearing capacity of URC reinforced concrete anisotropic columns is not apparent. However, because stirrups and tie bars will form a tie effect" with the concrete in the core area, the smaller the spacing between stirrups and tie bars is, the development of diagonal cracks can be inhibited, the strength degradation is slow, and the curve is flat.
(4) With the decrease in slenderness ratio, specimens' ultimate bearing capacity and ductility have improved significantly. It can be seen that the slenderness ratio is an essential factor affecting

the seismic performance of URC special-shaped columns, and the slenderness ratio should be strictly controlled in practical engineering.

6 OUTLOOK

The research data (theoretical analysis and experimental research) on URC columns are very scarce, and there are many limitations in finite element simulation, such as the slip between reinforcement and concrete, the setting of friction coefficient, etc. Therefore, this paper still needs to rely on the actual test research to study the integrity and local characteristics of components and provide a theoretical basis for the practical application of engineering.

REFERENCES

Diao B & Li S.C (2010). Experimental study on low cycle repeated loading of reinforced concrete L-section column frame. J. Building structure (6), 4.

GB 50010-2010, design code for concrete structure, China building industry press,2010. (in Chinese)

Huang C.K.& Qu F.L (2008). Experimental study on mechanical behavior of reinforced concrete special-shaped columns under low cyclic load. J. Building Structure 38 (11), 3.

JGJ 149-2017, technical specification for concrete special-shaped column structure, China Building Industry Press.

Li F. Q (2021). Seismic performance of U-shaped concrete shear wall. D.

Li J.Y & Wang X.Y (2021). Finite element analysis of u-shaped reinforced concrete column. J. IOP Conference Series: Earth and Environmental Science 768 (1).

Teng Z.C.& Zhao T.J (2022). Numerical analysis of seismic performance of T-shaped RC special-shaped column.

Zhang R & Su X.Z (2006). Experimental study on seismic performance of vertical prestressed reinforced concrete columns. J. Journal of Tongji University: Natural Science 34 (12), 5.

Zhao J. & Cai G. & Larbi A. S. (2018). Hysteretic behavior of steel fiber RC coupled shear walls under cyclic loads: Experimental study and modeling. J. Engineering Structures 156, 92–104.

Frontiers of Civil Engineering and Disaster Prevention and Control – Yang & Rahman (Eds)
© 2023 The Author(s), ISBN: 978-1-032-31200-2

Current application status of red mud in construction and structural materials

Jiahang Tian*
Department of Materials, Wuhan University of Technology, Wuhan, China

ABSTRACT: China is a big producer of alumina. Red mud is the residual solid waste in the aluminum processing industry after refining aluminum oxide from bauxite. Its comprehensive utilization, modified treatment, and efficient environmental resource recovery have become increasingly concerning hot spots and difficult problems in the alumina industry. The application of red mud in construction and structural materials is an important way of red mud recycling and reuse.

1 PRODUCTION OF RED MUD

Red mud is a kind of strong alkaline solid waste residue with high water content left after refining alumina from bauxite in the aluminum processing industry. Its comprehensive utilization has

Table 1. Analysis of different chemical components of red mud.

Component (%)	SiO_2	Al_2O_3	CaO	Fe_2O_3	TiO_2	MgO	Na_2O	K_2O
Bayer red mud	12~18	18~22	2~15	23~30	1~4	–	6~8	0.8
Sintering red mud	22~25	6~8	38~46	6~10	3~4	–	2~3	0.08
Joint red mud	20	8~9	44~47	8	6~7	–	2.8~3	0.5

Figure 1. Red mud.

*Corresponding Author: 504641662@qq.com

DOI 10.1201/9781003308577-69

become a global problem. Many usable components in red mud cannot be reasonably used, resulting in a serious waste of secondary resources. Red mud is the solid waste produced by the alumina industry, and the red mud produced is also different due to the different alumina production processes. In bauxite with high aluminum content, smelting aluminum by the Bayer method, the red mud is called Bayer red mud (Khairul 2019). In bauxite with low aluminum content, it adopts the sintering method and the Bayer method for smelting aluminum, and the red mud is called sintering red mud or joint red mud.

2 NATURE OF THE RED MUD

2.1 Physical properties of the red mud

The physical properties of red mud are as follows: particle diameter of $0.088 \sim 0.25$ mm, specific gravity of $2.7 \sim 2.9$, bearing capacity of $0.8 \sim 1.0$, melting point of 1200 to $1250°C$.

2.2 The hydraulic properties of the red mud

The hydraulic properties of red mud mainly include permeability, disintegrative, and expansion. It is mainly controlled by the physical composition of red mud and its evolution after stacking.

2.2.1 Water holding characteristics
Red mud has not only large water content, but also water-holding characteristics. Its water holding capacity is up to $79.03 \sim 93.20\%$, especially the water precipitation during vibration is still $5 \sim 14.93\%$. This means that when red mud vibrates, it changes its structure and deteriorates engineering performance. At the same time, it reveals that the red mud in the red mud storage yard has accumulated dozens of meters deep, which can still retain the water for decades and is difficult to consolidate, showing the soft plastic-flow plastic silt state, low strength, and high compressibility.

2.2.2 Disintegrative
Distilled water was added to the experiment with little change. After the addition of 5% HCl and H_2SO_4 in the experiment, the samples changed in color after 24 h and collapsed, but did not disintegrate. Over time, the sample surface was precipitated with abundant white salt and significantly hardened as a result of recrystallization and cementation of $CaCO_3$ and $NaSiO_3$.

2.2.3 Contraction and expansion
Although the red mud has high porosity and high water content, it does not shrink after a dry water loss, indicating that the high water content is not the result of hydrophilic minerals. At the same time, the red mud is not expansive. With increasing desiccation, the surface appears to have an obvious hardening phenomenon; a large number of white salts precipitate on the surface and cemented. The shrinkage test showed that the moisture content of the soil samples decreased after 45 d of air drying, but the soil samples did not shrink.

2.2.4 Liquefaction potential
Since the newly accumulated red mud has high water content, its value is mostly greater than the liquid limit. In addition, powder and sand are hydrophobic aragonite and calcite. Therefore, the newly accumulated red mud may cause liquefaction under vibration.

2.2.5 Corrosive characteristics
Red mud contains a large amount of alkaline material, which can corrode aluminum and form $NaAlO_2$ under the action of NaOH. SiO_2 is very stable in both neutral and acidic media but has a high solubility in an aqueous solution of pH> 11. Therefore, direct contact with the red mud has an obvious burning feeling and can corrode the skin.

2.3 Physiochemical properties of red mud

The physicochemical characteristics of red mud mainly include two indexes: cation exchange and specific surface area.

2.3.1 Cation interchange

The cation exchange amount of red mud is generally high, and the change amount is large. The cationic exchange amount of red mud is smaller than 578.1 me/kg and larger than 207.9 me/kg, mostly 250 me / kg $\sim$ 300 me / kg. This value is higher than expansion soil and kaolin and lower than illite and montmorillonite, indicating that the exchange capacity of red mud is unstable.

2.3.2 Specific area

The specific surface area of red clay (the size of the specific surface area reflects the dispersion and mineral lattice structure of clay minerals) is relatively high, with a large value of 186.9 m^2 / g and a smaller value of 64.09 m^2 / g (Bhat 2002; Shi 2020). The change range is very large, indicating the significant difference in the mineral dispersion and lattice structure of red mud.

Red mud has a large cation exchange amount and specific surface area, which is related to the strong alkalinity of red mud, proving that its physicochemical properties are active (Wang 2012, 2004). The disintegration test of the red mud in different solutions of H_2O, HCl (15%), and H_2SO_4 found that the red mud did not disintegrate, but after the interaction with the pore water solution, which will affect the engineering performance of the red mud to some extent.

2.4 Compressibility of red mud

Since red mud is a soft silty material with high porosity and high water content, the consolidation characteristics of red mud under accumulation conditions have become one of the problems that must be considered in the study of red mud discharge and treatment. In addition, due to the special chemical and mineral composition of red mud, as well as the discontinuity of stacking site and filling time, the surface is often dry, and the soluble salt from liquid (pore solution) to solid-state, with the conversion of aragonite to calcite (increased $CaCO_3$ crystallization degree) and the cementation of goethite, gibbsite, hydrophane and sodium silicate (Na_2SiO_3), significantly reducing the compressibility of long-term aged red mud.

2.5 Strength characteristics of red mud

Red mud has high moisture content and low density, but the shear strength is not low, and the internal friction angle is 20 $\sim$ 30. The reasons for the high shear strength of red mud are as follows: first, the high content of coarse powder and fine sand greater than 0.01 mm, the recrystallization of red mud mineral phase from aragonite to calcite, and the bonding of SiO_2, Al_2O_3 and Fe_2O_3; second, there is a strong water segregation phenomenon in the shear process, namely the conversion from weak bound water to free water.

2.6 Durability of red mud

During aging, drying, and hardening, large amounts of free oxides also undergo dehydration and recrystallization, such as SiO_2 colloids into hydrophane and sodium silicate, Al_2O_3 colloids into gibbsite, and Fe_2O_3 colloids into goethite or aqueous hematite. The content of free SiO_2, Al_2O_3, and Fe_2O_3 in red mud is high (about 25%$\sim$30%), which plays an important role in the strength hardening of red mud, and this bond is irreversible and will not soften by repeated dry and wet red mud, so it has high durability.

3 HARM OF RED MUD

According to statistics, since the industrial production of aluminum in 1888, the global aluminum industry has discharged more than 4.5 billion tons of red mud, and now the annual emissions are more than 170 million tons. At present, only about 10% of the global red mud is comprehensively utilized, while China's overall utilization rate is less than 4%. Such a large amount of red mud not only increases the cost of red mud treatment in alumina production enterprises and causes serious environmental pollution, but also leads to a great waste of resources.

At present, most of the red mud of Chinese alumina production enterprises is disposed of by dam construction and storage. During the process of red mud stacking, its surface often forms a layer of white "frost," which is the soluble alkali precipitated from the red mud. Because sodium hydroxide is added in the production process of alumina, the hydroxide, carbonate, aluminates, and so on existing in the red mud make it very highly alkaline. When the red mud is not removed, "frost" is used as building materials. Due to the high content of alkali, we must reduce the ratio of red mud in raw materials to ensure the quality of building materials, so that a large number of red muds cannot be used effectively. The impact of red mud on the environment is mainly reflected in the following aspects:

3.1 *Pollution of air*

The granularity of the red mud is very different due to the different production processes. When the red mud is dehydrated and weathered, the adhesion of the surface layer becomes worse, which is easy to cause dust pollution. And dry red mud is easy to form dust, causing an increase in fine particulate concentration. Due to the impact of wind erosion dust, red mud will reduce the visibility of ash storage fields, destroy the ecological environment, and cause serious pollution.

3.2 *Pollution of the groundwater*

The water pollution caused by red mud is manifested in two aspects: on the one hand, the infiltration of red mud leachate will increase the water hardness of underground water and even cause more serious arsenic, chromium, and other elements pollution of water (Lyberopulu 1995); on the other hand, the ash is discharged directly into the water body, forming sediment, suspended matter, soluble matter and so on, causing pollution (Liu 2015; Renforth 2012). The alkali content of red mud attached to the liquid is very high. Red mud storage area downstream of the groundwater is the main object affected by the red mud in seepage control measures of red mud yard near the high basicity seeped into the underground or into surface water, the pH of water increases, leading to groundwater total hardness and higher pH value. There are more than groundwater water quality standards of class I phenomenon. The total hardness of groundwater is nearly 1600mg/ L (2.53 times exceeding the standard), and the pH value is 11.2 (the standard is 6.5~8.5). Fluoride in red mud is another major pollutant in water pollution.

3.3 *Effects on building surfaces and soil*

Red mud is alkaline. In the humid air, red mud does not erode the surface of the building. The suspended particles landing on the ground alkaline the soil, pollute the soil surface and affect the planting and grazing. Red mud and its attachments have a strong salinization effect on the underground clay layer and can change the structure and chemical composition of the underground clay layer. The intensity and alkalinization of the soil will disturb the normal physiological activities of the plant roots and affect the absorption of nutrients. Most plants are not suitable for survival in the soil stacked with red mud. Therefore, it is almost impossible for the red mud-accumulated soil to be reclaimed and planted with plants (Martin 2013; Yang 2018).

3.4 Radioactivity of red mud

Because of the mineral components contained in some red mud raw mines, the radium, polonium, and other radioactive elements and toxic substances contained will cause harm to people and animals near the stacking yard. It will also cause inevitable radioactive harm to the environment (Piga 1993; Yang 2013).

4 APPLICATION OF RED MUD IN CONSTRUCTION AND STRUCTURAL MATERIALS

4.1 Use red mud as a PVC filler

Red mud can also be used for the PVC filler, and adding a certain amount of red mud to the PVC can improve the flame retardant performance, aging resistance, and thermal stability energy of the PVC, which can prolong the life of plastic products. Building materials can also be produced using the flame retardant properties of red mud. PVC with a certain amount of red mud last two to three times longer than the average PVC. Ordinary PVC will lose its strength completely after eight years in the open air, but red mud polyfluoroethylene plastic products can maintain good mechanical properties, without finding any aging phenomenon. In recent years, the application of red mud in the plastic industry has continuously made new progress.

4.2 Coagulant is prepared using red mud

Aluminum ferrosilicate is a new type of inorganic polymer coagulant, which has both electric neutralization and adsorption frame and bridging effect. Due to its characteristics of low price, wide source of raw materials, good coagulation effect, and fewer residues in the treated water, it has become a hot topic of inorganic polymer coagulants at home and abroad. The method of preparing polyaluminum ferric silicate (PSAF) coagulant by leaching red mud with dilute sulfuric acid and mixing it with polysilicic acid under atmospheric pressure and oxygen was reported in the literature. Compared with the polymeric ferric sulfate coagulant, the removal rate of CODc and chroma increased by about 20% and 28%, and the removal rate of suspended matter increased by about 10%. Therefore, the coagulant is an inorganic polymer coagulant worthy of attention and has a better treatment effect.

4.3 Red mud is used as a pavement base material

The use of the accumulated sintering method of red mud to develop high-grade road base materials is a promising comprehensive utilization technology for large-scale consumption of red mud. Based on the test and research basis of red mud road materials, a road about 15 m wide and about 4 km long was built in Luocun Town, Zichuan District, Zibo City, Shandong Province. After on-site drilling sampling by the Transportation Bureau, it shows that the red mud road base has reached the strength requirements of the stable soil of the lime industry waste residue and the highway, creating a good demonstration effect for the promotion of the comprehensive utilization technology of red mud.

4.4 Using red mud to produce bricks

Using red mud as the main raw material can produce a variety of bricks, such as steam-free burning brick, coal ash brick, black granular material decorative brick, and ceramic glazed brick. Among them, glaze brick prepared by sintering red mud is taken as an example. This method takes red mud as the main raw material, takes the traditional needed ceramic raw materials, supplemented by clay and silicon materials. It not only reduces the cost of raw materials, but also makes a great contribution to environmental protection. In the building materials industry, red mud can also produce glass and plastic packing. However, in the application of red mud, it must be noted that the red mud itself contains alkali fluid, and some red mud also contains radioactive elements, which directly harm human health.

4.5 *Use red mud to produce various cement*

Red mud is most commonly used in the production of building materials, as it can produce various types of cement (Liu 2011; Tsakiridis 2004). Russia uses Bayer red mud to produce cement. The ratio of raw red mud can be up to 14%. Japan Mitsui Aluminum Company works in cooperation with the cement plant to use red mud as the iron raw material with cement raw material. Each ton of cement clinker can use 5~20 kg of red mud. Volkhov, Achin, and Karev aluminum oxide plants in Russia use red mud as raw material to produce cement. The experiment on two components of limestone and red mud has been carried out. A total of 629~795 kg of red mud can be used per ton of cement, opening up an effective way for comprehensive utilization of sintering red mud.

An aluminum plant in China has studied the comprehensive utilization of red mud in the early stage of its construction and built a large cement plant in the early 1960s for comprehensively using red mud. In addition, sintering red mud is used to produce ordinary silicate cement. The average annual ratio of red mud in cement raw material is 20%~38.5%, the utilization amount of red mud in cement is 200~400kg / t, and the comprehensive utilization rate of producing red mud reaches 30%~55%. Compared with the general cement plant products, the cement produced, in addition to the low compressive strength, other performances are equal to or better than the general cement, especially the folding strength, early compressive strength, and sulfate erosion coefficient are particularly obvious. The product cost was reduced by 15% more than in the conventional cement plants.

5 CURRENT SHORTAGE OF RED MUD APPLICATION

5.1 *Lack of a large amount of red mud consumption and industrial competitiveness of the key technologies*

Red mud has the characteristics of strong alkaline, large specific surface area, various components, and embedding, which makes it difficult for its comprehensive utilization to learn from some mature processes, technologies, and equipment in other fields. A technical support system suitable for efficient utilization and large-scale promotion has not been formed in China.

5.2 *Lack of corresponding standards, the product market recognition is low*

At present, part of the comprehensive utilization of red mud products has been developed. However, due to the lack of national standards or industrial standards support, materials such as red mud can only refer to similar product standards and receive low market recognition, resulting in limited product application and difficulty in the large-scale promotion.

5.3 *Lack of targeted support policies*

Regarding the current fiscal and tax preferential policies in China, the strong alkaline red mud is not fully considered, making the comprehensive utilization far more difficult than that of other industrial waste residues. This is accompanied by a lack of targeted supporting policies and a low enthusiasm of enterprises to use red mud.

5.4 *The attention to the comprehensive utilization of red mud needs to be improved*

The comprehensive utilization of red mud is the non-main business of alumina enterprises, which is at the end of the industry, and the economic benefits are poor. The environmental and safety risks of red mud storage are long-term and hidden, resulting in the lack of attention from enterprises and relevant departments.

6   RESEARCH OUTLOOK

Because a large number of red muds is not fully used and treated, it occupies a large amount of land for a long time, consumes a lot of storage yard construction and maintenance and management costs, and alkaline waste liquid pollutes the surface and groundwater sources, seriously damages the natural ecological environment, directly endangers people's health, and restricts the sustainable development of aluminum industry (Samal 2012). The comprehensive treatment and utilization of red mud have always been an urgent problem in the aluminum industry. With the enhancement of environmental awareness, the red mud problem has attracted wide attention at home and abroad. At present, many countries in the world are actively looking for countermeasures, and strive to obtain a more effective red mud treatment technology, so that the discharged red mud can be effectively used(Sutar 2014; Wu 2012).

7   CONCLUSION

Although more research has been done on the comprehensive utilization of red mud at home and abroad, some red mud-based products have also been developed. However, there is still a long way to go for the comprehensive treatment and utilization of red mud. We need to carry out deeper research work, comprehensively analyze the composition and nature of red mud, and expand the application field of red mud. Exploring the production method of high value-added red mud products is the main way to recover and use red mud in the future. The effective use of red mud will bring great promotion to the production of the aluminum industry and drive the benign development of other industrial fields, solve the problems of waste of resources and environmental deterioration, and bring huge economic and social benefits.

ACKNOWLEDGMENTS

This work was financially supported by the National innovation and entrepreneurship training program for college students (157).

REFERENCES

A review of soil heavy metal pollution from industrial and agricultural regions in China: Pollution and risk assessment[J]. Qianqi Yang, Zhiyuan Li, Xiaoning Lu, Qiannan Duan, Lei Huang, Jun Bi. Science of the Total Environment. 2018

Bioleaching of rare earth and radioactive elements from red mud using Penicillium tricolor RM-10[J]. Yang Qu, Bin Lian. Bioresource Technology. 2013

Contaminant mobility and carbon sequestration downstream of the Ajka (Hungary) red mud spill: The effects of gypsum dosing[J]. P. Renforth, W. M. Mayes, A. P. Jarvis, I. T. Burke, D.A.C. Manning, K. Gruiz. Science of the Total Environment. 2012

Immobilization of beryllium in solid waste (red-mud) by fixation and vitrification[J]. P.N. Bhat, D.K. Ghosh, M.V.M. Desai. Waste Management. 2002 (5)

Low-cost red mud modified graphitic carbon nitride for the removal of organic pollutants in wastewater by the synergistic effect of adsorption and photocatalysis[J]. Weilong Shi, Hongji Ren, Xiliu Huang, Mingyang Li, Yubin Tang, Feng Guo. Separation and Purification Technology. 2020 (C)

Metallurgical process for valuable elements recovery from red mud—A review[J]. Zhaobo Liu, Hongxu Li. Hydrometallurgy. 2015

Physical and Chemical Properties of Sintering Red Mud and Bayer Red Mud and the Implications for Beneficial Utilization[J]. Ping Wang, Dong-Yan Liu. Materials. 2012 (10)

Progress of Red Mud Utilization: An Overview[J]. Harekrushna Sutar, Subash Chandra Mishra, Santosh Kumar Sahoo, Ananta Prasad Chakraverty, Himanshu Sekhar Maharana. American Chemical Science Journal. 2014 (3)

Proposal for resources, utilization and processes of red mud in India — A review[J]. Sneha Samal, Ajoy K. Ray, Amitava Bandopadhyay. International Journal of Mineral Processing. 2012

Recovering metals from red mud generated during alumina production[J]. Luigi Piga, Fausto Pochetti, Luisa Stoppa. JOM. 1993 (11)

Red mud addition to the raw meal for the production of Portland cement clinker[J]. P.E. Tsakiridis, S. Agatzini-Leonardou, P. Oustadakis. Journal of Hazardous Materials. 2004 (1)

Removal of dyes from aqueous solution using fly ash and red mud[J]. Shaobin Wang, Y. Boyjoo, A. Choueib, Z.H. Zhu. Water Research. 2004 (1)

Selective separation and determination of scandium from yttrium and lanthanides in red mud by a combined ion-exchange/solvent extraction method[J]. M. Ochsenkühn-Petropulu, Th. Lyberopulu, G. Parissakis. Analytica Chimica Acta. 1995 (1)

Spontaneous vegetation encroachment upon bauxite residue (red mud) as an indicator and facilitator of in situ remediation processes.[J]. Santini Talitha C, Fey Martin V. Environmental Science & Technology. 2013 (21)

Stockpiling and Comprehensive Utilization of Red Mud Research Progress[J]. Chuan-Sheng Wu, Dong-Yan Liu. Materials. 2012 (7)

The composition, recycling, and utilization of Bayer red mud[J]. M.A. Khairul, Jafar Zanganeh, Behdad Moghtaderi. Resources, Conservation & Recycling. 2019

Utilization of red mud in cement production: a review[J]. Xiaoming Liu, Na Zhang. Waste Management & Research. 2011 (10)

Frontiers of Civil Engineering and Disaster Prevention and Control – Yang & Rahman (Eds)
© 2023 The Author(s), ISBN: 978-1-032-31200-2

Study on preparation and key performance of C35 machine-made sand pavement concrete in Foshan first ring road

Bing Qiu*
Foshan Transportation Science and Technology Co., Ltd, Guangdong, China

Shuangxi Yu*
Foshan Jianying Development Co., Ltd, Guangdong, China

Zhuojie Yu* & Zhongyun Chen*
Foshan Highway and Bridge Engineering Monitoring Station Co., Ltd, Guangdong, China

Xi Chen*, Putao Song* & Jing Wang*
China Academy of Building Research, Beijing, China

ABSTRACT: In this section, the influences of water-binder ratio, the total amount of cementitious materials, sand ratio, and different fly ash content on the long-term performance and durability of C35 machine-made sand pavement concrete, such as workability, mechanical properties, electric flux, dry shrinkage, early crack resistance and so on, are studied. It is found that when the water-binder ratio is 0.37, the total amount of cementitious materials is 382 kg/m^3, the sand ratio is 37%, and the fly ash content is 15%. The performance indexes of concrete are as follows: the 7-d compressive strength is 37.2 MPa, the 28-day compressive strength is 46.6 MPa, the flexural strength is 5.3 MPa, and the elastic modulus is 32.6 Gpa. The 28 d shrinkage rate is 283×10^{-6}, the total cracking area per unit area is 664 mm^2/m^2, and the 56 d electric flux is 1415 C. The slump is 215 mm, the expansion degree is 550 mm, and the emptying time of the inverted slump cone is 12 s.

1 INTRODUCTION

Unlike ordinary river sand, machine-made sand is artificial aggregate, which has natural defects such as coarse particle size, poor gradation, high fineness modulus, and high stone powder content. In the process of concrete preparation with machine-made sand, there are some characteristics such as high viscosity, poor fluidity, easy bleeding, poor workability, etc. (Chen et al. 2021; Chai & Zou 2021; Ruan & Li 2020; Sun et al. 2021; Xie 2020; Yu 2020), which are also problems in the popularization and application of machine-made sand concrete that seriously restrict the application of machine-made sand concrete in road and bridge engineering. However, unlike river sand (Ding et al. 2019), machine-made sand is an aggregate formed by mechanical processing. The material properties of machine-made sand can be adjusted by adjusting processing parameters, and its performance indexes are adjustable and controllable. It is found that through reasonable processing control, the surface roughness of machine-made sand particles can be reduced, the particle gradation of machine-made sand can be optimized, the stone powder content and silt content of machine-made sand can be reasonably controlled, and the natural defects of machine-made sand can be made up as much as possible. At the same time, in the process of preparing machine-made sand concrete, through reasonable mix proportion optimization and optimal adjustment of additives (Dong et al. 2020; Qiao et al. 2020; Yan & Xie 2020), the viscosity of machine-made sand concrete can be greatly reduced, the workability of concrete can be improved, the strength and durability

*Corresponding Authors: 13902598673@139.com, ya_ci@163.com, 1660793982@qq.com, 1273220100@qq.com, 202022212014115@gs.zzu.edu.cn, song-pu-tao@163.com and wangking3007@126

     DOI 10.1201/9781003308577-70

of concrete can be enhanced, and the comprehensive performance of machine-made sand concrete can be comparable to or even better than that of ordinary natural river sand concrete.

In this paper, taking the C35 concrete reinforced in the West Extension of Foshan First Ring Road as the engineering background, the preparation and application technology of C35 machine-made sand pavement concrete are studied. The effects of water-binder ratio, the total amount of cementitious materials, sand ratio, and different fly ash content on the long-term performance and durability of C35 machine-made sand pavement concrete, such as workability, mechanical properties, electric flux, dry shrinkage, early crack resistance, are studied, and the optimal key mix ratio parameters of C35 machine-made sand pavement concrete are put forward, so as to be the pavement concrete of Foshan First Ring Road Project.

2 RAW MATERIALS

Cement: P.O52.5 cement produced by Guangdong Taini Cement Factory, with 3-day strength of 32.2MPa and 28-day strength of 59.5MP. Fly ash: Grade I fly ash produced by Qingyuan Power Plant in Guangdong, with ignition loss of 3.3%, 45μm sieve residue of 4.5%, and water demand ratio of 95%. Mineral powder: S95 mineral powder produced in a certain place in Guangdong: 78.2% in total activity in 7 days, 95.3% in total activity in 28 days, and 97.2% in fluidity. Fine aggregate: machine-made granite sand, MB value 0.5%, stone powder content 10%, fineness modulus 2.8%. Coarse aggregate: 5~20mm of continuously graded crushed stone, 5% of crushed stone needle flake content, 1400 Kg/m^3 of loose bulk density, 40.4% of void ratio, and 0.5% of silt content. Admixtures: Admixtures are polycarboxylate high-performance water reducing agents produced by an admixture factory in Jiangsu, with a solid content of 35% and a water-reducing rate of 34%.

3 TEST

3.1 *Performance requirements*

3.1.1 *Mechanical properties*
The mechanical properties of concrete meet the requirements of specifications and design documents, as shown in Table 1:

Table 1. Mechanical performance index of concrete.

compressive strength	rupture strength	modulus of elasticity
7d≥35MPa; 28d≥43.5MPa	28d rupture strength≥5MPa	28d modulus of elasticity≥31GPa

3.1.2 *Long-term performance and durability*
From the aspect of improving the quality of manufactured sand concrete, this paper refers to the "Design Code for Durability of Concrete Structures in Highway Engineering" (Ministry of Transport of the People's Republic of China. 2019)(JTG/T 3310-2019) and the existing research results and experience in China, and puts forward the requirements of mixing indexes from the aspect of concrete electric flux, which is as follows:

(1) The electric flux of the standard 56d concrete specimen (environmental action grade 50 years, III D) is ≤1500 C.
(2) At the same time, the early cracking performance, drying shrinkage, and carbonation resistance of concrete are relatively optimal.

3.1.3 *The workability of concrete is the best.*

Table 2. Performance index of concrete.

Slumps	Divergence	Emptying time of inverted slump cone	Gas content
(200±20)mm	≥ 500mm	≤ 15s	3%~4%

3.2 *Mixture ratio*

Considering the environmental conditions and on-site construction technology of the West Extension Project of the First Ring Road, combined with past design experience and test verification, it is preliminarily estimated that the range of test mix ratio parameters is as follows: a water-binder ratio of 0.35~0.39, water consumption of 134kg/m3~149kg/m^3, a sand ratio of 34%~40% and fly ash content of 10%~20%. See Table 3 for a specific test mix ratio.

Table 3. Test mix ratio.

No.	Water-binder ratio	Sand ratio (%)	Fly ash ratio (%)	Total amount of cementitious materials (kg/m^3)	Single dosage of concrete raw materials(kg/m^3)					
					Cement	Fly ash	Crushed stone	Machine-made sand	Mixing water	Admixture
A1	0.37	37	15	382	325	57	1210	710	142	6.12
A2	0.35	37	15	406	345	61	1194	701	142	6.90
A3	0.39	37	15	364	309	55	1220	717	142	6.92
B1	0.37	37	15	402	342	60	1192	700	149	6.43
B2	0.37	37	15	362	308	54	1227	721	134	6.52
C1	0.37	34	15	382	325	57	1267	653	142	6.12
C2	0.37	40	15	382	325	57	1152	768	142	6.12
D1	0.37	37	10	382	344	38	1210	710	142	6.12
D2	0.37	37	20	382	306	76	1210	710	142	6.12

4 RESULTS AND ANALYSIS

4.1 *Influence of water-binder ratio*

According to the reference mixture ratio, the influence of water-binder ratio change on concrete mixture performance and basic physical and mechanical properties is studied. As the total amount of cementitious materials is more sensitive to the shrinkage and cracking performance of concrete, the adjustment of the water-binder ratio is based on the constant water consumption. The change of water-binder ratio is controlled by adjusting the dosage of cementitious materials, and the dosage of water reducing agent is appropriately increased or decreased with the change of water-binder ratio to ensure that the workability of concrete mixtures in each test group is basically the same. The water-cement ratio is 0.35, 0.39, and 0.37, and the specific ratio is shown in Table 3.

4.1.1 *Performance of mixture*

The performance test results of concrete mixtures with different water-binder ratios are shown in Table 4.

Table 4. Performance of concrete mixture with different water-binder ratios.

No.	Water-binder ratio	Performance of mixture				
		slumps (mm)	expansion degree (mm)	Inverting cylinder (s)	air content (%)	Wet apparent density (kg/m³)
A1	0.37	215	550	12	3.6	2450
A2	0.35	210	550	14	3.2	2460
A3	0.39	215	560	10	3.8	2430

The test results show that (Table 4): concrete with different water-binder ratios can achieve roughly the same slump and expansion degree, but the viscosity of flow resistance is different. With the increase in the water-binder ratio, the pouring time of concrete decreases, and the viscosity decreases. Apparent density is inversely proportional to the water-binder ratio of concrete, which is mainly related to the increase of the proportion of cement slurry with lower density in concrete composition materials. Generally speaking, the performance of the mixture of each test group meets the preparation requirements and is in good condition, which is beneficial to pumping construction.

4.1.2 Mechanical properties

Test results of mechanical properties of concrete with different water-binder ratios are shown in Table 5.

Table 5. Mechanical properties of concrete with different water-binder ratios.

No.	Water-binder ratio	7d compressive strength (MPa)	28d compressive strength (MPa)	Bending strength (MPa)	Elastic module (GPa)
A1	0.37	37.2	46.6	5.3	32.6
A2	0.35	40.1	49.1	5.5	33.5
A3	0.39	35.1	44.3	5.0	31.2

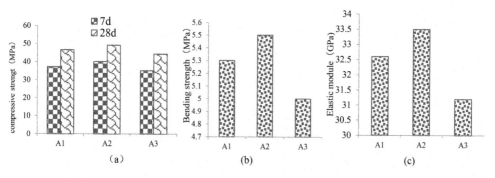

Figure 1. ((a) Compressive strength of concrete with different water-binder ratios; (b) Flexural strength of concrete with different water-binder ratios; (c) Elastic modulus of concrete with different water-binder ratio.)).

The test results show that (Table 5): the compressive strength, flexural strength, and elastic modulus of concrete are inversely proportional to the water-binder ratio. When the water-binder ratio is 0.39, the surplus of the mechanical properties of concrete is smaller than the required value. Considering all kinds of potential fluctuations in the construction process, the lower water-binder ratio (0.37) should be selected in actual construction to ensure the construction quality of concrete on site.

4.1.3 Durability

The durability test results of machine-made sand concrete with different water-binder ratios are shown in Table 6.

Table 6. Durability of concrete with different water-binder ratios.

No.	Water-binder ratio	28d shrinkage $(\times 10^{-6})$	Total cracking area per unit area (mm^2/m^2)	56d electric flux (C)
A1	0.37	283	664	1415
A2	0.35	304	682	1386
A3	0.39	296	676	1519

The test results show that (Table 6):

(1) Drying shrinkage: With the increase in the water-binder ratio, the 28d drying shrinkage of concrete firstly decreases and then increases, and the water-binder ratio is 0.37, which is the lowest. With the increase in the water-binder ratio, the slurry-bone ratio of concrete decreases, and the synergistic effect of the two causes the drying shrinkage of concrete to decrease at first and then increase.

(2) Early cracking resistance: With the increase in the water-binder ratio, the total cracking area per unit area of concrete first decreases and then increases, and the lowest water-binder ratio is 0.37, which should be chosen to prevent early cracking of concrete.

(3) Electric flux: When the water-binder ratio is 0.35, the electric flux of concrete is the lowest. With the increase in the water-binder ratio, the electric flux of concrete tends to increase.

4.2 The influence of the total amount of cementitious materials

The total amount of cementitious materials affects the slurry-bone ratio of concrete and then greatly influences the workability, strength, and durability of the concrete mixture. The influence of the change of the total amount of cementitious materials on various properties of concrete is studied with reference to the benchmark mixture ratio, and the total amount of cementitious materials is taken as a single variable. In the process of mixing ratio adjustment, the amount of admixture should be appropriately increased or decreased with the change in the number of cementitious materials to ensure that the working performance of the concrete in the test group is basically the same. The total amount of cementitious materials is 362kg/m^3, 382kg/m^3 and 402kg/m^3. See Table 3 for specific proportions.

4.2.1 Performance of mixture

See Table 7 for the performance test results of concrete mixtures with different cementitious materials.

Table 7. Performance of concrete mixture with different total cementitious materials.

No.	Total amount of cementitious materials (kg/m^3)	Performance of mixture				
		Slumps (mm)	Expansion degree (mm)	Inverting cylinder (s)	Air content (%)	Wet apparent density (kg/m^3)
A1	382	215	550	12	3.6	2450
B1	402	225	590	10	4.1	2420
B2	362	190	510	25	3.1	2460

The test results show (Table 7) that with the increase of the total amount of cementitious materials, the slump, expansion degree, and air content of concrete increase, the fluidity increases, the pouring time and wet apparent density of concrete decrease, and the viscosity decreases. The air content of test group B1 and the pouring time of mixture ratio of group B2 do not meet the preparation requirements, which indicates that the low dosage of cementitious materials is not conducive to pumping construction, and the performance of the concrete mixture of group A1 meets the preparation requirements, so it is recommended that the total amount of cementitious materials be 382kg/m³.

4.2.2 Mechanical properties

The test results for the mechanical properties of concrete with different cementitious materials are shown in Table 8.

Table 8. Mechanical properties of concrete with different total cementitious materials.

No.	Total amount of cementitious materials (kg/m³)	7d compressive strength (MPa)	28d compressive strength (MPa)	Bending strength (MPa)	Elastic module (GPa)
A1	382	37.2	46.6	5.4	32.6
B1	402	37.6	46.3	5.4	32.3
B2	362	36.1	45.5	5.2	32.5

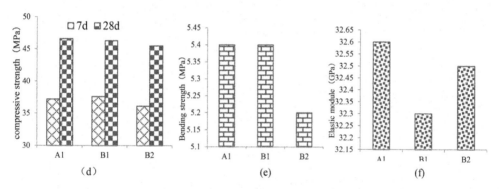

Figure 2. ((d)Compressive strength of concrete with different dosages of cementitious materials; (e) Flexural strength of concrete with different dosages of cementitious materials; (f) Elastic modulus of concrete with different dosages of cementitious materials)).

The test results show that (Table 8) with the increase of the total amount of cementitious materials, the 7th-day compressive strength of concrete gradually increases, the 28th-day compressive strength of concrete first increases and then slightly decreases, the flexural strength of concrete first slightly increases and then remains unchanged, and the elastic modulus of concrete increases and then decreases. Group A1 has the best performance in all mechanical properties of concrete except 7d compressive strength. This shows an optimal dosage of cementitious material in concrete, and under the optimal dosage, the cementitious material system has the best gradation and the lowest internal porosity (Jin, Yuan, 2017).

4.2.3 Durability

The test results for the durability of concrete with different total cementitious materials are shown in Table 9.

Table 9. Durability of concrete with different cementitious materials.

No.	Total amount of cementitious materials (kg/m³)	28d shrinkage ($\times 10^{-6}$)	Total cracking area per unit area (mm²/m²)	56d electric flux (C)
A1	382	283	664	1415
B1	402	326	689	1489
B2	362	276	656	1422

The test results show that (Table 9):

(1) Drying shrinkage: With the increase in the total amount of cementitious materials, the 28d drying shrinkage of concrete gradually increases, and the total amount of cementitious materials is 362kg/m³, which is the lowest. With the increase of cementitious materials, the cement-aggregate ratio of concrete increases, and the dry shrinkage of concrete increases. Considering the performance of the concrete mixture, B2 group concrete has the lowest drying shrinkage at 28d, but the viscosity of the concrete mixture is the highest, so it is not suitable for use.
(2) Early cracking resistance: With the increase in the total amount of cementitious materials, the total cracking area per unit area of concrete gradually increases, and the total amount of cementitious materials is 362kg/m³, which is the lowest. Considering the prevention of early cracking of the concrete, the water-binder ratio of 0.37 should be selected, but considering the workability, the viscosity of the mixture is the highest, which is not conducive to the control of cracking resistance of concrete pouring entities.
(3) Electric flux: When the total amount of cementitious materials is 382kg/m³, the electric flux of concrete is the lowest. With the increase of the total amount of cementitious materials, the electric flux of concrete first decreases and then increases, but the concrete corresponding to the water-binder ratio of each group all meets the requirement of standard curing for 56 days with electric flux ≤ 1500 C.

4.3 Influence of the sand ratio

Sand ratio affects the stacking and filling of aggregate particles in the concrete system, and then has a great influence on the workability, concrete strength, and durability of the concrete mixture. The influence of the change of sand ratio on various properties of concrete is studied with reference to the benchmark mixture ratio, and the sand ratio is taken as a single variable. In the process of mixing ratio adjustment, with the change of sand ratio, the dosage of concrete admixture remains unchanged. The contrast sand ratios are 34%, 37%, and 40%. See Table 3 for specific mixing ratios.

4.3.1 Performance of mixture
The performance test results of concrete mixtures with different sand ratios are shown in Table 10.

Table 10. Performance of concrete mixtures with different sand ratios.

No.	Sand coarse aggregate ratio (%)	Slump (mm)	Expansion degree (mm)	Inverting time (s)	Gas content (%)	Apparent density (kg/m³)
A1	37	215	550	12	3.6	2450
C1	34	210	550	15	3.0	2460
C2	40	210	520	21	3.7	2440

The test results show that (Table 10): with the increase of sand ratio, the slump of concrete first increases and then decreases, the expansion degree of concrete remains unchanged at first and then decreases, the pouring time of concrete first decreases and then increases, the air content of concrete increases and the wet apparent density decreases. When the sand ratio is 40%, the pouring time does not meet the preparation requirements. When the sand ratio is 37%, the slump of concrete is larger, the viscosity is lower, and the performance of the mixture is the best. This shows that there is an optimal sand ratio in concrete for the workability of the mixture. This is because mortar plays a lubricating role among aggregates and reduces the friction between aggregates, so the fluidity of concrete increases with the increase of sand ratio within a certain range. However, because the machine-made sand contains a certain amount of stone powder (Pang, Li et al., 2021), the specific surface area of these powders is very large. With the increase in the sand ratio, the total amount of stone powder introduced by machine-made sand increases, and the total surface area of aggregate increases. Under the condition of a certain amount of cementitious materials, the amount of slurry wrapped by aggregate decreases, the lubrication effect decreases, and the fluidity of the mixture decreases. Therefore, when the sand ratio exceeds a certain range, the fluidity of concrete decreases with the increase in sand ratio.

4.3.2 *Mechanical properties*

The test results of the mechanical properties of concrete with different sand ratios are shown in Table 11.

Table 11.   Mechanical properties of concrete with different sand ratios.

No.	Sand coarse aggregate ratio (%)	7d compressive strength (MPa)	28d compressive strength (MPa)	Bending strength (MPa)	Elastic module (GPa)
A1	37	37.2	46.6	5.4	32.6
C1	34	36.9	46.1	5.3	32.5
C2	40	36.5	46.3	5.2	32.2

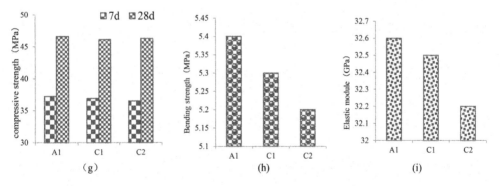

Figure 3.   ((g) Compressive strength of machine-made sand concrete with different sand ratios; (h) Flexural strength of machine-made sand concrete with different sand ratios; (i)Elastic modulus of machine-made sand concrete with different sand ratios)).

The test results show that (Table 11): with the increase of sand ratio, the 7th- and 28th-day compressive strength, flexural strength, and elastic modulus of concrete all increase first and then decrease. When the sand ratio is 37%, the mechanical properties of concrete are the best. This is because when the sand ratio is too low, the mortar in concrete and the stone powder in machine-made sand can't effectively fill the gaps between aggregates, which leads to a decrease in the compactness and strength of concrete. When the sand ratio is too high, with the same dosage of cementitious materials and an increase in the sand ratio, a dense skeleton structure cannot be formed, and the strength of concrete decreases. At the same time, the specific surface area of aggregate increases, and the slurry wrapped on the surface of aggregate becomes thinner, resulting in the decrease of the cohesive force between concrete aggregates and affecting the strength (Fan, Li et al., 2019).

4.3.3 *Durability*

Test results of the durability of concrete with different sand ratios are shown in Table 12.

Table 12.   Durability of concrete with different sand ratios.

No.	Sand coarse aggregate ratio (%)	28d shrinkage ($\times 10^{-6}$)	Total cracking area per unit area (mm^2/m^2)	56d electric flux (C)
A1	37	283	664	1415
C1	34	276	659	1459
C2	40	296	686	1442

The test results show that (Table 12):

(1) Drying shrinkage: With the increase of sand ratio, the 28d drying shrinkage of concrete gradually increases, and the concrete is the lowest when the sand ratio is 34%, which may be related to the introduction of stone powder content in machine-made sand. After the increase in the sand ratio, the content of stone powder introduced from machine-made sand increases, the actual slurry quantity increases, and the drying of concrete increases. Considering the performance of the concrete mixture, the pouring time of concrete with a sand ratio of 34% does not meet the requirements, and the workability of the concrete mixture is poor, so it should not be used.
(2) Early cracking resistance: With the increase in the sand ratio, the total cracking area per unit area of concrete gradually increases, and the sand ratio is the lowest when it is 34%. Considering the prevention of early cracking of the concrete, a lower sand ratio should be selected. Considering the performance of concrete mixture, the workability of concrete mixture with a sand ratio of 34% is poor, which is not conducive to the control of crack resistance of concrete pouring entity.
(3) Electric flux: With the increase of sand ratio, the electric flux of concrete first decreases and then increases, but the concrete corresponding to the water-binder ratio of each group all meet the requirement of 56d electric flux $\leq$ 1500C, and the electric flux of concrete is the lowest when the sand ratio is 37%.

4.4 *Influence of fly ash content*

The proportion of fly ash affects the stacking and filling of particles in the powder material of the concrete system, and then affects the workability of the concrete mixture, the composition of hydration products in concrete, and then the strength and durability of concrete. According to the reference mix ratio, the influence of the change of fly ash content on various properties of concrete is studied, and the fly ash content is a single variable. In the process of mixing ratio adjustment, the amount of admixture remains unchanged with the change in fly ash mixing ratio. The mixing ratios of different fly ash mixing ratios are shown in Table 3.

4.4.1 *Performance of mixture*

See Table 13 for the performance test results of concrete mixtures with different admixture proportions.

Table 13. Performance of concrete mixtures with different proportions of fly ash.

No.	Fly ash ratio (%)	Slump (mm)	Expansion degree (mm)	Inverting time (s)	Gas content (%)	Apparent density (kg/m^3)
A1	15	215	550	12	3.6	2450
D1	10	210	530	15	3.1	2460
D2	20	220	590	6	4.1	2430

The test results show (Table 13) that the air content of 20% fly ash does not meet the preparation requirements. With the increase of fly ash content, the slump and expansion of concrete increase, the workability is improved, the pouring time of concrete decreases, and the viscosity decreases. This is mainly due to the "shape effect" and "micro-aggregate effect" of fly ash, which reduces the internal voids of concrete and exerts its lubrication function among aggregates (Cui, 2021), which significantly improves the fluidity of concrete and reduces the viscosity. However, with the increase of fly ash content, the air content of concrete increases, and the wet apparent density decreases. Therefore, considering the control of the air content of concrete, it is necessary to control the fly ash content properly.

4.4.2 Mechanical properties
The test results of mechanical properties of concrete with different fly ash content are shown in Table 14.

Table 14. Mechanical properties of concrete with different fly ash content.

No.	Fly ash ratio (%)	7d compressive strength (MPa)	28d compressive strength (MPa)	Bending strength (MPa)	Elastic module (GPa)
A1	15	37.2	46.6	5.4	32.6
D1	10	37.2	46.4	5.3	32.4
D2	20	36.1	46.2	5.0	32.5

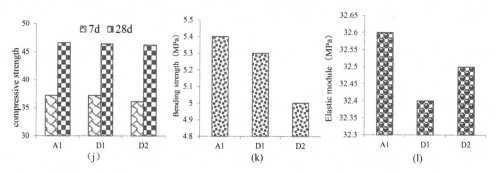

Figure 4. ((j)Compressive strength of machine-made sand concrete with different fly ash content; (k) Flexural strength of machine-made sand concrete with different fly ash content; (l) Elastic modulus of machine-made sand concrete with different fly ash content).

The test results show that (Table 14): With the increase of fly ash content, the 7th-day compressive strength of concrete remains unchanged and then decreases, while the 28th-day compressive strength, flexural strength, and elastic modulus of concrete all increase first and then decrease, while the flexural strength of concrete increases first and then decreases. When the content of fly ash is 15%, the mechanical properties of concrete are the best. The fly ash improves the compactness of concrete, optimizes the micro-morphology of the interface between slurry and aggregate, and improves the strength of concrete (Guo, 2021; Yang, Zhao, Zhou, 2021; Wu, He et al., 2021). Table 13 Test results show that with the increase of fly ash content, the air content of concrete increases, which affects the strength of concrete. The multiple effects of fly ash content on concrete lead to the strength of concrete first increasing and then decreasing after adding fly ash.

4.4.3 *Durability*

The durability test results of concrete with different fly ash content are shown in Table 15.

Table 15. Durability of concrete with different fly ash content.

No.	Fly ash ratio (%)	28d shrinkage ($\times 10^{-6}$)	Total cracking area per unit area (mm²/m²)	56d electric flux (C)
A1	15	283	664	1415
D1	10	295	692	1472
D2	20	298	671	1469

The test results show that (Table 15):

(1) Drying shrinkage: With the increase of fly ash content, the 28d drying shrinkage of concrete firstly decreases and then increases, and the lowest is found when the fly ash content is 15%.
(2) Early cracking resistance: With the increase of fly ash content, the total cracking area per unit area of concrete first decreases and then increases, and the lowest is when the fly ash content is 15%. Considering the prevention of early cracking of the concrete, 15% fly ash content should be selected.
(3) Electric flux: With the increase of fly ash content, the electric flux of concrete first decreases and then increases, and the electric flux of concrete is the lowest when the fly ash content is 15%. Table 13 Test results show that with the increase of fly ash content, the air content of concrete increases, which leads to the increase in the number of air holes in concrete and the increase of the electric flux of concrete. The multiple effects of fly ash content on concrete lead to the electric flux of concrete first decreasing and then increasing after adding fly ash.

To sum up, considering the influence of various mix ratio parameters on concrete performance, mix ratio A1 (water-binder ratio of 0.37, total cementitious material of 382kg/m³, the sand ratio of 37%, and fly ash content of 15% has the best performance.

5 CONCLUSION

(1) Control the water consumption unchanged and change the dosage of cementitious materials. The influence of water-binder ratio change on concrete mixture performance and basic physical and mechanical properties is studied, and the optimal water-binder ratio is 0.37.
(2) Studying the influence of the total amount of cementitious materials on the slurry-bone ratio of concrete, and then the performance of concrete mixture and various basic physical and mechanical properties, the optimal dosage of cementitious materials is 382kg/m³.
(3) Studying the influence of sand ratio on the packing of aggregate particles in the concrete system, and then on the performance of concrete mixture and basic physical and mechanical properties, the optimal sand ratio is 37%.
(4) Studying the influence of fly ash ratio on the stacking and filling of particles of powder material in the concrete system, and then on the performance of concrete mixture and basic physical and mechanical properties, the optimal fly ash content is 15%.

(5) In the process of preparing concrete with machine-made sand, there are some characteristics such as high viscosity, poor fluidity, easy bleeding, and poor workability. Through reasonable process control, the particle size distribution of manufactured sand can be optimized. In the process of preparing machine-made sand concrete, the viscosity of machine-made sand concrete can be greatly reduced, the workability of concrete can be improved, the strength and durability of concrete can be enhanced, and the comprehensive performance of machine-made sand concrete can be equal to or even better than that of ordinary natural river sand concrete. According to this paper, we should explore the appropriate proportion of higher-strength concrete with machine-made sand in the next step.

REFERENCES

Chai Tianhong, Zou Xiaoping. 2021, Discussion on the problems and application of machine-made sand concrete [J]. Jiangxi Building Materials, 23–24+28.

Chen Jindong, Zhu Wen, Huang Yibin. 2021, Characteristics of sand making with wet rod mill and its application in concrete [J]. New Building Materials, 30–34.

Cui Yongfeng. 2021, Study on the influence of fly ash content on the performance of machine-made sand concrete [J]. Fujian Building Materials,16–17+68.

Ding Qingjun, Peng Cheng Kang Yan, et al. 2019, Effect of fine aggregate on performance of ultra-high performance concrete [J]. Silicate Bulletin, 488–494.

Dong Rui, Xue Changwu, et al. 2020, Study on the preparation and performance of machine-made sand cement concrete [J]. Shanxi Communications Technology, 51–54+58.

Fan Jinpeng, Li Weihong, et al. 2019, Influence of material factors on the performance of machine-made sand concrete [J]. Concrete,152–155+160.

Guo Dongfeng. 2021, Study on the influence of fly ash content on the durability of machine-made sand concrete [J]. Comprehensive utilization of fly ash,111–115.

Jin Qiaolan, Yuan Chengfang. 2017, Influence of material factors on compressive strength of C40 machine-made sand concrete [J]. People's Yellow River, 129–132.

Ministry of Transport of the People's Republic of China. 2019.Technical specification for construction of highway bridges and culverts: JTG/T 3310–2019 [S]. Beijing: People's Communications Press.

Pang Jianrong, Li Wenjie, et al. 2021, Study on the influence of stone powder on the performance of recycled concrete with machine-made sand [J]. Heilongjiang Science, 12(18):98–99.

Qiao Donghua, He Fei, et al. 2020, Preparation and quality control of machine-made sand self-compacting concrete [J]. Highway, 276–282.

Ruan Yuhe, Li Hongbin. 2020, Application of machine-made sand high-performance concrete in large precast box girder [J]. Guangdong Civil and Architecture, 76–80.

Sun Xinghai, Liu Ze, et al. 2021.Study on micro-morphology and mortar properties of different manufactured sand [J]. Concrete and Cement Products.

Wu Yinfang, He Honglong, et al. 2021,35(04), Experimental study on the influence of fly ash on pumping performance of machine-made sand self-compacting concrete [J]. Comprehensive utilization of fly ash, 83–87.

Xie Huawei.2020, Measures to improve the surface quality of C50 machine-made sand T-beam concrete and its engineering application [J]. Guangdong Civil and Architecture, 113–115.

Yan Wei, Xie Bo. 2020, Study on preparation of machine-made sand self-compacting concrete based on orthogonal test [J]. Building Materials World, 14–17+46.

Yang Zezheng, Zhao Qingxin, Zhou Meiru. 2021, Effect of pretreatment of low-quality fly ash on mortar performance and microstructure [J]. Journal of Wuhan University of Technology, 70–74+85.

Yu zhang. 2020, Study on performance and mix design of machine-made sand concrete with composite binder system [D]. Beijing: Beijing Jiaotong University.

Frontiers of Civil Engineering and Disaster Prevention and Control – Yang & Rahman (Eds)
© 2023 The Author(s), ISBN: 978-1-032-31200-2

Preparation of integral waterproof concrete with addition of alkyl alkoxy silane emulsion

Li Li*

Department of Architectural Engineering, Dalian Vocational & Technical College, Dalian, Liaoning Province, China

ABSTRACT: Based on the analysis of the overall waterproof concrete material test design points, this paper uses the test to analyze the compressive strength, waterproof effect, water freeze resistance, flexural tensile strength, microscopic performance, anti-scour performance, and fatigue performance of waterproof concrete mixed with alkyl alkoxy silane emulsion. By studying the measures of strengthening the quality management of raw materials, controlling the proportion of alkyl alkoxy silane emulsion, and strengthening the control of the mixing process, this paper aims to give full play to the application effect of concrete and improve the waterproof performance of the structure after construction.

1 INTRODUCTION

As a common material during the construction of building engineering at this stage, the construction quality of concrete will also directly affect the stability and service life of the project. Alkyl alkoxy silane emulsion can be added to the preparation of waterproof concrete to improve the properties of concrete and continuously improve the impermeability of waterproof concrete. This paper analyzes the performance of waterproof concrete mixed with alkyl alkoxy silane emulsion to demonstrate the application value of alkyl alkoxy silane emulsion.

2 KEY POINTS OF EXPERIMENTAL DESIGN FOR INTEGRAL WATERPROOF CONCRETE MATERIALS

2.1 *Raw material and mix proportion*

In the overall waterproof concrete material test, we use P.O42.5 cement, and the basic performance indexes are shown in Table 1. Coarse aggregate is selected from granite macadam with a continuous gradation of 5~25 mm in particle size. The fine aggregate is river sand with a maximum particle size of 5 mm and a fineness modulus of 2.9. The mixing water is ordinary tap water, and the admixture is a polycarboxylic acid superplasticizer. Protectosil MH50 alkyl alkoxy silane produced by EVONIK Company of Germany was selected in the experiment, which contains 50% active ingredients and is a water-soluble suspension emulsion of silane colloidal particles. The water-cement ratio of

Table 1. Basic performance index of P.O42.5 cement.

C3A (%)	Alkali content (%)	Total chloride ion (%)	Fineness (m^2/kg)	Loss on ignition (%)
12.32	0.886	0.1824	362.32	2.51

*Corresponding Author: 2013511215@dlvtc.edu.cn

DOI 10.1201/9781003308577-71

concrete specimens is 0.5, and the mixing water consumption takes into account the moisture in silane. The mixing ratio and the corresponding number of each concrete are shown in Table 2 (Wang 2021).

Table 2. Concrete proportion (Unit of measurement, kg/m^3).

No.	Cement	Coarse aggregate	Fine aggregate	Mixing water	Water reducing agent	Silane emulsion
A0	330	650	1250	160.3	1.5	0
A1	330	650	1250	158.3	1.5	6.4
A2	330	650	1250	156.3	1.5	9.6
A3	330	650	1250	153.3	1.5	12.8
A4	330	650	1250	150.5	1.5	19.2

2.2 Sample preparation and treatment

We add the silane emulsion into the mixing water to mix thoroughly and then pour it into the concrete mixture. After being evenly stirred, it is put into a 100 mm × 100 mm × 100mm steel mold for vibrating molding. In order not to affect the waterproof effect of silane, the mold release agent is not used during pouring. 48h after molding, dismantle the mold, and then put the specimen in the standard curing room (20±3)°C, and cure it with a relative humidity greater than 90%. The compressive strength of some specimens was tested at 7d and 28d ages. The rest of the specimens were taken out after curing for 14 days and then placed in the indoor environment of the laboratory at (20±3)°C, with a relative humidity of about 60% (Liu 2021).

3 PERFORMANCE ANALYSIS OF INTEGRAL WATERPROOF CONCRETE

3.1 Compressive strength

In order to improve the intuitiveness of the analysis results, waterproof concrete with different amounts of silane emulsion needs to be made according to the requirements. After 7d, 14d, and 28d, the samples are taken out and put into the compressive strength test platform, the relevant data are counted, and the corresponding graphs are drawn by computer software (Han 2021). According to the statistical results of big data, it can be known that with the increase of test pressure, the deformation of concrete with different amounts of silane emulsion will increase continuously, and its overall fluctuation trend will keep high similarity. From the point of view of mechanical properties, with the increase of silane emulsion content, the compressive strength of the structure will also decrease continuously. According to the statistical results of Figure 1, when the addition of silane emulsion is below 3%, the concrete strength hardly fluctuates. When the amount of silane emulsion is 3%-5%, the concrete strength decreases by 10%-20%. When the addition of silane exceeds 6%, the strength of concrete decreases by more than 20%. It can be seen that when the addition of silane emulsion is less than 3%, the strength of concrete remains relatively stable.

3.2 Waterproof effect

In the test of the waterproof effect, it is necessary to make waterproof concrete with different amounts of silane emulsion according to the requirements. After curing for 14 days, weigh the sample weight and record the initial data at this time, immerse the weighed sample in the water for 72 hours, dry the surface moisture every 24 hours, then weigh, count the relevant data and make a chart. According to the data obtained, it can be known that the waterproof effect of concrete with

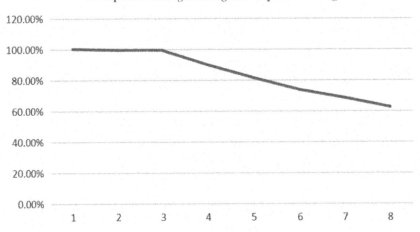

Figure 1. The compressive strength test results.

different silane emulsion content will decrease with the increase of immersion time, and there is a certain similarity in the curve fluctuation trend. According to the statistical results of big data, when the amount of silane emulsion is less than 3%, the total water seepage of concrete is least at 72h, and the water seepage growth curve is relatively gentle. When the amount of silane emulsion is 3%-5%, the change speed of the total water seepage of concrete at 72h accelerates, and the growth rate of the water seepage growth curve starts to increase after 24h. When the addition of silane exceeds 6%, the total water seepage of concrete changes greatly in 72 hours, and the impermeability decreases with the increase of time. Therefore, when the amount of silane emulsion is less than 3%, the waterproof effect of concrete remains relatively stable (Guo 2020).

3.3 Water freeze resistance

In the test of water freeze resistance, it is necessary to make waterproof concrete with different amounts of silane emulsion as required. After curing for 24 days, the sample was immersed in water for 72 hours. After being saturated with water, it was put into a freeze-thaw box, and the freeze-thaw environment and parameter settings met the corresponding standard requirements. The total number of freeze-thaw cycles is 100 times, with an interval of 25 cycles. Then, we take out the specimen, clean up the surface sediment and moisture, check the external damage, weigh its quality, and use information technology to count the relevant data and organize it into a chart. According to the obtained data, it can be known that the water freeze performance of concrete with different amounts of silane emulsion is different. According to the statistical results in Figure 2, when the amount of silane emulsion is less than 3%, the external damage to concrete is not obvious, and the quality change is less. When the amount of silane emulsion is 3%-5%, there is obvious damage outside the concrete, and the mass reduction is 2%-3%. When the amount of silane added exceeds 6%, the exterior of concrete is seriously damaged, and the quality decreases by more than 5%. Therefore, when the amount of silane emulsion is below 3%, the water freeze performance of concrete is in a relatively stable state (Ji 2020).

3.4 Flexural tensile strength

In the test of flexural tensile strength, waterproof concrete with different amounts of silane emulsion should be made according to the requirements. After curing for 36 days, the surface stains should be

Mass reduction

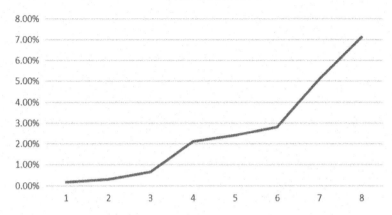

Figure 2.   Analysis diagram of water freeze resistance.

cleaned up, and 1/3 of the side of the concrete specimen should be marked with a marker pen. Then, the pressure side should be placed on the operating table, and the loading rate should be controlled at 50mm/min. After the specimen is damaged, the critical value of the damage state should be recorded, and the flexural tensile strength value of the specimen can be obtained according to the calculation formula. Based on the big data technology, the statistical results are sorted out. According to the feedback data in Figure 3, it can be known that the changing trend of waterproof concrete with different amounts of silane emulsion is basically the same. When the amount of silane emulsion is less than 3%, the flexural and tensile strength of waterproof concrete with the amount of silane emulsion is the highest, which is 10%-15% higher than the standard. When the amount of silane emulsion is 3%-5%, the flexural and tensile strength of waterproof concrete with the amount of silane emulsion is close to the standard. When the amount of silane emulsion exceeds 6%, the flexural and tensile strength of waterproof concrete with the amount of silane emulsion decreases rapidly, which is lower than the corresponding standard (Zhu 2020). Therefore, when the amount of silane emulsion is below 3%, the stability of silane emulsion concrete is the highest, and it has good adaptability.

Ratio of flexural tensile strength to standard value

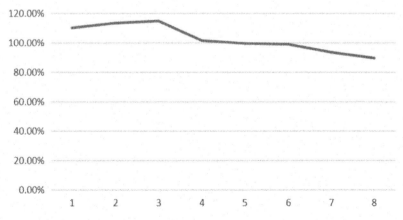

Figure 3.   Statistical results of flexural tensile strength.

3.5 Microscopic performance

In the microscopic performance test, it is necessary to make waterproof concrete with different amounts of silane emulsion according to the requirements. After curing for 24 days, clean up the surface stains and then put it under an electron microscope for observation. The scanning times of the microscope used are 500 times and 3000 times and then show the corresponding scanning pictures. According to the statistical results, the microscopic performance of waterproof concrete with different amounts of silane emulsion are different. When the amount of silane emulsion is less than 3%, the uniformity of waterproof concrete with the amount of silane emulsion is high, and there is almost no "hugging" phenomenon. The crystals obtained by the hydration reaction will cover the coarse aggregate with good compactness. When the amount of silane emulsion is 3%-5%, the amount of silane emulsion waterproof concrete begins to be unevenly distributed, and the phenomenon of "hugging" begins to appear locally. The crystals obtained by hydration reaction are unevenly distributed, and the density is locally uneven. When the amount of silane emulsion exceeds 6%, the distribution of waterproof concrete with the amount of silane emulsion fluctuates greatly, and the phenomenon of "hugging" is serious. Moreover, the crystal volume obtained by hydration reaction is large, the structure begins to appear to have serious loose problems, and the overall density is in an uneven state. It can be seen that the stability of silane emulsion concrete is the highest when the amount of silane emulsion is below 3% (Shuai 2020).

3.6 Anti-scour performance

In the test of erosion resistance, waterproof concrete with different silane emulsion content should be made according to the requirements. After curing for 28 days, we put it in a drying box for drying, put it into a washing cylinder for fixing after weighing, and added clean water into the washing cylinder. The clean water needs to be less than 10cm from the top of the sample. Subsequently, the sample is washed 100 times, and then the sample is taken out and dried and weighed, and the difference before and after washing is the loss of quality. At the same time, we count the erosion marks on the sample surface and use big data technology to assist the finishing process. According to the statistical results, when the amount of silane emulsion is less than 3%, the surface of the waterproof concrete sample with silane emulsion addition has no obvious scouring trace, and the quality of scouring loss is less than 1%. When the amount of silane emulsion is 3%-5%, the surface of the waterproof concrete sample with the amount of silane emulsion appears with a small number of scouring marks, and the quality of scouring loss is 2%-4%. When the amount of silane emulsion exceeds 6%, the surface of the waterproof concrete sample with the amount of silane emulsion appears as obvious erosion marks, and the erosion loss quality exceeds 5% (Yue 2019). Therefore, when the addition of silane emulsion is below 3%, the obtained silane emulsion concrete has the strongest stability.

3.7 Fatigue performance

In the test of fatigue performance, it is also necessary to make waterproof concrete with different amounts of silane emulsion. After curing for 28 days, we clean up the surface stains, mark 1/3 of the side of the concrete specimen with a marker pen, place it in the fatigue testing machine, and fix the specimen with tools. Pre-press the specimen with 0.2 times stress intensity for 2min, then apply Haversine wave (frequency is 10Hz) at the marked points, sort out these data using big data technology, and calculate related data using fatigue equation. According to the statistical results, cracks will appear in the waterproof concrete samples with silane emulsion addition with the increase of load, and these cracks will continue to expand under the long-term load. After reaching a certain critical value, the cracks will be in a state of unstable expansion, and finally, the sample penetration or big crack disease will appear. In this process, when the addition of silane emulsion is less than 3%, the increasing speed and changing frequency of the obtained silane emulsion concrete is slow, and it has stronger stability in the application (Li 2019).

4 MEASURES TO IMPROVE THE APPLICATION EFFECT OF INTEGRAL WATERPROOF CONCRETE

4.1 Strengthen the quality management of raw materials

Strengthening the quality management of raw materials can lay a good foundation for improving the quality of concrete, thus improving the waterproof effect of concrete. In the concrete practice, firstly, the basic conditions such as hydrological conditions, precipitation conditions, and geological conditions of the area to be operated are sorted out to determine the best proportion of waterproof concrete and raw material parameters around the mixture ratio. Its contents include strength, gradation, fineness parameters, and particle size (Ren 2018). Second, before the raw materials enter the site, we need to carry out a quality inspection as required, and materials that do not meet the requirements are not allowed to enter the construction site. Moreover, in the process of promoting construction activities, it is also necessary to do a good job of quality sampling and temporary inspection to improve the quality compliance of raw materials used and ensure the quality of concrete mixing.

4.2 Control the proportion of alkyl alkoxy silane emulsion

Controlling the proportion of alkyl alkoxy silane emulsion can give full play to the application effect of silane emulsion and improve the application performance of concrete. From the practical point of view, first, it is necessary to sort out the basic situation of the site, and understand the concrete layout area, stratum water content, soil erosion state, etc., so as to serve as a reference for the addition ratio of silane emulsion. Second, the experiment on the addition ratio of silane emulsion is carried out in the laboratory. The most suitable mixture ratio is determined according to the evaluation results of various performance parameters to ensure that the performance of silane emulsion can be fully developed and exerted (Xiao 2019). Third, when mixing on site, it is also necessary to adjust the proportion according to the external temperature and environmental humidity to ensure that the mixing results can meet the requirements of building codes.

4.3 Strengthen the mixing process control

Strengthening the control of the mixing process can ensure the uniformity of concrete mixing results and the reliability of concrete pouring results. In the concrete practice, first, the material addition error is controlled. Based on the past construction experience, the single addition error of materials should be controlled within 2% to avoid the negative impact caused by excessive addition errors of materials. Second, the material mixing process needs to be dry-mixed for 40s, and after the material is evenly mixed, water is added for 2min to ensure the uniformity of the mixed material. Third, control the temperature of the mixing water. For example, when mixing at a high temperature in summer, it is necessary to use ice water to avoid premature concrete condensation caused by too high water temperature. During the construction at low temperatures in winter, warm water will be used for mixing to avoid the freezing of concrete due to low temperature (Tang 2013).

5 CONCLUSION

To sum up, strengthening the quality management of raw materials can lay a good foundation for improving the quality of concrete, controlling the proportion of alkyl alkoxy silane emulsion can give full play to the application effect of silane emulsion, and strengthening the control of mixing process can ensure the uniformity of concrete mixing results. The application of alkyl alkoxy silane emulsion in waterproof concrete can improve the properties of concrete, and it is of positive significance to enhance the waterproof property of the structure and prolong the service life of the concrete structure.

ACKNOWLEDGMENTS

This work was supported by the Scientific Research Innovation Fund Project of Dalian Vocational & Technical College(NO.DZ2018CXJJ01); Annual Scientific Research Innovation Platform of Dalian Vocational & Technical College in 2019 (NO.2019XJPTYF1-1); Research Backbone Project of Dalian Vocational & Technical College in 2019; Scientific Research Funding Project of Liaoning Education Department in 2020 (NO.JYT202005); Education Science Planning Project of Liaoning Province in 2020 (NO.JG20EB037); Liaoning Modern Distance Education Society's 14th Five-Year Plan 2021 Annual Project (Research on Training Mode of Modern Apprenticeship Talents in Construction Engineering Technology Specialty.NO.2021XH-31); Subject of Dalian Academy of Social Sciences (Research Center) in 2021(NO.2021dlsky125); Liaoning Economic and Social Development Project in 2022; 2021 Subject of China Society of Higher Education.

REFERENCES

Guo Jufu (2020). Performance Test and Construction Control of Waterproof Concrete in Green Building Basement. Railway Construction Technology, (12):73—76.
Han Jianwei (2021). Experimental Study on the Influence of Nano-modified Waterproof Agent on the Performance of Underground Rigid Waterproof Concrete. Installation, (01):70–72.
Ji Xiankun, Xu Ke (2020). Waterproof Concrete, Structural Self-waterproofing, Rigid Waterproofing and Engineering Application. China Building Waterproofing, (10):49–57.
Li Bing, Guo Rongxin, Yan Feng (2019). Study on Self-healing Performance of Permeable Crystalline Waterproof Concrete with Water-cement Ratio. Non-metallic Mines,42(02):84–86.
Liu Quan (2021). Performance Analysis and Application of FS102 Waterproof Compacting Agent. Construction Machinery & Maintenance, (02):156–157.
Ren Hui, Zhou Xianping, Tan Zhongsheng, Liu Qiang (2018). Study on the Influence of Coastal Environment on Self-waterproofing Performance of Lining Concrete Structure. Tunnel Construction,38(S2):98–103.
Shuai Guo (2020). Experimental Study on Preparation and Impermeability of Cellulose Fiber Waterproof Concrete for Subway Station. Anhui University of Science and Technology.
Tang Tiejun (2013). Experimental Study on Impermeability of Basalt Fiber Waterproof Concrete. Traffic Engineering & Technology for National Defence,11(04):25–27.
Wang Chao (2021). Waterproof Method for Improving Overall Performance of Concrete Structure. Popular Standardization,(12):138–140.
Xiao Fei (2019). Study on Preparation and Performance of High-Performance Anti-crack Waterproof Concrete. Shandong Province, Shandong Provincial Institute of Architectural Science, (01).
Yue Changqing, Sun Liying (2019). Experimental Analysis and Research on Action Mechanism and Performance of Concrete Waterproofing Agent in Waterproof Concrete. Construction Technology,48–52.
Zhu Yinhu, Meng Qingchao, Xu Wen (2020). Study on Performance of Anti-cracking Self-waterproof Concrete and Its Application in Nanjing Metro Line 10. China Building Waterproofing, (06):34–38.

Frontiers of Civil Engineering and Disaster Prevention and
Control – Yang & Rahman (Eds)
© 2023 The Author(s), ISBN: 978-1-032-31200-2

Preparation of low shrinkage geopolymer concrete and its application in seawall engineering

Kexin Zhao
South China University of Technology, Guangzhou, China

ABSTRACT: In recent years, the scale of China's marine engineering construction has increased year by year, and the demand and performance requirements of concrete are also getting higher and higher. Geopolymer concrete has higher stability and stronger corrosion resistance than ordinary silicate concrete. It is more suitable for application in marine engineering, and its fast hardening and early strength characteristics are more suitable for rapid construction at tidal water level seawalls. However, the self-shrinkage of geopolymer concrete is relatively large, and there is a risk of cracking during construction. In this paper, low-shrinkage geopolymer concrete was prepared, the shrinkage deformation characteristics and mechanical properties of geopolymer concrete were studied, and the engineering application of geopolymer was carried out in the seawall of Guangdong Province. The research results show that the self-shrinkage of the improved geopolymer concrete was greatly reduced, the temperature rise and the temperature difference between inside and outside were also smaller, and the risk of cracking was lower. The engineering application of geopolymer was carried out in the seawall of Guangdong Province of China. The self-shrinkage of the improved geopolymer concrete was greatly reduced. Geopolymer concrete had a small temperature rise and temperature difference between inside and outside, and the risk of cracking was low. The characteristics of rapid hardening and early strength of geopolymer concrete were more suitable for the rapid construction of seawalls at tidal water levels compared with ordinary cement concrete.

1 INTRODUCTION

Most of the marine projects are far from the land. There is a shortage of materials and fresh water during the construction process, and the construction is greatly affected by the tidal effect. During the service process, the buildings will suffer from harmful ion corrosion in seawater, as well as the influence of typhoons and waves (Sun et al. 2021; Su 2016; Wang et al. 2015). Therefore, according to marine engineering construction and service characteristics, a cementitious material with excellent seawater corrosion resistance is urgently needed. Geopolymers (Kozub et al. 2021; Li et al. 2021; Ukrainczyk 2021) are a new class of inorganic polymers obtained from burning clay and industrial waste residues as the main raw materials through appropriate technological treatment and chemical reactions. Compared with ordinary cement-based materials, geopolymers have excellent properties, such as hard and fast setting, early strength, high strength, and corrosion resistance. They can be used for rapid construction by tidal intermittent, and sea sand can be used as aggregate from local materials (Gc et al. 2021; Valente et al. 2007).

The hydration products of geopolymers are stable, the slurry structure is dense, and they have excellent resistance to seawater erosion (Meicke & Paasch. 2012; Zhou et al. 2020). However, due to the large early volume shrinkage of geopolymers, its wide application is restricted (Grassl et al. 2010; Huang & Liang, 1996; Rozière et al. 2007). The chemical shrinkage and autogenous shrinkage of geopolymers are much larger than those of Portland cement hardened bodies. It is pointed out that the slag base substrate is about two times that of Portland cement slurry (Li et al. 2017). At present, there are three main measures to reduce the volume shrinkage of geopolymers at home and abroad: (1) Reduce the shrinkage by adjusting the ratio of the raw materials of the geopolymer and the activator. The shrinkage reduction effect is more obvious, and the shrinkage reduction effect is gradually enhanced with the increase of the modulus of water glass (Bakharev 2005). (2) Use the shrinkage reducer to reduce the volume shrinkage of the geopolymer because it can reduce the water in the pores of the hardened slurry, surface tension, resulting in a decrease in capillary pressure in the pore structure, thereby reducing the shrinkage of the geopolymer (Palacios & Puertas 2005). (3) Add an expansion agent to the geopolymer to compensate for the shrinkage (Jia et al. 2018). (4) Use high-temperature steam curing to reduce the shrinkage of geopolymers (Noushini et al. 2018). The above reduction measures have limited effect, and further systematic research is needed.

In this paper, the mix ratio of low-shrinkage geopolymer slurry and concrete was designed, and the low-shrinkage geopolymer concrete was applied to the actual seawall project. The seawall panels were poured with geopolymer concrete and ordinary Portland concrete. The working performance, compressive strength, internal temperature rise, and strain of ordinary Portland cement concrete and geopolymer concrete in a real seawater environment were compared to provide technical support for the engineering application of geopolymer concrete.

2 ABOUT THIS EXPERIMENT

2.1 *Raw materials for the experiment*

The geopolymer cementitious material used in the test was mainly composed of Shaogang granulated blast furnace slag, Huangpu Power Plant fly ash, Aiken silica fume, activator, and retarder enhancer (Shaogang S95 slag; Huangpu Power Plant Class II fly ash; Silica fume: a mixture of solid sodium silicate and sodium carbonate with a modulus of 1.4 Activator: retarder enhancer = 70:12:5:8:5). The chemical composition and physical and mechanical properties of the geopolymer cementitious materials are shown in Table 1 and Table 2, respectively. In addition, P·II 42.5 cement produced by Zhujiang Cement Plant and granulated blast furnace slag (S95) produced by Shaoguan Iron and Steel Group were used. The chemical composition of blast furnace slag is shown in Table 3.

Table 1. Chemical composition of geopolymer binder/wt%.

SiO_2	Al_2O_3	Fe_2O_3	TiO_2	CaO	MgO	SO_3	P_2O_5	K_2O	Na_2O	Loss
30.56	19.56	2.22	0.85	34.40	3.23	1.2 3	0.05	2.09	4.88	0.93

Table 2. Physical and mechanical properties of geopolymer binder.

Standard consistency water consumption	Specific surface area/(m²/kg)	Density/(g/cm³)	Flexural strength/MPa			Compressive strength/MPa		
			3 d	7 d	28 d	3 d	7 d	28 d
0.224	375	2.91	4.5	6.5	7.8	30.6	35.2	50.2

Table 3. Chemical composition of GBFS /wt%.

SiO_2	Al_2O_3	Fe_2O_3	TiO_2	CaO	MgO	SO_3	P_2O_5	K_2O	Na_2O	Loss
35.07	12.15	0.32	0.74	37.08	11.27	1.19	–	0.49	0.25	−0.61

The active MgO used was prepared by calcining magnesite. The main mineral component of magnesite is $MgCO_3$; the loss on ignition exceeded 50% (Table 4), the decomposition temperature was about 650°C, and it could be completely decomposed at 700°C. The magnesite that was ground to a certain fineness was heated to 800°C, 900°C, and 100°C, respectively, (the heating rate was 10°C/min) for 1 h, and cooled in the furnace to prepare three kinds of MgO powders with activities marked by 60s, 150s, and 220s. The chemical activity of MgO was measured by the citric acid discoloration method (YBT 4019-2006) according to the time of reaction discoloration. The shorter the time, the higher the activity. Referring to the GBMF 19-2017 standard, MgO was divided based on the reaction time. 0-100s was fast-type MgO, which was high-activity MgO; 100-200s was medium-speed MgO, which was medium-activity MgO; 200-300s is low-speed type MgO, namely low activity MgO.

Compared with the geopolymer cementitious material powder, the prepared MgO had a larger specific surface area and smaller particle size than the geopolymer cementitious material (D50 is about 10μm), as shown in Table 5. After being mixed with geopolymer cementitious material, MgO could be uniformly filled into the pores of geopolymer cementitious material to achieve close packing of particles to create favorable conditions for rapid and uniform compensation of shrinkage.

Table 4. Chemical composition and characteristics of magnesite.

	Chemical components/wt%						Proportion by weight	Average particle size/μm
SiO_2	MgO	CaO	K_2O	Na_2O	Al_2O_3	Loss		
0.14	46.94	0.70	0	0	0.10	51.72	3.16	11.96

Table 5. Properties of MgO powders prepared by different calcination processes.

MgO activity	Processing Temperature/°C	Active Time/s	Specific Surface Area/(m^2/g)	D_{50}/μm
High activity	800	60	37.42	9.05
Moderate activity	900	150	32.52	12.59
Low activity	1000	220	20.92	13.41

2.2 Cementitious material paste mix ratio

The idea of the mix ratio design was to first take the pure slurry as the research object, determine the active mix and dosage of MgO, and then formulate geopolymer concrete. In view of the large shrinkage of the geopolymer slurry, active MgO was proposed to compensate for the volume shrinkage of the geopolymer slurry. We changed the content of MgO and the activity of MgO, mixed MgO with different activities, and then mixed them evenly with the geopolymer cementing material, added water, and stirred evenly according to the water-to-binder ratio of 0.38, and controlled the fluidity to be 180 ± 20 mm. For the polymer pulp, the specific ratio is shown in Table 6.

Table 6. Summary of mix proportions of slurry.

No.	Comparison Group	4%	6%	8%	12%	60s	150s	220s	60s+150s	150s+220s	60s+220s
Cementitious Material/g	100	96	94	92	88	94	94	94	94	94	94
MgO (60 s)/g	–	4	6	8	12	6	–	–	3	–	3
MgO (150 s)/g	–	–	–	–	–	–	6	–	3	3	–
MgO (220 s)/g	–	–	–	–	–	–	–	6	–	3	3
Water/g	38	38	38	38	38	38	38	38	38	38	38

2.3 Experiment method

2.3.1 Autogenous shrinkage

Autogenous shrinkage was tested according to the method modified by Gao et al. (Jia et al. 2018). When testing the self-shrinkage, the slurry was sealed in the bellows; the volume deformation was converted from the bellows to the length deformation; the length deformation was converted into a voltage signal by the displacement sensor and then converted into a digital signal by the analog-to-digital converter and transmitted to the computer. Finally, the real-time recording of data was conducted with self-designed software.

2.3.2 Compressive strength

According to GB/T 50081-2019 "Standards for Test Methods for Physical and Mechanical Properties of Concrete," the Italian Matest C088-01 universal testing machine was used to compress a 150 mm cube at a loading rate of 0.5 MPa/s to determine the cube of the sample and the compressive strength (FCU). Three cubes were tested per group to ensure that the data dispersion did not exceed 15% (the strength of each specimen did not exceed 15% of the median), and the average was calculated.

3 TEST RESULTS AND ANALYSIS

3.1 Influence of MgO on Autogenous Shrinkage of Cementitious Paste

The autogenous shrinkage of the geopolymer slurry was relatively large, and the autogenous shrinkage was as high as 1500μm/m in 1d, and the autogenous shrinkage still increases after 7d, as shown in Figure 1(a). With the increase of MgO content, the autogenous shrinkage of the geopolymer slurry gradually decreased, and the autogenous shrinkage of the slurry after 14 d became smooth. When the content of MgO is 4%, the increasing trend of autogenous shrinkage in the later period (after 35 days) increased slightly; that is, the ability of MgO to compensate for autogenous shrinkage of geopolymer slurry was not significant. When the content of MgO was 6%, compared with the 3d and 7d autogenous shrinkage (1237μm/m, 2490μm/m) of undoped MgO geopolymer slurry were reduced by 65% and 72%, respectively. When the content of MgO was 12%, and the age was 3d and 7d, the autogenous shrinkage of the geopolymer slurry was reduced by 71% and 78%, respectively, indicating that MgO can effectively compensate for the autogenous shrinkage of the geopolymer slurry. In addition, when the content of MgO is less than 6%, the autogenous shrinkage of the geopolymer slurry at each age decreases significantly with the increase of the content of MgO. When the content of MgO exceeds 6%, the degree of compensation for autogenous shrinkage decreases. It could be seen from Figure 1(b) that the autogenous shrinkage of the MgO-added geopolymer slurry was significantly reduced. When using high-activity MgO (reaction discoloration time is 60 s), the self-shrinkage of the hardened slurry was smaller in the early stage (0~14d), but the self-shrinkage of the slurry after 14d was larger, which was roughly the same as the shrinkage level when MgO incorporation was added for 150s. This indicates the high activity of MgO can effectively compensate for the early autogenous shrinkage of the geopolymer slurry. When the low activity MgO (reaction discoloration time was 220s) was used, the hardening (0-13d) had a large self-shrinkage, but it could effectively compensate for the late self-shrinkage of the geopolymer slurry. MgO with different activities was incorporated into the geopolymer slurry, and the autogenous shrinkage of the slurry is shown in Figure 1(c). When using MgO with the 60s and 220s activities, both the early autogenous shrinkage and the late autogenous shrinkage of the hardened paste were smaller. By using MgO in combination, the expansion stage of different active MgO could be fully utilized so that the volume shrinkage of the hardened slurry of geopolymer could be compensated in different stages.

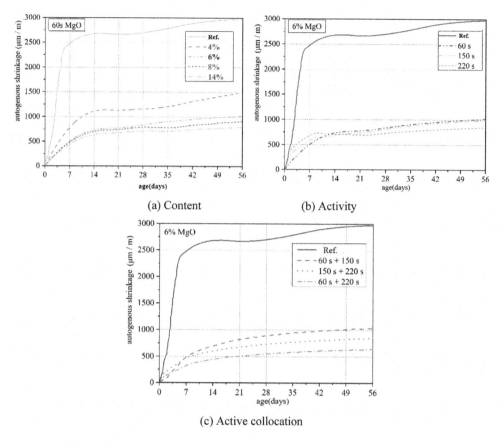

(a) Content

(b) Activity

(c) Active collocation

Figure 1.   Effect of MgO on the autogenous shrinkage of geopolymer pastes.

3.2   *Influence of MgO on the compressive strength of the cementitious paste*

It could be seen from Figure 2(a) that with the increase of MgO content, the mortar strength of the geopolymer in the early stage (before 7d) was improved, and the 28d mortar strength showed a trend of first increasing and then decreasing. Compared with the geopolymer without MgO, the compressive strength of the geopolymers 3d, 7d, and 28d with MgO content of 6% increased by 24.0%, 30.2%, and 17.1%, respectively. Compared with the geopolymer with 8% MgO content, when the MgO content is 12%, the compressive strength of its 28-day pure slurry shrinks; that is, the excessive MgO content is detrimental to the strength of the geopolymer and should be controlled. The content of MgO is 4-8%.

Figure 2(b) showed the results of the compressive strength of mortar mixed with different active MgO geopolymers. When MgO with 60s activity was incorporated, the compressive strength of 3d and 28d increased by 19.4% and 12.5%, respectively. When MgO with 150s activity was incorporated, the compressive strength of 3d and 28d increased by 14.1% and 10.1%, respectively. When MgO was added, the compressive strengths of 3d and 28d mortars increased by 10.5% and 19.7%, respectively. High activity MgO (60s) was beneficial to improving the early (3d) compressive strength of geopolymer mortar, and low activity MgO (220s) was beneficial to the development of geopolymer mortar's late (28d) compressive strength.

When the 60s and 220s active MgO were used in combination, the compressive strength of the mortar was significantly improved, and the 3d and 28d compressive strengths of the geopolymer mortar without MgO were increased by 18.1% and 15.3%, respectively (Figure 2(c)). Compared

545

with the geopolymer doped with 60s MgO, the early strength of MgO with 60s and 220s activity was slightly lower, but the 28d strength was significantly improved. The early strength of the geopolymer of MgO combined with MgO was significantly improved and had little effect on the 28-d strength. That is to say, the appropriate combination of different active MgO could improve the strength of the geopolymer mortar, and the strength development of each age was more coordinated.

Based on the results of MgO compensating for the volume shrinkage of the geopolymer slurry and the strength of the mortar, the geopolymer could be effectively reduced by adding 6% MgO (60 s: 220s=1:1) to the geopolymer (G1). The volume of the slurry shrinks and the modified geopolymer cementitious material with higher strength was obtained.

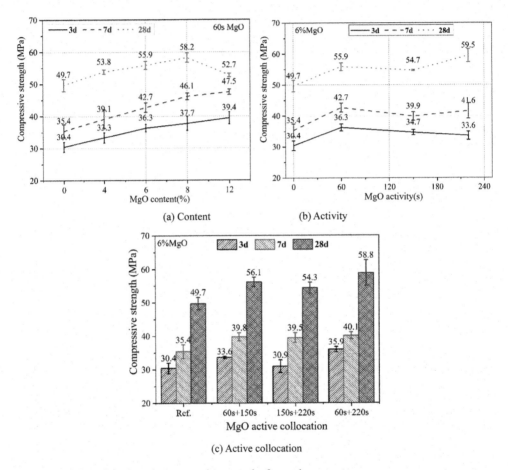

(a) Content

(b) Activity

(c) Active collocation

Figure 2. Effect of MgO on the compressive strength of geopolymer pastes.

4 APPLICATION OF GEOPOLYMER CONCRETE IN SEAWALL ENGINEERING

This paper selected a new seawall of 1.2 km, and the seawall panels were poured with geopolymer concrete. Two separation joints (10m×2m) were selected for sampling quality inspection of pouring, monitoring of concrete structure temperature and strain, and on-site concrete core-pulling test to verify that geopolymer concrete could be used in seawall engineering.

4.1 Raw materials and mix ratio

The environmental category of a seawall reinforcement project that meets the standards was category 3a. According to the requirements of the "Code for Design of Concrete Structures" (GB50010-2010) (Gao et al. 2014), concrete with a strength grade of C35 or above should be used. Therefore, the application section of the project adopted a strength grade of C35(Modified Geopolymer Concrete, G35). The raw materials of concrete were natural sea sand from Huilai and limestone crushed stone of 5–31.5mm. The physical properties of crushed stone and sea sand are shown in Tables 7 and 8. The mixing water was tap water, which was used when preparing ordinary concrete (C35). The mix ratio of the prepared geopolymer concrete is shown in Table 9, and the physical and mechanical properties of the concrete are shown in Table 10.

Table 7. Physical properties of aggregate.

Needle-like and flaky particle content/%	Crushing Index/%	Bulk density/(kg/m^3)	Apparent density/(kg/m^3)	Porosity/%
7.5	9.1	1560	2750	43.3

Table 8. Physical properties of sea sand.

Bulk density/(kg/m^3)	Apparent density/(kg/m^3)	Fineness modulus	Mud content/%
1520	2590	2.5	0.2

Table 9. Mix proportions of geopolymer and Portland cement concretes used in seawall construction.

No.	Water-cement ratio	Cementitious material/(kg/m^3)	Sea sand/(kg/m^3)	Gravel/(kg/m^3)	Water reducer/(kg/m^3)	Water/ (kg/m^3)
C35	0.40	440	760	1050	7.92	176
G35	0.40	440	720	1000	0	176

4.2 Core pulling strength

Twenty-five days after the seawall concrete slab was poured, the geopolymer concrete slab and the ordinary concrete slab were core-pulled at different positions. Three groups of nine core samples were taken out, cut and ground, soaked in water for two days, and taken out for one day. Then the core compressive strength test was carried out. The core-like surface structure of the geopolymer concrete seawall panel was dense, and no aggregate exposure and honeycomb voids were found. Both concretes achieved design strength levels, as shown in Figure 3. The 28-day compressive strength of the geopolymer concrete core samples was 48.1-57.0MPa, and the average compressive strength was 52.8 MPa. Compared with the geopolymer concrete core sample, the core sample structure of ordinary concrete was also dense and had no apparent defects, and the average 28 days compressive strength of the concrete core sample was 48.8 MPa, which was relatively low.

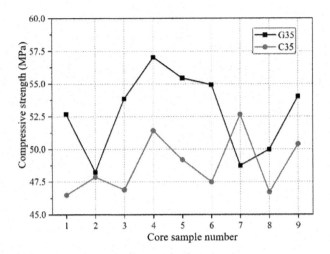

Figure 3.   Compressive strength of core-drilling samples of geopolymer and Portland cement concrete.

4.3   *Temperature monitoring*

4.3.1   *Measuring point layout*

The layout of the measuring points for monitoring the temperature and deformation of geopolymer concrete and ordinary cement concrete in the engineering test section is shown in Figure 4. The length of the seawall panel along the slope was 5 meters, and the length of the structural seam was 10 meters. The measuring points were arranged with sensors in four sections: in section 1, three temperature sensors (T1, T2, and T3) were arranged at the upper, middle, and lower positions, respectively. Two temperature sensors (T4 and T5) were arranged in the left and right positions, respectively.

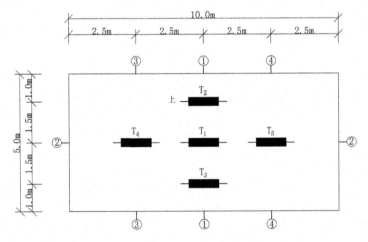

Figure 4.   Location of temperature in seawall geopolymer concrete panel.

4.3.2   *Monitoring results*

The air temperature at the location of the test section was recorded from the time the concrete was poured. During the engineering test section, the temperature was relatively high (24~34°C), and the temperature difference between morning and evening was large. The highest temperature

548

was 38.5°C, and the lowest temperature was 17.6°C. The temperature monitoring results showed that (Table 10), the temperature of the geopolymer concrete increased rapidly at each measuring point within the initial 10 h, and the temperature reached a peak at about 30 h (the maximum temperature of each measuring point is 43~49°C). It reached stability at about 120 hours of age, and then fluctuated due to the influence of temperature. In section ①, the temperature of the middle measuring point was the highest, the bottom was second, and the upper part was the lowest. In section ②, the temperature of the three measuring points was not much different, the temperature of the middle measuring point was the highest, and the temperature measuring points on the left and right sides were slightly lower.

The temperature of the measuring point of geopolymer concrete was lower than that of the corresponding measuring point of ordinary concrete, and the highest temperature of the measuring point T1 was 7°C lower. The ordinary concrete measuring point temperature was similar to that of the geopolymer concrete. In other words, the temperature of the middle measuring point was the highest, the left and right and upper and lower temperature measuring points were lower, the temperature of the upper temperature measuring point was the lowest, and the temperature of each measuring point was in the range of 46~53°C.

According to the requirements of "Construction Specifications for Mass Concrete" (GB 50496-2018) [20], the temperature rise inside the concrete mold should not be greater than 50°C, and the temperature difference between the inside and outside should not be greater than 25°C. The temperature monitoring results showed that compared with ordinary cement concrete, geopolymer concrete had a smaller temperature rise and temperature difference between inside and outside and a lower risk of cracking.

Table 10. Maximum temperature of sea wall geopolymer concrete during casting and monitoring stages.

	T1/°C	T2/°C	T3/°C	T4/°C	T5/°C
G35	48.75	43.52	/	45.95	48.35
C35	52.96	47.95	48.32	47.95	49.35

5 CONCLUSION

(1) MgO in the geopolymer slurry could compensate for the volume shrinkage of the geopolymer. The combination of low activity (220 s) and high activity (60 s) MgO could improve the early and late volume shrinkage of geopolymers. Incorporating 6% MgO into the geopolymer (60 s : 220 s = 1 : 1) could obtain geopolymers with less volume shrinkage and meet practical engineering needs.

(2) The geopolymer concrete prepared from sea sand was applied to a 1.2-kilometer new sea-wall. The engineering test results showed that the geopolymer concrete had high compressive strength, with an average value of 50.1 MPa. The maximum temperature rise of geopolymer concrete was 48°C, 6°C lower than that of ordinary concrete, and the geopolymer concrete had a lower risk of cracking.

REFERENCES

Bakharev T. Geopolymeric materials prepared using Class F fly ash and elevated temperature curing[J]. Cement and Concrete Research, 2005, 35(6): 1224–1232.

E Rozière, Granger S, Turcry P, et al. Influence of paste volume on shrinkage cracking and fracture properties of self-compacting concrete[J]. Cement & Concrete Composites, 2007, 29(8):626–636.

Gao P, Zhang T, Luo R, et al. Improvement of autogenous shrinkage measurement for cement paste at a very early age: Corrugated tube method using non-contact sensors[J]. Construction and Building Materials, 2014, 55: 57–62.

Gc A, Mm A, Pvb C, et al. Synthesis of clay geopolymers using olive pomace fly ash as an alternative activator. Influence of the additional commercial alkaline activator used[J]. Journal of Materials Research and Technology, 2021, 12:1762–1776.

Grassl P, Hong S W, Buenfeld N R. Influence of aggregate size and volume fraction on shrinkage induced micro-cracking of concrete and mortar[J]. Cement & Concrete Research, 2010, 40(1):85–93.

Huang Y J, Liang C M. Volume shrinkage characteristics in the cure of low-shrink unsaturated polyester resins[J]. Polymer, 1996, 37(3):401–412.

Jia Z, Yang Y, Yang L, et al. Hydration products, internal relative humidity, and drying shrinkage of alkali-activated slag mortar with expansion agents[J]. Construction and Building Materials, 2018, 158: 198–207.

Kozub B, Bazan P, D Mierzwiński, et al. Fly-Ash-Based Geopolymers Reinforced by Melamine Fibers[J]. Materials, 2021, 14(2):400.

Li H, Chi Z, Yan B, et al. An innovative wood-chip-framework substrate used as slow-release carbon source to treat high-strength nitrogen wastewater[J]. Journal of Environmental Sciences, 2017, 51(001):275–283.

Li O H, Yun-Ming L, Cheng-Yong H, et al. Evaluation of the Effect of Silica Fume on Amorphous Fly Ash Geopolymers Exposed to Elevated Temperature[J]. Magnetochemistry, 2021, 7(1):9.

Meicke S, Paasch R. Seawater lubricated polymer journal bearings for use in wave energy converters[J]. Renewable Energy, 2012, 39(1):463–470.

Noushini A, Castel A, Gilbert R I. Creep and shrinkage of synthetic fiber reinforced geopolymer concrete[J]. Magazine of Concrete Research, 2018, 71(20): 1–43.

Palacios M, Puertas F. Effect of superplasticizer and shrinkage-reducing admixtures on alkali-activated slag pastes and mortars[J]. Cement and Concrete Research, 2005, 35(7): 1358–1367.

Su G. Influence of Mn on the Corrosion Behaviour of Medium Manganese Steels in a Simulated Seawater Environment[J]. International Journal of Electrochemical Science, 2016:9447–9461.

Sun Y, K Wang, Zhong X, et al. Assess the Typhoon-driven Extreme Wave Conditions in Manila Bay through Numerical Simulation and Statistical Analysis[J]. Applied Ocean Research, 2021, 109(3):102565.

Ukrainczyk N. Simple Model for Alkali Leaching from Geopolymers: Effects of Raw Materials and Acetic Acid Concentration on Apparent Diffusion Coefficient[J]. Materials, 2021, 14(6):1425.

Valente M, Sambucci M, Sibai A. Geopolymers vs. Cement Matrix Materials: How Nanofiller Can Help a Sustainability Approach for Smart Construction Applications-A Review[J]. Nanomaterials, 2021, 11(8):2007.

Wang W, Li X B, Wang J, et al. Influence of biofilms growth on corrosion potential of metals immersed in seawater[J]. Materials & Corrosion, 2015, 55(1):30–35.

Zhou S, Zhou S, Zhang J, et al. Relationship between Moisture Transportation, Efflorescence and Structure Degradation in Fly Ash/Slag Geopolymer[J]. Materials, 2020, 13(23):5550.

Frontiers of Civil Engineering and Disaster Prevention and
Control – Yang & Rahman (Eds)
© 2023 The Author(s), ISBN: 978-1-032-31200-2

Study of cement mortar uniformity testing methods

Yue Chen*, Jinming Wang*, Caiying Sun, Lei Liu & Mengxi Zhang
School of Information Engineering, Inner Mongolia University of Science and Technology, Baotou, China

ABSTRACT: Cement mortar is an important material used in the construction industry, and its uniformity detection is the key index for judging the quality of cement mortar. To measure the uniformity of cement mortar, ABAQUS software is used for simulation design. The ultrasonic sound attenuation method is selected to measure the uniformity of five lines on the same plane. The uniformity of cement mortar is measured by comparing the difference in receiving amplitude through the relationship between uniformity, density, and ultrasonic receiving amplitude. According to the simulation results, this method can conveniently and accurately measure the uniformity of cement mortar quantitatively. A certain threshold can be set for qualitative judgment in the actual measurement.

1 INTRODUCTION

As a significant contributor of the national economy and a pillar industry, the construction industry is developing rapidly, and the scale of output value is expanding. Cement mortar is an important industrial material used during construction, but the mixing process often leads to decreased strength and consistency of cement mortar due to uneven mixing, which affects the quality of the building.

In recent years, most of the ultrasonic sound attenuation methods are used in the process of slurry detection. In 2016, Wei Lu et al., based on the exploration of the characteristic quantities of ultrasonic sound velocity, sound pressure, and sound intensity, analyzed the transmission characteristics of ultrasonic waves on the surface of the solid-liquid two-phase medium and the fluctuation law of ultrasonic waves in the medium, based on which the attenuation mechanism of ultrasonic waves in a non-uniform medium was studied in depth (Wei 2016). In 2019, Zhou Shengyou et al. measured oil density by measuring the acoustic impedance and sound velocity of oil simultaneously (Zhou 2019). In 2020, Guo Xinyi compared the application of several common slurry density measurement methods for wet desulfurization systems, analyzed the basic principles and advantages and disadvantages of each density measurement method, and made suggestions on the selection of ultrasonic slurry density measurement methods (Guo 2020).

This paper defines the concept of cement mortar uniformity from the principle of uniformity testing, gives a set of feasible testing methods, and uses ABAQUS software to simulate the design and testing of the testing system. The results show that the method has accurate testing results and few interfering factors and can be employed to test the quality of cement mortar on construction projects.

*Corresponding Authors: 1249479695@qq.com and yb15661510224@163.com

DOI 10.1201/9781003308577-73

2 TESTING PRINCIPLE

For the detection of cement mortar uniformity, we need to detect the density of cement mortar first and then through the consistency of density, we can determine the uniformity of cement mortar.

Ultrasonic detection of cement mortar density speed of sound method, acoustic attenuation method, acoustic impedance method three, and acoustic impedance method are mostly used to measure the hardness and strength of the medium, rarely used in life. The acoustic velocity method is used to obtain the number of parameters by measuring the speed of sound of the signal propagating in the medium, the density of cement mortar is not obvious concerning the speed of sound. The acoustic attenuation method involves the use of ultrasonic waves in the medium to propagate its ability to increase with the propagation distance and reduce the principle of detecting density. This paper is a better fit as the final choice of the acoustic attenuation method (Wang 2014).

Both scattering attenuation and absorption attenuation of acoustic waves obey the exponential law, i.e., there is a relationship between the amplitude of the echo and the propagation distance as follows.

$$U = U_0 e^{-\beta x}$$

U is the initial sound intensity, β is the attenuation coefficient of sound waves, and x is the propagation distance.

$$2\beta = N\pi a^2 \left[\frac{4}{9} k^4 a^4 + \frac{3}{4} \pi ka \frac{s(\sigma - 1)^2}{\sigma^2 + (\sigma + \tau^2)^2} \right]$$

β is the attenuation coefficient of the sound wave; k is the wave number of the ultrasonic wave; w is the angular frequency; a is the radius of the cement mortar particles; v is the viscosity rate of the liquid; N is the number of cement mortar particles per unit volume; σ is the density ratio of the cement mortar density to that of the clear water.

The uniformity of the cement mortar refers to the consistency of the density of the measured medium in different detection channels. In this paper, the uniformity of the cement mortar is calculated based on the variation of the amplitude in different detection channels. The received amplitude and uniformity conversion formula is as follows (Chen 2019):

$$\overline{X} = \frac{X_1 + X_2 + X_3 + X_4 + X_5}{5}$$

$$S = \sqrt{\frac{(X_1 - \overline{X})^2 + (X_2 - \overline{X})^2 + (X_3 - \overline{X})^2 + (X_4 - \overline{X})^2 + (X_5 - \overline{X})^2}{10 - 1}}$$

$$CV = \frac{S}{\overline{X}} \times 100\%$$

$$M = 1 - CV$$

$\overline{X}$ is the mean, S is the standard deviation, CV is the coefficient of variation, and M is the uniformity.

In this paper, cement mortar uniformity testing is based on ultrasonic waves propagating in the same medium under the same conditions and at the same distance with the same acoustic attenuation coefficient and the same magnitude of received amplitude.

3 SYSTEM DESIGN

3.1 Scheme design

In this system design, a cement mortar with a weight ratio of cement: sand: water = 1:3:0.65, a mortar density of 2000 kg/m³, a medium sand particle size of 0.3 mm, and a mortar strength of M30 is used as an example to design.

When designing the cement mortar uniformity detection system, the first step is to determine the sensor spacing. Very large sensor spacing will result in a weak received waveform, and no amplitude signal change will be visible; very small will result in a superimposed ultrasonic received waveform graph, affecting judgment. Considering the actual detection needs, a spacing of 40 cm was chosen as the measurement stroke.

The ABAQUS/standard module enables kinetic display analysis, and the finite element software ABAQUS is used to establish a cement mortar model for finite element analysis. A cement mortar model with length, width, and height of 50 cm, 30 cm, and 50 cm, respectively, with properties of density $\rho = 2000 \text{kg} \cdot \text{m}^{-3}$, modulus of elasticity E = 21,800 MPa, and Poisson's ratio $\mu = 0.28$. At the same frequency, sensors are placed at different heights, with a 40 cm interval at the same level of the corresponding sensor. The receiving probe and the transmitting probe is placed, as shown in Figure 1. The grid cell length is 3 mm, the sampling interval is $2.5e^{-7}$s, using structured meshing, and the cell type is hexahedral cell C3D8. The modeling image is shown in Figure 2.

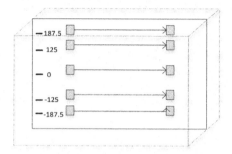

Figure 1. Positioning of transceiver probes.

Figure 2. Cement mortar modeling diagram.

3.2 Detection signal selection

Ultrasonic frequency selection needs to be determined through simulation. The density of cement mortar is concentrated in the range of 1800 kg/m³–2600 kg/m³, and three densities of 1800 kg/m³, 2000 kg/m³, and 2600 kg/m³ are selected for simulation. At position 0 of the cement mortar model, the frequency of the ultrasonic waves was changed, the other conditions of the mortar were unchanged, and the simulation experiments were carried out using 100 kHz–800 kHz, respectively, and the waveforms received by the receiving probe were similar at different densities, as shown in Figure 3.

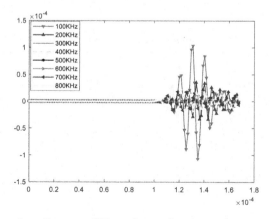

Figure 3. Receiving waveform diagram at different frequencies.

3.3 *Judgement of test results*

The solution used in this paper uses the test on uniformity in the same plane on 5 test lines. The test results are considered as the uniformity of cement mixing on the test surface. The difference between the uniformity value and the benchmark value gives a qualitative judgment of the test results. The benchmark value for uniformity in this paper is set at 90%, which can be adjusted according to actual requirements.

4 EXPERIMENTS AND ANALYSIS OF RESULTS

4.1 *Influence of density on detection results*

The ultrasonic frequency was set at 100 kHz, and the densities were chosen to be 1800 kg/m^3, 2000 kg/m^3, and 2600 kg/m^3, respectively. The simulation results of uniform mortar at different densities are shown in Table 1.

Table 1. Simulation results with different densities of homogeneous mortar.

	187.5	125	0	−125	−187.5	$\overline{X}$	S	CV	M
1800 kg/m^3	1.21E-04	1.19E-04	1.18E-04	1.16E-04	1.23E-04	1.19E-04	1.6E-06	1.34%	98.66%
2000 kg/m^3	1.13E-04	1.07E-04	1.08E-04	1.08E-04	1.13E-04	1.10E-04	1.91E-06	1.74%	98.26%
2600 kg/m^3	1.03E-04	1.01E-04	9.80E-05	9.36E-05	1.04E-04	1.00E-04	2.86E-06	2.86%	97.14%

The table shows the relationship between the density of cement mortar and the amplitude of the emitted and received amplitude signal. When the ultrasonic frequency and other characteristics are constant, i.e., the density of cement mortar increases, the sound amplitude received by the terminal transducer increases. The uniformity is stable at over 95%.

In the same model, a layered design simulation is carried out. Option 1: Set the cement mortar density to 2000 kg/m^3 and 2600 kg/m^3 in two layers, and place the transmitting and receiving probes in five ultrasonic probes for simulation. The simulation schematic is shown in Figure 4. Option 2: Divide the cement mortar into three layers of water, cement mortar, and sand, and carry out the same operation simulation. The simulation schematic is shown in Figure 5.

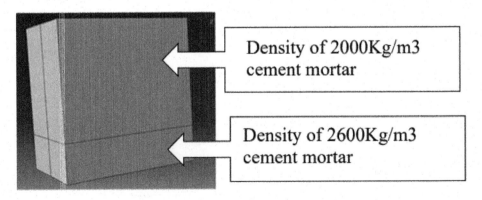

Figure 4. Simulation diagram of Option 1.

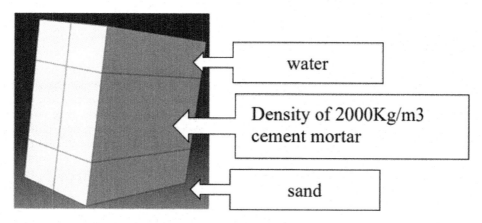

Figure 5.   Schematic diagram of Option 2 simulation.

The simulation results for scenarios one and two are shown in Table 2.

Table 2.   Simulation results under different layering conditions.

	187.5	125	0	−125	−187.5	$\overline{X}$	S	CV	M
Homogeneous mortar	1.13E-04	1.07E-04	1.08E-04	1.08E-04	1.13E-04	1.10E-04	1.91E-06	1.74%	98.26%
Option 1	6.22E-05	6.92E-05	6.21E-05	7.20E-05	9.64E-05	7.24E-05	1.15E-05	15.95%	84.05%
Option 2	0	4.00E-07	2.21E-06	4.31E-06	9.00E-06	3.18E-06	2.99E-06	94.18%	5.82%

4.2   The influence of temperature on test results

For cement mortars applied in construction projects, the environmental conditions involved in summer and winter vary considerably, with temperatures varying from below zero to tens of degrees. And the density of cement mortar varies with temperature.

As the acoustic impedance changes with temperature, the density of the cement mortar also changes. Thus, to exclude the disturbing influence of temperature on the detection results, the other parameters are kept constant, and only the temperature is allowed to vary. Therefore, the ultrasonic frequency f was set to 100 kHz during the simulation, and four sets of simulations were done for temperatures starting from −20°C and ending at 40°C with an interval of 20°C, as shown in Table 3.

Table 3.   Simulation results at different temperatures.

	187.5	125	0	−125	−187.5	$\overline{X}$	S	CV	M
40°C	0.000366	0.000377	0.000353	0.000359	0.000368	0.000365	8.27E-06	2.27%	97.73%
20°C	0.000283	0.000268	0.000277	0.000274	0.000296	0.00028	9.44E-06	3.37%	96.63%
0°C	0.000235	0.000216	0.000218	0.000228	0.000222	0.000224	6.89E-06	3.08%	96.92%
−20°C	0.000163	0.000159	0.000156	0.000175	0.00016	0.000163	6.67E-06	4.10%	95.90%

Observing Table 3, it can be obtained that although the cement mortar temperature varies with climatic temperature, its temperature variation has less impact on the homogeneity test results, which all remain above 90% homogeneity. Therefore, the temperature factor can be disregarded while designing the method for testing the uniformity of cement mortars.

5 CONCLUSIONS

By conducting simulations, the following conclusions can be drawn.

1) At the same density, the cement mortar uniformity is stable above 90%. When the cement mortar is not mixed uniformly and the overall density of the cement mortar changes, the cement mortar uniformity is below 90%. When impurities appear in the cement mortar, the mortar density changes locally, and the uniformity also changes significantly below 90%. Therefore, the uniformity of the cement mortar can be detected by the change in density.
2) The higher the temperature, the greater is the acoustic attenuation and the smaller the reception amplitude. Simulations were carried out at different temperatures with small changes in uniformity, and the uniformity was kept above 90%, so the effect of temperature on the test results can be ignored.
3) From the simulation results, it can be seen that in a homogeneous medium, the uniformity reaches above 90%, while in a non-homogeneous medium, the uniformity is below 90%. Therefore, the homogeneity of cement mortar can be tested using the scheme described in this paper, and a qualitative judgment can be derived.

REFERENCES

Chen Hui, Peng Junjian. Exploration of the use of different sizes of cuvettes in testing mixing uniformity by methyl violet method[J]. Feed and Animal Husbandry, 2019(12):42–45.

Guo Xinyi. Application of ultrasonic density meter in slurry density measurement of desulfurization system [J]. Science and Technology Innovation, 2020(01):27–28.

Wang Dan yang. Analysis of ultrasonic mud propagation attenuation characteristics [D]. Shenyang University of Technology, 2014.

Wei Lu. Research on ultrasonic slurry particle size and concentration detection system [D]. Yanshan University, 2016.

Zhou Sheng You, Xiong Gang, Shi Yong Gang. Research on the technique of measuring oil density by ultrasonic multiple echo reflection method[J]. Applied Acoustics, 2019,38(03):392–396.

Frontiers of Civil Engineering and Disaster Prevention and
Control – Yang & Rahman (Eds)
© 2023 The Author(s), ISBN: 978-1-032-31200-2

Research on mechanical properties of arch-pylon cable-stayed bridge

Wen-gang Ma*, Shi-xiang Hu, Ling Cong & Yu-qin Zhu
Nanjing Institute of Technology, Nanjing, China

ABSTRACT: The arch pylon cable-stayed bridge possesses the advantages of both arch and cable-stayed bridges, which exhibits not only reasonable mechanical properties, but also a good landscape effect. The properties of arch-pylon cable-stayed bridges with different alignments of arch axis were analyzed in this paper. Then, the mechanical properties of the bridge with different alignments and the corresponding applicable condition were put forward based on comparative analysis. Finally, the advantages and disadvantages of arch-pylon cable-stayed bridges with different alignments of the arch axis were described. The elliptical arch-pylon is a type of reasonable alignment that could be recommended for an arch-pylon cable-stayed bridge.

1 INTRODUCTION

The arch pylon cable-stayed bridge combines the advantage of the arch bridge and the cable-stayed bridge. This type of bridge embodies the beauty of the arch bridge and easily integrates into the natural environment and becomes a part of the natural scenery. At the same time, it also shows the slenderness and adaptability of the cable-stayed bridge.

The arch pylon cable-stayed bridge has been widely built since it appeared in public view for the first time in 1996, which was located in front of the new passenger terminal of Tokyo International Airport (Haneda Airport) named Haneda SkyArch (Shiomi & Nakamura 1996). The bridge is used to connect the original facilities of the airport and the post-expansion part. It requires the bridge to have a landmark function. In addition, because it is located around the airport, it has strict requirements on the height of the building. Since then, lots of this kind of bridge have been constructed, and corresponding research has been reported. The design of the Huashan bridge(arch-pylon) was described (Liu Hou-jun et al.) (Liu et al. 2020). The dynamic characteristics and seismic performance of double steel arch cable-stayed bridge with cable damage was researched (Huang Hua et al.) (Huang et al. 2021). The vibration mitigation and isolation for the cable-stayed bridge with double arch pylons were studied (Li Zhi-qiang et al.) (Li et al. 2018). The arch pylon cable-stayed bridges with different alignments of the arch axis show different mechanical characteristics and aesthetic considerations of the designer.

This paper presents the kinds of alignment of the arch axis, which reflect the apparently different mechanical properties based on the corresponding alignment, and also indicate the aesthetic design of the landmark building. Based on different surrounding environments and aesthetic considerations, designers adopt different alignments of the arch axis to make the arch pylon cable-stayed bridge present a colorful landscape effect. Summarizing the cable-stayed bridge with arch pylon, the alignment of the arch axis could be divided into four categories: the circular-arc arch-pylon, the parabola-shape arch-pylon, and the ellipse arch-pylon, the curve (circular-arc or parabola-shape) + linear arch-pylon. The alignment of the arch axis is adopted in most cable-stayed with arch-pylon based on the information of constructed projection or under construction.

---

*Corresponding Author: morgan@njit.edu.cn

DOI 10.1201/9781003308577-74

2 THE CIRCULAR-ARC ARCH-PYLON

The Haneda Skyarch was the representative of a circular-arc arch-pylon cable-stayed bridge (Figure 1). The arch span is 160 m with two separated decks apart from 80m. The distance between two arch springing must be much larger than the width of a deck if the alignment of the arch axis is adopted, as shown in Figure 2. Assuming the span of the arch is close to the dimension of the cross-section of the deck, there will be two questions that need to be solved. First, clearance between the outside of the deck and the bottom of the arch is not enough for the vehicle passing through because of the two mutual perpendiculars. Moreover, the space between cable and deck also will be affected no matter whether the single cable plane or double cable plane is employed. The dimension of the cross-section of the arch must be increased, which would lead to a larger dead-load in the event of adopting a larger span of arch, and the tied bar needs to be increased, which aims to balance the thrust of arch springing, as shown in Figure 2.

Therefore, the cable-stayed bridge with circular-arc arch-pylon is usually built in scenic spots or airports, with a small rise-span ratio and much smaller live-load.

Figure 1.   Completed bridge.   Figure 2.   Alignment of arch axis and cross arrangement.

3 THE PARABOLA-SHAPE ARCH-PYLON

The cable-stayed bridge adopting a parabola-shaped arch pylon could also be found in the completed projection, such as the Mau-Lo Hsi cable-stayed bridge located in Taiwan province of China, the Miho Museum bridge of Japan, and the Zhivopisny bridge striding the Moskva River, etc.

Compared with the circular-arc arch pylon, the alignment of the parabola-shaped arch pylon possesses much more variables by changing the alignment to suit the projection requirement.

However, the clearance below the cable could not be utilized completely if the alignment of the parabola shape is much flat. As shown in Figures 3 and 4, the Mau-Lo Hsi cable-stayed bridge only set four anchoring points to resolve insufficient clearance.

Figure 3.   Mau-Lo Hsi cable-stayed bridge.   Figure 4.   Mau-Lo Hsi cable-stayed bridge.

The Miho Museum Cable-Stayed Bridge (Figure 5(a)) is another famous landscape bridge located in the mountain nature reserve in Japan. The bridge, which is120m long, is used for pedestrians and cars. One end is connected to a 200-meter-long tunnel, and the other end is connected to the museum. The main arch pylon is 19 meters high and inclined 30 degrees. Owing to the excellent local rock formations, one end of the stay cable is anchored to the main beam, and the other end is anchored to the tunnel opening. This bridge has attracted much attention because of its perfect integration with the surrounding environment and won the "Outstanding Structure Award" issued by the International Bridge and Structural Engineering Association in 2002.

The Miho Museum Cable-Stayed Bridge also employs a parabola-shaped arch pylon with a 7.5m width, as shown in Figure 5(b). To eliminate the negative effects of cables, the additional triangle cross-section was applied, which was composed of one steel bar and two tense cables at the anchor point (Figure 5(c)).

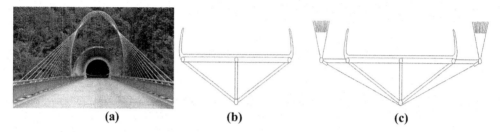

(a)　　　　　　　　　　**(b)**　　　　　　　　　　**(c)**

Figure 5.　Miho Museum cable-stayed bridge.

The Zhivopisny (Figure 6), located in Russia, also uses a secondary parabolic arch pylon. The bridge deck is 44 meters wide, and the height of the pylon is 102 meters, with the distance between the skewbacks of 138m. The skewbacks are set on both banks of the river, which would not affect the navigation under the bridge and could effectively reduce the cost of the foundation.

Figure 6.　Zhivopisny Bridge.

The above analysis shows that using a parabolic arch axis has certain limitations. If the pylon height is small and the distance between the skewbacks is small, there will be a problem of insufficient clearance of the bridge deck. If a high-order parabola is used, the height of the pylon increases sharply when the section of the beam widens. But the secondary parabola is a good choice, which meets the needs of the project and improves the landscape effect of the bridge structure when the distance between skewbacks is much larger than the main girder section.

4 THE ELLIPTICAL ARCH-PYLON

Owing to much more adaptability, the elliptical arch pylon is the most commonly adopted in arch-pylon cable-stayed bridge. The Zhi-Jiang Bridge (Figure 7) is a typical case that stretches across the Qian-tang River in Hangzhou (Cao et al. 2016).

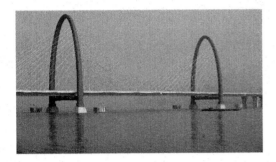

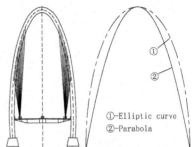

①-Elliptic curve
②-Parabola

Figure 7.   Zhi-Jiang Bridge.   Figure 8.   Comparison of ellipse and parabola.

The advantage of the elliptical arch pylon is more obvious by comparing the alignment of the parabola-shaped and elliptical arch pylon. As shown in Figure 8, the angle between the pylon and bridge deck is much bigger in an elliptical arch pylon than parabola-shape pylon with the same width of deck and height of the pylon, which means the cable has much fewer influences on the clearance of the bridge deck, especially when the deck is close to the minor axis of the ellipse. The adverse effect could be ignored while the first anchoring point between cable and pylon has a certain distance (Figure 8). Furthermore, the elliptical arch-pylon bridge possesses a larger longitudinal angle between cable and deck, decreasing the horizontal force at the skewback.

5 THE CURVE (CIRCULAR-ARC OR ELLIPTICAL) + LINEAR ARCH-PYLON

Based on aesthetic consideration, some designers take the curve (circular-arc or elliptical) + line as the alignment of the arch-pylon. According to different inclination angles of line, the alignment of the arch-pylon could be classified as curve + inward straight line (Figure 9, Figure 10) and curve + outward straight line (Figure 11). The former also could be subdivided into two types according to different anchoring points, which lead to quite different mechanical behavior.

The alignment of the elliptical segment + arch segment + inward straight line has been adopted by San-Hao Bridge (Figure 9). The stay cables were arranged along the elliptical section near to end, the whole arch segment, and the partial straight-line segment. Owing to the arrangement of stay cables, the horizontal component of the stay cable has balanced the corresponding axial force of the pylon, but the non-stayed cable segment is no longer mainly bearing the compression, but a bending moment, so the arch can't give full play to its advantage of pressure.

The Tai-Jiang Bridge (Figure 10) adopts the alignment of arch segment + transition segment + inward straight line, which is seemingly similar to the San-Hao Bridge as far as shape, but the mechanical characteristics are completely different due to the different arrangement form of stay cables. The non-traditional arrangement of the cable plane was adopted based on aesthetic consideration. One end of the cable is anchored at the highest point on the pylon. The other end is anchored at the nearest place on the horizontal center of the bridge deck (the farthest place according to traditional arrangement). Such arrangement would be bound to affect the clearance of the deck. There are two ways to resolve the aforementioned problems: one is to increase the height of the pylon, and the other is to anchor the stay cable at the tower top as much as possible.

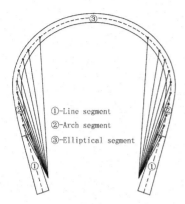

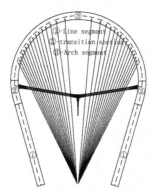

Figure 9. San-Hao Bridge.   Figure 10. Tai-Jiang Bridge.   Figure 11. MHH Bridge.

The Tai-Jiang Bridge employs the second method, namely anchoring the stay cable at the top of the arc segment; meanwhile, the transition segment + inward straight line was adopted for the remainder due to the small distance between skewback. The advantage of compression of the arch could be given full play under load with the alignment of arch segment + transition segment + inward straight line, but the horizontal force of skewback can't be balanced due to the inward straight line. In order to solve the imbalance, the transverse cable was set, and the vertical cable was also employed, which connected the main girder at one end and the transverse cable at another end, and the balance of horizontal force of skewback could be implemented by adjusting the force of vertical cable.

The MHH Bridge (Figure 11), located in the USA, introduces the alignment of a parabolic+ outward straight line (Scott et al. 2009). The Santiago Calatrava Bridge, built in Reggio Emilia of Italy, also adopted a similar alignment of arch axis: elliptical + outward straight line (Stefania 2011). Although the curve form is different, the force form and design purpose are the same. These two bridges also adopt a similar arrangement form of cable as the Tai-Jiang Bridge (namely, anchor the stay cable at the top of the arc segment). The same problem that the stayed-cable affects the clearance of the deck will exist if the arch pylon is low. Therefore, heightening the pylon not only solves the clearance, but also promotes an aesthetic effect. The bending moment would be very large at the non-stayed cable segment as the component of cable force is not distributed along the axis if the whole pylon adopts a curve. So, the above-mentioned two bridges employ the straight line instead of the curve at the non-stayed cable segment and ensure the direction of the component of cable force is consistent with the straight line to decrease the bending moment effectively.

6 CONCLUSIONS

Based on the above analysis in this paper, the following main conclusions are obtained:

1. The arch-pylon cable-stayed bridge, as an ancient and modern bridge style, combines the merit of a cable-stayed bridge and arch bridge, which not only unfold gentler configuration but also supply sufficient traffic capacity.
2. The clearance would be affected, and some necessary measures should be employed when the alignment of the arch axis adopts the circular arc or parabola shape.
3. The elliptical arch-pylon is a type of reasonable alignment of the arch axis based on the comparative analysis of various alignments of the arch axis.
4. The alignment of curve + inward straight line possesses some unreasonable mechanical reaction, and the curve + outward straight line could be recommended.

ACKNOWLEDGMENTS

The project was supported by the Scientific Research Foundation for the High-level Personnel of Nanjing Institute of Technology (Grant No. YKJ201984) and the Open Research Fund of NJIT Institute of Industrial Economy and Innovation Management (No. JGKC202004 and No. JGKC202006) and the Natural Science Foundation of the Higher Education Institutions of Jiangsu Province, China (No. 20KJB580002).

REFERENCES

Cao Yi-shan, Xu Qiang, Li Ke-bin (2016). Experimental study on model of long-span cable-stayed bridge with arch-shaped steel pylon[J]. Journal of Highway and Transportation Research and Development, 33(5):73–77.

Huang Hua, Bai Bao, Zhou Wen-jie, et al. (2021). Analysis of dynamic characteristic and seismic performance of V-shape double steel arch cable-stayed bridge with cable damage[J]. Journal of Chang'an university, 41(2):114–124.

Li Zhi-qiang, Wang Hao-ming, Wang Da-qian. Study of vibration mitigation and isolation for long-span cable-stayed bridge with double arch pylons based on friction pendulum bearings[J]. World Bridges, 2018, 46(6): 36–40.

Liu Hou-jun, Zheng Guo-fu, Lu Qin-fen (2020). Design of main bridge of Huashan Bridge in Nanjing[J]. Bridge construction, 50(S2):82–92.

Masaki Shiomi, Yoshio Nakamura (1996). Aesthetic Design of Haneda Sky Arch[J]. Transportation Research Record: Journal of the Transportation Research Board, (1549):35–41.

Scott Lomax, Charles E. Quade (2009). Margaret Hunt Hill (Woodall Rodgers) Bridge. Structures 2009: Don't Mess with Structural Engineers, Texas, 164–173.

Stefania Palaoro (2011). Arch Bridge: design-construction-perception. University of Trento, Ph.D.

Frontiers of Civil Engineering and Disaster Prevention and
Control – Yang & Rahman (Eds)
© 2023 The Author(s), ISBN: 978-1-032-31200-2

Research on test method and influencing factors of adhesion between asphalt and aggregate

Yongzhen Li, Shengjie Zhou & Ning Zhao*
Shandong Transportation Institute, Shandong Jinan, P.R. China

Qingtao Zhang
Shandong Hi-speed Company Limited., Shandong Jinan, P.R. China

ABSTRACT: In this paper, the water immersion method, image processing method, and contact angle method are used to study the adhesion between different asphalt and aggregate to provide a quantitative, accurate, and rapid method to predict the water stability of asphalt mixture. The correlation analysis between the adhesion grade, peeling rate, adhesion work, and asphalt mixture freeze-thaw splitting test is carried out to explore the influence relationship between the adhesion and asphalt aggregate index. The results show that the water immersion method with image processing method and contact angle method can accurately and quantitatively characterize the adhesion between the base asphalt and the aggregate, which can be used as the test method to evaluate the water stability of the base asphalt mixture. However, there are still some disadvantages in the evaluation of the adhesion between modified asphalt and aggregate, which means further research and development of a wider range of methods are still needed. In conclusion, the higher the SiO_2 content of aggregate, the greater the spalling rate of asphalt on the aggregate surface, the smaller the adhesion work, and the worse the adhesion. The viscosity of asphalt at 135°C has a certain influence on the adhesion, but not as much as the chemical properties of aggregate.

1 INTRODUCTION

Asphalt pavement has become the most important form of pavement in my country due to its excellent performance, easy maintenance, and good flatness. The performance of asphalt pavement is closely related to the performance of asphalt mixture (Zhou 2015). The various properties of asphalt and aggregates and the adhesion between them determine the various properties of asphalt mixtures. The current specifications (Shen et al. 2004) has clear testing methods and evaluation standards for asphalt and aggregate technical indicators. Although there is a testing method for asphalt-aggregate adhesion, it is a semi-quantitative test based on human observation and judgment, and the quantification is not accurate (Liu 2021; Wang 2018; Zhang 2021). The main methods for characterizing the adhesion between asphalt and aggregates are the boiling water immersion method, water immersion Marshall test, and asphalt mixture freeze-thaw splitting test. In actual engineering applications, the boiled water immersion method has poor relevance to the actual engineering and is too different from the actual force of the asphalt mixture (Fu et al. 2021; Zhao et al. 2021)[0]. Engineers all use the boiled water immersion method as a simple reference and do not use it as a key indicator to predict the performance of the asphalt mixture. Therefore, in order to better characterize the adhesion characteristics of asphalt-aggregate and integrate and guide actual engineering, the use of high-tech experimental evaluation methods to develop a rapid evaluation method for the adhesion characteristics of asphalt-aggregate has a great impact on the quality of

---

*Corresponding Author: liyz314@163.com

the project (Industry Standard-Traffic. Highway Engineering Asphalt and Asphalt Mixture Test Regulations (Industry Standard-Traffic 2011)[0]. Assurance is of great significance.

This paper uses the water immersion method combined with the image processing method and contact angle method to study the adhesion between different asphalt and aggregates, analyze the correlation between adhesion grade, peeling rate, adhesion work, and other indexes of asphalt mixture freeze-thaw splitting test, and explore the influence relationship between adhesion and asphalt and aggregate indexes. It is expected to provide an adhesion test method that can quantitatively accurately and quickly predict the water stability of asphalt mixtures for practical engineering applications.

2 RAW MATERIALS AND TEST METHODS

2.1 *Raw materials*

Four kinds of asphalt are selected, namely No. 50 asphalt, No. 70 asphalt, No. 90 asphalt, and SBS modified asphalt. The technical indicators of the four asphalts are shown in Table 1.

Table 1. Asphalt technical indicators.

Projects	50#	70#	90#	SBS
Penetration 25°C, 5s, 100g (0.1mm)	51	69	84	57
Softening Point (R&B) (°C)	49.0	48.0	45.0	88.0
135°C Viscosity (Pa·s)	0.45	0.40	0.38	1.6

According to the specification, SiO_2 content is mainly used to judge the acidity and alkalinity of aggregates. Aggregates with SiO_2 content less than 52% are called alkaline aggregates, and aggregates with SiO_2 content between 52% and 65% are called neutral aggregates. Aggregates with SiO_2 content greater than 65% are called acid aggregates. Three types of aggregates with different chemical properties are selected, namely limestone, basalt, and granite, respectively. The content of SiO_2 in the three aggregates is shown in Table 2. Limestone is alkaline, basalt is neutral, and granite is acidic.

Table 2. SiO_2 content of three aggregates.

Aggregate type	SiO_2 /%
Limestone	15.50
Basalt	53.99
Granite	72.77

2.2 *Experiment method*

2.2.1 *Water immersion*

The water immersion test adopted the method in literature (Industry Standard-Traffic. Highway Engineering Asphalt and Asphalt Mixture Test Regulations. (Industry Standard-Traffic 2011). The image collection of the test results of the water immersion method and a certain technical processing method was used to obtain the exfoliation rate of the asphalt on the aggregate. The specific image collection and processing methods refer to the research method of Fan Liang (Fan et al. 2011).

2.2.2 Contact angle test

The contact angle method adopted the lying drop method and the DSA 100 contact angle measuring instrument from KRUSS, Germany. The test method and data processing method refer to the test method of Liu Yanjun (Liu & Zhang 2017), and the adhesion work between asphalt and aggregate was calculated.

2.2.3 Freeze-thaw split test

Freeze-thaw split test method in reference (Industry Standard-Traffic. Highway Engineering Asphalt and Asphalt Mixture Test Regulations. (Industry Standard-Traffic 2011).

3 ADHESION TEST RESULTS AND DISCUSSION

3.1 Evaluation of asphalt-aggregate adhesion by water immersion method combined with image processing method

Using the water immersion method, one was to visually judge the adhesion level of asphalt and aggregate by three technicians, and the other was to use image processing technology to calculate the peeling rate of asphalt on the surface of the aggregate. The results are shown in Table 3.

Table 3. Adhesion of asphalt to different aggregates.

Project	Visually judge the adhesion level			Image processing method to calculate the peeling rate		
	Limestone	Basalt	Granite	Limestone	Basalt	Granite
50#	4	3	3	1.7%	11.0%	22.4%
70#	4	3	3	2.1%	25.3%	34.2%
90#	5	4	2	0.7%	14.0%	33.5%
SBS	5	4	4	0.2%	1.4%	5.7%

It could be seen from Table 1 that limestone and basalt have the same adhesion levels as SBS and No. 90 asphalt when using the visual inspection method, and the adhesion levels of No. 50 and No. 70 asphalts were the same as those of the three aggregates. When the image processing method was used, each asphalt had the same adhesion levels as the exfoliation rate of the material had a clear value, which can accurately and quantitatively characterize the adhesion of various asphalt and aggregate. This demonstrated that the visual judgment method could not well compare the adhesion difference between the same kind of aggregate and different asphalt. The water immersion method combined with the image processing method was used to obtain the peeling rate of the asphalt on the aggregate surface, which could accurately and quantitatively characterize the adhesion between asphalt and aggregate.

3.2 Evaluation of asphalt-aggregate adhesion by contact angle method

Using the lying drop method, the contact angles of four asphalts and three aggregates were measured, and the adhesion work was calculated to evaluate the adhesion of asphalt to aggregates quantitatively. The results are shown in Table 4.

From Table 2, under the same asphalt condition, it could be found that the adhesion results of different aggregates were limestone>basalt>granite. Under the same aggregate condition, the adhesion result was 50#>70#>90#>SBS. The adhesion work reflected the work required to separate the asphalt from the aggregate surface to produce a new interface. Generally, the greater the adhesion work, the better the adhesion between these two (Wei 2008). The higher the asphalt grade, the better

Table 4. The contact angle and adhesion work of asphalt and aggregate.

Asphalt and aggregate types	Adhesion work/mJ·m^{-2}		
	Limestone	Basalt	granite
50#	47.5	46.8	46.5
70#	46.2	45.8	45.6
90#	45.8	45.4	45.2
SBS	42.6	40.7	40.3

the wettability of the asphalt on the surface of the aggregate, the less work required for separation, and the worse the adhesion. The adhesion work data can quantitatively characterize the excellent adhesion between the asphalt and the aggregate.

3.3 Correlation analysis with a freeze-thaw split test

Analyze the linear correlation between the adhesion grade, peeling rate, and adhesion power obtained in the test and the corresponding asphalt and aggregate freeze-thaw splitting strength ratio (TSR) of the asphalt mixture. The equations and correlation coefficients obtained are shown in Table 5.

Table 5. Correlation analysis between three adhesion indexes and freeze-thaw splitting strength ratio (TSR).

Asphalt type	Adhesion level		Peeling rate		Adhesion work	
	Linear equation	Correlation coefficient	Linear equation	Correlation coefficient R^2	Linear equation	Correlation coefficient R^2
50#	y=7.25x+58.4	R^2=0.57	y=-0.53x+88.78	R^2=0.98	y=9.98x-385.54	R^2=0.87
70#	y=15.15x+34.9	R^2=0.75	y=0.59x+97.49	R^2=0.94	y=32x-1381.9	R^2=0.94
90#	y=4.09x+67.5	R^2=0.93	y =-0.39x+88.64	R^2=0.97	y=4.04x-1010.6	R^2=0.98
SBS	y=0.8x+86.8	R^2=0.02	y=-1.1x+92.91	R^2=0.67	y=0.32x+ 76.736	R^2=0.04

It can be seen from Table 5 that for the matrix asphalt, the linear correlation coefficient of the ratio of the peeling rate and the adhesion work to the freeze-thaw splitting strength is far greater than the adhesion grade. It can be concluded that the peeling rate and the adhesion function are appropriate ways to characterize the adhesion of asphalt-aggregate. The two adhesion test methods of the immersion + image processing method and the contact angle method have a high correlation with the water stability of the asphalt mixture. The water immersion + image processing method and contact angle method can be used to predict the pros and cons of the asphalt mixture's water stability.

For SBS modified asphalt, the linear correlation coefficients of adhesion grade, peeling rate, adhesion work index, and freeze-thaw splitting strength ratio are relatively small, indicating that the three test methods have certain drawbacks in evaluating the adhesion of modified asphalt to aggregates. As the adhesion between asphalt and aggregate is related to the physical and chemical properties of asphalt and aggregate, the evaluation of the adhesion between SBS modified asphalt and aggregate requires research on the asphalt components, aggregate chemistry, and interfacial adhesion mechanisms.

4 ANALYSIS OF INFLUENCING FACTORS OF ADHESION

4.1 Influence of aggregate SiO₂ content

Analyze the influence of aggregate SiO_2 content on the results of the two adhesion tests. Take the aggregate SiO_2 content as the x-axis, and use the exfoliation rate and adhesion work as the y-axis. Linear fitting is used to obtain the linear equation and correlation coefficient, as shown in Table 6.

Table 6. Linear equation of exfoliation rate, adhesion work, and aggregate SiO_2 content

Label	Exfoliation rate and aggregate		Adhesion power and aggregate	
50#	y = 0.34x - 4.58	$R^2 = 0.94$	y = -0.02x + 47.75	$R^2 = 0.9997$
70#	y = 0.57x - 6.39	$R^2 = 0.997$	y = -0.01x + 46.34	$R^2 = 0.996$
90#	y = 0.54x - 9.55	$R^2 = 0.91$	y = -0.01x + 45.90	$R^2 = 0.9995$
SBS	y = 0.09x - 1.64	$R^2 = 0.76$	y = -0.04x + 43.18	$R^2 = 0.97$

It can be seen from Figure 1 and Table 4 that the higher the SiO_2 content of the aggregate, the greater the exfoliation rate of the asphalt on the surface of the aggregate. The correlation coefficients of the linear equation between the exfoliation rate of the different asphalt on the surface of the aggregate and the SiO_2 content of the aggregate are, respectively: No. 50 pitch 0.94, No. 70 pitch 0.997, No. 90 pitch 0.91, SBS pitch 0.76.

It has been shown that the aggregate SiO_2 content has a high correlation with the adhesion test results, and the correlation coefficient is basically greater than 0.9. The aggregate SiO_2 content reflects the acid-based chemical properties of the aggregate, indicating that the adhesion and adhesion between the asphalt and the aggregate are related to each other. The acid-based chemical properties of the aggregate have a good correlation.

4.2 Influence of asphalt viscosity at 135°C

The influence of asphalt 135°C on the results of the two adhesion tests is analyzed. The viscosity of asphalt at 135°C is taken as the x-axis, and the peeling rate and adhesion work as the y-axis. Linear fitting is used to obtain the linear equation and correlation coefficient, as shown in Table 7.

Table 7. Linear equation of peeling rate, adhesion work, and viscosity content of asphalt at 135°C.

Type of material	Spalling rate and 135°C viscosity of asphalt		Linear equation between adhesion power and asphalt viscosity at 135°C	
Limestone	y = -1.03x + 1.89	$R^2 = 0.51$	y = -3.17x + 47.75	$R^2 = 0.84$
Basalt	y = -13.10x + 22.18	$R^2 = 0.63$	y = -4.37x + 47.76	$R^2 = 0.93$
Granite	y = -20.85x + 38.68	$R^2 = 0.87$	y = -4.50x + 47.60	$R^2 = 0.95$

It can be seen from Figure 3 and Table 5 that the correlation coefficients between the spalling rate and the viscosity of asphalt at 135°C are limestone 0.51, basalt 0.63, and granite 0.87. As a whole, the viscosity of asphalt increases, and the adhesion to aggregates is better.

As shown in Table 7, the correlation coefficients between the adhesion work and the viscosity of asphalt at 135°C are limestone 0.84, basalt 0.93, and granite 0.95.

It can be concluded that the correlation between the work of adhesion and the viscosity of asphalt at 135°C is greater than the peeling rate, and the correlation between the peeling rate and the viscosity of asphalt at 135°C is poor. This shows that the viscosity of asphalt at 135°C has a certain effect on the adhesion, but not as good as the chemical properties of aggregate.

5 CONCLUSION

Based on the results and discussions presented above, the conclusions are obtained as below:

(1) The two indexes of peeling rate and adhesion work can accurately and quantitatively characterize the adhesion of different asphalts and aggregates.
(2) The water immersion method combined with the image processing method and the contact angle method can accurately and quantitatively characterize the adhesion between the base asphalt and the aggregate and can be used as a test method to evaluate the water stability of the base asphalt mixture.
(3) The two methods of water immersion + image processing method and contact angle method have certain drawbacks when evaluating the adhesion of modified asphalt and aggregates. Further research and development of methods with a wider scope of applications are still needed.
(4) The higher the SiO_2 content of the aggregate, the greater the exfoliation rate of the asphalt on the surface of the aggregate, the lower the adhesion work, and the worse the adhesion. The viscosity of asphalt at 135°C has a certain effect on the adhesion, but it is not as good as the chemical properties of the aggregate.

REFERENCES

Fan Liang, Zhang Yuzhen, Wang Lin. Preliminary Study on Image Analysis Method of Asphalt Adhesion Test[J]. Highway, 2011(12): 151–154.

Fu Jun, Li Zhongjie, Yang Rongqi, Jia Dawei, Ding Qingjun Analysis of nondominant structural integrity failure characteristics of asphalt pavement [J] Journal of Chang'an University (NATURAL SCIENCE EDITION), 2021,41 (04): 11–20

Industry Standard-Traffic. Highway Engineering Asphalt and Asphalt Mixture Test Regulations. [S], 2011.

Liu Hongcheng, Tang Xiaodan, Kan Tao, Qi Guangzhi Study on the influence of different anti-stripping agents on the pavement performance of granite asphalt mixture [J] Sino foreign highway, 2021,41 (04): 366–370

Liu Yanjun, Zhang Yuzhen. Wetting performance and adhesion work of different types of asphalt and aggregates[J]. Journal of Petrochemical Universities, 2017, 30(3): 1–8.

Shen Jinan, Li Fupu, Chen Jing. JTG F40-2004. Technical Specification for Highway Asphalt Pavement Construction. [S]. Beijing: People's Communications Publishing House, 2004.

Wang Ge. Evaluation of Asphalt-Aggregate Interface Bonding Performance and Influencing Factors[D]. Shandong Jianzhu University, 2018.

Wei Jianming. Research on the surface free energy of asphalt and aggregates and the diffusion of water in asphalt[D]. China University of Petroleum, 2008.

Zhang Xu Quantitative evaluation and analysis of attenuation law of asphalt stone interface adhesion effect during asphalt mixture service [J] Highway, 2021,66 (07): 267–270

Zhao Wenkun, Feng Hao, Zhu Yujie Experimental study on the influence of granite manufactured sand on the water stability of asphalt mixture [J] Sino foreign highway, 2021,41 (03): 343–347

Zhou Lan. Research on Performance Evaluation and Prediction of Expressway Asphalt Pavement[D]. Southeast University, 2015.

*Frontiers of Civil Engineering and Disaster Prevention and
Control – Yang & Rahman (Eds)
© 2023 The Author(s), ISBN: 978-1-032-31200-2*

Research on mechanical properties of recycled broken ceramic tile concrete

Chen Shan, Liang Xiaoguang* & Huang He
College of Civil Engineering and Architecture, Nanning University, Nanning, Guangxi, China

ABSTRACT: Recycled broken tile concrete was prepared with different broken tile replacement rates. Compared with ordinary concrete, the effect of different replacement rates on the workability and compressive strength of concrete was studied. The results show that: compared with ordinary concrete, recycled broken tile concrete has higher compressive strength. Under the effect of the same water-cement ratio, different replacement rates have different effects on the strength of concrete. In the test, increasing the replacement rate of broken tiles can effectively increase the compressive strength of recycled tile concrete test blocks. The best replacement rate in the test is 66%. In the case of the same broken tile replacement rate, if the water-cement ratio is reduced, the compressive strength of the recycled tile concrete test block can be improved. To effectively use broken tiles as recycled materials, the replacement rate of broken tiles can be increased, while the water-cement ratio in concrete can be reduced, and the strength of concrete test blocks can be improved. Recycled broken tile concrete has good advantages in material utilization and compressive strength, and it is expected to be fully used in engineering practice.

1 INTRODUCTION

With the development of my country's economic construction, various buildings have been increasing, and many buildings have undergone frequent demolition and reconstruction. While the demand for concrete is increasing, how to recycle and reuse waste building materials has become a hot topic now. Since the 1990s, China has become the world's number one ceramic producer (Hou & Zeng 2005). As a decoration material, ceramic tiles are frequently replaced in demolition and reassembly. Broken tiles are a kind of non-degradable solid waste material. Burying it out not only wastes resources, but also pollutes the environment. If it can be recycled as aggregates of recycled concrete for reuse, it can turn waste into treasure and save resources (Khaloo et al. 2008). Luan (Luan et al., 2015) focuses on the production of building materials from ceramic waste and the classification of ceramic waste, which is widely used in construction engineering. Liu (Liu et al. 2014) has studied that mixing waste ceramics will cause the fluidity of concrete to deteriorate, but the toughness is better than that of ordinary concrete. Pei (Pei et al. 2017)studied that replacing natural aggregate by waste broken ceramic tile in the same proportion can improve the strength of concrete and reduce the shrinkage rate. Cheng (Cheng et al. 2013) studied the shrinkage of ceramic tile coarse aggregate concrete, and the results showed that ceramic tile recycled coarse and fine aggregates can improve the shrinkage performance of concrete than natural gravel aggregates. The research of Qiao (Qiao et al. 2019) shows that different tile replacement rates have a greater impact on the freeze-thaw cycle resistance of recycled concrete. With the increase in the replacement rate, the freeze-thaw cycle resistance of recycled concrete decreases drastically. Binici (Binici 2002) studied the chloride ion permeability, abrasion resistance, and workability of recycled concrete with different substitution rates than ordinary concrete. Nie (Nie et al. 2014) prepared recycled

---

*Corresponding Author: 292019782@qq.com

concrete mortar by grinding waste ceramic tile into grits to replace natural sand and studied the strength change law of recycled masonry mortar under different conditions. Bignozzi (Bignozzi & Andrea 2012) replaces natural gravel with 100% waste tiles, and the prepared waste tile coarse aggregate concrete has reduced compressive strength, split tensile strength, and flexural strength by 3.8%, 18.2%, and 6%, respectively, while having lower Tension-compression ratio and good working performance. Meanwhile, Wajeeha Mahmood (Wajeeha et al. 2021) considered the durability comparison of recycled concrete and ordinary concrete under the same water-cement ratio of 0.43 and different replacement rates of recycled aggregate. Correia (Correia et al. 2006) also proved that discarded ceramic tiles used as coarse aggregate instead of gravel have little impact on the mechanical properties of concrete and have economic advantages. Based on the research results of previous scholars, this paper uses broken ceramic tiles as recycled aggregates, replaces the broken stone aggregates in ordinary concrete with different replacement rates, and prepares recycled broken ceramic tile concrete. The issue of tile recycling and reuse provides a basic basis.

2 EXPERIMENT

To study the mechanical properties of recycled broken tile concrete, this paper aims at a water-cement ratio of 1.25 and designs three sets of tests for research and a set of ordinary concrete as a comparative test. Among them, the recycled concrete coarse aggregate uses broken tile blocks with a particle size within 40 mm and has not been sieved. The sand used for mixing is natural river sand. The cement is Conch brand ordinary Portland cement, the strength grade is 32.5 MPa, and the measured strength is 41.0 MPa. The water used for mixing is the domestic water used in Nanning City. No additives are used to prepare concrete test blocks; the test block size is $150 \times 150 \times 150$ mm^3.

In this experiment, a group of ordinary concrete B1 groups with the same mix ratio was prepared as the comparison group. Groups A1-A3 use different proportions of broken tiles as coarse aggregate instead of crushed stones without screening treatment under the same material ratio. Among them, Group A1 has a 33% replacement rate, Group A2 has a 66% replacement rate, and Group A3 has a 100% replacement rate. Group A4 is a concrete test block prepared with a water-cement ratio of 1.00 and a broken tile replacement rate of 33%. It is used to compare Group A1 at the same replacement rate and different water-cement ratios. After calculation, the initial mix ratio of concrete material used is shown in Table 1.

Table 1.   Material parameter of preliminary mix ratio (kg).

Group	Water	Cement	Sand	Gravel	Broken tiles
A1	4.00	3.20	14.74	16.12	7.94
A2	4.00	3.20	14.74	8.18	15.88
A3	4.00	3.20	14.74	0.00	24.06
A4	4.00	4.00	14.44	15.79	7.77
B1 (Comparison group)	4.00	3.20	15.50	25.30	0

Because the method of adding additives is not used to prepare concrete, the water ash is relatively large, and the tiles are not subjected to water immersion treatment, resulting in serious workability that does not meet the test requirements in the actual operation process, as shown in Figure 1. After repeated adjustments to the workability of concrete, the final material usage was determined, as shown in Table 2.

(a)   (b)

Figure 1. Concrete mixing test photos.

Table 2. Material parameters after adjustment and workability (kg).

Group	Water	Cement	Sand	Gravel	Broken tiles
A1	5.10	4.08	14.74	16.12	7.94
A2	6.60	5.28	16.24	8.18	15.88
A3	6.70	5.36	14.74	0.00	24.06
A4	5.10	5.10	14.44	15.79	7.77
B1 (Comparison group)	4.60	3.68	15.50	25.30	0

3 RESULTS AND DISCUSSION

In this experiment, the YAW-3000 pressure testing machine in the laboratory of the College of Civil Engineering and Architecture of Nanning University was used to carry out the loading test of each group of test blocks in turn. And record the load peak value of each group of specimens at the time of failure, and the maximum load value obtained is shown in Table 3.

Table 3. Maximum load value of each group of Specimens (kN).

Group	Specimens 1	Specimens 2	Specimens 3	Average value
A1	221.2	228.5	210.8	220.2
A2	238.4	236.7	240.9	238.7
A3	244.0	252.8	269.1	255.3
A4	279.6	266.2	273.6	273.1
B1(Comparison group)	165.6	178.7	173.8	172.7

Because the curing age exceeds the prescribed 28 days, to better compare with the specifications, the data is converted according to the requirements of the specifications, and the calculated 28-day compressive strength value is shown in Table 4.

It can be seen from Table 4 that when the water-cement ratio of the comparative group B1 is 1.25, the measured maximum strength is 7.9MPa and the minimum strength is 7.4MPa, which basically meets the requirements of the ordinary concrete mix ratio design regulations. The results show that

571

Table 4. Compressive strength value of each group of specimens (MPa).

Group	Specimens 1	Specimens 2	Specimens 3	Average value	Equivalent 28-day intensity
A1	9.8	10.2	9.4	9.8	7.7
A2	10.6	10.5	10.7	10.6	8.8
A3	10.8	11.2	12.0	11.3	9.0
A4	12.4	11.8	12.2	12.1	10.0
B1 (Comparison group)	7.4	7.9	7.7	7.7	6.1

the water-cement ratio is too large and the water consumption is large, which causes the cement slurry to have more pores in the test block due to the existence of free water during the setting and hardening process, so the compressive strength is low.

Group A1 has a maximum strength of 10.2 MPa, a minimum strength of 9.4 MPa, and an average strength of 9.8 MPa, which is equivalent to a 28-day strength of 7.7 MPa. In the case of the same ratio of other materials, group A1 replaced the broken stones in group B1 with 33% of broken tiles. The final measured compressive strength value of 7.7MPa is higher than that of group B1 of 6.1MPa, and the strength increase rate is 26.2%. The reason may be that dry, broken tiles are used. In the case of the same water consumption, because the tiles have more pores, they will absorb part of the water in the mixing water. This leads to the fact that the water that contacts the cement for hydration reaction is less than the added water. At the same time, the actual water-cement ratio is smaller than the originally designed water-cement ratio. Therefore, when the water-cement ratio is too large, broken tiles absorb water, which indirectly reduces the water consumption and reduces the water-cement ratio, which is beneficial to improving the strength of concrete.

In the same way, compared with Groups B1, A2, A3, and B1 replaced the gravel in ordinary concrete with 66% and 100% replacement rates, respectively. The 28-day compressive strength value has been greatly improved by 44.3% and 47.5%. However, when the replacement rate changed from 66% to 100%, the intensity growth increased from 44.3% to 47.5%, and the growth rate slowed down. The reason for this result may be that when the replacement rate is 66%, the capacity of broken tiles to absorb water is close to saturation, and the water-cement ratio in cement concrete remains basically unchanged. When the replacement rate becomes 100%, the water-cement ratio in the concrete is basically the same as when the original replacement rate was 66%. Therefore, although the test was carried out with two different replacement rates, the compressive strength was basically close.

Comparing the A4 group and the A1 group, under the same broken tile replacement rate, the compressive strength of the A4 group with smaller water ash is significantly higher than the compressive strength of the A1 group by 29.9%. This shows that in the study of the mechanical properties of recycled broken tile concrete, the water-cement ratio also plays a decisive role in the compressive strength of concrete. The smaller the water-cement ratio, the higher the compressive strength of recycled broken tile concrete.

Comparing group A4 with group B1, while the replacement rate of broken tiles increased from 0% to 33%, the water-cement ratio decreased from 1.25 to 1.00. The results show that the compressive strength value of the A4 group is 63.9% higher than the compressive strength value of the B1 group. A conclusion can be drawn from this: we can prepare a set of ideal recycled broken tile concrete test blocks to achieve a better comprehensive mechanical property while increasing the replacement rate of broken tiles, appropriately reducing the water-cement ratio. At the same time, to meet the requirements of the workability of the concrete mixture, admixtures such as water-reducing agents can be added when mixing the concrete to improve the workability and meet the construction requirements.

(a)   (b)

Figure 2. Photo of compression test.

As shown in Figure 2, when the test block is subjected to the compressive strength test, the outer skin is easy to peel off and form flake-like fragments. This is because the strength of the cement mortar on the surface and corners of the test block is low, and it is exposed to greater external pressure. When the maximum stress that can be resisted is exceeded, the failure is given priority. After the crack appears on the surface of the test block, the maximum load value of about 20kN is continued to increase, and the test piece is completely destroyed.

(a)   (b)

Figure 3. Photo of the damaged section of the specimen.

As shown in Figure 3, it can be seen from the broken section of the test block that because the broken tile is a sheet-like material and one side is a smooth surface, the bonding force is weak during the contact and hardening process with the cement slurry, so when external force acts, damage occurs first from this part. And there are a lot of exposed broken tiles in the damaged section without cement grout. The surface of the broken concrete block is smooth, which is one reason for the low compressive strength of the recycled broken ceramic tile concrete test block. In addition, there are many tiny pores in the cement mortar, which are caused by the excessively large water-cement ratio. Therefore, in the later preparation of the test block, an admixture can be considered to effectively control the water-cement ratio and improve the compressive strength of the recycled tile concrete test block.

4 CONCLUSION

(1) Under the same water-cement ratio, different replacement rates have different effects on strength. Increasing the replacement rate of broken tiles is beneficial to increase the compressive strength of recycled tile concrete test blocks, and the best replacement rate is 66%.
(2) In the case of the same replacement rate, reducing the water-cement ratio is beneficial to increase the compressive strength of the recycled ceramic tile concrete test block.
(3) In order to achieve the purpose of comprehensive utilization, it is possible to reduce the water-cement ratio and increase the strength of the concrete test block while increasing the replacement rate of broken tiles.

ACKNOWLEDGMENTS

This work is financially supported by the Professor Cultivation Project of Nanning University (2018JSGC14), the Scientific Research Project of Nanning University (2018XJ26), the Basic Research Ability Improvement Project for Young and Middle-aged Teachers in Guangxi Universities (2021KY1808), and the Nanning Excellent Young Scientist Program and Guangxi Beibu Gulf Economic Zone Major Talent Program (RC20190208).

REFERENCES

Bignozzi M C, Andrea Saccani. Ceramic waste as aggregate and supplementary cementing material: A combined action to contrast alkali-silica reaction (ASR). Cement and Concrete Composites, 2012,34(10):1141–1148.

Binici H. Strength of concrete made from crushed concrete coarse aggregate. Materials Journal.2002,1(8):283–288.

Cheng Y H, Huang F, Li Y Z. Experimental study on shrinkage performance of waste ceramic aggregate concrete. Highway, 2013(08):258–262.

Correia J R, Brito J D, Pereira A S. Effects on concrete durability of using ceramic aggregates. Materials & Structures,2006,39(2):169–177.

Hou L G, Zeng L K. Current situation of comprehensive utilization of ceramic waste. China Ceramic Industry, 2005(04): 41–44.

Khaloo A R, Dehestani M, Rahmatabadi P. Mechanical properties of concrete containing a high volume of tire–rubber particles. Waste Management,2008,28(12).

Liu J H, Gu X S, Liu F L. Experimental study on the performance of waste ceramic recycled mixed sand concrete[J]. Construction Technology, 2014, 43(09): 52–54+59.

Luan X F, Cao Y N, Xiao L H, Peng H J, et al. Application Progress of ceramic waste in building materials. Materials Report, 2015,29(13): 145–150.

Nie Y H, Liu X H, Peng Y, Yang X L, et al. Study the Strength Regularity of Recycled Masonry Mortar Produced with Recycled Ceramic Tiles. Applied Mechanics and Materials, 2014, 2972(507): 378–382.

Pei Q W, Wang G X, Shen Y, Zhu M Q, et al. Effect of recycled aggregate of waste ceramic tile on mortar and concrete performance. Silicate Bulletin, 2017,36(03): 797–802.

Qiao H X, Peng K, Chen K F, Guan L J, et al. Freeze-thaw cycle resistance and reliability analysis of recycled concrete from waste tile aggregates. Functional Materials, 2019, 50(07): 7139–7144+7151.

Wajeeha M, Asad R K, Tehmina A. Mechanical and Durability Properties of Concrete Containing Recycled Concrete Aggregates. Iranian Journal of Science and Technology, Transactions of Civil Engineering, 2021: 1–20.

Frontiers of Civil Engineering and Disaster Prevention and
Control – Yang & Rahman (Eds)
© 2023 The Author(s), ISBN: 978-1-032-31200-2

Experimental study on mechanical properties of steel fiber reinforced concrete

Jiuyang Li, Zhenwei Wang, Li Chen, Jinpeng Guo & Guangchao Hu
Changchun Institute of Engineering, Changchun, China

ABSTRACT: In order to further explore the mechanical properties of end-type low-volume steel fiber reinforced concrete and determine its optimal incorporation, five groups of specimens with a steel fiber content of 0%, 0.2%, 0.4%, 0.5%, and 0.7% were designed in this paper, and the mechanical properties of steel fiber reinforced concrete were tested. The research shows that with the increase of steel fiber content from 0% to 0.7%, the improvement of compressive strength of steel fiber reinforced concrete gradually tends to be stable, while the improvement of splitting tensile strength increases linearly and the improvement is obvious. Based on the experimental data, this paper further fitted the relationship between the relative values of the compressive strength and the splitting tensile strength of the steel fiber reinforced concrete (i.e., the ratio of the compressive strength and the splitting tensile strength of the steel fiber reinforced concrete to the compressive strength and the splitting tensile strength of the plain concrete) and the steel fiber content, and obtained the corresponding curve equation. Finally, according to the fitting curve and the characteristic equation, the optimal content range of the steel fiber was obtained.

1   INTRODUCTION

In the new era, construction projects tend to be green and intelligent, and the development of construction materials pays more attention to high performance. Concrete, as the main building material, has developed into many types. From the most common mixture of sand, cement, and stone to the gradual addition of fiber and other functional materials, fiber-reinforced concrete materials with improved mechanical properties are obtained (Wu 2017). Steel Fiber Reinforced Concrete (SFRC) is a new composite material formed by the disorderly distribution of steel fiber as reinforcement material in the concrete matrix with concrete as the matrix. The main function of steel fiber in concrete is to hinder the expansion of microcracks in concrete and block the occurrence and development of macro cracks (Song 2016). Therefore, compared with ordinary concrete, concrete with different steel fiber content has different effects and effects on the improvement of deformation capacity, toughness, tensile, shear, bending, and crack resistance. It is of great significance to explore the optimal dosage of steel fiber and obtain the optimal comprehensive performance of steel fiber coagulation.

2   RAW MATERIALS AND MIXTURE RATIO

The cement used in this experiment was P. O. 42.5 grade cement of a local brand. The coarse aggregate was crushed stone, and the particle size was 5–20 mm. The fine aggregate specification is grade II medium sand. Water reducing agent is high-performance water reducing agent. Steel fiber adopts an end hook type, a length of 35mm, a diameter of 0.75mm, an aspect ratio of 47, and tensile strength is 1000-1800 MPa. In order to study the effect of steel fiber content on the properties of concrete, four groups of different steel fiber content and one group of reference

DOI 10.1201/9781003308577-77

Table 1. Proportion of steel fiber reinforced concrete.

Numbering	Volume fraction of steel fiber/%	Water/ kg	Cement/ kg	Sand/ kg	Stone/ kg	Water cement ratio	Water reducer dosage/%
A	0	184	307	804	1155	0.6	0.75
B	0.2	184	307	804	1155	0.6	0.75
C	0.4	184	307	804	1155	0.6	0.75
D	0.5	184	307	804	1155	0.6	0.75
E	0.7	184	307	804	1155	0.6	0.75

concrete were designed according to C30 concrete in this experiment, and a 0.75% water reducer was added. The mix ratio of test concrete is shown in Table 1.

3 TEST METHODS

Test requirements: three test blocks for each group. According to the test method standard of physical and mechanical properties of concrete (GB/T 50081-2019), the test block size for concrete compressive strength and splitting tensile strength test is 150 mm × 150 mm × 150 mm; Splitting tensile test and compressive test of steel fiber reinforced concrete after curing 28 days (GB/T 50081 – 2019).

According to the test method of "Standard Test Method for Physical and Mechanical Properties of Concrete" (GB/T 50081 – 2019), the test block was put into a pressure testing machine and loaded according to the standard test method. The cube compressive strength and splitting tensile strength of steel fiber reinforced concrete were measured. Because the test block is vibrated on the horizontal vibration table, it is easy to directly cause a large difference in the distribution of steel fibers in the horizontal and vertical directions. Therefore, when loading, the side of the specimen is used as the bearing surface of the compressive strength test, and the upper and lower parts of the specimen cannot bear the pressure (GB/T 50081 – 2019). When the specimen is close to failure, the throttle of the testing machine should be adjusted in time to load slowly until the specimen is damaged, and the failure load should be recorded in time. The cubic compressive strength of steel fiber reinforced concrete can be calculated by (1):

$$f_{cc} = \frac{F}{A}$$

In the formula: f_{cc}—Concrete cube specimen compressive strength (MPa), the calculation results should be accurate to 0.1 MPa;
F—Failure load of the specimen (N)
A—Pressure area of the specimen (mm^2)
Calculation formula of splitting tensile strength of steel fiber reinforced concrete:

$$f_{ts} = \frac{2F}{\pi A} = 0.637 \frac{F}{A}$$

In the formula: f_{ts}—Concrete splitting tensile strength (MPa)
F—Failure load of the specimen (N)
A—Pressure area of the specimen (mm^2)
According to the requirements of tensile strength in 'Standard Test Method for Physical and Mechanical Properties of Concrete' (GB/T 50081-2019), specimens are made and loaded to obtain tensile strength.

4 EXPERIMENTAL RESULTS AND ANALYSIS

According to the strength calculation method of "steel fiber reinforced concrete" (JG/T 472 – 2015), the test strength of three concrete blocks in each group was calculated as the compressive strength value and the splitting tensile strength value by calculating the average value, so as to obtain the corresponding strength value of steel fiber reinforced concrete (JG/T465 – 2019). The experimental data of steel fiber reinforced concrete and the content of steel fiber are set as the vertical and horizontal coordinates, respectively. Not only the broken lines of the improvement degree of compressive strength and splitting tensile strength of steel fiber reinforced concrete are drawn and analyzed, but also the fitting curves of the relationship between the relative values of compressive strength and splitting strength and the content of steel fiber are drawn and analyzed.

4.1 Effect of steel fiber on compressive strength of concrete

It can be seen from Figure 1 that when the volume fraction of steel fiber is 0.2%–0.7%, the improvement of the compressive strength of steel fiber reinforced concrete relative to that of plain concrete increases with the increase of the volume fraction of steel fiber. When the volume fraction of steel fiber is 0.7%, the maximum improvement is not more than 15%. So, the compressive strength increases with the change in steel fiber content, but the increase is not obvious (Niu et al. 2019).

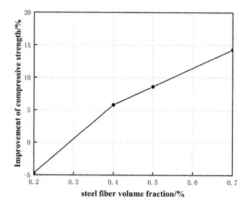

Figure 1.   Improvement of compressive strength.

The compressive strength of concrete with different steel fiber contents was compared with that of plain concrete, and origin software was used for data fitting to obtain the compressive strength relationship image of steel fiber concrete relative to plain concrete (Xie etal. 2020). The ratio of compressive strength of steel fiber reinforced concrete to plain concrete and the content of steel fiber are set as the vertical and horizontal coordinates, respectively, and the relevant curves are plotted for analysis, as shown in Figure 2.

The relationship between the ratio of the compressive strength of steel fiber reinforced concrete to the compressive strength of plain concrete and the content of steel fiber in Figure 2 is further analyzed, and the influence of different content of steel fiber on the relative compressive strength of concrete is explored by linear fitting (Chang et al. 2020). Equation (3) is obtained.

$$\frac{f_{cc}}{f_c} = 0.99887 - 0.78366x + 3.39794 \times x^2 - 2.84783 \times x^3$$
$$R^2 = 0.9858$$

According to Equation (3), the relationship between the relative value of compressive strength (the ratio of compressive strength of steel fiber reinforced concrete and plain concrete) and steel

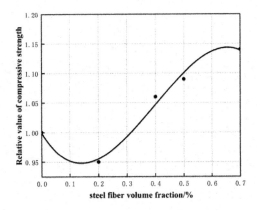

Figure 2. Relationship between steel fiber content and relative compressive strength.

fiber content is a cubic function equation. It can be seen from Figure 2 that when the steel fiber content is 0.2%~0.7%, this function shows a monotonically increasing trend and reaches a maximum value when the steel fiber content is 0.7%. By analyzing the compressive strength of the steel fiber content (0.2%~0.7%), it is obvious that the optimal content of steel fiber is 0.6%~0.7%. The value of R2 in Equation (3) is very close to 1, indicating that Equation (3) is highly fitted. Therefore, the equation can be used as a general formula for the compressive strength of steel fiber reinforced concrete. When the two factors of plain concrete compressive strength, steel fiber concrete compressive strength, and steel fiber content are known, the value of another unknown factor can be calculated, and the applicable scope is that the steel fiber content is not more than 0.7%.

4.2 Effect of steel fiber on splitting tensile strength of concrete

It can be seen from Figure 3 that when the steel fiber content is 0.2%~0.7%, the improvement of splitting tensile strength of concrete increases to different degrees with the increase of steel fiber content. When the steel fiber content is 0.7%, the improvement of the splitting tensile strength of concrete increases by nearly 35%. It can be seen that the different content of steel fiber has an obvious influence on splitting tensile strength.

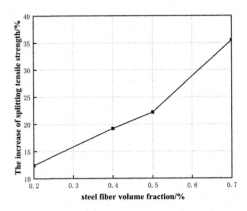

Figure 3. Increase in splitting tensile strength.

The splitting tensile strength of concrete with different content of steel fiber is compared with plain concrete. The ratios of splitting tensile strength of steel fiber concrete to plain concrete and the content of steel fiber are set to vertical and horizontal coordinates, respectively, and the relative value of splitting tensile strength of concrete is fitted by function, as shown in Figure 4.

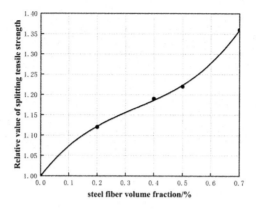

Figure 4. Relative value of splitting tensile strength.

The relationship between splitting tensile strength of steel fiber reinforced concrete and compressive strength of plain concrete and steel fiber content is shown in Equation (4).

$$\frac{f_{fts}}{f_{ts}} = 0.99966 + 0.89348 \times x - 1.77085 \times x^2 + 1.75561 \times x^3$$
$$R^2 = 0.9996 \tag{1}$$

According to Equation (4), the relationship between the relative value of splitting tensile strength of concrete (the ratio of splitting tensile strength of steel fiber concrete to splitting tensile strength of plain concrete) and steel fiber content is a cubic function equation. It can be seen from Figure 4 that the relative value of concrete splitting tensile strength increases with the increase of steel fiber content, although the increase has a slow and rapid throw is a monotonically increasing trend. It can be clearly seen that this function curve reaches the maximum when the steel fiber content is 0.7%. Steel fiber content (0.2%~0.7%) on splitting tensile strength analysis, the optimal content of steel fiber is 0.7%. Because the difference between the value of R2 and 1 in Equation (4) is very close to 0.0004, indicating that the fitting degree of Equation (4) is almost the same. Therefore, Equation (3) can be used as a general formula for the relative value of concrete splitting tensile strength. When two values of plain coagulation, splitting tensile strength of steel fiber reinforced concrete, and steel fiber content are known, another unknown factor can be calculated, and the applicable scope is that the steel fiber content is not more than 0.7%.

5 CONCLUSIONS

(1) When the content of end hooked steel fiber is in the range of 0.2%~0.7%, the increase of compressive strength of concrete is not obvious, and the maximum increase of compressive strength is only close to 15%. The relative value of compressive strength by fitting analysis shows that the optimal volume content of steel fiber is 0.6%~0.7%.
(2) When the end hook type steel fiber content is in the range of 0.2%~0.7%, the increase of splitting tensile strength of concrete is in the range of 13%~%, and the increase is the largest when the steel fiber content is 0.7%. In general, the splitting tensile strength of concrete is significantly affected by the steel fiber content. By fitting and analyzing the relative value of splitting tensile strength, the optimal volume content of steel fiber is 0.7%.

REFERENCES

Chang H, Shen P, Gu F G. Experimental study on mechanical properties of steel fiber reinforced concrete [J] Concrete, 2020(04):67–69.

GB/T 50081-2019, Standard Test Method for Physical and Mechanical Properties of Concrete [s]. Beijing: Construction Industry Press, 2019.

JG/T465 – 2019, steel fiber reinforced concrete [s]. Beijing: Construction Industry Press, 2018.

Niu L L, Zhang S P Wei Y X. Effect of Steel Fiber Content on Mechanical Properties of Concrete [J]. Concrete and Cement Products 2019 (3):52–54.

Song H Y. Distribution of steel fiber in concrete matrix and its relationship with toughness [D]. Dalian University of Technology,2016.

Wu H J. Development and Utilization of Intelligent Concrete [J]. Green Building Materials, 2017(10):79.

Xie Y F, Sheng M, Lv Y F. Experimental study on mechanical properties of steel fiber reinforced concrete [J]. Sichuan Cement, 2020(07):14–15.

Frontiers of Civil Engineering and Disaster Prevention and
Control – Yang & Rahman (Eds)
© 2023 The Author(s), ISBN: 978-1-032-31200-2

Influence of type of machine-made sand and content of stone powder on creep property of C60 concrete

Min Zhou, Tengyu Yang, Lixian Guo & Bing Qiu
Foshan Transportation Science and Technology Co., Ltd, China

Putao Song
China Academy of Building Research, China

Liangbo Wang
Foshan Road and Bridge Supervision Station Co., Ltd, China

Jing Wang
China Academy of Building Research, China

ABSTRACT: C60 manufactured sand concrete was prepared with different kinds of manufactured sand with different content of limestone powder. The influence of the types of manufactured sand (granite, sandstone, limestone pebble) and the content of limestone powder of granite manufactured sand (5%, 7%, 10%, 15%) on the creep of concrete at 150d were studied. The results show that the 150d creep of sandstone manufactured sand concrete ($33 \times 10^{-6} \text{MPa}^{-1}$) is the largest, the granite manufactured sand concrete is the second ($22 \times 10^{-6} \text{MPa}^{-1}$), and the limestone pebble concrete is the smallest ($21 \times 10^{-6} \text{MPa}^{-1}$). The creep degree of manufactured sand concrete increases first and then decreases slowly with the increase of stone powder content. Relatively speaking, when the content of machine-made sand and stone powder is 5%, the 150d creep degree ($21 \times 10^{-6} \text{MPa}^{-1}$) is the lowest, and the 150d creep degree ($24 \times 10^{-6} \text{MPa}^{-1}$) is the highest when the stone powder content is 10%. Generally speaking, based on meeting the requirements of compressive strength, the creep ($21 \times 10^{-6} \text{MPa}^{-1}$) of C60 manufactured sand concrete with 5% stone powder content is the lowest.

1 INTRODUCTION

River sand is a non-renewable resource. In recent years, the excessive use of natural river sand in domestic large-scale concrete projects such as high-speed railways, dams, and super large bridges makes the river sand resource increasingly scarce and has the momentum of depletion. It has been found that (Wang & Gao 2019; Wang 2018; Zhao et al. 2019; Zheng 2019), as the best substitute for river sand, machine-made sand has been gradually applied to industrial and civil construction, municipal engineering and other engineering fields, as well as pier body, bearing platform, pile foundation and other substructures in large projects such as high-speed railway and a super large bridge. However, due to the different properties of machine-made sandstone and stone powder content, machine-made sand concrete has certain differences in creep and other long-term properties. The results of Li Xinggui (Li et al. 2002) show that the shrinkage of concrete with 16% and 12% stone powder increases by 12.8% and 4.8%, respectively, compared with the concrete with 3% stone powder. When the content of stone powder is more than 12%, the dry shrinkage increases rapidly. According to Chen Zhao wen (Chen 2001), the dry shrinkage rate of machine-made sand concrete increases with the increase of the content of the stone powder. When the content of stone powder increases from 12% to 21%, the dry shrinkage rate increases by about 1% for every 1% increase in the content of the stone powder. The study of Li Jian (Li et al. 2001) shows that the early dry shrinkage of stone chip concrete is larger than that of river sand concrete, mainly because the stone chip gradation is not as good as river sand. According to Tahir Celik (Celik & Mara 1996), dry shrinkage increases with the increase of stone powder content when the content of stone powder is less than 10% and decreases when the content of stone powder is more than 10%. Zhou Min (Zhou et al. 2021) used granite and limestone instead of sandstone to prepare C60 concrete and found that the 90d creep of granite C60 concrete and limestone C60 concrete decreased by 47% and 53%, respectively, compared with sandstone C60 concrete. Therefore, the mechanical sandstone powder content and aggregate lithology greatly influence the drying shrinkage and creep properties of concrete.

For high-strength concrete with a strength grade of C60 and above, especially high-strength pre-stressed concrete, the changes in machine-made sandstone and stone powder content have greater differences in the effects on concrete creep and other long-term properties. The potential differences and uncertainties in the application of machine-made sand in high-strength and high-strength pre-stressed concrete limit its use in high-strength and high-strength pre-stressed concrete. In addition to the restrictions on the application of machine-made sand in pre-stressed engineering in current domestic and foreign standards and specifications, the application of machine-made sand in high-strength concrete, especially pre-stressed high-strength concrete, is very few (Lu et al. 2019; Zhuang 2016).

Based on Beijiang Bridge in Foshan of China, C60 concrete for a pre-stressed box girder with satisfactory performance is prepared using bulk materials such as machine-made sand around the project. The effects of machine-made sand type and stone powder content on the creep performance of C60 machine-made sand concrete for pre-stressed box girder are studied to provide a reference for the application of machine-made sand in the pre-stressed beam.

2 RAW MATERIALS

1) Cement: P.O 52.5 cement, produced by Guangdong Taini Cement Plant, with 3-day strength of 32.2 MPa and 28-day strength of 59.5 MPa, the performance of cement is tested in accordance with the test method specified in the current standard GB 175.2). Fly ash: Grade I fly ash, produced by Guangdong Qing yuan Power Plant, with a loss on ignition of 3.3%, 45 μm sieve residue is 4.5%, density is 2.3 g/cm^3, water demand ratio of 95%. The performance of fly ash is tested according to the test method stipulated in the current Standard GB/T 1596. 3). Slag powder: S95 granulated slag powder, produced in a place in Guangdong, with a 7-day activity index of 78.2%, 28-day activity index of 95.3%, fluidity ratio of 97.2%, density of 3.1 g/cm^3, and specific surface area of 420 m^2/kg. 4). Fine aggregate: sandstone manufactured sand (SY), with a stone powder content of 7%, MB value of 0.5 % and fineness modulus of 2.8, water absorption rate of 2.4%, crushing value of 23.2%, needle flake content of 16%, and firmness of 22.7%. Limestone pebble manufactured sand (LS) has a stone powder content of 7%, MB value of 0.5%, fineness modulus of 2.9, water absorption rate of 1.9%, crushing value of 4%, needle flake content of 21%, firmness of 12.3%. Granite manufactured sand (Hg): with an MB value of 0.8%, water absorption rate of 18.8%, crushing value of 11.5%, needle flake content of 10%, and firmness of 12.3%. The content of stone powder is 0%, 5%, 7%, 10% and 15%, and the corresponding fineness modulus is 3.0%, 2.8%, 2.8%, 2.7% and 2.6% respectively. The fine aggregate performance is tested according to the test method specified in the current standard GB/T 14684. The chemical composition of sandstone is mainly quartz, followed by Albite, calcite, and Muscovite. Calcite is the main chemical composition of limestone, followed by quartz and dolomite. The chemical composition of granite is mainly quartz, followed by Albite, nickel chloride, and calcite.

5) Coarse aggregate: 5~20 mm continuously graded crushed stone, with 5% needle flake content, 1400 Kg/m^3 loose bulk density, 40.4% porosity, and 0.5% mud content. The performance of coarse aggregate is tested in accordance with the test method specified in the current standard GB/T

14685.6). Admixture: the admixture is a polycarboxylic acid high-performance water reducer produced by an admixture factory in Jiangsu, with a solid content of 35% and water reduction rate of 34%.

3 MIX PROPORTIONS AND TEST METHODS

According to the project requirements, the concrete shall have good pumping performance, the initial slump of the concrete shall be 220 ± 20 mm, and the C60 manufactured sand concrete shall meet the preparation strength requirements of the corresponding strength grade, with 7d compressive strength $\geq$ 54MPa and 28d compressive strength$\geq$69.9MPa. In order to ensure 7d tensioning, the 7d elastic modulus of concrete is $\geq 2.9 \times 104$ MPa.

The mix design of C60 machine-made sand concrete was prepared according to standard JGJ 55-2011. The mix proportion of C60 machine-made sand concrete with different types and stone powder content is shown in Table 1. As shown in Table 1, numbers 2-A, 2-B, and 2-C are the concrete mix ratios of granite machine-made sand, limestone pebble machine-made sand, and limestone gravel machine-made sand with a stone powder content of 7%, respectively. It is used to compare the influence of different kinds of manufactured sand on the properties of manufactured sand concrete. 2-2A, 2-A, 2-2B, and 2-2C are the concrete mix proportion of machine-made granite sand with 5%, 7%, 10%, and 15% stone powder, respectively, in order to compare the influence of different stone powder content on the performance of machine-made sand concrete.

Firstly, the effects of the type of manufactured sand and the content of stone powder on the properties and mechanical properties of the C60 manufactured sand concrete mixture are studied. The properties of the C60 machine-made sand concrete mixture include concrete slump, expansion, inverted slump, and emptying time (inverted time). The mechanical properties of C60 manufactured sand concrete include compressive strength (7d, 28d) and elastic modulus (7d). The slump, expansion, inverted slump, and emptying time (inverted time) performance test of C60 manufactured sand concrete mixture shall be carried out in accordance with the current national standard GB/T 50080-2016, and the mechanical performance test shall be carried out in accordance with the current national standard GB/T 50081-2019.

Table 1. Mix Proportion of C60 Machine-made Sand Concrete with Different Types and Different Stone Powder Content.

NO	Consumption of concrete raw materials (kg/m^3)							Type of manufactured sand	Stone powder content of manufactured sand %
	Cement	Fly ash	Slag powder	Manufactured sand	Coarse aggregate	Water	Admixture		
2-A	367.5	73.5	49.0	657	1118	147	7.35	HG	7
2-B	367.5	73.5	49.0	657	1118	147	7.35	SY	7
2-C	367.5	73.5	49.0	657	1118	147	7.35	LS	7
2-2A	367.5	73.5	49.0	657	1118	147	7.35	HG	5
2-2B	367.5	73.5	49.0	657	1118	147	7.35	HG	10
2-2C	367.5	73.5	49.0	657	1118	147	7.35	HG	15

4 RESULTS

4.1 *Workability and mechanical properties of the mixture*

The test results of workability and mechanical properties of C60 manufactured sand concrete with different types of manufactured sand and stone powder content are shown in Table 2.

Table 2. Workability and mechanical properties of C60 manufactured sand concrete mixture.

NO	Slump (mm)	Slump flow (mm)	Inverted time (s)	Compressive strength (MPa)		7d elastic modulus (GPa)
				7d	28d	
2-A	230	580	12	67.10	81.63	3.68
2-B	200	450	30	58.13	69.10	3.08
2-C	240	650	8	56.73	66.20	3.36
2-2A	225	500	15	65.30	74.50	3.45
2-2B	240	620	22	66.87	76.50	3.65
2-2C	230	590	35	64.93	70.50	3.60

It can be seen from Table 2 that except that the 28d compressive strength of the 2-c sandstone machine-made sand group does not meet the preparation requirements, the 7d and 28d compressive strength and 7d elastic modulus of all the groups meet the project requirements. On the basis that the basic material properties of manufactured sand (the content of stone powder is 7%, fineness modulus is 2.8) and mix proportion parameters are basically consistent, the slump and slump flow of limestone pebble manufactured sand concrete are the largest, followed by granite manufactured sand concrete, and the slump and the slump flow of sandstone manufactured sand concrete are the smallest. The inverted time of limestone pebble machine-made sand concrete is the shortest, followed by granite machine-made sand concrete, and sandstone machine-made sand concrete is the longest. Granite machine-made sand concrete has the highest compressive strength at 7d and 28d, followed by sandstone machine-made sand concrete, and limestone pebble machine-made sand concrete has the lowest compressive strength at 7d and 28d. The 7d elastic modulus of granite manufactured sand concrete is the highest, followed by limestone pebble manufactured sand concrete, and the 7d elastic modulus of sandstone manufactured sand concrete is the lowest.

Limestone pebble machine-made sand particles have relatively round morphology, fewer edges and corners than machine-made sand particles, small friction resistance between machine-made sand particles, and low adhesion between machine-made sand and slurry. Under the same slurry bone ratio, limestone pebble machine-made sand can obtain better fluidity and lower viscosity, so its slump and slump flow are the largest, and the inverted time is the lowest. The rounder surface morphology reduces the bite force between machine-made sand particles and the adhesive force between machine-made sand and slurry, resulting in relatively lower compressive strength and elastic modulus. In the range of stone powder content of 5%~15%, with the increase of random sand powder content, the slump and slump flow of concrete first increase and then decrease. When the stone powder content is 10%, the slump and the slump flow of concrete are the largest, and when the stone powder content is 5%, the slump and the slump flow of concrete are the smallest. The concrete inverted time first decreases and then increases. The inverted time is the shortest when the stone powder content is 7%, and the inverted time is the longest when the stone powder content is 15%. When the content of stone powder is 10%, the slump and the slump flow of concrete are the largest, and when the content of stone powder is 5%, the slump and the slump flow of concrete are the smallest. The 7d and 28d compressive strength of concrete increases first and then decreases. When the content of machine-made sand and stone powder is 7%, the 7d and 28d compressive strengths of concrete are the largest, and when the content of stone powder is 15%, the 7d and 28d compressive strengths of concrete are the smallest. The 7d elastic modulus of concrete increases first and then decreases. When the content of machine-made sand and stone powder is 7%, the 7d elastic modulus of concrete is the largest, and when the content of stone powder is 5%, the 7d elastic modulus of concrete is the smallest.

In recent years, stone powder has been applied to concrete in the form of admixture, and the current standard JGJ/T 385-2015 also includes stone powder in the category of composite admixture. For machine-made sand concrete, the increase of an appropriate amount of stone powder can increase the number of slurries in machine-made sand concrete (the number of composite admixtures and cementitious materials increases), improve the slurry bone ratio, and improve the fluidity of concrete. However, the constant actual water consumption will reduce the actual water binder ratio of concrete (the number of composite admixtures and cementitious materials increases) and increase the viscosity of concrete. The fluidity decreases, so the fluidity of concrete increases in the range of appropriate stone powder content (5%~7%), and the fluidity of concrete decreases when the stone powder content exceeds a certain amount. Within a certain range (5%~7%), the filling effect of stone powder on the voids of machine-made sand improves the compactness of concrete, and the strength and elastic modulus of concrete increase. After exceeding a certain range (5%~7%), the consumption of excessive stone powder on cement paste (cement paste wrapping stone powder particles) reduces the number of cement paste involved in cementation and weakens the cementation force. Concrete strength and elastic modulus decrease (Sun 2019; Wang 2008; Xie et al. 2019; Zhang et al. 2011).

4.2 Creep degree

The creep and drying shrinkage test methods of C60 manufactured sand concrete with different types of manufactured sand and stone powder content at the same age shall be carried out in accordance with the current standard GB/T 50082-2009. The concrete test block for the creep test shall be cured for 7d after the formwork is removed and then moved to the creep room to continue drying and curing until 28d. The creep value of concrete at different ages shall be tested after loading according to the load specified in GB/T 50082. The creep degree of concrete at different ages shall be calculated according to the formula specified in the standard. The test ages were 1d, 3d, 7d, 14d, 28d, 45d, 60d, 90d, and 150d, respectively. The creep test results of C60 manufactured sand concrete with different types of manufactured sand and stone powder content are shown in Table 3.

Table 3. Creep degree of C60 machine-made sand concrete.

Curing age (d)	Creep degree ($\times 10^{-6} \mathrm{MPa}^{-1}$)					
	2-1A	2-1B	2-1C	2-2A	2-2B	2-2C
1	5	9	5	5	6	6
3	7	11	7	7	8	7
7	9	15	9	9	10	9
14	11	18	12	11	13	12
28	13	22	14	14	16	15
45	15	24	15	15	17	17
60	17	25	16	16	18	18
90	20	32	20	20	23	21
150	22	33	21	21	24	23

According to the comparison of 2-1A, 2-1B, and 2-1C, when using sandstone manufactured sand, the 150d creep of C60 manufactured sand concrete is the largest, followed by granite manufactured sand, and the 150d creep of C60 manufactured sand concrete prepared with limestone pebble manufactured sand is the lowest. In general, the creep degree of manufactured sand concrete in each group increased the fastest from 0 to 14 days, the growth rate of manufactured sand concrete creep degree slowed down after 14 days, and the creep degree of manufactured sand concrete in

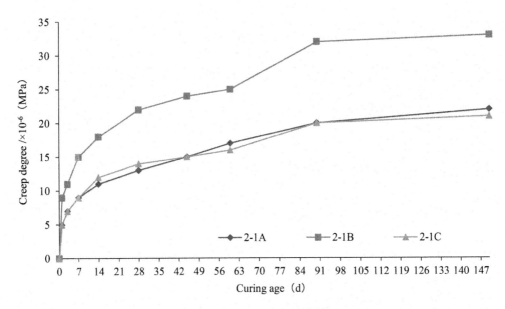

Figure 1. Effect of type of manufactured sand on creep degree of C60 concrete.

each group almost did not increase after 90 days. The maximum 150d creep of sandstone machine-made sand C60 concrete is $33 \times 10^{-6} MPa^{-1}$, compared with the 150d creep results of sandstone manufactured sand C60 concrete, the 150d creep of granite manufactured sand C60 concrete is reduced by 33%, and the 150d creep of limestone pebble manufactured sand C60 concrete is reduced by 36%. The elastic modulus of aggregate is the key factor affecting the creep degree of concrete. Relatively speaking, the elastic modulus of limestone pebble is the lowest, followed by granite and sandstone. Under the same load, the aggregate with a higher elastic modulus has the smallest elastic deformation. The concrete prepared based on the same mix proportion parameters and other raw materials has the smallest deformation and the lowest creep. Therefore, the concrete prepared with limestone pebble machine-made sand has the lowest creep, followed by granite and sandstone machine-made sand (Zhang et al. 2017; 2016).

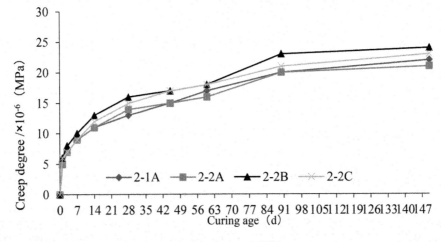

Figure 2. Effect of stone powder content on creep degree of C60 concrete.

According to the comparison of 2-1a, 2-2a, 2-2b, and 2-2c, with the increase of stone powder content, the creep of C60 concrete first increases and then decreases slowly. Relatively speaking, the creep of 150d is the lowest when the content of manufactured sandstone powder is 5%, and the creep of 150d is the highest when the content of stone powder is 10%. The creep of each group of C60 manufactured sand concrete increases fastest within 14 days, slows down after 14 days, and almost no longer increases after 90 days. When the stone powder content is 10%, the maximum creep of C60 concrete for 150d is $24 \times 10^{-6} \text{MPa}^{-1}$. Compared with the creep degree, when the stone powder content is 5%, 7%, and 15%, the creep degree of C60 manufactured sand concrete for 150 days is reduced by 13%, 8%, and 4%, respectively. The influence of stone powder content on the creep degree of machine-made sand concrete is mainly manifested in three aspects (Tang 2012; Zheng 2014; Zhao 2009): 1) when the stone powder content is low (within 5%), with the increase of stone powder (admixture), the actual water binder ratio of machine-made sand concrete decreases and the creep decreases. 2) The "micro-aggregate effect" of an appropriate amount of stone powder (less than 5%) optimizes the pore structure in the concrete, reduces the through holes and "evaporated water," and reduces the shrinkage and creep caused by it. 3) When the content of stone powder exceeds a certain value (5%), the demand for excessive stone powder content for cement paste reduces the binding force between paste and aggregate, and then has a negative effect on the reduction of concrete creep. The comprehensive effect of the above factors leads to the creep degree of machine-made sand concrete increasing first and then decreasing slowly with the increase of stone powder content.

In general, the influence of the type of machine-made sand on the creep of C60 machine-made sand concrete is greater than the content of machine-made sand and stone powder. When preparing concrete with machine-made sand, we should pay more attention to the influence of the type of machine-made sand on the creep performance of concrete. According to the test results of mixture performance and mechanical properties of C60 manufactured sand concrete, on the basis of ensuring that the compressive strength meets the requirements, the creep degree of C60 manufactured sand concrete prepared with 5% granite manufactured sand is the lowest in 150d, and the creep degree is $21 \times 10^{-6} \text{ MPa}^{-1}$.

5 CONCLUSIONS

1) Different types of machine-made sand C60 concrete have different mixture properties and mechanical properties. Relatively speaking, limestone pebble machine-made sand concrete has the best performance, granite machine-made sand concrete takes the second place, sandstone machine-made sand concrete has the worst performance, granite machine-made sand concrete has the best mechanical performance, and sandstone machine-made sand concrete takes the second place, The mechanical properties of limestone pebble manufactured sand concrete are the worst.

2) With the increase of random sand and stone powder content, the concrete slump and slump flow first increase and then decrease, and the concrete inverted time first decreases and then increases. When the stone powder content is 10%, the concrete slump and slump flow are the largest, the inverted time is the shortest when the stone powder content is 7%, the 7d and 28d compressive strength and 7d elastic modulus first increase, and then decrease, and when the machine-made sand and stone powder content is 7%, the concrete 7d. The 28d compressive strength and 7d elastic modulus are the largest, the 7d compressive strength and the 28d compressive strength of concrete are the smallest when the stone powder content is 15%, and the 7d elastic modulus of concrete is the smallest when the stone powder content is 5%.

3) The creep degree of C60 concrete with different types of machine-made sand is different. When using machine-made sandstone sand, the 150d creep degree of C60 machine-made sand concrete is the largest, followed by machine-made granite sand, and the 150d creep degree of C60 machine-made sand concrete prepared with limestone pebble machine-made sand is the lowest.

4) With the increase of stone powder content, the creep degree of C60 concrete increases first and then decreases slowly. Relatively speaking, when the content of machine-made sand and stone powder is 5%, the creep degree of 150d is the lowest, and when the content of stone powder is 10%, the creep degree of 150d is the highest.

5) The influence of manufactured sand type on the creep degree of C60 manufactured sand concrete is greater than that of manufactured sandstone powder content. Based on ensuring that the compressive strength meets the requirements, the creep degree of C60 manufactured sand concrete prepared with 5% granite manufactured sand is the lowest in 150d, and the creep degree is $21 \times 10^{-6} \text{MPa}^{-1}$.

REFERENCES

Chen Zhaowen. Test and study of the properties of artificial-sand concrete with high content of stone powder for Mianhuatan Hydropower Station[J]. Water Power,2001(7)32:35+66.

Li Jian, Hie Youjun, Liu Baoju, et al. Mechanical properties of concrete with limestone chips[J]. Concrete,2001(07):23–25+11.

Li Xing-gui, Zhang Heng-quan, Chen Xiao-yue. Experimental study on dry-shrinkage property of fully graded concrete with high powder-content artificial sand[J]. Journal of Hehai University (Natural Sciences),2002(4)37:40.

Lu Jialin, Lan Cong, Chen Jing, et al. Research and application of C50 pre-stressed concrete prepared by machine-made sand [J]. New Building Materials, 2019,46(08): 80–82+135.

Sun Shengwei. Influence of stone powder content on performance of C30 and C50 manufactured sand concrete [J]. Journal of Water Resources and Building Engineering, 2019,17(03): 160–63.

Tahir Celik, Khaled Marar. Effects of Crushed Stone Dust 0n Some Properties of Concrete [J]. Cement and Concrete Research. 1996, 26(7):1121~1130.

Tang Jing. Study on shrinkage and durability of artificial sand concrete [D]. Chongqing University, 2012.

Wang Chengqi, Gao Hailao. Experimental study on application of ultra-high pumping machine-made sand concrete [J]. Concrete, 2019(06): 112–14.

Wang Jianjun. Research on the Application of Machine-made Sand and Stone Powder in Commercial Concrete in Shanxi Jinzhong Area [D]. Lanzhou University of Technology, 2018.

Wang Jiliang. Study on the influence of machine-made sand characteristics on concrete performance and its mechanism [D]. Wuhan University of Technology, 2008.

Xie Kaizhong, Wang Hongwei, Xiao Jie, et al. Test on the influence of stone powder content on mechanical properties of manufactured sand concrete [J]. journal of architectural science and engineering, 2019,36(05): 31–38.

Zhang Lihua, Liu Laibao, Zhou Yongsheng, et al. Influence of stone powder content on mechanical properties and microstructure of manufactured sand concrete [J]. Concrete and Cement Products, 2011(12): 22-26.

Zhang Rulin, Chen Yuqian, Liu Shutong, et al. Study on the influence of sand-gravel powder content on concrete performance [J]. Concrete, 2016(03): 84-85.

Zhang Yingxue, Lou Zongke, Zhang Zhen, et al. Prediction of the influence of aggregate size on elastic modulus and bond strength of concrete interface transition zone [J]. Concrete, 2017(07): 710+14.

Zhao Jinhui. Study on Creep Performance of C60 High-Performance Concrete [D]. Central South University, 2009.

Zhao Youming, Han Zili, LI Huajian, et al. Application status and existing problems of machine-made sand concrete in railway engineering in China [J]. China Railway, 2019(08): 1–7.

Zheng Ke. Research on Application of Machine-made Sand in High-grade Concrete Structure of Expressway Bridge [D]. South China University of Technology, 2019.

Zheng Yi. Study on long-term deformation performance of limestone machine-made sand concrete [D]. Huazhong University of Science and Technology, 2014.

Zhou Min, Yang, Tengyu, Qiu Bing, et al. Influence of different types and quality of coarse aggregate on creep property of C60 concrete [J]. New Building Materials. 2021(08) Page:21–24.

Zhuang Hongbin. Research on Application of Machine-made Sand in Bridge High-grade Concrete [D]. Dalian University of Technology, 2016.

Frontiers of Civil Engineering and Disaster Prevention and Control – Yang & Rahman (Eds)
© 2023 The Author(s), ISBN: 978-1-032-31200-2

Influence of two-stage heat treatment on the morphology of mesophase pitch

Jun Li & Dong Liu*

State Key Laboratory of Heavy Oil Processing, China University of Petroleum, Qingdao, Shandong, China

ABSTRACT: The comparative study of different two-stage heat treatments was performed to illustrate the effect of improving the yield and optical texture of the mesophase pitch. In such two-stage term treatments, the pressured condensation stage promotes the conversion of raw oil to mesogen molecules. Subsequently, the vacuum condensation stage removes the non-mesogen molecules, reducing the steric hindrance and thus inducing the rapid formation of mesophase pitch. The optical structure of the product was characterized by polarized light microscopy. The results show that the mesophase pitch with a large flow domain structure and high anisotropic content was obtained at a reaction temperature of 310°C and a vacuum degree of 0.06 MPa for 30 min in the second stage.

1 INTRODUCTION

Mesophase pitch, also known as anisotropic pitch, is petroleum residue, coal tar, pure aromatic compounds, and other organic compounds that undergo a series of continuous reactions, including bond breaking, dehydrogenation, condensation, and other reactions, resulting in the gradual formation of liquid crystal compounds containing a considerable number of anisotropic phases in the process of heat treatment between 350~550°C (Marsh 1973). Mesophase pitch is solid at room temperature and has optical anisotropy when observed under a polarized light microscope. In addition, mesophase pitch has a regularly oriented macromolecular lamellar structure inside, which is easily graphitized when treated at high temperature and is typical of easily graphitized carbon. It was found that the mesophase pitch with 100% anisotropic content has a H/C atomic ratio between 0.35~0.5, contains 15~20% volatile components, has a density of $1.3~1.5 g/cm^3$, and its average molecular weight is around 2000, which is about 4~5 times higher than the raw material (Kim et al. 1993). Mesophase pitch belongs to the nematic liquid crystal, mainly composed of planar aromatic macromolecules, which are highly oriented under the action of its unique π-electron cloud (Wang 2005), so that the mesophase pitch has various properties of crystals and liquids, such as crystalline physical properties of heat, light, electricity, and magnetism of crystals, and physical properties of liquids such as rheology, viscosity, and deformation. Mesophase pitch can be used as raw material to prepare many high-performance carbon materials: such as ultra-high modulus mesophase pitch-based carbon fibers (Yue, Liu, Vakili), needle coke (Wang & Eser 2007), mesocarbon microbeads (Gong et al. 2021), composite materials, fluorocarbon materials, carbon foams, carbon sheets, and films, mesophase pitch-based electrode materials, high-temperature lubricants. Among them, ultra-high modulus mesophase pitch-based carbon fibers, compared with polyacrylonitrile-based (PAN-based) carbon fibers, have higher thermal conductivity and tensile modulus and a smaller

*Corresponding Author: liudong@upc.edu.cn

DOI 10.1201/9781003308577-79

coefficient of thermal expansion. So, they are in great demand in aviation, aerospace, and military fields.

This study carried out the heat treatment of aromatic-rich oils using a two-stage thermal method. The effect of different treatment conditions on the product yield and optical structure was investigated to illustrate the effect of the removal of non-mesocrystalline molecules on the reaction carbonization behavior.

2 EXPERIMENTAL SECTION

2.1 *Feedstock*

The raw material used in this study is a rich aromatic component in hydrogenated tail oil, named HCTO-2, the basic properties of the raw oil analysis are listed in Table 1. HCTO-2 is less dense and more viscous, with a H/C ratio of 1.35, indicating a lower degree of condensation and a certain amount of aliphatic structure. The four-component analysis shows that HCTO-2 is rich in aromatics with an aromatic fraction of 63.94 wt% and does not contain asphaltenes. In summary, HCTO-2 is an ideal raw material for preparing mesophase pitch.

Table 1.   Basic properties of raw material HCTO-2.

Properties		HCTO-2
Density $(20°C)/g·cm^{-3}$		0.9944
Viscosity $(100°C)/mm^2·s^{-1}$		9.90
Micro carbon residues/wt%		0.19
Number average molecular weight (M_n)		383
Elemental composition/wt%	C	87.93
	H	9.94
	H/C	1.35
Four components (SARA) analysis/wt%	Saturates	22.26
	Aromatics	63.94
	Resins	13.69
	Asphaltene	–

2.2 *Thermal treatment*

Two-stage heat treatment experiments were conducted in a 500ml stainless steel high-pressure reactor with a PID program control. In the thermal polycondensation stage, 180g of HCTO-2 was reacted under a nitrogen atmosphere at 440°C and 3MPa for 6 h to obtain the carbonaceous intermediate product. Afterward, turn on the vacuum pump and reduce the pressure to the required level within 10 min, and the reaction is continued for 30 min at a specific temperature. After the reaction, the reactor was quickly cooled to room temperature to obtain the mesophase pitch, named MP-T (°C)-P (MPa), where T (°C) and P (MPa) represent the temperature and vacuum degree of vacuum polycondensation stage, respectively.

2.3 *Samples characterization*

The polarized optical microphotographs of the mesophase samples were characterized by XP-4030 metallographic microscopy.

3 RESULTS AND DISCUSSION

3.1 *Effect of the vacuum temperature on the preparation of mesophase pitch by two-stage heat treatment*

The yields of the products at different vacuum temperatures are shown in Figure 1. As can be seen from Figure 1, the higher the temperature of the vacuum treatment, the lower the yield of the product. The decrease in temperature has an inhibitory effect on the free radical reaction, so that the intermediate does not have enough temperature to undergo thermal cleavage to form small molecules, thus allowing more light components to be present in the system, resulting in higher product yields.

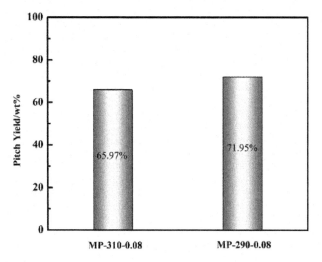

Figure 1. Effect of vacuum temperature on product yield.

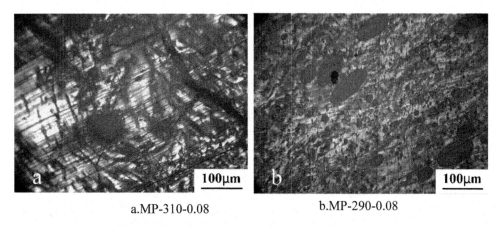

a.MP-310-0.08   b.MP-290-0.08

Figure 2. Effect of vacuum temperature on product optical structure.

Figure 2 shows the polarized structure of the products under different vacuum temperatures. From the polarized photos, it can be clearly observed that when the evacuation temperature is 310°C, the mesophase pitch is a small flow domain structure with an optical size of around 50–100 μm. However, when the vacuum temperature is 290°C, the isotropic content increases significantly, and

591

the optical structure deteriorates. The higher vacuum temperature causes the thermal cleavage reaction and thermal polycondensation of the products to be violent, which makes the small molecule removal system more effective, reducing the steric hindrance to facilitate the insertion and longitudinal stacking of the mesogen molecules. Consequently, the optical texture of the mesophase pitch shows a flow domain type.

3.2 *Effect of the vacuum degree on the preparation of mesophase pitch by two-stage heat treatment*

The carbonaceous yields of the products under different vacuum degrees are presented in Figure 3. The carbonaceous yield decreased from 65.97 wt% at 0.08 MPa to 51.33 wt% at 0.06 Mpa. The reduction of vacuum degree promotes the liquid-phase carbonization reaction causing more alkyl side chains or naphthene structures in the intermediate products to crack and escape from the liquid-phase system along with the gas stream, resulting in lower yields of the products.

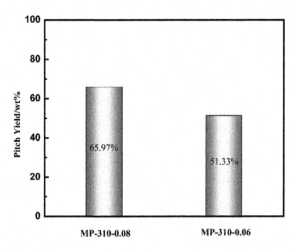

Figure 3.   Effect of vacuum degree on product yield.

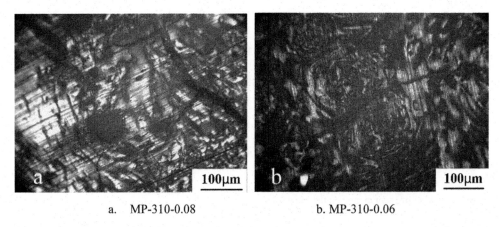

a.   MP-310-0.08   b. MP-310-0.06

Figure 4.   Effect of vacuum degree on product optical structure.

It can be seen from Figure 4 that the optical structure size of the product under the vacuum degree of 0.06 MPa is larger than 200 μm and exhibits a flow domain, which has a certain development compared with the product under the vacuum degree of 0.08 MPa. The reduction of the vacuum led to a further increase in the viscosity of the reaction system and complete removal of the non-mesogen molecules, which increased the degree of condensation of the products and increased the planarity of the aromatic lamellae, resulting in the formation of more long-range ordered structures, which in turn exhibited better optical textures.

4 CONCLUSION

Based on the results and discussions presented above, the conclusions are obtained as below:

(1) The above analysis shows that a large flow domain mesophase pitch with high anisotropy content can be obtained at 310°C and 0.06 MPa for 30 min.
(2) The low temperature and high vacuum degree do not allow the complete removal of non-mesogen molecules from the system, which not only affects the fluidity of the system in the middle and later stages of the reaction but also affects the flatness of the already formed aromatic molecules, resulting in the degradation of the optical structure quality.

ACKNOWLEDGMENTS

This work was financially supported by the National Natural Science Foundation of China (22108309).

REFERENCES

Gong X, Lou B, Yu R, et al. Carbonization of mesocarbon microbeads prepared from mesophase pitch with different anisotropic contents and their application in lithium-ion batteries[J]. Fuel Processing Technology, 2021, 217: 106832.

Kim C J, Ryu S K, Rhee B S. Properties of coal tar pitch-based mesophase separated by high-temperature centrifugation[J]. Carbon, 1993, 31(5): 833–838.

Marsh H. Carbonization and liquid-crystal (mesophase) development: Part 1. The significance of the mesophase during carbonization of coking coals[J]. Fuel, 1973, 52(3): 205–212.

Wang G, Eser S. Molecular composition of the high-boiling components of needle coke feedstocks and mesophase development[J]. Energy & fuels, 2007, 21(6): 3563–3572.

Wang G. Molecular composition of needle coke feedstocks and mesophase development during carbonization[J]. The Pennsylvania State University, 2005.

Yue Z, Liu C, Vakili A. Solvated mesophase pitch-based carbon fibers: thermal-oxidative stabilization of the spun fiber[J]. Journal of Materials Science, 2017, 52(13): 8176–8187.

Frontiers of Civil Engineering and Disaster Prevention and Control – Yang & Rahman (Eds)
© 2023 The Author(s), ISBN: 978-1-032-31200-2

Preparation of isotropic pitch with high spinnability through the co-carbonization of aromatic-rich distillate oil and polyvinyl alcohol

Luning Chai & Dong Liu*

State Key Laboratory of Heavy Oil Processing, China University of Petroleum, Qingdao, Shandong, China

ABSTRACT: An isotropic pitch with high spinnability was prepared by co-carbonization of aromatic-rich distillate oil (HCTO) and polyvinyl alcohol (PVA). The chemical structure of isotropic pitch and mechanical properties of carbon fibers were characterized by elemental analysis, Fourier transforms infrared spectra (FT-IR), tensile strength tester, etc. The results showed that the pitch yield and spinnability of isotropic pitch increased with the increase of PVA content at a certain reaction temperature. The addition of PVA led to an increase in the aliphatic structure of pitch molecules. The tensile strength of carbon fibers increased first and then decreased with the increase of PVA content.

1 INTRODUCTION

Isotropic pitch-based carbon fibers (IPCFs) have attracted extensive attention due to their wide source of raw materials, low production costs, and simple production processes (Mochida et al. 2000; Liedtke & Hüttinger 1996). However, the tensile strength and elastic modulus of most IPCFs were only o.5-1.0GPa and 30-40GPa, which could not meet the application needs of many specialized fields such as transportation, aerospace, and military hardware. Therefore, how to obtain IPCFs with excellent mechanical properties and low cost has become the pursuit of many researchers.

In order to obtain isotropic pitch with high spinnability, several methods, such as air blowing (Mochida et al. 1990), thermal treatment (Zhang et al. 2013), bromination-dehydrobromination polymerization, and adding chemical additives (Li et al. 2020; Shimanoe et al. 2020), have been used to modify the chemical structure of isotropic pitch. Although we could obtain IPCFs with good mechanical properties from these isotropic pitches, the harsh operating conditions result in higher carbon fiber production costs.

This study selected polyvinyl alcohol (PVA) and an aromatic-rich distillate oil as raw materials to prepare an isotropic pitch with smoothed reaction conditions. The effect of different PVA additions on the chemical structural composition of the isotropic pitch was studied. Several modern characterization techniques were used to analyze the molecular chemical structure of isotropic pitch to demine the optimal addition amount of PVA.

---

*Corresponding Author: liudong@upc.edu.cn

DOI 10.1201/9781003308577-80

2 EXPERIMENTAL SECTION

2.1 *Raw materials*

The aromatic-rich distillate oil was provided by Qilu Petrochemical Industries Co. Sinopec. The PVA was provided by Shanghai Macklin Biochemical Co. Ltd. The basic properties of the aromatic-rich distillate oil are listed in Table 1.

Table 1.   Properties of HCTO.

Items	Results
Density (20°C)/g·cm^{-3}	0.9944
Carbon residues/$wt\%$	19.0
Average molecular weight (VPO)	383
C/$wt\%$	87.93
H/$wt\%$	9.94
N/$wt\%$	0.61
S/$wt\%$	0.44
Aromatic carbon content (C_A)/$wt\%$	54.0
Naphthenic carbon content (C_N)/$wt\%$	23.9
Paraffinic carbon content (C_P)/$wt\%$	22.1
Saturates/$wt\%$	22.3
Aromatics/$wt\%$	63.9
Resins/$wt\%$	13.8
Asphaltenes/$wt\%$	0

2.2 *Preparation of isotropic pitch*

First, the HCTO and PVA were added to the stainless-steel reactor according to a certain proportion. Then, the atmosphere was replaced by high-purity nitrogen and repeated three times. The co-carbonization reaction was carried out at 390°C for 8 h. According to the PVA addition amount, the isotropic pitch was named H-PVA-X, where X represents the additional amount of PVA.

2.3 *Preparation of carbon fibers*

The melt spinning process was used to prepare green carbon fibers. Before spinning, 50g isotropic pitch was placed into the barrel, and then raising the temperature to 50°C above the softening point, the isotropic pitch was extruded by nitrogen. In order to avoid the melting of pitch fibers during high-temperature carbonization, the pitch fibers were stabilized in a horizontal tube furnace at 280°C for 1 h. And then, the stabilized fibers were carbonized at 1000°C for 30 min under an argon atmosphere.

2.4 *Samples characterization*

PerkinElmer 2400 series elemental analyzer was used to determine the composition of raw oil and isotropic pitches. According to the national standard GB/T4507-84, the softening point of isotropic pitches was determined by the needle penetration method. NEXUS infrared spectrometer was used for FT-IR characterization. The surface structure characteristics of carbon fibers were observed by field emission electron microscopy (JSM-6700F). The mechanical properties of carbon fibers were determined by the single-filament drawing method according to the national standard (ASTM D4018-2011).

Table 2.   Physical characteristics of the raw oil and spinnable pitches.

Sample	Ultimate analysis/%					C/H (Atomic ratio)	Group composition/%			Softening point/°C
	C	H	S	N	O		HS	HI-TS	TI	
H-PVA0	92.74	4.83	1.33	0.21	0.89	1.6	5.57	30.66	63.77	226
H-PVA5	92.58	4.97	1.31	0.24	0.87	1.55	5.85	40.71	53.44	229
H-PVA10	92.01	5.54	1.37	0.18	0.85	1.38	7.71	57.07	35.22	235
H-PVA15	91.87	5.98	1.41	0.21	0.81	1.28	8.85	62.21	28.94	237

Table 2 indicates the elemental composition of the pitch samples. The oxygen content of H-PVA0, H-PVA5, H-PVA10, and H-PVA15 were 0.89, 0.87, 0.85, and 0.81, respectively. Compared with the raw oil HCTO, the oxygen content of the pitch samples was slightly lower. The C/H atomic ratio of H-PVA0 was 1.6, a growth of about 0.86 compared with that of raw material (0.74), and the C/H atomic ratios of H-PVA5, H-PVA10, and H-PVA15 were 1.55, 1.38, 1.28, respectively, which indicate that the pyrolysis fractions of PVA engendered evidently influence on the chemical composition of the pitches. It can be speculated that the addition of PVA could change the reaction process to reduce the aromaticity of pyrolysis products.

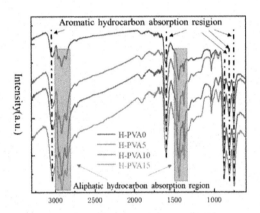

Figure 1.   FT-IR curves of H-PVA0, H-PVA5, H-PVA10 and H-PVA15.

In order to elucidate the structural features of isotropic pitches, FT-IR spectra are demonstrated in Figure 1. The absorption peak at 750 cm^{-1} corresponds to the cata-configuration CH structure in the pitch molecule; the absorption peak distributions at 880 cm^{-1} and 824 cm^{-1} correspond to the adjacent and isolated CH structures in the peri-condensation of the molecule. The absorption peak at 3040 cm^{-1} is attributed to the stretching vibration of the CH single bond in the aromatic structure. The absorption peak in the region of 2800~2980 cm^{-1} corresponds to the CH stretching vibration in aliphatic hydrocarbons. According to the changing trend of absorption peaks in the curve, it was found that with the increase of PVA addition, the content of aliphatic components in isotropic pitches increases significantly. The hydrogen ratios of the cata- and peri- condensation of pitch molecule gradually decreased. The aromaticity index (I_{Ar}) of the four isotropic pitches could be calculated by function: formula, $I_{Ar} = A_{(3150-2990)}/[A_{(3150-2990)} + A_{(2990-2880)}]$, where $A_{(3150-2990)}$ corresponds to the C-H stretching vibration of the aromatic component, and $A_{(2990-2800)}$ corresponds to the C-H stretching vibration of the aliphatic components [6]. The results showed that H-PVA0 had the highest aromatic index of 0.568, while that of H-PVA15 was 0.47, and the aromatic index of H-PVA5 and H-PVA10 was 0.534 and 0.501, respectively.

Table 3. Chemical structure of the spinnable pitches examined by ^{13}C-NMR.

Sample	Aliphatic (%)			Aromatic (%)	
	CH_3^a	CH_2^b	C_{chain}^c	$Car_{1,2}^d$	$Car_{1,3}^e$
H-PVA0	10.37	1.9	1.04	63.91	22.78
H-PVA5	8.33	6.06	1.22	62.87	21.52
H-PVA10	5.04	10.54	1.67	60.94	21.81
H-PVA15	4.27	15.54	1.79	57.76	20.64

a. Methyl carbons, b. Methylene carbons, c. CH_2 carbon in bridge/hydroaromatic structures, d. Catacondensed aromatic carbons, e. Pericondensed aromatic carbons.

The information on the chemical structures of the isotropic pitches was summarized in Table 3. The content of aromatic components gradually decreased in the order of H-PVA0>H-PVA5>H-PVA10>H-PVA15, with the increase of PVA content. The number of aliphatic chains gradually increases from H-PVA0 to H-PVA15. The change in the chemical structure of isotropic pitch molecules could improve the spinnability of the pitch.

Table 4. Spinnability of spinnable pitches.

Samples	Spinning temperature (°C)	Softenning point (°C)	Winding speed (rpm)	Diameter (μm)	Breakage number (in/3min)
H-PVA0	277	226	760	14	6
H-PVA5	281	229	820	10	3
H-PVA10	284	235	880	9.8	2
H-PVA15	289	237	960	9.2	0

Table 4 summarizes the spinnability of the four isotropic pitches. During the spinning process, the H-PVA broke six times at a winding speed of 760 rpm, which showed the poorest spinnability. The diameter of carbon fiber derived from H-PVA0 was 14 μm. The isotropic pitch prepared from HCTO co-carbonization with 15% PVA exhibited better spinnability than H-PVA5 and H-PVA10. The H-PVA15 was almost no instances of breakage, even at the severe winding speed of 96-rpm. The average diameter of carbon fiber derived from H-PVA15 was 9.2 μm. The results confirmed the spinnability of the isotropic pitch was gradually increased with the increase of PVA addition content. The improvement in spinnability of the pitch was a result of the change in the molecular chemical structure of isotropic pitches.

Figure 2 shows the curves of mechanical properties of carbon fibers with additional content of PVA. The results showed that under the same carbonization conditions, the tensile strength of carbon fibers first increased and then decreased with the increase of PVA content. When the PVA content was 5%, the tensile strength of carbon fiber was 1200 MPa. Based on the above analysis, adding an appropriate amount of PVA (5 wt%) in the co-carbonization reaction system could significantly improve the chemical structure of isotropic pitch molecules and obtain carbon fibers with excellent mechanical properties.

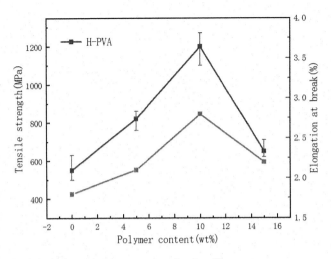

Figure 2.　Mechanical properties vs. polymer content of carbon fibers.

4 CONCLUSION

Isotropic pitches were formed by the co-carbonization of aromatic-rich distillate oil and PVA. The pitch yield and spinnability of isotropic pitch gradually increased with the increase of PVA content. Compared to H-PVA0, H-PVA5, H-PVA10, and H-PVA15 contained a higher proportion of aliphatic side chains and relatively few condensed aromatic molecular compositions. The tensile strength of carbon fibers increased first and then decreased with the increase of PVA content. The appropriate addition amount of PVA should be 5% (wt%).

ACKNOWLEDGMENTS

This work was financially supported by the National Natural Science Foundation of China (22108309).

REFERENCES

Li L, Lin X C, Zhang Y K. (2020) Characteristics of the mesophase and needle coke derived from the blended coal tar and biomass tar pitch. Journal Of Analytical and Applied Pyrolysis, 150: 104889.

Liedtke V, Hüttinger K J. (1996) Mesophase pitches a matrix precursor of carbon fiber reinforced carbon: I mesophase pitch preparation and characterization. Carbon, 34: 1057–1066.

Mochida I, Korai Y, Ku C-H. (2000) Chemistry of synthesis, structure, preparation, and application of aromatic-derived mesophase pitch. Carbon, 38: 305–328.

Mochida I, Toroshi H, Koral Y. (1990) A structural study on the stabilization and enhancement of mesophase pitch fiber. Journal of Materials Science, 25: 76–82.

Shimanoe H, Mashio T, Nakabayashi K. (2020) Manufacturing spinnable mesophase pitch using direct coal extracted fraction and its derived mesophase pitch-based carbon fiber. Carbon, 158: 922–929.

Zhang Bo, Song Huaihe, Chen Xiaohong. (2013) Transformation of Lewis acid during the carbonization and graphitization of mesophase pitches. Journal of Analytical and Applied Pyrolysis, 104: 433–440.

*Frontiers of Civil Engineering and Disaster Prevention and
Control – Yang & Rahman (Eds)*
© *2023 The Author(s), ISBN: 978-1-032-31200-2*

Simplified calculation of lateral seismic resistance of shield tunnel in soil-rock combination stratum

Jinhua Shang & Yanan Dong
Jinan Rail Transit Group Co., Ltd Jinan, P.R. China

Kejin Li
China Railway 14th Bureau Group Co. Ltd. Jinan, P.R. China

Guangbiao Shao*
School of Civil Engineering, Shandong Jianzhu University Jinan, P.R. China
*Key Laboratory of Building Structural Retrofitting and Underground Space Engineering, Ministry of
Education Jinan, P.R. China*

ABSTRACT: The typical soil-rock combination stratum in the Jinan area is used as the research
object, and the MIDAS/GTS NX finite element software is used to establish the two-dimensional
finite element stratum-structure model for the tunnel and geotechnical body. First, the seismic
response analysis is carried out separately for the soil-rock combination stratum and the common
soil site to obtain the horizontal acceleration results of the two sites. Then, the reaction acceleration
method is used to input the site horizontal acceleration results into the two-dimensional stratum
structure. The shear force and bending moment of the tunnel in the two types of strata are calculated
in the model, and the force characteristics of the shield tunnel in the two types of strata are analyzed.
Due to the difference in stiffness between the soil and rock formations, the shear force and bending
moment of the shield tunnel under transverse ground shaking are abruptly changed at the soil-rock
interface, which has a significant impact on the seismic performance of the tunnel. Research shows
that the effect of the soil-rock combination on the tunnel force under lateral earthquake is much
greater than that of a homogeneous soil layer, so the soil-rock interface should be considered the
weak location in the seismic design of the tunnel.

1 INTRODUCTION

In recent years, with the rapid development of China's urbanization, the problem of urban land
shortage has become more and more prominent, while with the increasing number of private cars,
urban congestion is extremely serious. In order to solve the problem of road land tension and urban
traffic congestion, people began to develop and utilize underground space, and the subway, as an
important part of urban rail transit, has been developed rapidly.

At present, most of the underground structures are located in soil or rock strata, but there is less
research on the underground structures located in soil-rock combination strata. Therefore, it is of
strong theoretical significance and engineering application value to carry out research on lateral
seismic resistance of shield tunnels under soil-rock combination stratum.

The current calculation methods for seismic analysis of underground structures at home and
abroad include the seismic coefficient method (Liu et al. 2010), free-field deformation method
(Hashash et al. 2001), flexibility coefficient method (Wang 1993), reaction displacement method

---

*Corresponding Author: sgbou@126.com

DOI 10.1201/9781003308577-81

(Kazuhiko 1994), reaction acceleration method (Tateishi 2005), pushover analysis method for underground structures (Liu et al. 2008), and nonlinear dynamic time method (Wang 2007). Among them, the pushover analysis method and nonlinear dynamic time method for underground structures have the characteristics of high professional knowledge requirement for operators and large calculation workload, so they are less used in practical engineering. The remaining five are simplified seismic analysis methods with easy operation and small calculation workload compared with the first two methods, so simplified seismic analysis methods for underground structures are widely used in practical engineering. Since the seismic coefficient method, free field deformation method, flexibility coefficient method, reaction displacement method, and other common simplified seismic analysis methods indirectly consider the interaction between the underground structure and the surrounding soil, and it is difficult to guarantee the calculation accuracy. In contrast, the reaction acceleration method can accurately respond to the interaction between the underground structure and the surrounding soil, which establishes a two-dimensional overall finite element model of the stratum-structure, and firstly, through the site seismic response analysis to obtain the acceleration function of the stratum at the most unfavorable moment, and then input the acceleration function into the calculation model for the solution to obtain the dynamic response law of the subsurface structure. Li Xinxing (Li 2015) used the reaction acceleration method for the seismic design of underground stations and their enclosure structures and verified the applicability of reaction acceleration in the seismic design of complex structures. Yang Yaqin et al. (Yang et al. 2016) used the reaction acceleration method to compare the calculation results with the reaction displacement method and verified that the calculation results of the reaction acceleration method are more reasonable and accurate. Xu Kunpeng et al. (Xu et al. 2020) compared the calculation results of the reaction acceleration method with the Xu Kunpeng et al. (Xu et al. 2020) compared the results of the reaction acceleration method with the results of the dynamic time analysis and verified the accuracy of the solution of the reaction acceleration method. Therefore, in this paper, the reaction acceleration method will be used for the simplified calculation of transverse seismic resistance of shield tunnels in the soil-rock combination stratum.

2 PROJECT OVERVIEW

Taking Jinan Metro Line 2 as the engineering background, the shield tunnel is located in the soil-rock combination stratum, the overlying soil layer of the tunnel is 11.8m thick, the upper part is miscellaneous fill and powdered clay layer, the lower part is medium weathered limestone, the miscellaneous fill is 2m thick, and the powdered clay is 13m thick. The shield tunnel is located in the stratum section, as shown in Figure 1.

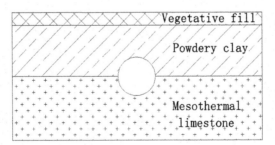

Figure 1.   Shield tunnel in the stratigraphic section.

Due to the large difference in stiffness between soil and rock layers, this not only brings certain difficulties to the subway construction, but more importantly, there is still a lack of understanding of the earthquake damage mechanism and seismic performance of subway underground structures, and the soil-rock combination stratum is more uncertain and challenging to the ground vibration response mechanism of subway and other underground structures in this type of stratum due to the effect of the embedded solid constraint of rocks and ground vibration propagation speed.

3 ONE-DIMENSIONAL SEISMIC RESPONSE ANALYSIS OF SOIL-ROCK COMBINATION SITES

3.1 *Calculated profile and parameters*

According to the calculation steps of the reaction acceleration method, firstly, a 1D seismic response analysis of the site is required to obtain the ground acceleration results at the most unfavorable moment and then input them into the two-dimensional integral finite element model of the stratum structure to solve the force characteristics of the underground structure.

The free-field vibration calculation of transverse seismic action is carried out for the Jinan soil-rock combination stratum, and the free-field vibration situation is compared and analyzed with that of the ordinary soil stratum site conditions. The geological conditions of the soil-rock combination site and the common soil stratum site are shown in Figure 2.

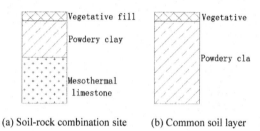

(a) Soil-rock combination site   (b) Common soil layer

Figure 2.   Site geological conditions.

According to the geological survey report and seismic safety evaluation report of the project site, the geodynamic parameters of the site foundation are obtained. The values of dynamic elastic modulus, Poisson's ratio, density, and shear wave velocity of the geotechnical body are shown in Table 1. The relationship between geotechnical modulus ratio and shear strain and the relationship between damping ratio and shear strain are shown in Table 2.

Table 1.   Table of foundation geodynamic parameters.

Material	E_d//MPa	ν	$\gamma/(kN{\cdot}m^{-3})$	$V_S/(m{\cdot}s^{-1})$
Vegetative fill	0.125	0.40	18.0	154.3
Powdery clay	0.15	0.35	19.0	194.6
Mesothermal limestone	5	0.25	22.0	900.6

3.2 *Seismic response analysis*

To consider the nonlinearity of the geotechnical body, an equivalent linearization method (Hu 2003; Jin & Kong 2004; Li 2014) was used to analyze the one-dimensional ground seismic response of the engineering site. The method uses an equivalent nonlinear viscoelastic model, which converts the nonlinear problem into a linear problem by replacing the shear modulus and damping ratio that vary with the strain amplitude with an equivalent shear modulus and damping ratio, and finally solves the linear fluctuation equation in the frequency domain.

The El Centro horizontal real seismic recording wave is selected as the input ground shaking. The seismic wave holding time is 53.46s, the peak acceleration reaches the maximum at 11.46s, and the maximum peak acceleration is 0.2142g. The El Centro wave acceleration time curve is shown in Figure 3.

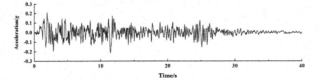

Figure 3.   El-Centro wave time history curve.

The seismic response analysis was carried out for the combined soil-rock strata and soil strata by inputting El Centro seismic waves and adjusting the peak acceleration to 0.10g to obtain the horizontal acceleration results of the strata at the most unfavorable moment, as shown in Figure 4.

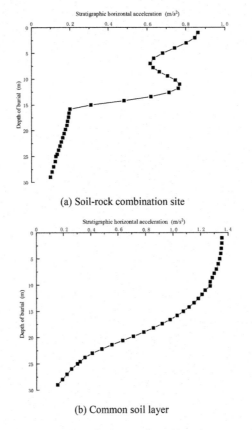

(a) Soil-rock combination site

(b) Common soil layer

Figure 4.   Ground level acceleration results at the most unfavorable moment.

According to the results of horizontal acceleration of strata, the peak acceleration of seismic waves in both strata increases with decreasing depth of burial, and there is an amplification effect in the process of seismic wave transmission from bedrock input to the surface. In the soil-rock combination stratum, the peak horizontal acceleration changes abruptly from 0.10g to 0.8g, while in the ordinary soil stratum, the horizontal acceleration varies uniformly with the burial depth.

4 SIMPLIFIED CALCULATION OF LATERAL SEISMIC RESISTANCE OF SHIELD TUNNEL IN SOIL-ROCK COMBINATION STRATUM

4.1 *Computational models*

The reaction acceleration method is used for calculation, and a two-dimensional stratigraphic-structural integral finite element model is established. The model is 80 m long and 30 m high, and the thickness of the overlying soil layer of the tunnel is 11.8 m. The soil, rock, and tunnel are all linear elastic models, with the geotechnical body as a plane strain unit and the tunnel as a beam unit. The tunnel is equated into a homogeneous circle with an outer diameter of 3.2m, an inner diameter of 2.9m, and a concrete strength of C50 for the tube piece. A fixed boundary is used at the bottom of the model, and horizontal sliding supports are used on both sides. In order to truly reflect the interaction between soil, rock, and structure, the contact unit-Goodman unit is established between soil and rock, between soil and structure, and between rock and structure. The finite element model is shown in Figure 5.

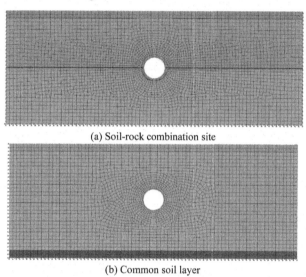

(a) Soil-rock combination site

(b) Common soil layer

Figure 5.   Finite element model.

4.2 *Calculation results*

The horizontal acceleration function is obtained from the seismic response analysis of the site and input into the calculation model. The results of shear force and a bending moment of the shield tunnel under the two stratigraphic conditions are calculated and shown in Figures 6 to 7.

From the results of the tunnel shear calculation, it can be seen that the shear results of the tunnel in the rock part of the soil-rock combination are significantly smaller than those in the soil part of the tunnel, and the tunnel shear changes abruptly at the soil-rock interface, and the abrupt change is the maximum value of the shear force, which is 580.44kN; the shear results in the normal soil layer are symmetrically arranged, and the maximum value of the shear force occurs at the lower-left 45° of the tunnel, with the maximum value of the shear force is 74.76kN.

From the results of the tunnel moment calculation, it can be seen that the tunnel moment in the rock part of the soil-rock combination is significantly smaller than that in the soil part of the tunnel, the tunnel moment at the soil-rock interface, and the sudden change is the maximum value of the moment, the maximum value of the moment is 460.32kN·m. The ordinary soil moment results are symmetrically arranged, the maximum value of the moment occurs at the bottom of the tunnel arch position, and the maximum value of the moment is 79.20kN·m.

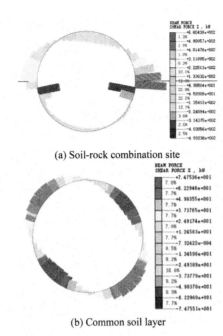

(a) Soil-rock combination site

(b) Common soil layer

Figure 6.   Shear calculation results.

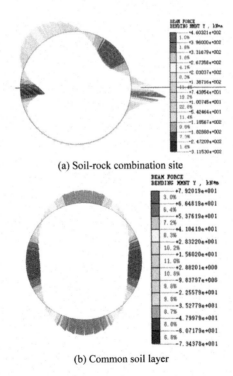

(a) Soil-rock combination site

(b) Common soil layer

Figure 7.   Bending moment calculation results.

5 CONCLUSION

Simplified calculations of lateral seismic resistance for shield tunnels in soil-rock combination formations lead to the response law of soil-rock combination formations and the force characteristics of shield tunnels. The conclusions are summarized as follows.

(1) In the soil-rock combination strata, the horizontal acceleration of the strata will change abruptly at the soil-rock intersection.
(2) In the soil-rock combination, the shear force and bending moment of the shield tunnel will change abruptly at the soil-rock interface, and the value of the abrupt change is the maximum value of the structural force, so the soil-rock interface should be considered as the seismic weak position when designing the tunnel.

FUNDING PROJECT

Jinan Science and Technology Development Program (NO.201807005); Shandong Provincial Department of Transportation Science and Technology Plan Project (2019B11); Shandong Provincial Key Research and Development Program (NO.2019JZZY010428).

AUTHOR

Corresponding author: SHAO Guangbiao (1978-), Male, Ph.D., Engaged in research on excavation and support technology of urban underground engineering and disaster prevention and reduction of underground engineering. Email: sgbou@126.com

REFERENCES

Hashash Y M A, Hook J J, Schmidt B, et al. Seismic design and analysis of underground structures[J]. Tunneling and Underground Space Technology, 2001, 16(4):247–293.
Hu, Luxian. Tutorial on Seismic Safety Evaluation [M]. Beijing: Earthquake Press, 2003.
Jin Xing, Kong Ge. Nonlinear analysis of seismic response in horizontally stratified sites[J]. Earthquake Engineering and Engineering Vibration, 2004, 24(3):38–43.
Kazuhiko Kawashima. Seismic Design of Underground Structures [M]. Japan: Kashima Publishing, 1994.
Li Lingling. Analysis of nonlinear seismic response of underground structures[D]. Tianjin: Tianjin University, 2014.
Li Xinxing. Seismic design of underground stations considering enclosure structure by reactive acceleration method[J]. Underground Engineering and Tunneling, 2015(03):1–5+50.
Liu Jingbo, Liu Xiangqing, Li Bin. Pushover analysis method for seismic analysis and design of underground structures[J]. Journal of Civil Engineering, 2008, 41(4):73–80.
Liu Jingbo, Wang Wenhui, Zhao Dongdong. Review of seismic design analysis methods for cross-section of underground structures[J]. Construction Technology, 2010, 39(06):91–95.
Tateishi A. Study on seismic analysis methods in the cross-section of underground structures using static finite element method[J]. Structural Engineering /Earthquake Engineering, 2005, 22(1):41–54.
Wang Guo-Bo. Research on the theory and method of calculating three-dimensional seismic response of soft ground railway station structure[D]. Shanghai: Tongji University, 2007.
Wang J N. Seismic design of tunnels: a simple state-of-the-art design approach[M]. New York: Parsons Brinkerho Quade and Douglas Inc, 1993.
Xu Kunpeng, Jing Liping, Bin Jia. Comparative analysis of forced reaction displacement method and reaction acceleration method for underground structures[J]. Journal of Earthquake Engineering, 2020, 42(04): 967–972.
Yang YAG, Zhang CHUNJIN, Wang GUOBO. Application of reaction acceleration method in seismic analysis of underground structures[J]. Roadbed Engineering, 2016(03):144–147.

Frontiers of Civil Engineering and Disaster Prevention and Control – Yang & Rahman (Eds)
© *2023 The Author(s), ISBN: 978-1-032-31200-2*

Influence of coal gangue coarse aggregate on the basic mechanical behavior of concrete

Hanqing Liu* & Guoliang Bai*
School of Civil Engineering, Xi'an University of Architecture & Technology, Xi'an, PR China;
Key Lab of Structural Engineering and Earthquake Resistance, Ministry of Education (XAUAT), Xi'an, P.R. China

ABSTRACT: This paper analyzes the physical and mechanical properties of coal gangue coarse aggregate, and the mechanical properties of concrete with coal gangue as coarse aggregate were studied. The test results show that the crushing index of coal gangue coarse aggregate meets the requirements for the III grade crushed stone, which indicates that coal gangue coarse aggregate can be used to produce medium-strength concrete. Based on the test results, the relationship between the strength of coal gangue coarse aggregate concrete and the water-cement ratio is established, and the conversion relationship between the cubic compressive strength and the splitting strength is obtained by regression. The conversion relationship between the cube compressive strength and the axial compressive strength can be calculated according to the formula in GB 50010-2010.

1 INTRODUCTION

Coal gangue is a solid waste discharged during coal mining (GB/T 29162, 2012). The traditional accumulation treatment of coal gangue will occupy a large amount of land and will also pollute the environment around the mining area in the long-term accumulation process (Li & Wang 2019). At present, sustainable development has become a new development concept, and how to use coal gangue in China has attracted wide attention.

Existing research shows that coal gangue is a kind of rock with similar physical and chemical composition to natural aggregate. Compared with natural aggregate and recycled aggregate, coal gangue can also be used as aggregate (Chen 1994). At present, some research has been carried out on CGC. Duan et al. (2014) analyzed the effect of basic properties of coal gangue aggregate on the mechanical properties of CGC. The test results show that the CGC could achieve the maximum strength under the condition of coal gangue-cement matrix strength being harmonious with the cement matrix strength. Liu et al. (2020) studied the elastic modulus of CGC through experiments and gave the applicable calculation model, which provided a reference for the design of CGC. Guo et al. (Guo & Zhu 2011) designed multiple sets of specimens to study the durability of CGC and finally gave the mix proportion of coal gangue concrete with good durability. Taking water-cement ratio, coal gangue replacement ratio, and carbonation time as variables, Yi et al. (2017) studied the durability of CGC and gave the carbonation model of CGC.

This paper mainly carries out the following research work. Firstly, the physical and mechanical properties of coal gangue coarse aggregate are analyzed. After that, several groups of tests on the mechanical properties of CGC were designed. Through the test, the failure mode and failure mechanism of CGC were studied, and the effect of the water-cement ratio on the basic mechanical properties of coal gangue concrete with a 100% replacement ratio was studied. Finally, through the regression analysis of the experimental results, the strength conversion relationship suitable for CGC was established, which provides a reference for the engineering application of CGC.

---

*Corresponding Authors: hqliu@xauat.edu.cn and guoliangbai@126.com

DOI 10.1201/9781003308577-82

2 EXPERIMENTAL PROGRAM

2.1 *Materials*

The cement uses 42.5 MPa ordinary Portland cement. The fine aggregate is ground yellow sand from northern Shaanxi, with a fineness modulus of 1.1. Coarse aggregates were divided into two categories: (1) coal gangue coarse aggregates (CGCA); (2) natural coarse aggregates (NCA). To master the characteristics of coal gangue coarse aggregate characteristics, the apparent density, bulk density, crushing index, and water absorption were evaluated according to GB/T 14685-2011 (GB/T 14685, 2011), and the test results are shown in Table 1. It can also be seen from the crushing index of coarse aggregate that the crushing index of CGCA is smaller than that of NCA, which meets the requirements for the III grade crushed stone ($\leq$30%) and is suitable for producing medium-strength concrete.

Table 1.   Physical performance indexes of aggregate.

Coarse aggregate	Grading (mm)	Apparent density (kg/m^3)	Bulk density (kg/m^3)	Crushing index (%)	Water absorption (%)
CGCA	5-31.5	2400	1350	21.0	5.7
NCA	5-31.5	2656	1590	10.7	0.8

2.2 *Mixture proportions*

To investigate the mechanical properties of CGC, seven groups of CGC with different water-cement ratios were designed in this test. Due to the high-water absorption of coal gangue, the coarse aggregate of coal gangue was pre-wetted before the test. According to JGJ 55-2011 (JGJ 55, 2011), eight groups of concrete with different mix proportions were designed, as shown in Table 2.

Table 2.   Mix ratio of concrete.

Mix	r (%)	w/c	Mixture proportions (kg/m^3)					
			Cement	Water	Sand	NCA	CGCA	Water reducer
CGC1	100	0.65	210	136.5	661	0	1342	1.05
CGC2	100	0.60	325	195.0	604	0	1226	0.00
CGC3	100	0.55	280	154.0	632	0	1284	2.80
CGC4	100	0.50	320	160.0	617	0	1253	1.60
CGC5	100	0.45	433	195.0	568	0	1154	0.00
CGC6	100	0.40	341	136.5	618	0	1254	3.41
CGC7	100	0.35	390	136.5	602	0	1222	3.90
NC	0	0.40	488	195.0	550	1117	0	0.00

According to the requirements of GB/T 50081-2002 (GB/T 50081, 2003), two types of specimens were designed to test the mechanical properties of concrete. The number and size of test specimens used for different tests are presented in Table 3. Each group of specimens was cured under the same condition, and the test was carried out after 28 days of curing.

2.3 *Test loading*

The test loading was carried out using the WAW-1000 universal testing machine. The mechanical properties of CGC were evaluated according to the method recommended by GB/T 50081-2002 (GB/T 50081, 2003). The elastic modulus and Poisson's ratio of CGC are obtained by pasting transverse and vertical strain gauges on the specimen surface.

Table 3. Specimen details for each mix.

Test	Number of specimens	Dimensions of test specimens for concrete
Cube compressive strength	3	Cube – 150 mm× 150 mm × 150 mm
Prism compressive strength	3	Prism – 150 mm × 150 mm × 300 mm
Split tensile strength	3	Cube – 150 mm× 150 mm × 150 mm
Modulus of elasticity	3	Prism – 150 mm × 150 mm × 300 mm
Poisson's ratio	3	Prism – 150 mm × 150 mm × 300 mm

3 RESULTS AND DISCUSSION

The test results are shown in Table 4, and the analysis of test failure phenomena and test results is shown in the following.

Table 4. Test results of mechanical properties.

Mixtures	f_{cu}^a(MPa)	f_c^b(MPa)	f_{ts}^c(MPa)	Elastic modulus (GPa)	Poisson's ratio
CGC1	15.7	11.9	0.98	7.42	0.236
CGC2	18.5	13.4	1.13	9.25	0.148
CGC3	17.0	12.3	1.09	8.67	0.166
CGC4	20.8	15.2	1.31	9.73	0.227
CGC5	23.3	18.3	1.45	11.25	0.199
CGC6	26.4	18.6	1.48	11.07	0.182
CGC7	28.7	20.3	1.67	11.81	0.217
NC	44.2	37.5	3.27	35.18	0.202

$^a f_{cu}$: cube compressive strength; $^b f_c$: axial compressive strength; $^c f_{ts}$: splitting strength.

3.1 *Failure mode*

3.1.1 *Cube compressive strength*

After the specimen was loaded, the upper and lower ends of the specimen were restrained by the loading plate, and the transverse deformation was small, while the middle part had the largest transverse expansion deformation. At the early stage before the 30% ultimate load, the specimen was in the elastic stage, and there were no obvious cracks on the specimen surface. With the increase of load, a vertical crack appeared in the middle of the height of the specimen, near the side surface, and then extended upward and downward, gradually turning to the corner of the specimen to form an x-shape crack, as shown in Figure 1.

(a) NC   (b) CGC1

Figure 1. Fracture propagation morphology of Typical Specimens.

3.1.2 *Axial compressive strength*

At the beginning of loading, the deformation of the specimen was small, and the specimen was in the elastic deformation stage. With the increase of vertical load, the first vertical crack appeared on the surface of the middle part of the specimen. After that, the load continued to increase, and a number of discontinuous longitudinal cracks appeared on the specimen successively and continued to extend, expand and connect, and gradually through the full section. When the specimen finally failed, a macroscopic oblique crack was formed on the surface. The surface crack propagation morphology of the CGC specimen is basically consistent with that of the NC specimen, as shown in Figure 2.

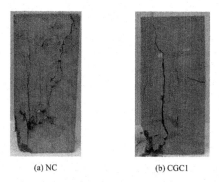

(a) NC   (b) CGC1

Figure 2.　Fracture propagation geometry of Typical Specimens.

3.1.3 *Split tensile strength*

The process of test loading can be divided into two stages. When the load was less than the ultimate load, the deformation of the specimen was not significant, while when the load reached the peak load, the sample was quickly split into two halves from the middle, and the damage was very sudden. Figure 3 shows the splitting section shape of each specimen.

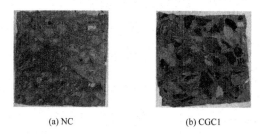

(a) NC   (b) CGC1

Figure 3.　Cross-sectional shape.

It can be seen from Figure 3 that due to the existence of NCA, NC can effectively inhibit the expansion of cracks. The destruction of NC is caused by insufficient bonding strength between the aggregate and mortar, so the split section is uneven. For CGC, the existence of CGCA cannot effectively prevent the expansion of cracks, resulting in cracks directly passing through the CGCA, so the splitting section is very flat. Through the analysis of the failure modes of the two types of concrete, it can be found that different from the failure mode of NC, when CGC is damaged, in addition to the bond failure between CGCA and cement mortar, there is also the failure of CGCA.

3.2 *Strength regression*

From the test results in Table 4, the water-cement ratio greatly influences the strength of CGC. The strength of CGC increases with the decrease of the water-cement ratio, and the relationship between them is approximately inverse. Therefore, $y = a/x + b$ was used as the mathematical model to regress the relationship between strength and water-cement ratio. Finally, the regression curve is shown in Figure 4, and the corresponding regression equation is shown in Eq. (1).

$$f_{cu} = \frac{10}{w/c} + 0.42 \qquad (1)$$

where: f_{cu} is the cube compressive strength, w/c is the water-cement ratio.

The fitting curve in Figure 4 is consistent with the experimental results, and $R^2 = 0.95$, indicating that the regression results are reasonable. Therefore, in engineering applications, Eq. (1) can be used as an estimation formula for the strength of CGC with a replacement ratio of 100%.

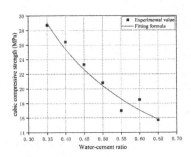

Figure 4.   Intensity regression curve of CGC.   Figure 5.   Splitting strength of CGC.

3.3 Axial compressive strength

In the design of the concrete structure, axial compressive strength is mainly used. According to GB 50010-2010 (GB 50010, 2010), the calculation formula of axial compressive strength (fc) is given:

$$f_c = 0.76 f_{cu} \qquad (2)$$

The theoretical results of prism compressive strength of CGC calculated by Eq. (2) are given in Table 5. The maximum relative error between the experimental results and the theoretical results is −6.9%, which indicates that the result calculated by Eq. (2) is more accurate. Therefore, it is suggested to use Eq. (2) to calculate the prism compressive strength of CGC with a 100% replacement ratio.

Table 5.   Experimental and theoretical results of prism compressive strength of CGC.

Groups	Prism compressive strength (MPa)		Relative error (%)
	Experimental results	Theoretical results	
CGC1	11.9	11.9	0.0
CGC2	13.4	14.1	−5.0
CGC3	12.3	12.9	−4.7
CGC4	15.2	15.8	−3.8
CGC5	18.3	17.7	3.4
CGC6	21.2	20.1	5.5
CGC7	20.3	21.8	−6.9

3.4 Splitting tensile strength

The formula for calculating the splitting strength of NC given in GB 50010-2010 is as follows:

$$f_{ts} = A f_{cu}^{B} \qquad (3)$$

Where f_{ts} is the splitting tensile strength, and A and B are regression parameters. For ordinary concrete, A=0.19 and B=0.75.

The test results show that the regression parameters A=0.19 and B=0.75 of NC are no longer applicable to CGC, so the conversion relationship between splitting strength and cubic compressive strength will be re-regressed based on Eq. (3). Finally, A=0.135 and B=0.75 were obtained through nonlinear regression:

$$f_{ts} = 0.135 f_{cu}^{0.75} \tag{4}$$

The regression curve is presented in Figure 5. The regression results of Eq. (4) are in good agreement with the test results, indicating that Eq. (4) can be used to calculate the splitting strength of CGC with a 100% replacement ratio.

4 CONCLUSIONS

Based on the results and discussions presented above, the conclusions are obtained as below:

(1) Although the physical and mechanical properties of coal gangue coarse aggregate are poor compared with natural coarse aggregate, the crushing index of coal gangue coarse aggregate meets the requirements for the III grade crushed stone, which indicates that coal gangue coarse aggregate can be used to produce medium-strength concrete.
(2) The formula for calculating the strength of coal gangue coarse aggregate concrete under different water-cement ratios is given, and the conversion relationship between cubic compressive strength and splitting tensile strength of coal gangue coarse aggregate concrete is established.
(3) The conversion relationship between the axial compressive strength and the cubic compressive strength in the "Specification for Concrete Structure Design" (GB 50010-2010) is also applicable to coal gangue coarse aggregate concrete.

ACKNOWLEDGEMENTS

This study was funded by the Shaanxi Provincial Science and Technology Plan Achievement Promotion Project (2020CGHJ-017) and the Key Laboratory Project of the Shaanxi Provincial Department of Education (20JS071).

REFERENCES

B. P. Chen, Strength of gangue concrete, Industrial Construction, 7(1994) 29–32.

C. Yi, H. Q. Ma, H. G. Zhu, et al., Investigation on Anti-carbonation Performance of Coal Gangue Coarse Aggregate Concrete, Journal of Building Materials. 05 (2017) 131–137.

GB 50010, Code for Design of Concrete Structures, Chinese Building Construction Publishing Press, Beijing, 2010.

GB/T 14685, Pebble and crushed stone for construction, Standards Press of China, Beijing, China, 2011.

GB/T 29162, Classification of gangue, Standards Press of China, Beijing, China, 2012.

GB/T 50081, Standard for test method of mechanical properties on ordinary concrete, Chinese Building Construction Publishing Press, Beijing, 2003.

H. Q. Liu, Q. Xu, Q. H. Wang, et al., Prediction of the elastic modulus of concrete with spontaneous-combustion and rock coal gangue aggregates, Structures, 28(2020)774–785.

J. M. Guo, L. L. Zhu, Experimental Research on Durabilities of Coal Gangue Concrete, Advanced Materials Research. 306–307 (2011) 1569–1575.

J. Y. Li, J. M. Wang, Comprehensive utilization and environmental risks of coal gangue: A review, J. Clean Prod. 239 (2019) 117946.

JGJ 55, Specification for mix proportion design of ordinary concrete, Chinese Building Construction Publishing Press, Beijing, 2011.

X. M. Duan, J. W. Xia, F. Z. Yang, et al., Experimental research on the influence of coal gangue aggregate behavior on the mechanical properties of concrete, Industrial Construction. 3 (2014) 114–118.

Frontiers of Civil Engineering and Disaster Prevention and Control – Yang & Rahman (Eds)
© 2023 The Author(s), ISBN: 978-1-032-31200-2

Design and application of formwork platform for the north side tower of Wujiagang Yangtze River bridge

Kaiqiang Wang, Deng Yang*, Xiaosheng Liu, Leilei Zhu & Yong Zhou
China Construction Third Engineering Bureau Co., Ltd., Wuhan, China

ABSTRACT: To solve the low efficiency, multi-fulcrums, and poor safety in cast-in-place concrete bridge tower climbing construction, the double formwork cycle construction and the corner climbing were proposed, and a new formwork was designed for the Wujiagang Yangtze River Bridge project in Yichang, which was successfully applied after verification of structural loading capacity by finite element analysis. Under specific conditions, the construction period of the double formwork cycle construction process is shorter than that of a single section in traditional formwork by one day. It adopts a supporting and lifting system with large stroke and high loading capacity arranged at four corners for continuous construction and avoiding high-altitude demolition and modification of formwork. The frame is characterized by integrity, automatic contraction, high loading capacity, and small demolition and modification.

1 INTRODUCTION

There have been a series of research and application of new formworks (Zhang et al. 2022) to construct super high-rise buildings worldwide. For example, China Construction Third Engineering Bureau (Pan et al. 2017) designed an integrated construction platform for the Shenyang Baoneng Global Financial Center Project, reducing the construction period and saving costs. Zhang Hao from Tsinghua University (Zhang et al. 2022) discussed the development status and trend of the construction platform through a literature review and interview survey. However, there is no research and application on the new formwork of bridge tower construction, which still focuses on hydraulic climbing formwork (Huang 2019; Yin & Wang 2018; Yin 2015). With the complex construction environment, industrialized construction technology, and intelligent operation (Zhang 2018), there are three problems in the traditional climbing formwork technology: First, sequential operation leads to the construction efficiency limited by the concrete age. Second, the climbing formwork fulcrum, with a low loading capacity and a large number, is greatly limited by the transverse members and protruding structures. It often needs high-altitude demolition and modification with high safety risk. Third, the climbing formwork, elevator, and other equipment have poor adaptability, inconvenient access, and potential safety hazards. Therefore, China Construction Third Engineering Bureau designed and applied a new formwork on the north main tower of the Wujiagang Yangtze River Bridge for the above problems.

The mainline of the Wujiagang Yangtze River Bridge (Miao 2020) is 2.813 km long. As shown in Figure 1, the main span length and the bridge width of the main bridge are 1 160 m and 31.5 m, respectively, which is a suspension bridge with an orthotropic plate and steel box girder; the rise span ratio of the main cable in the middle span of the main bridge is 1/9, and the rise is 128.889 m. The transverse center distance between the two main cables is 26.5 m. The south bank adopts gravity anchorage, while the north bank adopts tunnel anchorage. Moreover, vertical bearings, longitudinal

---

*Corresponding Author: 949569531@qq.com

DOI 10.1201/9781003308577-83

limit bearings, and transverse wind-resistant bearings are set at the upstream and downstream of the crossings between the main tower and the main beam, and hydraulic dampers are set between the longitudinal end of the main beam and the main tower.

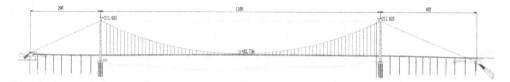

Figure 1. Elevation of Wujiagang Yangtze river bridge.

As shown in Figure 2, the main tower is composed of pile foundations, a platform, tower columns upstream and downstream, upper and lower beams, and a steel truss. The tower column adopts a single box rectangular, and the tower top is 7 m × 5.5 m, the bottom tower section is 10 m × 7 m, the corner fillet radius is 100 mm, and the facade has protruding decorative strips. The upper and lower beams are box section cast-in-place prestressed rebar concrete structures, with the upper beam 7 m high and 6.6 m wide, with the lower beam 9 m high and 8.5 m wide. The steel truss comprises seven sections, eight closure sections, and seven embedded sections connected with the upper beam and tower column by PBL shear connectors and shear studs.

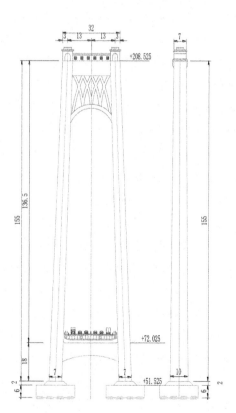

Figure 2. Elevation of main tower of Wujiagang Yangtze river bridge.

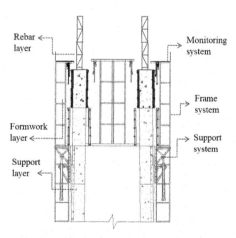

Figure 3. Formwork composition.

2 FORMWORK DESIGN

2.1 Composition and principle of formwork

The main tower construction formwork is shown in Figure 3, composed of support, frame, formwork, and monitoring systems. The steel frame is divided into the rebar layer, the formwork layer, and the support layer according to the functions of each layer. The rebar layer is used for rebar binding; the formwork layer has four layers with two construction sections for the recycling of two formworks; there are four supporting layers, of which the upper two layers are the formwork load-bearing layer, and the lower two layers are used for the removal of supporting parts. The plane layout is shown in Figure 4. The steel frame is composed of four L-shaped units on the plane. The slidable walkway plates form a channel between the units to be automatically contracted with the contraction of the bridge tower section.

The supporting and lifting system of the formwork, including the upper support frame, lower support frame, lifting cylinder, load-bearing parts, and embedded bolts, is shown in Figure 5. The load-bearing parts are also used as corner fillet formwork. In the construction state, the load path is that the steel frame transmits the upper load to the upper support frame, the upper support frame to the load-bearing parts, and the load-bearing parts to the M64 embedded bolts. In the lifting state, the load path is that the steel frame transmits the upper load to the upper support frame, the upper support frame transmits the horizontal force to the load-bearing parts, the vertical load to the lifting cylinder, the lifting cylinder to the lower support frame, the lower support frame to the load-bearing parts, and the load-bearing parts to the M56 embedded bolts. The Wujiagang Yangtze River Bridge adopts four lifting cylinders, with a maximum stroke of 3700 mm, maximum lifting force of 100 t, and maximum lifting pressure of 26 MPa. Each cylinder is equipped with an independent hydraulic pump station and an external cable transducer. The cable transducers read the cylinder stroke in real-time and feed it back to the lifting control system. The control system adjusts the cylinder speed through the proportional valve to synchronize multiple cylinders. If the stroke difference of the cylinder exceeds 5 mm, the system will automatically stop and give an alarm. It is needed to continue lifting after the cause is identified and the alarm is removed.

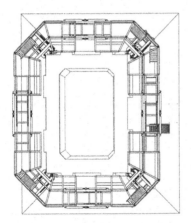

Figure 4. Plan view of formwork.

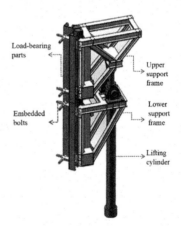

Figure 5. Composition of support system.

2.2 Analysis and calculation of formwork

The overall finite element model of the formwork is established, as shown in Figure 6, composed of a support frame, connecting rod, and conventional steel structure frame. Beam elements are used to simulate all members. The two ends of all connecting rods are hinged, simulating the pin connection by releasing the beam bending moment. Besides, the calculation refers to the engineering standard for hydraulic climbing formwork (JGJT195-2018) and other relevant standards, which shows that

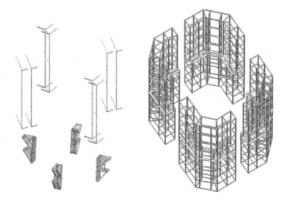

Figure 6. Model analysis.

under various working conditions, the maximum combined stress is 147 MPa and the maximum shear stress is 85 MPa, which is less than the designed strength of Q235 steel. The calculating results of the support reaction show that the maximum lifting load of a single cylinder is 52 t, and the maximum lifting pressure of the cylinder is about 8.2 MPa, which is consistent with the measured lifting pressure of the hydraulic system of 7.1 MPa ~ 8.8 MPa. In addition, the analysis results of the action effect of the steel frame cooperative connecting rod are shown in Figures 7 and 8. If the four frame units are connected as a whole using the connecting rod, their maximum combined stress under the same working condition is reduced from 176 MPa to 147 MPa.

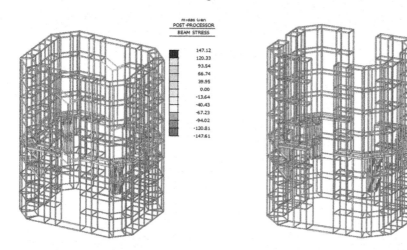

Figure 7.   Stress results with connecting rod model. Figure 8.   Stress results without connecting rod model.

3   FORMWORK APPLICATION

3.1   *Climbing planning*

The main tower of the Wujiagang Yangtze River Bridge is 155 m, divided into 36 sections, 1.5 m + 4 × 4.5 m+3.0 m+28 × 4.5 m + 3.0 m + 3.5 m. The first two and last two sections do not require climbing construction, and the climbing is done only in sections of 4.5 m and 3.0 m. The standard load-bearing part is 2.25 m, and the non-standard load-bearing part is 0.75 m, forming two combinations of 2.25 + 2.25 = 4.5 and 2.25 + 0.75 = 3, which meet the construction and climbing of different sections.

3.2 *Formwork installation*

After the completion of the bridge tower platform, the bearing parts shall be used as the corner formwork of the tower column from the first section. Positioning support is required while installing the first section, and the installation of other sections only shall be connected to the lower load-bearing parts. M56 embedded bolts are installed before formwork closing. After the construction of the third section, the load-bearing parts begin to be recycled: top layer installation – formwork pouring – platform lifting – bottom layer removal, and cleaning.

The main construction points of load-bearing parts are as follows: 1) it is needed to remeasure the top elevation of load-bearing parts after each concrete pouring. For large height differences, steel plates shall be added between the load-bearing parts of the next section for leveling to avoid error accumulation. 2) the track surface at the joint should be checked if it is flush after jointing the load-bearing parts and the next section. 3) it is required to move the bottom load-bearing parts to the top layer timely after completing the Formwork lifting. First, it is necessary to remove the connecting bolts between the left and right load-bearing parts and then remove the embedded bolts of the load-bearing parts in pieces. Crossing the pair while assembling the load-bearing parts before hoisting them to the top floor is not allowed.

The temporary supports are used for the construction of 1 ~ 3 sections. First, the hanging claw shall be manually flattened before hoisting the support frame when the concrete strength meets the requirements. The hanging claw shall be turned upright using a tower crane and fixed at temporary support, consistent with the installation posture. Second, the support frame is lifted manually to the top of the load-bearing part aligning the slide block with the slideway and lowering it smoothly. Finally, the hanging claw can be loosened after hovering above 20 cm from the predetermined hanging shoe. It is essential to confirm that the hanging claw is flattened and lower to the steel wire rope for unloading. A single support frame is installed, as shown in Figure 9. As shown in Figure 10, the steel frame is assembled into a truss in the factory. The truss shall be assembled into frame units on-site and hoisted by units to form a platform layer by layer for rebar and concrete construction.

Figure 9.　Supporting system installation.

Figure 10.　Frame installation.

4 BRIDGE TOWER CONSTRUCTION

4.1 *Layout of tower crane and elevator*

The height of the formwork for the north main tower of the Wujiagang Yangtze River Bridge is up to 20 m, which significantly impacts the layout and type selection of the tower crane, especially the layout of the attached wall. In general, the minimum hoisting clearance of bridge tower construction, controlled by the overall installation condition of the rigid skeleton, shall be at least more than 8.0 m. Minimum hoisting clearance = maximum cantilever height – attached wall spacing – Formwork height. Whether the attached wall spacing meets the requirements can be calculated. If necessary, the attached wall spacing can be appropriately reduced. For example, for the upstream tower crane STT 293, the attached wall spacing is 21 m; for the downstream tower crane STT 200, the attached wall spacing is 18 m. The guide of the construction elevator is obliquely arranged along the tower column. The guide is 6.0 m away from the attached wall, whose length remains unchanged. With the continuous contraction of the tower column section, the relative position between the elevator and the platform constantly changes. A sliding channel shall be set up under the platform to adjust the distance between the elevator and the platform after each lifting, forming a safe passage of the formwork. As shown in Figure 11, the construction elevator can directly access the hanger under the platform.

Figure 11. Real scene of tower crane, elevator, and formwork.

4.2 *Standard section construction of tower column*

As shown in Figure 12, the formwork of the north main tower of the Wujiagang Yangtze River Bridge is equipped with two sets of formwork A and B. The recycling process is as follows.

Step 1: Pour concrete on formwork B. Install rigid skeleton and rebar on the top layer, remove formwork A simultaneously. Return to the main platform, clean and brush oil.
Step 2: Formwork A lifts a section along with the platform, as high as Formwork B.
Step 3: Lift the formwork A to the layer to be poured using the electric hoist, close the formwork, and pour concrete.

Repeating the above three steps is the circular construction process of double formwork. In theory, the removal and cleaning of formwork shall be carried out simultaneously with the rebar binding before formwork closing to improve construction efficiency. According to the actual construction period statistics, the rebar binding efficiency is low, generally three days, which restricts the work efficiency of the double formwork process. The rebar binding construction of some sections takes two days, and that of a single section takes five days. The double formwork process can save one day for a single section.

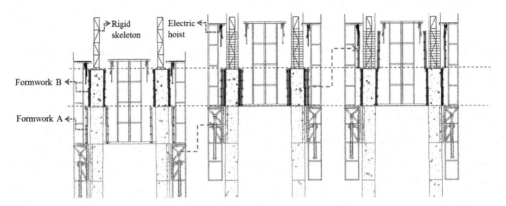

Figure 12.   Circular construction process of double formwork.

4.3   *Prestressed beam construction*

The upper and lower beams of the main tower are constructed asynchronously (Wang et al. 2021). In order to prevent the root concrete damage caused by the excessive free height of the inclined tower column, the concrete of section six of the tower column shall not be poured before the pretension of the lower beam is completed, so the formwork cannot be lifted to avoid the construction support of the lower beam. When installing the formwork, installing the lower hanger in conflict with the support can be suspended, as shown in Figure 13.

4.4   *Construction of decorative strips and embedded section of steel truss*

In the traditional climbing formwork, the protruding decorative strips affect the layout of the formwork fulcrum, the embedded section of the steel truss obstructs the normal formwork lifting, and the high-altitude demolition and modification increase the operational risk and the workload of temporary measures. However, the angular line climbing of the north main tower formwork of the Wujiagang Yangtze River Bridge avoids the decorative strips and the embedded section of the steel truss. As shown in Figure 14, the lifting construction does not need demolition and modification but turning over the edge flap, with high efficiency, strong integrity, and high safety.

Figure 13.   Construction of lower beam.

Figure 14.   Climbing through embedded section.

5 CONCLUSION

According to the construction of the north side tower of the Wujiagang Yangtze River Bridge, the following conclusions can be drawn. First, the double formwork cycle construction process has the efficiency of synchronous construction of multi construction. When it takes two days for the rebar binding of a 4.5 m section, the shortest work time of a single standard section is five days. To this end, one day is shortened compared with the single section work time of the traditional formwork under the same conditions. Secondly, similar bridge tower construction adopts corner support, making the formwork avoid the large embedded parts and decorative structures on the rebar surface. Then, the steel structure frame on corner support is connected as a whole by cooperative connecting rods between adjacent units, effectively improving the formwork's loading capacity. The design and application results show that the formwork of the north side tower of the Wujiagang Yangtze River Bridge has a high reference for similar bridge construction and has a high market value with further cost reduction.

ACKNOWLEDGMENTS

Research Project of China State Construction Engineering Corporation (CSCEC-2017-Z-19).

REFERENCES

Huang Z W, Design and Construction of Hydraulic Self-Climbing Formwork for Thin-Walled Hollow High-Rise Piers of Mountainous Highway Bridge. World Bridges, 2019 47 (03): pages 10–14.

Miao R C, Study of Key Seismic Design Techniques for Wujiagang Yangtze River Bridge in Yichang [J] Bridge Construction, 2020,50 (S2): 36–40.

Pan C L, Quan W B, Zhang W S, Wang G F, Chen S M. Technical of the Equipment Integrated Platform of High-rise Building [J] Construction Technology, 2017,46 (16): 1-4 + 17.

Wang H Z, Cao Z J, Lan Q P, Study on Internal Force Distribution Mode of Lower Beam of Wujiagang Yangtze River Bridge in Yichang [J] Journal of China & Foreign Highway, 2021,41 (03): 135-140 DOI: 10.14048/j.issn. 1671-2579.2021.03.028.

Yin H M, Wang H M, Construction Technology of Hydraulic Climbing Formwork in Equal Section Hollow Pier of Bridge. Highway, 2018 63 (10): PP. 131–134.

Yin Z J, Construction Technique of Using Hydraulic Climbing Formwork to Pylons of Huanggang Changjiang River Rail-cum-Road Bridge. World Bridges, 2015 43 (01): pages 18–22.

Zhang H, Ma L, Tian S C, Guo H L. Critical Construction Scenarios, Elements, and Development Paths for Intelligent Construction Platforms [J / OL] Journal of Tsinghua University (Science and Technology): 1-6 [2022-01-06] DOI:10.16511/j.cnki. qhdxxb. 2022.22.004.

Zhang K, Development and Expectation of Construction Technology of Super High-rise Building. Construction Technology, 2018 47 (06): pages 13–18 + 93.

Frontiers of Civil Engineering and Disaster Prevention and Control – Yang & Rahman (Eds)
© 2023 The Author(s), ISBN: 978-1-032-31200-2

Research on monitoring scheme for closure construction of rail-cum-road steel truss girder cable-stayed bridge

Songbao Cai, Junlong Zhou, Wenbin Geng & Weiliang Qiang
China Construction Sixth Engineering Bureau Co., Ltd, Tianjin, China

ABSTRACT: As one of the important processes in the construction of cable-stayed bridges, the closure of the bridge structure is a key link in the system transformation. The construction quality of the closure processes will directly affect the internal force and the alignment of the completed bridge. Taking the Hongyancun Jialing River Bridge as an example, this paper establishes a spatial model of the whole bridge and conducts a numerical analysis of the bridge's completion stage by using finite element software, then arranges the construction monitoring points of the main tower and key components of the steel truss girder according to the results of the numerical analysis. Therefore, it can be regarded as a guide for the construction of the closing section of the bridge to ensure that the completed bridge state meets the design requirements.

1 INTRODUCTION

The main method of construction monitoring is to use the intelligent sensors arranged in the key parts of the bridge structure to obtain the load and structural response data during the bridge construction process in real-time and guide the bridge construction process. In terms of monitoring technology, the current domestic research mainly covers fiber optic sensor monitoring technology, computer vision monitoring technology, wireless sensor monitoring technology, etc. Lu (Lu et al. 2019) used externally attached long gauge fiber grating strain sensors to obtain the dynamic monitoring data of Wanlongshan Bridge and studied the dynamic and quasi-static strain separation method of long gauge length strain gauges for large-span rigid frame bridges under vehicle loads. Hou (Hou & Wu 2019) designed a low-cost sensor system that can be used for bridge displacement monitoring, using low-power wireless communication technology to collect displacement data, and realizing the analysis and visualization of data in a remote web interface. Ye (Ye et al. 2019) proposed an improved stepped liquid level sensing system for bridge dynamic deflection monitoring. The system uses stepped pipelines instead of straight pipelines to eliminate the interference of inclination angles on the monitoring system and improve dynamic deflection monitoring precision.

In this paper, taking the Hongyancun Jialing River Bridge as the engineering background, finite element software is used to establish the spatial model of the whole bridge, and the static analysis of the bridge is carried out. According to the calculation results, the construction monitoring points of the bridge tower and the key components of the steel truss girder are arranged as the basis for the construction of the bridge.

2 PROJECT OVERVIEW

The Hongyancun Jialing River Bridge is a Rail-cum-Road Steel Truss Girder Cable-Stayed Bridge with high-low towers and double cable planes. The span of the bridge is arranged as 91.4+138.6+375+120=725m. The overall layout of the bridge is shown in Figure 1.

DOI 10.1201/9781003308577-84

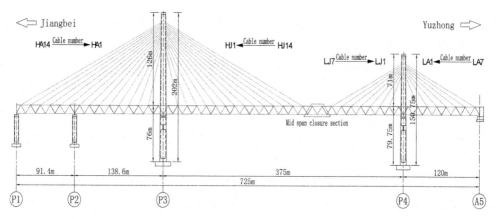

Figure 1. Overall layout of the bridge.

The main tower is a portal frame structure composed of the top, middle, and bottom tower columns and beams. The total height of the P3 tower is 202m, and the total height of the P4 tower is 150.75m. The main girder is a continuous steel truss girder that consists of a double main truss of the same height. The main truss is a triangular truss, the width of the truss is 28.2m, and the height of the truss is 11.163m.

The steel truss girder is divided into 49 segments. The length of the segment from the P1 pier to the P3 tower is 15.4m, and the length of the segment from the P3 tower to the A5 abutment is 15m. The standard segment steel beam includes five components: the top chord, the bottom chord, the top deck, the bottom deck, and the inclined web member. The top deck and the bottom deck are all orthotropic steel decks.

3 ARRANGEMENT PRINCIPLES OF MONITORING POINTS AND CALCULATION METHOD

3.1 *Arrangement principles of monitoring points*

Since the previous procedures of the closure construction have been completed, we only focus on the arrangement of the construction monitoring points of the procedures following the closure construction and before the procedure of the second-phase load, such as the auxiliary facilities of the bridge. The arrangement of the monitoring points should be able to accurately reflect the actual stress of the bridge structure, the alignment of the main girder, and other information after the closure to ensure the state of the bridge after the closure meets the design requirements.

3.2 *Calculation method*

The bridge is modeled by Midas Civil. The pylons, top and bottom chords of the main girder, web members, and beams are all simulated by beam elements, and the stay cables are simulated by cable elements. The finite element model of the full-bridge is shown in Figure 2. The simulation analysis includes calculating the cable force, deformation, and stress of the completed bridge before the procedure of the second-phase load, such as the auxiliary facilities of the bridge.

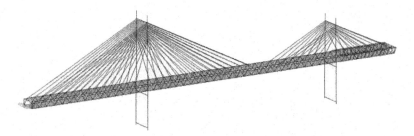

Figure 2.   Finite element model of the cridge.

4   MONITORING SCHEME AND MONITORING POINT ARRANGEMENT

4.1   *Temperature monitoring*

Temperature monitoring is the basis of the closure construction since it has an impact on key parameters such as the width of the closure opening, the elevation of the main girder, and the cable force (Tu et al. 2013). The difference between the actual temperature of the environment and the design reference temperature during closure construction will cause thermal stress and deformation of the steel truss girder and cables, and then the width of the closure-opening, the elevation, and the angle of the bridge deck on both sides are changed. This means higher accuracy for closure construction. Meanwhile, the difference in temperature between the top and bottom decks will also cause different responses in the top chord and bottom chord because the bridge is a highway-railway dual-purpose bridge. Therefore, we should obtain the system temperature and the difference in temperature between the top and bottom decks to provide better protection for construction.

A total of 8 sections are set up in the longitudinal direction of the bridge for temperature monitoring, and 16 wireless thermometers are arranged on the top and bottom decks of each section for monitoring, as shown in Figure 3.

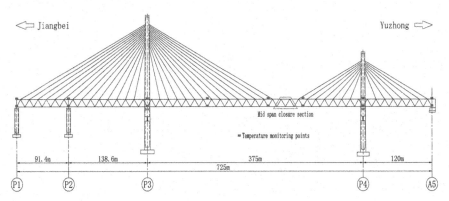

Figure 3.   Arrangement of temperature monitoring points of the bridge.

4.2   *Displacement monitoring of the main tower*

The horizontal displacement of the tower is a comprehensive reflection of asymmetric load, cable force, temperature, and other factors (Tian & Qi, 2012; Wu et al. 2011). The bridge tower is a reinforced concrete structure with a high tower height. During the erection of the steel beam, the displacement of the tower has been in a state of change. In order to ensure that the top displacement of the tower is within the monitoring requirements, a displacement sensor is set up on the top of the tower to measure the top displacement data of the tower at any time.

The reflective mode is used as the observation point on the top of each main tower, and the polar coordinate method of the total station is used for monitoring. Multiple monitoring rounds are carried out during the monitoring. Figure 4 shows the calculation results of the top displacement of the main tower after closure.

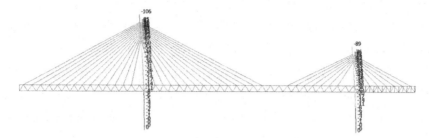

Figure 4.   Displacement of the main tower (mm).

It can be seen from the calculation results that the top displacement value of the tall tower after closure is 106mm, and the top displacement value of the low tower is 89mm.

4.3   Cable force monitoring

The cable is the main force-bearing member of the cable-stayed bridge. The cable force not only affects the stress state of the main beam, but also affects the linear shape of the main beam after the bridge is completed (Tian & Qi 2012; Wu et al. 2011). In order to accurately obtain the variation of the cable force during the closure process, a pressure sensor is installed in each cable anchor area to measure the cable force. During the construction process, it is necessary to measure the cable force several times to check whether the actual cable force is consistent with the calculated cable force.

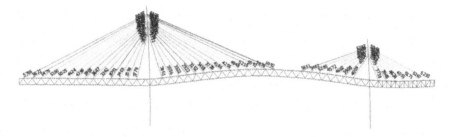

Figure 5.   Result of the cable force (kN).

It can be seen from the calculation results that the cable forces of the outermost stay cables HA14, HJ14, LA7, and LJ7 are the largest after the closure. The closer to the main tower, the cable force of the cable decreases, which is in line with the designed cable force, and the value is smaller than the trend of the designed cable force. This is because the cable force of the completed bridge given by the designer is the final cable force of the completed bridge after applying the pressure on both sides and the second-phase load of the bridge, and the design requirements can be met after the weight block, and the second-phase load is constructed in the later stage, and the cables are properly adjusted.

4.4   Stress monitoring of steel truss girder

Whether the internal force of the main beam after closure is consistent with the calculation means the quality of the bridge construction or not, the internal force of the main beam will redistribute due to different structural systems before and after the closure of the bridge. Strain sensors arranged at the key nodes of the main girder according to the calculation results are important to obtain the stress variations of the key nodes before and after the closure of the bridge (Tian & Qi 2012; Wu et al. 2011). The stress results of the top and bottom chords of the main girder after closure are shown in Figures 6 and 7.

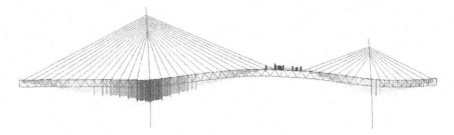

Figure 6.   Top chord stress of the steel truss girder (MPa).

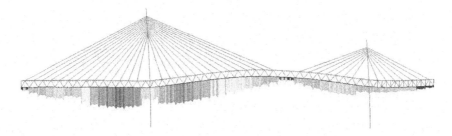

Figure 7.   Bottom chord stress of the steel truss girder (MPa).

It can be seen from the calculation results that the maximum stress of the top chord of the main girder after closure is 204 MPa, the maximum stress of the bottom chord is 162 MPa, and the stress of the chord nearby the main towers is relatively large. According to the calculation results, measuring points are arranged on the top, bottom chord, and inclined web members of the main truss, respectively, to monitor the stress of the main girder after closure. The layout of stress monitoring points is shown in Figure 8.

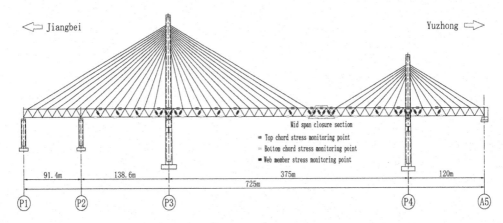

Figure 8.   Arrangement of stress monitoring points of the steel truss girder.

5 CONCLUSION

During the closure construction monitoring process of the Hongyancun Jialing River Bridge, an accurate closure construction process was accomplished by determining the monitoring targets such as the displacement of the main tower, the cable force, the alignment and the stress of the steel truss girder, and formulating a reasonable monitoring scheme, therefore ensured the precision and quality of the closure construction process, and lay the foundation for the realization of the target state of the bridge. The monitoring data during the closure process and the monitoring data such as the alignment of the main girder and the displacement of the main tower are in good agreement with the theoretical calculation results, which indicate that the construction process of the bridge is standardized and the construction quality is well controlled.

REFERENCES

Hou, S. T., Wu, G., "A low-cost IoT-based wireless sensor system for bridge displacement Measuring," Smart Materials and Structures, 28(8):085047 (2019).

Lu, H. X., Gao, Z. C., Wu, B. T., Zhou, Z. W., "Dynamic and quasi-static signal separation method for bridges under moving loads based on long-gauge FBG strain Measuring," Journal of Low-Frequency Noise, Vibration and Active Control, 38(2):388–402 (2019).

Tian, J., Qi, Y., "Construction Monitoring of Chaobaihe Extradosed Cable-stayed Bridge" Construction Techniques, 41(366):22–27 (2012).

Tu, G. Y., Yan, D. H., Chen, C. S., Yi, Z. P., "Construction Techniques for Closure of Central Span of Jingyue Changjiang River Highway Bridge," Bridge Construction, 43(4):105–109 (2013).

Wu, Y. H., Yue, Q., Zhu, L. M., Huang, X. H., "Control Measures for Closure Construction of Jintang Primary Fairway Bridge," Construction Techniques, 40(3):15–17 (2011).

Ye, X. J., Sun, Z., Cai, X., Mei, L., "An improved step-type liquid level sensing system for bridge structural dynamic deflection Measuring," Sensors, 19:2155 (2019).

Frontiers of Civil Engineering and Disaster Prevention and
Control – Yang & Rahman (Eds)
© 2023 The Author(s), ISBN: 978-1-032-31200-2

Construction monitoring technology and application of interzone tunnel construction in urban metro

Jingyi Zhang*
Department of Civil Engineering, Shenyang Urban Construction University, Shenyang, P.R. China

ABSTRACT: Aiming at the influence of the interzone tunnel construction in the urban metro on new projects and existing cable structures, this paper discusses the construction monitoring technology and application of the shield method in the interzone tunnel combined with the construction monitoring of Guang-heng Interzone Tunnel of Guangzhou Metro Line 12. Construction monitoring items include surface settlement, building settlement, building tilt, vault settlement, vertical displacement of the arch bottom, horizontal displacement of the pipe section, gap convergence of pipe section, pipe settlement, and displacement. Through monitoring point data collection and processing, the cumulative maximum value, change rate, and corresponding monitoring point location are obtained, and compared with the control value to judge the construction state. Site inspection includes construction conditions, pipe deformation, surrounding environment, and monitoring facilities. Through the comprehensive analysis and judgment of site inspection, combined with the monitoring data of abnormal areas and construction conditions, the potential accident or accident symptoms can be found in time.

1 INTRODUCTION

As a form of railway transportation, the urban metro refers to the urban rail transit system, which mainly operates underground. It is a fast and large-volume urban rail transit mode with electric traction developed from the traditional surface railway system (Zhang et al. 2010). Urban metro has many advantages, including that the underground railway lines set up transfer stations for convenient passenger transfer. The underground railway is built below the ground, which can save ground space and resist the adverse impact of rain and snow weather on travel. The underground railway transportation system can reduce automobile noise and exhaust emission pollution to the environment. In addition, the urban traffic becomes fast and convenient because of the stable driving speed and not overlapping with the ground road. At present, metro engineering is an important means to solve the problem of urban traffic congestion as an important part of urban rail transit.

Compared with other traffic engineering, there are many non-unpredictable problems in engineering design and construction in the special geotechnical geological environment. It is almost impossible to accurately predict the dynamic response of the tunnel and surrounding environment in the construction process. In order to ensure the safety of the metro construction process, an effective method is to strengthen monitoring in the construction process based on displacement monitoring data and site inspection information to optimize or modify the design and construction scheme. In short, whether it is to guide the construction or ensure the safety of construction, construction monitoring is an important role (Xie 2018). Metro construction safety monitoring has become an indispensable part of the current construction throughout the whole process of construction.

---

*Corresponding Author: dq_zjy@syucu.edu.cn

DOI 10.1201/9781003308577-85

2 METRO CONSTRUCTION MONITORING CONTENTS

Metro construction monitoring refers to continuously observed displacement, tilt, pressure, internal force, cracks, uplift of the basement, and changes in the underlying water level of the surrounding rock, support structure, ground, surrounding buildings, underground facilities, other objects with instruments, equipment, and other means during metro construction. And the observation results are analyzed and fed back. Monitoring data is not only the direct feedback on engineering safety, but also an important basis for guiding the subsequent design and construction

The monitoring items in metro construction should be reasonably determined according to the characteristics of the monitoring object, the monitoring grade of the project, the influence zone of the project, the requirements of design and construction, and reflect the changing characteristics and safety state of the monitoring object (Hallowell et al. 2013; Lai et al. 2013). These projects are determined comprehensively according to geological conditions, surrounding environment, tunnel buried depth, section size, excavation method, and design requirements, which are generally divided into testing items and selected testing items. Testing items refer to the project that should be monitored daily for the stability and construction safety of the engineering support structure, surrounding environment, and surrounding rock king body in the construction process. Selected testing items refer to the monitoring items carried out in local sections or parts for the special needs of design, construction, and research.

2.1 Horizontal displacement monitoring

For the station, when the open excavation method or the cover excavation method is adopted, the monitoring items include the supporting surface (wall), the horizontal displacement of the top of the slope, and the horizontal displacement of the column structure. For the interval, the monitoring item includes the horizontal displacement of the segment structure when the shield construction method is adopted. For the surrounding environment monitoring of the project, the monitoring items include the horizontal displacement of the building (structure) and the horizontal displacement of the underground pipeline, and the horizontal displacement of the tunnel structure should be monitored when the existing urban rail transit is around.

Horizontal displacement monitoring points are divided into three types: datum point, working base point, and deformation monitoring point. The datum point is generally located outside the scope of construction influence, used to check and restore the reliability of the working base point, the working base point is arranged in a stable place around the foundation pit, and the deformation monitoring point is arranged in accordance with the requirements of the monitoring design document. The stability of the head control network constituted with datum and work base points should be checked before each monitoring. Free stations or set up on the working base points are used for horizontal displacement monitoring, and the polar coordinate method is used to observe the monitoring points.

2.2 Vertical displacement monitoring

For the station, the monitoring items include the retaining pile (wall), the vertical displacement of the top of the slope, the vertical displacement of the column structure, the surface settlement, the vertical displacement of the soil layer, and the uplift of the bottom of the pit when the open excavation method or the cover excavation method is adopted. For the interval, the monitoring items include vertical displacement of segment structure, surface settlement, and vertical displacement of the soil layer in shield construction (Ming 2014). For engineering environment monitoring, monitoring projects including vertical displacement of building, underground pipeline, subgrade and pavement, the bridge pier. Differential settlement of bridge pier, when the urban rail transit is surrounding, should also be monitoring of tunnel and the vertical displacement of track structure and differential settlement deformation of tunnel structure joints.

Although there are many kinds of vertical displacement monitoring projects, the monitoring method is simple and flexible, and the monitoring cost is low (Zhang & Erchu 2015). It is the main monitoring form in construction safety monitoring and the most important monitoring data source in metro construction safety monitoring. Vertical displacement can be monitored by geometric leveling, electromagnetic wave ranging triangular elevation, and static leveling. The vertical displacement monitoring is usually carried out by the second-level survey. When other methods are adopted, the relevant requirements of the second-level survey should also be met. For the layered solid displacement, the layered settlement instrument is used to monitor the vertical displacement, and the nozzle of the layered settlement pipe is used as the monitoring point. After the formal implementation of the monitoring work, the initial values of all the monitoring points of each project are collected three times as required, and the average value is taken after qualification.

2.3 Site inspections

The deformation of the project is always forewarned, and some of the warnings are not only reflected in the microscopic monitoring data deformation, but also reflected in the macroscopic external performance (Yu 2013; Zhen & Liu 2008). Therefore, in the construction of metro engineering, regular monitoring and patrol should be carried out to find out the cause of deformation intuitively, find out the hidden trouble directly, find out the situation and ensure safety.

The inspection includes shield structure, support structure, surrounding environment, contact channel, and monitoring facilities (Zhang et al. 2011). Inspection is mainly by visual inspection; the contents of the inspection should be checked and filled in form item by item and saved the image data by taking photos of the existing cracks, ground subsidence, and other conditions to fill in the inspection report after the inspection. Inspection work generally includes construction conditions, support structure, surrounding environment, monitoring equipment, construction units, supervision units, and third-party monitoring units should conduct inspections according to the geological conditions of the excavation face, the supporting structure system, and the surrounding environment of the foundation pit or tunnel, At the same time, supervision units are required to inspect the construction technology and equipment, construction organization and management and operation conditions (Qiu et al. 2012). In addition to the formulation of a unified inspection table before the inspection, it is also necessary to statistics on the sources of danger along all lines to ensure the uninterrupted inspection work during the construction period.

3 APPLICATION OF METRO CONSTRUCTION MONITORING

3.1 Project overview

Guangzhou Metro Line 12 Guang-heng Interzone Tunnel from Guangyuan Xincun station to Hengfu Road Station, about 74 buildings were monitored along the line, Xiatang West Road runs through DN325 steel gas pipe with burial depth of 1.4M, concrete DN1000 rain pipe with burial depth of 4.17m, concrete DN800 water supply pipe with burial depth of 3.4m and concrete DN500 sewage pipe with burial depth of 1.65m, among which the minimum net distance between the concrete DN1000 rain pipe and the interval is about 8.7m, halfway through Lu-hu fault zone and Guang-chong fault zone.

The engineering geology of the section is complex, and the fault zone, long-distance hard rock section (998m full section hard rock), and karst development area are crossed successively. The surrounding environment is complex and diverse, and it has passed through 66 buildings (24 underpass buildings and 42 side-pass buildings), a military railway, Guangzhou University of Traditional Chinese Medicine, Guihuagang Primary School, one civil air defense passage, Guangzhou Sculpture Park, Guangzhou Luhu Golf Course, and Donghao Yong pile foundation. Among them, there are 12 ϕ1m piles in the Donghaochung pile foundation, which are 11.5cm away from the tunnel vertically. Since the mileage is DK25+179.100, it forms a stacking section with

Guangzhou Metro Line 11 (Line 12 crosses Line 11). The length of the stacking section is 50mm in total, and the minimum vertical net distance between the two lines is 2.73m. Therefore, it is necessary to carry out comprehensive construction monitoring during the construction process of the Guang-heng Interzone Tunnel.

3.2 Construction monitoring

(1) Setting up datum points and monitoring points

Datum points are the basis of the site surrounding environment monitoring, the selection of reference point must ensure that the point is solid and stable, has good visibility conditions, is conducive to long-term preservation and observation of signs, the elevation control network consists of datum and working base points. Seven datum points of leveling work are buried in the Gungheng Construction Interval, three datum points are buried in the initial section, respectively, and four datum points are buried in the leveling work near Jinlu Mountain Villa. Seven datum elevation data were collected, respectively, and the datum results were checked and accepted by supervision.

In order to complete the data collection of monitoring projects in the process of metro construction, the monitoring points of surface settlement, building settlement, building tilt, pipeline settlement, vault settlement, vertical displacement of the arch bottom, horizontal displacement of the segment, and segment clearance convergence are laid out, and their initial values are collected. The statistics of the number of buried monitoring points of monitoring projects, as shown in Table 1.

Table 1. Statistical table of monitoring points quantity of monitoring items.

Monitoring Items	Number of monitoring sites	Actual number of monitoring points	Number of monitoring points cannot be arranged
Surface settlement	338	325	13 (Need to coordinate)
Building settlement	130	130	0
Building tilt	20	20	0
Vault settlement	30	30	0
Vertical displacement of the arch bottom	30	30	0
Horizontal displacement of segment	60	60	0
Segment clearance convergence	30	30	0
Pipeline settlement and displacement	32	32	0

(2) Monitoring project control analysis

The control values of monitoring items are divided into deformation monitoring control values and mechanical monitoring control values according to the nature of monitoring items. The deformation monitoring control value should include the accumulated change value and change rate value of deformation monitoring data. The control value of mechanical monitoring should include the maximum value and minimum value of mechanical monitoring data. The control value of monitoring items should be determined according to geological engineering conditions, engineering design parameters, engineering monitoring level and local engineering experience, etc. When there is no local experience, international standards can be implemented.

In order to ensure the timeliness of monitoring data, monitoring information must be timely fed back to all units. During construction monitoring, all monitoring data shall be calculated and analyzed after completion of each monitoring work, and relevant analysis data, corresponding charts, and inspection information shall be timely delivered to all parties concerned for analysis and use in combination with the project overview and construction progress. The statistics of construction monitoring results on a certain day in the Guangheng Interzone Tunnel are shown in Table 2.

Table 2. Statistical table of construction monitoring results in Guangheng Interzone Tunnel.

Monitoring Items	Maximum change of this time		Cumulative maximum change			Engineering control value	
	Point position	Numeric value	Point position	Numeric value	Change rate	Variation per day	Cumulative value
Surface settlement	DBCZ630	0.37mm	DBCY630	−6.31mm	0.06mm/d	±3mm	+10, −20mm
Building settlement	JGC130	0.16mm	JGC078	−14.71mm	0.14mm/d	±3mm	±30mm
Building tilt	JGQ012Y	−0.078‰	JGQ020X	−0.382‰	0.016‰	/	±2‰
Vault settlement	GGCZ019	0.26mm	GGCY018	0.26mm	0.19mm/d	±2mm	±20mm
Vertical displacement of arch bottom	GDCY019	0.11mm	GDCY019	0.18mm	0.11mm/d	±2mm	±20mm
Segment clearance convergence	GDCZ016	−0.30mm	GDCZ017	−0.30 mm	−0.2mm/d	±3mm	±12.8mm
Pipeline settlement and displacement	GXC203	−0.34mm	GXC205	−4.13mm	−0.02mm/d	±2mm	±10mm

According to this monitoring data, the maximum cumulative change in the monitoring data of surface settlement is −6.31mm, the change rate is 0.03mm/d, the maximum cumulative change in the monitoring data of building (structure) settlement is −14.71mm, and the change rate is 0.14mm/d, in the monitoring data of building tilt, the maximum cumulative change is −0.382‰ and the change rate is 0.016‰, in the monitoring data of vault settlement, the maximum cumulative change is 0.26mm, and the change rate is 0.19mm/d, in the monitoring data of vertical displacement of the arch bottom, the maximum cumulative change is 0.18mm, and the change rate is 0.11mm/d, in the segment clearance convergence monitoring data, the maximum cumulative change is 0.30mm and the change rate is 0.2mm/d, in the monitoring data of pipeline settlement and displacement, the maximum cumulative change is −4.13mm and the change rate is −0.02mm/d. The monitoring values of each monitoring item are all within the control values. Among them, the maximum cumulative change point of surface settlement is point DBCY630, and the maximum cumulative change is −6.31mm (the warning value is 16mm). The maximum settlement rate point of DBCZ630 is 0.37mm/d, which requires continuous monitoring of this position. The cumulative changes of other monitoring parameters in the interval are small.

3.3 Construction site inspection

In the construction process, according to the inspection plan, the construction progress, inspection should be done timely, and make detailed inspection records. Any anomalies detected during site inspections must be taken seriously. Combined with the monitoring data of abnormal areas and construction conditions, comprehensive analysis and judgment are made to timely discover the potential accident or symptoms and timely start the plan.

The shield construction method is adopted in Guangheng Interzone Tunnel. Inspection items during the construction period are divided into construction conditions, segment deformation, surrounding environment, and detection facilities. The daily inspection and monitoring report is filled in according to the inspection content. The inspection situation of the construction site on a certain day in the Guangheng Interzone Tunnel is shown in Figure 1.

The inspection of the construction includes the soil reinforcement at the beginning of the shield and the receiving end, the shield driving position, the time and location of shield machine shutdown, the warehouse opening, the shutdown and closure of the excavation face, the hinged seal or shield tail seal water surge, the contact passage openings and other openings. The construction inspection report shows that the left line of shield tunneling was normal on that day, the left line was tunneling 760 rings in total, and the face course was 26+206.8. The cumulative driving of the right line is 694 rings, the mileage of the palm surface is 26+306.2, and the shield tail seal is normal.

| (a) Normal shield tail | (b) Normal surrounding buildings |
| (c) Monitoring points are normal | (d) Surrounding surface is normal |

Figure 1.   The inspection situation of the construction site in Guangheng Interzone Tunnel.

The inspection content of segment deformation includes inspection of whether there is damage, cracking, misplacement of the segment, segment water leakage, and other special circumstances, etc. There is no abnormal segment on the day of on-site inspection.

The surrounding environment inspection includes the deformation, crack location, number and width of construction (structure) structures, bridge piers or beams, existing rail transit structures and so on, the location, size, and the number of concrete spalling, and whether the facilities can be used normally, the water accumulation and seepage of underground structures, the surrounding underground pipeline deformation suddenly increased or appeared cracks, water leakage, the air leakage, and other situations, the variation of water level of rivers and lakes, whether there are vortices and bubbles on the water surface and their locations and ranges, the width, depth, number, and development trend of dike cracks, and so on. Excavation, stowage, piling, and other activities around the project may affect the safety of the project. There is no abnormality of the building (structure), no new cracks, and no subsidence, uplift, or slurry on the surrounding pavement or surface through site inspection.

Monitoring facilities inspection includes the intact condition and protection of datum points, the monitoring points, and the integrity and protection of components. The inspection found that the datum and monitoring points are intact.

4   FUTURE RESEARCH PROSPECT

Based on the construction monitoring of the Guangheng Section of Guangzhou Metro Line 12, this paper discusses the application of shield construction monitoring in the construction of the metro section from the aspects of the burying of reference points and monitoring points, monitoring items and control, and site inspection through the combination of manual and automatic equipment to timely control the deformation degree and regularity. However, the manual method is generally adopted in the monitoring project data processing and analysis, and there is a certain time difference between the data processing and management. With the development of construction information technology, monitoring data processing and management can be combined to build an information management system for subway construction monitoring. With the help of the Internet and computer technology, real-time data collection, processing, analysis, and feedback are realized to assist in the optimization of construction schemes and emergency management, and the informatization level of subway construction monitoring can be improved.

5 CONCLUSIONS

This paper introduces the application of metro construction monitoring technology in practical engineering from the aspects of base point and monitoring point burying, monitoring items and control, and site inspection.

Daily monitoring items of the interval include surface settlement, building settlement, building tilt, vault settlement, vertical displacement of the arch bottom, horizontal displacement of the segment, segment clearance convergence, pipeline settlement, and displacement. Through data collection and processing, the accumulated maximum value, change rate, and corresponding monitoring points are obtained, and compared with the control value to judge the construction status.

Site inspection includes construction conditions, segment deformation, surrounding environment, and monitoring facilities. Through comprehensive analysis and judgment of site inspection combined with monitoring data of abnormal areas and construction conditions, potential accident hazards or symptoms can be found in time.

ACKNOWLEDGMENT

This work was supported by the Construction Project of the University-level Characteristic Major in 2019: Road, Bridge, and River Crossing Project ([2019] No.58, Urban Construction University of China).

REFERENCES

Hallowell, M. R., Hinze, J. W., Baud, K. C., & Wehle, A. Proactive Construction Safety Control: Measuring, Monitoring, and Responding to Safety Leading Indicators[J]. Journal of Construction Engineering and Management, 2013, 139(10):04013010.

Lai, H. B., De -Ying, M. A., Wang, G. X., Zhao, Y. B., & Song, J. Research on Deformation Monitoring of Meilan Airport Underground Station on Hainan Eastern-ring Railway[J]. High-Speed Railway Technology, 2013.

Ming L I. Application of Automatic Monitoring Technology in Construction of Shield-bored Metro Tunnel Crossing Underneath High-speed Railway in Urban Area[J]. Tunnel Construction, 2014, 34(4):368–373.

Qiu D, Huang H, Song D S. Deformation Monitoring and Prediction Technique of Existing Subway Tunnel: A Case Study of Guangzhou Subway in China[J]. Journal of the Korean Society of Surveying Geodesy Photogrammetry & Cartography, 2012, 30(6_2):623–629.

Xie R. Research on Data Processing and Management Technology of Subway Construction Safety Monitoring[J]. Jiangxi Building Materials, 2018.

Xin-Mei Y U. On deformation inspection in deep foundation pit construction at subway stations[J]. Shanxi Architecture, 2013, 154(5):1272–1276.

Zhang J T, Erchu C O. Application of Monitoring and Control Technology in Urban Metro Shield Construction[J]. Equipment Manufacturing Technology, 2015.

Zhang Z, Kong N, Shen F, et al. Scheme and Analysis of Deformation Monitoring for Subway [J]. Journal of Geomatics, 2010.

Zhang, Z. L., Shen, F. F., Kong, N., & Ling-Yan, L. U. A method for datum points choice of deformation monitoring networks in subway tunnels [J]. Science of Surveying and Mapping, 2011.

Zhen C, Liu J F. Monitoring Technology Application of Information Construction in Beijing Metro Line 10[J]. Construction Technology, 2008.

*Frontiers of Civil Engineering and Disaster Prevention and
Control – Yang & Rahman (Eds)*
© 2023 The Author(s), ISBN: 978-1-032-31200-2

Key technologies for the design and rapid construction of steel cofferdams at Guojiatuo Yangtze River bridge

Wenbin Geng* & Xiaomin Liu
China Construction Sixth Engineering Bureau Limited, Tianjin; China

Dianyong Wang, Renliang Li & Chuandong Liu
China Construction Bridge Company Limited, Chongqing, China

ABSTRACT: Guojiatuo Yangtze River Bridge is a double tower three-span continuous steel truss girder suspension bridge with a main span of 720 m. the South Main Pier Foundation is located in the Yangtze river shoal area with complex hydrogeological conditions. The construction scheme of cofferdam before platform and pile foundation is adopted for the bored pile foundation. The cofferdam is a dumbbell double-wall structure with a size of 73.6 m × 29.4 m × 30.0 m, a wall thickness of 2.0 m, a total weight of about 2550 t, four layers of internal support, and a bottom sealing concrete thickness of 4.5 m. The finite element software is used to calculate the deformation and stress of the steel cofferdam under the cofferdam control condition and the anti-floating safety factor, and the maximum tensile stress of the cofferdam bottom sealing concrete after pumping. The calculation results meet the requirements. Considering the factors affecting the construction site, the cofferdam adopts the rapid manufacturing and assembly construction technology of vertical sections and horizontal blocks. During the construction process, the deformation and stress of the key mechanical parts of the cofferdam structure are monitored. The results show that the structural state of the cofferdam during construction and operation is good, safe, and reliable, its structural design meets the requirements of on-site construction, and the cofferdam adopts the bridge position for rapid construction, which is convenient to improve the construction quality and risk control.

1   INTRODUCTION

In recent years, the construction of bridges in China has been changing rapidly, especially across deepwater basins such as rivers, lakes, and seas, where world-renowned achievements (Zhong et al. 2021) have been made in the construction of large span bridges. The construction of pile foundations and bearings in deep water areas often encounters natural phenomena such as floods, rivers, and surges, making construction difficult, with high safety risks and long construction cycles. With the continuous development of construction technology, steel cofferdams have become the main choice for deepwater foundation construction due to their high strength, adaptability to water depth, and recyclable materials. The main forms of cofferdams are locking pile cofferdams, single-walled steel cofferdams, and double-walled steel cofferdams (Wang 2016; Wu 2020). At present, researchers at home and abroad have conducted in-depth research on deepwater steel cofferdam structures from structural design, construction techniques, and monitoring tests and have achieved certain results (Benmebare et al. 2005; Gui & Han 2009; Hsiung 2009; Huang & Zhao 2018; Zuo et al. 2019; Zhou et al. 2020).

As a water-retaining and supporting structure for deepwater foundation construction, the rationality of its structural design will directly affect the safety and economy of deepwater foundation

---

*Corresponding Author: 1592577163@qq.com

DOI 10.1201/9781003308577-86

construction. In order to ensure that the strength, stiffness, and stability of steel cofferdams meet the actual engineering requirements during the construction of deepwater bridge foundations, special design (Xu et al. 2016) of steel cofferdams is required in conjunction with the actual project water depth, flow rate, and construction conditions.

2 PROJECT OVERVIEW

The main bridge of the Guo-jia-tuo Yangtze River Bridge is a single-hole suspension double tower three-span continuous steel truss girder suspension bridge with a span of 253.3 m + 720 m + 253.3 m (Guo 2017) (see Figure 1). The main pier foundation of the bridge tower on the South Bank of the main bridge adopts 34 pieces $\varphi 3.0$m bored pile group pile foundation, and the bearing platform is a dumbbell-shaped embedded structure (see Figure 2). The bearing platform size is 69.6 m (length) $\times$ 25.4 m (width) $\times$ 6.0 m (high), the design bottom elevation of the bearing platform is 156.726m, the design top elevation is 162.726m, and the two ends of the bearing platform are a $\varphi 25.4$ m circular structure, connected in the middle by a tie beam with a width of 17 m.

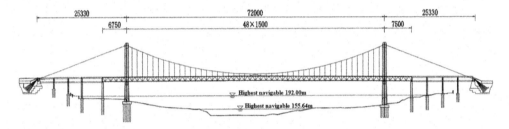

Figure 1.   Elevation view of the main bridge of Guo-jia-tuo Changjiang River Bridge.

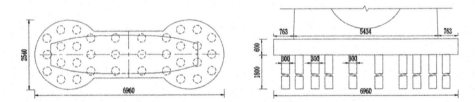

Figure 2.   Foundation structure of south pylon pier.

The Guo-jia-tuo Yangtze River Bridge is located in a curved river section with a wide and narrow planform, and the water volume and level are greatly influenced by the river. The main pier on the south bank is located in a shallow area of the Yangtze River, where the hydrogeological conditions are complex, with a thin cover of soil and locally exposed sloping rocks, a short dry period, and a fast flow of water during the storage period. The bridge abutment is constructed by cofferdam. With further consideration of construction conditions, duration, economy, etc., the construction plan (Xiao 2019) of fast assembling cofferdam at the bridge position first and then the platform and pile foundation is proposed, and a temporary processing plant is set up for the cofferdam next to the bridge position and processing the cofferdam in pieces, and lifting it to the bridge position for assembling after the processing is completed (Qin 2018; Zhang 2019). The method with fast construction, high accuracy of installation, and a guarantee of structural safety factors can provide a reference for similar projects.

3 COFFERDAM DESIGN AND VERIFICATION

3.1 *Cofferdam design*

The tower pier foundation cofferdam of the south bank bridge is designed as a dumbbell-shaped double-wall steel boxed cofferdam (Jin 2018; Hu et al. 2021, 2019; Xu 2021) (see Figure 3), with a plane size of 73.6 m × 29.4m, 30.0m high and 2.0m apart. It is divided into 28 compartments with a total weight of about 2550 t. The top elevation of the cofferdam is +182.226 m, and the bottom elevation is +152.226 m. the cofferdam soil is sealed with 4.5 m thick C30 underwater concrete (Xu & Wang 2018), and the bottom elevation of the sealing concrete is +152.226 m.

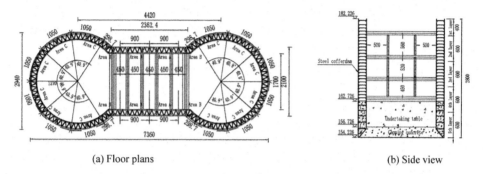

(a) Floor plans   (b) Side view

Figure 3.   Structural arrangement of the double-walled steel jacketed box cofferdam at the south main pier.

The structure of the cofferdam consists of vertical stiffening ribs, horizontal trusses, horizontal ring plates, vertical ring plates, and internal and external panels. The cofferdam is constructed using a rapid fabrication and assembly process (Yan & Wang 2019; Zhao et al. 2011; Zhang et al. 2020) in vertical sections and horizontal blocks. The cofferdam is vertically divided into five layers, and each layer is divided into three different types of standard blocks, including six straight-line section standard blocks a, four corner standard blocks B and twelve curve standard blocks C, with a total of twenty blocks. The cofferdams are spliced with 15cm splicing plates. The steel cofferdam is set with four layers in the straight-line section, with six supports in each layer. The thickness of the double-wall panel is 8 mm. The ring plate is 16 mm thick and adopts two sizes. The size of ring plate one is 300 mm × 16 mm, and the size of ring plate two is 250 mm × 16 mm. The vertical rib adopts ∠100 mm × 10 mm angle steel. The horizontal, diagonal bracing adopts ∠125 mm × 12 mm angle steel, reinforced locally, double. The thickness of the compartment board is 16 mm, and vertical angle steel stiffeners are used on it. The double-wall steel cofferdam is made of Q235 steel. The allowable stress of compression, bending, and tensile resistance of the steel is 215 MPa, and the allowable stress of shear resistance is 125 MPa.

3.2 *Structural checks*

Midas software was used to establish the finite element analysis model of the double-walled steel cofferdam. Solid units were used for the concrete filled in the wall slab and the sealed bottom concrete, stiffened slab units for the panel, truss units for the internal support, and beam units for the rest, and displacement constraints were designed at the bottom sealed bottom concrete. The results of the deformation and stress calculations are as follows.

(1) Overall deformation. The maximum overall deformation of the steel cofferdam under controlled conditions was 11.4mm, and the maximum deformation occurred in the vicinity of the circular area between the two inner supports in the middle near the straight area (see Figure 4).

(2) Combined panel and vertical rib stresses. The maximum value of the combined stresses in the steel weir panels and vertical ribs under controlled conditions was 84.2MPa, and the maximum

deformation occurred at the junction of the sealing bottom concrete and the steel weir wall slab (see Figure 5).

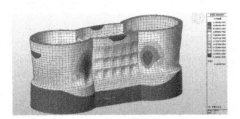
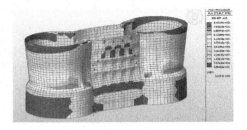

Figure 4. Overall deformation.

Figure 5. Combined stress of panel and vertical rib.

(3) Combined stresses in the horizontal ring plate. The maximum value of the combined stress in the horizontal ring plate of the steel cofferdam under controlled conditions was 100MPa, and the maximum deformation occurred at the junction of the sealing bottom concrete and the steel cofferdam wall plate (see Figure 6).

(4) Combined stresses in the horizontal truss. The maximum value of the combined stress in the horizontal ring plate of the steel cofferdam under the control condition is 155.5MPa, and the maximum deformation occurs at the intersection of the sealing bottom concrete and the steel cofferdam wall plate (see Figure 7).

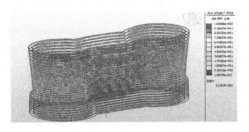
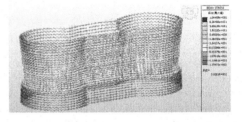

Figure 6. Combined stress of horizontal ring plate.

Figure 7. Combined stress of horizontal truss.

3.3 Undercover concrete checks

(1) Floatation stability

The annual flood level at the bridge site was considered to be181.48 m under natural conditions, and the overall floatation stability was carried out after the bottom sealing concrete had been poured and the water had been pumped into the cofferdam. The bonding force between the concrete and the steel shoring is taken as 170kN/m^2, the effective thickness of the bottom sealing concrete is taken as 4.5 m, and the net area of the inner wall of the steel cofferdam is A=1726 m^2. The calculation process is shown below (Liu et al. 2021).

Self-weight of steel cofferdam G_1:$G_1 = 1726 \times 4.5 \times 23 = 33000KN$

Self-weight of capped concrete G_2:$G_2 = 1726 \times 4.5 \times 24 = 186408KN$

Self-weight of concrete in pressurized bins G_3:$G_3 = 358 \times 10.5 \times 24 = 90216KN$

Self-weight of pressurized water G_4:$G_4 = 358 \times 15 \times 10 = 53700KN$

Adhesion between sealing concrete and steel shoring G_5:$G_5 = 10.36 \times 4.5 \times 150 \times 34 = 237762KN$

Downward force G:$G = G_1 + G_2 + G_3 + G_4 + G_5 = 601086KN$

Water buoyancy F_f:$F_f = \rho g V_v = 10 \times 30 \times 1726 = 517800KN$

Floatation stability factor K=G/F=1.16>1.05, the seal thickness meets the floatation requirements.

(2) Calculation of tensile stresses in sealed bottom concrete

The allowable tensile stress of C30 sealed concrete is known to be 1.04Mpa.

The upward uniform load on the encapsulated concrete is:

$$q = \gamma_w h_w - \gamma_c h_c = 182 \text{kN} / \text{m}^2$$

The calculations can be simplified using one-way plates:

$$M = 1 / 8 q l^2 = 2238.9 \text{kN} \cdot \text{m}$$

$$\sigma_{Max} = M / W_x = 0.66 MPa < [\sigma] = 1.04 MPa$$

The maximum tensile stress in the sealed bottom concrete meets the specification requirements.

4 KEY CONSTRUCTION TECHNIQUES

The key construction techniques for the steel cofferdam bridge assembly include the fabrication of the steel cofferdam in pieces, excavation of the lying bed, assembly of the first section, blocking and protection, installation of the steel shoring, construction of the concrete for the bottom sealing, cofferdam joint height and installation of the internal support.

4.1 *Excavation of foundation pits*

Before the excavation of the foundation pit, a cut-off ditch (width 0.6 m, depth 0.5 m) should be set up around the top of the pit at a position m from the edge line 2 of the slope. The excavation is carried out by an excavator according to the slope1: the slope is 0.5 released, the slope is trimmed in time during the excavation, and the slope is protected by a layer 10 of C20 concrete sprayed on the surface of the slope, the excavated soil is transported to the designated dumping ground by the dump truck. The rock stratum shall be excavated by machinery (crushing hammer + excavator) according to the slope of 1:0.3 and shall be constructed orderly according to the characteristics of vertical layering and plane blocking.

4.2 *Fabrication and transport of cofferdams in blocks*

A temporary processing plant with a hardened area of about 3000 m² is set on the downstream bank of the tower pier of the South Bridge to manufacture the steel cofferdam in sections and blocks (see Figure 8). Two 50-t gantry cranes are arranged in the processing plant to be responsible for the material transportation in the processing plant, and a 150-t truck crane is configured to assist the turnover work during the processing of the unit block of the steel cofferdam. After the unit block is manufactured and accepted, it is transferred to the storage area by the gantry crane for temporary storage. A 260-t crawler crane is used to transport it to the construction site and assemble the steel cofferdam. Before the standard block is made, the performance of raw materials and various technical indexes shall be sampled and rechecked according to the provisions of relevant standards. During the manufacturing process, the accurate blanking of profiles shall be ensured, the defective profiles shall be corrected or replaced in time, and the pre-welding, welding, and post-welding inspection shall be carried out according to the welding process required by design. After fabrication, the molding inspection, correction, acceptance, pre-assembly, and storage shall be carried out in time.

4.3 *Bottom section cofferdam assembly*

To ensure that the elevation of the first steel cofferdam is level when assembled, steel pads were installed under the bottom section of the cofferdam, and a total station was used to place the

Figure 8.   Field processing of unit block.

pad elevation and plane position to ensure that the elevation error between the steel pads met the design requirements. The first layer of the cofferdam was assembled symmetrically from the upstream corner B to the straight section A and the curved section C. The steel cofferdam was assembled with a large volume of work and a large number of units, and the standard blocks were numbered in advance according to the pre-assembly sequence. In order to ensure the stability and safety of the cofferdam during lifting and assembling, internal and external temporary supports were set up on the inside and outside of the cofferdam, respectively, in the order of cofferdam assembling, and the plane position and verticality of the cofferdam unit blocks were checked using measuring instruments. In order to ensure the rigid connection of the steel cofferdam sections and the quality of the overall assembly, the key stressing welds were tested by ultrasonic flaw detection according to the design requirements, and the rest of the welds were tested for tightness by paraffin penetration. Once the first layer of the steel cofferdam has been assembled, the temporary support of the cofferdam can be removed.

4.4   Concrete sealing of the cofferdam

After the pier position of the first section of the cofferdam is assembled, the gap between the inner and outer baffles of the cofferdam and the riverbed surface shall be sealed with sandbags. After sealing, the outer side of the cofferdam shall continue to use the rubble to press the bottom. During the excavation of the foundation pit, the rubble soil shall be backfilled to the height of the surrounding riverbed. After observing and confirming that the sealing effect is good, the underwater concrete bottom sealing construction shall be carried out. The cofferdam bottom sealing adopts the construction technology of multi-point underwater pouring and one-time concrete forming. Pouring conduit adopts the $\varphi300$ card type quick loading vertical conduit is arranged with concrete distribution points in quincunx shape according to the spacing of 4.5 m. each set of the conduit is equipped with adjustment joints with different lengths of 2 m, 1.5 m, and 1 m, which is convenient for the lifting and removal of the conduit when sealing the bottom, so as to control the buried depth of the conduit. Before installing and using the conduit, the water pressure and water tightness test shall be conducted first. The pressure test strength shall be 1.3 times the head pressure. During the assembly, the conduit shall be numbered and butted to determine the length and installation and splicing sequence of the conduit. The pouring sequence of bottom sealing concrete shall be carried out point by point from low to high, and the ball cutting pouring shall be carried out point by point from the middle to around, and the initially buried depth of the conduit shall not be less than 1.0 m. The total thickness of the bottom sealing concrete is 4.5 m. when the pouring height of the bottom sealing concrete is 4.3 m, the remaining 0.2 m is used as the leveling layer on the bottom surface of the bearing platform after pumping. When the strength of the standard test block under the same conditions of the bottom sealing concrete reaches the design value, pump water in the cofferdam, pour the leveling layer by the concrete pump truck dry method, and set the apron slope, drainage ditch, and sump on the top surface of the bottom sealing concrete according to the design, and arrange pumping equipment.

4.5 *Cofferdam joint height and internal support*

The assembly sequence and welding process of cofferdam heightening are basically the same as that of the first layer of steel cofferdam, and the installation of internal support is carried out with the heightening of the cofferdam. As the main bearing component in the horizontal direction of the double-wall steel cofferdam, the construction quality of the inner support is related to the structural safety of the steel sheet pile cofferdam. In order to improve the overall stability of the structure, 24 (four layers and six layers per layer) are set in the straight section of the cofferdam, with the model of $\varphi820\times12$ mm steel pipe internal support, $\varphi377\times6$ mm connection system is set between the steel pipe internal support on the same floor. In order to improve the construction efficiency and the construction quality of the on-site assembly of the internal support structure, the design size shall be retested before the internal support processing and welded and assembled in the bridge location processing plant. The inner support is assembled from the middle to both sides, transported to the construction site by 260 t crawler crane, placed on the concave corbel welded on the inner wall, and finally welded with the side plate as a whole, and other inner supports are installed in turn.

5 COFFERDAM MONITORING

Under the condition that the steel cofferdam meets the requirements of the design specifications, considering the unfavorable working conditions such as the complex structure of the steel cofferdam itself, pumping in flood season, and being located in the key protection zone of the ecological area, in order to prevent large deformation or stress in the cofferdam during flood season construction and further ensure the safety of steel cofferdam construction and operation stage, It is necessary to monitor the key mechanical parts of the steel cofferdam and the surrounding hydrological environment. The layout of monitoring points during cofferdam construction: eight vibrating wire surface strain gauges are arranged on the horizontal ring of the outer wall plate 10 m away from the bottom of the cofferdam, and five vibrating wire surface strain gauges are arranged on the horizontal axis of the support midpoint in the steel pipe. According to the monitoring data, the maximum deformation of each measuring point of the cofferdam wall plate is 8.4 mm during the process from the pumping of the cofferdam bottom to the pouring of the bearing platform. The maximum deformation of each measuring point of the inner support is 10.5mm, the measured value (maximum value) of each unfavorable position is less than the theoretical calculation value, and the overall stress of the cofferdam is stable and in a safe state.

6 COFFERDAM MONITORING

The construction scheme of cofferdam before platform and pile foundation is adopted for the tower pier foundation of the South Bridge of Chongqing Guo-jia-tuo Yangtze River Bridge. Aiming at the Yangtze river shoal area with complex hydrogeological conditions at the bridge site, the dumbbell double-wall steel boxed cofferdam adopts the rapid manufacturing and assembly construction technology of vertical section and horizontal block through scheme comparison. A temporary processing plant is set at the bridge site to manufacture and assemble the cofferdam in blocks without a floating platform, large water transportation, and hoisting machinery and equipment, which has remarkable economic benefits, effectively avoids the important problems of difficult positioning control and high safety risk of large tonnage and large-size cofferdam, so as to improve the construction quality and risk control.

During the flood season in 2018, floods above the warning water level broke out in many tributaries of the Yangtze River, and the maximum inflow peak flow of the Three Gorges Reservoir reached 60000 cubic meters per second. During the flood season and cushion cap construction, the cofferdam structure was in good condition, and the monitoring results of the cofferdam were within the controllable range of safety indicators, which fully verified the rationality of the design and construction of the steel cofferdam. It can provide a reference for pier foundation construction in a similar environment.

AUTHOR

Geng Wenbin (1990), male, engineer, engaged in structural analysis and construction control of large-span bridges.

ACKNOWLEDGEMENTS

This work is supported by CSCEC (Grant Nos. CSCEC-2020-Z-21).

REFERENCES

Benmebare K, N, S, et al. Numerical studies of seepage failure of sand within a cofferdam[J]. Computers & Geotechnics, 2005, 32(4).

China Merchants Bureau Chongqing Transportation Research and Design Institute Co. Construction drawing design of Guo-jia-tuo Yangtze River Bridge [Z]. Chongqing, 2017.

Gui M W, Han K K. An investigation on a failed double-wall cofferdam during construction[J]. Engineering Failure Analysis, 2009, 16(1):421–432.

Hsiung B. A case study on the behavior of a deep excavation in sand[J]. Computers & Geotechnics, 2009, 36(4):665–675.

Hu Hao, He Peng, Yang Xuexiang. Key technologies for the construction of the Wuxiu Yangtze River Highway Bridge steel jacket cofferdam[J]. World Bridges, 2019, 47(5):23–26.

Hu, Haibo, Zhou, Yadong, Zhang, Jing. Key technology for sealing the bottom of the steel cofferdam of the north main pier of Wuhan Qingshan Yangtze River Highway Bridge[J]. Bridge Construction,2021,51(01):8–13.

Huang Xiaojian, Zhao Xu. Monitoring analysis of double-row sheet pile cofferdam [J]. Engineering Quality,2018, 36(4):58–63.

Jin Hongyan. Construction technology for sinking steel jacket box cofferdams of Wuhan Qingshan Yangtze River Highway Bridge[J]. Bridge construction,2018,48(02):10–15.

Liu Xiaomin, Zhang Qiang, Li Fei, Yang Jinyong. Study on the construction technology of deepwater foundation with weir before pile in bare rock area[J]. Construction Technology, 2021, 50(7):31–34.

Qin Yonggang. Key technologies for the construction of tower foundations for the Dongting Lake Bridge of the Meng Hua Railway[J]. World Bridges, 2018, 46(3):23–26.

Wang Jun. Research on the construction technology scheme of locking pile cofferdam in shallow bare rock area[J]. Railway Construction, 2016, 000(012):33–36.

Wu Lingzheng. Design and construction of cofferdams for non-navigable bore bridges in the east flood relief area of the Shenzhong Channel[J]. World Bridges, 2020, 48(5):21–25.

Xiao Shibo. Key technology of cofferdam construction in deepwater of Pingtan Strait Public-Railway Bridge[J]. Bridge construction,2019,49(2):13–16.

Xu F, Li S C, Zhang Q Q, et al. Analysis and Design Implications on Stability of Cofferdam Subjected to Water Wave Action[J]. Marine Geotechnology, 2016,34(2):181–187.

Xu Qili. Design of cofferdams for bridge tower piers of the Pingtan Strait Public-Railway Bridge navigable pore bridge[J]. Bridge Construction,2021,51(01):115–120.

Xu Xin, Wang Tongmin. Steel cofferdam sealing technology for the tower pier of No4. 1 bridge at Yichang Xiangxi River[J]. Bridge Construction,2018, 48(2): 105–110.

Yan J, Wang Yiping. Construction technology of large double-walled steel cofferdams for the south tower of Chibi Yangtze River Highway Bridge[J]. World Bridges, 2019, 49(1):20–25.

Zhang Chengran, Zhu Bing, Zhang Zhen, Zhao Yujia. Construction technology of combined cofferdam for pier No3. 1 of Lingang Yangtze River Public-Railway Bridge[J]. Bridge Construction,2020, 50(4):101–106.

Zhang Yanhe. Construction technology of steel cofferdams in deepwater shallow overburden areas[J]. World Bridges, 2019, 47(2):28–31.

Zhao Xingzhai, Zhang Binbin, Liu Bei. Double-walled steel cofferdam construction technology for Xiangfan Hanjiang Fifth Bridge in layers and blocks [J]. Railway Construction, 2011, 51(12):31–32.

Zhong Qi, Zheng Chunyu, Wei Kang, Mu Xubiao, Zhang Peng. Study on the selection and design of steel cofferdam types in the complex geological deepwater environment[J]. Highway, 2021, 10(10):220–225.

Zhou Xinya, Liu Changyou, Qian Youwei. Key technologies for the design and construction of deepwater foundation ultra-long sheet pile cofferdams[J]. World Bridges, 2020, 48(2):20–24.

Zuo L, Fu J, Ji Rongxian, et al. Design and field test study of self-balancing combined steel cofferdam under deep water conditions[J]. South-North Water Diversion and Water Conservancy Science and Technology, 2019(3):139–146.

*Frontiers of Civil Engineering and Disaster Prevention and
Control – Yang & Rahman (Eds)*
© 2023 The Author(s), ISBN: 978-1-032-31200-2

Experimental study on torsional behavior of concrete-filled steel tubes

Jucan Dong
Shenzhen Expressway Engineering Consultant Co., Ltd, Shenzhen, China

Zhenwei Lin*
College of Civil Engineering, Yango University, Fuzhou, China

Yiyan Chen, Qingxiong Wu & Chao Zhang
College of Civil Engineering, Fuzhou University, Fuzhou, China

Xinmin Nie & Zhaojie Tong
Shenzhen Municipal Design & Research Institute Co., Ltd, Shenzhen, China

ABSTRACT: To study the contribution of concrete in the tubes to the torsional stiffness of concrete-filled steel tube (CFST) members, torsion tests of five groups of CFST members and corresponding hollow steel tube members were performed, and the finite element model was verified based on the different steel ratios and diameter-to-thickness ratios. The test results show that the hollow steel tubes will be destroyed due to plastic hinges, but the CFST members have no obvious buckling phenomenon. The parameter analysis was carried out by finite element analysis (FEA), which considered steel ratio, steel yield strength, and concrete strength. The results show that the steel ratio greatly influences the torsional stiffness of CFSTs, while the concrete strength and steel strength have minimal effect on the torsional stiffness. Finally, considering the contribution of the concrete in the steel tubes to the torsional stiffness, the superposition method of linear stiffness was used to determine the regression formula of the torsional stiffness coefficient of concrete.

1 INTRODUCTION

Concrete-filled steel tubes (CFSTs) are widely used as compressive members in the superstructure of arch bridges and substructures of bridges in China. At the same time, CFST members are also increasingly popularized and enriched as part of composite beam systems. Based on a composite truss girder bridge with CFSTs and a composite box girder bridge with corrugated steel webs, a composite box girder with corrugated steel webs and trusses is proposed by Yiyan Chen (2008), which is shown in Figure 1. It consists of two bottom CFSTs and a top concrete slab connected by corrugated steel webs, and the two bottom CFSTs use a truss connection. This new composite structure can reduce the weight of the main girder and improve the structural crack resistance, mechanical performance, integrity, and material utilization of the structure. To study the torsional behavior of composite box girders with corrugated steel webs and trusses, the torsion mechanism of concrete slabs, and corrugated steel webs, CFSTs need to be understood. However, there are few studies on the torsional stiffness of CFSTs. The study on the torsional stiffness of CFSTs can be helpful in understanding the torsional behavior of the novel composite beams.

Domestic and foreign scholars have conducted some different research on the torsional behavior of steel tubes. Beck and Kiyomiya (2003) conducted an experimental study on the torsional behavior of CFSTs, corresponding hollow steel tubes, and plain concrete columns. The results show that

---

*Corresponding Author: linzhenwei2005@126.com

DOI 10.1201/9781003308577-87

the CFST members present ductile failure and do not appear local buckling or other failure modes similar to hollow steel tubes and plain concrete columns. An Gong (1989) conducts an experimental study of CFST short columns under the action of torsional load and the action of compressive and torsional loads, while Jing Zhou (1990) conducts an experimental study of CFST long columns under the action of the compressive and torsional loads. The results show that the interface between the steel tubes and the concrete in the tubes is well-bonded, and no relative slippage occurs. Based on the plastic theory, the calculation of torsional bearing capacity is deduced. Linhai Han et al. (1995) conduct an experimental study on the torsional behavior of CFST members. The load-deformation curves of the test beams are analyzed by FEA, and a simplified calculation formula for the torsional bearing capacity of CFSTs is proposed. Yiwei Chen (2003) conducts experiments and research on CFSTs under cyclic torsions. The torsional stiffness of CFSTs subjected to multiple loads is analyzed, and the formula of the torsional bearing capacity is given. Baochun Chen (2008) conduct experimental studies on the torsional behavior of CFSTs, corresponding hollow steel tubes, as well as the plain concrete columns. The results show that the deformation of the steel tubes and the concrete in the steel tubes are coordinated. Due to the interaction between the steel tubes and the concrete inside the tubes, the torsional bearing capacity of CFSTs is greater than the linear addition of the steel tube bearing capacity and the concrete bearing capacity. Guohuang Yao (2006) uses ABAQUS software to study the torsional behavior of CFSTs and analyzes the mechanics characteristics of each stage. The formula of CFST torsional bearing capacity is obtained by parameter analysis and regression method.

It can be seen that the majority of the above results are based on experimental research combined with analytical methods to obtain the formula for the torsional bearing capacity of CFSTs. In general, the study on the torsional stiffness formula is rare. Moreover, the steel ratios of the members are mainly in the range of 0.04–0.15. However, the steel ratios in the composite girder bridges are much larger.

In order to study the torsional stiffness of CFSTs, ten specimens with a length of 3.6m were tested. The steel ratios of CFSTs are mainly in the range of 0.1–0.3, which are much larger than that of CFST arch structures. Based on the steel ratios on different sections, the torsional tests and FEA were carried out. Considering the contribution of the concrete in the steel tubes to the torsional stiffness, the superposition method of linear stiffness was used to determine the regression formula of the torsional stiffness coefficient of concrete.

2 TEST OVERVIEW

2.1 *Specimen design*

The section steel ratios of CFST members in arch bridges are between 0.04 and 0.15. When CFST members are used for composite girder bridges, CFSTs require larger steel ratios. Table 1 presents the parameters for 11 composite box girder bridges with the bottom steel tubes filled with concrete. It can be seen that the diameter-to-thickness ratios of CFSTs are between 10 and 40. Meanwhile, the corresponding section steel ratios are between 0.1 and 0.5, mainly in the range of 0.1 to 0.3, which is much larger than that of the CFST arch structures. Therefore, according to the parameters collected in the literature, the test specimens had steel ratios between 0.1 and 0.3. The influence of steel ratios on the torsional stiffness of CFSTs was investigated. The load-torsion curves in the whole torsion test process were analyzed, and the torsional stiffness of CFSTs was defined.

According to the test site and the effective transmission of the torque during the torsion test, the length of the test specimens was 3.6m, and the torsion test of CFSTs was carried out. In this test, five sets of CFST specimens with a length of 3.6 m were made. The steel tubes were made of Q235c, and the concrete inside the tubes was C30. The size and material properties of CFST specimens are shown in Table 2. CFST represents concrete-filled steel tube; ST represents hollow steel tube. The specimen design is shown in Figure 1.

Table 1. Parameters of CFST composite bridges.

Project	Diameter d(mm)	Thickness t(mm)	Diameter-to-thickness ratio(d/t)	Steel ratio $\rho_s(A_s/A_c)$
Xiangjia Dam bridge	700	20	35.0	0.12
Wanzhou Yangtze river bridge	500	16	31.3	0.14
Chongqing Wanzhou Daohegou bridge	325	8	40.6	0.11
Shanghai Yunlian road pedestrian bridge	402	14	28.7	0.16
Ganhaizi bridge	813	28	29.0	0.15
Lully viaduct	508	36	14.1	0.36
Dättwil bridge	508	50	10.2	0.55
Aarwangen interstate bridge	406	36	11.3	0.48
Maluanshan park viaduct	720	20	36	0.12
Beijing-Hong Kong-Macau express service area footbridge	500	24	20.8	0.22
Daxigou pipeline bridge	600	15	40	0.11
Xiaoshixi pipeline bridge	600	15	40	0.11
Maupre bridge	610	20	30.5	0.15

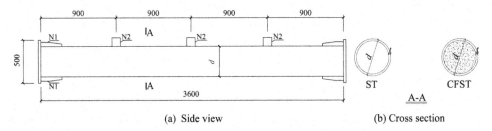

(a) Side view   (b) Cross section

Figure 1. CFST specimen design (Unit: mm).

Table 2. Size and material properties of CFST specimens.

Specimen number	$(d \times t \times l)$ mm	Diameter to thickness ratio(d/t)	Steel ratio(A_s/A_c)	E_s (10^5MPa)	E_c (10^4MPa)	f_y (MPa)	f_{cu} (MPa)
CFST-1	$\Phi70 \times 4 \times 3600$	17.50	0.275	2.05	3.05	240	30.1
ST-1	$\Phi70 \times 4 \times 3600$	17.50	——	2.05	3.05	240	30.1
CFST-2	$\Phi114 \times 4.5 \times 3600$	22.80	0.202	2.05	3.05	235	30.1
ST-2	$\Phi114 \times 4.5 \times 3600$	22.80	——	2.05	3.05	235	30.1
CFST-3	$\Phi180 \times 6.5 \times 3600$	25.71	0.176	2.05	3.05	246	30.1
ST-3	$\Phi180 \times 6.5 \times 3600$	25.71	——	2.05	3.05	246	30.1
CFST-4	$\Phi240 \times 8 \times 3600$	30.63	0.145	2.05	3.05	242	30.1
ST-4	$\Phi240 \times 8 \times 3600$	30.63	——	2.05	3.05	242	30.1
CFST-5	$\Phi351 \times 10 \times 3600$	35.10	0.124	2.05	3.05	238	30.1
ST-5	$\Phi351 \times 10 \times 3600$	35.10	——	2.05	3.05	238	30.1

2.2 Measuring point arrangement and loading device

To understand the mechanical properties of CFST specimens under the torsion, the measuring points were arranged on the three sections of the specimens (l/2, l/4, and 3l/4, l is the length of the test specimen). Four strain test points were evenly arranged along the circumferential direction on the steel tube surface, and each measuring point was arranged by a three-direction strain gauge, as shown in Figure 2. CFST specimens were welded and poured on-site, as shown in Figure 3. The test was carried out at the Experimental Teaching Center of the School of Civil Engineering of Fuzhou University. The test was carried out using a torsion test machine, and Figure 4 shows the machine, which consists of a fixing device, a loading device, and a reaction force test device. To be more specific, a thick steel plate is installed in one end of the specimen and then locked in the fixing device such that the rotation of this end can be restrained. The other end is embedded into a fan-shaped plate in the loading device. A steel wire rope is wound along the arc surface of the fan-shaped plate. A hydraulic jack pulls the steel wire rope during torsional tests, and then the fan-shaped plate rotates, generating a torsional load for the test specimen. One end of the specimen was embedded in the fixing device, and the other end was embedded in the fan-shaped plate. Several steel wires were arranged in the groove on the circumference of the fan-shaped plate, and the steel wires were connected to the jack through the pulley on the anchor. The jack pulled the steel wires, and the pulley turned to rotate the fan-shaped plate to apply torque to the specimen. Therefore, the torsion was loaded.

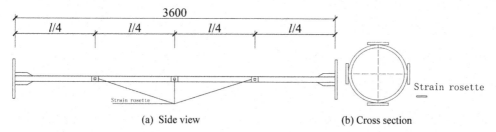

(a) Side view   (b) Cross section

Figure 2.   CFST measuring point layout (Unit: mm).

Figure 3.   CFST specimen.   Figure 4.   CFST torsion test machine.

To measure the torsion angle of the specimens, a concentric steel cylinder suspended with heavy hammers was welded to the torsion arm, and the torsion angle was measured and checked by displacement meters and an inclinometer installed on the loading end of the torsion testing machine. The test load was applied step by step. Each stage load was 1/10 (elastic stage) or 1/15 (elastoplastic stage) of the estimated ultimate load. The load of each stage was held for about 3~5min. Before the specimen was destroyed, slow continuous loading was performed, and the data was collected until the specimen was destroyed.

2.3   Test results and analysis

As shown in Figure 5(a), for the CFST specimen, a slight creaking sound occurred during the loading process. This is mainly caused by the cracking of the concrete inside the steel tube and the cracking of the local surface of the steel tube. There was no dent, bulge, or fracture in the outer steel tube. As shown in Figure 5(b), for the hollow steel tube specimen, the surface of the steel tube produced a 45°C slip line during the loading process. Meanwhile, the steel tube showed local dent and bulge deformation, and it was slightly flat or even dislocated from the side view. Local buckling occurred on the surface of the steel tube when the specimen was broken.

Comparing the test phenomenon and test results of CFSTs with the corresponding hollow steel tube, the following conclusions can be obtained. When the torsional load acted on the hollow steel tube specimen, the angle between the principal tensile stress direction and the principal compressive stress direction of the specimen along the axis of the steel tube was about 45°. The hollow steel tube produced oblique deformation whose direction was the same as the principal stress. Finally, the hollow steel tube was damaged due to the plastic hinge. However, the CFST specimens had no obvious buckling under the torsional load and had better plasticity and stability.

 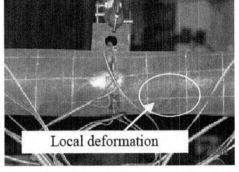

(a) CFST specimen   (b) Hollow steel tube specimen

Figure 5.   Failure mode of specimens.

Figure 6 shows torque-torsion angle curves of a CFST specimen and a hollow steel tube specimen with a diameter $d = 70$ mm. It can be seen that curves and the test process can be divided into three stages under the torsional load:

(1) Elastic stage (O-A): For the CFST specimen, the steel tube and the core concrete all worked and had no interaction, and the curves were straight lines. At point A', the core concrete would crack with increased torque. For the hollow steel tube, the force state was a relatively simple shear state in a circumferential direction. Point A in Figure 6 is the critical point at which the steel of the specimen entered the elastoplastic stage from the elastic stage.

(2) Elastoplastic stage (A-B): For the CFST specimen, the core concrete gradually cracked with the increase of torque. The deformation of the core concrete was greater than the deformation of the steel tube. And the interaction between the core concrete and the steel tube occurred, leading them to a complex stress state dominated by shearing in both directions. Point B in Figure 6 was the critical point at which the steel of the CFST and ST specimens entered the plastic stage from the elastoplastic stage, at which the steel tubes began to yield.

(3) Plastic stage (B-C): For the CFST specimen, a large number of micro-cracks appeared in the concrete inside the steel tube. However, the torsional bearing capacity of the specimen continued to grow due to the constraint of concrete on the deformation and buckling of the steel tube and the constraint of the steel tube on the concrete. The hollow steel tube specimen showed plastic deformation with the increase of the torsional load. Both specimens showed good plasticity.

The torque-torsion angle curve of each specimen was measured by the test, and the torsional stiffness can be calculated by the following formula:

$$GJ = T/\theta \tag{1}$$

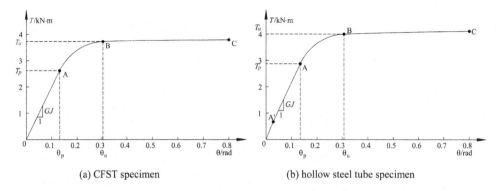

(a) CFST specimen   (b) hollow steel tube specimen

Figure 6. Torque- torsion angle curves of specimens (T/θ).

Before the steel tube reached the proportional limit (point A in Figure 6), the CFST specimen had higher torsional stiffness than the corresponding hollow steel tube specimen due to the support of the concrete inside the steel tube. The slash slope of point A is taken as the torsional stiffness of the two sets of specimens. The torsional stiffness values of all the specimens are shown in Table 3. And $k1$ in equation (2) is defined as the torsional stiffness ratio of each CFST specimen to the corresponding hollow steel tube specimen (i.e., the torsional stiffness increase coefficient of CFST):

$$k_1 = \frac{K_t}{K_{ht}} = \frac{(GJ)_{sc}}{G_s J_s} \qquad (2)$$

Table 3 lists the values k_1 of all the specimens. The value k_1 of each specimen is between 1.1 and 1.2. It can be seen that the steel ratio ρ_s is inversely proportional to k_1. The main reason is that when the steel ratio ρ_s increases, the support effect of the concrete inside the steel tube is reduced, so that k_1 is reduced.

Table 3. Torsional stiffness comparison results.

No.	ST-1	CFST-1	ST-2	CFST-2	ST-3	CFST-3	ST-4	CFST-4	ST-5	CFST-5
$GJ(kNm^2)$	69.2	76.8	388	436.9	2180.4	2480	6754.2	7950	25204.6	30011.6
$\rho_s(A_s/A_c)$	—	0.275	—	0.202	—	0.176	—	0.145	—	0.124
k_1		1.11		1.126		1.137		1.177		1.191

Figure 7 shows the torque-maximum shear strain curves of CFST specimens and the hollow steel tube specimens. It can be seen that the torque-maximum shear strain of each specimen developed linearly at the initial stage of torsion. At the elastoplastic stage, the torque-maximum shear strain of each specimen developed nonlinearly. The CFST specimens exhibited uniform torsional deformation. For the hollow steel tube specimens, it could be observed that the surface of the specimens had obvious local deformation and other phenomena. On the whole, the ductility of the CFST specimens and the hollow steel tube specimens were good, and the torque-maximum shear strain curves did not show any descent stage.

3 FINITE ELEMENT CALCULATIONS

3.1 Finite element modeling

The ABAQUS finite element software was used to analyze the torsional behavior of CFST members and hollow steel tube members. Among them, the constitutive relation of plain concrete adopted

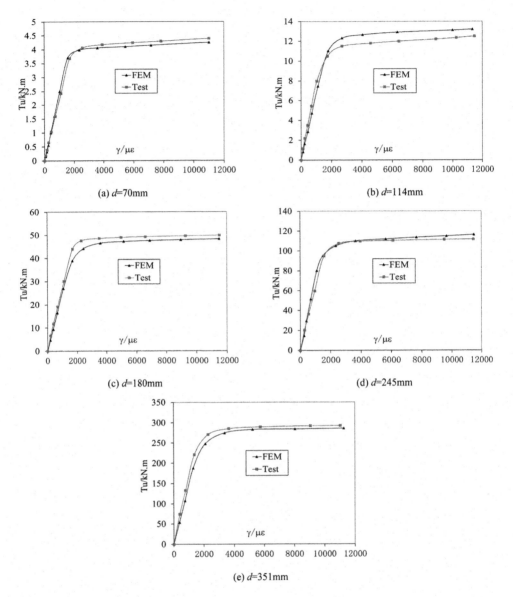

Figure 7. Torque-maximum shear strain curves.

the model that was established by Han et al. (2007), as shown in Figure 8(a). The constitutive relationship of concrete inside the steel tubes was determined by the following formula:

$$y = \begin{cases} 2x - x^2 & (x \leq 1) \\ \dfrac{x}{\beta_0(x-1)^{\eta_0}+x} & (x > 1) \end{cases} \tag{3}$$

$$\beta_0 = f_{ck}^{0.1}/(1.2\sqrt{1+\theta}) \tag{4}$$

With $x=\varepsilon/\varepsilon_0$, $y=\sigma/\sigma_0$ $\sigma_0=f_{ck}$, $\varepsilon_0=[(1300+12.5\times f_{ck}+800\times\theta^{0.2}]\times10^{-6}$, $\eta_0=1.6+1.5/x$, $\theta=A_sf_y/A_cf_{ck}$.

648

The steel used the constitutive model with two polylines, as shown in Figure 8(b). The elastic modulus and Poisson's ratio of the steel were 2.06×10^{11} Pa and 0.3, respectively.

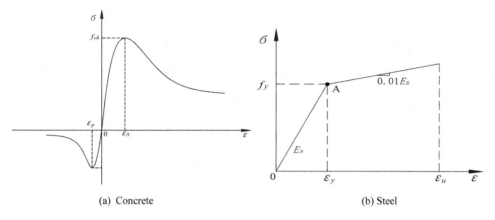

(a) Concrete   (b) Steel

Figure 8. Constitutive relationships.

Considering that the element mesh would be twisted during the torsion simulation of CFST members, this paper used the three-dimensional solid element (C3D8R) provided by ABAQUS to simulate the steel tubes and core concrete in the numerical analysis of CFST torsion resistance. When the mesh was twisted, and the calculation did not converge, the steel tubes and concrete were replaced by a two-dimensional element (C3D20R). The assumption of a complete bond without slip was applied to the interface between the steel tubes and the core concrete, and the interface was simulated by the common node method that the displacement of two materials was coordinated.

3.2 Comparison of calculation results with experimental results

The torsional nonlinear numerical analysis was carried out for each specimen. Figure 9 and Figure 10 show the torque-torsion angle curves and the torque (T)-the maximum shear strain (ρ_s) curves of CFST with diameter $d=70$mm, respectively. The finite element was in good agreement with the experimental results.

The torsional stiffness (K_{te}) of CFSTs measured by the experiment and the torsional stiffness (K_{tc}) of CFSTs calculated by the finite element model are summarized in Table 4. The maximum difference in torsional stiffness is 5.8%. It can be seen that the finite element calculation results are in good agreement with the experimental results.

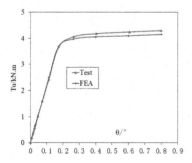

Figure 9. Torque-torsion angle comparison curves.

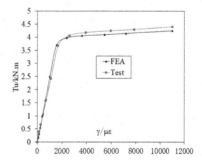

Figure 10. Torque-shear strain curves.

Table 4. Comparison of torsional stiffness of specimens.

No.	Specimen number	$K_{te}(\text{kN·m}^2)$	$K_{tc}(\text{kN·m}^2)$	K_{te}/K_{tc}
1	CFST-1	76.8	79.1	0.971
2	ST-1	69.2	71.8	0.964
3	CFST-2	436.9	417.9	1.045
4	ST-2	388	368.3	1.053
5	CFST-3	2480	2420.9	1.024
6	ST-3	2180.4	2115.4	1.031
7	CFST-4	7950	7663.8	1.037
8	ST-4	6754.2	6634.5	1.018
9	CFST-5	30011.6	29430.6	1.023
10	ST-5	25204.6	23914.2	1.058

3.3 *Parameter analysis*

By using the finite element model of CFST with diameter d=70mm, different parameters were selected for torsional stiffness parameter analysis: concrete strength (C30~C60), steel strength (Q235~Q420), and steel ratio (ρ_s =0.1~ 0.3). Other basic parameters are steel tube diameter d=70mm, length of specimen l= 3600mm, the thickness of the steel tube t = 4mm, concrete C30, and steel Q235.

Figure 11 shows the influence of various parameters on the torsional stiffness increase coefficient k_1 of CFSTs. It can be seen that the steel ratio ρ_s is inversely proportional to k_1 and has the greatest influence on k_1. Because the reduction of the steel ratio ρ_s will increase the support effect of concrete on the steel tubes, the increase of the torsional stiffness is more obvious.

4 SIMPLIFIED CALCULATION FORMULAS FOR TORSIONAL STIFFNESS

Through the above test and parameter analysis, the torsional stiffness increase coefficient of the CFSTs is mainly related to the steel ratio ρ_s. Therefore, considering the contribution of the concrete in the steel tube to the torsional stiffness in the form of linear stiffness superposition, the calculation formula of the torsional stiffness of CFSTs is:

$$K_t = k_1 G_s J_s = G_s J_s + k_2 G_c J_c \tag{5}$$

Figure 12 shows the relationship between the torsional stiffness and the steel ratio ρ_s of 24 specimens, including both the test value in this paper and the finite element value in the literature. Through regression analysis, the relationship between the steel ratio ρ_s is obtained. The calculated results are well fitted to the experimental results, so that the formula k_2 is:

$$k_2 = 0.901\rho_s + 0.170 \tag{6}$$

In this paper, the concrete inside the steel tube is equivalent to a thin-walled steel tube with the same torsional stiffness and then superimposed on the thickness of the steel tube, which is:

$$J_{sc} = \frac{G_s J_s + k_2 G_c J_c}{G_s} \tag{7}$$

$$\delta_{sc} = \frac{d}{2} - \frac{1}{2}\sqrt[4]{d^4 - J_{sc}\frac{32}{\pi}} \tag{8}$$

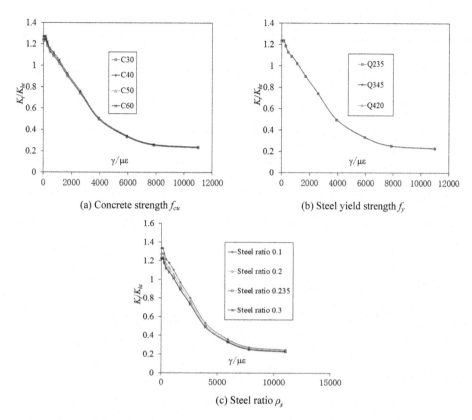

(a) Concrete strength f_{cu}

(b) Steel yield strength f_y

(c) Steel ratio ρ_s

Figure 11. k_1—γ relationship curves.

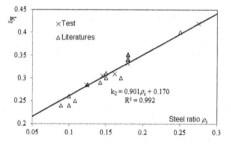

Figure 12. k_2—ρ_s relationship curve.

5 CONCLUSIONS

Based on the results and discussions presented above, the conclusions are obtained as below:

(1) Under the action of torsional load, the hollow steel tubes would be destroyed by the plastic hinge, and the CFST members had no obvious buckling phenomenon and had superior plasticity and stability.

(2) Comparing the test results of CFST torsion angle, maximum shear strain, and torsional stiffness with the corresponding finite element model calculations, the test curves agreed well with the finite element calculation curves.

(3) The steel ratio has a great influence on the torsional stiffness of CFSTs. The steel ratio is inversely proportional to the torsional stiffness, while the strength of the concrete and the strength of the steel has little influence on the torsional stiffness.

(4) Considering the contribution of the concrete inside the steel tubes to the torsional stiffness, the superposition method of linear stiffness was used to determine the regression formula of the torsional stiffness coefficient of concrete inside the steel tubes.

ACKNOWLEDGMENTS

This research is funded by the National Natural Science Foundation of China (Project No. E51508102). Natural Science Foundation of Fujian Province (Project No. 2019J05128). Natural Science Foundation of Fujian Province (Project No. 2019J01233).

REFERENCES

An Gong (1989). Study on Concrete-filled Steel Tubular Short Columns under Compression and Torsion. Beijing: Beijing Institute of Civil Engineering and Architecture.

Beck, J., & Kiyomiya, O. (2003). Fundamental pure torsional properties of concrete-filled circular steel tubes. Doboku Gakkai Ronbunshu, 2003(739), 285–296.

Baochun Chen (2007). Concrete-Filled Steel Tube Arch. Beijing: China Communications Press.

Chen, Y., Dong, J., & Xu, T. (2018). Composite box girder with corrugated steel webs and trusses–A new type of bridge structure. Engineering Structures, 166, 354–362.

Chen, B. C., & LI, X. H. (2008). Experimental study on restricted torsion for concrete-filled steel tube (single circular). Journal of Fuzhou University (Natural Science Edition), 5.

Guohuang Yao (2006). Research on Behavior of Concrete-Filled Steel Tubes Subjected to Complicated Loading States. Fuzhou: Fuzhou University.

Jing Zhou (1990). The Experimental Research on the Concrete Filled Steel Tubular Slender Column under Combined Compression and Torsion[D]. Beijing: Beijing Architecture and Civil Engineering Institute.

Linhai Han, Shantong Zhong (1995). The Studies of Pure Torsion Problem for Concrete Filled Steel Tube[J]. Industrial Construction, 25(1):7-13.

Yiwei Chen (2003). Study on Shape Factors and Torsional Toughness Behavior of Concrete Filled Steel Tubular Columns. Taiwan: Central University.

*Frontiers of Civil Engineering and Disaster Prevention and
Control – Yang & Rahman (Eds)*
© 2023 The Author(s), ISBN: 978-1-032-31200-2

Design and analysis of variable angle and inclined tower crane

Kun Zhang, Hui Wang, Kaiqiang Wang, Baiben Chen*, Yong Zhou, Dongdong Mu &
Xiaolin Fang
*Department of Engineering Technology Research, China Construction Third Engineering Bureau Co.,
Wuhan, China*

ABSTRACT: Considering the problem that the efficiency of traditional tower cranes decreases
and the safety risk increases with the increase of height during the construction of a bridge tower
with an inclined surface, a variable angle, and inclined tower crane is designed and developed,
based on the principle of variable angle action. Part of the tower body of the tower crane can be
deformed with the inclined surface of the structure to shorten the anchorage length and achieve
the purpose of construction efficiency. Through the structural design and analysis of the new tower
crane, the following conclusions are drawn: Based on the deformation principle of cooperative
work, the variable angle and inclined tower crane act simultaneously by relying on the tower crane
transformation unit and adjustable anchorage to achieve the purpose of tower body deformation.
The operation of the new tower crane depends on the simultaneous contraction of two adjustable
anchorages that drive the transformation units to rotate around the hinge point in the middle,
forming a cooperative deformation system. Through the comparative analysis of vertical state and
inclined state, it is found that the stress, displacement, and stability of tower cranes meet the safety
and use requirements.

1 INTRODUCTION

The shapes of high-rise buildings, bridge towers, cooling towers, and other high-rise structures
have different styles due to the structural, functional, and aesthetic requirements, most of which
have inclined and curved facades. In the construction process of these inclined or curved building
structures, the application of a self-climbing tower crane is essential. It undertakes the functions
of vertical and horizontal transportation for construction materials and local components. It also
provides reliable lifting conditions for structural construction.

However, due to the limitation of the lifting weight and lifting range of the tower cranes, the
arrangement scheme and quantity need to be carefully considered for the construction of a high-rise
structure. Taking the bridge tower as an example, the current arrangement of tower cranes used in
bridge tower construction mainly includes three ways: (1) one large tower crane is arranged on the
side of a single tower limb; (2) one middle tower crane is arranged between two tower limbs; (3)
two middle tower cranes are arranged on the side of both tower limbs. The typical arrangements
are shown in Figure 1.

There are some problems in the actual construction of bridge towers using the above traditional
tower crane layout schemes, especially in the case of high bridge towers and large slope shapes.
Such as Sutong Bridge (Zhang et al. 2009), Hangzhou Bay bridge (Li & Wen 2008), and Shanghai
Tongjiang railway Yangtze River Bridge (Li 2019), the bridge tower is high and has a certain
inclination angle, which elongates the attachment length the tower crane, increases the local stress

---

*Corresponding Author: cbb0505@163.com

| (a) 1st arrangement scheme | (b) 2nd arrangement scheme | (c) 3rd arrangement scheme |

Figure 1.   The typical arrangement scheme of tower cranes for bridge tower construction.

of the structure and impacts the tower crane with potential safety hazards (Mohamed et al. 2021). To solve the above problems, scholars have carried out some research work. Still, the existing research is mainly focusing on the aspects of optimization of tower crane layout position, selection of tower crane model, and local strengthening of tower crane structure (Al-Fadhli & Khorshid 2012; Chen et al. 2021), without considering the expansion of tower crane's function.

Based on the above problems and focusing on expanding the functional properties of the tower crane, a new type of tower crane – "variable angle and inclined tower crane" was developed, which breaks the technical bottleneck of the existing tower crane industry by analyzing and studying the design idea and structural stress of the new equipment. Moreover, it provides equipment guarantee and technical support for the applicability and efficiency of tower cranes in the field of inclined building construction.

2   DESIGN OF VARIABLE ANGLE AND INCLINED TOWER CRANE

2.1   *Overall design idea*

Compared with the traditional tower crane, the upper structure of the variable angle inclined tower crane remains unchanged. The main difference is that the lower structure of the tower body of new equipment can be tilted at a certain angle according to the structural shape. It adapts to the inclined plane on the side of the building to reduce the attachment length and makes the tower body closer to the building. The hoisting range and efficiency of the tower crane is improved. During the whole process, the top of the tower body remains vertical.

2.2   *Structure composition*

TC6013-8 tower crane produced by Joinhand Construction Machinery CO. Ltd. was used to design. The jib of the tower crane is 50 m long, the maximum lifting capacity is 8 tons, the rated lifting torque is 1000 kN.m, and the maximum lifting torque is 1104 kN.m. The free height of the outrigger fixed tower crane is 45 m, the maximum attached height is 150 m, and the total power of the motor is 42.4 kW.

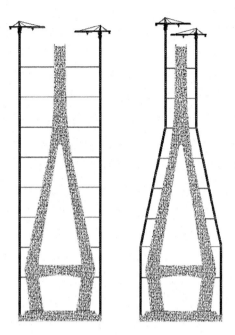

Figure 2.   Overall design idea of variable angle and inclined tower crane.

Variable angle and inclined tower crane is mainly composed of a standard unit, a transformation unit, adjustable anchorage, normal anchorage, and superstructure of the tower crane, as shown in Figure 3.

(1) The tower body of the tower crane is mainly composed of the traditional tower crane standard units, which are the main bearing component of the variable angle and inclined tower crane.
(2) According to the needs, the transformation units shall be arranged at the part of the tower crane body where the angle needs to be changed. The overall dimension and interface of the transfer units are consistent with the ordinary standard units. The transformation units can be installed by jacking the tower crane's standard unit.
(3) In the tower body, where the tower crane does not need to change the angle, normal anchorage can be installed according to the application needs to realize the height lifting and construction of the tower crane.
(4) Adjustable anchorage is installed at each transformation unit, and the length and angle of adjustable anchorage can be adjusted. By adjusting adjustable anchorages, the transformation unit can rotate a certain angle around the hinge joint to realize inclining a certain angle for the tower body.
(5) The superstructure of variable angle and inclined tower crane is consistent with that of conventional tower crane, including rotating part, jib, counterweight, hook, cab, counterweight, etc. Their functions are consistent with that of the conventional tower crane.

2.3 *Working mechanism*

The installation process of variable angle and inclined tower cranes is completely consistent with the operation process of a conventional tower crane. After the anchorage of the tower crane is installed, the tower body under the following anchorage can be transformed into an inclined state.

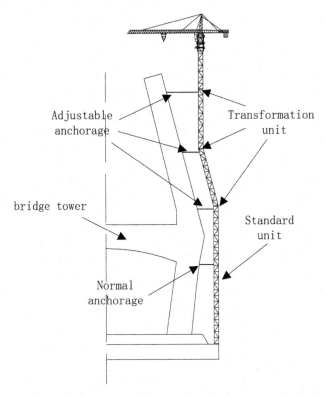

Figure 3. Structure composition of variable angle and inclined tower crane.

First, unlock the locking devices of penultimate and antepenultimate transformation units, which leads to the simultaneous contraction of tailender and penultimate adjustable anchorages. Under the contraction of these two anchorages, the penultimate and antepenultimate transformation units rotate at a certain angle. However, the tower body above the penultimate anchorage always remains vertical. When the angle transformation of the tower body under the penultimate anchorage is completed, the transformation units are locked to complete the process of transforming the penultimate transformation unit into an inclined state.

2.4 *Design of key components*

(1) Transformation unit

The upper and lower structure of the transformation unit is composed of the main limb, horizontal brace, inclined rod, etc. Between the upper and lower parts of the transfer joint, one end of the structure is hinged, and the other is connected through the sliding mechanism, fine adjustment, and limit mechanism.

The transformation unit can be regarded as a standard unit of the tower crane that can change the angle. The envelope, overall dimension, and interface of the transformation unit are consistent with those of the conventional tower crane standard unit. The transformation unit can be converted from a vertical state to a fixed inclined state or from an inclined state to a vertical state, as shown in Figure 4.

(2) Adjustable anchorage

Each adjustable anchorage needs to shrink twice in the whole process. The designed adjustable anchorage mechanism is shown in Figure 5. The brace of adjustable anchorage is designed to be a component of fixed length and rotatable angle, and the attachment frame is designed to be slidable. The sliding is driven by hydraulic cylinders arranged on both sides.

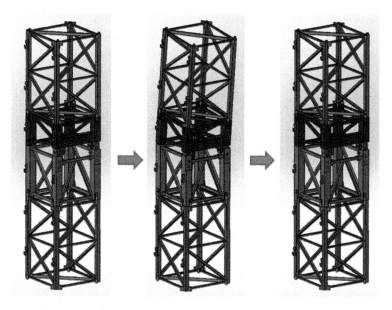

Figure 4. Angle change of transformation unit.

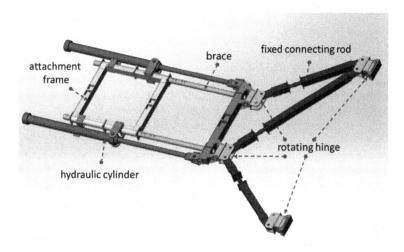

Figure 5. Adjustable anchorage Designed.

3 ANALYSIS OF VARIABLE ANGLE AND INCLINED TOWER CRANE

To analyze the feasibility of the overall operation mechanism of variable angle and inclined tower crane, changing angle of the inclined tower crane is set to 6°. Based on the TC6013 tower crane, the standard units at the bottom and 20 m height of the tower crane body are replaced with transformation units. At the same time, two adjustable anchorages are set, which are located at a height of 21.125 m and 41 m of the tower body, and the free height of the tower body is 30 m. The stress state of the whole and local members of the structure in the vertical state and inclined state are analyzed. The analysis diagram is shown in Figure 6.

657

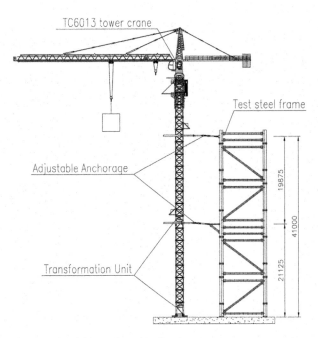

Figure 6.   Analysis diagram of variable angle and inclined tower crane.

3.1   *Establishment of finite element model*

Software Midas civil 2019 is used to establish the overall model of the tower crane and the test steel frame. The vertical support rod and transfer joint of the tower crane are simulated with Q345 steel; the inclined rod, anchorages of the tower crane, and test steel frame are simulated with Q235 steel. For both steel frames, the elastic modulus is 2.06×10^5 MPa, Poisson's ratio is 0.31, linear expansion coefficient is 1.2×10^{-5}, and unit weight is 78.5 kn/m^3.

The $135 \times 135 \times 12$ mm hollow rectangular section was used in the main vertical pole of the standard unit of the tower crane, the $80 \times 80 \times 5$ mm hollow rectangular section was used in the inclined web member, the $50 \times 50 \times 4$ mm hollow rectangular section was used in the crossbar, and d76 $\times$ 5 mm hollow circular section was used in a transverse circular pipe. And the $250 \times 250 \times 12$ mm hollow rectangular section was used in the main vertical pole of the transformation unit of the tower crane; the $80 \times 80 \times 5$ mm hollow rectangular section was used in the inclined web member.

The bottom of the tower crane and the test steel frame are simulated by a six-way full consolidation constraint of general constraint, the top boom of the tower crane is transformed into the load, the top standard unit is formed as a whole by rigid connection, and the anchorage and test steel frame are simulated by the rigid connection of elastic connection. A total of 548 nodes and 1167 beam elements are established. The finite element model is shown in Figure 7.

3.2   *Analytical load and working case*

The analytical load includes self-weight and external load. The self-weight is simulated by the gravity coefficient, which is 1, and the external load at each construction process is simulated in the form of concentrated force and bending moment.

According to the model, the additional applied vertical force is set as 40t, which is equally distributed to the four limb vertical rods; the bending moment is 2430 kN.m, which is equally distributed to 4-limb vertical rods; The horizontal force is 85 kN, which is applied to the nodes on both sides of the outermost attachment.

According to the different swing angles of the boom, the typical position in the boom swing half-circle is selected as the static analysis condition, i.e., 0°, 45°, 90°, 135°, and 180°.

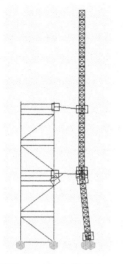 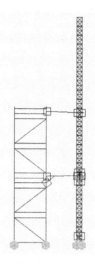

(1) Tower crane inclined state   (2) Tower crane vertical state

Figure 7. Finite element model.

3.3 Structural analysis results

(1) Structural stress analysis

The stress analysis of the main structure of the tower crane is carried out. The structural stress state under various working conditions is shown in Table 1, "+" is positive stress, and "−" is negative stress.

Table 1. Structural stress analysis (MPa).

Maximum stress	0°	45°	90°	135°	180°
Inclined state	−90.5	103	−123.5	−153.4	−127.1
Vertical state	−93.5	109.1	−108.6	−135.1	−109.3

According to the above stress analysis, the maximum stress of the tower crane in the inclined state is 153.4 MPa, the maximum stress of the tower crane in the vertical state is 135.1 MPa, and the stress of the tower crane in the inclined state increases by 13.5%, and the structure is safe.

(2) Structural displacement analysis

The horizontal displacement of the tower crane is analyzed. The horizontal displacement state under various working conditions is shown in Table 2.

Table 2. Structural displacement analysis (mm).

Maximum displacement	0°	45°	90°	135°	180°
Inclined state	142.6	166.3	185.3	173.5	155.8
Vertical state	142.8	172.9	175.3	179.8	155.8

According to the above displacement analysis, the maximum displacement of the top of the tower crane under the inclined state is 188 mm; in the vertical state, the maximum displacement of the top of the tower crane is 199.6 mm; the displacement of the tower top is small.

(3) Structural stability analysis

The overall stability analysis of the overall model of the tower crane and test frame is carried out. The weight is taken as a constant load, and the other loads are variable loads. The analysis results are shown in Table 3.

Table 3. Structural stability analysis.

Stability coefficient	0°	45°	90°	135°	180°
Inclined state	14.2	14.2	14	13.7	13.6
Vertical state	13.8	13.8	13.6	13.4	13.2

In the above analysis, the minimum buckling mode in the inclined state is 13.6, and the minimum buckling mode in the vertical state is 13.2. The structure is stable.

4 CONCLUSION

Based on the design of the variable angle action principle, a variable angle and inclined tower crane was designed and developed. Part of the tower body can deform with the structural inclined plane, to improve the construction efficiency and reduce the safety risk. Through the structural design and stress analysis of the new tower crane, the following conclusions are drawn:

(1) Based on the deformation principle of cooperative work, the variable angle and inclined tower crane act simultaneously by relying on the tower crane transformation unit and adjustable anchorage to achieve the purpose of tower body deformation.

(2) When the new tower crane works, it depends on the simultaneous contraction of two adjustable anchorages to drive the transformation units to rotate around the hinge point in the middle, forming a cooperative deformation system.

(3) Through the comparative analysis of vertical state and inclined state, the stress, displacement, and stability of tower crane structures meet safety and use requirements.

(4) Based on the working principle of variable angle, the deformation of multi-section towers with different inclination angles can be expanded for the better adaptation to more special-shaped bridge tower structures.

REFERENCES

Al-Fadhli A, Khorshid E. Payload oscillation control of tower crane using smooth command input. Journal of Vibration and Control. December 2021.

Chen Y., Zeng Q., Zheng X.Z., etc., "Safety supervision of tower crane operation on construction sites: An evolutionary game analysis," Safety Science. Papers 105578 (2021).

Li J.T., "Key Construction Techniques for Pylons of Main Navigational Channel Bridge of Hutong Changjiang River Bridge." Bridge Construction, Papers 46(6), 1–5 (2019). (in Chinese)

Li W.Z., Wen Z.Q., "Construction of Main Pylon of South Navigable Bridge of Hangzhou Bay Sea-crossing Bridge." China Harbour Engineering, Papers 4, 50–54 (2008). (in Chinese)

Mohamed H.E.O., Nabil B.K., Saiful I., etc., "A Smart Tower Crane to Mitigate Turbulent Wind Loads," Structural Engineering International. Papers 31, 18–29 (2021).

Zhang H., Zhang Y.T., You X.P., "Construction Control for Tower of Sutong Bridge." Highway, Papers, 3, 16–20 (2009). (in Chinese)

Frontiers of Civil Engineering and Disaster Prevention and
Control – Yang & Rahman (Eds)
© 2023 The Author(s), ISBN: 978-1-032-31200-2

Research on detection of architectural ceramic ornamentation based on template fractal

Xiaoping Zhou & Yanmin Liu*
College of Architecture and Civil Engineering, Xiamen Institute of Technology, Xiamen Fujian, China

Xiaobo Lian
Fine Art and Design College, Quanzhou Normal University, Quanzhou, China

Huiting Guo
School of animation design, Hoseo University, South Chung Ching Road, South Korea

Sangyoung Lee
TongMyong University, South District, Busan Metropolitan City, South Korea

ABSTRACT: With exquisite craftsmanship, rich tool types, and decorative features, ancient ceramics vividly show the essence of traditional Chinese culture. They have important historical, artistic, and scientific research values. In this paper, based on template fractal, the detection of architectural ceramic decoration is studied. The edge detection and image enhancement methods of architectural ceramic decoration and fractal art application in ceramic decoration pattern design are introduced. The composition of two-dimensional continuous pattern fractal art in ceramic decoration pattern design is introduced. In characteristic analysis, it is found that the research point of fractal art still has some limitations in the application of ceramic decorative pattern design. The application of two-way and four-way continuous splicing of architectural ceramic products requires that the pattern must meet the seamless splicing process of the product application. The raw materials, different molding and decoration processes, and different firing temperatures should be selected according to local conditions to create rich shapes and decorative images feature.

1 INTRODUCTION

In the 21st century, the concept of the human home has undergone fundamental changes (Chen 2011). People pay more and more attention to the role of porcelain in beautifying and decorating the home environment. Ceramics are defined as hard substances formed by a series of physical and chemical reactions after formulation, crushing, and molding with clay and other minerals as raw materials and firing at high temperatures. They include daily-use and building ceramics (Xu & Zack 2012). The main purpose of image enhancement is to improve the visual effect of the image and improve the definition of the image (Liu & Liang 2016). Chinese ancient ceramics have a clear origin, broad system, and wide-spreading range. They are important material evidence and footnotes in the inheritance of Chinese cultural heritage, forming a large system of ceramic technology and culture inherited in one continuous line and diverse (Liang & Liu 2016). In pattern design, the self-similarity of a fractal can be used to construct a variety of artistic patterns with arbitrary high-resolution structures (Kang 2016). Computer graphics has a very broad prospect with the development and progress of science and technology, and the fractal map directly drawn by computers has become a new source of pattern design (Pan 2018). The combination of fractal theory, computer, and art have derived fractal art with traditional aesthetic standards. It not only has basic aesthetic standards, such as symmetry and balance, rhythm, change, and unity, but also

---

*Corresponding Author: zhouxiaoping925@163.com

has its unique characteristics, such as the high self-similarity between part and whole (Sun 2016). There are many kinds of daily-use artistic ceramics, which have high practicability and partial appreciation, and are a model of the combination of industry and technology (Ren & Zack 2016). In the production of building ceramics, patterns and stripes are mainly finished by the cloth machine and pressed in the molding workshop (Zack 2018). The fractal pattern breaks the traditional pattern generation mode and provides new design inspiration and pattern sources for designers (Zack 2017). The appearance of white porcelain is one of the five milestones in China's ceramic development history, which provides a material basis for the prosperity of blue-and-white porcelain, multicolored porcelain, and pastel porcelain in later generations. Especially, the appearance of Xingyao Sui Dynasty transparent white porcelain has aroused great concern in academic circles, and its exquisite manufacturing technology is breathtaking. At present, architectural ceramic enterprises mainly rely on manual inspection to detect patterns and stripes, that is, to see whether there are defects with naked eyes. The inspection efficiency is low, the accuracy is not high, and it takes a lot of time, which has some limitations.

In the design of ceramic decorative patterns of a single pattern, two-way continuous patterns, and four-way continuous patterns. Combine fractal theory with ceramic decorative pattern design, guide the research of ceramic decorative pattern design innovation, create more inspiration for ceramic decorative pattern designers and expand the thinking of designers. Edge detection and image enhancement are often combined in image processing technology, because both image enhancement and edge detection process gray images. Before that, people could not imagine that the mathematical formula could produce infinitely fine and novel patterns after repeated iterations by using the computer. It provides a theoretical basis and a preliminary practical basis for new ceramic decorative pattern design practice, and provides theoretical and creative guidance for ceramic decorative pattern meter and ceramic art education. The research and application of architectural ceramic decoration detection based on template fractal are one of the brilliant achievements of China's ceramic science and art. This unique glaze decoration technique has been widely used in Jizhou kiln, Duandian kiln, Qiongyao kiln, and Changzhou kiln.

2   EDGE DETECTION AND IMAGE ENHANCEMENT METHODS IN RESEARCH

2.1   *Edge detection method*

In recent years, the fractal graph has been widely used, and it has outstanding performance in decoration design, architectural design, and simulation design. The theme and forms of decorative patterns are ever-changing, such as decorative architectural patterns, which can be used in ceramic decorative pattern design after changes, and the patterns in silk can also be used in its ceramics. The input and output of the ceramic edge detection system are shown in Figure 1 below.

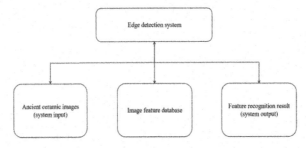

Figure 1.   Input and output of ceramic edge detection system.

To further expand the application scope of fractal art graphics in ceramic decorative pattern design, we should discuss the morphological characteristics of fractal art graphics, such as harmony, balance, and symmetry, and the composition characteristics of ceramic decorative pattern, such as nesting, proportion, and segmentation. In terms of artistic processing, it has preliminarily had the composition characteristics of unity, contrast, and symmetry and has created a variety of composition forms such as individual patterns and continuous patterns. Edge detection and image enhancement are very important in the fields of architectural ceramic decoration detection based on template fractal, numerical image segmentation, target region recognition, region shape extraction, and other image analysis. There are many forms of edge detection, most of which are defined by the first-order differential form and a few second-order differential forms, and the fast convolution function is mainly used to realize the calculation. Like many designers who make ceramic decorative patterns believe that fractal art only belongs to the school of mathematical formulas, those creators should have a certain mathematical foundation, such as matrix transformation, computer graphics, etc. The generation of the architectural ceramic pattern needs to go through the steps of conception, sketch design, refinement, and color filling before finally generating a pattern. At present, the mainstream identification methods of porcelain are traditional "visual identification" based on expert experience and "scientific identification" based on the support of modern science and technology. For example, a way to apply fractal graphics to ceramic decorative pattern design is to select different fractal graphics and different ceramic shapes for direct collage combination and apply 3D software to directly collage the generated fractal graphics on objects while seldom considering some basic pattern composition rules. In the application process of architectural ceramic products, there are a large number of application examples in which the same product is continuously distributed in two or four directions. Seamless splicing of patterns in the process of product tiling is an unavoidable problem in the process of pattern design. Using edge detection and image enhancement algorithms in digital image processing technology to analyze the image of architectural ceramics, we can quickly check the stripe, edge direction, and outline direction of architectural ceramics to judge whether the product is defective or qualified and achieve the purpose of rapid analysis.

2.2 Image enhancement method

The so-called image enhancement is to improve the picture quality by enhancing and highlighting the important content on the picture and suppressing the unimportant content simultaneously. When making a fractal graph with the target graph as a fractal element, you can choose any image you want to make. Png format is best, and only the required parts are reserved. The generated fractal graph will be more natural and harmonious and achieve excellent visual performance, which is suitable for patterns in various fields such as textile fabrics, decorations, and illustrations. The structure of the image enhancement system is shown in Figure 2 below.

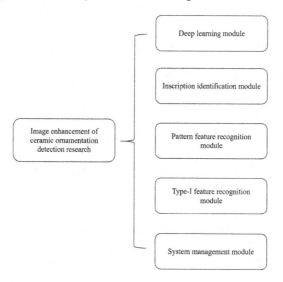

Figure 2.   Division of image enhancement function modules.

Image enhancement is commonly used in contour tracking, connected domain marking, object detection, etc. It only pays attention to the contour and other information of the image, pays less attention to other information, and there are only two elements, 0 and 1 (non-black or white). It is a special gray image. As a frontier, the application of fractal theory in ceramic decorative patterns is less than that in other aspects. Gray transformation enhancement is the basis of image enhancement, which can increase the dynamic range of the image, expand the contrast, and make the image features clear. Compared with other art forms, porcelain pattern design is affected by many factors such as textile technology and performance to form a unique expression, but in the form, it follows a certain law of formal beauty. The RGB model describes the general digital image; that is, each pixel of the digital image contains three quantities (R, G, b). The gray image is the simplified description of the RGB image, and the binary image is the simplified (contour) description of the gray image. The transformation path is from color image to gray image and then to binary image. Binary images can be understood as the smallest set containing only color image contour information. Image enhancement can apply fractal technology in two-dimensional discrete space to generate a fractal pattern based on template control. The composition of the pattern depends on the selection and design of the template. If the template itself is symmetrical (including two-way symmetry and four-way symmetry), the generated fractal pattern has the ability of seamless splicing in two-way or four-way continuous tiling. Compared with general gray images, binary images lack detail features, but can completely describe outline features. Using binary images to describe ancient ceramic features can reduce related interference factors, extraction, and recognition rate of lifter features. The application of fractal art in ceramic decorative patterns plays an important role in the application research of ceramic decorative pattern art design as an application in the early development stage. However, when the fractal theory develops to a certain stage, the application of fractal theory as a methodology in ceramic decorative pattern art design is not limited to simple direct addition, but requires innovative design and creation, just like other disciplines.

3 FEASIBILITY ANALYSIS OF FRACTAL ART APPLIED IN CERAMIC DECORATIVE PATTERN DESIGN

3.1 Application of fractal art in ceramic decorative pattern design

The road to combining art with science seems to be very far away and difficult. One of the most important reasons is the problem of means of expression. Overlay processing is to combine the same pattern or different patterns through certain changes to produce a new pattern effect and make the pattern produce a new appearance. The fractal theory provides a way to describe some unsmooth and irregular geometric shapes in nature and nonlinear systems that can't be described by traditional Euclidean geometry. Due to the particularity of this detection research, the related requirements of artificial intelligence-assisted ancient ceramic identification and the common characteristics of common application systems are fully considered in the design process, as shown in Figure 3 below.

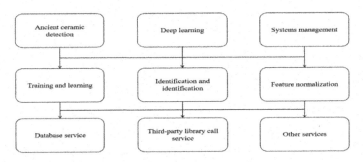

Figure 3.   Overall design of detection research.

Through fractal iteration, the graphics with unimaginable effects can be generated by simple mathematical formulas. The template is a macroscopic generalization of the pattern. The template is usually a pattern much smaller than the target pattern. The composition of data in the template determines the overall characteristics of the final pattern. Overlapping the same patterns generally does not change the original shape, layout, and arrangement. When it is necessary to detect the edge of the image, the simplest method is to calculate the absolute value of each pixel f, and finally make a threshold. Roberts operator is based on this idea. The operator is:

$$R(i, j) = \sqrt{(i,j) - f(i+1, j+1)^2 + (f(i,j+1) - f(i+1, j))} \tag{1}$$

The third layer of the usual pattern detection system is the data layer, and the bottom layer of the system needs not only database services, but also deep learning services and MATLAB language call services. Therefore, these are integrated into the service layer to replace the traditional data layer to provide services for the business logic layer. Taking the template as the initial pattern template, the Kronecker operation is performed on each data and template of the initial pattern by using the fractal iterative algorithm to enrich the template details; After reaching the specified number of iterations, the final template is converted into a pattern according to the specified tone. The main functions of the system are all aspects of artificial intelligence recognition of ancient ceramics. The main functions are three feature recognition modules: pattern recognition, device type, and inscription recognition image feature recognition module, other deep learning ancient ceramic image feature extraction, data storage, and other modules. As an important branch of nonlinear science, it is an effective mathematical model and tool to accurately describe natural and social phenomena. Fractal is different from the rigid feeling brought by symmetry in Euclidean geometry, which enhances the symmetry between local and whole in a large range. A Fractal is a complex figure generated by simple formula through iteration. Fractal uses its unique means to solve the relationship between whole and part. At present, there is no unified ancient ceramics data information storage and sharing system, and some existing ancient ceramics management systems have imperfect data information storage n56], or only some types of ancient ceramics data information can be stored and retrieved. The harmony of fractal graphics is a kind of mathematical harmony. Every shape change and the color transition of plum is a natural flow, without a stiff feeling.

3.2 *Analysis of the composition characteristics of two-dimensional continuous pattern fractal art in ceramic decorative pattern design*

Through a large number of different types and compositions of two-way continuous ceramic decorative patterns, it can be seen that the two continuous patterns created by fractals not only reflect the morphological characteristics of the traditional two continuous ceramic decorative patterns, such as symmetry and balance, change and unity, proportion and balance. Moreover, it also reflects the fractal graphic characteristics of dynamic change, local similarity to the whole, certainty, and randomness of fractal art graphics. The main function of the two-way continuous pattern fractal technology is to receive the ancient ceramic image, carry out the necessary preprocessing work on the image, and then compare and analyze the main features of the ancient ceramic image. The main features include decorative patterns, instrument types, and inscriptions, and the recognition results are given, as shown in Figure 4 below.

Grayscale transformation mainly consists of a linear transformation, piecewise linear transformation, etc. The unified web system realizes the integration of ancient ceramic image feature recognition, including the extraction, quantification, and recognition of the image features of ancient ceramic type, decoration, and inscription [N5W6Q], and gives complete recognition rate analysis results. The noise in the picture can be removed by low-pass filtering. Using the high-pass filtering method, high-frequency signals such as edges can be enhanced to make blurred pictures clear. The representative algorithms in the spatial method include the local averaging method and

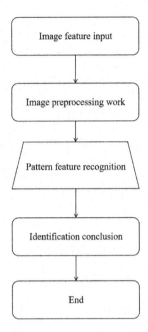

Figure 4.　Main function flow chart of binary, continuous pattern fractal technology.

median filtering method, which can be used to remove or weaken noise and are described as follows by the formula:

$$g(x, y) = f(x, y) * h(x, y) \qquad (2)$$

The database management module is the processing of formatting, storing, and extracting the decorative data, instrument data, and inscription data of ancient ceramics. According to the actual situation of the system, various data such as recognition time and image size are collected to facilitate the formation of corresponding data sets in the later stage. In the production of building ceramics, the image of building ceramics is directly generated by shooting, and then the image is directly transmitted to the computer, which is analyzed and detected by MATLAB software. Rapid detection of the pattern of architectural ceramics can find the problems in production in the shortest time, make it processed quickly, and minimize the production cost. For example, the linear transformation assumes the gray range [a, b] of the image f (x, y), and the gray range of the transformed image g (x, y) is linearly extended to [C, D], and its mathematical expression is as follows:

$$g(x, y) = c + \frac{d - c}{b - a}[(f(x, y) - a)] \qquad (3)$$

In the process of building a ceramic pattern generation algorithm based on template fractal, the pattern generated by the algorithm is symmetrical as long as the symmetry of the template is kept in the template selection or design, which can meet the seamless splicing of building ceramic patterns applying this pattern in the construction industry and ensure the application effect of products. The unrecognized image relearning module is a relearning module proposed for the system that can't give the recognition conclusion (the system doesn't contain the identification problem of ancient ceramics with such data). The system administrator screens the unidentified ancient ceramics data accumulated in the system, screens out the corresponding high-quality data, sorts them out, forms a data set, and then submits them to the deep learning module (EasyDL) for training and learning.

4 CONCLUSIONS

Study the extraction and recognition methods of image features of ancient ceramic types, patterns, and inscriptions, verify the implementation methods, and summarize and analyze the experimental results. As the concepts of dynamic change, whole and part, order and disorder, simplicity and complexity, finite and infinite, quantitative and qualitative change, certainty and randomness in scientific fractal theory, the philosophical connotation is profound and incisive. There are a lot of patterns reflecting fractal thought and fractal characteristics in ceramic decorative patterns. These patterns illustrate the relationship between fractal and ceramic decorative patterns. Through the application of edge detection, image enhancement, and other algorithms in digital image processing technology in decorative pattern detection in building ceramics production, it is found that the pattern style, edge direction, and outline of the experimental image after image processing are obviously enhanced, which is very helpful to identify decorative patterns and can greatly improve the detection efficiency in building ceramics production. China's ceramic decorative patterns are rich in types and widely used. Applying computer graphics to the research of fractal theory will be the development trend of fractal theory in the future. Its application in mechanical vibration, chemical engineering, material science, computer visualization, and other fields is extensive and deep, which provides a new scientific epistemology and methodology for the research of natural science and social science.

ACKNOWLEDGMENT

This article is funded by "Research on image restoration of ancient buildings in southern Fujian based on the convolutional neural network" (Project No. JAT210564) and is funded by "Exploration and practice of the double question exchange mechanism between high-caliber teachers and industry talents in civil architecture," Document No.: Fujian Jiaogao (2020) No. 4 (Fujian Provincial Department of Education).

REFERENCES

Chen Kaiyu. Application research of ceramic decorative patterns based on fractal theory [D]. Jingdezhen Ceramic Institute, 2011.

Kang Yang. Exploring the application of traditional decorative patterns in ceramic design [J]. Tomorrow's Fashion, 2016(11):1.

Liang Jinfeng, Liu Zijian. Research on digital protection method of ceramic decoration in Yaozhou kiln [J]. Ceramics, 2016(1):5.

Liu Zijian, Liang Jinfeng. Study on ceramic decoration of Yaozhou kiln based on template fractal [J]. Packaging Engineering, 2016, 37(4):5.

Pan Lu. On the characteristics of the times of ancient ceramic patterns [J]. Tomorrow's Fashion, 2018(20):1.

Ren Yuan, Zack Zhang. On the role of traditional decorative patterns in ceramic design [J]. Art and Technology, 2016(7):1.

Sun Ping. Study on the application of auspicious decoration in ceramic art [J]. Modern Decoration: Theory, 2016(3):2.

Xu Yanli, Zack Zhang. On the application of fractal technology in ceramic decorative pattern design [J]. Foshan Ceramics, 2012(3):3.

Zack Zhang. Research on the influence of traditional ceramic decoration in Jingdezhen ceramic culture industry [J]. Good Parents, 2017(54):1.

Zack Zhang. Study on the value of local traditional ceramic decoration in Jingdezhen creative culture industry [J]. Tomorrow's Fashion, 2018(16):1.

Frontiers of Civil Engineering and Disaster Prevention and
Control – Yang & Rahman (Eds)
© 2023 The Author(s), ISBN: 978-1-032-31200-2

Study on narrowing effect of high concrete face rockfill dam in narrow valley

Lu Xi*, Wang Wei* & Wang Jiayuan
PowerChina Northwest Engineering Corporation Limited, Xi'an, Shaanxi, China

ABSTRACT: With the development of dam engineering technology, building high concrete face rockfill dams in narrow river valleys under unfavorable geological conditions will be a trend. This paper discusses the problems existing in concrete face rockfill dams in narrow valleys, and different valley width-height ratio schemes are assumed. This paper analyzes the deformation of embankment, stress and stress level of the embankment, deformation of face slab, stress of face slab, displacement of joint, etc. The effective engineering measures to eliminate the narrowing effect of the valley are summarized, which can be used as a reference for similar projects.

1 INTRODUCTION

Generally, the concrete arch dam in a narrow river valley is a proper dam type, but it is not suitable for constructing the arch dam in a narrow river valley carved by some deep unloading rock mass or weak rock mass. The concrete face rockfill dam (CFRD) has become a dam type with obvious advantages, which will be increasingly adopted in the hydropower development in northwest and southwest China. As early as 2013, Yang Zeyan et al. (Yang et al. 2013) proposed that high CFRDs built in narrow river valleys are difficult problems and challenges for future development. In recent years, adequate attention has been paid to this technical concern, and various research has been carried out. For example, Wang Junli (Wang 2020) summed up the experiences and lessons taken from the design and construction of multiple CFRDs, including technologies adopted to construct dams on high steep bank slopes in a narrow valley. Shi Zhongle et al. (Shi et al. 2016) pointed out that the steep and narrow topography in the gorge area increases the difficulty of control for dam deformation and face slab cracking and fundamentals of dam deformation and face slab cracking as well as countermeasures have also been discussed. Yan Yongpu et al. (Yan et al. 2017) proposed pertinent major engineering measures based on several typical engineering examples of CFRDs built in narrow river valleys.

In short, for a high CFRD built in a narrow river valley, the deformation gradient of the embankment is likely too high, resulting in the destruction of the waterstop structure and face slab due to rheological characteristics of rockfill and the narrowing effect of the river valley. It is commonly argued that the valley effect can be nearly eliminated where the width-height ratio (W/H) of the river valley is above 2.5. In other words, the narrow valley effect is disregarded. Table 1 shows some CFRDs with a width-height ratio of less than 2.0 in China and abroad. Therefore, effective engineering measures shall be taken to control dam deformation and eliminate adverse effects caused by narrow valley topography.

---

*Corresponding Author: wwtiger@126.com

                    DOI 10.1201/9781003308577-90

Table 1. High CFRDs built in narrow river valleys.

No.	Project	Location	Height (m)	Width (m)	W/H	Rockfill material
1	Yang Qu	China	191	146	0.8	Limestone
2	Golillas	Colombia	125	109	0.87	Gravel
3	Acembe	Venezuela	162	146	0.9	Gravel
4	Long Shou Two	China	146.5	191	1.13	Diabase
5	Hou Zi Yan	China	223.5	284	1.27	Rhyolite
6	Ma Er Dang	China	211	308	1.46	Monzonite
7	Bai Yun	China	120	200	1.67	Sandstone, shale
8	Gu Shui	China	310	540	1.74	Sandstone, basalt
9	Anchicaya	Colombia	140	260	1.86	Hornblende, diorite
10	Jiang Ping He	China	219	414	1.89	Moraine conglomerate
11	Cethana	Australia	110	213	1.94	Quartzite

2 PROBLEMS OF CFRD BUILT IN NARROW RIVER VALLEY

The CFRD built in the narrow river valley is quite different from the one built in the wide river valley because of the narrowing effect of the river valley, resulting in the following problems:

(1) Effect on embankment deformation

The rockfill built in the narrow river valley has an arch effect, which restrains the deformation rate of the rockfill in the early stage. The arch effect will gradually disappear in a later stage. The deformation of rockfill increases greatly with the increase of the embankment height and the increase of the water pressure after the impoundment (Zhang & Wang 2017). The arch effect of CFRD refers to the scenario where the settlement deformation of the embankment in the middle of the riverbed is larger than that of the embankment on both sides of the dam abutment, and the gravity of the embankment is transferred to the dam abutment on both banks. As a result, the internal stress of rockfill decreases, the compaction is not easy to improve during construction, and the rheological phenomenon will be increasingly apparent with time due to the deformation modulus of rockfill embankment that is much lower than that of the mountain body on both banks (O et al. 2021; Wang et al. 2006; Zhang et al. 2017). The arch effect is more intense with the increase of dam height and higher settlement difference between the rockfill and mountains on the two sides of the valley.

(2) Effect on concrete face slab

On the whole, the stress and deformation of the embankment and the face slab in a narrow valley are lower than those built in a wide valley, but the deformation of the narrow valley in the later stage is larger. The rockfill stress and modulus of the lower face slab below the dam abutment of the river bed dam section decrease, leading to the deformation of the face slab in this area that increases obviously after the reservoir impoundment (Deng et al. 2015). As the proportion of the settlement after completion to the total settlement of the CFRD built in the narrow valley is much larger than that built in a wide valley, structural cracks are more likely to appear in the face slab of the CFRD built in the narrow river valley, compared with the wide valley CFRD.

(3) Effect on embankment stress

The stress and deformation of the dam embankment are complicated due to the complex topography of the dam site area, such as the twisted gullies, scarps, and funnel-shaped valleys of the main river bed along the dam axis. To the narrow valley topography, dam deformation and stress have a strong spatial effect. The extreme value of principal stress is not located in the middle of the river bed, but in the dam abutment.

(4) Possible damage caused by narrowing effect

The disappearance of the arch effect of the embankment will increase the deformation of the embankment, and the deformation along the dam axis from both banks to the riverbed will be larger due to the larger deformation gradient along the dam axis, and the axial tensile stress of the faceplate will be larger. The engineering practice shows that in the narrow river valley area, the rockfill may slip along both sides of river banks, which may lead to the failure of the concrete face slab and the shear deformation of the peripheral joints (Yang & Zhan 2007). Therefore, the failure of the high CFRD built in the narrow river valley shall be prevented.

3 ANALYSIS AND RESEARCH ON NARROWING EFFECT OF NARROW RIVER VALLEY

The narrowing effect of the high CFRD built in a narrow valley area will lead to a decrease in the deformation of the embankment during the early stage of the construction of a CFRD and the increase of the deformation of the embankment during the later stage of the construction of the CFRD, and the uneven deformation of the embankment will occur because of the large deformation gradient of the rockfill on the steep bank slope, causing cracks on face slab or break waterstop joints. In this paper, the width-height ratio of the river valley is assumed to be five different schemes (Figure 1) for a CFRD of a hydropower station in Qinghai Province, and the degree of narrow effect of the river valley and its influence on the embankment, face slab, and joints are analyzed using 3-D finite element calculation, a study on the engineering measures to reduce the valley narrowing effect is carried out. The 3-D model of the dam is shown in Figure 2, the calculation parameters are shown in Table 2, and the calculation results are shown in Table 3.

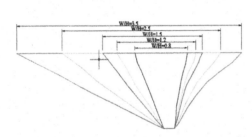

Figure 1. Schematic diagram of the valley effect study.

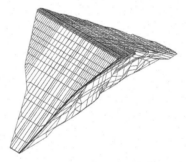

Figure 2. 3-D model of the dam (W/H 1.5:1).

Table 2. Static calculation parameters of dam body.

Item	Density ρ_d (g/cm³)	c (kPa)	Model parameters Φ_0 (°)	$\Delta\Phi$ (°)	k –	n –	R_f –	c_d (%)	n_d –	R_d –
Cushion zone	2.20	0	57.3	9.66	1100.0	0.42	0.56	0.40	0.50	0.47
Transition zone	2.18	0	58.7	10.90	1300.5	0.35	0.60	0.26	0.68	0.49
Main rockfill	2.15	0	55.7	10.30	1300.0	0.32	0.77	0.40	0.78	0.66
Secondary Rockfill	2.15	0	57.4	11.71	970.0	0.35	0.61	0.56	0.60	0.53
Downstream rockfill	2.15	0	55.7	10.30	1300.0	0.32	0.77	0.40	0.78	0.66

Note: 1. Case 1-As built period, Case 2-Impoundment period. 2. The deformation along the river is positive to the downstream and negative to the upstream. The axial deformation is positive to the

Table 3. Summary of characteristic values of stress and deformation with different W/H.

		Item	3.5	2.5	1.5	1.2	0.8
Embankment	Case 1	Deformation along river (cm)	22.4/-18.5	19.3/-16.4	14.7/-13.6	11.7/-10.1	7.3/-6.8
		Axial deformation (cm)	15.8/-14.6	15.0/-11.5	10.6/-6.5	10.3/-5.8	7.8/-2.8
		Settlement (cm)	104.3	95.8	93.8	91.5	87.3
		Major principal stress (MPa)	2.98	2.87	2.85	2.43	2.08
		Minor principal stress (MPa)	1.41	1.37	1.35	1.12	0.94
	Case 2	Deformation along river (cm)	23.4/-12.3	19.9/-10.6	15.0/-9.0	12.0/-5.9	7.7/-3.9
		Axial deformation (cm)	16.2/-14.8	15.4/-11.8	10.8/-6.7	10.5/-5.9	8.2/-2.9
		Settlement (cm)	107.6	98.7	97.0	94.3	90.3
		Major principal stress (MPa)	0.52	0.48	0.47	0.46	0.44
		Minor principal stress (MPa)	3.09	2.95	2.93	2.50	2.19
		Deformation along river (cm)	1.48	1.41	1.39	1.16	0.97
Face slab	Case 2	Deflection (cm)	32.1	30.6	29.9	29.6	29.0
		Axial deformation (cm)	3.8/-3.7	3.7/-2.8	3.6/-1.9	3.3/-1.3	2.8/-0.9
		Axial stress (MPa)	12.07/-2.20	11.86/-2.08	9.11/-1.82	9.05/-1.73	7.26/-1.66
		Stress along slope (MPa)	12.02/-1.28	11.53/-1.17	10.24/-0.96	8.30/-0.83	7.93/-0.41
Joints	Case 2		Dislocation (mm)	11.4	13.7	23.3	25.1	30.0
		Peripheric joint	Settlement (mm)	12.6	18.0	24.9	25.7	37.4
			Opening (mm)	6.2	9.0	11.0	11.9	12.8
		Vertical joint	Opening (mm)	4.3	4.9	6.4	7.8	8.0

right bank and negative to the left bank, the compressive stress is positive, and the tensile stress is negative.

3.1 *Deformation of embankment*

Figure 3 reflects the maximum deformation during the impoundment period of the embankment with different W/H ratios. As the ratio of W/H of the valley changes from 3.5 to 0.8, the settlement of the dam, the horizontal deformation along the river, and the axial deformation of the dam all show a trend of decreasing gradually. For example, the settlement rates of the dam are 0.52%, 0.48%, 0.47%, 0.46%, 0.44% respectively during the impoundment period. This is mainly due to the narrowing effect of the valley, which causes the arch effect in the rockfill, and the arch effect exists not only in the vertical direction, but also in the horizontal direction. In both directions, the settlement and horizontal deformation of the embankment show a decreasing trend. From the axial displacement of the embankment and the narrow valley and asymmetric valley left and right bank of the dam, axial deformation asymmetry is more obvious. If the ratio of the maximum settlement of the embankment to the length of the dam axis is defined as the deformation dip rate, the deformation dip rate of the embankment is 0.15%, 0.19%, 0.32%, 0.38%, 0.55% for the W/H ratio of 3.5, 2.5, 1.5, 1.2 and 0.8 respectively, although the deformation of narrow valley CFRD is relatively small, the gradient of deformation is relatively large, and the deformation is actually inhomogeneous. Certainly, with the lapse of time and the gradual disappearance of the arch effect, the large deformation in the later stage of the embankment is unfavorable to the operation of the embankment and the safety of the face slab.

3.2 *Stress and stress level of the embankment*

Figure 4 shows the principal stress of the embankment with different W/H ratios. When the ratio decreases from 3.5 to 0.8, the major and minor principal stress of the embankment shows a decreasing trend during the period of impoundment and construction. For example, the major stress decreases from 3.09MPA to 2.19MPa during the impoundment period, and the $\sigma_1/\gamma_s H$ at the bottom of the dam is 0.70, 0.67, 0.63, 0.55, and 0.47, respectively; this is because as the valley becomes narrower, the arch effect of the valley is getting stronger, and the major principal stress at the bottom of the dam of the steep, narrow valley is much less than the load of the overlying soil column. In addition, the restriction of bank slope hinders the transfer of self-weight load to the bottom of the dam, the stress level of the contact part between the embankment and bank slope will be increasingly higher, and the local stress level of the contact part between the embankment and bank slope reaches 0.90, which approximates to the plastic limit.

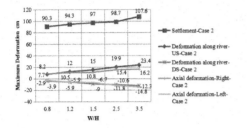

Figure 3. Maximum deformation of embankment with different W/H ratios.

Figure 4. Maximum stress of embankment with different W/H ratios.

3.3 Deformation of face slab

Figure 5 shows the extreme values of axial deformation and deflection of the face slab of a dam with different ratios of W/H. The maximum deflection decreases from 32.1cm to 29.0cm, and the maximum deformation from the left bank to the right bank decreases from 3.8cm to 2.8cm, with a decreasing rate of 26.3%, the maximum deformation from the right bank to the left bank decreases from 3.7cm to 0.9cm with a decreasing rate of 75.7%. As the right bank slope of the valley is steeper, the deformation of the right face slab pointing to the left bank decreases obviously with the narrowing of the valley, and the asymmetry of the axial deformation of the left and right face slab is more obvious.

3.4 Deformation of face slab

Figure 6 shows the maximum axial stress and the maximum stress along the slope at different W/H ratios during the impoundment period. From 3.5 to 0.8 of W/H, the maximum axial compressive stress decreases from 12.02MPa to 7.26MPa, the maximum axial tensile stress decreases from 2.20MPa to 1.66MPa, the maximum longitudinal compressive stress decreases from 12.07MPa to 7.93MPa, the maximum along-slope tensile stress decreases from 1.28MPa to 0.41MPa, and the area of the tensile stress along-slope extends from near the top of the river bed to both sides and the bottom. It is shown that with the narrowing of the valley, the axial compressive and tensile stresses of the face slab are increasingly lower, and the compressive and tensile stresses along the slope are getting lower too.

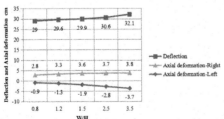
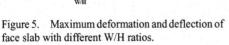
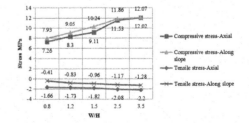

Figure 5. Maximum deformation and deflection of face slab with different W/H ratios.

Figure 6. Maximum stress of face slab with different W/H ratios.

672

3.5 Deformation of joints

Figure 7 shows the maximum deformation of peripheric joints and vertical joints at different W/H ratios. From 3.5 to 0.8 of W/H, the maximum settlement of peripheric joints increases from 12.6mm to 37.4mm, the maximum opening from 6.2mm to 12.8mm, the maximum dislocation from 11.4mm to 30.0mm, and the maximum opening of vertical joints from 4.3mm to 8mm. With the narrowing of the valley, the dislocation deformation, the subsidence deformation, and the opening deformation of the peripheric joints increase, especially the dislocation deformation and the subsidence deformation at the steep slope increase obviously, and the opening deformation of the vertical joints also increases with the narrowing of the valley.

4 ENGINEERING MEASURES TO ELIMINATE NARROWING EFFECT OF NARROW RIVER VALLEY

4.1 Higher compaction standard

If the compaction standard of a CFRD is raised, the deformation of the embankment, face slab, and joints will be reduced, which can effectively control the deformation of the embankment and reduce the influence of the valley narrowing effect on the deformation. According to statistics, the porosity of upstream rockfill materials ranges from 19% to 21%, as is proved by many high CFRD projects.

The dam height of a hydropower station in Qinghai is 212.5m, and the ratio of W/H is about 1.5. Through sensitivity analysis, the maximum settlement of the embankment under different compaction standards is shown in Figure 8. The settlement rate of the embankment is 0.63%, 0.55%, 0.50%, 0.48%, and 0.46%, respectively, the porosity is reduced from 21% to 19%, the maximum settlement of the embankment is reduced from 133.8cm to 98.2cm, and it is reduced by 26.6%, the settlement deformation of the embankment decreases obviously.

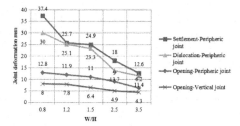

Figure 7. Maximum deformation of joints with different W/H ratios.

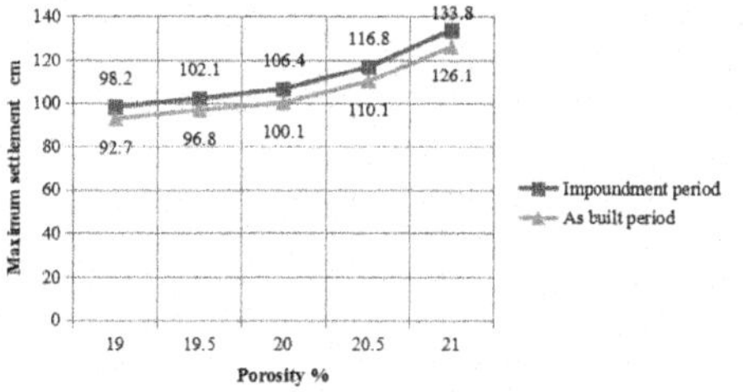

Figure 8. Maximum settlement of embankment under different compaction standards.

4.2 Effective construction technique

In general, the method of the balanced rise of upstream and downstream is required for rock filling of the CFRD of the narrow river valley, which is an effective construction measure to reduce differential settlement of different rockfill materials upstream and downstream. It can ensure the uniform settlement of the embankment to the utmost extent and reduce the unfavorable effects such as excessive deformation of waterstop joints caused by differential settlement of the narrow valley. In addition, to ensure the quality of dam rockfilling, a GPS high-precision real-time process monitoring system can be used during dam rockfilling. It can monitor the vibration roller compaction variable, the walking track, and the walking speed in a real-time manner and perform dam rockfilling construction automation control. While greatly reducing the labor intensity of construction and improving efficiency, a good visual interface is conducive to effective control of construction quality.

4.3 Improving valley shape

In order to control the adverse effect of valley shape, the slope of the lower reaches of the plinth is made into a smooth continuous surface by cutting the slope or steep bank slope, followed by placement of concrete to avoid the abrupt change of the steep slope so as to reduce the settlement deformation of the rockfill during the operation period and adverse effect on the stress and deformation of the face slab in the later period.

Moreover, if the width of the valley in the riverbed is extremely narrow, dry lean concrete can be rolled back to widen the width of the bottom of the valley and improve the stress condition of the face slab. It is also a desirable method to reduce the deformation and failure of peripheric joints and vertical joints due to the valley arch effect.

4.4 Adding modulus-improvement zone

It is possible to set up a horizontal modulus-improvement zone of a certain width on the steep bank slope on both sides of the river (DL/T 5016, 2011; Ge & Deng 2020). The porosity control standard can be increased by 1 to 2 percent, the deformation gradient between the foundation bank slope and the rockfill, and the deformation of the dam and the joints can be reduced.

In the example of a hydropower station in Qinghai mentioned above, the materials and grain sizes of the modulus-improvement zone are as same as the transition zone's, the horizontal thickness is 15m ~ 20m, and the two sides of the dam are arranged on the steep bank slope above the dam axis, the porosity of the bank slope modulus-improvement zone is 18%, compared with 19.5% in the rockfill zone, the porosity is increased by 7.7%. The results show that the deformation of the embankment, face slab, and joints decreases gradually after setting up the bank slope modulus-improvement zone, and the deformation and stress of embankment and face slab decrease, but the deformation gradient of embankment at bank slope decreases obviously, which, in particular, is more obvious at a higher elevation, with a decline rate of about 17%, joint deformation is reduced, especially the peripheric joint displacement deformation, which reduces obviously, by about 10%.

4.5 Proper timing of face slab placement

Attention shall be focused on pre-settlement measures before the construction of the face slab for the high CFRD built in the narrow river valley to ensure that the embankment will not have large deformation after the construction of the face slab. Two indexes can be used for quantifying the pre-settlement of the dam: one is the pre-settlement time control index. Before each stage of the construction of the face slab, there should be a period allowing pre-settlement for not less than three months for the lower part of the face slab; the other one is the pre-settlement convergence control index, i.e., settlement deformation rate of rockfill at the lower part of face slab that tends to converge. The settlement monitoring curve passes the inflection point and tends to be smooth, and monthly settlement deformation is not higher than 2mm ~ 5mm.

5 CONCLUSION

It is obvious that a narrow valley, more or less, has a narrowing effect on CFRD, which will cause some unfavorable effects. Modulus-improvement zone placed along the bank can alleviate this effect, but cannot reduce dam deformation. It is necessary to take such measures as the increase of compaction standards, application of effective construction technology, improvement of valley shape, and selection of proper face slab-pouring time for effective prevention of narrow valley effect on the embankment, face slab, and joints. In general, there will be no technical constraints for the construction of CFRD built in narrow valleys if appropriate measures are taken.

REFERENCES

Deng Gang, Wang Xiaogang, Wen Yanfeng, Yu Shu, Chen Rui. Study on conceptualization method of deformation pattern and horizontal breakage of face slab of concrete faced rockfill dam [J]. Journal of Hydraulic Engineering, 2015,46(04): 396–404.

DL/T 5016-2011, Design specification for concrete face rockfill dams [s].

Ge Fangyong, Deng Chengjin. Application of cemented sand and gravel in the design of the mold-enhancing area of Dashixia CFRD [J]. Northwest Hydropower, 2020(04): 78–81.

O Bo, Yuan Lina, Xiang Guoxing. Influence of rheology on stress and deformation of high face rockfill dam in narrow valleys [J]. Water Conservancy Construction and Management, 2021,41(04): 10–15.

Shi Zhongle, Ding Yutang, Xiong Guowen. Investigation on construction and research progress for concrete face rockfill dams in a narrow valley in China. Jiangsu Water Resources, 2016(11): 50–53.

Wang Hui, Chang Xiao-lin, Zhou Wei. Effects of rockfill rheological on deformation and stress of high concrete faced rockfill dam [J]. Rock and Soil Mechanics, 2006,27(S1): 85–89.

Wang Junli. Design technology Progress and safety concept prospect of concrete face rockfill dam [J]. Northwest Hydropower, 2020(01): 11–18.

Yan Yongpu, Sun Baoping, Dang Lincai. Study on engineering design and countermeasure of high concrete face rockfill dam in narrow valley [J]. Water power, 2017,43(02): 35–39.

Yang Zeyan and Zhan Zhenggang. Research and application of construction techniques for concrete face slab rockfill dam with height about 200m in narrow valley region [J]. Advances in Science and Technology of Water Resources, 2007(05): 33–37.

Yang Zeyan, Jiang Guocheng, Zhou Jianping, Xu Zeping, Sun Yongjuan, Wang Fuqiang. Development of modern concrete face rockfill dam [c]/Hydropower 2013-proceedings of the 2013 annual conference of the China Association of large dams and the 3rd International Symposium on Rockfill Dams, 2013: 568–582.

Zhang Shuyu, Duan Bin, Tang Maoying, Li Peng, Wang Guanqi. Rheological behavior and three-dimensional finite element analysis of ultra-high rockfill dam materials [C]/earth-rockfill dam technology 2017, 2017: 273–282.

Zhang Xiaomei, Wang Junli. Key technologies for deformation control of high face rockfill dam in narrow valleys [c]/earth-rock dam technology 2017, 2017: 59–68.

675

*Frontiers of Civil Engineering and Disaster Prevention and
Control – Yang & Rahman (Eds)*
© 2023 The Author(s), ISBN: 978-1-032-31200-2

Study on supporting effect of steel tube and HAT composite steel sheet pile

Bing Du*
CCCC First Highway Consultants Co., Ltd., Xi'an, China

Shou-Chen Jing*
Geotechnical Research Institute, Hohai University, Jiangsu Nanjing, China
Jiangsu Zhusen Architectural Design Co., Ltd. Xuzhou Branch, Jiangsu Xuzhou, China

Hao-Yu Fang, Jian Zhang & Tu-Gen Feng
Geotechnical Research Institute, Hohai University, Jiangsu Nanjing, China

ABSTRACT: A new type of composite support structure is a cross combination of steel pipe and Hat type steel sheet pile, PHC support structure for short. It has the advantages of large section modulus, stiffness, and a good steel sheet pile sealing effect. Using ABAQUS to carry out a three-dimensional finite element simulation of the foundation pit excavation. From the dimensions of steel pipe, Hat steel sheet pile, and other aspects of the excavation supporting effect analysis.

1 INTRODUCTION

In order to meet people's various needs, such as transportation and living services, it is a trend to develop underground areas in urban centers where land resources are scarce. The foundation pit is an indispensable part of underground space engineering. Due to the complexity of geological conditions and construction conditions, various foundation pit supporting methods have been proposed. Steel sheet pile has been widely used because of their high construction efficiency, good impermeability, economy and durability, and little environmental pollution (Sun & Zhang 2019). However, the traditional U-shaped steel sheet pile is easy to cause excessive deformation of the foundation pit due to its small section modulus and insufficient bending stiffness, which affects the engineering safety. As shown in Figure 1, it is a new type of combined support structure that is used by crossing steel pipes and Hat steel sheet piles. This kind of structure not only has high bending rigidity but also has water and soil retaining capacities. At present, there is little research on the influencing factors of the supporting effect of this structure. This paper analyzes the influence of these parameters on the deformation and internal force of the structure by changing the supporting structure parameters.

Figure 1.   A new combined supporting structure of steel pipe +Hat steel sheet pile.

Many factors affect the support effect. For example, appropriately increasing the diameter of the steel pipe can effectively reduce the horizontal displacement of the pile body (Xu 2019), while increasing the thickness of the steel pipe can improve the stiffness of the support structure and reduce

---

*Corresponding Authors: dubing222@qq.com and 913140315@qq.com

DOI 10.1201/9781003308577-91

the lateral displacement of the pile body. Increasing the depth of the embedding, the displacement of the top of the pile may increase or decrease, which is related to the position of the center of gravity of the pile and the overturning of the pile (Zhong 2014). At the same time, the increase in the pile length will also cause a change in the internal force of the pile. The friction force is transmitted between the piles through the action of the lock, so that the combined steel sheet piles can be stressed together. During the modeling analysis, the friction coefficient of the contact surface of the lock is also different. It will lead to a change in the stiffness of the combined pile body (Shen et al. 2014), which will affect the lateral displacement and internal force of the pile body after the excavation of the foundation pit. Based on a tunnel excavation project in Jinan, Zhang Guoyu (Zhang 2019) compared the difference between the numerical simulation structure and the theoretical calculation structure of single/double-row steel pipe support and carried out a three-dimensional numerical simulation to analyze the single/double-row steel sheet pile support. Based on the difference in the effect, the influence of the spacing of the row piles, the diameter of the steel pipe, and the wall thickness of the steel pipe on the support structure of the tunnel excavation was discussed, and the value of the spacing of the steel pipe piles was optimized from the perspective of "cost-support effect." Li Zhi (Li 2017) took a foundation pit project in Beijing as an example, carried out the design of steel pipe piles, and studied the support effect of steel pipes with different diameters through physical model tests. In this paper, the three-dimensional numerical simulation method will be used to analyze the influence of various factors on the supporting effect of composite steel sheet piles to provide suggestions and references for a specific construction.

2 NUMERICAL SIMULATION OF FOUNDATION PIT EXCAVATION

In this section, the Abaqus finite element model is established, and the parameters are adjusted to simulate the influence of different influencing factors on the horizontal displacement and bending moment of the supporting structure.

2.1 Project overview

Taking a foundation pit project in Ma'anshan, Anhui, as an example, the excavation range of the foundation pit is 50 m long, 30 m wide, and the excavation depth is 6 m. The basic parameter information of the soil layer is shown in Table 1:

Table 1. Basic parameters of soil layer after simplification.

Soil layer name	Soil weight (kN/m³)	Void ratio	Cohesion (kPa)	Internal friction angle φ (°)	Compression modulus E_s (Mpa)	Thickness (m)
Miscellaneous fill	18.0	0.970	10.0	10.0	4.30	1.74
Plain fill	18.6	0.950	12.0	12.6	4.62	1.50
Silt mixed with silt	19.0	0.811	13.2	14.2	7.94	16.20
Silt Silty Clay	17.6	1.103	12.3	8.9	3.60	15.40

Considering that the excavation area of the foundation pit is large and there are buildings around it, the combined supporting structure plus internal supporting system is adopted as the supporting scheme in this project. Single support is adopted, and the support is located at a depth of 2.5 m in the foundation pit. The heap load of construction and driving load of surrounding roads is 20 kPa.

2.2 Finite element model establishment

The soil adopts the M-C model. The steel pipes, Hat-type steel sheet piles, and concrete inner supports in the PHC composite support system are all simulated by elastic materials. The steel pipe diameter is 426 mm, thickness 10 mm, length 17 m, Hat-type steel sheet pile effective width 700 mm, height 240 mm, thickness 7.5 mm, length 17 m, the elastic modulus of supporting structure 210 GPa, Poisson's ratio 0.3, concrete inner support is a rectangle of 0.8 m × 0.8 m, using C30 concrete with an elastic modulus of 30 GPa and a Poisson's ratio of 0.167. The solid element (C3D8R) is used to simulate the soil. The foundation pit shall be excavated twice, with the internal support set after excavation for 3 meters and then excavation for 3 meters. The soil model of the foundation pit and its mesh division are shown in Figure 2.

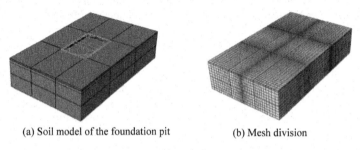

(a) Soil model of the foundation pit   (b) Mesh division

Figure 2.   Soil model and mesh division diagram of foundation pit.

The steel pipe and the Hat steel sheet pile are connected through the lock, and friction is established between the locks. The normal direction is hard to contact, the tangential direction is a penalty function, and the friction coefficient is set. The steel pipe and Hat-type steel sheet piles are simulated with shell elements (S4R). The internal support system and column pile are simulated by the beam element. The grid division of steel pipe and Hat steel sheet pile is shown in Figure 3.

Figure 3.   Grid division diagram of steel pipe and Hat steel sheet pile.

2.3 Model loads and boundary conditions

Apply gravity load to the soil and support system, and apply 20 kPa stacking a load on the top surface of the soil. Because 1.5 m of grading is carried out before the excavation of the foundation pit, the grading soil is equivalent to a uniform load of 27 kPa and is applied to the ground. On the top surface of the soil, the combined steel sheet piles are embedded in the soil and bound to the inner support at a depth of 1 m to limit the displacement in the z-direction of the bottom of the soil model and to limit the horizontal displacement of the soil around the foundation pit.

3 INFLUENCE OF SUPPORT STRUCTURE PARAMETERS

3.1 *Influence of steel pipe diameter*

For the modeling analysis of steel pipe diameter, keep the thickness of steel pipe unchanged at 10 mm, and take the diameters of 380 mm, 430 mm, 480 mm, and 530 mm, respectively. The horizontal displacement curves of the pile body of steel pipes with different diameters are shown in Figure 4. It can be seen that with the increase of the diameter of the steel pipe, the lateral displacement of the pile body decreases, and the horizontal displacement of the pile top decreases more obviously. This is because when the diameter of the steel pipe becomes larger, and the remaining support conditions remain unchanged, the section modulus of the support structure becomes larger, and the flexural rigidity also increases accordingly, which reduces the horizontal displacement of the support structure. The influence of the diameter on the horizontal displacement of the pile body is greater in the part above the excavation surface and has little effect on the part embedded in the soil below the excavation surface.

The curve of the maximum bending moment of steel pipes with different diameters is shown in Figure 5. It can be seen that with the increase of the diameter of the steel pipe, the maximum bending moment of the steel pipe is also increasing, but the increase rate decreases with the increase of the diameter. The maximum bending moment of the pile first decreases and then tends to be flat and slightly increased. This may be due to the increase in the diameter of the steel pipe. Without considering the change of the neutral axis, the stiffness of the steel pipe in the PHC combined steel sheet pile supporting structure increases, while the stiffness of the Hat-type steel sheet pile remains unchanged. The stiffness increases, the load internal force is redistributed, and the proportion of the steel pipe to bear the load becomes larger, which becomes the main force-bearing member of the entire supporting structure. When the diameter increases to a certain extent, the maximum bending moment of the Hat-type steel sheet pile no longer decreases with the increase of the diameter, but increases somewhat.

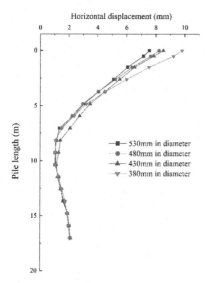

Figure 4. Horizontal displacement curves of steel pipe piles with different diameters.

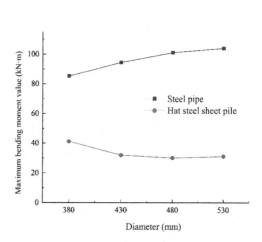

Figure 5. Variation curve of maximum bending moment of steel pipes with different diameters.

3.2 *Influence of embedded depth*

The embedded depth of the support structure is an important parameter in the design of the foundation pit support. The insufficient depth of the support structure will affect the safety of the foundation pit envelope structure. In the three-dimensional numerical simulation, the depths of soil penetration are taken as 1.2 H, 1.4 H, 1.6 H, 1.8 H, and 2.0 H, respectively (H is the excavation depth of the foundation pit). Figure 6 and Figure 7 show the change curves of the horizontal displacement and the maximum bending moment of the pile under different embedding depths of the supporting structure.

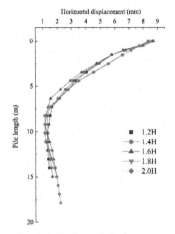

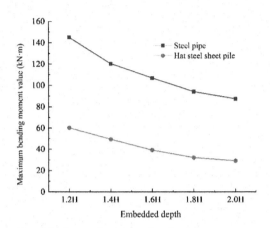

Figure 6. Horizontal displacement curve with different embedded depths.

Figure 7. Curve of maximum bending moment with different embedded depths.

It can be seen from Figure 6 that when the embedded depth is 1.6 H, the horizontal displacement of the pile top reaches the maximum value, and the horizontal displacement of the pile body can be reduced by appropriately increasing/decreasing the embedded depth of the supporting structure. In a certain range, increasing or decreasing the embedded depth has a certain control effect on the horizontal displacement of the pile body. With the increase of embedded depth, the horizontal displacement of the pile bottom is gradually increasing. When the embedded depth of the structure is lengthened, the bending rigidity of the supporting structure decreases, and the partition effect on the inside and outside of the foundation pit weakens, which shows the horizontal displacement at the bottom of the wall gradually decreases. When the embedded depth reaches 2.0 H, the influence of the embedded depth changes on the horizontal displacement of the pile decreases.

It can be seen from Figure 7 that with the increase of embedded depth of supporting structure, the maximum bending moment of steel pipe and Hat steel sheet pile gradually decreases with the increase of embedded depth of pile body. The bending moment value of steel pipe is always greater than that of steel sheet pile. When the embedded depth increases from 1.2 H to 1.8 H, the maximum bending moment value decreases obviously. Still, from the above, it can be seen that increasing the embedded depth can reduce the internal force of the supporting structure and control the horizontal displacement of the pile top after more than 1.6 h. However, attention should be paid to the cost problem caused by increasing the pile length, so as to obtain a better supporting effect.

4 CONCLUSION

This paper expounds on some factors that affect the supporting effect of composite steel sheet piles, including the diameter of steel pipes and the embedded depth of the supporting structure. According to these influencing factors, Abaqus three-dimensional numerical simulation with different values is carried out, and the influence of different parameter values on the horizontal displacement and internal force bending moment of the supporting structure is analyzed. Based on the results and discussions presented above, the conclusions are obtained as below:

(1) The diameter of the steel pipe and the embedded depth of the supporting structure have a great influence on the horizontal displacement and maximum bending moment of the supporting structure;

(2) The embedded depth of the pile can be controlled within the range of 1.8 h. When the horizontal displacement is strictly required, the embedded depth can be further increased to reduce the bending moment and horizontal displacement further.

REFERENCES

Li Zhi. Model test and numerical simulation of steel pipe pile foundation pit support stability [D]. China University of Mining and Technology, 2017. (in Chinese)

Shen Yupeng, Wang Huihuang, Jing Peng, Tian Yahu, Fu Xiaoyan. Orthogonal Analysis of Influencing Factors of Foundation Pit Support Near Existing Lines [J]. Chinese Journal of Transportation Engineering, 2014, 14(02): 14–20. (in Chinese)

Sun Guangli, Zhang Yining. Application of Larsen steel sheet pile in foundation pit support engineering [J]. Sichuan Cement, 2019(08):121. (in Chinese)

Xu Jie. Numerical simulation analysis of support structure properties during foundation pit excavation in coastal reclamation soft soil area [D]. Zhejiang Ocean University, 2019. (in Chinese)

Zhang Guoyu. Application research of steel pipe pile support technology in shallowly buried tunnel engineering in complex urban environment [D]. Shandong Jianzhu University, 2019. (in Chinese)

Zhong Lin. Analysis of key influencing factors of deformation of multi-level foundation pit supporting structure [J]. Fujian Architecture, 2014(05): 84–86+80. (in Chinese)

Author index